Francis C. Moon
Applied Dynamics

Related Titles

Meriam, J. L.

Engineering Mechanics – Statics 6e with Engineering Mechanics Statics – Solving Statics Problems in Maple 6e Set

660 pages
2006
Hardcover
ISBN: 978-0-470-10221-3

Nayfeh, A. H., Balachandran, B.

Applied Nonlinear Dynamics

Analytical, Computational, and Experimental Methods

700 pages
1995
Hardcover
ISBN: 978-0-471-59348-5

Heard, W. B.

Rigid Body Mechanics

Mathematics, Physics and Applications

262 pages with 43 figures
2006
Softcover
ISBN: 978-3-527-40620-3

Banerjee, S.

Dynamics for Engineers

294 pages
2005
Softcover
ISBN: 978-0-470-86844-7

Tongue, B. H., Sheppard, S. D.

Dynamics

Analysis and Design of Systems in Motion

Approx. 560 pages
Hardcover
ISBN: 978-0-471-40198-8

Francis C. Moon

Applied Dynamics

With Applications to Multibody and Mechatronic Systems

Second, completely revised edition

WILEY-VCH Verlag GmbH & Co. KGaA

The Author

Prof. Dr. Francis C. Moon
Ithaca, NY
USA
E-mail: fcm3@cornell.edu

All books published by **Wiley-VCH** are carefully produced. Nevertheless, authors, editors, and publisher do not warrant the information contained in these books, including this book, to be free of errors. Readers are advised to keep in mind that statements, data, illustrations, procedural details or other items may inadvertently be inaccurate.

Library of Congress Card No.:
applied for

British Library Cataloguing-in-Publication Data
A catalogue record for this book is available from the British Library.

Bibliographic information published by the Deutsche Nationalbibliothek
Die Deutsche Nationalbibliothek lists this publication in the Deutsche Nationalbibliografie; detailed bibliographic data are available in the Internet at <http://dnb.d-nb.de>.

© 2008 WILEY-VCH Verlag GmbH & Co. KGaA, Weinheim

All rights reserved (including those of translation into other languages). No part of this book may be reproduced in any form – by photoprinting, microfilm, or any other means – nor transmitted or translated into a machine language without written permission from the publishers. Registered names, trademarks, etc. used in this book, even when not specifically marked as such, are not to be considered unprotected by law.

Typesetting   VTEX, Litauen
Printing   betz-druck GmbH, Darmstadt
Binding   Litges & Dopf GmbH, Heppenheim

Printed in the Federal Republic of Germany
Printed on acid-free paper

ISBN: 978-3-527-40751-4

Contents

Preface *IX*
Preface to the Second Edition *XIII*

1 **Dynamic Phenomena, Design and Failures** *1*
1.1 Introduction *1*
1.2 What's New in Dynamics? *2*
1.3 Dynamic Failures *15*
1.4 Basic Paradigms in Dynamics *19*
1.5 Coupled and Complex Dynamic Phenomena *32*
1.6 Dynamics and Design *33*
1.7 History of Dynamics Principles *35*
1.8 Modern Physics of Dynamics and Gravity *37*

2 **Basic Principles of Dynamics** *41*
2.1 Introduction *41*
2.2 Kinematics *41*
2.3 Equilibrium and Virtual Work *47*
2.4 Systems of Particles *50*
2.5 Rigid Bodies *60*
2.6 D'Alembert's Principle *64*
2.7 The Principle of Virtual Power *67*
Homework Problems *68*

3 **Kinematics** *77*
3.1 Introduction *77*
3.2 Angular Velocity *80*
3.3 Matrix Representation of Angular Velocity *83*
3.4 Kinematics Relative to Moving Coordinate Frames *84*
3.5 Constraints and Jacobians *88*
3.6 Finite Motions *91*

3.7   Transformation Matrices for General Rigid-body Motion   *98*
3.8   Kinematic Mechanisms   *103*
      Homework Problems   *124*

4   Principles of D'Alembert, Lagrange's Equations, and Virtual Power   *139*
4.1   Introduction   *139*
4.2   D'Alembert's Principle   *143*
4.3   Lagrange's Equations   *151*
4.4   The Method of Virtual Power   *169*
4.5   Nonholonomic Constraints: Lagrange Multipliers   *182*
4.6   Variational Principles in Dynamics: Hamilton's Principle   *188*
      Homework Problems   *192*

5   Rigid Body Dynamics   *211*
5.1   Introduction   *211*
5.2   Kinematics of Rigid Bodies   *214*
5.3   Newton–Euler Equations of Motion   *224*
5.4   Lagrange's Equations for a Rigid Body   *248*
5.5   Principle of Virtual Power for a Rigid Body   *260*
5.6   Nonholonomic Rigid Body Problems   *278*
      Homework Problems   *285*

6   Introduction to Robotics and Multibody Dynamics   *305*
6.1   Introduction   *305*
6.2   Direct Newton–Euler Method Using Graph Theory   *309*
6.3   Kinematics   *315*
6.4   Equations of Motion: Lagrange's Equations and Virtual Power Method   *319*
6.5   Inverse Problems   *339*
6.6   PD Control of Robotic Machines   *346*
6.7   Impact Problems   *351*
      Homework Problems   *366*

7   Orbital and Satellite Dynamics   *383*
7.1   Introduction   *383*
7.2   Central-force Dynamics   *384*
7.3   Two-body Problems   *396*
7.4   Gravity Force of Extended Bodies   *398*
7.5   Rigid-body Satellite Dynamics   *403*
7.6   Control Moment Gyros   *414*
7.7   Tethered Satellites   *417*
      Homework Problems   *421*

8 Electromechanical Dynamics: An Introduction to Mechatronics *435*

8.1 Introduction and Applications *435*
8.2 Electric and Magnetic Forces *438*
8.3 Electromechanical Material Properties *446*
8.4 Dynamic Principles of Electromagnetics *455*
8.5 Lagrange's Equations for Magnetic and Electric Systems *458*
8.6 Applications *470*
8.7 Control Dynamics *481*
Homework Problems *488*

9 Introduction to Nonlinear and Chaotic Dynamics *501*

9.1 Introduction *501*
9.2 Nonlinear Resonance *503*
9.3 The Undamped Pendulum: Phase-Plane Motions *507*
9.4 Self-Excited Oscillations: Limit Cycles *511*
9.5 Flows and Maps: Poincaré Sections *514*
9.6 Complex Dynamics in Rigid Body Applications *524*
Homework Problems *537*

Appendix A. Second Moments of Mass for Selected Geometric Objects *545*

Appendix B. Commercial Dynamic Analysis and Simulation Software Codes *549*

References *555*

Index *561*

Preface

The modern post industrial era has ushered in a new set of dynamics problems and applications, such as robotic systems, high-speed maneuverable aircraft, microelectromechanical systems, space-craft dynamics, magnetic bearings, active suspension in automobiles, and 500-kph magnetically levitated trains. Up until the 1950s, engineers generally dealt with dynamic effects in machines and structures from a quasi-static point of view or not at all. In the last quarter of this century, however, incorporation of dynamic forces in design has become necessary, as new materials have permitted higher loads, speeds, and temperatures, resulting in more lightweight and optimally designed dynamical devices.

One success of the computer revolution in the field of dynamics has been the codification of analysis tools in linear dynamical systems. Codes are now available to accurately predict natural frequencies and mode shapes of complicated structures and machines. This has pushed the frontiers of dynamical analysis into nonlinear dynamics and multibody systems, and coupled field dynamical problems such as electromagneto-dynamics, fluid-structural dynamics, and intelligent control of machine–structure interactions. In Europe and Japan the combined field of dynamics, control, and computer science is called *Mechatronics*.

So why another textbook in dynamics? This book is written to fill a gap between elementary dynamics textbooks taught at the sophomore level, such as Meriam, Beers, Johnson, etc., and advanced theoretical books, such as Goldstein, Guckenheimer, and Holmes, etc., taught at the advanced graduate level. The focus of this book is on modern applied problems and new tools for analysis.

The goals of this textbook are:

- To illustrate the phenomena and applications of modern dynamics through interesting examples without excessive mathematical abstraction.
- To introduce the student to a clear statement of the principles of dynamics in the context of modern analytical and computational methods.
- To introduce modern methods of virtual velocities or principle of virtual power as developed by Jourdain, Kane, and others through clear illustrative examples.
- To develop educated intuition about advanced dynamics phenomena.
- To integrate modeling, derivation of equations, and solution of equations as much as possible.

Applied Dynamics. With Applications to Multibody and Mechatronic Systems. Second Edition. Francis C. Moon
Copyright © 2008 WILEY-VCH Verlag GmbH & Co. KGaA, Weinheim
ISBN: 978-3-527-40751-4

- To provide an introduction to applications related to robotics, mechatronics, aerospace dynamics, multibody machine dynamics, and nonlinear dynamics.

The level chosen for this text is at the undergraduate senior, masters degree, or first-year graduate level, although an honors junior class should have no difficulty with the material.

Most parts of this book have been used in an Intermediate Dynamics course taught at Cornell University by the author over the course of a decade. Students who have taken the course have included mechanical and civil engineers, theoretical mechanics, and applied physics students. The levels have ranged from seniors, master of engineering, to Ph.D.-level students. The core of the material (Chapters 2–6) can be taught in a one-semester course or in two quarter-system courses. In recent years the author has taught the course using *MATLAB* and *MATHEMATICA*. Students are asked to write programs to automatically derive Lagrange's equations or to obtain a time history of the motion. Also term projects have been used to teach the course in which students have 4–5 weeks to analyze and write a report about a specific dynamics application, often using the class notes as a launching platform to venture into more advanced material as given in the list of references and advanced texts.

Both vector and matrix methods are presented in the book. Experience has shown that students easily master Lagrange's equations, but still struggle with the three-dimensional vector dynamics introduced in elementary courses in dynamics. Thus the author has kept a strong element of kinematics in the book.

An important feature of this book is the development of the *method of virtual velocities* based on the *principle of virtual power*. Virtual power ideas go back many centuries, but were formalized in dynamics by Jourdain at the turn of this century. In the 1960s Professor Thomas Kane of Stanford University developed a related formalism to derive equations of motion using virtual velocities. However, in Europe and Germany a less formal, more direct use of the principle of virtual power has been used in textbooks and software. This book presents the less formal approach. The method is not only simpler in many cases than Lagrange's equations, but is sometimes more suited to solving multibody problems using computer methods.

The second half of the book reflects the author's research interests, especially in magneto-mechanical dynamics and nonlinear phenomena. There are an increasing number of electromechanical applications, and there are few pedagogical treatments of the derivation of equations, of motion in such systems. In the last two decades, research in dynamics has shown that deriving equations of motion does not always give one intuitive knowledge of the dynamical phenomena that is embodied in them. The chapter on nonlinear dynamics is included to review some of the important phenomena associated with the nonlinear equations of rigid-body dynamics.

Of course, modern problems in dynamics are sometimes closely linked with control. A brief mention of control issues is discussed in Chapter 1, and a few of the problems in Chapter 8 incorporate feedback control forces. However, the subject is too broad to squeeze into this text.

Finally, the pedagogical style of writing emphasizes phenomena and applications. Thus the book is intended for a broad spectrum of students, not just the most advanced, although the advanced student with develop dynamical intuition with this book. The goals here are to motivate, develop educated intuition in dynamics, and to develop confidence to use modern software tools and programs that solve dynamics problems.

A final word to the Lecturer. Professors wishing to use this book as a text may obtain a copy of solved selected homework problems from the Author, care of Mechanical and Aerospace Engineering, Upson Hall, Cornell University, Ithaca, NY 14853.

Acknowledgements

I wish to acknowledge the help of some of my former and present graduate students who have provided criticism, computations and have checked some of the solutions. In particular, I thank Dr. Matt Davies (National Institute of Standards and Technology), Dr. Erin Catto (Parametrics Corporation), Dr. Mark Johnson (General Electric), Mani Thothadri, Tamas Kolmar-Nagy and Duncan Calloway.

Finally, I am indebted to the skilled typing of Ms. Debbie De Camillo and the drawing of the figures by Ms. Teresa Howley. The help and patience of the Wiley staff in this project is also acknowledged, especially Greg Franklin, John Falcone, and Lisa Van Horn.

Ithaca, New York *Francis C. Moon*
April 1998

Preface to the Second Edition

In planning for the second edition of *Applied Dynamics* I initially thought that the addition of 80 new homework problems would be sufficient. However in the decade since the first edition, many new areas of applications had developed and I spent an extra year adding new material to the text. In particular I have added more discussion of electric forces at the MEMS and nanometer scale such as Casimir forces in recognition of the advances in nanotechnology. At the macro-scale I have added new material on the kinematics and dynamics of robotic machines. Also since the 1990s, the teaching of design synthesis of machines and robots has expanded at many universities and I have included new material on kinematics of mechanisms such as centroids, Jacobians and mobility criteria.

In addition to new applications, I have tried to keep the dynamics fundamentals at a basic level recognizing that there are many excellent texts in advanced dynamics, robotics and multi-body dynamics. With multi-body codes becoming more accessible to both the student and practicing engineer, it is important more than ever to develop a sense of "educated intuition" about dynamic phenomena so that one can exercise good judgment when using these codes.

In the development of my own dynamics intuition I must recognize my early teachers especially at Pratt Institute in Brooklyn and Cornell University. At Pratt Professor Irving H. Shames (now retired from the University of Buffalo) taught me vector dynamics and Professor Charles Mischke (now retired from Iowa State University) taught me dynamics of machines. At Cornell I was fortunate to take a course in classical advanced dynamics from Professor Phillip Morrison before he went to MIT. I have also been influenced by the classic text of Goldstein, and the elementary texts of Den Hartog and Meriam. Recently I have used the robotics text of Craig and the German dynamics text of Friedrich Pfeiffer (retired from the Technical University of Munich). I have also benefited from discussions with Cornell dynamics and design faculty Hod Lipson, Ephahim Garcia, Mason Peck, Richard Rand and Andy Ruina.

In my early studies of dynamics decades ago, I largely bypassed the study of kinematics of machines instead devoting efforts into techniques to solve applied differential equations. In the last decade however I discovered the geometric kinematics work of Franz Reuleaux (1876) and have recently published a historical book on kinematics: *The Machines of Leonardo da Vinci and Franz Reuleaux* (Moon, 2007).

Applied Dynamics. With Applications to Multibody and Mechatronic Systems. Second Edition. Francis C. Moon
Copyright © 2008 WILEY-VCH Verlag GmbH & Co. KGaA, Weinheim
ISBN: 978-3-527-40751-4

Some of the material on kinematics has filtered into the writing of this second edition.

Finally I want to acknowledge the contributions of the many students I have taught in the last 30 years at Cornell. In particular I wish to thank graduate students Scott Adams and Anthony Johnson for ideas in electro-dynamics. I also want to thank recent students in my Applied Dynamics classes for their new homework problems ideas among them; Justin Atchison, Michele Carpenter, Jonathon Hiller, Joanna Hinks, Matthew Kosakowski, Dan Periad Ephram Rubin, Danelle Schrader, Yused Syed Joseph Shoer, Adal Tecleab, and Noah Zych. Scott Schriffies deserves special thanks for updating the MATLAB problems in the text.

I also want to recognize the encouragement and patience of my editors at Wiley–VCH, in particular Dr. Christoph v. Friedeburg and Regina Hagen in Berlin.

Ithaca, New York, 2008 *Francis Moon*

1
Dynamic Phenomena, Design and Failures

1.1
Introduction

At the end of the first decade of the new millennium, technical headlines describe new technologies in which dynamics plays an important and even a crucial role. Increasingly the role of dynamics in design is coupled to feedback control technology. However this does not diminish its importance for the engineer. In transportation technology in the first decade of the new century we have witnessed the flight of the world's largest airliner, the Airbus 380, the new French high-speed train TGV has reached test speeds over 500 kph (300 mph) and the Germans and the Chinese have built the world's first passenger service MAGLEV train in Shanghai with a regular speed of 420 kph. In some cities one can often see a squadron of tourists on two-wheel transporters called 'Segway' machines, inverted pendulums controlled with gyro sensors. In energy technology we have seen the development of thousands of wind turbines in both Europe and the United States with blade diameters reaching 80 meters or more and power generators of over five megawatts.

In mechatronics technology, robotics has seen a renewal with both walking machines as well as semi-robotic home cleaning machines in the ROOMBA vacuum systems. At the micro-scale, MEMS technology is now applied to optical displays. At the nanometer scale, we have seen molecular machines in NEMS technology research. In biomechanics there has been advances in control augmented prostheses as well as robot assisted surgery in the popular da Vinci robot surgical system. From the nano- to the macro-scale from biotech to energy and transportation systems, the understanding of the principles of dynamics continues to play a crucial role in technology more that four centuries after Newton's *Principia*.

For the scientist, dynamics is both familiar and fascinating; its subject matter is close to everyday experience, yet it still holds surprises and mysteries. To the engineer, dynamics is a tool to predict forces and to design machines. But dynamics also embodies problems that can destroy and damage both the machines and the people that use them. These many facets of dynamics are the subject of this book.

In the study of dynamics the student either learns how to predict motions from a given set of forces or how to calculate the resultant forces from the motions. Force

Applied Dynamics. With Applications to Multibody and Mechatronic Systems. Second Edition. Francis C. Moon
Copyright © 2008 WILEY-VCH Verlag GmbH & Co. KGaA, Weinheim
ISBN: 978-3-527-40751-4

is the interaction of one body with another or the interaction of one part of a body with another, interactions that can deform, damage, and destroy the machines the engineer designs and builds. Yet before one can acquire the skills to successfully predict and design, most students must build a catalog of dynamic phenomena and then slowly begin to formalize the ways of modeling these phenomena with mathematics.

This catalog building begins early with play: swings, bicycles, sports. Then these experiences are formalized in laboratory experiments or careful observation of dynamic events. Finally constructs such as time, space, velocity are introduced with which one begins to learn the mathematical laws and techniques to calculate and simulate. Although the main focus of this book is on model building, analysis, and calculation, the student is encouraged to re-experience the phenomena through play and informal experiment. We take for granted the importance of play in forming our early concepts of dynamics. The late Nobel laureate Richard Feynman told about visualizing early concepts of his theory of quantum electrodynamics by observing the wobble of empty food plates that students were playfully throwing like Frisbees in the cafeteria at Cornell University. I often encourage my students to buy dynamic toys such as gyros and spinning tops, and when no one is looking to play with some of the dynamic toys of their younger siblings.

While avoidance of excessive forces, stresses, and motions is a prime interest of the student as engineer, to the student as scientist the challenge and joy of understanding the complexities of the subject often inspires the long hours of mathematical study required to master the subject. Although this text is directed toward technical mastery of dynamics, the student should not lose sight of the intellectual fascination of the subject.

We began this text by reviewing the motivation for studying dynamics at the end of the first decade of the 21st century. This includes looking at new applications as well as reviewing the relation between dynamics and failure in modern engineering. We also present a catalog of some of the basic phenomena of dynamics and their simplest mathematical models. A more general review of the principles of dynamics is presented in Chapter 2. The development of a working knowledge of the principles of dynamics is then given in Chapters 3, 4, and 5.

It is assumed that the reader has had elementary courses in the dynamics of particles and rigid bodies as well as a basic working knowledge of vector calculus and ordinary differential equations.

1.2
What's New in Dynamics?

What could be new in a field that celebrated its tricentennial in 1986, the anniversary of Newton's *Principia*[1]? Ten years earlier one could have said "not much". But

---

[1] Isaac Newton (1642–1727) occupied the Lucasian Chair of Mathematics at Cambridge University. Newton invented the calculus, along with Leibniz. In 1687 he published his *Philosophiæ Naturalis*

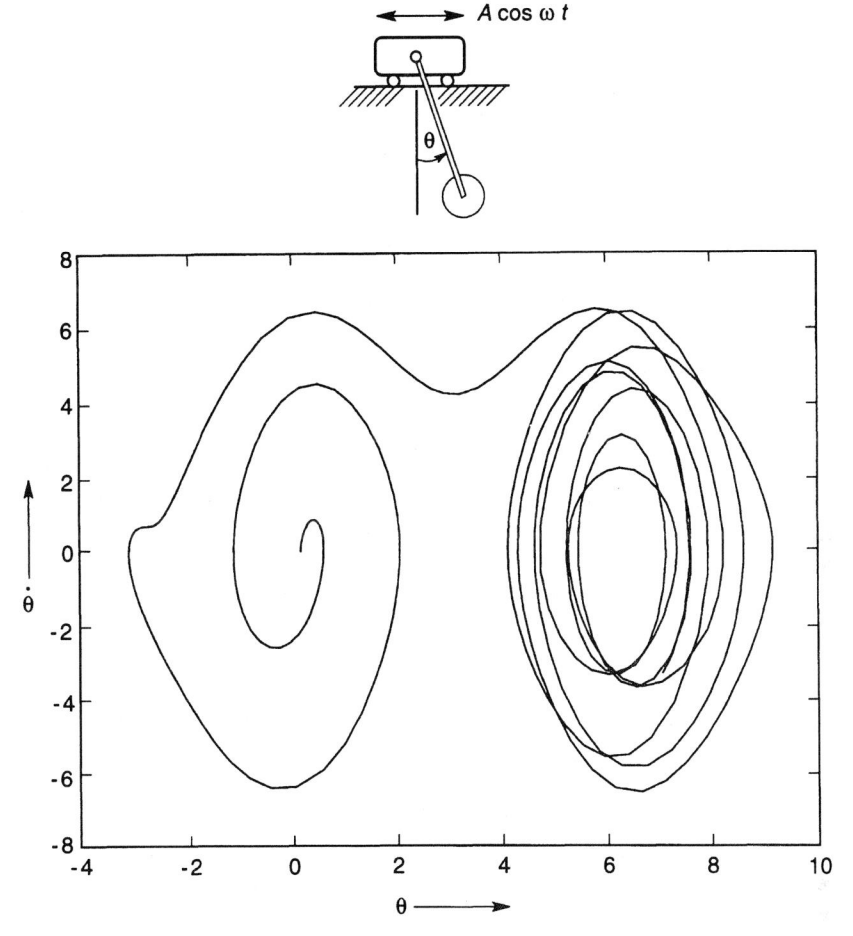

Fig. 1.1 Phase-plane trajectory of the periodically forced pendulum with a small amount of viscous damping showing chaotic-like vibrations.

in the last quarter of the twentieth century, the field of dynamics has developed a new vigor inspired by new discoveries, new methods of formulation, new computational and experimental tools, and new applications.

New Discoveries

The most widely recognized discovery in dynamics in the last twenty years has been *chaotic dynamics*. Chaos theory is a branch of nonlinear dynamics. Out of new theoretical ideas have come experimental evidence for deterministic chaos –

Principia Mathematica in which he presented his famous laws of motion as well as his law of gravity. The Lucasian Chair is now occupied by Stephen Hawking whose new theories of gravity seek to replace those of Newton and Einstein (see Hawking, 1988).

motions whose time histories are sensitive to initial conditions. Such motions resemble random dynamics but originate from nonrandom deterministic systems. An example is shown in Fig. 1.1 for the forced motion of a simple pendulum. The input motion is periodic, but the output motion is a complex pattern of rotary and oscillatory motions. These ideas had their origin in the work of Henri Poincare (ca. 1905), but their full meaning in dynamics took nearly a century of mathematical study and the advance of computers to become clear. Mechanical systems that are now known to exhibit chaotic dynamics are:

- Buckled structures under dynamic loading
- Impact problems
- Friction devices
- Fluid-structure problems
- Magnetomechanical systems
- Forced pendular-type problems

An introduction to chaotic and nonlinear dynamics is given in Chapter 9. (See also Moon, 1992 or Guckenheimer and Holmes, 1983.)

New Methods of Formulation of Dynamic Models

The formulation of mathematical models for simulation and analysis is a major part of the practice of dynamics. Until the early 1960s the two principal methods were the Newton–Euler force approach, taught in the first two years of engineering training, and Lagrange's equations based on energy functions. Since the 1960s however, two methods have been rediscovered. These are D'Alembert's principle of *virtual work* and Jourdain's principle of *virtual power*. The former is essentially an extension of the principle of virtual work to dynamics, while the latter principle goes back to the time of Aristotle but was presented in mathematical form by Philip E. B. Jourdain in 1907. The virtual power or virtual velocity method was rediscovered by Professor Thomas Kane of Stanford University in the 1960s. He extended this method to rigid bodies and developed it into a formal structure sometimes referred to as Kanes' equations (see, e.g., Kane and Levinson, 1985, as well as Chapters 4 and 5 in this book).

New Applications

In rigid-body dynamics, two dominant new areas of application have been multibody dynamics and mechatronics. Multibody dynamics involves formulation of models for connected rigid or near rigid bodies such as a robot manipulator arm shown in Fig. 1.2. Mechatronics involves the introduction of electromagnetic actuators and feedback forces sometimes incorporating computer intelligence.

Multibody Problems
The most common multibody system is the bicycle. The four principal rigid-body components are two wheels, a frame holding a seat, and a fork steering bar. Pedal-

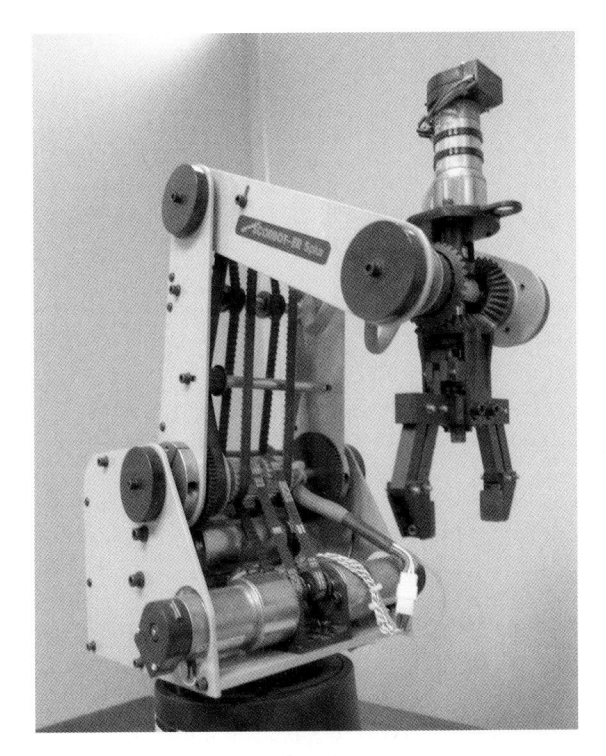

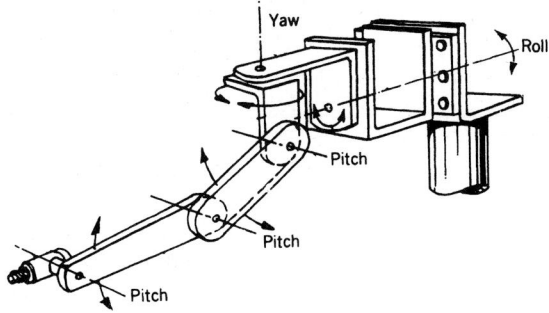

Fig. 1.2 Serial link, robot manipulator arm with gripper.

sprocket, drive chain, and gearing add dozens of other rigid-body subcomponents. And, of course, the rider can sometimes be viewed as a multilinked system. In spite of the ubiquitous nature of the bicycle and the recent improvements in the so-called mountain bike, very little hard dynamic knowledge is known about this system apart from empirical trials and observations. Adding to the complexity, is the rolling constraint, which adds so-called *nonholonomic* constraints (Chapters 4 and 5).

The subject of multilink kinematics is an old field with a long history. The collection of linkage systems by F. Reuleaux in Germany for teaching kinematics in the nineteenth Century has yet to be rivaled in this century. A subset of these mod-

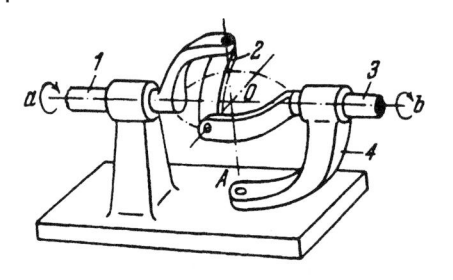

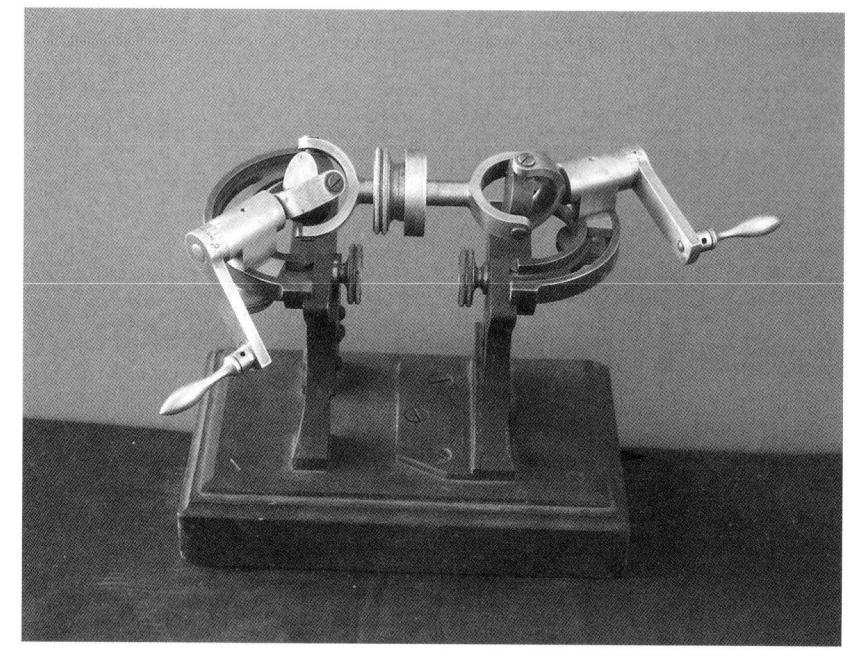

Fig. 1.3 *Top*: Four-bar spherical mechanism (from Artobolevsky, 1979). *Bottom*: Photo of double universal joint from the Cornell Collection of Reuleaux Kinematic Models (Model P-2).

els is in the Kinematic Collection of Cornell University. The Reuleaux models can be viewed on the Cornell website KMODDL: Kinematic Models for Design Digital Library. And the beautiful catalog of kinematic mechanisms by the Russian Artobolsky (see the References section) shows the great variety of devices for creating motion in a mechanical system (Fig. 1.3). See Erdman (1993) for a review of modern kinematics.

One of the world's most ubiquitous simple multibody mechanisms is the slider crank (Fig. 1.4) used in internal combustion engines. Earlier dynamic analysis of the forces in such systems used graphical methods. Today there are powerful commercial and research codes to analyze and graphically simulate the motions of complicated dynamic systems (see Schiehlen, 1990).

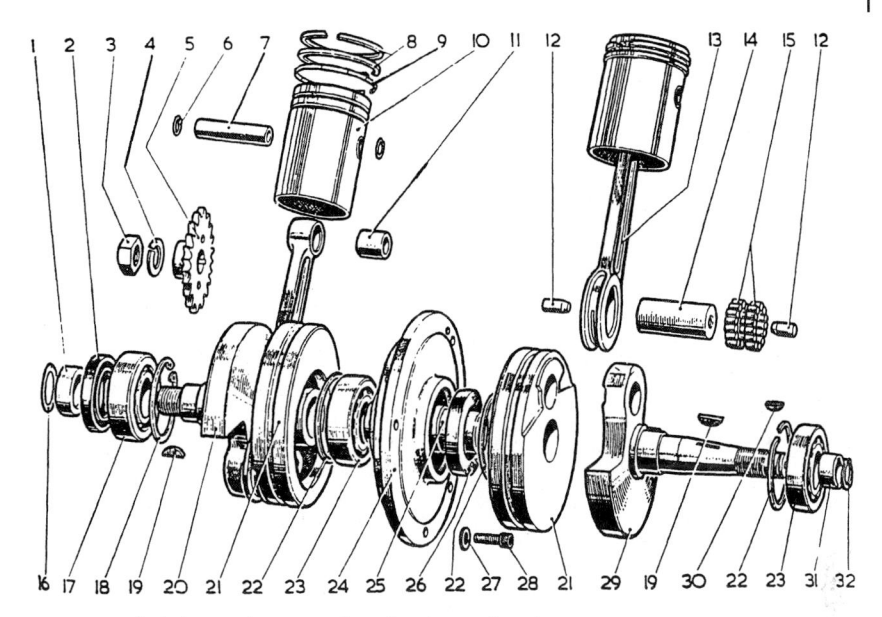

Fig. 1.4 Exploded view of an internal combustion engine piston and crankshaft assembly. (From *Two-Stroke Engines*, Roy Bacon, Osprey Publ. Ltd, London.)

One of the recent tools has been the use of *graph theory* to represent the *connectedness* topology of the multibody system. Fig. 1.5 shows chain, closed-loop, and tree-type multibody geometries. The classic serial link mechanism (Fig. 1.6) can be an open chain, a closed chain or a loop structure if it comes in contact with its base environment. Some of the most difficult dynamics problems in these systems are friction, impact, and rolling contracts, which still require more research. Practical examples of multibody systems include:

- Robotic devices
- Door and latch mechanisms in aircraft design, jigs, and fixtures
- Train, tandem trucks, and buses
- Docking of space vehicles
- Solar panel deployment on satellites
- Helicopters
- Biomechanical systems: walking and mobility prosthetics

Modern texts in multibody dynamics include Dankowicz (2004), Huston (1990), Shabana (1989), Roberson and Schwertassek (1988), and Wittenburg (1977, 2007).

Multibody Dynamics and Computer Game Programming
Research issues in multibody dynamics can be found in the journal *Multibody Systems Dynamics*. Besides the development of efficient computational algorithms, modeling of 3D impact and friction is an open item as well as integration with

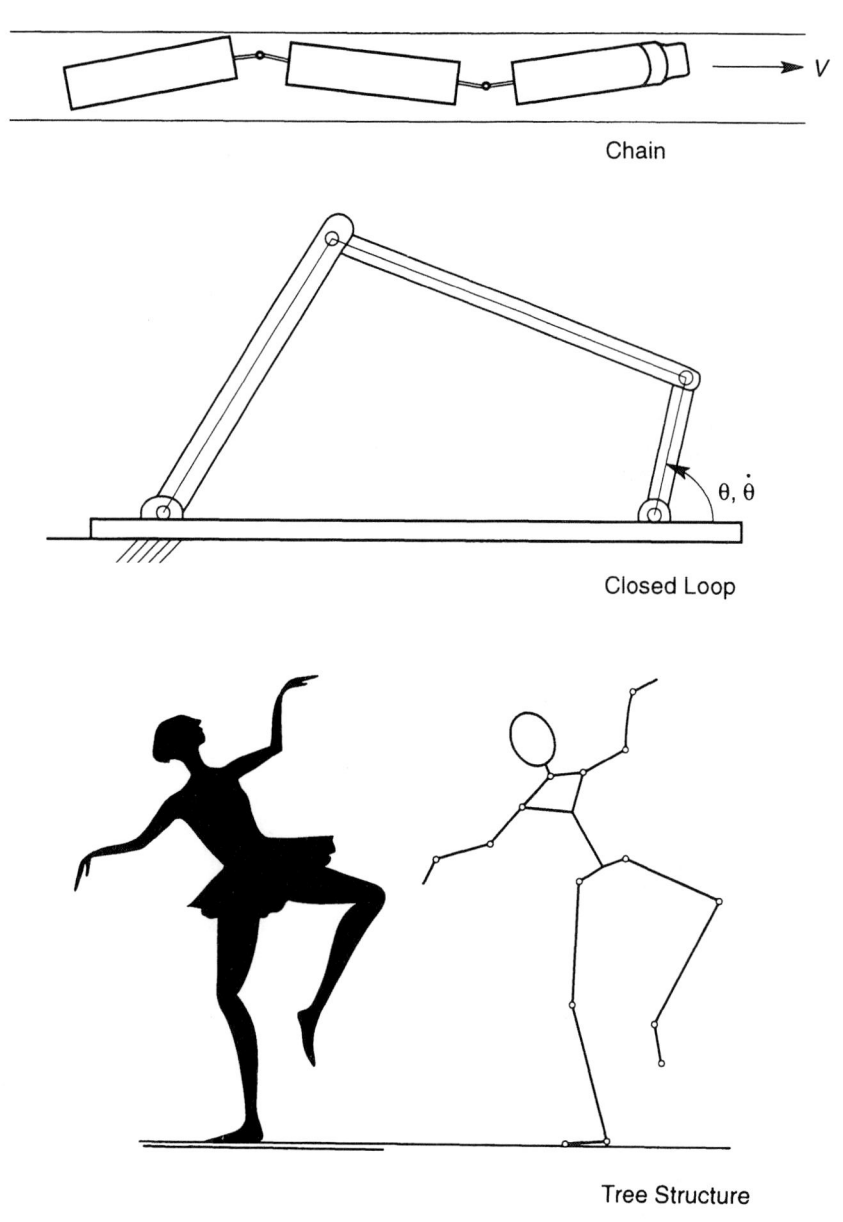

Fig. 1.5 Linked multibody systems: *top*: chain structure; *middle*: closed-loop structure; *bottom*: tree structure.

finite element codes to handle elastic and plastic deformations and vibrations in systems of solid body machines. Collision detection is also area of research.

Another research community that has developed efficient computational methods for multibody systems has been the computer and video games area. Here the goal of video game programming is to develop super fast visual images of solid

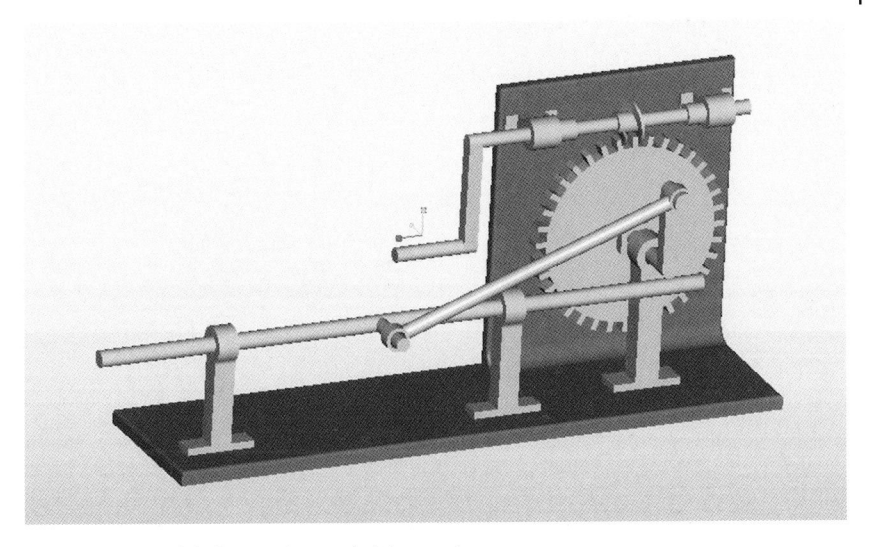

Fig. 1.6 CAD model of worm-drive and slider-crank mechanisms of Leonardo da Vinci, *Codex Madrid*. (See also Moon, 2007.)

object dynamics without regard to accuracy of the physics. However, the visual imagery developed by this group is very impressive and employs not only dynamics modeling but also light and shadow effects. The collision of both convex and nonconvex bodies is of interest as well as damage simulation. The Computer Graphics Lab at Stanford University has some very impressive visuals on its website for example. An early paper on realistic animation of rigid bodies is by Hahn (1988). One of the venues for presentation of this research and dynamic animation products is the SIGGRAPH conference whose proceedings are often published in the journal *Computer Graphics*.

Mechatronics
Other names are used for this class of problems, including controlled machines, smart machines and structures, and intelligent machines. However the term, *mechatronics* is in use in Europe and Japan where mechatronic devices such as magnetic bearings and automated cameras and video equipment have been pioneered. One of the large-scale applications of mechatronics is the Mag-Lev train shown in Fig. 1.7 that has been developed in Japan and Germany. Active magnetic bearings for large and small rotary machines such as gas pipeline pumps and machine tool spindles have been pioneered in Switzerland, France, and Japan (see, e.g., Moon, 1994). In the United States a significant amount of research and development has been directed toward microelectromechanical systems or MEMS (Fig. 1.8).

The distinguishing feature of most of these systems compared to traditional controlled machines, has been the incorporation of *sensing*, *actuation*, and *intelligence* in producing and controlling motion in machines and structures. Thus the engineer is called on to integrate control and intelligence into the mechanical design from

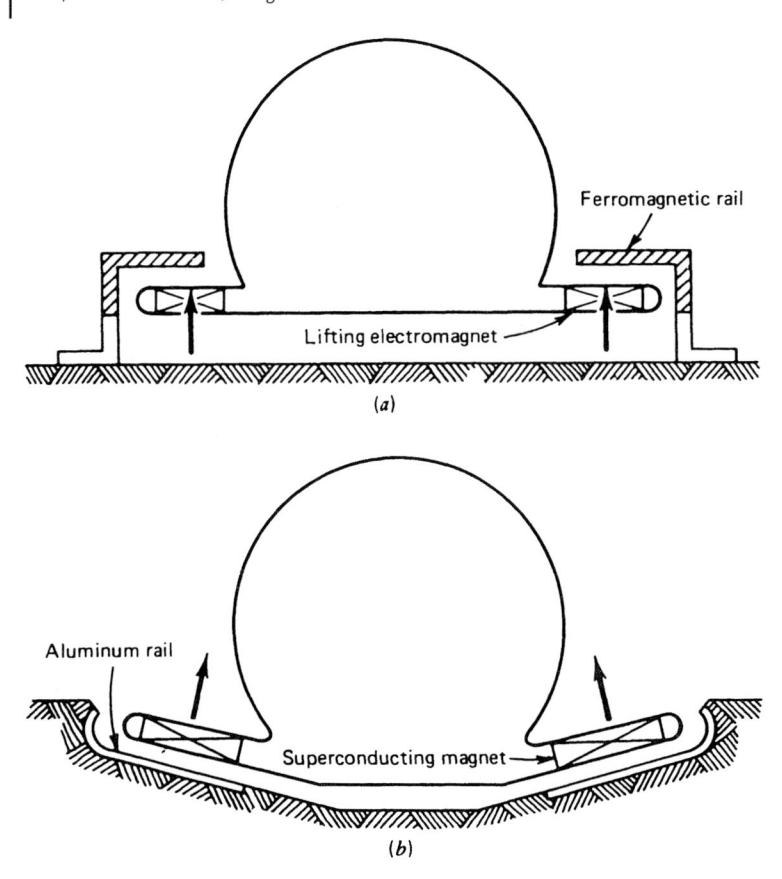

Fig. 1.7 Sketch of two Mag-Lev systems. (*a*) Electromagnetic levitation (EML), which requires electronic feedback control. (*b*) Electrodynamic levitation using superconducting magnets. (From Moon, 1994.)

the very beginning and not as an add-on after the machine is designed. For example, the control motions at the joints of serial robot arms will effect the dynamics of manipulation and must be optimized along with the rest of the system.

MEMS Technology
In the first decade of the new millennium, micro-electromechanical technology or MEMS, has made further advances. Initially employed as micro-accelerometers for automobile airbag crash sensors, there are now a host of applications of this processing technology. Initially implemented in silicon, these micron sized devices can now be made in silicon, polymers and metals. The applications include, automotive accelerometers, micro-gyroscopes, pressure sensors for automobile tires as well as blood pressure devices, digital micro-mirror displays, optical switching devices for communications networks, inkjet printer heads, microfluidics, and bio-

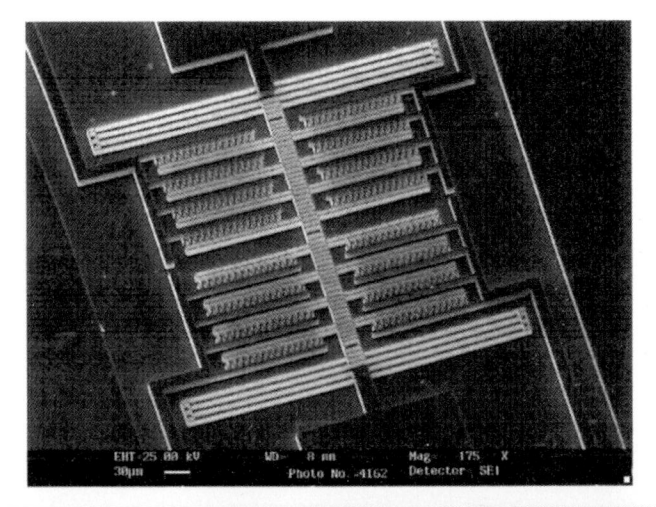

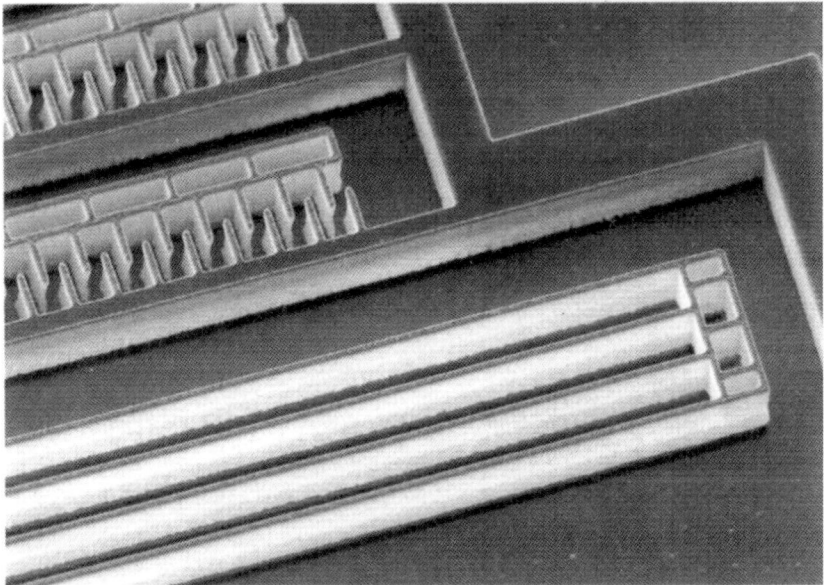

Fig. 1.8 Microelectromechanical (MEMS) accelerometer photoetched from a silicon substrate. See also Chapter 8, Fig. 8.24. (Courtesy Adams, 1996.)

medical applications such as lab-on-a-chip diagnostic devices. Accelerometer, gyro and optical micro-mirror displays involve dynamics design.

One of the principal actuator concepts is a multi-fingered array called a comb-drive electro-actuator as shown in Fig. 1.8. This device acts as a variable capacitor. In Chapter 8 we shall describe how one can determine the generalized electric force from the change of capacitance with the separation of the fingers of the combs.

1 Dynamic Phenomena, Design and Failures

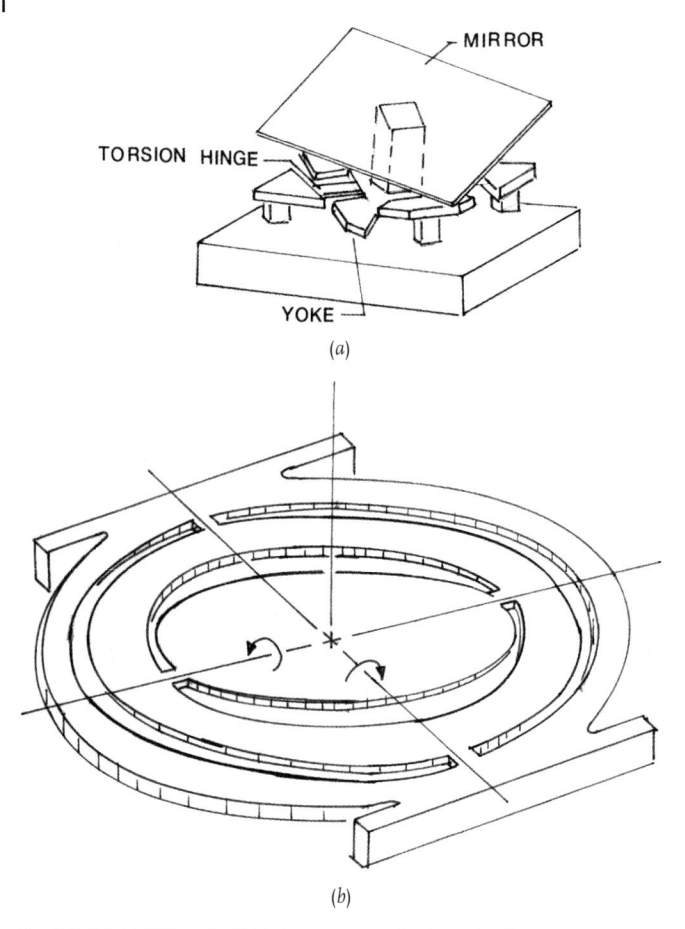

Fig. 1.9 (a) MEMS scale digital mirror compliant mechanism.
(b) MEMS scale mirror on a double compliant gimbal mechanism.

Another dynamic MEMS design element is a digital optical display shown as a schematic in Fig. 1.9a. There are both single and double degrees of freedom. Chips of arrays of such mirrors, 17 μm square, have been made by Texas Instruments with 1280 × 1820 mirror pixels. Electrostatic forces create a torque on the mirror yoke that rotates the mirror +10 or −10 degrees, deflecting a laser beam into or out of a light projection device. These components are known as DMD for *Digital Micro-mirror Devices* and are part of larger systems called DLP for *Digital Light Processing* systems now appearing in flat panel TV screens. Because the mirrors rotate until they hit a stop, dynamically these devices are nonlinear in their equations of motion.

Another sketch of a digital micro-mirror is shown in Fig. 1.9b with a two degree of freedom yoke or gimbals system. Unlike rigid body gyro or universal joint mechanisms, these micro-'mechanisms' are actually very flexible structures called compliant mechanisms. The mirror part of the structures in both Fig. 1.9a and Fig. 1.9b can be modeled dynamically however as rigid body components.

Nanomachines and Nanomotors
Beginning in the 1980's engineers began to make machine-like structures out of silicon at the scale of microns (10^{-6} m). In 1987, K. Eric Dressler published a book, *Engines of Creation*, that envisioned the manipulation of molecular structures at the nano scale (10^{-9} m) inspired by the Nobel Prize winner Richard Feynman's quote about the richness of science at smaller and smaller scales of physics. In the last decade, billions of dollars and euros have been spent on nanotechnology and the question arises as to whether there are problems in the products and processes of this technology that require a knowledge of classical dynamics?

On the one hand at the nanoscale level, many of the physical processes are governed by the laws of quantum mechanics. The interactions involve forces between electron shells of molecules as well as interchange of photons and phonons. However because of the complexities of calculating quantum processes at the molecular level, physicists and chemists have often used the classical limit of quantum physics, using equivalent force potentials and simulations that use the classical laws of Newtonian dynamics. Because of the large number of degrees of freedom in molecular dynamics simulation, there has been a search for so-called low-order models based on classical multi-body dynamics simulation codes. Thus there appears to be a role for the use of classical dynamics at the nano and molecular dynamics scale.

However there are important differences between using dynamics concepts at the molecular or micron scales and macro-size problems. At such small scales, surface forces dominate over inertia effects. Also in a chemical soup of trillions of molecules and atoms, random forces are often dominant in transmitting energy to a group of atoms and molecules that one seeks to organize. Another difference at these dimensions is the very concept of *nanomachine* which some have termed *nanites*. While some have defined 'machine' so broadly that anything that is useful is a machine, in the classical definition we envision a group of structures that move relative to one another. At the classical macro scale these structures are often modeled as rigid bodies. However at the nanometer scale, one must view these structures as molecular structures and chains that can 'bend' as well as rotate and translate relative to other 'soft' chemical structures.

One such soft 'machine' involves a ring of molecules that can translate along a linear chain of other molecules. This 'soft nanomotor' was developed at the University of Edinburgh, in Scotland, and uses sunlight or photons to convert the energy into reciprocating motion of the two molecular chains relative to each other. It is notable that the Scottish research lab is near to where James Watt made his seminal improvements to the macro scale steam engine around 1780.

Nanotechnology research can be divided into biological and non-bio applications. This requires dynamicists trained in classical mechanics to study the basics of molecular biology as well as quantum mechanics if they are to potentially contribute to this new field.

So at this writing there appears to be scientific and technological models of nanomachines that might require the use of classical dynamic modeling concepts coupled to constructs of statistical physics, quantum mechanics and chemical structures.

New Methods of Computation

The development of microprocessors, workstations, and computer graphics has resulted in a plethora of codes and software to solve particle, rigid body, and coupled elastic body dynamics problems (see Appendix B and Schiehlen, 1990). An example is shown in Fig. 1.9. Often, however, the methods of formulation and solution of dynamic problems are not transparent to the user who cannot assess the limitation of such codes.

These new codes allow one to specify geometry, materials, and the type of joints and constraints. The derivation of equations of motion is hidden from the user. The integrated motions are presented in the form of computer-generated graphs and movie frame simulation of the moving objects. Two of the major workstation codes in use are *DADS* and *ADAMS* and another for the PC is *SOLID WORKS*; an example is shown in Fig. 1.6.

An intermediate level code for dynamic analysis is MATLAB. Here the user must provide the equations of motion and MATLAB will numerically integrate and display the results graphically. A few examples of the use of MATLAB are given in this text.

In addition to these numerical codes, however, software to perform symbolic mathematical operations have appeared, such as *MACSYMA*, *MAPLE*, and *MATHEMATICA*. A few examples of the use of these codes to derive equations of motion are given in this text.

New Experimental Tools

In the last two decades the ability of the dynamicist and the engineer to observe and measure dynamical systems has been greatly enhanced by new sensors, signal processing, and actuator devices for control.

Sensing and Measurement Systems

The two most important developments have been optical sensing systems and video recording. These include:

- Optical follower cameras for large motions
- Optical velocimeter
- Fiber-optic proximity systems for small motions
- Optical encoders for rotary and linear motions
- Holographic methods for elastic deformation
- High-speed framing cameras
- X-Ray methods
- Portable video scanning cameras

Signal Processing Methods
There are four basic signal-processing systems for storing data from dynamical systems; *analog* and *digital oscilloscopes, dedicated signal analysis electronics,* and *computer hardware,* such as hard drives. These devices are either directly connected to sensing electronics or a buffer is used, such as a data-acquisition system. Many measurement systems use a direct input to a computer or microprocessor where the data are analyzed in software.

One of the important advances in dynamic measurement has been the invention of the fast Fourier transform chip (FFT). This allows on-line calculation of dynamic Fourier transforms or spectra, which in the past required post-data-acquisition processing.

System Identification: Linear Systems
This field has seen the development of software tools to allow the construction of a linear mathematical model (usually differential or difference equations) from a digitally sampled dynamic data record. In some systems where the data are sampled at different spatial locations on the machine or structure, the spatial eigenmodes can also be reconstructed. In control systems that use model-based feedback, algorithms such as Kalman filters are used to obtain an approximate model of the system under control.

System Identification: Nonlinear Systems
Methods for nonlinear dynamical systems are under continual development. In recent years the development of Chaos Theory (Chapter 9) has produced an evolving methodology based on constructing the topological and geometric features of the dynamical object in phase space. New techniques based on time series sampling of dynamic records such as *fractal dimension* and *false nearest neighbors,* allow us to place a bound on the dimension of the phase space (see Moon, 1992; Abarbanel, 1996).

Other methods, such as *Poincare maps* (Chapter 9) and *bifurcation diagrams,* have been used to characterize the nature of motions in nonlinear systems.

1.3
Dynamic Failures

While dynamic analysis in engineering is often used to create motions in physical systems, in many cases unwanted dynamic failures are to be avoided. Such failures include:

- Large deflections
- Fatigue
- Motion-induced fracture
- Dynamic instability, e.g., flutter, chatter, wheel shimmy
- Impact-induced local damage

- Dynamic buckling in structures
- Motion-included noise
- Instability about a steady motion, e.g., wheel on rails
- Thermal heating due to dynamic friction

A few of the most important failure modes are fatigue, dynamic instability, and noise.

Fatigue

Fatigue is an insidious type of failure in machines and structures because it does not result from either large forces or excessive stresses. Instead, it is the result of many small oscillations whereby microcracks in the material grow to form a large failure-producing crack. In some materials the failure may occur only after millions of cycles. An example of a fatigue crack is shown in Fig. 1.10. Fatigue sometimes occurs near so-called "stress risers," corners, holes, etc. But fatigue can also occur through repeated dynamic contact such as in gear teeth in transmissions or rolling problems in rail-wheel systems. A standard plot of dynamic stress vs. number of cycles to failure is shown in Fig. 1.11 for aluminum and titanium. In the case of steel there sometimes exists a dynamic stress below which cracks don't grown, a so-called "fatigue limit stress." In the case of aluminum, however, a limit doesn't exist, so any level of repeated stress will eventually lead to failure. Fatigue is one reason why the study of small vibrations is such an important subgroup of dynamics. (See, e.g., Frost et al., 1974.)

Dynamic Instabilities

Many machines are designed to operate in a steady dynamic state characterized by a constant velocity, such as an aircraft or constant rotation as in turbines and generators. A satellite pointing toward the earth involves both steady translation and rotation. One classic failure mode in these machines is a significant departure from steady state. Examples include:

- Wing flutter
- Machine tool chatter
- Vehicle skidding, wheel shimmy
- Train derailment
- Uncontrollable spin of aircraft
- Electro-magnetic launcher dynamics

In many of these systems, a linear stability analysis can be performed. As an example consider the lateral-yaw motions of a Mag-Lev vehicle moving along a guideway, shown in Fig. 1.12 (see Chapter 8, also Moon, 1994). The lateral motion of the center of mass, u, and the yaw, ϕ, can be modeled as two coupled oscillators:

1.3 Dynamic Failures

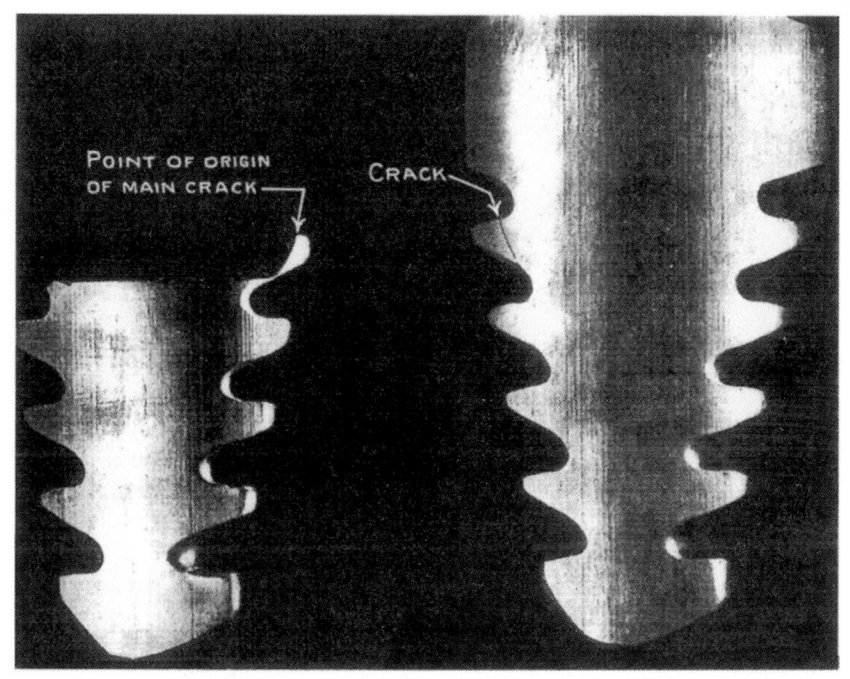

Fig. 1.10 Fatigue failure at the root of a turbine blade. (From Hutchings and Unterweiser, 1981.)

$$\begin{bmatrix} m & 0 \\ 0 & I \end{bmatrix} \begin{bmatrix} \ddot{u} \\ \ddot{\phi} \end{bmatrix} + \begin{bmatrix} d_{11} & d_{12} \\ d_{21} & d_{22} \end{bmatrix} \begin{bmatrix} \dot{u} \\ \dot{\phi} \end{bmatrix} + \begin{bmatrix} k_{11} & k_{12} \\ k_{21} & k_{22} \end{bmatrix} \begin{bmatrix} u \\ \phi \end{bmatrix} = 0 \qquad (1.1)$$

where some of the constants depend on the steady velocity parameter, V_0. In linear stability analysis, one looks for a solution of the form

$$\begin{bmatrix} u \\ \phi \end{bmatrix} = \begin{bmatrix} A \\ B \end{bmatrix} e^{st} \qquad (1.2)$$

This problem results in a so-called *eigenvalue problem*. In general there might be two such solutions, i.e., two sets of $\{A_i, B_i, s_i\}$ where $s_1 = f(V_0), s_2 = g(V_0)$. These values can be complex, i.e., $s_i = \alpha_i + i\beta_i$.

When $\alpha_i = 0$, we say the mode is oscillatory; $\alpha_i < 0, \beta_i \neq 0$, the system is damped oscillatory, and $\alpha_i > 0$, the system is unstable. Two classes of instability can be observed in many problems: $\beta_i = 0, \alpha_i > 0$; this case is called *divergence, buckling*, or a *static instability*. The case $\beta_i \neq 0, \alpha_i > 0$ is called *"flutter"* or a *dynamic instability*. In nonlinear dynamics, this oscillation instability is called a *Hopf bifurcation* (see Chapter 9).

In either case, the motion cannot become unbounded, and eventually will be limited by either a sudden failure, e.g., a bearing seizure, or will evolve to another steady state motion determined by the nonlinear effects in the problem (see Chapter 9).

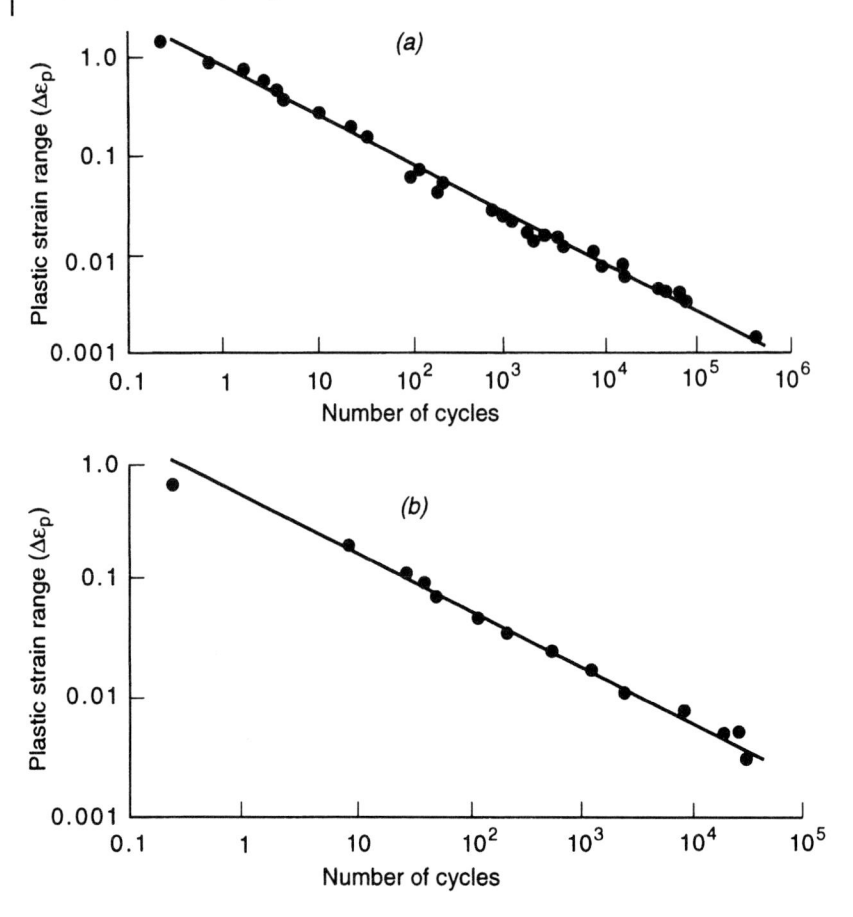

Fig. 1.11 S-N Curve: Plastic strain vs. cycles to failure. *Top*: aluminum, commercially pure; *bottom*: titanium, commercially pure. (From Frost et al., 1974.)

Dynamic Generation of Noise

Although stresses and motions can be small in dynamical machines, excessive noise can produce an unacceptable product as in the case of automobiles, printing machines, dishwashers, and components for submarines. Two dynamic problems shown in Fig. 1.13 can produce noise: resonance and impact. In the impact problem a multibody system undergoes periodic, random, or chaotic impact, generating structure-borne noise, which then couples to some resonant part of the machine structure to create airborne noise. Resonance generation of noise is often associated with rotating components. Other noise-generation mechanisms are rolling, friction and fluid-structure interaction.

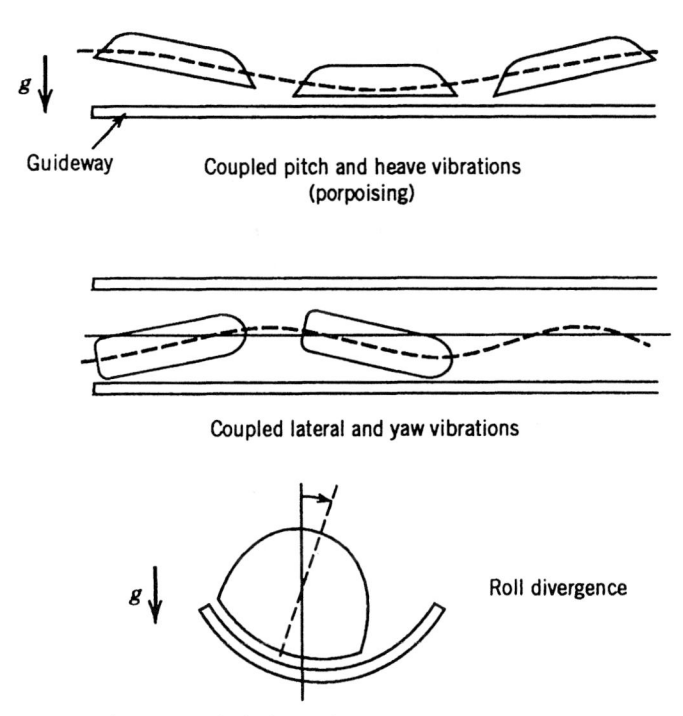

Fig. 1.12 Three types of vehicle instabilities in a Mag-Lev vehicle: *top*: porpoising (coupled pitch and heave); *middle*: snaking (coupled lateral and yaw); *bottom*: roll divergence. (From Moon, 1994.)

Avoidance of Dynamic Failure

The engineer has a number of options to prevent dynamic failure in a machine or structure. A short list would include the following:

- Understand the dynamics *before* the design becomes a product, using simulation tools or experiments
- Choose materials with enhanced properties to resist fatigue, fracture or wear, or choose higher damping materials to minimize resonance
- Use passive damping
- Use active control
- Use internal diagnostics, sensors, limit switches, etc., to detect imminent failure and avoid catastrophe

1.4
Basic Paradigms in Dynamics

The general principles of dynamics used in this text stem from Newton's law of linear momentum and Euler's extension of this principle to angular momentum.

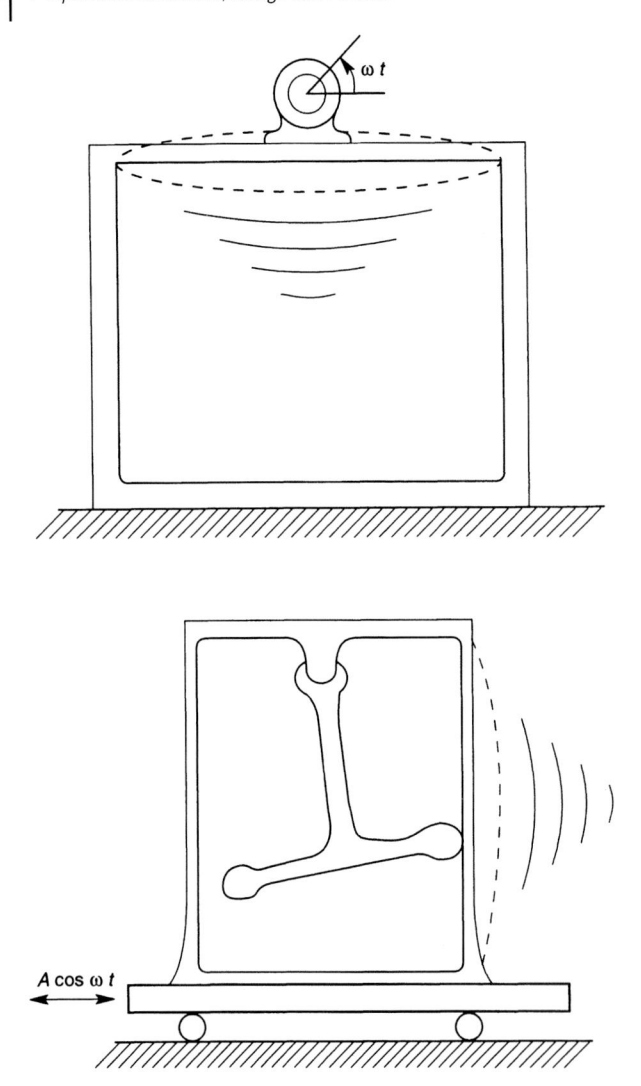

Fig. 1.13 Sketches of vibration induced noise sources. *Top*: resonance; *bottom*: impact.

In this text we address the classical dynamics of particles, systems of particles, and rigid bodies. The Newton–Euler principles can be codified into two mathematical equations

$$\mathbf{F} = m\frac{d\mathbf{V}_c}{dt} \tag{1.3a}$$

$$\mathbf{M}_c = \frac{d}{dt}\mathbf{H}_c \tag{1.3b}$$

where the subscript denotes the velocity of the center of mass $\mathbf{V}_c$; the moment of the forces about the center of mass $\mathbf{M}_c$; and the angular momentum about the cen-

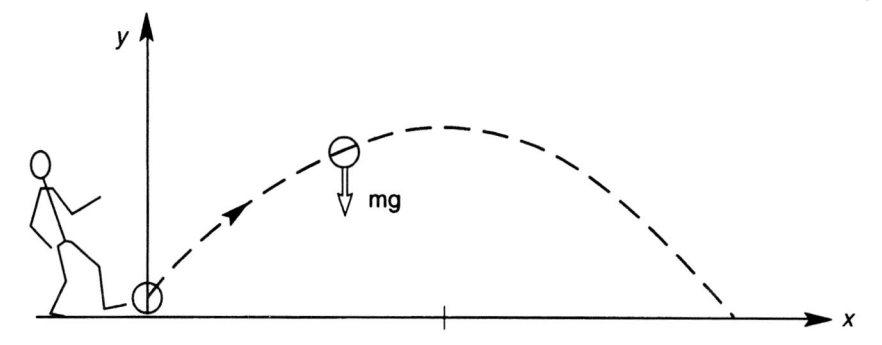

Fig. 1.14 Parabolic path of a projectile under gravity without aerodynamic drag.

ter of mass $\mathbf{H}_c$. However, most students do not learn dynamics by starting with these vector differential equations. Instead the novitiate begins with simple examples and simple mathematical models. A few of these simple models are reviewed in this section;

1. Motion under a constant force
2. The pendulum
3. The linear oscillator and resonance
4. Gyroscopic motion
5. Central-force motion

Motion Under a Constant Force

Motion under a constant force is a classic problem studied by Galileo Galilei (1564–1642) and by every high school student. If the constant force is gravity, then the equation of motion is given by (see Fig. 1.14):

$$m\dot{\mathbf{v}} = -mg\mathbf{e}_y \tag{1.4}$$

(In this text an overdot will indicate a total derivative with respect to time.) This equation can be separated into two scalar equations (for planar motion) and integrated in time to obtain:

$$v_y = -gt + v_{oy}$$
$$v_x = v_{0x}$$
$$x = x_0 + v_{ox}t$$
$$y = y_0 + v_{oy}t - \frac{1}{2}gt^2 \tag{1.5}$$

The motion takes the form of a parabola, as shown in Fig. 1.14. The addition of air-drag to this problem can significantly alter these results as any football, soccer, tennis, or baseball player knows.

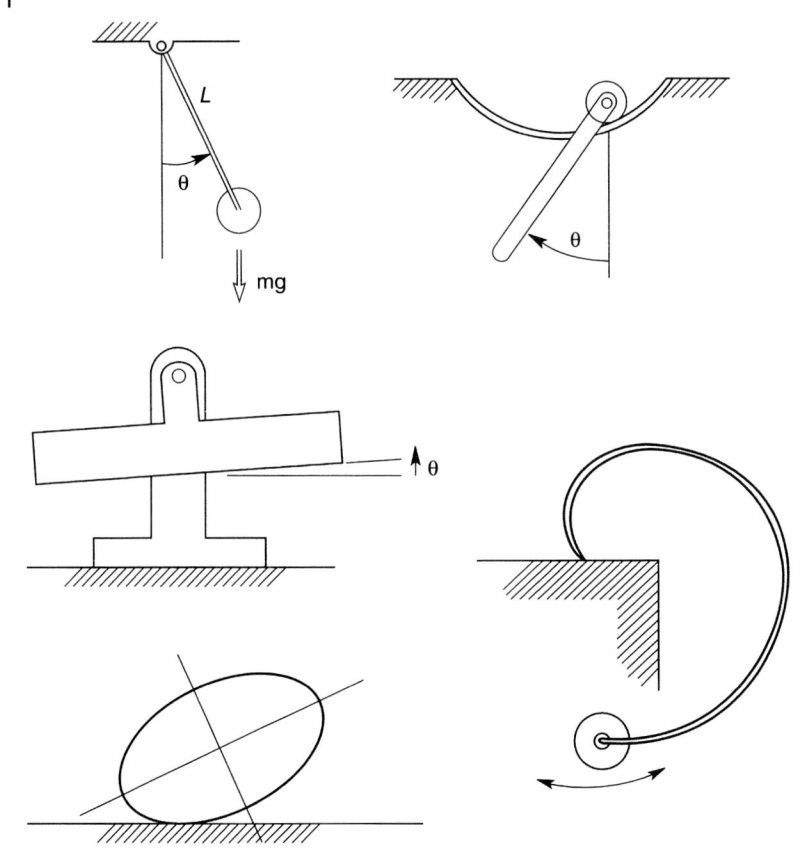

Fig. 1.15 Pendula-type motions.

Pendulum

The planar pendulum (top left Fig. 1.15) is a single-link, single-degree-of-freedom device with circular motion under the force of gravity. The human arm locked at the elbow is an example of pendulum system, albeit without the planar constraint. The basic equation of motion involves taking the cross product of Newton's law to obtain a planar version of the law of angular momentum (1.3b)

$$\dot{H}_z = M_z \tag{1.6}$$

where H_z is the angular momentum of the mass distribution of the link and M_z is the moment of the gravity force. For a link with a mass concentration at a distance L from the joint, the equation of motion takes the form

$$\ddot{\theta} + (g/L)\sin\theta = 0 \tag{1.7}$$

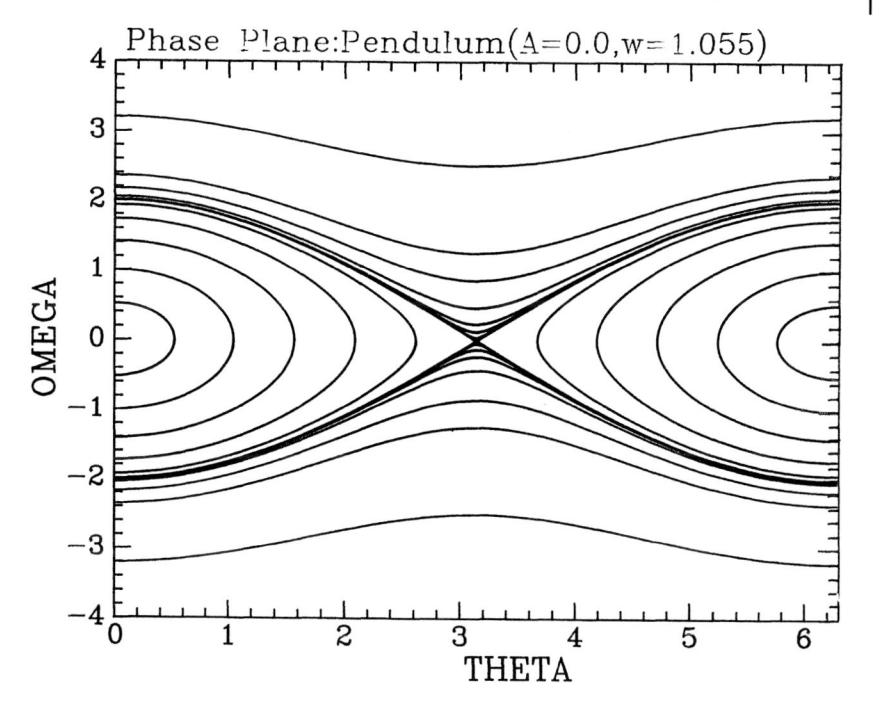

Fig. 1.16 Phase-plane trajectories of the pendulum. Enlarge with a copy machine, cut out, and join $\theta = 0, 2\pi$ to form a cylindrical phase space. Time flow for Omega > 0, left to right.

This is a nonlinear ordinary differential equation. The nature of the solution can be seen by looking at the phase plane $(\theta, \dot{\theta})$ in Fig. 1.16. There are oscillatory notions about $\theta = 0$, and for initial conditions outside the so-called separatrix curve in Fig. 1.16, the motions are circulatory, either clockwise or counterclockwise. The solution near $\theta = \pi$, is statically unstable. When the angle is small we have

$$\ddot{\theta} + \frac{g}{L}\theta = 0 \tag{1.8}$$

and the solution is given by

$$\theta(t) = A \cos \omega_0 t + B \sin \omega_0 t \tag{1.9}$$

where $\omega_0^2 = g/L$. The natural frequency in cycles per second is given by

$$f = \frac{1}{2\pi}\sqrt{g/L} \tag{1.10}$$

and we can see that the longer the length L, the slower the oscillation. In pendulum or gravity forced systems, the natural frequencies do not depend on the mass.

The Linear Oscillator: Resonance

Many flexible structures, such as a cantilevered beam, offer a resistance to motion that can be modeled as a linear reaction force or spring force. A classic dynamics paradigm is the single-degree-of-freedom motion of a mass connected to a linear spring and viscous damper (Fig. 1.17). This could be a model for a simple structure under base motion, as in an earthquake:

$$m(\ddot{x} + \ddot{b}(t)) = -kx - c\dot{x} \tag{1.11}$$

When the base motion is sinusoidal, the standard form of the equation of motion becomes

$$\ddot{x} + 2\gamma\dot{x} + \omega_0^2 x = A \cos \Omega t \tag{1.12}$$

where $\omega_0^2 = k/m$, $\gamma = c/2m$.

When the damping and forcing are both zero, $\gamma = 0$, $A = 0$, there are two sinusoidal solutions 90° out of phase:

$$x = C_1 \cos \omega_0 t + C_2 \sin \omega_0 t \tag{1.13}$$

The frequency of oscillation in cycles per second is given by, $\omega_0/2\pi$. When the damping is small, the frequency in radians per second is given by $(k/m)^{1/2}$, where we assume in this book standard international units of meter, kilogram, seconds, so that k is in Newton's per meter and m is in kilograms.

When small damping is present, the motion becomes a damped sinusoidal motion

$$x(t) = e^{-\gamma t}[C_1 \cos \omega_1 t + C_2 \sin \omega_1 t] \tag{1.14}$$

where

$$\omega_1^2 = \omega_0^2 - \gamma^2 \tag{1.15}$$

In the study of ordinary differential equations this solution is called a *homogeneous solution*.

When forcing is present, $A \neq 0$ in (1.4), one can show that the *particular solution* is also of the same form as the driving signal but with a phase shift. The maximum amplitude of the motion exhibits a classic resonance effect shown in Fig. 1.18 for small damping. Thus when the driving frequency is close to the natural frequency ω_1 or ω_0, large amplitude motions can occur that may lead to failure.

Phase Plane

Another way of viewing the motion of an oscillation is the phase plane. For the unforced, damped oscillation we rewrite the equation of motion (1.12) with $A = 0$ as two first-order equations

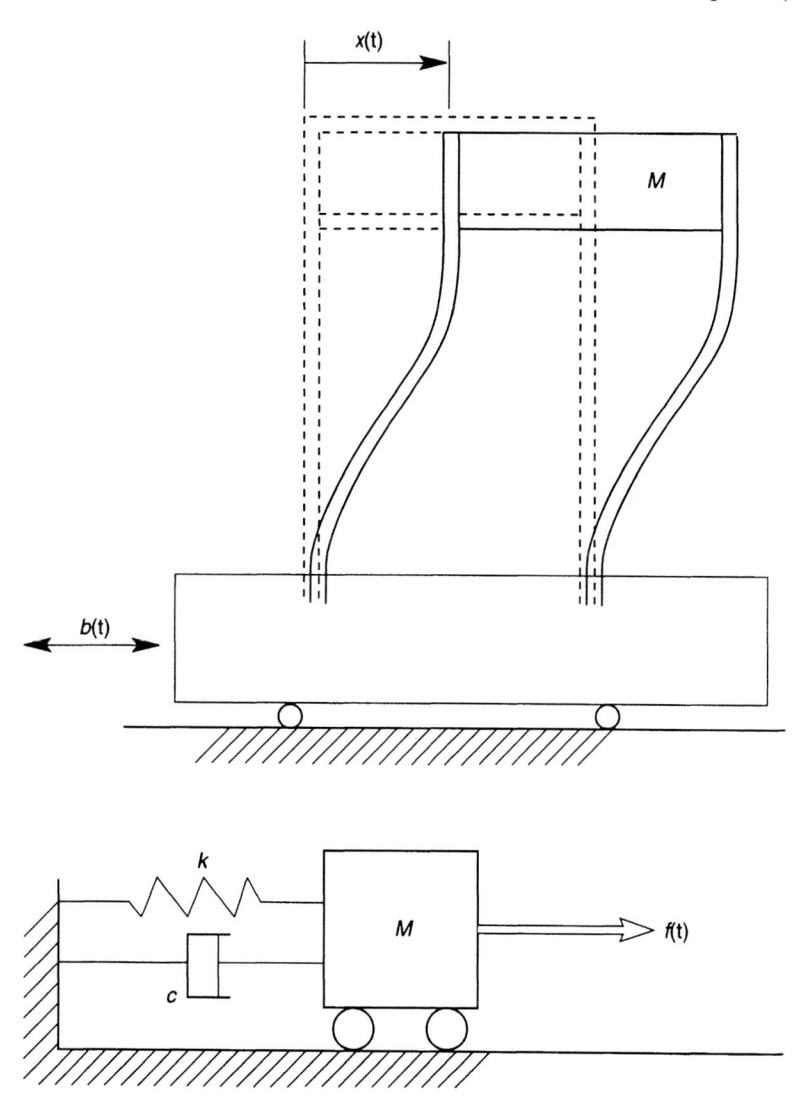

Fig. 1.17 *Top*: Model for earthquake excitation of a structure. *Bottom*: Forced vibration of a spring-mass oscillator.

$$\dot{x} = v$$
$$\dot{v} = -2\gamma v - \omega_0^2 x \tag{1.16}$$

Thus we can define a *state vector* $[x, v]$, and $[\dot{x}, \dot{v}]$ becomes the rate of change of that state vector. When damping is zero, $\gamma = 0$, the trajectories of motion become ellipses in this state space or phase plane as shown in Fig. 1.19a. When damping is present, this motion becomes a damped spiral (Fig. 1.19b). The phase space representation is especially useful for analysis of nonlinear systems (Chapter 9).

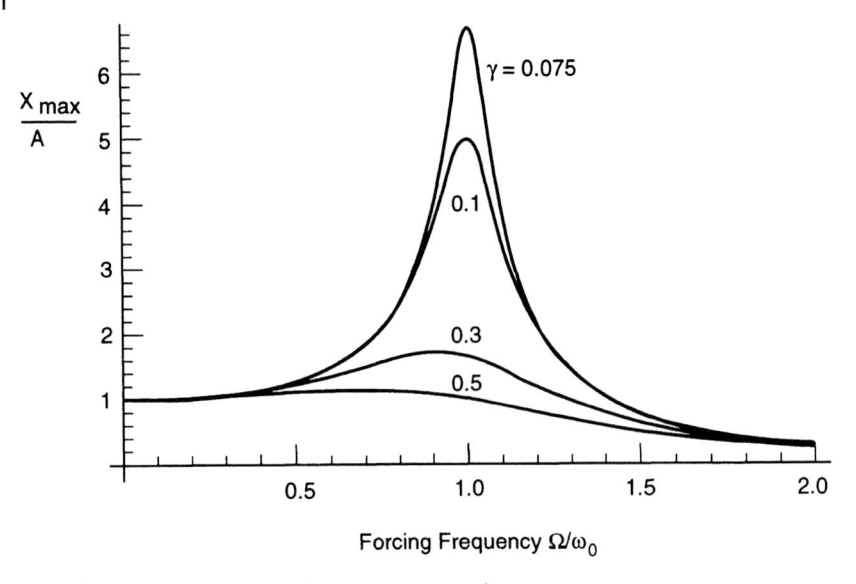

Fig. 1.18 Resonance curves for a forced-spring-mass-damper oscillator, for different damping ratios, γ.

Gyroscopic Motion

When a rigid body is in pure rotation about a point, the equation of motion relates the change in angular momentum, $\dot{\mathbf{H}}$, to the applied force moment $\mathbf{M}$, or

$$\mathbf{M} = \dot{\mathbf{H}} \tag{1.17}$$

The example of a cylindrically symmetric body with steady rotation about two of its principal axes illustrates the simplest example of the gyroscopic effect. Referring to Fig. 1.20, the rotation vector can be written in the moving axis frame $\boldsymbol{\omega} = \dot{\phi}\mathbf{e}_\phi + \dot{\psi}\mathbf{e}_\psi$. The rotation about the body symmetry axis is called the *spin*, $\dot{\phi}$, and the rotation about the vertical axis, $\dot{\psi}$, is called the *precession*. The angular-momentum vector $\mathbf{H}$ can be shown to lie in the x-z plane where $\mathbf{e}_x = \mathbf{e}_\phi, \mathbf{e}_z = \mathbf{e}_\psi$ (Chapter 5), i.e.,

$$\mathbf{H} = I_x \dot{\phi} \mathbf{e}_x + I_z \dot{\psi} \mathbf{e}_z \tag{1.18}$$

It is shown in Chapter 5 that $\dot{\mathbf{H}} = \dot{\psi}\mathbf{e}_z \times \mathbf{H}$.

Thus the angular-momentum vector changes in a direction normal to the x-z plane, i.e., the moment must have a y component, $\mathbf{M} = M_y \mathbf{e}_y$. It can be shown that Eq. (1.17) then becomes, $M_y = \omega_z H_x$, or

$$M_y = I_x \dot{\psi} \dot{\phi} \tag{1.19}$$

Thus we have a *troika* relationship; the y component of $\mathbf{M}$ is equal to the product of the z component of $\boldsymbol{\omega}$ and the x component of $\mathbf{H}$. This gyroscopic troika is revealed

Zero Damping

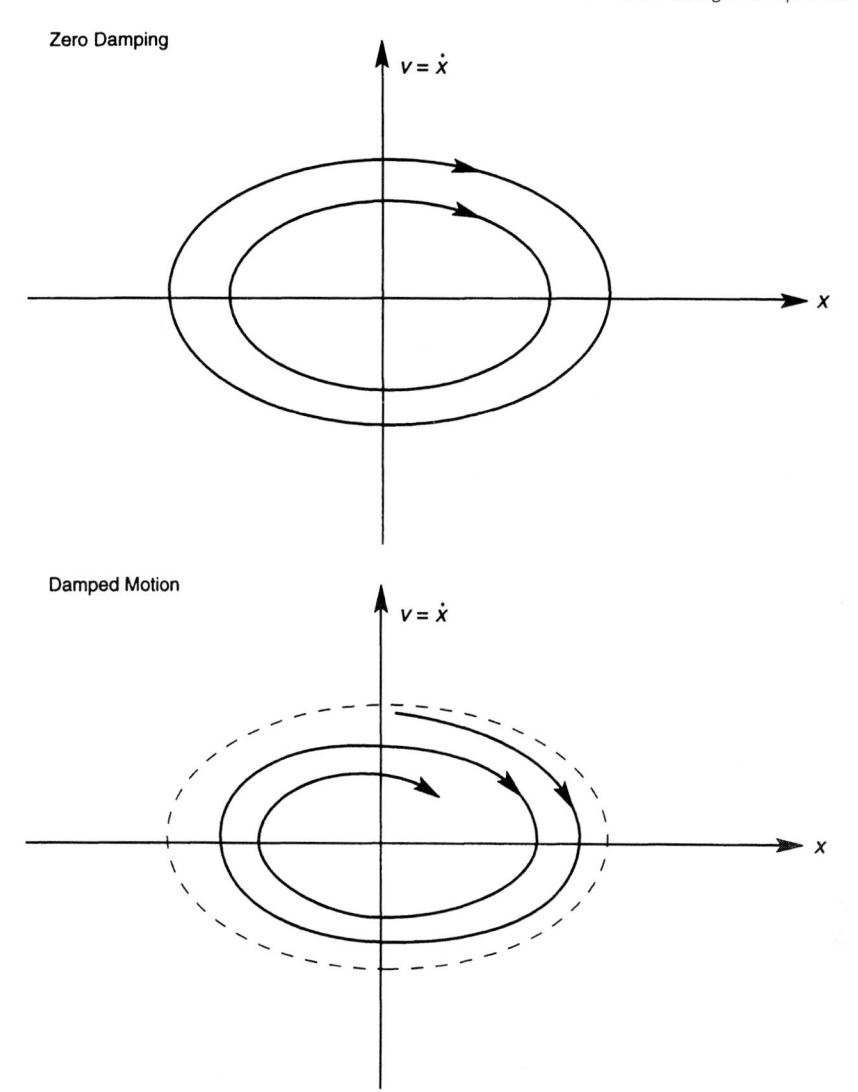

Damped Motion

Fig. 1.19 Phase-plane trajectories for the unforced, spring-mass oscillator. *Top*: without damping; *bottom*: with small damping.

in the remarkable motion of a top or gyroscopic in which the moment is created by the force of gravity, i.e., $M_y = mgL$. The spin rate $\dot{\phi}$, and the precession rate $\dot{\psi}$ are related by the hyperbolic relation.

$$\dot{\psi} = \frac{mgL}{I_x \dot{\phi}} \tag{1.20}$$

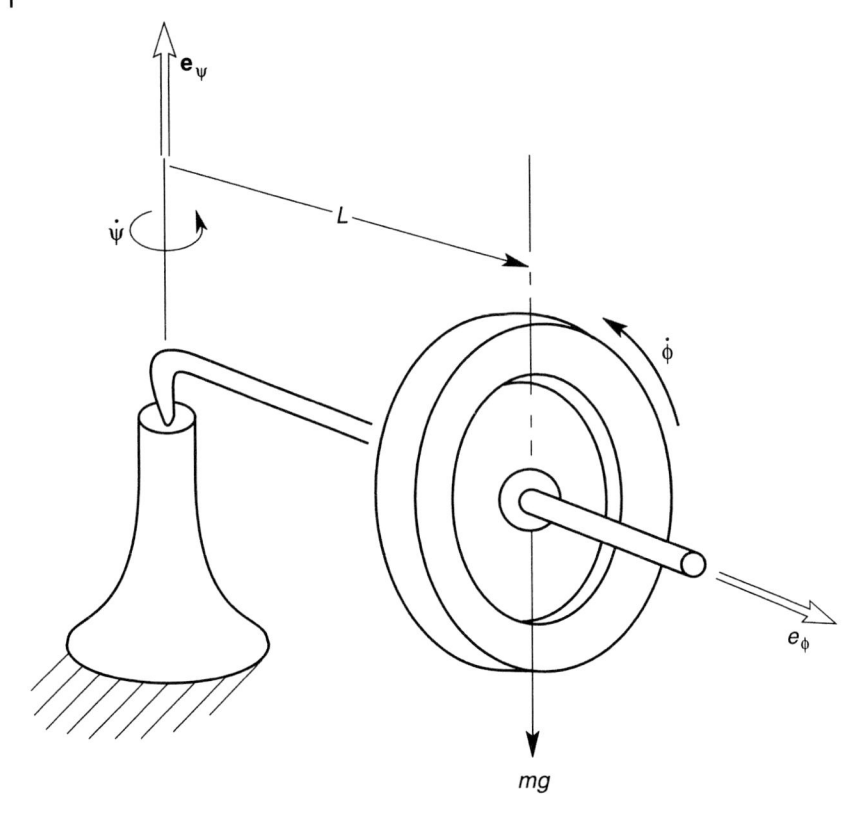

Fig. 1.20 Gyroscopic motion of a spinning wheel.

The reader is encouraged to buy a small toy gyro and demonstrate this effect. The gyroscopic moment is present whenever rotating bodies are turned about two principal axes such as a bicycle wheel turning left or right. One application shown in Fig. 1.21 is the retraction of a landing wheel on an aircraft wing.

Central-force Motion

The first dynamics example the student often studies is circular motion of a particle under a constant radial force. One learns that the acceleration is v^2/r and the tension is given by $T = mv^2/r$. Although the velocity is tangential to the circle, the change in the velocity vector is in the radial direction pointing toward the center.

The second example of central-force motion that the student encounters is orbital motion about a gravitational center (Fig. 1.22). This problem goes back to the work of Johannes Kepler (1571–1630) who "discovered" the elliptical orbits of orbiting bodies, and Sir Isaac Newton (1642–1722), who developed Kepler's three laws into his mathematical laws of motion and law of gravitation. The classical

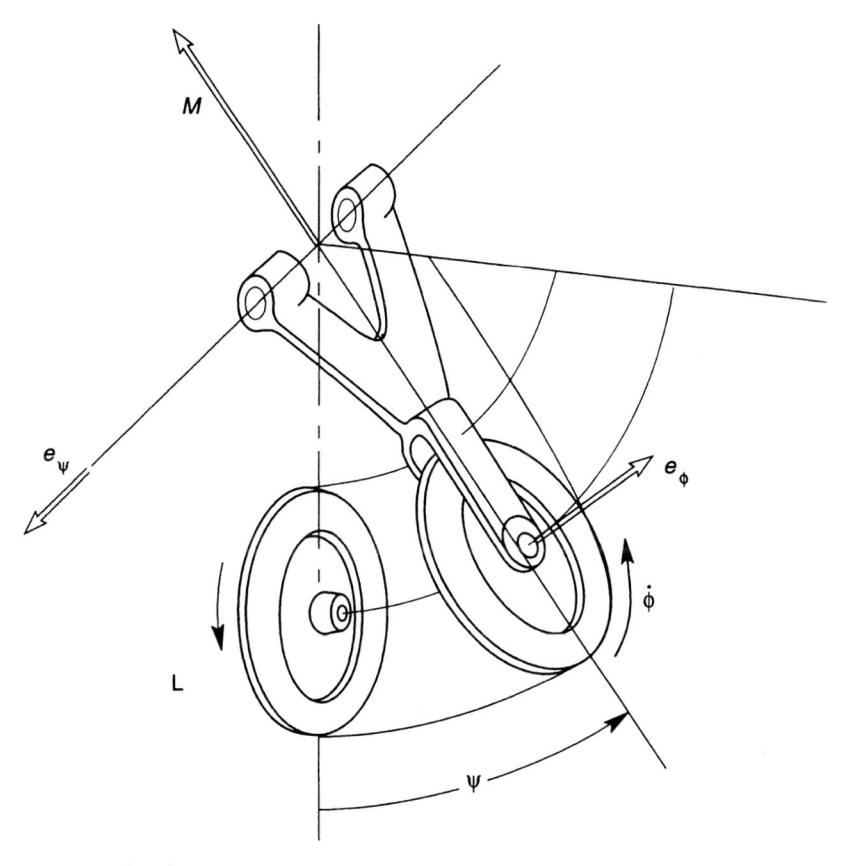

Fig. 1.21 Induced gyroscopic moment, **M**, on a retracting aircraft wheel.

Newtonian force law of gravity assumes an inverse-square law between bodies of masses m_1, m_2:

$$F = \frac{Gm_1 m_2}{r_{12}^2} \tag{1.21}$$

where r_{12} is the radial distance between the two gravitational centers of m_1, m_2. The complete problem is discussed in Chapter 7. However, one result is easy to see, namely the conservation of angular momentum. For all radial-force laws, the total circumferential acceleration must be zero, since there is no force in this direction. Using polar coordinates the circumferential acceleration becomes

$$a_\theta = r\ddot{\theta} + 2\dot{r}\dot{\theta} = 0 \tag{1.22}$$

or $(r^2 \dot{\theta})$ is constant in time.

Taking the cross product of Newton's law (1.3a) and the radial vector, one also finds that

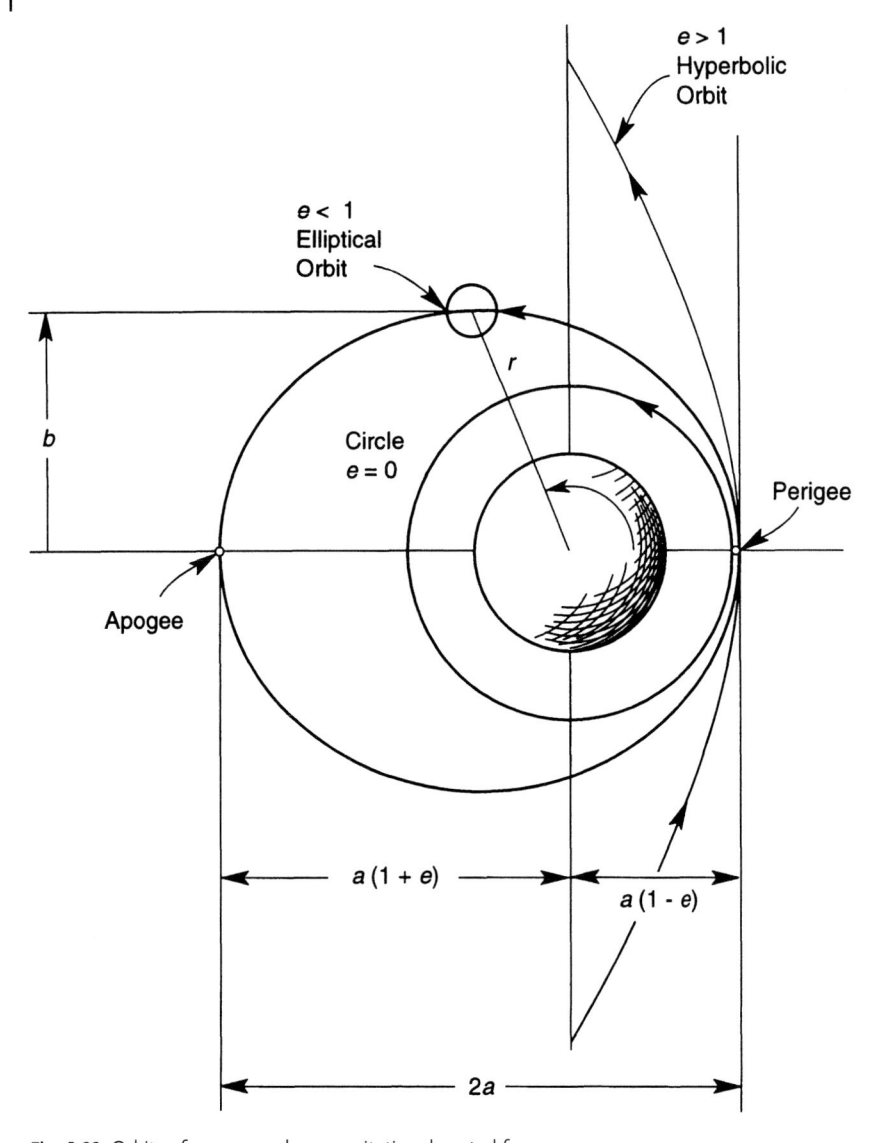

Fig. 1.22 Orbits of a mass under a gravitational central force.

$$\frac{d}{dt}(mr^2\dot{\theta}) = \frac{d}{dt}(H_z) = 0 \tag{1.23}$$

This says that the angular momentum, **H**, is conserved. Writing $\dot{\theta} = \omega$, we see that the angular velocity for all central-force motions is inversely proportional to the square of the radius:

$$\omega = \frac{H_z}{mr^2} \tag{1.24}$$

The bounded motion under gravity is either a circle or an ellipse (Fig. 1.22). When m_1 is much larger than m_2, the equation for a circular orbit is obtained by equating the radial acceleration $\omega^2 r$ to the gravitational force (1.21) per unit mass:

$$\omega^3 = Gm_1/r^3$$

or since $v = r\omega$,

$$v = (Gm_1/r)^{1/2} \tag{1.25}$$

For a low earth orbit, this velocity is on the order of 8000 m/s. The period is around 90 min (see also Table 7.1).

Electromagnetic Forces

There are basically two types of electromagnetic effects in dynamics modeling; direct electric or magnetic forces that appear in Newton's equations of motion or indirect deformations such as piezoelectric effects that appear in stress–strain and voltage–dielectric relations.

Direct Effects

Electromagnetic forces act between masses separated in space. To represent the effect of one charged or current carrying mass on another such mass, the concepts of electric field **E**, and magnetic field **B**, were introduced. For example if a mass m_1 carries a charge Q_1, the force produced by a second charge Q_2, that produces a field $\mathbf{E}_2$, is given by

$$\mathbf{F}_1 = Q_1 \mathbf{E}_2 \tag{1.26}$$

In a similar way if a mass m_1 carries a current loop with current I_1, the effect of the magnetic field produced by another current circuit I_2, or other source of magnetic field $\mathbf{B}_2$ is given by the integral

$$\mathbf{F}_1 = \oint I_1 \times \mathbf{B}_2 \, ds_1 \tag{1.27}$$

where the integral is carried around the circuit of current I_1. One can see that if the magnetic field $\mathbf{B}_2$, is uniform, i.e. its gradient in the vicinity of m_1 is zero, then the magnetic force is zero if we assume that the circuit loops back on itself. On the other hand, one can show that there will still be a magnetic couple $\mathbf{M}_1$, produced by a uniform field $\mathbf{B}_2$, on a closed current circuit given by

$$\mathbf{M}_1 = \mathbf{m}_1 \times \mathbf{B}_2 \tag{1.28}$$

where

$$\mathbf{m}_1 = I_1 \oint \mathbf{r} \times \mathbf{ds}_1$$

The student should not worry about the details of these formulas at this stage, though electromagnetic forces can be quite intimidating at first. In Chapter 8 we shall outline a method of calculating electromagnetic forces using electric and magnetic energy functions, that will in many cases simplify the determination of electromagnetic forces and moments.

In general, electric forces on charges are small except at very small scales such as in MEMS or NEMS problems. In general, magnetic forces are more effective in large scale problems such as magnetic levitation of trains or magnetic launchers.

There are two other principal sources of direct electromagnetic forces; forces on electric polarized materials and forces on magnetized materials. These are discussed in Chapter 8.

1.5
Coupled and Complex Dynamic Phenomena

Coupled Fields

Some of the most exciting problems in applied dynamics arise from the coupling of dynamics to other fields of science. Some of these problem areas are listed below:

- Fluid mechanics and dynamics, e.g.,
    - Aircraft dynamics
    - Submarine, wave-ship motions
    - Flutter of flexible structures
    - Turbine or pump dynamics
- Electromechanical dynamics, e.g.,
    - Levitated vehicles and trains
    - Magnetic bearings
    - Linear actuators
    - Electro-magnetic launchers
    - Sensors, accelerometer
    - Motors, generators
    - Microelectromechanical systems (MEMS)
- Control systems, e.g.,
    - Robotic systems
    - Active vehicle control
- Orbital dynamics, e.g.,
    - Spacecraft altitude stability
    - Vibrations of flexible structures in space
    - Deployment dynamics of space structures
- Biomechanical dynamics, e.g.,
    - Walking machines
    - Animal locomotion
    - Wheelchair stability
    - Prosthetic design

- Dynamics in thermal and energy systems, e.g.,
    Combustion machines
    Thermoelastic effects
    Wind energy machines

Another set of problems that provide challenges for the next generation of dynamicists involve material and geometric complications. Some of these problems are listed below.

Complex Dynamics

- Material nonlinearities, e.g.,
    Friction forces
    Impact problems
    Fracture and fatigue
    Elastoplastic forces, inelasticity
    Cutting forces in material processing
- Geometric nonlinearities, e.g.,
    Unimodal contact problems
    Gaps or play in mechanisms
- Nonconservative forces, e.g.,
    Fluid forces
    Nonlinear damping
- Nonholonomic problems
    Rolling
    Wheeled vehicles
- Many-body systems, e.g.,
    Linked systems with dozens of connected components
    Transmission systems
- Large motions of flexible bodies, e.g.,
    Flexible robotic manipulators
    Flexible vehicles and satellites
    Tethered satellites
    Towed vehicles

1.6
Dynamics and Design

Until the advent of modern computers, dynamicists and vibration engineers were ambulance chasers in engineering design. They arrived on the scene after the system was designed, built, and deployed in service, usually called in to solve some post design problem or failure. One of the most famous examples was the dynamic failure of the suspension bridge over the Tacoma Narrows gorge in the State of Washington. At the time, civil engineers did not often consider dynamics in design. In recent decades we have witnessed incorporation of dynamics into the design of structures in earthquake-prone areas.

In the area of machine design, kinematic principles were employed well over a century ago. However, the operation of machines in off-kinematic behavior due to machine flexibility, power limits, etc., has only recently been incorporated in design with the advent of multipurpose dynamics codes. Still many manufacturing machines are designed initially for steady state, and then only in the second stage of the design process checked for vibration modes and transient behavior using finite-element codes.

Many flight vehicles and wheeled vehicles are initially designed for steady-motion operation, and only after the initial design is completed will dynamic and vibration behavior be addressed. Exceptions are often in high performance flight vehicles where flutter of wings and control surfaces sometimes must be considered early in order to avoid limitations on operational speed.

The one area where dynamics analysis has been central to initial design has been in robotics. In earlier stages of the design, a kinematic approach is used, followed by dynamic and control aspects of the design. With the need for high-speed manipulators, or mobile robotics, dynamic effects will take center stage in optimization and choice of design parameters.

Dynamic Design Synthesis

In recent decades several design synthesis methodologies have evolved for dynamic systems. There are traditional *design optimization methods* where one can distribute mass in the machine element to optimize vibration modes. In many control problems one doesn't have exact measures of component inertias and other dynamics related properties. In these cases there has evolved a number of *parameter estimation methods* to determine geometric and dynamic properties of the system from limited experimental data. These techniques have been useful in estimating dynamic models of space satellites for example.

In the area of robotics, there are so-called *evolutionary, genetic synthesis methods* that can evolve the design of a robot, optimized to accomplish a given set of design constraints. One of the principal proponents of this methodology is Prof. Hod Lipson of Cornell University. For example if one desires to create a crawling robot that will move from one domain to another and meet several design goals, evolutionary design codes look at a large selection of geometric or topological configurations. For the robot crawler, one parameter is the number of legs and joints. The algorithm runs dynamic simulations and then changes the design parameters in a way that mimics that of natural evolutionary genetics. (See, e.g., Lipson, 2005.) In Lipson's lab these same codes translate the design into rapid prototype instructions and the robot is realized in plastic and literally 'walks' out of the processing machine. Thus we can expect in the future that the traditional role of dynamic analysis as the last element of design may be reversed.

1.7 History of Dynamics Principles

Pre-Newtonian Era

Although classical dynamics is often referred to as Newtonian mechanics, the development of the principles of dynamics took place over a millennium from Greek antiquity to the 17th century and the time of Newton. Ideas of mechanics can be traced to the classical writings of ancient Greek philosophers and mathematicians such as Aristotle, Archimedes and Hero. One of the sources for history of dynamics is the work of Dugas (1988). In his book, Dugas attributed the seeds of virtual velocities or virtual power to the Peripatetic school of Aristotle (4th C. B.C.E.). The Greeks were fascinated with machines and sought to explain the properties of the lever and screw, with primitive notions of energy or work. Kinematic concepts of velocity, acceleration and rotations had to wait until the work of both Arab scholars of the middle ages and the so-called Schoolmen of the late middle ages in Paris and Oxford. For example, William Heytesbury of Oxford is credited by Dugas with defining the concept of acceleration in the 14th century. These Church schools also developed methods of formal reasoning as well as provided a theological basis for Christian Europe to study the natural world.

In the Renaissance, important contributions to dynamics were made to dynamics by Galileo Galilei (1564–1642), especially in his book *Two New Sciences* (1648). At this time motivation for the study of dynamics came from astronomical observations of the motions of the planets. It was on the astronomical observations of Tycho Brahe (1546–1601) and the deduction of the planetary laws of motion by Johannes Kepler (1571–1630) that Newton relied on when he began his deliberations on the mathematical principles of dynamics.

Newtonian Era

Isaac Newton (1642–1727) provided three pillars of rational science in the late 17th century; the calculus, the law of universal gravitation and equations of dynamics of particles relating forces and accelerations. However he was not alone in this endeavor. Both Robert Hooke (1635–1703) and Gottfried Leibniz (1646–1716) were also working on related problems. Leibniz had independently developed his own version of the calculus and Robert Hooke is reported to have also discovered the inverse square law to explain Kepler's laws of planetary motions. In 1686, Newton published his monumental work the *Principia*, in which he postulated the inverse square law of gravitation, defined the concept of force and wrote down his famous equations of motion, which required the primitive concepts of the calculus.

Post Newtonian Era

The next important contributor was the Swiss mathematician Leonard Euler (1707–1783). He effectively extended Newton's laws to the dynamics of rigid bodies. In his

Theoria motus corporum solidorum seu rigidorum (1760) [theory of the motion of rigid bodies], Euler defined the nature of a rigid body, its center of mass and moments of inertia. He also showed how to divide the rigid body dynamics problem into equations for motion of the center of mass, and equations for rotations about the center of mass, using what we now call Euler angles.

Other important players in the 18th century were the Swiss family Bernoulli, Jean, Jacques and Daniel. Jean Bernoulli (1667–1748) extended the idea of virtual work in 1717 and the idea of virtual velocities. In 1743 and 1758, Jean D'Alembert (1717–1783) published his *Traite de dynamique*, in which he developed his principles of dynamics.

In the late 18th century and continuing into the early 19th century, important mathematical formulations of dynamics occurred, beginning with the work of the Italian Joseph Lagrange (1736–1813) in Paris and Berlin, and culminating with the Irish mathematician William Hamilton (1805–1865) in the 19th century. Lagrange published his *Traite de mecanique analytique* in 1788 and taught calculus to engineering students at the Ecole Polytechnique in Paris. Poisson further developed Lagrange's famous equations of motion in 1809. In 1835, Hamilton continued to extend Lagrange's variational methods of dynamics that eventually became known as Hamilton's equations of dynamics.

Oddly, although this was the age of machines, the focus of the mathematical theories was not on the dynamics of constrained bodies as in a machine system but on planetary motions. There were notable exceptions such as James Clerk Maxwell's analysis of the stability of steam engine control systems.

Important contributions to the kinematics of machines was made by Robert Willis (1800–1875) at Cambridge and Franz Reuleaux (1829–1905) at Berlin. Both developed the idea that a machine was the integrated sum of basic kinematic components and that there were a basic set of mechanisms. Reuleaux's goal was to develop inverse methods for synthesis of mechanisms and machines; i.e. given the desired motion, find the mechanism to effect these motions. His principle work was *The Kinematics of Machinery* published originally in German in 1875 and quickly translated into English, French and Italian. He is known today as the *Father of Kinematics of Machines*. A visual history of kinematic mechanisms may be found on the website called *KMODDL: Kinematic Models for Design Digital Library* (http://kmoddl.library.cornell.edu). Readers interested in the history of kinematics for machines from the time of Leonardo da Vinci to the 20th century should consult the Author's recent book *The Machines of Leonardo da Vinci and Franz Reuleaux* (Moon, 2007).

Other contributors to the kinematics of constrained rigid bodies in the 19th century were Rankine, Alexander Kennedy, Burmester, Grübler as well as mathematicians such as Chebyshev and Sylvester. The development of a unified theory of kinematics and dynamics of constrained bodies such as machines would wait until the first half of the 20th century as in the textbooks of Den Hartog and Timoshenko.

Aside from general theories of kinematics and dynamics, there were many contributors to technical dynamics such as Huygens and his clocks. He not only built the first pendulum controlled escapement clock, but also developed a theory for it before Newton had published his *Principia*. There was also a separate track of development of the gyroscope in the 19th century as well as the mathematical theory aircraft stability in the early 20th century.

Special technical problems, so-called nonholonomic problems, such as rolling bodies, cannot be easily handled with Lagrange's equations and new formulations of the laws of dynamics for such problems were developed at the end of the 19th century as in the work of Jourdain (1909). In addition the development of mathematical methods for electromagnetic dynamics by James Clerk Maxwell (1831–1879) and others as well as the theories of elasticity and fluid mechanics provided mathematical methods that were exploited by practitioners of dynamics. Thus when the student of dynamics uses modern computational codes such as ADAMS or SOLID WORKS to analyze some dynamics problem, she or he might reflect on the centuries long, evolutionary history embodied in these modern tools.

1.8
Modern Physics of Dynamics and Gravity

The physics of dynamics has undergone a number of revolutions from the time of Aristotle (300 B.C.), Newton (1686), Einstein (1909), to the modern quest for a quantum theory of gravity and a unification of all the mass and force laws. For all of the applications described in this book, the Newtonian ideas of dynamics and gravitation will suffice. This theory is based on the assumption of a universal time unit, independent of motion, and the use of a reference system in the "fixed" stars against which all accelerations are measured. The theory also assumes that gravitational forces are universal, proportional to the product of masses of each particle pair, and change instantaneously when one particle moves with respect to the other. This assumption is sometimes called "action at a distance" and implies an infinite force propagation speed. Implicit in this theory is that mass is intrinsic to particles and does not change with the velocity.

However, all students of introductory physics know that this theory is challenged when distances are great, when the velocity is near the speed of light and when distances become smaller than the size of the atom. The great scientific revolutions of *relativity* and *quantum mechanics* at the beginning of this century, and the new world of particle physics and the *theory of everything* at the end of this century have dramatically changed the Newtonian structure of dynamics. So while the continued use of this structure for engineering problems is sufficient, the intellectual excitement of these new theories challenges all dynamicists to become aware of the limitations of our dynamical ideas and tools.

The first great assault on Newtonian dynamics came at the beginning of this century by Albert Einstein (1879–1955) in his special theory of relativity. This theory was based on observations of the independence of the speed of light to the frame of reference, as well as theoretical work on the equations of electromagnetics as formulated by J. C. Maxwell in 1865. As most high school students learn, this theory resulted in the notion of the change of length and time scales with velocity, as well as the equivalence of energy and mass summarized in the famous equation $E = mc^2$, where c is the velocity of light.

However, Einstein's general theory of relativity shed new light on gravitation. In this theory the four-dimensional world of space and time is warped by the presence of objects possessing mass. It is this deformation of space–time that "causes" one mass to be attracted to another. Thus the "force of gravity" is replaced by a geometric warping of space–time. His theory also dealt a blow to the "action at a distance" feature of the Newtonian view of gravity by including the notion of *gravitational waves*. That is, any changes in the relative position of two masses are not felt instantaneously, but travel at the speed of light. Observations of pulsars by Joseph Taylor and Russell Hulse of Princeton in 1974 provided evidence for gravitation waves, which won them a Nobel Prize in 1993. However, measurements of gravitational waves on earth have been more difficult despite substantial ongoing projects by several physics groups.

The second major development in classical dynamics was in quantum mechanics, which was developed by several physicists in the first quarter of the twentieth century and was embodied in the famous equation of Schrödinger (1926). Instead of assigning a position vector to describe the motion of a particle, a wave function representing the probability of a particle's position in space was posited. Out of this theory came the now famous *uncertainty principle* of Heisenberg (1927). This principle says that one cannot measure both the particle's position and velocity with infinite accuracy.

In the last quarter of the twentieth century there has been a bewildering display of discoveries in particle physics that has challenged the imagination of those trained in classical dynamics. Mass particles have been replaced by atomic nuclei (protons and neutrons), which have been replaced by sub-nuclear particles: *quarks* and *leptons*. Physicists now classify the forces of nature into four groups: *electromagnetic*, *gravity*, *weak nuclear*, and *strong nuclear forces*, each transmitted by another group of particles called *bosons*. In fact forces are now viewed as an exchange of particle-like messenger objects. For example, electromagnetic forces between particles are realized by an exchange of massless *photons* each traveling with the speed of light. Gravitational forces are imagined to be an exchange of gravity quanta, called *gravitons*, again traveling at the speed of light. Thus the force of the earth on a satellite is viewed as a continual exchange of gravity quanta between mass particles in the earth and mass in the satellite. These theories were developed in the period 1974–84. To date, however, experimental evidence for the graviton has not been found.

The most recent theories of mass, forces, and gravity attempt to incorporate all known forces of nature into one grand theory called the theory of everything (TOE). One class of TOE replaces mass particles with stringlike objects in ten dimensions. Hence the term *string* or *superstring theory*.

What prompted these diversions in this book is the observation that the continued mystery of the nature of classic concepts of mass, force, and gravity has become one of the principal thrusts of modern physics. The reader wishing to gain further insight into these ideas and discoveries should read some popular books, such as those of Hawking (1988), Zee (1989), Weinberg (1993), Lederman (1993), and Thorne (1994) referenced at the end of this book.

2
Basic Principles of Dynamics

2.1
Introduction

It is assumed that the student has had an introduction to dynamics in a college physics course as well as a one-semester course covering kinematics and dynamics of particles and rigid bodies at the sophomore or junior level. Sometimes it is useful to take a large view of a field. In Chapter 1 we have summarized some of the phenomena and paradigms of dynamics without emphasizing the basic principles. Here we catalog the various mathematical definitions of dynamical measures along with the basic principles and equations that embody Newton's laws of dynamics. Additional concepts are introduced later, such as Lagrange's equations and virtual power method. In later chapters we revisit some of the basic principles in greater detail.

2.2
Kinematics

The teaching of kinematics has evolved in the last century. Prior to the 1950s, use of scalar components of velocities and accelerations were mainly used and graphical methods of solution were taught. Beginning with the 1950s a vector approach was used that de-emphasized the scalar components. In recent years the use of matrices has been introduced, especially to describe rotations. This approach is most useful when using the digital computer to formulate and solve problems.

In this text we primarily use a vector approach to do the model development and analysis. The matrix form of certain equations is included since it is used quite extensively in modern computer codes in robotics and multibody dynamics as well as in the simulation code MATLAB.

Classical books on dynamics usually omit kinematics of mechanisms and machines. However there is a large literature as well as active interest in the kinematics of systems of constrained bodies. In Chapter 3 we shall review a small sample of this rich field. The classic text of Reuleaux (1876) has been republished

Applied Dynamics. With Applications to Multibody and Mechatronic Systems. Second Edition. Francis C. Moon
Copyright © 2008 WILEY-VCH Verlag GmbH & Co. KGaA, Weinheim
ISBN: 978-3-527-40751-4

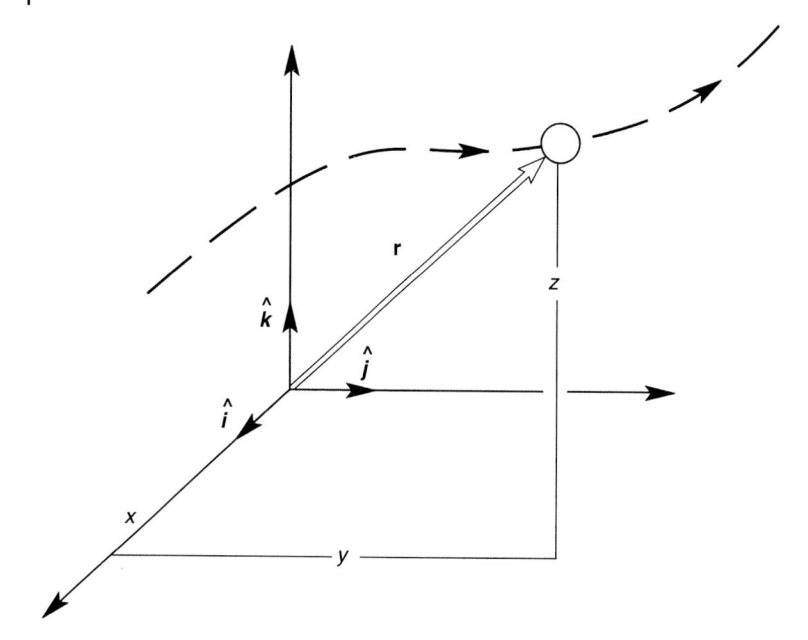

Fig. 2.1 Position vector, Cartesian components.

by Dover Publications in 1963. A contemporary textbook with computer and web-based material is Erdman et al. (2001). Readers are also encouraged to visit the Cornell website named KMODDL: Kinematic Models for Design Digital Library at http://kmoddl.library.cornell.edu for descriptions, tutorials and videos of several hundred kinematic mechanisms.

Position Vector

Kinematics is the mathematical description of motion. In dynamics we are concerned with position, velocity, and acceleration. The position of a particle is tracked by a vector whose origin begins at the origin of a reference frame (Fig. 2.1) and ends at the particle. The position vector is often projected on a reference frame with three mutually orthogonal unit vectors or basis vectors.

$$\mathbf{r} = r_1 \mathbf{e}_1 + r_2 \mathbf{e}_2 + r_3 \mathbf{e}_3 \tag{2.1}$$

When a Cartesian frame is used, the three components are often written as a column vector in matrix notation, i.e., $\mathbf{r} = [x, y, z]^T$, and the three basis vectors are written as $\{\mathbf{e}_1, \mathbf{e}_2, \mathbf{e}_3\} = \{\mathbf{i}, \mathbf{j}, \mathbf{k}\}$. (The superscript '$T$' on the square bracket indicates a matrix transpose. Thus a 1×3 row matrix becomes a 3×1 column matrix.) However, in cylindrical coordinates (Fig. 2.2) only two basis vectors are used.

$$\mathbf{r} = \rho \mathbf{e}_\rho + z \mathbf{e}_z \tag{2.2}$$

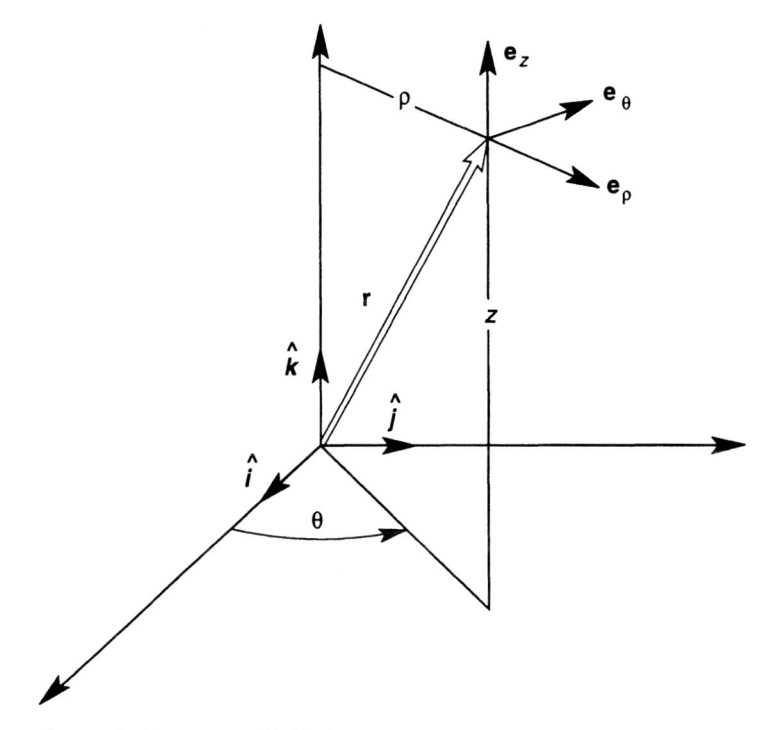

Fig. 2.2 Position vector, cylindrical components.

Here the other variable, the angle θ, is implicit since the radial basis vector $\mathbf{e}_\rho$ depends on θ.

Velocity

Using the basic definition of a derivative in vector calculus, the velocity vector of a moving particle is given by the derivative of the position vector,

$$\mathbf{v} = \frac{d\mathbf{r}}{dt} \qquad (2.3)$$

In rectangular coordinates this becomes

$$\mathbf{v} = \dot{x}\mathbf{i} + \dot{y}\mathbf{j} + \dot{z}\mathbf{k} \qquad (2.4)$$

where the derivative is taken in a reference frame where the basic vectors $(\mathbf{i}, \mathbf{j}, \mathbf{k})$ are assumed to be fixed.

In cylindrical coordinates one must account for the fact that $\mathbf{e}_\rho$ changes as $\theta(t)$ changes. One can show (see Chapter 3) that in these coordinates the velocity is given by differentiating (2.2) with respect to time

$$\mathbf{v} = \dot{\rho}\mathbf{e}_\rho + \rho\dot{\theta}\mathbf{e}_\theta + \dot{z}\mathbf{e}_z$$

or

$$v_\rho = \dot{\rho}, \qquad v_\theta = \rho\dot{\theta}, \qquad v_z = \dot{z} \qquad (2.5)$$

Acceleration

By extension of the ideas just discussed, the acceleration **a** is a vector quantity that is the time derivative of the velocity vector

$$\mathbf{a} = \frac{d\mathbf{v}}{dt} \qquad (2.6)$$

In Cartesian coordinates we have

$$\mathbf{a} = \ddot{x}\mathbf{i} + \ddot{y}\mathbf{j} + \ddot{z}\mathbf{k} \qquad (2.7)$$

and in cylindrical coordinates

$$\mathbf{a} = (\ddot{\rho} - \rho\dot{\theta}^2)\mathbf{e}_\rho + (\rho\ddot{\theta} + 2\dot{\rho}\dot{\theta})\mathbf{e}_\theta + \ddot{z}\mathbf{e}_z \qquad (2.8)$$

Here the term $-\rho\dot{\theta}^2\mathbf{e}_\rho$ is known as the *centripetal* acceleration and comes from the change of $\mathbf{e}_\theta$ in (2.5). The term $2\dot{\rho}\dot{\theta}\mathbf{e}_\theta$ is called the *Coriolis*[1] acceleration and originates from the time rate of change of ρ in v_θ and the time rate of change of $\mathbf{e}_\rho$.

Constraints: Jacobian

The motions of many components in machines are often limited by geometric constraints such as pistons in an engine cylinder or balls in a bearing race. This means that the number of degrees of freedom is less than three for a particle and less than six for a rigid body. In such problems it is usually convenient to replace the spatial position variables by generalized variables that are natural to the constraint. Two examples are shown in Fig. 2.3. In case (a) the particle is constrained to move along a curve whose length is given by $s(t)$. This means that the components of the position vector $\mathbf{r} = [x, y, z]^T$ are functions of s. In this case the velocity is given by

$$\mathbf{v} = \dot{\mathbf{r}} = \frac{d\mathbf{r}}{ds}\dot{s} = \left(\frac{dx}{ds}\mathbf{i} + \frac{dy}{ds}\mathbf{j} + \frac{dz}{ds}\mathbf{k}\right)\dot{s}$$

or

$$\mathbf{v} = \mathbf{J}\dot{s} \qquad (2.9)$$

The vector **J** is called a 3×1 *Jacobian*[2] matrix. Note that **J** is tangent to the path of the particle. The Jacobian relates the physical velocity **v** to the generalized velocity $\dot{s}$. An

---

1) Gaspard-Gustave de Coriolis (1792–1843), French engineer and mathematician.
2) K. G. J. Jacobi (1804–1851), German mathematician.

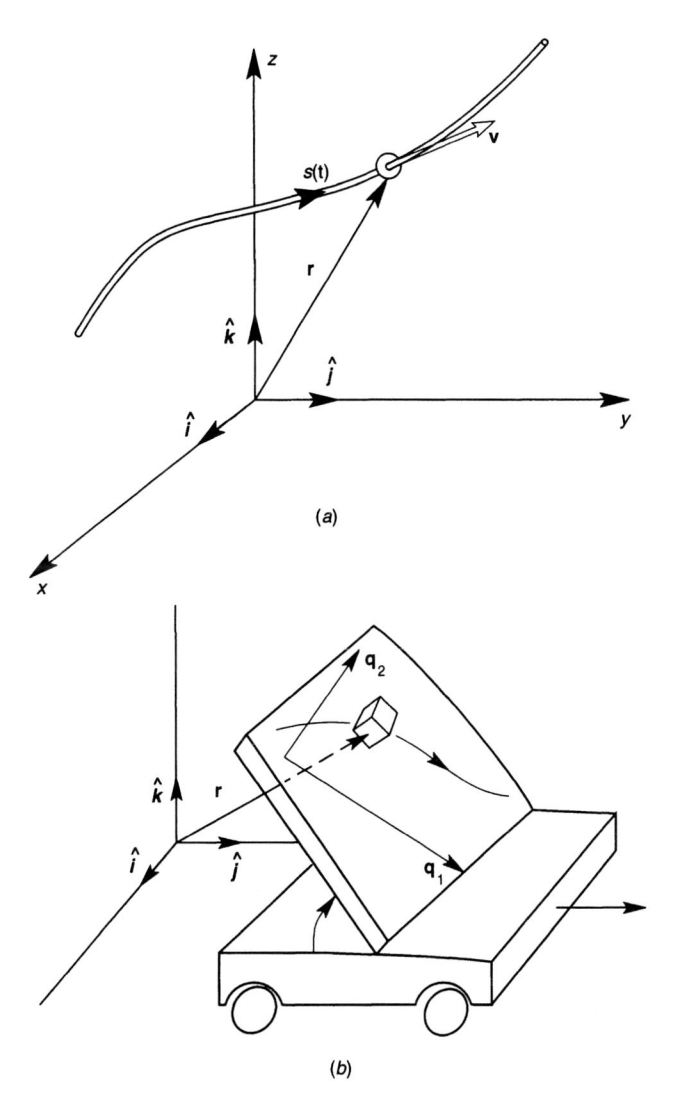

Fig. 2.3 Two examples of constraints: (a) motion on a curve, and (b) motion on a moving surface.

alternative representation is to choose one component as the independent variable, say x, then $\mathbf{r} = [x, y(x), z(x)]^T$ and the velocity is given by

$$\dot{\mathbf{v}} = \dot{x}\left(\mathbf{i} + \frac{dy}{dx}\mathbf{j} + \frac{dz}{dx}\mathbf{k}\right) \tag{2.10}$$

In the second example the particle is constrained to move on a surface that may itself be moving. In Fig. 2.3b we show two generalized coordinates $(q_1(t), q_2(t))$, embedded in the moving surface. In this case each of the components (x, y, z) is a function of (q_1, q_2) as well as an explicit function of time so that

$$\mathbf{v} = \sum_{i=1,2} \dot{q}_i \left(\frac{\partial x}{\partial q_i}\mathbf{i} + \frac{\partial y}{\partial q_i}\mathbf{j} + \frac{\partial z}{\partial q_i}\mathbf{k} \right) + \frac{\partial \mathbf{r}}{\partial t} \qquad (2.11)$$

In general, where $\mathbf{r} = \mathbf{r}(q_i, t)$

$$\mathbf{v} = \sum_i \frac{\partial \mathbf{r}}{\partial q_i}\dot{q}_i + \frac{\partial \mathbf{r}}{\partial t} \qquad (2.12)$$

or

$$\mathbf{v} = \mathbf{J}\dot{\mathbf{q}} + \frac{\partial \mathbf{r}}{\partial t}$$

The coefficient of $\dot{q}_i$ in Eq. (2.12) is called a Jacobian matrix, J. In the example of motion on a surface, this matrix is 3 rows by 2 columns, i.e.,

$$\mathbf{J} = \begin{bmatrix} \frac{\partial x}{\partial q_1} & \frac{\partial y}{\partial q_1} & \frac{\partial z}{\partial q_1} \\ \frac{\partial x}{\partial q_2} & \frac{\partial y}{\partial q_2} & \frac{\partial z}{\partial q_2} \end{bmatrix}^T \qquad (2.13)$$

Also $\dot{\mathbf{q}} = [\dot{q}_1, \dot{q}_2]^T$. Jacobian matrices embody information about geometric constraints and are very important in the study of robotic and other multibody machine dynamics.

Example 2.1 Kinematic Constraints in a Robot Arm. Consider the parallelogram mechanism shown in Fig. 2.4. This mechanism is the basis of the so-called MIT robot manipulator arm. (See, e.g., Craig, 2005.) This mechanism has $n = 6$ links and $r = 6$ revolute joints, counting the double joint at the base. The number of degrees of freedom of this planar robot arm is $F = 3(n-1) - 2r = 3$. The last joint before the gripper is called the wrist joint. It is useful for robotic path planning to be able to have geometric and kinematic relationships between the Cartesian positions of the wrist point and the two joint variables:

$$\mathbf{r}_w = (b\cos\theta_1 + c\cos\theta_2)\mathbf{e}_x + (b\sin\theta_1 - c\sin\theta_2)\mathbf{e}_y$$

$$\mathbf{v}_w = \mathbf{J}[\dot{\theta}_1, \dot{\theta}_2]^T$$

$$\mathbf{J} = \begin{bmatrix} -b\sin\theta_1 & -c\sin\theta_2 \\ b\cos\theta_1 & c\cos\theta_2 \end{bmatrix} \qquad (2.14)$$

There are positions of the mechanism where one loses one degree of freedom called *singularities*. These positions are found by setting the determinant of the Jacobian matrix (2.14) equal to zero. (See also Chapter 3.) For the MIT arm the singularities are given by

$$\det \mathbf{J} = bc\sin(\theta_1 + \theta_2) = 0$$

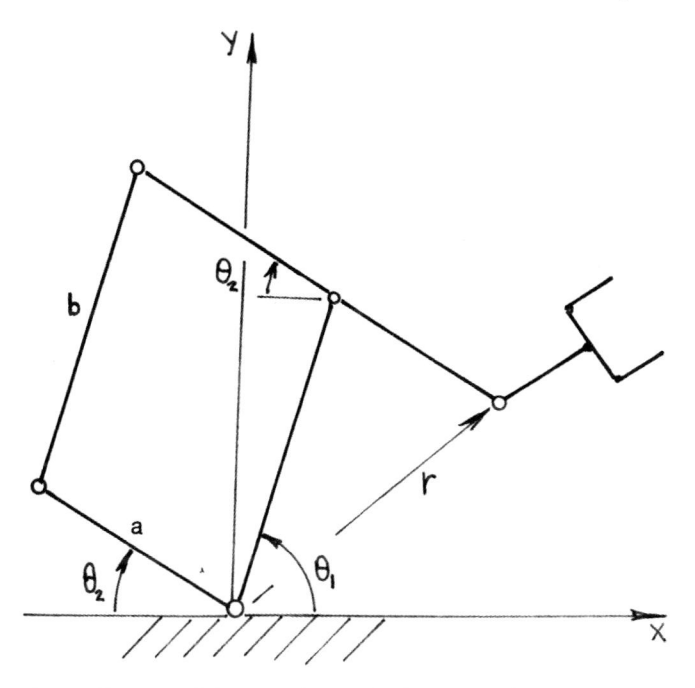

Fig. 2.4 Planar, parallelogram, robot manipulator arm with end-effecter. (See "MIT Arm" in Rosheim, 1994.)

or

$$\theta_1 + \theta_2 = 0, +\pi, -\pi$$

At these points, one cannot invert the velocity equation to find $[\theta_1, \theta_2]$ in terms of $[v_1, v_2]$.

2.3
Equilibrium and Virtual Work

In dynamics, equilibrium is the absence of acceleration, not the absence of motion. For a single particle this means that the sum of all the forces on the particle must be zero, i.e.,

$$\sum_i \mathbf{F}_i = 0 \tag{2.15}$$

This is a vector equation and as such we can project the forces onto any direction described by a unit vector, $\mathbf{e}$,

$$\sum_i \mathbf{F}_i \cdot \mathbf{e} = 0 \tag{2.16}$$

For a single particle we can obtain three independent scalar equations relating the forces acting on the body. In (2.16) **e** is arbitrary, but all the forces are involved. In the next section we show how to choose **e** to eliminate some of the forces from the equations of equilibrium.

Virtual Work

In the Newtonian view of dynamics, changes of velocity or acceleration are produced by a vector quantity that we call force. This vector approach appears very natural to modern students of dynamics. However, an alternative view of dynamics can be formulated using work and energy concepts, which is a scalar-based view of dynamics. This work–energy approach can be quite useful in solving many practical problems of both an equilibrium nature (the method of virtual work) and a dynamic nature (Lagrange's equation and the method of virtual power). These scalar methods can be derived without using vectors, but in this text we begin with the vector notation.

The methods derived from the principle of virtual work are most useful in problems in which there are workless constraints. For example, consider the particle shown in Fig. 2.5a that is constrained to move on a rigid surface. We write the forces on the particle as the sum of applied forces $\{\mathbf{F}_i^a\}$ and a constraint force $\mathbf{F}^c$ between the surface and the particle. We next imagine that the particle is allowed to move a small amount on the surface. This amount is denoted by $\delta \mathbf{r}$. This small vector is not arbitrary, but is tangential to the surface. If the surface normal is denoted by $\mathbf{n}$, then $\delta \mathbf{r} \cdot \mathbf{n} = 0$. In this case we say that the small test displacement $\delta \mathbf{r}$, called a *virtual displacement*, is consistent with the constraints.

If there is no sticking and no friction between the particle and the surface, we can write $\mathbf{F}^c = F^c \mathbf{n}$, where $\mathbf{n}$ is a unit vector, so that $\mathbf{F}^c \cdot \delta \mathbf{r} = 0$. The vector form of equilibrium is

$$\sum_i \mathbf{F}_i^a + \mathbf{F}^c = 0 \tag{2.17}$$

However, taking the projection onto $\delta \mathbf{r}$, we have

$$\sum_i \mathbf{F}_i^a \cdot \delta \mathbf{r} = 0 \tag{2.18}$$

In this equation we have eliminated the constraint force $\mathbf{F}^c$ and have a relation between the applied forces as projected onto the direction $\delta \mathbf{r}$. Since $\delta \mathbf{r}$ is arbitrary, we can choose two independent directions tangential to the surface and obtain two scalar equations of equilibrium of the applied forces. Defining the independent variables q_1, q_2, the small virtual displacement $\delta \mathbf{r}$ can be resolved into independent variations $\delta q_1, \delta q_2$ or

$$\delta \mathbf{r} = \sum_k \frac{\partial \mathbf{r}}{\partial q_k} \delta q_k$$

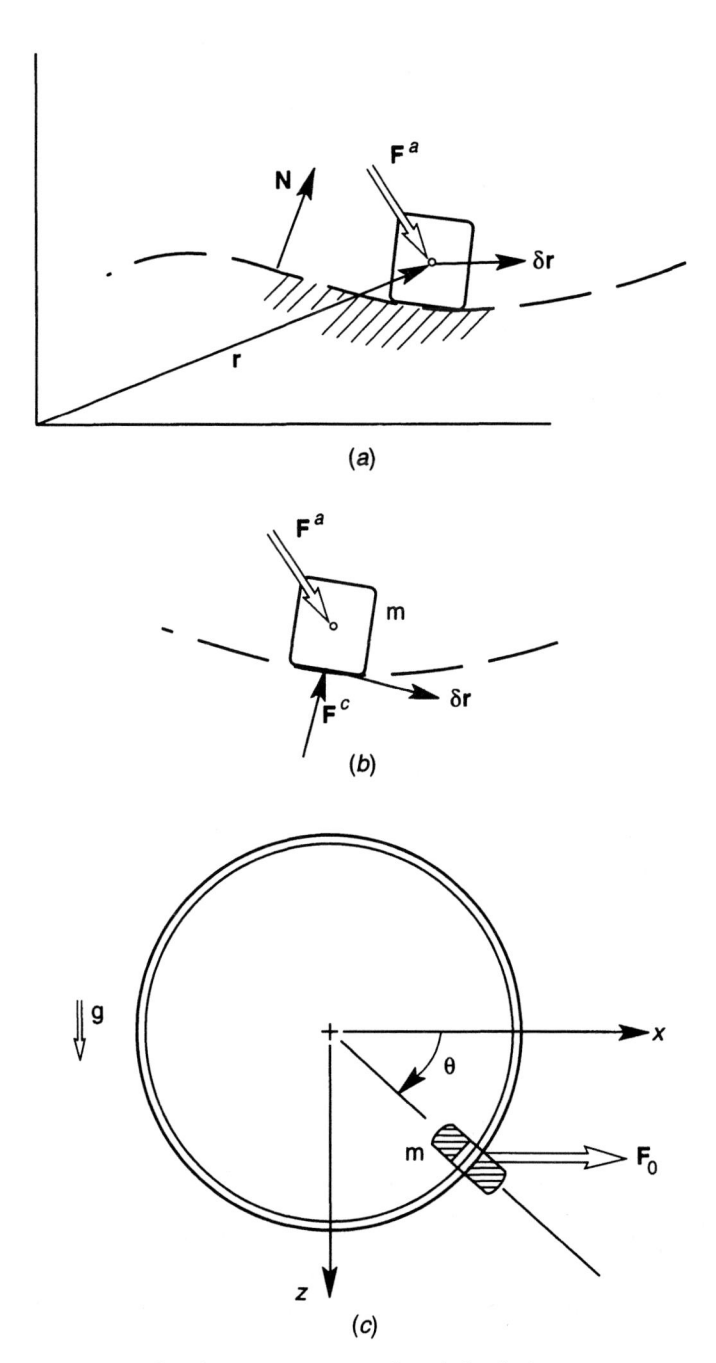

Fig. 2.5 Virtual work: (a) motion on a surface; (b) free-body diagram showing workless force $\mathbf{F}^c$. (c) Constrained motion of a mass on a circular ring under gravity.

And since these variations are assumed to be independent, we have

$$\sum_i \mathbf{F}_i^a \cdot \frac{\partial \mathbf{r}}{\partial q_k} = 0, \quad k = 1, 2 \tag{2.19}$$

Here we see again the role of the Jacobian matrix (2.13). The column vectors of J represent projection vectors in (2.19). Another statement of the principle of virtual work is the matrix equation below (see, e.g., Roberson and Schwertassek, 1988, p. 167).

$$J^T \mathbf{F}^c = 0$$

Example 2.2 Virtual Work. As an example of the virtual work method consider the particle on a circular ring as in Fig. 2.5c. For this example we assume that the horizontal force $\mathbf{F}_0$, is created by a spring force given by $F_0 = k(x_0 - x)$, where we assume that the reference position of the spring is greater than the radius R of the circular ring. Thus the applied force $\mathbf{F}^a$ and the constraint force $\mathbf{F}^c$ are given by

$$\mathbf{F}^a = F_0 \mathbf{e}_x + mg \mathbf{e}_z, \qquad \mathbf{F}^c = -N \mathbf{e}_R$$

Where the unit vectors $\{\mathbf{e}_R, \mathbf{e}_\theta\}$ are the planer polar coordinate basis vectors normal to and tangential to the circular ring. The virtual displacement of the mass on the ring must be in the direction of the circumferential unit vector; $\delta \mathbf{r} = R \delta \theta \mathbf{e}_\theta$. The statement of virtual work for equilibrium of forces on the mass is given by

$$\mathbf{F}^a \cdot \delta \mathbf{r} = 0 \tag{2.20}$$

$$F_0 (\mathbf{e}_x \cdot \mathbf{e}_\theta) + mg (\mathbf{e}_z \cdot \mathbf{e}_\theta) = 0$$

$$\frac{k x_0}{mg} \left[1 - \frac{R}{x_0} \cos \theta \right] \tan \theta = 1 \tag{2.21}$$

The expression for the equilibrium angle θ, where the gravity, spring and constraint forces are in balance, is a transcendental equation.

2.4
Systems of Particles

The classic equation of dynamics embodied in Newton's law applies strictly for a point mass. In the case of extended bodies, we have to determine which point the law of acceleration applies to, as different points can have different accelerations. In formulating the laws of dynamics for extended bodies, we use the artifact of assuming that fluid, solid, or gaseous bodies can be represented by a finite collection of point masses. This allows us to use summation operations instead of integrals. Almost all of the definitions that follow can be rewritten in integral form over mass densities. For the student who desires a more rigorous treatment of continuous

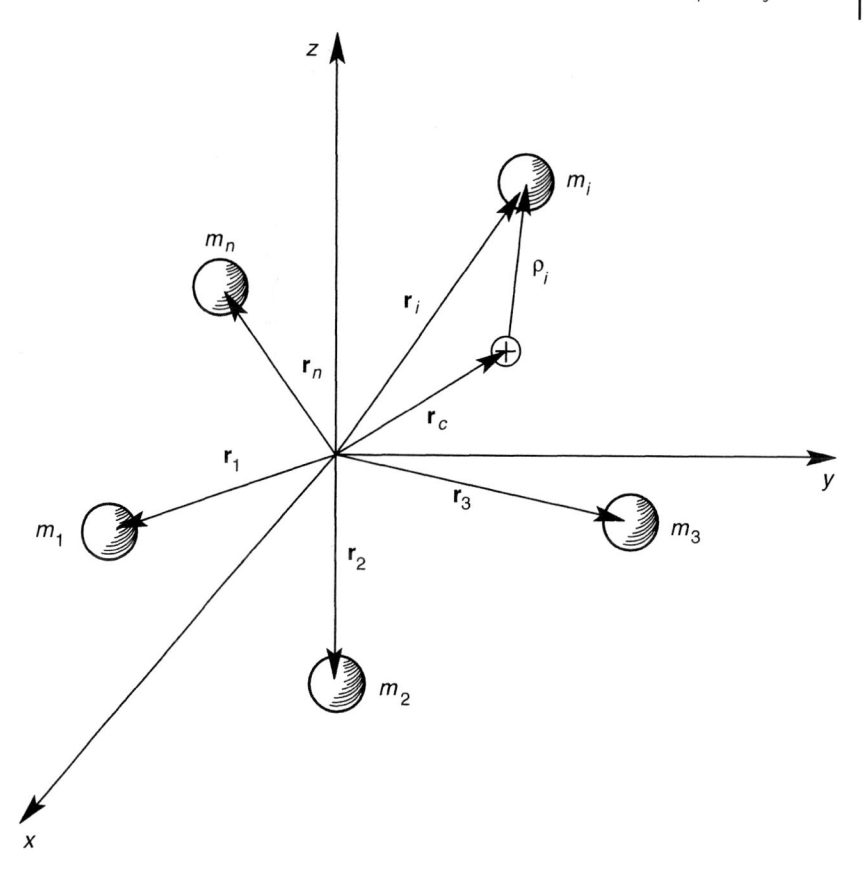

Fig. 2.6 System of particles.

matter dynamics, appropriate texts in solid, fluid and continuum mechanics are recommended.

Also, while some of the principles and equations are applicable to deformable mass systems, in this text, the extended bodies of interest will, in almost all cases, be rigid bodies.

We begin with several definitions that are based on the diagram in Fig. 2.6. Here we assume there exists a distribution of point masses $\{m_i\}$ whose positions are each described by position vectors $\{\mathbf{r}_i\}$. From this assumption the following definitions are given.

Center of Mass

The total mass given by

$$M = \sum m_i$$

The first moment of mass is called the *center of mass*:

$$\mathbf{r}_c = \frac{1}{M} \sum \mathbf{r}_i m_i \tag{2.22}$$

From this definition we can define relative position vectors $\{\boldsymbol{\rho}_i\}$ whose first moment of mass is zero, i.e.,

$$\boldsymbol{\rho}_i = \mathbf{r}_i - \mathbf{r}_c$$
$$\sum \boldsymbol{\rho}_i m_i = 0 \tag{2.23}$$

Linear Momentum

From the previous section on kinematics, the time rate of change of the position vectors is the velocity of each point mass, i.e., $\dot{\mathbf{r}} = \mathbf{v}$.

The linear momentum is then defined as

$$\mathbf{P} = \sum m_i \mathbf{v}_i \tag{2.24}$$

and by the preceding definition

$$\mathbf{P} = M \mathbf{v}_c$$

where $\mathbf{v}_c = \dot{\mathbf{r}}_c$

Angular Momentum

The linear momentum is defined with respect to a given reference frame. The angular momentum is the first moment of linear momentum with respect to a particular point. In the case where this point is the origin we define the angular momentum

$$\mathbf{H} = \sum \mathbf{r}_i \times m_i \mathbf{v}_i \tag{2.25}$$

We can also define an angular momentum measure with respect to a reference frame translating with the center of mass

$$\mathbf{H}_c = \sum \boldsymbol{\rho}_i \times m_i \dot{\boldsymbol{\rho}}_i \tag{2.26}$$

It is easy to show, using (2.23), that

$$\mathbf{H} = \mathbf{H}_c + \mathbf{r}_c \times M \mathbf{v}_c \tag{2.27}$$

Example 2.3 Angular Momentum for a System of Particles. To illustrate the definition of the angular momentum vector for a system of particles, consider the four particles in Fig. 2.7, with equal masses at the corners of a massless rectangular frame. We assume that the frame rotates about the diagonal axis of the frame. We

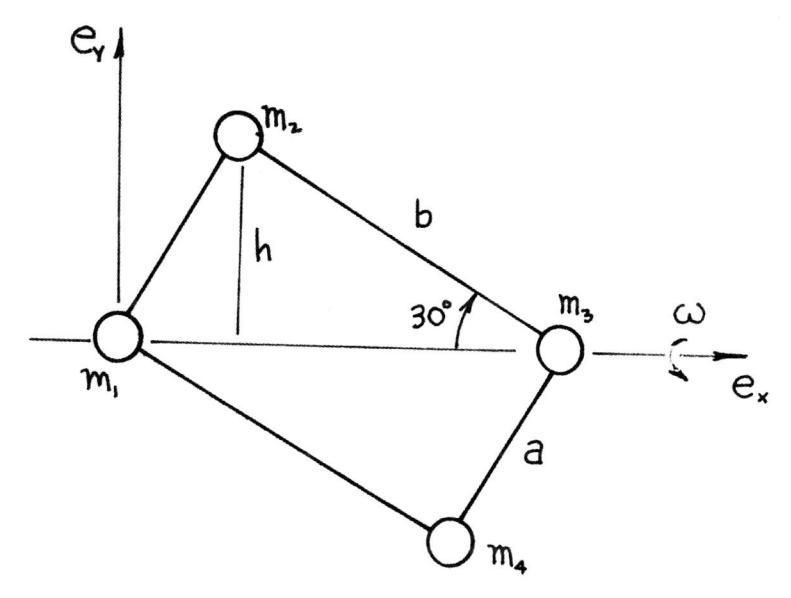

Fig. 2.7 Four-mass structure rotating about the x axis.

also align the x axis with the axis of rotation, and place the y axis in the frame such that the coordinate frame rotates with the mass-holder frame. With these definitions, we can evaluate the following vectors:

$$\boldsymbol{\omega} = \omega \mathbf{e}_x$$
$$\mathbf{r}_1 = 0, \qquad \mathbf{r}_2 = x_2 \mathbf{e}_x + h \mathbf{e}_y$$
$$\mathbf{r}_3 = \frac{2b}{3} \mathbf{e}_x, \qquad \mathbf{r}_4 = x_4 \mathbf{e}_x - h \mathbf{e}_y \tag{2.28}$$

where

$$x_2 = a \cos 60°, \qquad x_4 = b \cos 30°$$

The velocities of the particles at $\mathbf{r}_2$, $\mathbf{r}_4$ are each $\mathbf{v}_2 = \omega h \mathbf{e}_z$, $\mathbf{v}_4 = -\omega h \mathbf{e}_z$, $h = b \cos 30°$.

Using Eq. (2.25), we arrive at the expression for the angular momentum vector:

$$\mathbf{H} = \omega m b^2 \left[\frac{1}{2} \mathbf{e}_x + \frac{1}{2\sqrt{3}} \mathbf{e}_y \right] \tag{2.29}$$

This expression shows that the angular momentum vector $\mathbf{H}$ is not aligned with the axis of rotation and thus rotates with the mass frame. The rotation of $\mathbf{H}$ means that the direction of the change of angular momentum, $d\mathbf{H}/dt$, is normal to the plane of rotation. Thus according to (1.17) there must be a force moment $\mathbf{M}$ on the bearings normal to the plane of the particles. This makes sense if one looks at the two rotating masses as producing two offset centrifugal forces due to rotation and these centrifugal forces will produce a moment on the bearings normal to the plane of the particles.

Moment of Force

An important measure of the effect of point forces on extended bodies is the first moment of force defined by

$$\mathbf{M} = \sum \mathbf{r}_i \times \mathbf{F}_i$$

In this definition each mass is assumed to have an applied force $\mathbf{F}_i$ acting on it.

Laws of Linear and Angular Momentum

With the preceding definitions we can state the extension of Newton's laws for extended bodies as represented by a collection of N point masses $\{m_i;\ i = 1,\ldots N\}$. In stating these laws, we assume that, in addition to the forces applied to each mass from outside the system $\{\mathbf{F}_i\}$ there is another set of forces that act between pairs of masses represented by the set of vectors $\{\mathbf{f}_{ij};\ i \neq j\}$. We assume that these internal forces obey Newton's law of action and reaction or that $\mathbf{f}_{ij} = -\mathbf{f}_{ji}$. Thus for each mass particle we have Newton's law,

$$m_i \dot{\mathbf{v}}_i = \mathbf{F}_i + \sum_i \mathbf{f}_{ij} \tag{2.30}$$

If we sum these equations of motion over all the masses using (2.22) and (2.24), then we have Newton's law for the ensemble of particles

$$M\dot{\mathbf{v}}_c = \sum \mathbf{F}_i \tag{2.31}$$

This result tells us that in order to apply Newton's law to an extended body of particles, we must use the acceleration of the center of mass and the sum of all the external forces.

To obtain the law of angular momentum for a system of particles, we take the first moment of (2.30) and again sum over all the particles:

$$\sum \mathbf{r}_i \times m_i \dot{\mathbf{v}}_i = \sum \mathbf{r}_i \times \mathbf{F}_i + \sum_i \mathbf{r}_i \times \sum_j \mathbf{f}_{ij}$$

There are two more steps to obtain the standard form of this law. First, we recognize that on the left-hand side we can take the time derivative outside the summation and use (2.25):

$$\sum \mathbf{r}_i \times m_i \frac{d\mathbf{v}_i}{dt} = \frac{d}{dt} \sum \mathbf{r}_i \times m_i \mathbf{v}_i = \frac{d}{dt} \mathbf{H} \tag{2.32}$$

The second step is more controversial. Consider pairs of terms in the second expression on the right-hand side

$$\mathbf{r}_i \times \mathbf{f}_{ij} + \mathbf{r}_j \times \mathbf{f}_{ji} = (\mathbf{r}_i - \mathbf{r}_j) \times \mathbf{f}_{ij} \tag{2.33}$$

If we assume that the internal forces are central, that is, that they are directed along the line between the masses, the total moment of the internal forces is zero and we have the classic statement of the law of angular momentum

$$\frac{d}{dt}\mathbf{H} = \mathbf{M} \tag{2.34}$$

where the moment of the external forces $\mathbf{M}$ is defined in (2.28). However, Newton's third law $\mathbf{f}_{ij} = -\mathbf{f}_{ji}$ does not require that the forces be colinear, i.e., directed along the line between the masses $(\mathbf{r}_i - \mathbf{r}_j)$. In this case the pairs of internal forces would create couples that would contribute to the moment $\mathbf{M}$. Goldstein (1980), in the second edition of his book on classical mechanics, calls the assumption of central forces the strong form of the third law and the noncentral force law the weak form of the third law. There are classic force laws in electromagnetics in which the mutual forces between masses are not colinear. In this case, additional moment terms must be added in order to use the classic form of the angular momentum law (2.34).

Some authors have stated that deriving the law of angular momentum from Newton's laws for particles is artificial and that we should simply assume that the classic law of angular momentum for an extended body is an independent law of nature (see, e.g., Truesdell, 1968). In practice, we must require that both laws of linear and angular momentum (2.31), (2.34) be satisfied for every extended body.

The combination of both laws yields another form of the angular-momentum law. In particular, we use the relative coordinates $\boldsymbol{\rho}_i = \mathbf{r}_i - \mathbf{r}_c$ to rewrite the angular-momentum vector using (2.27)

$$\mathbf{H} = \mathbf{H}_c + \mathbf{r}_c + M\mathbf{v}_c$$

We also have the identity

$$\mathbf{M} = \mathbf{r}_c \times \sum \mathbf{F}_i + \mathbf{M}_c \tag{2.35}$$

where

$$\mathbf{M}_c = \sum \boldsymbol{\rho}_i \times \mathbf{F}_i \tag{2.36}$$

Here, the second term is the moment of the applied forces about the center of mass. The expressions (2.34), (2.27), (2.35) lead to the relation

$$\frac{d}{dt}\mathbf{H}_c = \mathbf{M}_c \tag{2.37}$$

The angular momentum $\mathbf{H}_c$, (2.26), is calculated by moments about the center of mass and from the relative velocities $\dot{\boldsymbol{\rho}}_i = \mathbf{v}_i - \mathbf{v}_c$. The time derivative is assumed to be taken in a nonrotating reference frame translating with the center of mass.

When the body is generated my a surface of revolution, as in a wheel or a flywheel (Fig. 1.20), it is customary to attach a coordinate system to the axis of revolution, say $\mathbf{e}_3$, and to write the angular velocity vector as a sum of two terms,

$$\omega = \dot{\phi}\mathbf{e}_3 + \Omega \times \mathbf{H}$$

where Ω is the angular velocity of the coordinate frame attached to the axis of revolution. The equation of motion relating the total force moment $\mathbf{M}$ and the change of angular momentum $d\mathbf{H}/dt$, then takes the form

$$\mathbf{M} = \left(\frac{d\mathbf{H}}{dt}\right)_{1\text{-}2\text{-}3} + \Omega \times \mathbf{H} \tag{2.38}$$

The first term in this expression is the change of $\mathbf{H}$ relative to the coordinate system $\{\mathbf{e}_1, \mathbf{e}_2, \mathbf{e}_3\}$ attached to the axis of revolution. For example in the case of the front wheel of a bicycle, it would be natural to attach the coordinate frame to the front fork of the bicycle. The first term in the above equation would represent the change in angular momentum due to the change of rotation speed of the wheel relative to the fork and in the second term, Ω would be the angular velocity of the bicycle front wheel fork.

There are other forms of the law of angular momentum that we do not discuss here. The form (2.37) is very important because the point about which the moments are taken moves with the body. This allows us to define another set of geometric quantities that measure the angular momentum per unit rate of rotation called "moments of inertia." We leave the discussion of moment of inertia to Section 2.5 and Chapter 5.

Example 2.4 Gyro Sensor. The expression for Euler's law of angular moment (2.34) can be applied to a gyro sensor shown in Fig. 2.8. A flywheel is made to rotate with a constant angular velocity $\dot{\phi}$, and is attached to a frame or gimbals that is fixed to a platform whose angular orientation about the vertical axis is Ω. One end of the shaft is attached to a revolute joint normal to the plane of the gimbal and the other end of the shaft is attached to a stiff spring of strength k, Newton's per displacement Δ. According to the expression (2.38), the force moment on the spinning rotor is given by

$$\mathbf{M} = \Omega \mathbf{e}_2 \times \mathbf{H} \tag{2.39}$$

where

$$\mathbf{H} = I_1 \dot{\phi} \mathbf{e}_1 + I_2 \Omega \mathbf{e}_2$$

$$\mathbf{M} = -I_1 \Omega \dot{\phi} \mathbf{e}_3$$

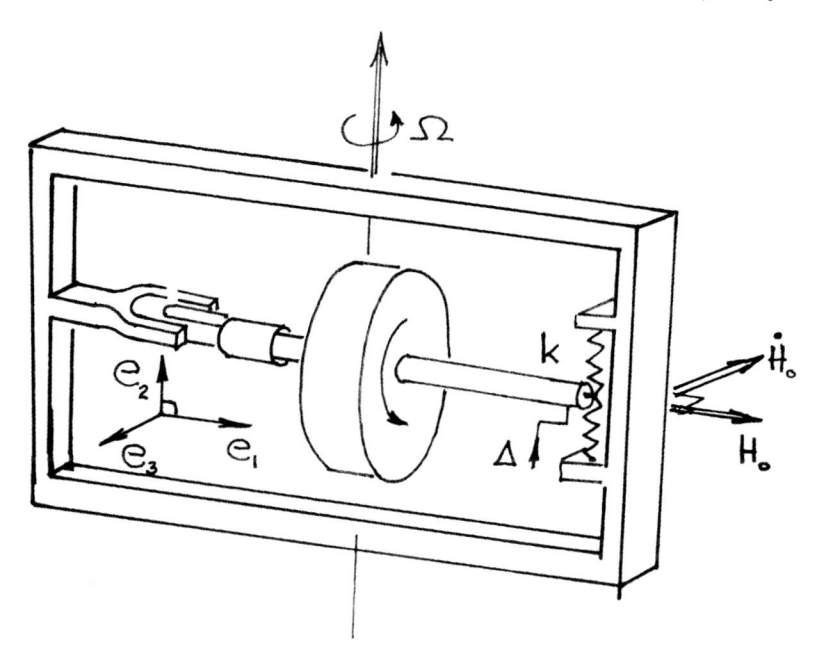

Fig. 2.8 Two-mass assembly on a massless semi-circle rolling on an inclined plane.

The force moment is created by the stiff spring and produces a moment about the rotary joint of $\mathbf{M} = -kL\Delta\mathbf{e}_3$. Thus measurement of the spring displacement is a measure of the angular velocity of the spinning rotor normal to its spinning axis:

$$\Omega = kL\Delta/I_1\dot{\phi} \tag{2.40}$$

Energy Principle

The laws of linear and angular momentum (2.31) and (2.37), are both vector equations. We can also obtain a scalar equation representing changes in kinetic and potential energy measures.

Kinetic Energy

The energy associated with motion is defined by

$$T = \frac{1}{2}\sum m_i \mathbf{v}_i \cdot \mathbf{v}_i \tag{2.41}$$

Suppose time-independent constraints, $\mathbf{r}_i(q_k(t))$, restrict the motion to a set of k generalized coordinates $\{q_k(t)\}$, where $k = 1, 2, \ldots k < 3N$. Then the kinetic energy assumes the form

$$T = \frac{1}{2}\sum_k\sum_\ell m_{k\ell}\dot{q}_k\dot{q}_\ell$$

where the ($K \times K$) mass matrix $m_{k\ell}$ is related to the Jacobian vectors $J_{ik} = \partial \mathbf{r}_i / \partial q_k$, $i = 1, 2, \ldots N$, i.e., $m_{k\ell} = \sum_i m_i J_{ik} J_{i\ell}$.

The equations of motion, Eq. (2.30), can be put into the form of an energy principle if we take the inner product of each of the terms in (2.30) with the particle velocities $\mathbf{v}_i$ and sum overall the particles:

$$\sum \mathbf{v}_i \cdot m_i \dot{\mathbf{v}}_i = \sum \mathbf{v}_i \cdot \mathbf{F}_i + \sum \sum \mathbf{v}_i \cdot \mathbf{f}_{ij}$$

Then, using the definition of kinetic energy (2.41), we can write

$$\frac{d}{dt} T = \sum \mathbf{v}_i \cdot \mathbf{F}_i + \sum \sum \mathbf{v}_i \cdot \mathbf{f}_{ij} \qquad (2.42)$$

This means that the rate of change of kinetic energy of the system is equal to the power of the external and internal forces.

Potential Energy

Some forces such as gravitational forces or forces due to nondissipative elastic deformation can be represented by a scalar potential energy function V, i.e.,

$$\mathbf{F} = -\nabla V(\mathbf{r}) \qquad (2.43)$$

The symbol ∇ is called the gradient operator. If we write the position vector in terms of Cartesian coordinates, $\mathbf{r} = x\mathbf{e}_1 + y\mathbf{e}_2 + z\mathbf{e}_3$, then

$$\mathbf{F} = -\left(\frac{\partial V}{\partial x} \mathbf{e}_1 + \frac{\partial V}{\partial y} \mathbf{e}_2 + \frac{\partial V}{\partial z} \mathbf{e}_3 \right) \qquad (2.44)$$

For a system of particles, the power becomes (using $\mathbf{v}_i = \dot{x}_i \mathbf{e}_1 + \dot{y}_i \mathbf{e}_2 + \dot{z}_i \mathbf{e}_3$)

$$\sum \mathbf{v}_i \cdot \mathbf{F}_i = -\sum \left(\frac{\partial V}{\partial x_i} \frac{dx_i}{dt} + \frac{\partial V}{\partial y_i} \frac{dy_i}{dt} + \frac{\partial V}{\partial z_i} \frac{dz_i}{dt} \right) = -\frac{dV}{dt}$$

This assumption changes the expression (2.42) into,

$$\frac{d}{dt}(T + V) = \sum \sum \mathbf{v}_i \cdot \mathbf{f}_{ij} \qquad (2.45)$$

If the system is a rigid body, then we can show that the right-hand side is zero. If the system is nondissipative and the internal forces can also be derived from a potential energy V_e representing elastic deformations, then we obtain a conservation-of-energy principle.

$$\frac{d}{dt}(T + V + V_e) = 0 \qquad (2.46)$$

Conservation Laws

There are three fundamental conservation laws for the dynamics of a system of particles:

Conservation of Linear Momentum

$$\sum \mathbf{F}_i = 0, \quad \text{implies } \mathbf{P} = \text{constant} \tag{2.47}$$

Conservation of Angular Momentum

$$\mathbf{M} = \sum \mathbf{r}_i \times \mathbf{F}_i = 0, \quad \text{implies } \mathbf{H} = \text{constant} \tag{2.48}$$

Conservation of Energy

$$\mathbf{F} = -\nabla V \text{ etc.}, \quad \text{implies } T + V + V_e = \text{constant} \tag{2.49}$$

The use of conservation laws can sometimes greatly simplify a problem. They are also useful to check the accuracy of numerical solutions.

Energy Method: Single Degree of Freedom Systems

In the case of zero dissipation, it is usually possible to obtain an equation of motion directly from the energy conservation law (2.49). This method is usually effective if one chooses an obvious generalized coordinate $q(t)$. For example, in a pendulum the angular position $\theta(t)$ shown in Fig. 1.15 is an obvious candidate for the generalized coordinate. And the kinetic energy takes the form $T = (1/2)I\dot{\theta}^2$. In general, the kinetic energy depends on both the generalized position and velocity. If T represents the kinetic energy of the system and V represents the potential energy we can write

$$T(q(t), \dot{q}(t)) + V(q(t)) = E_0$$

To obtain an equation of motion, differentiate the above equation with respect to time:

$$\frac{d}{dt}[T(\dot{q}, q) + V(q)] = 0$$

or

$$\frac{\partial T}{\partial \dot{q}}\frac{d\dot{q}}{dt} + \frac{\partial T}{\partial q}\frac{dq}{dt} + \frac{dV}{dt} = 0 \tag{2.50}$$

Example 2.5 Energy Method: MEMS Device. As an application of the conservation of energy, consider the MEMS device in Fig. 2.9, with a compliant rotary mechanism bent by the attraction between two equal and opposite charges of strength Q Coulombs. From elementary physics we know that the electric force between these charges is inversely proportional to the distance between them. Thus we can deduce that the electric energy function is given by

$$E = -\frac{1}{2}\frac{Q^2}{4\pi\varepsilon_0}\frac{1}{\rho} \tag{2.51}$$

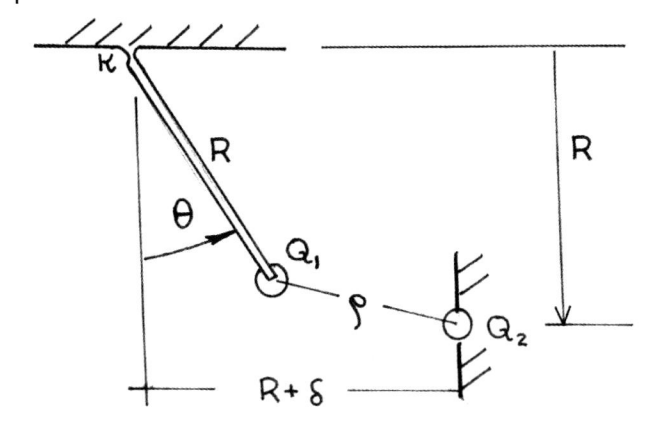

Fig. 2.9 MEMS compliant hinge lever with end mass and concentrated charge.

We assume that the movable charge is attached to a compliant cantilever that acts like a lever with a rotary spring of strength κ, and that the elastic energy function is given by

$$V = \frac{1}{2}\kappa\theta^2, \qquad T = \frac{1}{2}mR^2\dot{\theta}^2$$

Then by the conservation of energy, we can write a relationship relating the angular velocity and the bending angle:

$$\frac{1}{2}mR^2\dot{\theta}^2 + \frac{1}{2}\kappa\theta^2 - \frac{Q^2}{4\pi\varepsilon_0\rho^2} = E_0 \qquad (2.52)$$

where

$$\rho^2 = 2R^2(1 - \cos\theta) + (R + \delta)[(R + \delta) - 2R(R + \delta)\sin\theta]$$

In the above equation, E_0 is the electric energy when $\theta = \dot{\theta} = 0$.

2.5
Rigid Bodies

Angular Momentum and Moments of Inertia

The basic equations of linear and angular momentum for a system of particles (2.31) and (2.37) apply to rigid bodies and connected rigid bodies. In the case of a single rigid body (Fig. 2.10) we use the constraint that the relative velocity between two points in a rigid body, $\mathbf{r}_1, \mathbf{r}_2$ separated by a vector $\boldsymbol{\rho} = \mathbf{r}_2 - \mathbf{r}_1$, is given by

$$\dot{\boldsymbol{\rho}} = \boldsymbol{\omega} \times \boldsymbol{\rho} \qquad (2.53)$$

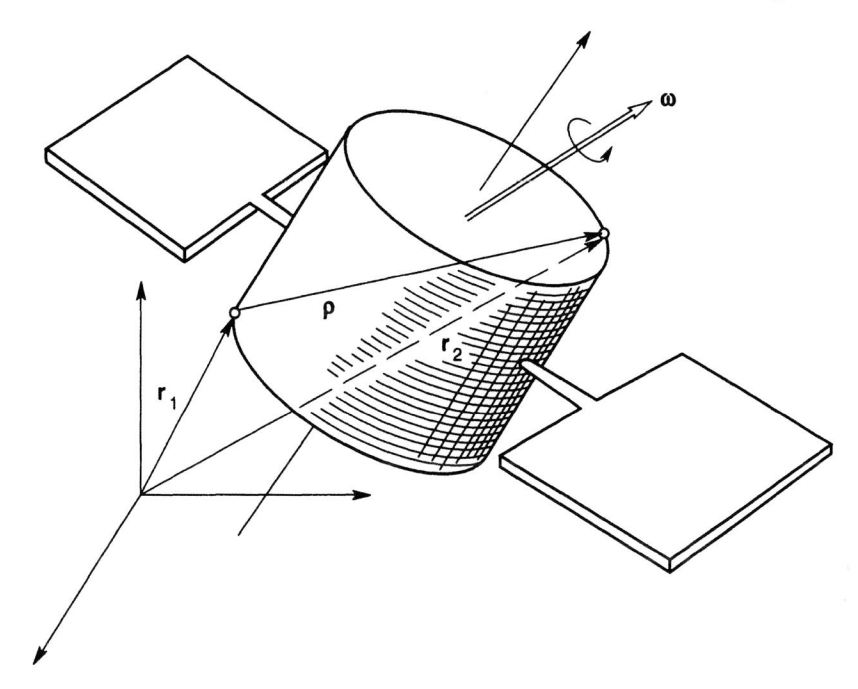

Fig. 2.10 Rigid body and rotation rate vector ω.

where ω is the rotation rate vector. The vector ω is a unique time-dependent vector that describes infinitesimal rotations of the body (see Chapters 3 and 5). Equation (2.53) can also be written in matrix form

$$\dot{\rho} = \bar{\omega}\rho \tag{2.54}$$

where

$$\bar{\omega} = \begin{bmatrix} 0 & -\omega_3 & \omega_2 \\ \omega_3 & 0 & -\omega_1 \\ -\omega_2 & \omega_1 & 0 \end{bmatrix}, \quad \rho = \begin{bmatrix} \rho_1 \\ \rho_2 \\ \rho_3 \end{bmatrix} \tag{2.55}$$

The subscripts refer to the components of the vectors ω, ρ referred to an orthogonal set of basic vectors. The ω vector is a property of the entire rigid body. This follows from *Euler's*[3] *theorem* (Chapter 3). The velocity of any point in the rigid body in pure rotation about a point can be represented as an instantaneous rotation about an axis with rate $|\omega|$ and direction of the axis of rotation given by the unit vector $\omega/|\omega|$. Thus every general motion of a rigid body is characterized by a translation of some point in the body (e.g., $\mathbf{v}_c$) and a rotation vector ω. The rotation vector is independent of the particular translation point chosen. With no constraints, $(\mathbf{v}_c, \omega)$ represents six state variables in addition to the three scalar positions, $\mathbf{r}_c$, and three angular variables representing the orientation of the body. Without constraints, the

---

[3] Leonard Euler (1707–1783), Swiss mathematician and physicist.

general motion of a rigid body is determined by the integration of *twelve* first-order differential equations.

With the recognition that $\boldsymbol{\omega}$ is a property of the entire rigid body, we can write the expression for the angular momentum separating the mass-distribution measures from the kinematic variables. In the case of a rigid system of particles in *pure rotation* about a fixed point, we have

$$\mathbf{H} = \sum \mathbf{r}_i \times (\boldsymbol{\omega} \times \mathbf{r}_i) m_i \tag{2.56}$$

This triple cross product can be rewritten using the identity

$$\mathbf{A} \times (\mathbf{B} \times \mathbf{C}) = (\mathbf{A} \cdot \mathbf{C})\mathbf{B} - (\mathbf{A} \cdot \mathbf{B})\mathbf{C}$$

or

$$\mathbf{H} = \sum m_i ((\mathbf{r}_i \cdot \mathbf{r}_i)\boldsymbol{\omega} - \mathbf{r}_i (\mathbf{r}_i \cdot \boldsymbol{\omega})) \tag{2.57}$$

This expression is usually written in matrix form as

$$\mathbf{H} = \mathbf{I}\,\boldsymbol{\omega} \tag{2.58}$$

Given an orthonormal set of basis vectors $\{\mathbf{e}_i\}$, the components of the symmetric matrix I are given by

$$I_{kj} = \sum_{i=1}^{N} m_i (\delta_{kj} r_i^2 - (\mathbf{r}_i \cdot \mathbf{e}_k)(\mathbf{r}_i \cdot \mathbf{e}_j)) \tag{2.59}$$

where $\mathbf{r}_i^2 = \mathbf{r}_i \cdot \mathbf{r}_i$, and $[\delta_{kj}]$ is the 3×3 identity matrix. (Note that the index i is summed over all the masses in the system, while k, j are summed over the three components of the position vector.)

In conventional Cartesian coordinates $\mathbf{r} = [x, y, z]^T$, we have

$$I_{11} = \sum_{i=1}^{N} m_i (y_i^2 + z_i^2)$$

$$I_{12} = -\sum_{i=1}^{N} m_i x_i y_i \tag{2.60}$$

The first term is called the *moment of inertia* about the x-axis (also called the *second moment of mass*), and the second term is called a *cross product of inertia*.

Using basic theorems from linear algebra, we can show that there exists an orthogonal set of basis vectors for which the cross products of inertia are zero. The remaining diagonal terms I_{11}, I_{22}, I_{33}, are called the *principal inertias* $\{I_1, I_2, I_3\}$. If we then write the components of I in the same principal coordinates, with unit vectors $(\mathbf{e}_1, \mathbf{e}_2, \mathbf{e}_3)$, the angular momentum vector takes a rather simple form

$$\mathbf{H} = I_1 \omega_1 \mathbf{e}_1 + I_2 \omega_2 \mathbf{e}_2 + I_3 \omega_3 \mathbf{e}_3 \tag{2.61}$$

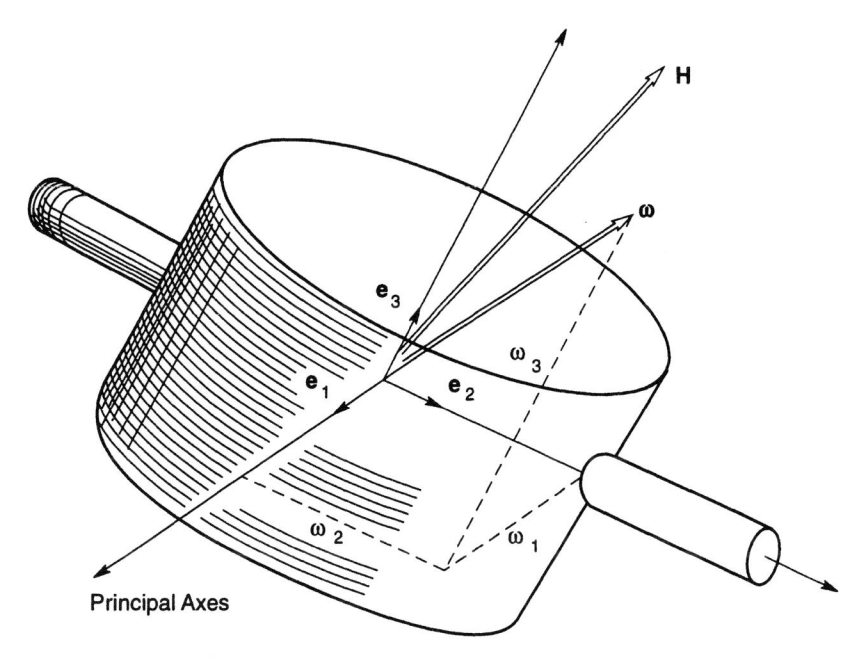

Fig. 2.11 Principal axes $\{e_1, e_2, e_3\}$ that move with the body.

In general, the vectors $\{e_i\}$ will move and rotate with the rigid body (Fig. 2.11), except in the cases of cylinders and spheres. What is confusing to many students is that $\mathbf{H}$ and $\dot{\mathbf{H}}$ represent kinematic measures of motion with respect to an inertial frame but are expressed in terms of a coordinate system that is not inertial (see Chapter 5 for further discussion).

For both translation and rotation of a rigid body, it is usually convenient to choose the center of mass as the translation point. In this case we use the angular-momentum equation about the center of mass and the expressions (2.59), (2.60) are understood to be calculated about $\mathbf{r}_c$. To calculate $\dot{\mathbf{H}}$, we must recognize that the $\{\dot{\mathbf{e}}_i\}$ are not zero and are given by

$$\dot{\mathbf{e}}_i = \boldsymbol{\omega} \times \mathbf{e}_i \tag{2.62}$$

This equation along with (2.37) and (2.61) yields the famous Euler's equations of motion for a rigid body.

$$I_1 \dot{\omega}_1 + (I_3 - I_2)\omega_2 \omega_3 = M_1 \tag{2.63a}$$

$$I_2 \dot{\omega}_2 + (I_1 - I_3)\omega_1 \omega_3 = M_2 \tag{2.63b}$$

$$I_3 \dot{\omega}_3 + (I_2 - I_1)\omega_1 \omega_2 = M_3 \tag{2.63c}$$

This set of first-order differential equations can be integrated once the components of the resulting applied moment vector are known as functions of time. These equations contain nonlinear terms which are sometimes called *gyroscopic* terms [see, e.g., (1.19)]. These terms are often responsible for some nonintuitive dynamics of spinning bodies [see (1.19), (1.20)] discussed in Chapters 1 and 5. In the case of rotation about a fixed axis, however, these nonlinear terms drop out and we simply have

$$I_1 \dot{\omega}_1 = M_1 \tag{2.64}$$

which is a form usually taught in high school or first year college physics courses in dynamics.

2.6
D'Alembert's Principle

The extension of the principle of virtual work (2.18) or (2.19) to dynamical problems is called D'Alembert's[4] principle and is covered in more detail in Chapter 4. It is not usually treated in a first course in dynamics. However, to complete our summary of basic equations of dynamics, it is included here for later reference.

The basic idea of the principle of virtual work is to project the forces onto directions in which the forces of constraint will drop out [see (2.18)]. These directions are chosen by imagining small displacements of the system in which the constraint forces do no work.

The same idea can be applied to Newton's law for a single particle of mass m, acted on by several active forces $\{\mathbf{F}^a_j\}$, and a constraint force $\mathbf{F}^c$. Following (2.18), (2.19) we seek directions $\boldsymbol{\beta}_i$ where

$$\left(\sum \mathbf{F}^a_j + \mathbf{F}^c - m\mathbf{a}\right) \cdot \boldsymbol{\beta}_i = 0 \tag{2.65}$$

and

$$\mathbf{F}^c \cdot \boldsymbol{\beta}_i = 0 \tag{2.66}$$

There are several methods for choosing the $\boldsymbol{\beta}_i$. In D'Alembert's method the inertia term $-m\mathbf{a}$ is treated as an effective force, and the principle of virtual work (2.18) is extended to the dynamics problem, i.e.,

$$\left(\sum \mathbf{F}^a_j - m\mathbf{a}\right) \cdot \delta\mathbf{r} = 0 \tag{2.67}$$

where $\delta\mathbf{r}$ does not violate the constraints. When $\delta\mathbf{r}$ is constrained by a surface (Fig. 2.12), we can choose two independent variables $\{q_1(t), q_2(t)\}$ and associated virtual displacements $\{\delta q_1, \delta q_2\}$ that lead to an equation similar to (2.19):

---

4) Jean Le Rond D'Alembert (1717–1783), French mathematician and mechanician. In 1743 he published *Traite' de dynamique* where his famous principle is developed which he had proposed in 1742.

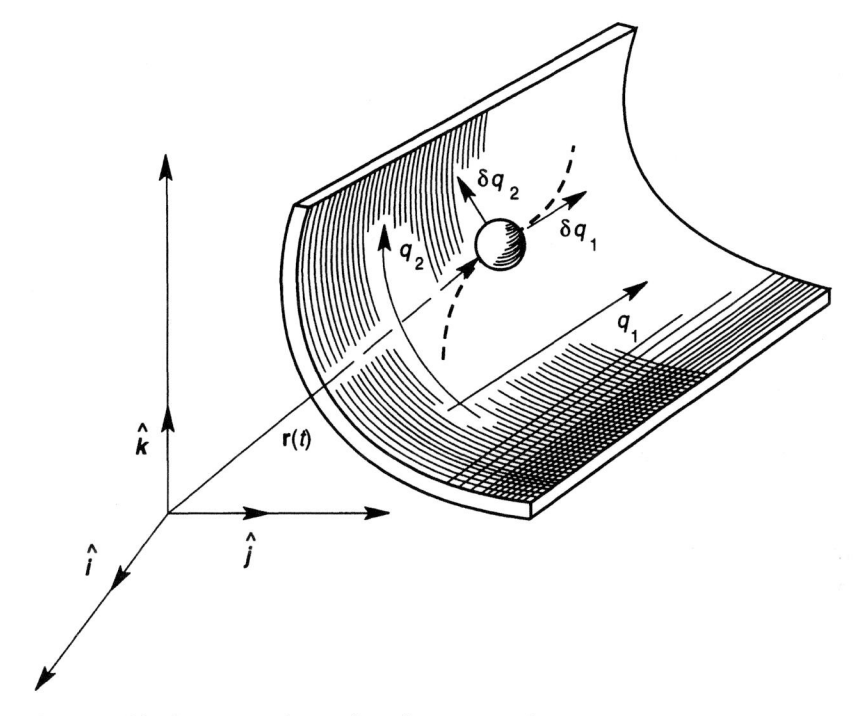

Fig. 2.12 D'Alembert's principle – surface of constraint with generalized coordinates $\{q_1(t), q_2(t)\}$.

$$\left(\sum \mathbf{F}_j^a - m\mathbf{a}\right) \cdot \frac{\partial \mathbf{r}}{\partial q_i} = 0 \tag{2.68}$$

Here the projection directions $\boldsymbol{\beta}_i$ are given by the Jacobian,

$$\boldsymbol{\beta}_i = \frac{\partial \mathbf{r}}{\partial q_i} \tag{2.69}$$

This requires that we express the constraint in the form $\mathbf{r} = \mathbf{r}(q_1, q_2, t)$. Such constraints are called holonomic. Some rolling constraints cannot be expressed in this form (see Chapter 4.)

D'Alembert's method can be directly applied to dynamic problems. It is also the basis for the derivation of another method, namely, *Lagrange's equations* (see Section 4.3).

Constraint Forces

The elimination of the constraint forces from the equations of motion is one advantage of D'Alembert's method, when one wants to determine the time history of the generalized coordinates. However in design, engineers often need to know the limits of the constraint forces such as the loads on the bearings. Constraint forces can be determined after the dynamic equations have been solved, by using the following equations:

$$\mathbf{F}_i = \mathbf{F}_i^a + \mathbf{F}_i^c$$

where $\mathbf{F}_i^a$ is the known applied or actuator force on the ith particle, and $\mathbf{F}_i^c$ is the unknown constraint force on the same particle.

$$\sum_{i=1}^{N} (\mathbf{F}_i^a - m_i \dot{\mathbf{v}}_i) \cdot \frac{\partial \mathbf{r}_i}{\partial q_k} = 0 \qquad (2.70)$$

After one solves for the velocities and accelerations, the constraint forces can be solved using the following equations:

$$\mathbf{F}_i^c = m_i \dot{\mathbf{v}}_i - \mathbf{F}_i^a$$

The constraint forces in this formulation are such that they do no work on the system, i.e.,

$$\sum_{i=1}^{N} \left(\mathbf{F}_i^c \cdot \frac{\partial \mathbf{r}_i}{\partial q_k} \right) = 0$$

Often in solving the dynamics problem, the generalized velocities are known and the particle velocities are given through a Jacobian relationship

$$\mathbf{v}_i = J_i \dot{Q}$$

where $\dot{Q} = [\dot{q}_1, \dot{q}_i, \ldots \dot{q}_M]^T$ is a column vector of generalized velocities. For example in the case of the pendulum, $q(t)$ might be the angle of the pendulum $\theta(t)$. In any case, the constraint forces can be determined from the equations

$$\mathbf{F}_i^c(t) = m_i \frac{d}{dt}[J_i \dot{Q}] - \mathbf{F}_i^a(t) \qquad (2.71)$$

Again in the case of a particle suspended on a pendulum of length L, as in Chapter 1, Fig. 1.15, the active force is the force of gravity, and the radial constraint force in the massless rod T is given by the centripetal acceleration forces less gravity, i.e.,

$$T = mL\dot{\theta}^2 + mg \cos \theta$$

where

$$\ddot{\theta}(t) + \frac{g}{L} \sin \theta(t) = 0$$

The time history of the angle of the pendulum is determined by the initial angle and initial angular velocity. Thus the dynamic constraint forces depend and both the applied forces as well as the initial conditions.

2.7
The Principle of Virtual Power

The second method of choosing independent projection directions $\{\boldsymbol{\beta}_i\}$ for Newton's equation is based on calculating the power of the forces and the effective inertia force under small changes in the velocity $\delta\mathbf{v}$. Again, the velocities are chosen consistent with the constraints, which in the absence of friction eliminates the constraint forces from Eq. (2.65), i.e.,

$$\left(\sum \mathbf{F}_j^a - m\mathbf{a}\right) \cdot \delta\mathbf{v} = 0 \tag{2.72}$$

This method has advantages over the D'Alembert method when the constraints depend on velocities as well as the position variables. For example, such constraints arise in problems of rolling of one body on another. This method has been codified in papers and books by T. R. Kane and co-workers of Stanford University and is sometimes known as Kane's equation. The method is also taught in European universities as *Jourdain's principle* based on a paper published in 1902.

Without further derivation or discussion, we state the simple form of this method when generalized coordinates $\{q_1, q_2\}$ are chosen along with generalized speeds $\{\dot{q}_1, \dot{q}_2\}$.

$$\left(\sum \mathbf{F}_j^a - m\mathbf{a}\right) \cdot \frac{\partial \mathbf{v}}{\partial \dot{q}_i} = 0, \quad i = 1, 2 \tag{2.73}$$

When the constraints in the problem are purely geometric with no velocity dependence (i.e., holonomic constraints), then the two methods are essentially the same since one can show that the Jacobian matrix can be calculated in two ways (see Chapter 4):

$$\frac{\partial \mathbf{r}}{\partial q_i} = \frac{\partial \mathbf{v}}{\partial \dot{q}_i} \tag{2.74}$$

where $\mathbf{v} = \dot{\mathbf{r}}$.

Example 2.6 Virtual Power Method. As an example of the virtual power method consider the massless cylinder in Fig. 2.13 with a concentrated mass. The cylinder rolls on the inclined plane without slipping. We assume that this rolling takes place without any loss of energy. The problem here is to determine the equilibrium angle θ of the mass on the cylinder at which the normal contact force, friction force and gravity force are all in equilibrium. One could use the direct form of Newton's law or virtual work. Instead we imagine that the cylinder is slowly in motion with an angular velocity $\dot{\theta}$. In (2.73) we set $\mathbf{a} = 0$. With a background in basic kinematics one can write the vector velocity of the mass and apply the principle of virtual velocity as follows:

$$\mathbf{v} = R\dot{\theta}\mathbf{e}_s + \rho\dot{\theta}\mathbf{e}_\theta$$
$$\frac{\partial \mathbf{v}}{\partial \dot{\theta}} = R\mathbf{e}_s + \rho\mathbf{e}_\theta$$

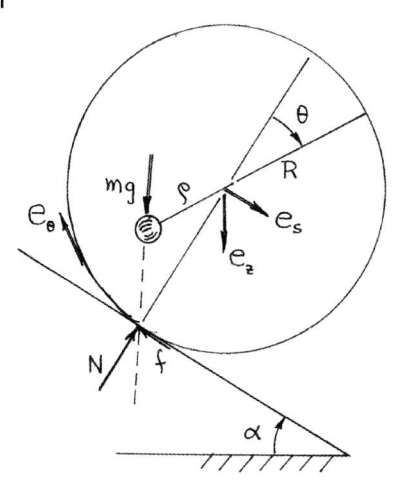

Fig. 2.13 Massless cylinder with point mass m on an inclined plane.

The equilibrium condition is given by

$$\mathbf{F}^a \cdot \frac{\partial \mathbf{v}}{\partial \dot{\theta}} = 0 \qquad (2.75)$$

where

$$\mathbf{F}^a = mg\mathbf{e}_z$$

This relation yields the following transcendental equation for the angle θ, as a function of the angle of the inclined plane α:

$$mg[\rho \sin(\theta + \alpha) - R \sin \alpha] = 0 \qquad (2.76)$$

It is interesting to note that the solution is independent of the mass m. As α increases, the cylinder rotates so that the gravity force goes through the contact point.

Homework Problems

2.1 (*Free Body Diagram*) Consider the rolling pendulum in Fig. 1.15 upper right-hand corner. It consists of two moving rigid bodies and a fixed grounded body. Draw a free-body diagram for the two moving masses and also show the forces on the grounded body. Assume there is no friction in the revolute joint.

2.2 (*Free Body Diagram*) The four-bar mechanism shown in Fig. 1.5 is a component in many machines such as the elliptical trainer exercise machine. Assume that three of the four rigid body links are moving and the lower link is grounded and that a driving torque is acting on the right-hand link on the lower revolute joint. Draw a free body diagram for each of the three moving links and include gravity forces acting at the center of each link.

2.3 (*Energy Concepts*) The force between two charges is proportional to the square of the distance between the charges [Eq. (1.21)]. The force is attractive when the charges are of opposite sign. Use Eq. (2.43) to derive a force potential $V(r)$ for the electric force.

2.4 (*Energy Concepts*) The force between two parallel wires carrying electric currents I_1, I_2, is proportional to the product of the currents and the distance between the wires. The force is repulsive when the currents are in opposite directions. Use Eq. (2.43) to derive a force potential $V(r)$ for the magnetic force between two wires.

2.5 Set $z = 0$ in Figs. 2.1 and 2.2. Use the fact that $x = r\cos\theta$, $y = r\sin\theta$ to derive expressions for velocity and acceleration in polar coordinates (2.5), (2.8).

2.6 A particle moves on a helical path on a cylinder of radius R and helical angle α (Fig. P2.6). Derive the Jacobian for the vector velocity **v** in terms of the angular rate about helical axis $\dot{\theta}$ [see (2.9)].

2.7 Consider the particle constrained to move on a circular path in a vertical plane as in Fig. 2.4c. Assume that gravity acts on the mass in the negative vertical direction and a constant force F_0 acts in the horizontal x-direction. Use the polar coordinates $z = R\sin\theta$, $x = R\cos\theta$.
  (a) Find the Jacobian matrix as in (2.19).
  (b) Use the virtual work equilibrium condition (2.19) to find the equilibrium angle given by $\tan\theta = mg/F_0$.

2.8 Four mass particles of masses $\{m, m, m, 2m\}$ are located, respectively, as follows: (1) at the origin; (2) along the x-axis at $x = a$; (3) along the y-axis at $y = b$; and (4) at the position $x = a, y = b, z = c$ (Fig. P2.8). Find the center-of-mass vector (2.22). Find the positions of each particle ρ_i relative to the center of mass. Sketch these vectors in an isometric view. Try writing a MATLAB program to draw a graphical sketch of the four vectors.

2.9 Suppose each of the masses in Problem 2.8 has a gravity force acting on them in the negative z-direction. Calculate the moment vector of the mass system about the origin. What is the component of the moment about the z-axis?

2.10 For the set of particles in Problem 2.8, find the moment of inertia matrix [Eqs. (2.59), (2.60)] with respect to the origin. Can you write a MATLAB program to find I_{ij} for any set of mass particles? (Answer: $I_{11} = m(3b^2 + 2c^2)$, $I_{22} = m(3a^2 + 2c^2)$, $I_{33} = 3m(b^2 + a^2)$, $I_{12} = -2mab$, $I_{13} = -2mac$, $I_{23} = -2mbc$.)

2.11 Consider the set of particles in Problem 2.8 as a rigid body and the rotation vector $\boldsymbol{\omega}$ is given by $\boldsymbol{\omega} = \omega_0 \cos\alpha\hat{\mathbf{i}} + \omega_0 \sin\alpha\hat{\mathbf{j}}$. Calculate the angular momentum vector (2.58). Are **H** and $\boldsymbol{\omega}$ parallel?

2.12 Suppose a rigid rectangular body is oriented with respect to a rectangular coordinate system such that one vertex is at the origin and three others are at $[A, 0, 0]$, $[0, B, 0]$, $[0, 0, C]$ (Fig. P2.12). Assume that the body rotates about an axis that goes through the origin and the vertex $[A, B, C]$. Find the matrix representation of the angular velocity $\bar{\boldsymbol{\omega}}$ as in (2.55).

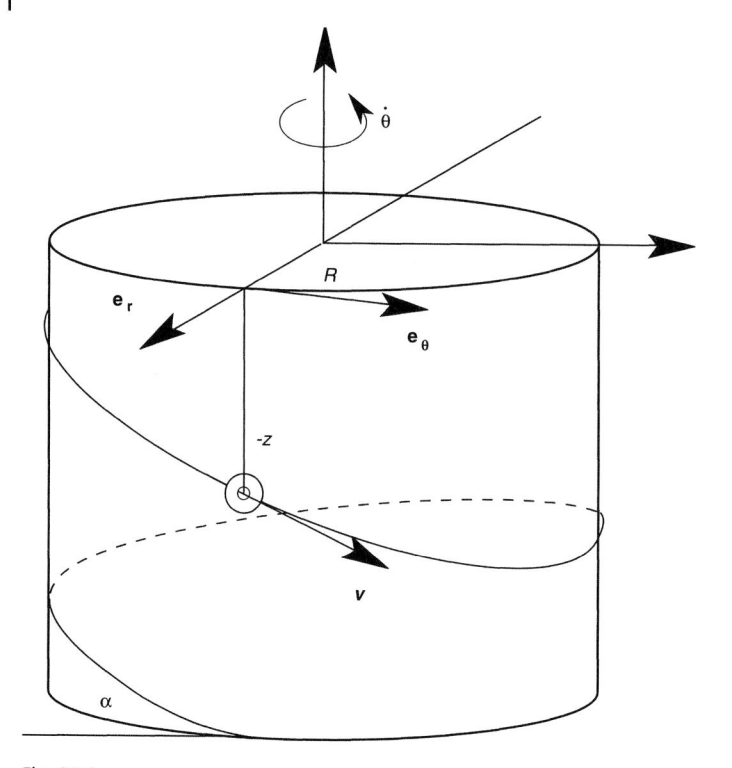

Fig. P2.6

2.13 Two equal masses rigidly connected by a link of length L move in a plane (Fig. P2.13). One mass is constrained to move in the horizontal direction located at $r = [x(t), 0, 0]^T$. The angular position of the second mass with respect to the vertical is $\theta(t)$. Choose generalized coordinates $q_1 = x(t)$, $q_2 = \theta(t)$ and show that the Jacobian matrix can be calculated in two ways as in (2.74).

2.14 Show that the potential energy function V defined by (2.43) for Newton's law of gravity is given by, $V = Gm_1m_2/r_{12}$. Use the conservation of angular momentum (1.23) or (1.24) and conservation of energy to relate the orbital radii and circumferential velocities at the apogee and perigee of an elliptic orbit of a small mass m_2 about a large mass m_1.

2.15 A mass is connected by two equal springs of stiffness k (Fig. P2.15). Initially the mass is at the origin and the two springs are stretched along the x-axis with an initial tension $F_0 = kd$. If the initial spring length is 'a' find an expression for the elastic energy function in (2.43) if the mass is moved transverse to the initial stretched configuration by an amount y. (*Hint*: Use Eq. (2.43); show that $\mathbf{F}(y) = 2ky(L-a)/L$, where $L = (y^2 + (a+d)^2)^{1/2}$; try solving the problem using a symbolic code such as *MATHEMATICA* or *MAPLE*.)

Homework Problems | 71

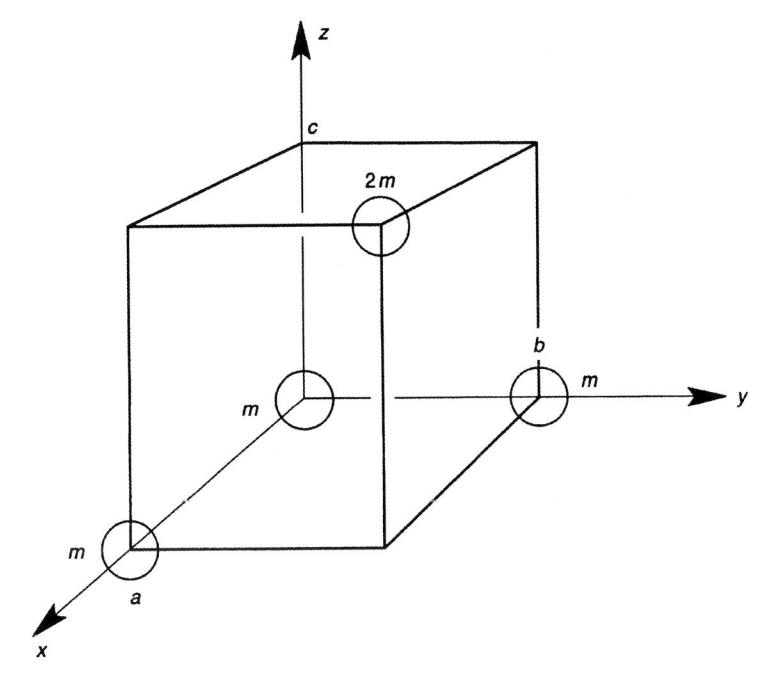

Fig. P2.8

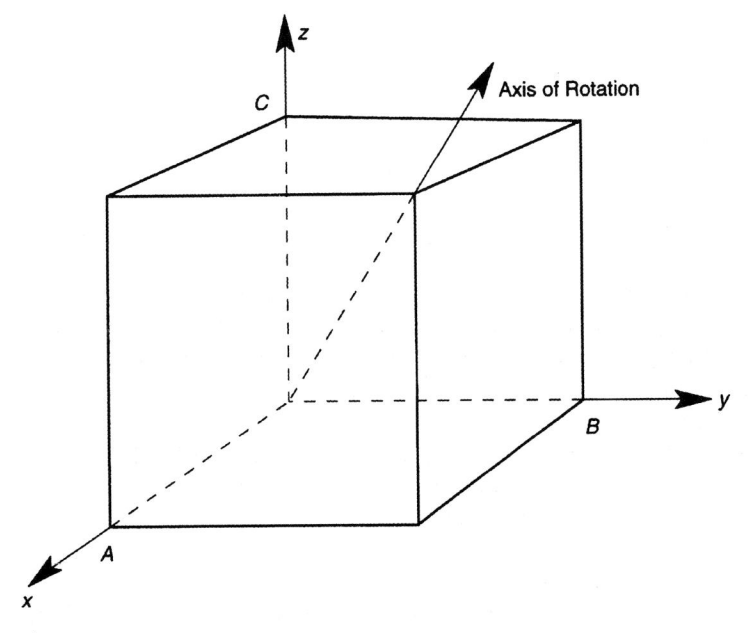

Fig. P2.12

2 Basic Principles of Dynamics

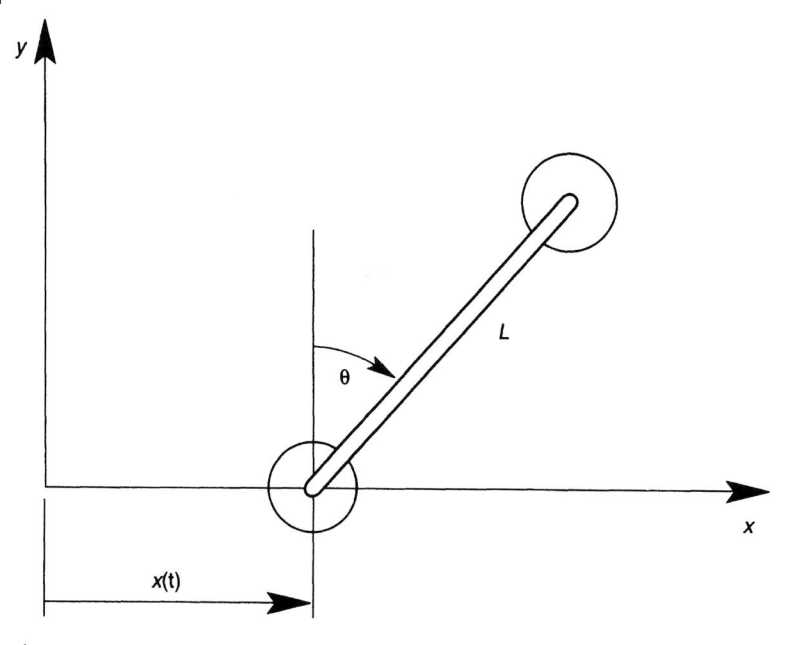

Fig. P2.13

A simple program in *MATHEMATICA* consists of the following input statements. (Here $k = 1$ in Problem 2.15.) This program plots $F[y]$ for $a = d = 1$, integrates $F[y]$ symbolically to get an energy function (Out[1]), differentiates the energy (Out[2]), and obtains a Taylor series of the force function (Out[3]). The symbol % indicates "previous statement" in the functions below.

```
L = Sqrt[y^2 + (a+d)^2]
F[y] = 2y(L-a)/L
a = 1
d = 1
F[y]
Plot[%, y, 0, 6]
Integrate[% %, y]
```
$$\text{Out}[1] = y^2 - 2\,\text{Sqrt}[4 + y^2]$$
```
Plot[%, y, 0, 6]
D[% %, y]
```
$$\text{Out}[2] = 2y - \frac{2y}{\text{Sqrt}[4 + y^2]}$$
```
Series[%, y, 0, 6]
```
$$\text{Out}[3] = y + \frac{y^3}{8} - \frac{3y^5}{128} + o[y]^7$$

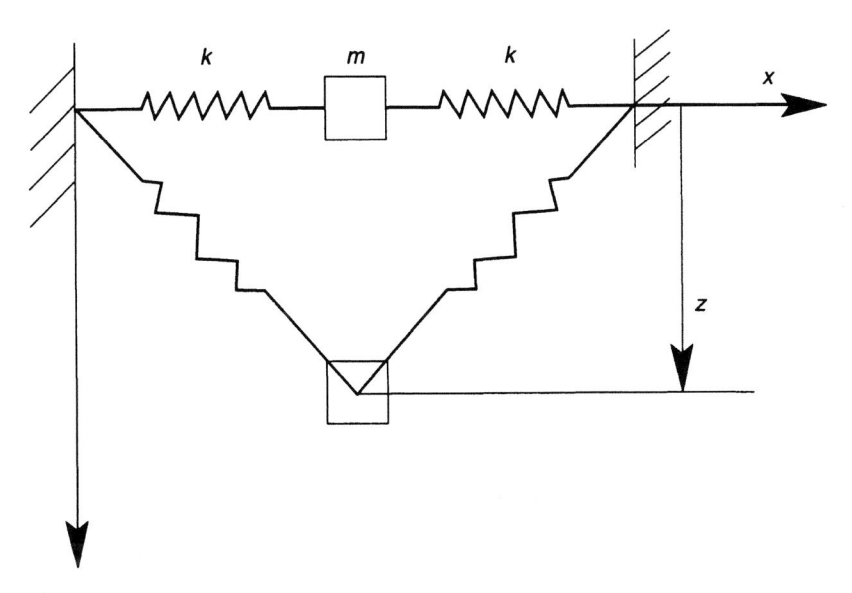

Fig. P2.15

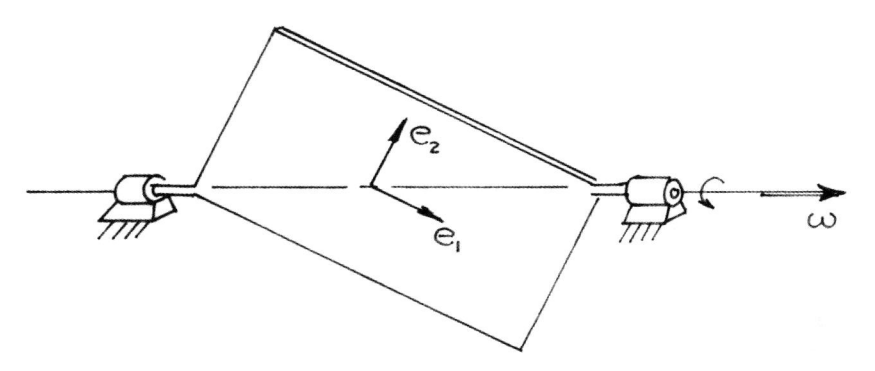

Fig. P2.16

2.16 (*Angular Momentum*) The idea of the change of angular momentum vector and applied moments captured in Eqs. (1.18), (1.19) for the spinning gyro wheel of Fig. 1.20, can be applied to the rotation of a thin rectangular plate about a diagonal axis through opposite corners as shown in Fig. P2.16. When the plate spins on the diagonal axis with constant speed, the angular momentum is off axis and rotates with the plate.
  (a) Use the Eqs. (2.61) and (2.40) to conclude that a force-moment must exist normal to the plate and that this moment is realized through equal and opposite bearing forces normal to the plate.
  (b) Draw a free-body diagram illustrating the conclusions of part (a).

2.17 (*Energy Concepts*) Consider the problem of a disc rolling down an inclined plane as shown in Fig. P2.17. When the rotary kinetic energy of the disc is

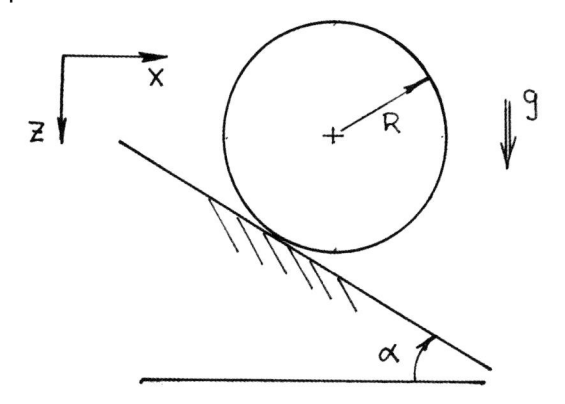

Fig. P2.17

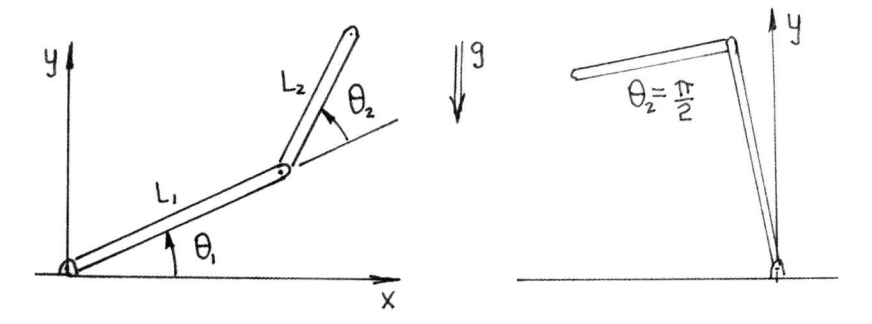

Fig. P2.18

neglected, we know from elementary physics that the velocity of the center of the mass is given by $\sqrt{(2gz)}$, where g is the gravity constant at the surface of the earth and z is the vertical drop. The additional kinetic energy due to rotation about the center of mass can be shown to be equal to the moment of inertia of the disc about the center of mass and one half the square of the angular velocity. Use Eq. (2.46) to derive the velocity of the center of mass with the correction for the rotary kinetic energy. Note that one must use the constraint between the angular velocity and the velocity of the center of mass of the disc.

2.18 (*Energy Concepts*) A two link arm shown in Fig. P2.18 is a model for the dynamics of a robot manipulator arm. Suppose the angle between the two arms is fixed at 90 degrees as shown and the inner arm is initially in the vertical position at zero velocity. Assuming that the arms fall under gravity with no frictional losses, use the conservation of energy principle to find the angular velocity of the constrained links when the inner arm is in the lower vertical position.

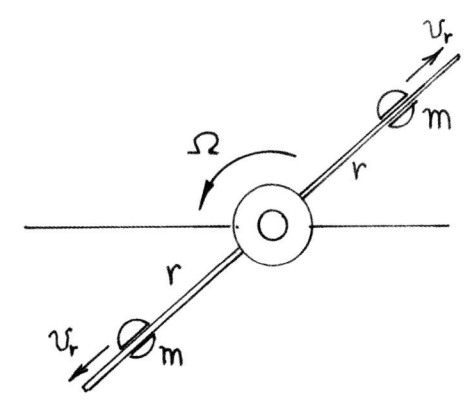

Fig. P2.20

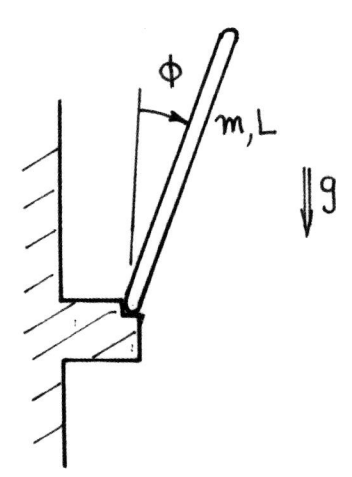

Fig. P2.21

2.19 (*Energy Concepts*) Do Problem 2.18 for the case when the angle between the two arms is arbitrary and fixed.

2.20 (*Conservation of Angular Momentum*) Two masses slide on frictionless radial arms as shown in Fig. P2.20. The arms are attached to a vertical shaft that is free to rotate about the vertical direction. Initially the arms rotate with angular velocity Ω_1, when the masses have zero radial velocity at position r_1. Because the masses are confined to the horizontal plane, gravity forces do no work. Use the principles of conservation of energy and angular momentum to find the radial velocity of the masses and the new angular velocity of the arms when the masses have slid from position r_1 to r_2.

2.21 (*Angular Momentum*) A rod of uniform mass of length L sits on a ledge and falls under the pull of gravity, and is assumed to act at the center of the rod (Fig. P2.21). Use the energy principle (2.49) to calculate the angular velocity of the rod when it reaches the horizontal position. Suppose at the horizontal position the constraint is removed and the rod is free to fall under gravity. Use the conservation of energy and angular momentum to describe the subsequent motion of the rod. Sketch the position at several times after loss of the constraint.

3
Kinematics

3.1
Introduction

For many students, kinematics is the most difficult element of applying dynamics to specific problems. Although kinematics is usually taught in an introductory course, the art of translating geometric and kinematic constraints into mathematics requires skill and practice in order to solve more advanced problems in dynamics. A brief introduction and review of kinematics was presented in the previous chapter. In this chapter we elaborate on the basic ideas and develop tools that are necessary to apply to problems.

An excellent review of modern kinematics as well as the historical background of the subject may be found in the book edited by Erdman (1993). A good tutorial textbook on kinematics may be found in Beatty (1986). A modern text on kinematic synthesis of mechanisms is the 4th edition of Erdman et al. (2001). The modern subject ranges from general mathematical principles to applications to specific classes of machines and mechanisms. The study of basic mechanisms such as the lever, wedge, wheel, and screw goes back to the Greeks, 300 B.C. The Greek word for both machine and mechanism is MHXANH or "mechane." The study of the principles of machines was undertaken by Aristotle (ca. 300 B.C.) He and his students formulated a principle of *virtual velocities* and the concept of equilibrium as related to energy or virtual work.

By the nineteenth century, the list of basic mechanisms used in the expanding industrial revolution grew to hundreds of linkages, gearing, and many exotic configurations. One of the first attempts to catalog and classify these mechanisms was by Franz Reuleaux (1875) in his *Theoretical Kinematics*, or as his translator Kennedy renamed it, *Kinematics of Machinery* (1876). Reuleaux built several hundred models with which to teach kinematics at the Technical University of Berlin.[1] (See Section 3.8, Fig. 3.18.) However by the 1960s, the use of mechanisms as a basis of teaching kinematics was disappearing from textbooks in favor of more generic mathematical formulas. Now at the end of the twentieth Century, there is a re-

---

1) One of the few remaining collections is at Cornell University.

Applied Dynamics. With Applications to Multibody and Mechatronic Systems. Second Edition. Francis C. Moon
Copyright © 2008 WILEY-VCH Verlag GmbH & Co. KGaA, Weinheim
ISBN: 978-3-527-40751-4

newed effort to apply computational and mathematical kinematics to real industrial mechanisms and machines, and there are now many codes and software to help the engineer design and analyze kinematic systems.

A modern catalog of kinematic devices may be found in the Russian two volume work of Artobolevsky (1979), which includes linkages, cams, ratcheting devices, clutches, parts feeders, positioning devices, indexing mechanisms, and aerospace-related mechanisms such as control linkages. For an on-line catalog of kinematic mechanisms, the reader is referred to the Cornell University website KMODDL – Kinematic Models for Design Digital Library (http://kmoddl.library.cornell.edu).

Kinematics is the Description of Motion

There are two types of kinematic tools: one to understand infinitesimal motions such as velocities and accelerations, and the other to describe finite motions of bodies.

The solution of dynamics problems generally involves both *kinematics* and *kinetics*. Kinematics is the process of describing motions; kinetics (or dynamics) involves determining the forces and moments or the resulting motions produced by these forces. The basic steps in the kinematic part of the problem are:

1. Define the degrees of freedom.
2. Assign reference frame.
3. Establish constraint relationships.
4. Relate velocities and angular velocities to constraints and geometry.
5. Calculate accelerations.

Another distinction is between kinematic and dynamic devices. In a *kinematic device* there are no free degrees of freedom. Once the motion of one link is given, then the motion of all the other linkages or connected parts are given. Gear transmissions are one example of such *zero-degree-of-freedom* devices. The dynamics problem here is to calculate the resulting forces in the mechanism from the imposed motions and acceleration.

A *dynamic device* has at least one degree of freedom whose motion is determined using both kinematic relations and the Newton–Euler laws of motion under applied forces. Usually devices with elastic elements fall into this class of problems. Also if a kinematic device violates a constraint, e.g., a wheel of a vehicle leaves the ground, then a new degree of freedom emerges and a kinematic problem can become a dynamic one.

Kinematics of Machines and Mechanisms

The description of motion of single particles and rigid bodies is usually introduced in the first course of dynamics. The next level of complexity is the description of motion of connected rigid bodies in mechanisms, machines and robotic devices. Kinematics of machines and mechanisms has a long mathematical history spanning over 150 years and the geometric description of mechanisms reaches back into the Renaissance and even classical antiquity of the Greek writers. (See Moon (2007) for a review of the history of machine kinematics from the time of Leonardo da Vinci to the late 19th century.) Although not as universally taught as in the past, the subject of kinematics of machines has seen a revival in its application to robotics. (See, e.g., Angeles, 1997; Ceccarelli, 2004.) Our goal in Section 3.8 is not to duplicate the material in advanced specialized books on kinematics of machines but to offer some of the flavor of the material.

A short list of modern applications of kinematic mechanisms includes the following:

Automobiles

> Engine components: pistons, crankshaft, cams, valves
> Gear transmission: planetary gears
> Rear axel: universal joint and differential, ball bearings
> Brake system
> Doors, hatches, hood

Aircraft

> Fuselage: passenger and cargo doors and hatches
> Landing gears
> Wings: flaps, control surfaces, ailerons
> Engine: bearings, gearing

Robotics

> Manipulator arm linkage
> End effecter, grippers
> Hydraulic actuators, pistons and valves
> Mobile robotic machines

Bio-engineering

> Artificial limbs: linkages and cable systems
> Artificial hands
> Joint replacements: bearings
> Mobile chairs: wheels, brakes

Space Technology

> Spacecraft antenna: folded structures
> Solar panels
> Control moment gyros, reaction wheels
> Planetary rover vehicles
> Space shuttle robotic arm

Manufacturing

> Machine tools: lathes, milling centers
> Assembly line components

Electronics and Computer Technology

> Cameras: lens focus mechanism
> Disc drives, microdrives
> Video recording and playback devices
> Computer printers: belt drive mechanisms

Food Production

> Farm machinery: plowing, seeding, harvesting
> Food packaging and bottling machines

Construction Machines

> Cranes: pulleys, cables
> Backhoes, loaders

In the past two decades, commercial software packages have appeared such as *Solid Works, Working Model, ADAMS, DADS*, that give the engineer new tools to animate, calculate dynamic motions as well as forces in complicated machines. Although these software tools free the engineer from learning arcane methods of analysis, it is still important for the student to gain an educated intuition about basic multi-body mechanisms such as the slider-crank, four-bar linkages and serial and parallel linkages used in robotics if for no other reason than to be able to recognize when codes produce realistic data and when they do not. There are many very good books on kinematics of mechanism design.

Finally the reader should note the distinction between analysis and synthesis of multibody machines. The focus of this book in on analysis. However the goal of the engineer is often to find a mechanical assemblage of bodies that will produce a given motion with force and torque constraints. The subject of kinematic synthesis is a difficult and ever evolving subject, using mathematical tools such as optimization theory that are beyond the scope of this book. Nonetheless it is the Author's goal to encourage the student in multi-body analysis and design to read further in the field of machine and mechatronic synthesis.

3.2
Angular Velocity

The concept of an angular velocity vector is one of the most important tools in the application of dynamics to rigid bodies and complex multibody problems. Although it is treated in elementary courses, the skilled use and understanding of angular velocity and acceleration usually takes further study and practice.

The derivation of an angular velocity vector can be obtained from *Euler's theorem* (see, e.g., Goldstein, 1980). This theorem states that the most general motion of a rigid body about a point is equivalent to a finite rotation about a unique axis. If we consider the rotation to be an infinitesimal angle $\Delta\theta$ in a time Δt, then the theorem says that the velocity of any point in a rigid body, with one fixed point, can be derived from the instantaneous circular motion about a fixed axis.

Referring to Fig. 3.1a, the mathematical form of the preceding theorem says that we can identify an axis with unit vector $\mathbf{n}$, and a vector

$$\boldsymbol{\omega} = \frac{d\theta}{dt}\mathbf{n} \tag{3.1}$$

such that the velocity of any point in the body is given by the cross-product operation

$$\mathbf{v} = \boldsymbol{\omega} \times \mathbf{r} \tag{3.2}$$

where $\mathbf{r}$ is a position vector from the fixed point to the moving body point. Looking down the axis at a plane normal to the axis (Fig. 3.1b), all the points are moving instantaneously in circles, with points further from the axis undergoing higher velocities proportional to the radius, i.e., $v = \rho\dot{\theta}$. In a general motion, however, both the axis of rotation and the angular rate can change in time.

Rate of Change of a Constant-length Vector

The preceding discussion of angular velocity and rigid-body rotation, leads us to a useful theorem in vector calculus concerning the rate of change of a constant-length vector.

The time derivative of a fixed length vector, $\mathbf{C}$, is given by the cross product of the rotation rate $\boldsymbol{\omega}$ and the vector, $\mathbf{C}$, i.e.,

$$\frac{d\mathbf{C}}{dt} = \boldsymbol{\omega} \times \mathbf{C} \tag{3.3}$$

Here we view $\mathbf{C}$ as if it were embedded in a rigid body rotating with $\boldsymbol{\omega}$ (Fig. 3.2). Further, we can write $\boldsymbol{\omega}$ with components parallel to and normal to $\mathbf{C}$, with $\boldsymbol{\omega} = \boldsymbol{\omega}_\parallel + \boldsymbol{\omega}_\perp$ and:

$$\frac{d}{dt}\mathbf{C} = \boldsymbol{\omega}_\perp \times \mathbf{C} \tag{3.4}$$

Thus a vector of constant length can only change due to rotation about an axis with a component normal to $\mathbf{C}$. This rule is very useful in calculating velocities and accelerations in moving coordinate frames.

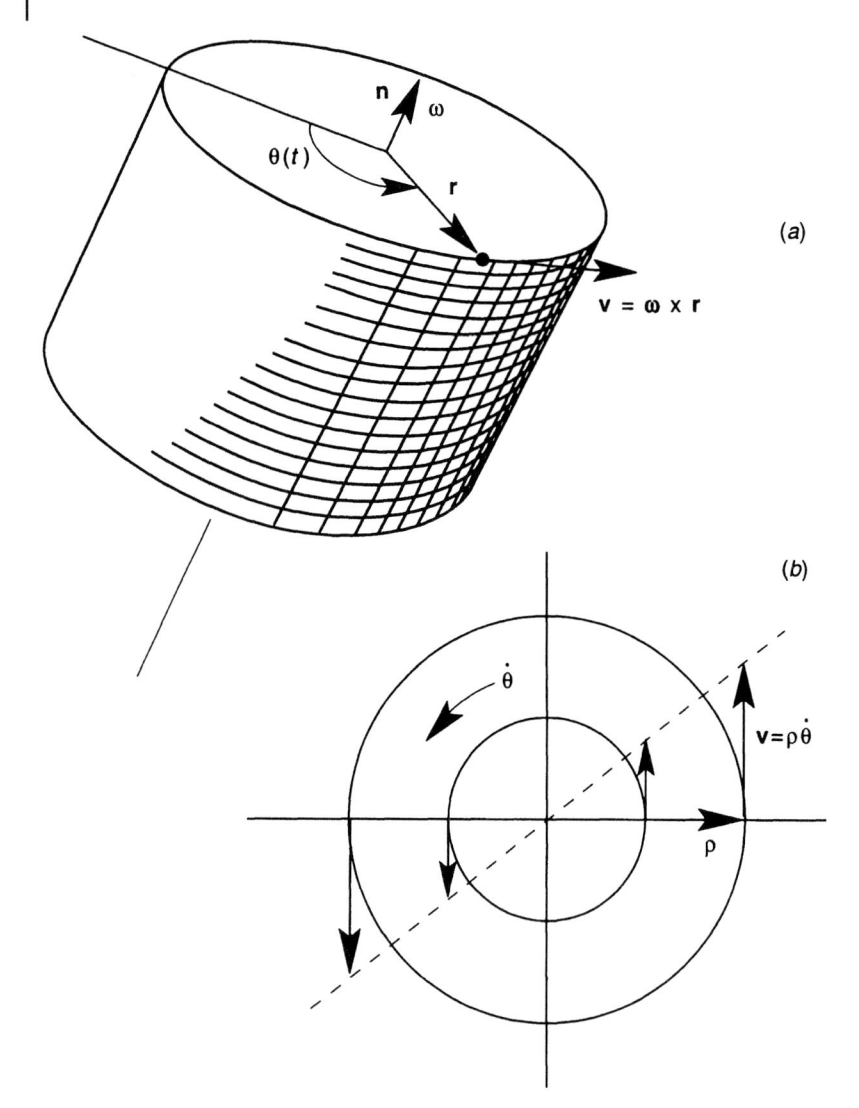

Fig. 3.1 (a) Rotation vector of a rigid body. (b) Circular motion about an axis normal to the plane.

Example 3.1. Suppose the airplane shown in Fig. 3.2 is undergoing a roll maneuver with rate $\dot{\phi}$, and a yaw or turn to the left with a rate $\dot{\psi}$. We want to find the relative velocity of a point on the horizontal stabilizer C with respect to the center of mass of the aircraft. The angular velocity vector, written with respect to the aircraft body axis $\{e_1, e_2, e_3\}$ is given by (e_2 directed from tail to nose, e_3 is the vertical yaw axis)

$$\boldsymbol{\omega} = \dot{\phi}\mathbf{e}_2 + \dot{\psi}\mathbf{e}_3 \tag{3.5}$$

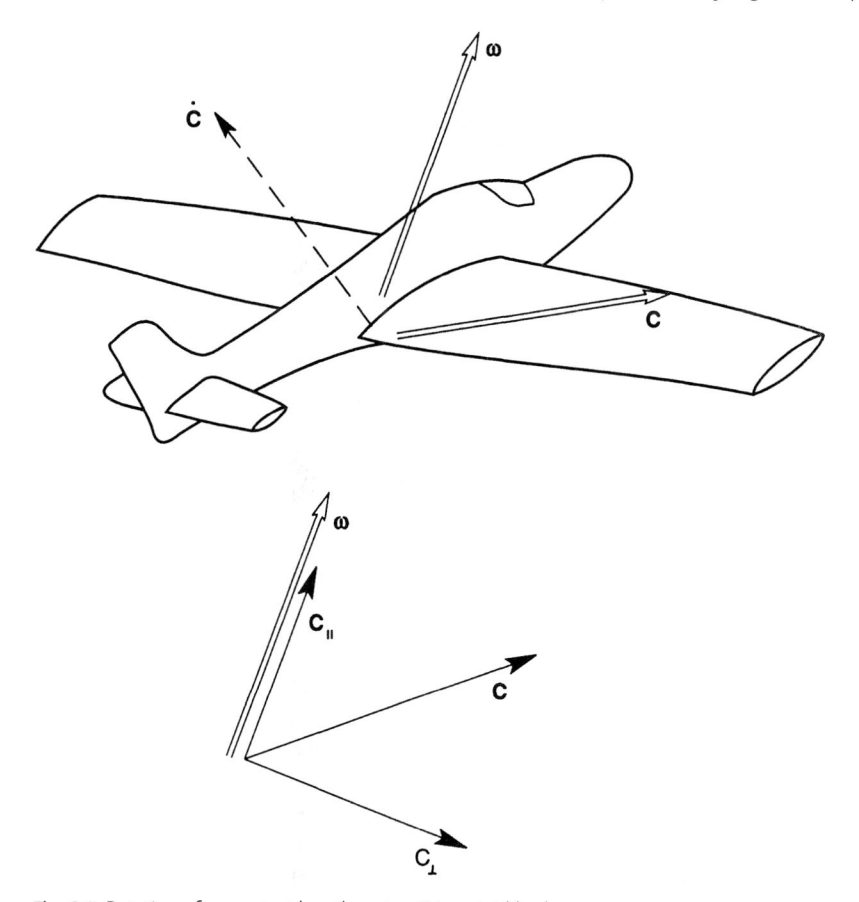

Fig. 3.2 Rotation of a constant length vector **C** in a rigid body.

Assuming that **C** lies in the plane of the wing, i.e., $\mathbf{C} = [b, a, 0]^T$, the relative velocity $\dot{\mathbf{C}}$ is given by

$$\dot{\mathbf{C}} = \boldsymbol{\omega} \times \mathbf{C} = -a\dot{\psi}\mathbf{e}_1 + b\dot{\psi}\mathbf{e}_2 - b\dot{\phi}\mathbf{e}_3 \tag{3.6}$$

3.3
Matrix Representation of Angular Velocity

It is easy to show that the cross product $\boldsymbol{\omega} \times \mathbf{C}$ in Eq. (3.3) can be represented by a matrix operation as in Eq. (2.48). (See also Beatty, 1986.) However, we must assume a right-handed orthogonal coordinate system in which the components of $\boldsymbol{\omega}$ and **C** and are given by

$$\boldsymbol{\omega} = \omega_1\mathbf{e}_1 + \omega_2\mathbf{e}_2 + \omega_3\mathbf{e}_3, \quad \text{or} \quad \boldsymbol{\omega} = [\omega_1, \omega_2, \omega_3]^T \tag{3.7a}$$

$$\mathbf{C} = C_1\mathbf{e}_1 + C_2\mathbf{e}_2 + C_3\mathbf{e}_3, \quad \text{or} \quad \mathbf{C} = [C_1, C_2, C_3]^T \tag{3.7b}$$

Then ($[\ldots]^T$ means transpose of the 1×3 row matrix into a 3×1 column matrix) (3.3) in matrix notation becomes,

$$\left[\left(\frac{d\mathbf{C}}{dt}\right)_1, \left(\frac{d\mathbf{C}}{dt}\right)_2, \left(\frac{d\mathbf{C}}{dt}\right)_3\right]^T = \begin{bmatrix} 0 & -\omega_3 & \omega_2 \\ \omega_3 & 0 & -\omega_1 \\ -\omega_2 & \omega_1 & 0 \end{bmatrix} \begin{bmatrix} C_1 \\ C_2 \\ C_3 \end{bmatrix} \quad (3.8)$$

This expression can also be derived from the finite rotation transformation discussed below. To use a compact notation we define the skew symmetric rotation matrix $\tilde{\omega}$ corresponding to the vector ω and write

$$\frac{d\mathbf{C}}{dt} = \tilde{\omega}\mathbf{C} \quad (3.9)$$

This product operator is different from either an inner product ($\cdot$) or cross product ($\times$). The product of the 3×3 matrix $\tilde{\omega}$ and the 3×1 vector $\mathbf{C}$ is understood to be a 3×1 vector, $\dot{\mathbf{C}}$.

The matrix representation of the cross product is very useful in computer calculations that are often based on matrix operations. You should always keep in mind, however, that the explicit form of $\tilde{\omega}$ implies a specific reference frame in which the components $[\omega_1, \omega_2, \omega_3]^T$ are calculated.

3.4
Kinematics Relative to Moving Coordinate Frames

In describing the motion of a body, we must distinguish between the coordinate frame in which the vector components are written, and the coordinate frame in which time derivatives are taken. A classic problem in kinematics is the description of motion in a reference frame moving relative to another frame (which we consider to be fixed). This problem is illustrated in Fig. 3.3. The fixed frame has an orthogonal set of basis vectors $\{\hat{\mathbf{i}}, \hat{\mathbf{j}}, \hat{\mathbf{k}}\}$. The moving frame has a position vector $\mathbf{R}$ to its origin, and a set of orthogonal basis vectors $\{\mathbf{e}_1, \mathbf{e}_2, \mathbf{e}_3\}$. The moving set of basis vectors acts as a rigid body and, therefore, has its own angular velocity, ω or $\tilde{\omega}$, which describes the rate of change of the angular orientation with respect to $\{\hat{\mathbf{i}}, \hat{\mathbf{j}}, \hat{\mathbf{k}}\}$.

We consider a motion trajectory $\mathbf{r}(t)$ and calculate $\dot{\mathbf{r}}(t)$, $\ddot{\mathbf{r}}(t)$ in terms of rates observed in the rotating coordinate system. To this end we introduce a relative position vector ρ,

$$\mathbf{r} = \mathbf{R} + \rho$$
$$\mathbf{v} = \frac{d\mathbf{r}}{dt} = \frac{d\mathbf{R}}{dt} + \frac{d\rho}{dt} \quad (3.10)$$

This leads to the formula

$$\mathbf{v} = \dot{\mathbf{R}} + \mathbf{v}_r + \omega \times \rho \quad (3.11)$$

3.4 Kinematics Relative to Moving Coordinate Frames

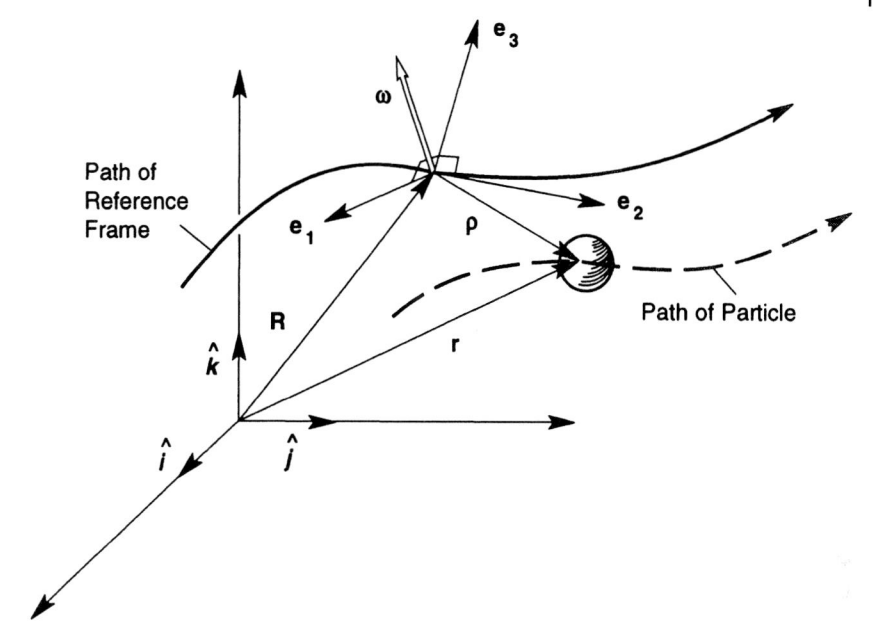

Fig. 3.3 Motion of a particle relative to a moving reference frame $\{e_1, e_2, e_3\}$.

where

$$\mathbf{v}_r = \left(\frac{d\boldsymbol{\rho}}{dt}\right)_{\text{rel}} \tag{3.12}$$

To see this we write $\dot{\boldsymbol{\rho}}$ in explicit form;

$$\boldsymbol{\rho} = \rho_1 \mathbf{e}_1 + \rho_2 \mathbf{e}_2 + \rho_3 \mathbf{e}_3$$

$$\frac{d\boldsymbol{\rho}}{dt} = (\dot{\rho}_1 \mathbf{e}_1 + \dot{\rho}_2 \mathbf{e}_2 + \dot{\rho}_3 \mathbf{e}_3) + \boldsymbol{\omega} \times \boldsymbol{\rho}$$

$$= \mathbf{v}_r + \boldsymbol{\omega} \times \boldsymbol{\rho} \tag{3.13}$$

The first term in (3.13) is the velocity as calculated in the moving reference coordinate system. The second term in (3.13) is the effect of the rate of rotation of the basis vectors and uses (3.3):

$$\dot{\mathbf{e}}_i = \boldsymbol{\omega} \times \mathbf{e}_i, \quad i = 1, 2, 3 \tag{3.14}$$

To obtain the expression for the acceleration we differentiate the velocity formula (3.11) with respect to time as observed in the fixed reference frame:

$$\mathbf{a} = \frac{d\mathbf{v}}{dt} = \ddot{\mathbf{R}} + \mathbf{a}_r + \boldsymbol{\omega} \times \mathbf{v}_r + \boldsymbol{\omega} \times \frac{d\boldsymbol{\rho}}{dt} + \frac{d\boldsymbol{\omega}}{dt} \times \boldsymbol{\rho} \tag{3.15}$$

The fourth term can be expanded using

$$\frac{d\boldsymbol{\rho}}{dt} = \mathbf{v}_r + \boldsymbol{\omega} \times \boldsymbol{\rho} \qquad (3.16)$$

The second and third terms come from the expression [see (3.13)]

$$\frac{d\mathbf{v}_r}{dt} = (\ddot{\rho}_1 \mathbf{e}_1 + \ddot{\rho}_2 \mathbf{e}_2 + \ddot{\rho}_3 \mathbf{e}_3) + \boldsymbol{\omega} \times \mathbf{v}_r \qquad (3.17)$$

The final formula takes the form

$$\mathbf{a} = \ddot{\mathbf{R}} + \mathbf{a}_r + \boldsymbol{\omega} \times (\boldsymbol{\omega} \times \boldsymbol{\rho}) + \dot{\boldsymbol{\omega}} \times \boldsymbol{\rho} + 2\boldsymbol{\omega} \times \mathbf{v}_r \qquad (3.18)$$

Before we blindly use this expression, careful attention should be paid to the implied derivatives. The derivatives of the scalar components $\ddot{\rho}_i$, $\dot{\rho}_i$ are rates observed in the moving reference. However, the rate $\dot{\boldsymbol{\omega}}$ is measured in the fixed system, although we could express the resulting vector components in either the $\{\hat{\mathbf{i}}, \hat{\mathbf{j}}, \hat{\mathbf{k}}\}$ or $\{\mathbf{e}_1, \mathbf{e}_2, \mathbf{e}_3\}$ bases or for that matter any convenient set of basis vectors.

The five terms in this expression have the following interpretation:

$\ddot{\mathbf{R}}$	Represents the acceleration of the moving reference point P.
$\mathbf{a}_r$	Is the acceleration of the object as observed in the moving reference.
$\boldsymbol{\omega} \times (\boldsymbol{\omega} \times \boldsymbol{\rho})$	Is a centripetal acceleration correction term for the local position vector $\boldsymbol{\rho}$ rotating with the moving frame angular velocity $\boldsymbol{\omega}$.
$\dot{\boldsymbol{\omega}} \times \boldsymbol{\rho}$	Is a correction term for the angular-acceleration vector $\dot{\boldsymbol{\omega}}$ of the moving references frame.
$2\boldsymbol{\omega} \times \mathbf{v}_r$	The Coriolis acceleration correction has two sources, both of which measure the rotation of the basis vectors of the moving reference frame.

Example 3.2 Satellite Solar Panel Deployment. The satellite shown in Fig. 3.4 has a steady spin Ω about the body fixed $\mathbf{e}_3$ axis. At the same time a solar panel arm simultaneously rotates about the $\mathbf{e}_2$ axis with a rate $\dot{\theta}$, and angular acceleration $\ddot{\theta} = 0$ and telescopes along the radial direction $\mathbf{e}_r$ with a steady rate $\dot{a} = \alpha$. In this example we wish to find the acceleration of the point P at the end of the solar panel relative to a nonrotating reference frame. We have several choices of intermediate reference frame. As an illustration let us choose a frame that rotates with the solar panel, which has a local coordinate system $\{\mathbf{e}_r, \mathbf{e}_\theta, \mathbf{e}_2\}$. In order to apply the five term acceleration expression (3.18), we define the following vectors:

$$\mathbf{R} = b\mathbf{e}_1, \; \boldsymbol{\rho} = (a(t) + c)\mathbf{e}_r$$

$$\boldsymbol{\omega} = \Omega \mathbf{e}_3 + \dot{\theta} \mathbf{e}_2 \qquad (3.19)$$

With these definitions, we can easily show that

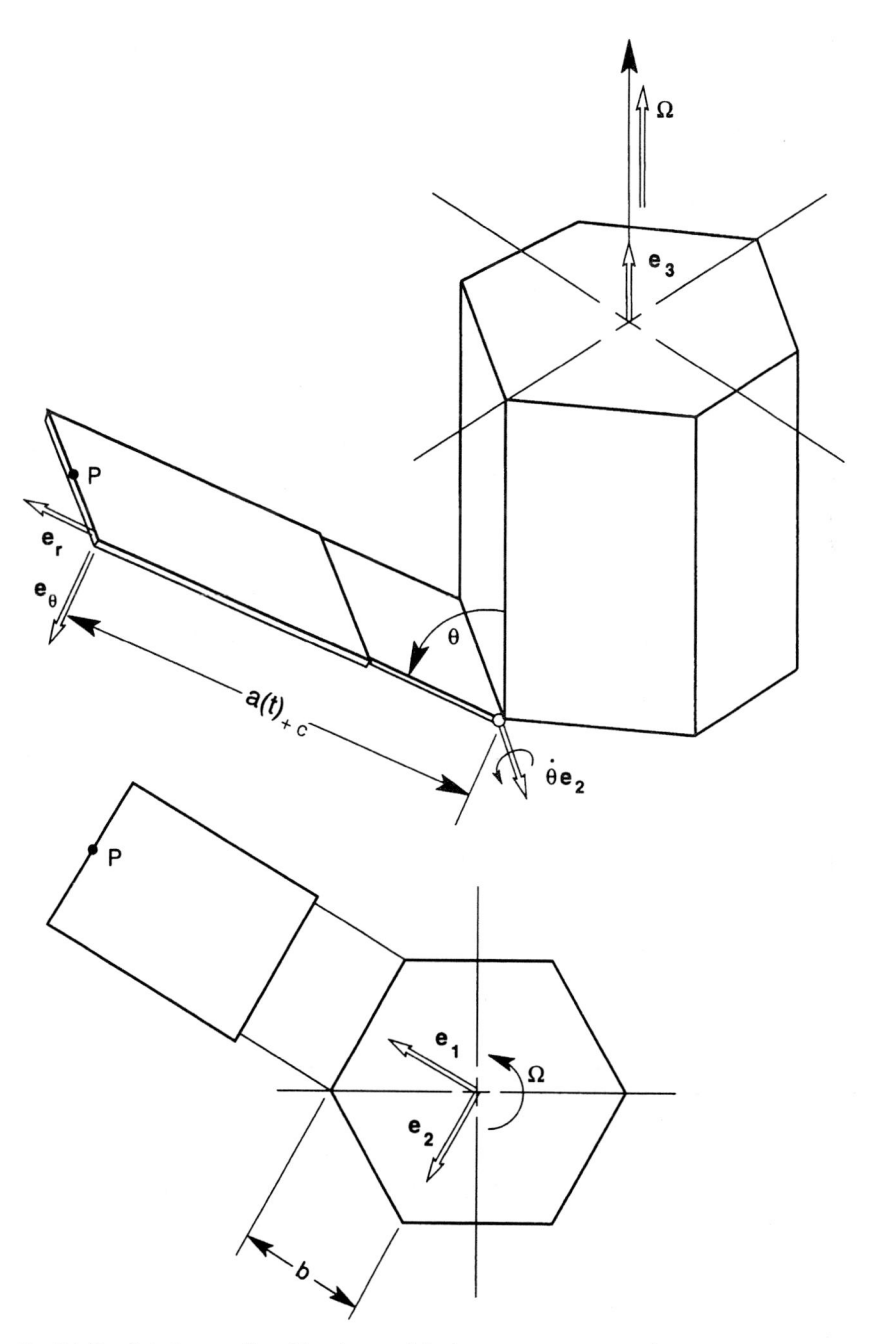

Fig. 3.4 *Top*: Rotating satellite with solar panel deployment. *Bottom*: View from top.

$$\mathbf{a}_r = \ddot{a}\mathbf{e}_r = 0, \qquad \mathbf{v}_r = \dot{a}\mathbf{e}_r = \alpha \mathbf{e}_r$$
$$\dot{\boldsymbol{\omega}} = \Omega \mathbf{e}_3 \times \dot{\theta}\mathbf{e}_2 = -\Omega\dot{\theta}\mathbf{e}_1 \tag{3.20}$$

This last term follows since the axis of rotation of $\dot{\theta}\mathbf{e}_2$ rotates with $\Omega\mathbf{e}_3$. With these results we can calculate the other four terms in (3.18)

$$\ddot{\mathbf{R}} = -b\Omega^2 \mathbf{e}_1$$
$$\dot{\boldsymbol{\omega}} \times \boldsymbol{\rho} = \Omega\dot{\theta}(a+c)\cos\theta \mathbf{e}_2$$
$$\boldsymbol{\omega} \times (\boldsymbol{\omega} \times \boldsymbol{\rho}) = -\Omega^2(a+c)\sin\theta \mathbf{e}_1 - \dot{\theta}^2(a+c)\mathbf{e}_r + \Omega\dot{\theta}(a+c)\cos\theta \mathbf{e}_2$$
$$2\boldsymbol{\omega} \times \mathbf{v}_r = 2\alpha(\Omega\sin\theta \mathbf{e}_2 + \dot{\theta}\mathbf{e}_\theta) \tag{3.21}$$

To express all quantities in one reference frame, we can use the transformation relation

$$\begin{bmatrix} \mathbf{e}_r \\ \mathbf{e}_\theta \end{bmatrix} = \begin{bmatrix} \cos\theta & \sin\theta \\ -\sin\theta & \cos\theta \end{bmatrix} \begin{bmatrix} \mathbf{e}_3 \\ \mathbf{e}_1 \end{bmatrix} \tag{3.22}$$

(Note that the expressions in the column "vectors" are unit vectors, not scalars.)

An alternate method to calculate $\mathbf{a}_p$ is to directly differentiate $\mathbf{r} = \mathbf{R} + \boldsymbol{\rho}$, i.e.,

$$\mathbf{v}_p = \left(b\Omega + (a+c)\Omega\sin\theta\right)\mathbf{e}_2 + (a+c)\dot{\theta}\mathbf{e}_\theta + \dot{a}\mathbf{e}_r \tag{3.23}$$

In order to calculate $\mathbf{a}_p$ from the preceding expression, we must note that $a(t), \theta(t)$ are time dependent and use expressions for $\dot{\mathbf{e}}_2, \dot{\mathbf{e}}_\theta, \dot{\mathbf{e}}_r$ for rotating unit vectors (3.3). It is left as an exercise to show that the two methods give the same result.

3.5
Constraints and Jacobians

A machine or a mechanism is defined by the way it is constrained to behave dynamically. The motion of a serial-link robot arm is limited by the constraints at each joint, and an all-terrain vehicle is supposed to operate with all four wheels rolling over the ground (Fig. 3.5). Although a set of N rigid bodies may have up to $6N$ degrees of freedom, in practice the actual number of degrees of freedom K is limited by the M constraints that define the machine or machine component, i.e., $K = 6N - M$.

A brief discussion of how constraints enter kinematic formulas was given in Chapter 2. In particular, it was shown that the sensitivity of the velocities of the bodies to the generalized velocities $\{\dot{q}_i\}$ was represented by a Jacobian matrix [see (2.13)]. The Jacobian matrices are an important part of D'Alembert's principle and the principle of virtual power. They allow you to project the forces and accelerations onto a reduced set of generalized equations that determine the dynamics of the system.

To illustrate how constraints come into kinematics, consider the example of the planar motion of a two-link chain represented by two masses, m_1, m_2 with position

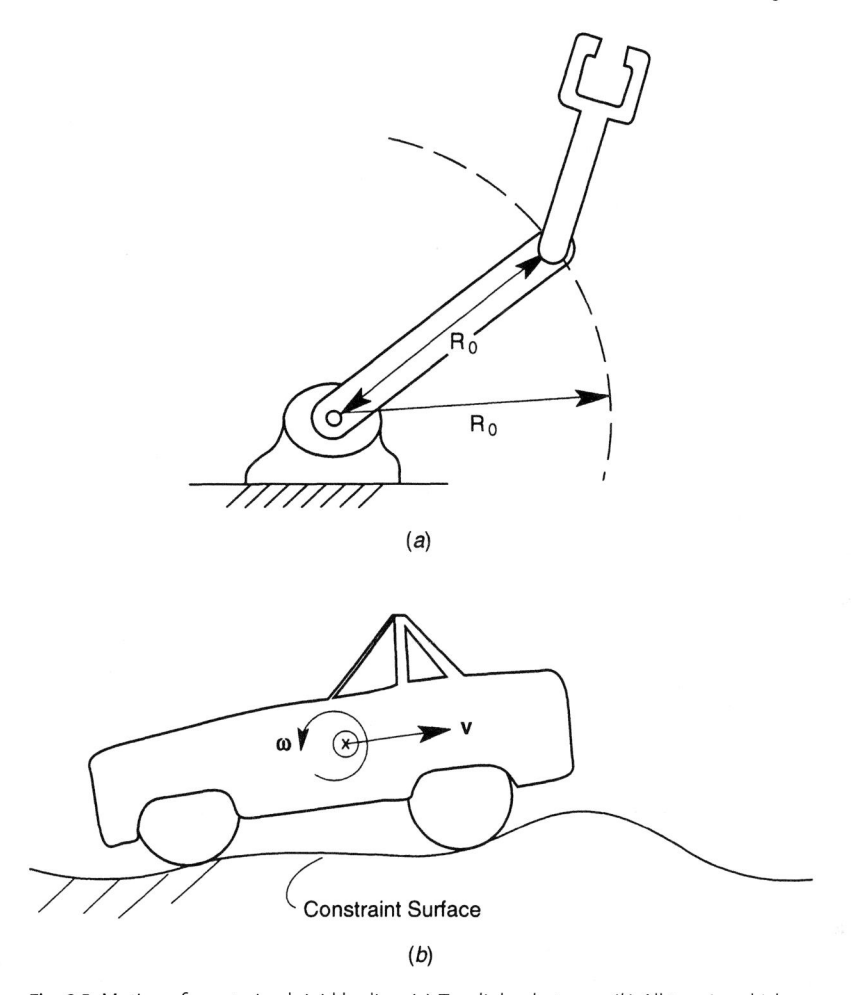

Fig. 3.5 Motion of constrained rigid bodies. (*a*) Two-link robot arm. (*b*) All-terrain vehicle.

vectors $\mathbf{r}_1, \mathbf{r}_2$. In general this system would have four degrees of freedom. There are two methods for dealing with constraints.

1. Use the constraint equations along with the equations of motion and solve simultaneously. This method generally involves opening up the system to expose the constraint forces.
2. Define a reduced set of independent coordinates and use constraint equations to eliminate coordinates, and constraint forces.

Example 3.3. In the example of the two-link mechanism shown in Fig. 3.6, we choose $(q_1, q_2) = (\theta_1, \theta_2)$ as the independent coordinates. The implicit form the constraint is $x_1^2 + y_1^2 = L_1^2$.

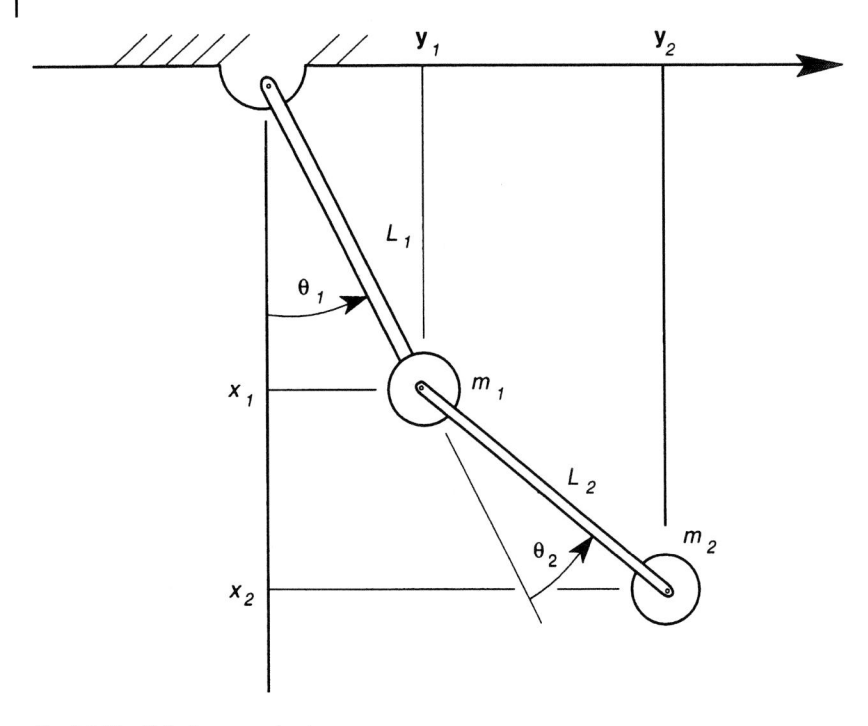

Fig. 3.6 Two-link planar mechanism.

The position-vector components, velocities, and accelerations are given by

$$\begin{cases} x_1 = L_1 \cos\theta_1 \\ y_1 = L_1 \sin\theta_1 \end{cases} \tag{3.24}$$

$$\begin{cases} \dot{x}_1 = -L_1 \dot{\theta}_1 \sin\theta_1 \\ \dot{y}_1 = L_1 \dot{\theta}_1 \cos\theta_1 \end{cases} \tag{3.25}$$

$$\begin{cases} \ddot{x}_1 = -L_1 \ddot{\theta}_1 \sin\theta_1 - L_1 \dot{\theta}_1^2 \cos\theta_1 \\ \ddot{y}_1 = L_1 \ddot{\theta}_1 \cos\theta_1 - L_1 \dot{\theta}_1^2 \sin\theta_1 \end{cases} \tag{3.26}$$

In the notation of (2.12) and (2.61),

$$[J_1] = \left[\frac{\partial \mathbf{r}_1}{\partial q_i}\right] = \begin{bmatrix} -L_1 \sin\theta_1 & 0 \\ L_1 \cos\theta_1 & 0 \end{bmatrix} \tag{3.27}$$

and

$$[\mathbf{v}_1] = [J_1][\dot{\mathbf{q}}]$$

where $[\dot{\mathbf{q}}] = [\dot{\theta}_1, \dot{\theta}_2]^T$.

Similar expressions follow for (x_2, y_2):

$$x_2 = x_1 + L_2 \cos(\theta_1 + \theta_2)$$
$$y_2 = y_1 + L_2 \sin(\theta_1 + \theta_2) \quad (3.28)$$

Then the kinematic constraint for $\mathbf{r}_2$ in matrix notation becomes

$$[v_2] = [J_2][\dot{q}] \quad (3.29)$$

where

$$[J_2]_{11} = -(L_1 \sin\theta_1 + L_2 \sin(\theta_1 + \theta_2))$$
$$[J_2]_{12} = -L_2 \sin(\theta_1 + \theta_2)$$
$$[J_2]_{21} = L_1 \cos\theta_1 + L_2 \cos(\theta_1 + \theta_2)$$
$$[J_2]_{22} = L_2 \cos(\theta_1 + \theta_2) \quad (3.30)$$

3.6
Finite Motions

Unlike structural dynamics, where the motions are generally small, mechanisms and machine components such as gears or robot manipulators undergo large motions. One of the major theorems of theoretical kinematics is that a general motion of a rigid body can be resolved into the translation of some point on the body and a finite rotation about this point. (Chasle's theorem is a generalization of Euler's theorem.) We begin with a discussion of the rotation of a rigid body about a point.

The mathematical problem in finite motions is to find a relation between a position vector of a point, P, before the motion $\mathbf{r}_P$ and the new position vector after the motion $\mathbf{r}'_P$ (Fig. 3.7). In matrix notation, we require a 3×3 transformation matrix relating $\mathbf{r}_p$ and $\mathbf{r}'_p$, i.e.,

$$\mathbf{r}'_p = \mathsf{T}\mathbf{r}_p \quad (3.31)$$

It is a theorem of theoretical kinematics attributed to Euler that this matrix is the same for any point in the rigid body. Thus we can drop the subscript on the position vector. Also the motion should be invertible so that, $\mathbf{r} = \mathsf{T}^{-1}\mathbf{r}'$, or that an inverse should exist. The other fact about T follows from the requirement that the length of $\mathbf{r}$ be unchanged after the motion. Using the matrix operation for an inner product ($\mathbf{r} \cdot \mathbf{r} = \mathbf{r}^T \mathbf{r}$), we have

$$\mathbf{r}^T \mathbf{r} = (\mathbf{r}')^T \mathbf{r}' = (\mathsf{T}\mathbf{r})^T \mathsf{T}\mathbf{r} = \mathbf{r}^T \mathsf{T}^T \mathsf{T}\mathbf{r} \quad (3.32)$$

so that the product $\mathsf{T}^T\mathsf{T}$ must be an identity matrix I

$$\mathsf{T}^T \mathsf{T} = \mathsf{I} \quad (3.33)$$

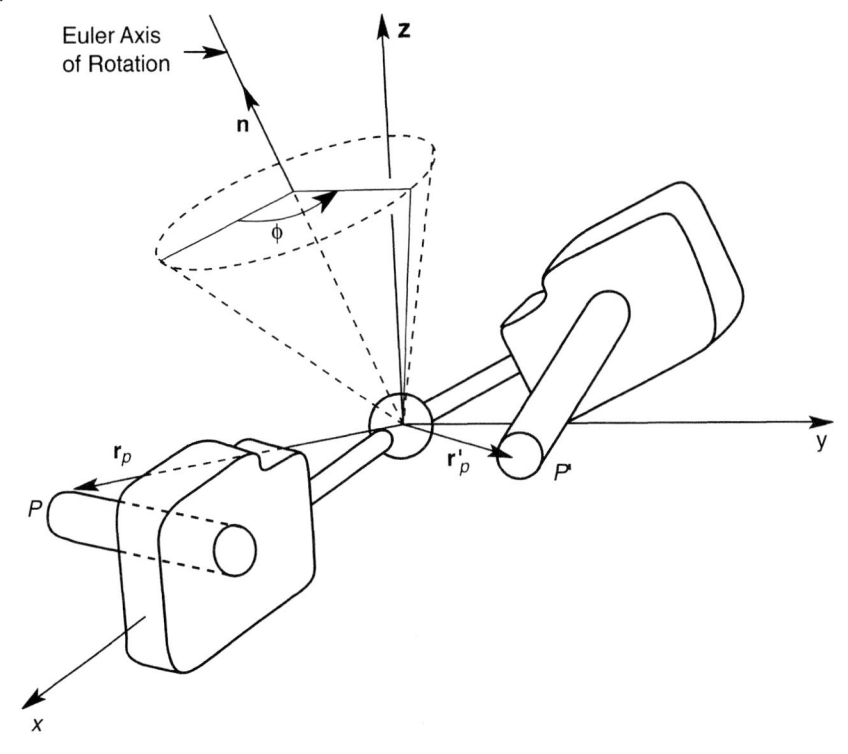

Fig. 3.7 Finite rotation of a rigid body.

This implies that T is an *orthogonal matrix*, with $T^T = T^{-1}$. Also the eigenvalues of T are unity, and the off-diagonal terms are skew symmetric, i.e., $T_{ij} = -T_{ji}, i \neq j$.

Another consequence of Euler's theorem is that any finite motion can be viewed as a rotation about a fixed axis. As a specific example, consider the rotation of the body in Fig. 3.8 about the z-axis through an angle ϕ. It is not difficult to find a transformation matrix, T, that relates the new coordinates $[x', y', z']$ to the old coordinates $[x, y, z]$

$$\begin{bmatrix} x' \\ y' \\ z' \end{bmatrix} = \begin{bmatrix} \cos\phi & -\sin\phi & 0 \\ \sin\phi & \cos\phi & 0 \\ 0 & 0 & 1 \end{bmatrix} \begin{bmatrix} x \\ y \\ z \end{bmatrix} \quad (3.34)$$

Here we denote the matrix by A_1,

$$A_1 = \begin{bmatrix} \cos\phi & -\sin\phi & 0 \\ \sin\phi & \cos\phi & 0 \\ 0 & 0 & 1 \end{bmatrix} \quad (3.35)$$

The inverse of the motion is to rotate the body through $-\phi$, so that $A_1^{-1} = A_1^T$, as is required. It is not difficult to show that a rotation about the y axis through an

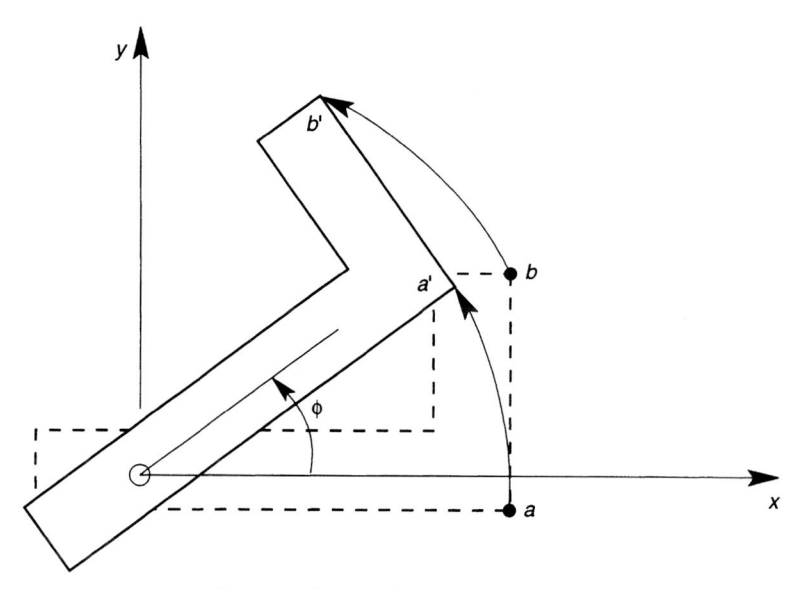

Fig. 3.8 Rotation transformation: finite rigid-body rotation, ϕ, about an axis normal to the plane.

angle θ is given by the matrix A_2:

$$A_2 = \begin{bmatrix} \cos\theta & 0 & \sin\theta \\ 0 & 1 & 0 \\ -\sin\theta & 0 & \cos\theta \end{bmatrix} \quad (3.36)$$

For a rotation about the x-axis one can show

$$A_3 = \begin{bmatrix} 1 & 0 & 0 \\ 0 & \cos\psi & -\sin\psi \\ 0 & \sin\psi & \cos\psi \end{bmatrix} \quad (3.37)$$

In general, however, these transformations do not commute, i.e.,

$$A_1 A_2 \neq A_2 A_1$$

This can be seen in Fig. 3.9, where an object is first rotated about the z-axis and then the y-axis and compared to the motion resulting from inverting the rotations. The object clearly ends up in different positions.

We can also find an expression for the new position vector $\mathbf{r}'$ in vector notation, if one specifies a unit vector $\mathbf{n}$ along the axis of rotation and angle of rotation ϕ (see Goldstein, 1980, p. 164,[2]) or Beatty, 1986) then

$$\mathbf{r}' = \mathbf{n}(\mathbf{n} \cdot \mathbf{r}) + [\mathbf{r} - \mathbf{n}(\mathbf{n} \cdot \mathbf{r})]\cos\phi + (\mathbf{n} \times \mathbf{r})\sin\phi \quad (3.38)$$

---

2) Note in Goldstein, ϕ is defined in the opposite sense to ϕ defined here.

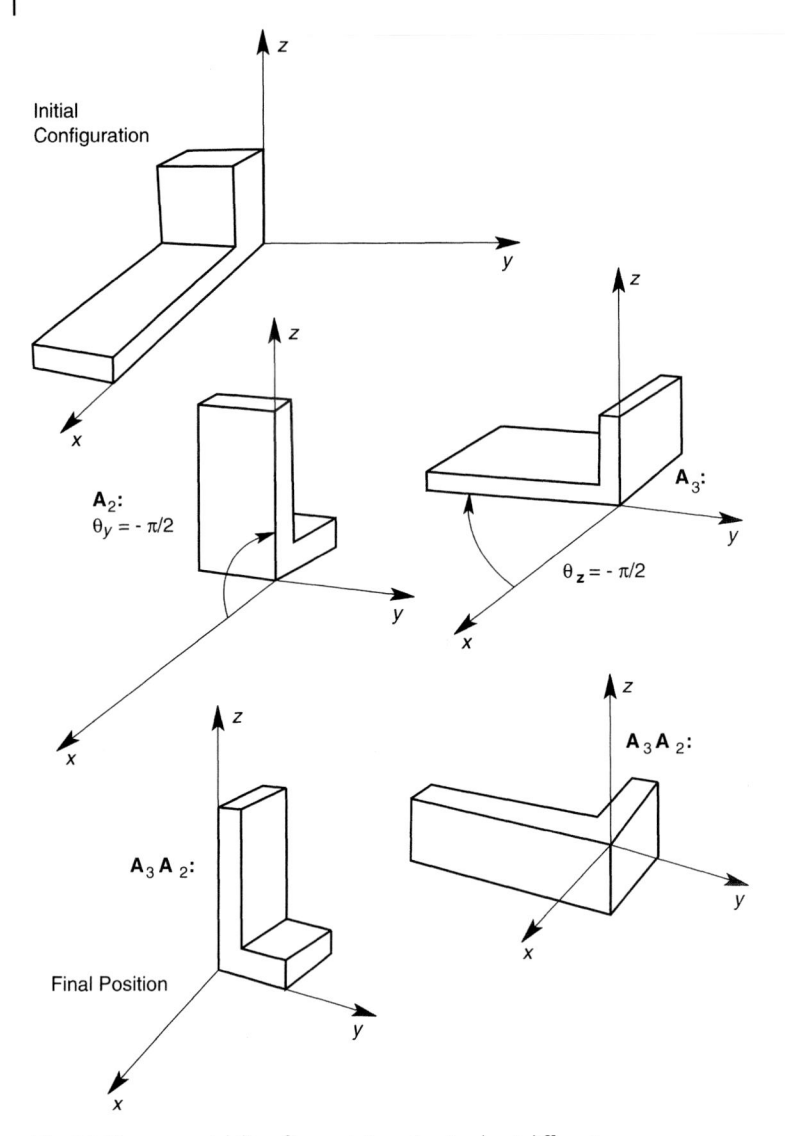

Fig. 3.9 Noncommutability of two rotations A_2, A_3 about different axes.

This expression has the advantage that it is independent of a specific coordinate reference.

Thus, in order to determine the finite motion of a rigid body about a point, three scalars are required; either three angles of rotation about the three basis axes (ϕ, θ, ψ) or as in the last expression $(\phi, n_x, n_y, n_z; \mathbf{n} \cdot \mathbf{n} = 1)$. The matrix representation, $\mathbf{r}' = \mathsf{T}\mathbf{r}$, (3.31) and the vector representation of a finite rotation (3.38) are related by

$$\mathsf{T}\mathbf{n} = \mathbf{n}, \qquad \text{Trace } \mathsf{T} = 1 + 2\cos\phi \tag{3.39}$$

For numerical codes, often so-called *Euler parameters* are used instead of $\{\phi, \mathbf{n}\}$. These four scalars $\{e_0, e_1, e_2, e_3\}$ are defined by (Goldstein, 1980, p. 165)

$$e_0 = \cos\frac{\phi}{2}$$
$$\mathbf{e} = \mathbf{n}\sin\frac{\phi}{2} \tag{3.40}$$

where

$$e_0^2 + \mathbf{e} \cdot \mathbf{e} = 1$$

For small angles of rotation, we can use the approximations $\cos\phi \sim 1$, $\sin\phi \sim \phi$. Thus, if we write $\phi \sim \omega_z \Delta t$, $\theta \sim \omega_y \Delta t$, we $\psi \sim \omega_x \Delta t$ one can show first that A_1, A_2, A_3 commute and that

$$\mathbf{r}' = \mathbf{r}(t + \Delta t) \simeq A_1 A_2 A_3 \mathbf{r}(t)$$

or

$$\mathbf{r}(t + \Delta t) = \begin{bmatrix} 1 & -\omega_z \Delta t & \omega_y \Delta t \\ \omega_z \Delta t & 1 & -\omega_x \Delta t \\ -\omega_y \Delta t & \omega_x \Delta t & 1 \end{bmatrix} \begin{bmatrix} x(t) \\ y(t) \\ z(t) \end{bmatrix}$$

Separating the transformation matrix into the identity matrix and a skew symmetric matrix, we obtain an expression for the velocity of a point in a rigid body undergoing rotation about some point.

$$\mathbf{v} = \lim_{\Delta t \to 0} \frac{\mathbf{r}(t + \Delta t) - \mathbf{r}(t)}{\Delta t} = \hat{\boldsymbol{\omega}} \mathbf{r}(t) \tag{3.41}$$

where

$$\hat{\boldsymbol{\omega}} = \begin{bmatrix} 0 & -\omega_z & \omega_y \\ \omega_z & 0 & -\omega_x \\ -\omega_y & \omega_x & 0 \end{bmatrix} \tag{3.42}$$

This is the same matrix representation of the rotation vector $\boldsymbol{\omega}$ described in (3.9).

A similar expression can be obtained from (3.38) where we write $\phi \sim \omega \Delta t$. Then

$$\frac{\mathbf{r}' - \mathbf{r}}{\Delta t} \sim (\mathbf{n} \times \mathbf{r})\omega \tag{3.43}$$

In the limit as $\Delta t \to 0$, we obtain the expression

$$\mathbf{v} = \boldsymbol{\omega} \times \mathbf{r}$$
$$\boldsymbol{\omega} = \omega \mathbf{n} \tag{3.44}$$

which is identical to (3.1) and (3.2).

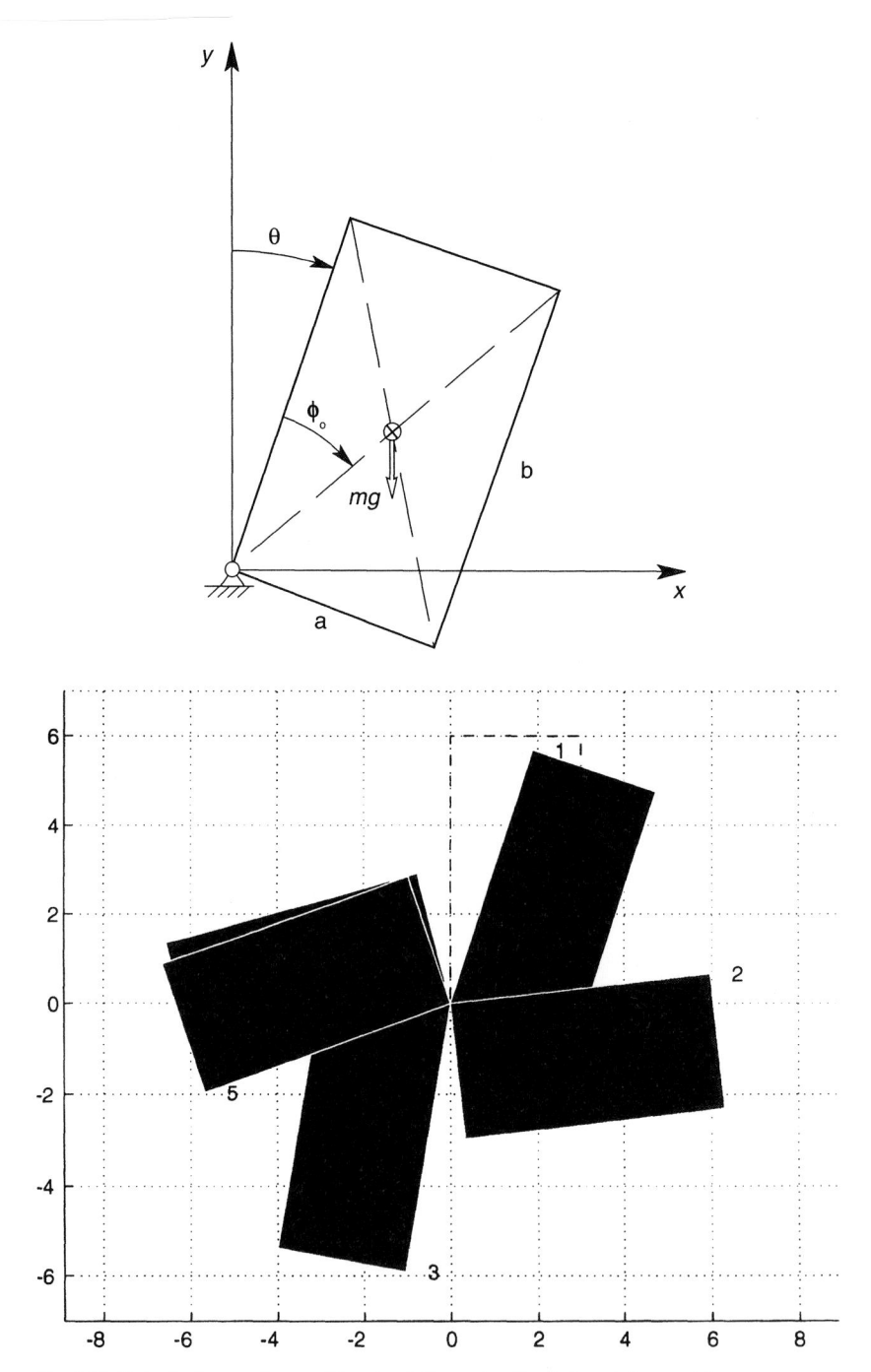

Fig. 3.10 Output of computer graphics simulation using the finite-rotation matrix.

The conclusion here is that finite rotations do not commute if there are rotations about different axes and that infinitesimal rotations do commute.

Example 3.4 MATLAB Computer Graphics Dynamic Simulation and Animation.
The finite rotation transformation is useful in designing a simple computer graphics program to simulate the dynamics of rigid-body motion. Consider the planar case of a hinged box shown in Fig. 3.10. The dynamics of the falling box are governed by the law of angular momentum

$$\mathbf{M} = \dot{\mathbf{H}}, \qquad \mathbf{H} = I\omega \mathbf{e}_z$$

The moment of the gravity force is $\mathbf{m} = mgL\sin(\theta + \phi_0)\mathbf{e}_z$. If a viscous torsional damper is added, the resulting second-order differential equation is (we define $L = 0.5[a^2 + b^2]^{\frac{1}{2}}$)

$$I\ddot{\theta} = mgL\sin(\theta + \phi_0) - C\dot{\theta}$$

For numerical integration, we rewrite this equation in the form of two first-order differential equations

$$\dot{\theta} = \omega$$
$$\dot{\omega} = \frac{mgL}{I}\sin(\theta + \phi_0) - \frac{C}{I}\omega \qquad (3.45)$$

The following program was written in MATLAB, which couples a simple Euler integration scheme to a plotting subroutine that replots a new picture of the box after a fixed time interval. We define MATLAB constants

$$e = mgL/I, \qquad c = C/I \qquad (3.46)$$

In the following, we retain the use of the Greek θ, ω, ϕ_0 in the MATLAB program for easier reading. You will have to define new variables, e.g., theta, omega, phi. Also the assignment of parameter values for dt, e, c, ϕ_0 must be added.

```
%      MATLAB Simulation and Animation Program for Falling Box
%      Define the vertices of the box
x = [0, a, a, 0, 0];
y = [0, 0, b, b, 0];
%      Define a matrix whose column vectors are the box vertices
r = [x; y];
%      Draw the box in the initial position
line(r(1, :), r(2, :), 'linestyle', '- -');
hold on
%      Integrate the equations of motion
θ = 0; ω = 0;
%      Do loop for new box graphic for
m = 1 : 5;
```

```
%       Do loop for elapsed time integration for
n = 1 : 20;
ω = ω + dt * e * sin(θ + φ_0) − dt * c * ω;
θ = θ + dt * ω;
%       end of inner loop
end
%       Rotate box graphic using finite rotation matrix
A = [cos(θ) sin(θ); − sin(θ) cos(θ)];
r1 = A * r;
x1 = r1(1, :); y1 = r1(2, :);
line(x1, y1, 'linestyle','– –'); axis ('equal')
end
```

The student can add color, shading, and a pause option to stop the action. Also to avoid distortion an equal line scale must be used in the horizontal and vertical scales. A good review of MATLAB is in the book *Getting Started with MATLAB7* by Pratap (2004).

3.7
Transformation Matrices for General Rigid-body Motion

One of the basic requirements of multibody kinematic analysis is to specify the location and angular orientation of one rigid body with respect to the reference frame of another rigid body. The general motion of one body relative to another involves translation of a point in the body (or an extended point) and a finite rotation about that point. For example, the displacement and rotation of the body about the z-axis shown in Fig. 3.11 can be described by a finite rotation of a typical vector in the body ρ and a translation vector **u**, i.e.,

$$\mathbf{r} = \mathbf{u} + \mathsf{A}\rho \tag{3.47}$$

where for rotation through an angle θ about the z-axis,

$$\mathsf{A} = \begin{bmatrix} \cos\theta & -\sin\theta & 0 \\ \sin\theta & \cos\theta & 0 \\ 0 & 0 & 1 \end{bmatrix} \tag{3.48}$$

Suppose **r** is expressed in terms of components $[x, y, z]^T$ with respect to the basis vectors $\{\hat{\mathbf{i}}, \hat{\mathbf{j}}, \hat{\mathbf{k}}\}$ and ρ has components $[\rho_x, \rho_y, \rho_z]^T$ with respect to the basic vectors $(\mathbf{e}_x, \mathbf{e}_y, \mathbf{e}_z)$, where this local basis triad is coincident with $(\hat{\mathbf{i}}, \hat{\mathbf{j}}, \hat{\mathbf{k}})$ when $\mathsf{A} = \mathsf{I}$. Then the new coordinates of the point in the body is given by the explicit form of (3.47):

$$\begin{bmatrix} x \\ y \\ z \end{bmatrix} = \begin{bmatrix} u_x \\ u_y \\ u_z \end{bmatrix} + \begin{bmatrix} \cos\theta & -\sin\theta & 0 \\ \sin\theta & \cos\theta & 0 \\ 0 & 0 & 1 \end{bmatrix} \begin{bmatrix} \rho_x \\ \rho_y \\ \rho_z \end{bmatrix} \tag{3.49}$$

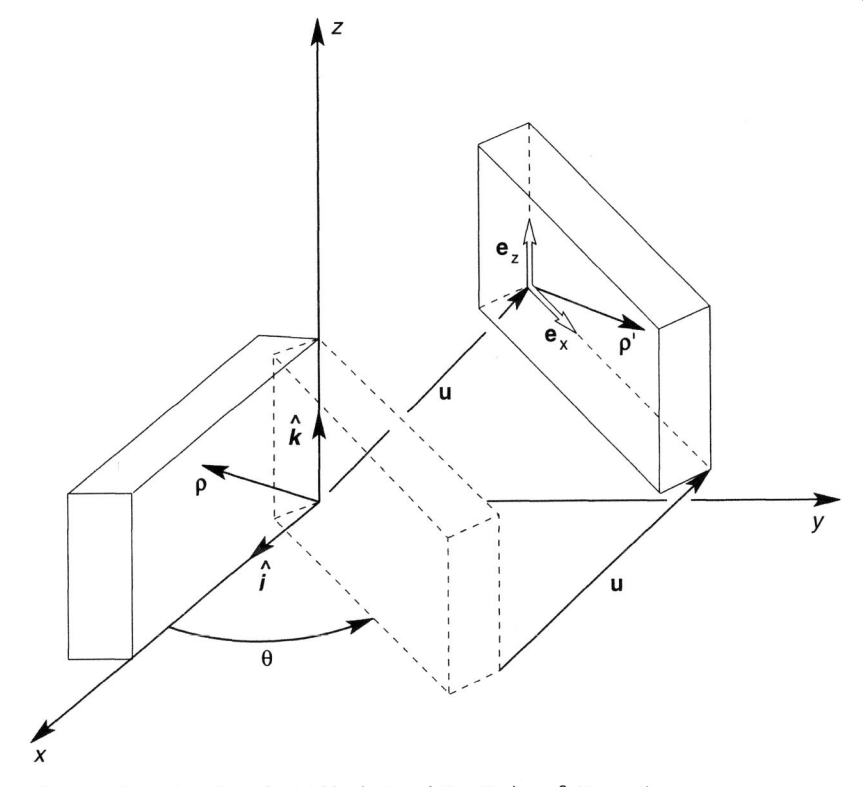

3.7 Transformation Matrices for General Rigid-body Motion

Fig. 3.11 General motion of a rigid body: translation **R** plus a finite rotation.

It is straightforward to show that both translation and rotation operations can be combined in one transformation matrix if we write $\mathbf{r}$, $\boldsymbol{\rho}$ in a fictitious four-dimensional space (see, e.g., Paul, 1981; Craig, 2005), i.e.,

$$\mathbf{r} = \begin{bmatrix} x \\ y \\ z \\ 1 \end{bmatrix}, \qquad \boldsymbol{\rho} = \begin{bmatrix} \rho_x \\ \rho_y \\ \rho_z \\ 1 \end{bmatrix} \tag{3.50}$$

Within this space one can show that

$$\mathbf{r} = \mathsf{T}\boldsymbol{\rho} \tag{3.51}$$

where

$$\mathsf{T} = \left[\begin{array}{c|c} \mathsf{A} & \begin{matrix} u_x \\ u_y \\ u_z \end{matrix} \\ \hline 0\ 0\ 0 & 1 \end{array} \right] \tag{3.52}$$

where A is the 3 × 3 rotation matrix. Using this notation, it is easy to show that a pure translation by a vector **u** is given by

$$T_1 = \begin{bmatrix} 1 & 0 & 0 & | & u_x \\ 0 & 1 & 0 & | & u_y \\ 0 & 0 & 1 & | & u_z \\ \hline 0 & 0 & 0 & | & 1 \end{bmatrix} \tag{3.53}$$

Also a pure rotation represented in a three-dimensional space by the orthogonal matrix A is written in 4 × 4 notation by

$$T_2 = \begin{bmatrix} & & & | & 0 \\ & A & & | & 0 \\ & & & | & 0 \\ \hline 0 & 0 & 0 & | & 1 \end{bmatrix} \tag{3.54}$$

Thus the *homogeneous transformation* (3.51) is given by

$$T = T_1 T_2 \tag{3.55}$$

This represents rotating the body first and then translating the body. Note that in general, the two operations do not commute, i.e.,

$$T_1 T_2 \neq T_2 T_1 \tag{3.56a}$$

Since

$$T_2 T_1 = \begin{bmatrix} & & & | & A_{xx}u_x + A_{xy}u_y + A_{xz}u_z \\ & A & & | & A_{yx}u_x + A_{yy}u_y + A_{yz}u_z \\ & & & | & A_{zx}u_x + A_{zy}u_y + A_{zz}u_z \\ \hline 0 & 0 & 0 & | & 1 \end{bmatrix} \tag{3.56b}$$

It is also straightforward to show that these 4 × 4 transformation matrices have an inverse, as can be shown by the following pair of equations:

$$T = \begin{bmatrix} n_x & m_x & k_x & | & u_x \\ n_y & m_y & k_y & | & u_y \\ n_z & m_z & k_z & | & u_z \\ \hline 0 & 0 & 0 & | & 1 \end{bmatrix} \tag{3.57}$$

$$T^{-1} = \begin{bmatrix} n_x & n_y & n_z & | & -\mathbf{u} \cdot \mathbf{n} \\ m_x & m_y & m_z & | & -\mathbf{u} \cdot \mathbf{m} \\ k_x & k_y & k_z & | & -\mathbf{u} \cdot \mathbf{k} \\ \hline 0 & 0 & 0 & | & 1 \end{bmatrix} \tag{3.58}$$

where $\mathbf{n} = [n_x, n_y, n_z]^T$ etc., and $\mathbf{n} \cdot \mathbf{n} = 1$, $\mathbf{m} \cdot \mathbf{m} = 1$, and $\mathbf{n} \cdot \mathbf{m} = 0$, etc. The 3 × 3 matrix is orthonormal and represents a rotation operation.

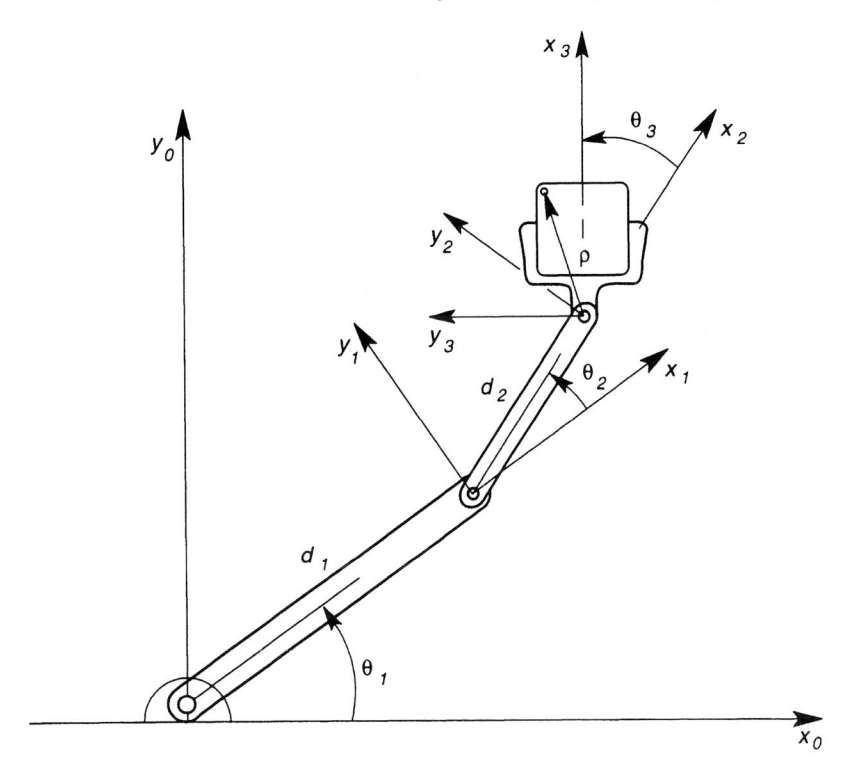

Fig. 3.12 Three-link planar manipulator.

Example 3.5 Three-link Planar Manipulator. Fig. 3.12 shows a device for positioning an object in a plane using three links with finite rotations $(\theta_1, \theta_2, \theta_3)$. It is usual in multibody problems to attach local coordinates to each body. In this example we define four sets of basis vectors with coordinates (x_0, y_0), (x_1, y_1), (x_2, y_2) and (x_3, y_3). The latter coordinates sit at the end of link number 3. Thus if $\rho = [x_3, y_3]^T$ represents the position of a point P in the movable object with respect to the gripper coordinates, its position relative to link number 2 is given by

$$\mathbf{r}_2 = \mathsf{T}_3 \rho \tag{3.59}$$

where

$$\mathsf{T}_3 = \begin{bmatrix} c_3 & -s_3 & 0 & | & 0 \\ s_3 & c_3 & 0 & | & 0 \\ 0 & 0 & 1 & | & 0 \\ \hline 0 & 0 & 0 & | & 1 \end{bmatrix} \tag{3.60}$$

where we have used the notation $c_3 = \cos\theta_3$, $s_3 = \sin\theta_3$. Then one can show that the position of a vector in link number 2 with respect to link number 1 involves a translation and a rotation, i.e.,

$$\mathbf{r}_1 = \mathsf{T}_2 \mathbf{r}_2$$

where

$$\mathsf{T}_2 = \begin{bmatrix} c_2 & -s_2 & 0 & | & d_2 c_2 \\ s_2 & c_2 & 0 & | & d_2 s_2 \\ 0 & 0 & 1 & | & 0 \\ \hline 0 & 0 & 0 & | & 1 \end{bmatrix} \quad (3.61)$$

The same form follows for link number 1 with the subscript 2 replaced by 1, i.e.,

$$\mathbf{r}_0 = \mathsf{T}_1 \mathbf{r}_1$$

$$\mathsf{T}_1 = \begin{bmatrix} c_1 & -s_1 & 0 & | & d_1 c_1 \\ s_1 & c_1 & 0 & | & d_1 s_1 \\ 0 & 0 & 1 & | & 0 \\ \hline 0 & 0 & 0 & | & 1 \end{bmatrix} \quad (3.62)$$

Thus the position of a point in the gripper object with respect to the base coordinates is given by

$$\begin{bmatrix} x_0 \\ y_0 \\ 0 \\ 1 \end{bmatrix} = \mathsf{T}_1 \mathsf{T}_2 \mathsf{T}_3 \begin{bmatrix} x_3 \\ y_3 \\ 0 \\ 1 \end{bmatrix} \quad (3.63)$$

For example, it is not difficult to show that the position of the point P in the base coordinates is,

$$\begin{aligned} y_0 &= d_1 \sin \theta_1 + d_2 \sin(\theta_1 + \theta_2) \\ &\quad + x_3 \sin(\theta_1 + \theta_2 + \theta_3) + y_3 \cos(\theta_1 + \theta_2 + \theta_3) \\ x_0 &= d_1 \cos \theta_1 + d_2 \cos(\theta_1 + \theta_2) \\ &\quad + x_3 \cos(\theta_1 + \theta_2 + \theta_3) + y_3 \sin(\theta_1 + \theta_2 + \theta_3) \end{aligned} \quad (3.64)$$

Screw Transformation

In a mechanical screw, the nut rotates and translates along a common axis with a geometric coupling of translation and rotation enforced by the helical screw. (See Fig. 3.15.) In mathematical kinematics one can define a generalized screw transformation as a finite rotation and a translation along the same axis where the axial translation and rotation angle are independent. In the 4×4 notation, this matrix is the product of a translation matrix and a simple rotation matrix as follows:

$$\mathbf{S}_Z(d_z, \theta_z) = \mathbf{R}_Z(\theta_z) \mathbf{D}_Z(d_z) \quad (3.65)$$

where

$$R_Z = \begin{bmatrix} \cos\theta_z & \sin\theta_z & 0 & 0 \\ \sin\theta_z & \cos\theta_z & 0 & 0 \\ 0 & 0 & 1 & 0 \\ 0 & 0 & 0 & 1 \end{bmatrix}$$

$$D_Z = \begin{bmatrix} 1 & 0 & 0 & 0 \\ 0 & 1 & 0 & 0 \\ 0 & 0 & 1 & d_z \\ 0 & 0 & 0 & 1 \end{bmatrix}$$

One of the unique properties of the screw transformation is that the translation and rotation about the same axis commute, i.e.

$$S_Z = R_Z D_Z = D_Z R_Z \qquad (3.66)$$

The screw transformation is especially useful in the kinematics of robotic machines. In a six degree of freedom manipulator arm for example, one can describe the position of a point in one link reference frame with respect to a neighboring link with two screw transformations. (See Chapter 6.)

If S represents the finite screw motion of coordinate frame A to frame B, then the position vector of a point in frame B can be related to the position vector in frame A by the matrix relation

$$r^A = S^A_B r^B \qquad (3.67)$$

We interpret this coordinate transformation as a translation of frame B along some axis relative to A and a rotation of frame B about the same axis. The vector r^B is the position vector to a geometric point measured with respect to frame B. Likewise r^A is the position vector to the same geometric point in space, but measured with respect to the coordinate frame A. We recall however that in the 4×4 matrix notation, the position vectors have a dummy 4th row:

$$r = [x \quad y \quad z \quad 1]^T$$

3.8
Kinematic Mechanisms

A pure kinematic mechanism is a collection of bodies with geometric constraints such that the motion of one or more bodies determines the motion of the rest of the bodies. In machines, mechanisms can have many functions, including

- Conversion of rotary to translation motion
- Conversion of translation to rotation
- Conversion of circular motion to oscillatory motion

- Conversion of circular motion about one axis to circular motion about another axis
- Ratcheting, indexing
- Produce similarity motions (pantographs)
- Produce dwell motions
- Stopping and breaking actions
- Clamping actions

Analysis of mechanisms can be divided into the kinematics of relative motion of *kinematic pairs*, or *joints* and the overall kinematics of the mechanism.

Examples of modern applications of mechanisms may be found in deployment of solar panels in satellites, aircraft door latches, VCR cassette receptacles, and windshield-wiper systems.

Variety of Mechanisms

Mechanisms are subassemblies of machines that are designed to produce a determinate motion under prescribed forces and torques. Robert Willis (1800–1875) and Franz Reuleaux (1829–1905) posited theories of machine synthesis that state that most machines can be decomposed into a finite set of simple or compound mechanisms. For example the internal combustion engine has a coordinated set of slider crank mechanisms, cam and pulley mechanisms, as well as pump mechanisms. Mechanisms can be pure kinematic devices whose motion is determined solely by the geometric constraints (Jacobian relationships) or dynamic devices such as escapements and engines. Dynamic machines require not only kinematic constraint equations but also Newton–Euler equations of motion, and perhaps control law equations to determine the motion and internal forces. For both pure kinematic or dynamic machines, the internal forces on the "rigid" elements as well as forces on the joints and bearings are determined by Newton–Euler equations of motion.

Although it is essential for students of machine design to know a variety of mechanisms in order to design a machine, very few formal courses exist at this time for the student to learn the nature of these sub-assemblies. The student of machine design is expected to acquire knowledge of a minimal set of mechanisms by either self-study or through experience. Today, many machine elements and mechanisms are modularized, i.e. are manufactured in a finite set of sizes and specific application classes, such as bearings, pumps, motors, engines, gear elements and complete gear transmissions.

Many types of mechanisms have been known and used since the Renaissance (Fig. 3.13*a*) and certainly from the late industrial era of the 19th century. (See for example the Author's book *The Machines of Leonardo da Vinci and Franz Reuleaux*, 2007, for a review of the history of machines and mechanisms.) However, new types of mechanisms are being invented and developed such as the harmonic drive or comb-drive mechanisms in MEMS devices as illustrated in Fig. 1.8. Microelectromechanical, micron scale gear mechanisms have also been fabricated in silicon as shown in Fig. 3.13*b*. The general classes of mechanisms are:

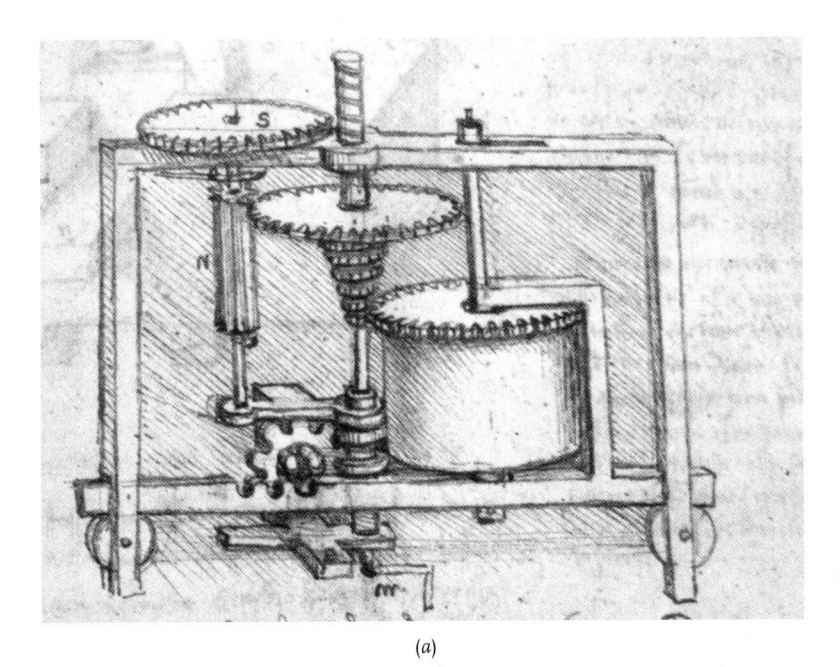

Fig. 3.13 (a) Speed-changing gear mechanism of Leonardo da Vinci (Codex Madrid, 1492). (b) Modern MEMS gear assembly made of micron sized silicon. (Courtesy Sandia National Lab.)

Linkages

Linkages are a collection of rigid bodies connected with kinematic joints such as revolute, prismatic, cylindrical or spherical bearings. The relative degrees of freedom between the rigid parts can be one [revolute, prismatic], two [cylindrical], or three [spherical bearing]. The overall degrees-of-freedom of the linkage depends on the number of grounded links [usually one link is grounded] and follows Grübler's criteria. (See below.) Linkages can undergo both planar motions or three-dimensional motions [e.g., universal joint]. Many robotic machines are made from jointed linkages.

A single degree of freedom, slider-crank linkage is shown in Fig. 3.14a from a 15th century drawing of Leonardo da Vinci. This is a pure kinematic mechanism. Motion of the worm crank determines the motion of the slider link. Modern robotic linkages may have from two or more degrees of freedom as shown in Fig. 3.14b in a passive walking machine from the laboratory of Professor Andy Ruina of Cornell University. In this device, the machine can mimic walking without control using principles of dynamic stability.

Geared Wheels

Gear systems transfer motion from one rotary axis to another through the contact of toothed wheels and helical contacts. For motion about parallel axes, Euler and other early theorists discovered that a smooth transfer of motion without accelerations can be accomplished if the shape of the gear teeth take the form of epicycloids or involute curves.

When the input and output motions occur about parallel axes, the toothed wheel pair is called *spur gears* and the smaller diameter gear is called a pinion. Transfer of motion about axes that intersect at 90 degrees are accomplished with *bevel gears*. When the axes are perpendicular but do not intersect a helical worm drive is used. A gear mechanism of Leonardo da Vinci is shown in Fig. 3.13a.

Pulley Systems

Belt and chain drives transfer motion from one rotary axis to another in a continuous way.

Ratchets

Ratchets often act as one-way devices that block motion in one direction while permitting motion in the opposite direction. They are analogs of diodes in electronic circuits and as such act as mechanical logic devices.

Escapements

Escapements are mechanical controllers that regulate the amount of energy into a machine. The classic escapement is found in mechanical clocks and often employs various versions of the ratchet. Escapements are also found in simple mechanical timing devices such as metronomes. However many functions of escapements have been replaced in mechatronic machines with microprocessors. The escapement is not a pure mechanism since its motion is governed by Newton–Euler

3.8 Kinematic Mechanisms | 107

(a)

(b)

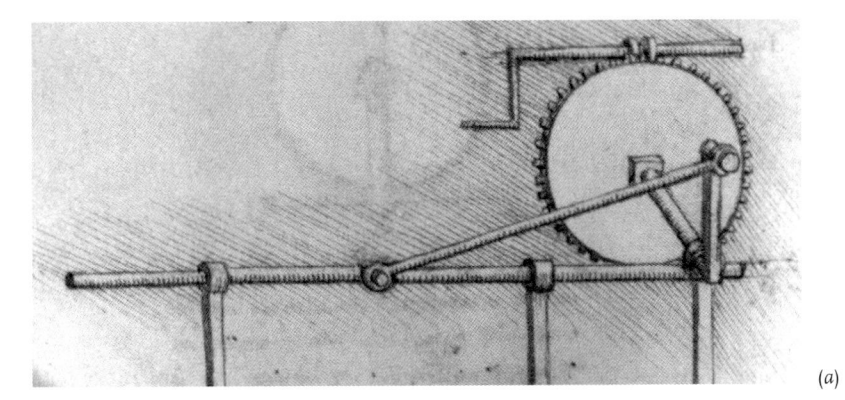

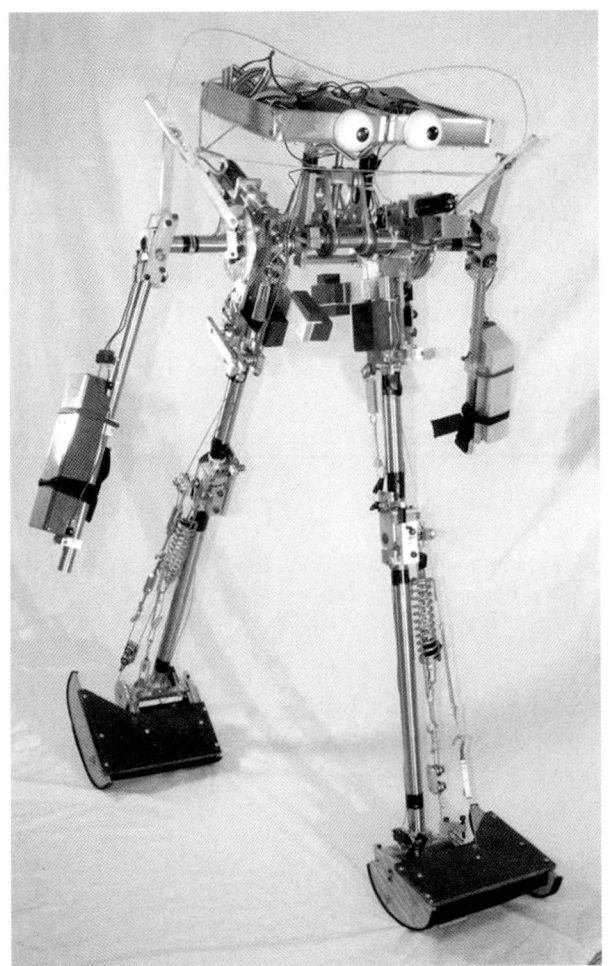

Fig. 3.14 (a) Slider-crank linkage and worm gear drawing of Leonardo da Vinci (Codex Madrid, 1492). (b) Passive walking linkage machine of Professor A. Ruina, Cornell University.

equations of motion. A discussion of clock escapement modeling may be found in a paper by the Author (Moon and Stiefel, 2006). Readers can also find ten escapements with videos of their motions on the Cornell KMODDL website.

Cams and Dwell Mechanisms

Cams have been used in internal combustion engines for over a century to control the fuel value opening and closing. Often they involve sliding motion of one part on another, as for example, the rotation of a noncircular closed curve against a movable piston-like surface. Cams sometimes require gravity or spring forces to maintain contact between the two sliding surfaces. However there are also positive-displacement cams such as the curved or Reuleaux triangle used in dwell mechanisms. These devices produce an oscillatory motion with a finite dwell, or dead time, in the period that is sometimes required in manufacturing processes. Readers can find a selection of positive displacement cam mechanisms on the Cornell KMODDL website.

Straight-line and Parallel Mechanisms

One of the challenges of mathematical kinematics in the 19th century was the search for a theorem to determine if an assembly of mechanical linkages could produce different classes of mathematical motions. Famous 19th century mathematicians such as Chebyshev and Sylvester worked on these problems. One famous example was the search for a mechanism to trace an exact straight line. James Watt had invented a multi-link mechanism to trace an approximate straight-line for his steam engines in order to convert linear motion of the piston into rocking motion of the *balancer* link. In 1867, a French engineer named Peaucellier (1864) invented an eight-link, planar linkage the drew an exact straight line and could also produce exact circular arcs of arbitrarily large radius of curvature (see Fig. 3.20). There are dozens of approximate and exact straight-line mechanisms available for the designer in the literature. Many can be found on the Cornell KMODDL website.

Another mathematical problem was to find mechanisms that would produce a motion of a body in pure translation so that lines would remain parallel during the motion. One example is shown below in one of Reuleaux's kinematic models in Fig. 3.18. Many other examples can be found on the KMODDL website.

Compliant Mechanisms

An important class of mechanisms, associated with MEMS technology, are structural devices that behave like kinematic mechanisms. In such devices material is removed to produce a resistance to bending such that the structural joint mimics a revolute kinematic joint. Such structures can be made out of the same material, usually silicon, without the necessity of designing kinematic bearings (see, e.g., Lobontiu, 2007; Pelesko and Bernstein, 2003).

Joints: Kinematic Pairs

In many mechanisms with rigid-body elements, the relative motion between bodies is constrained to one to five degrees of freedom. Several such kinematic pairs are shown in Fig. 3.15. The motion of one link relative to the other can be described

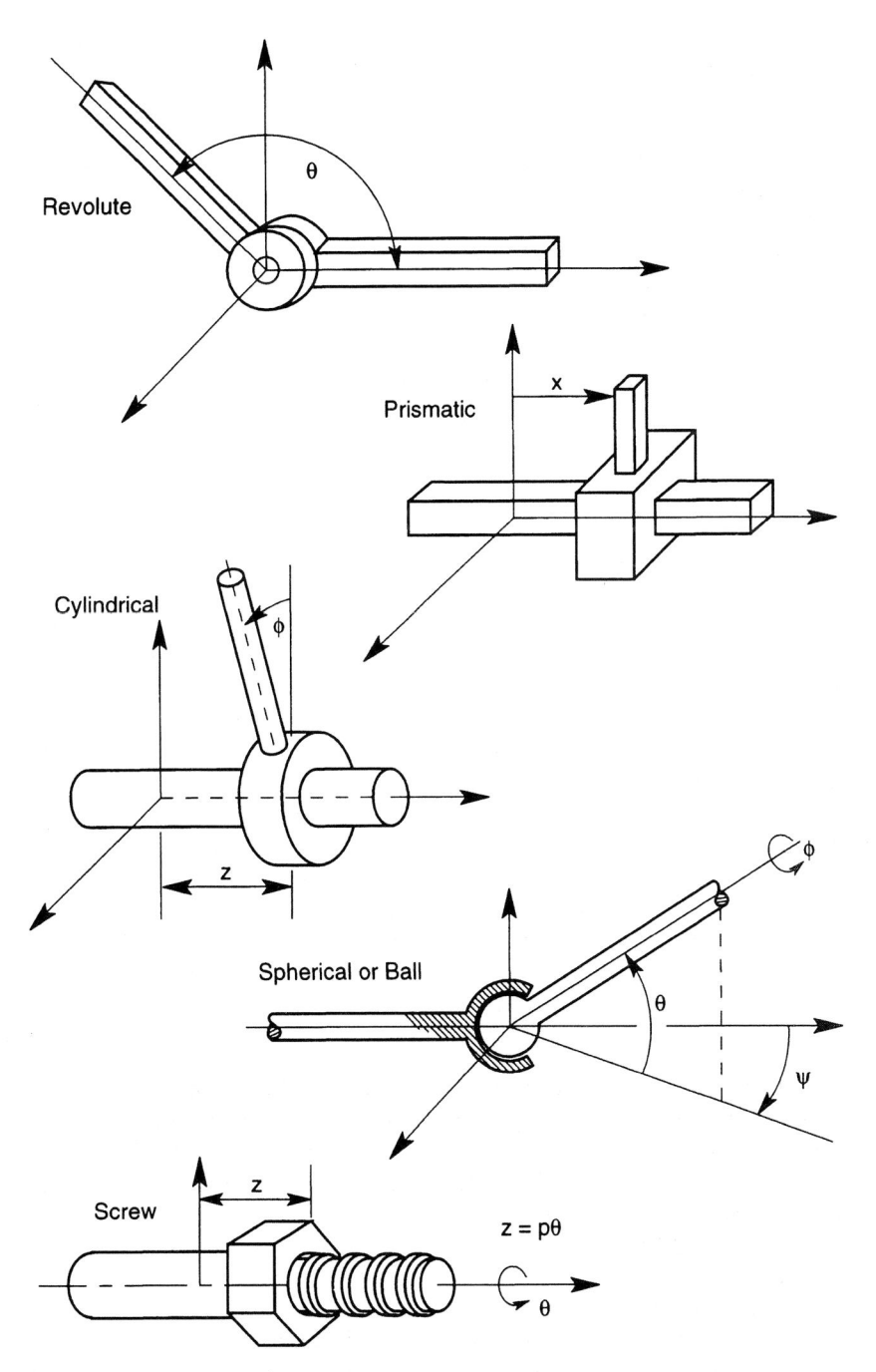

Fig. 3.15 Kinematic joint pairs.

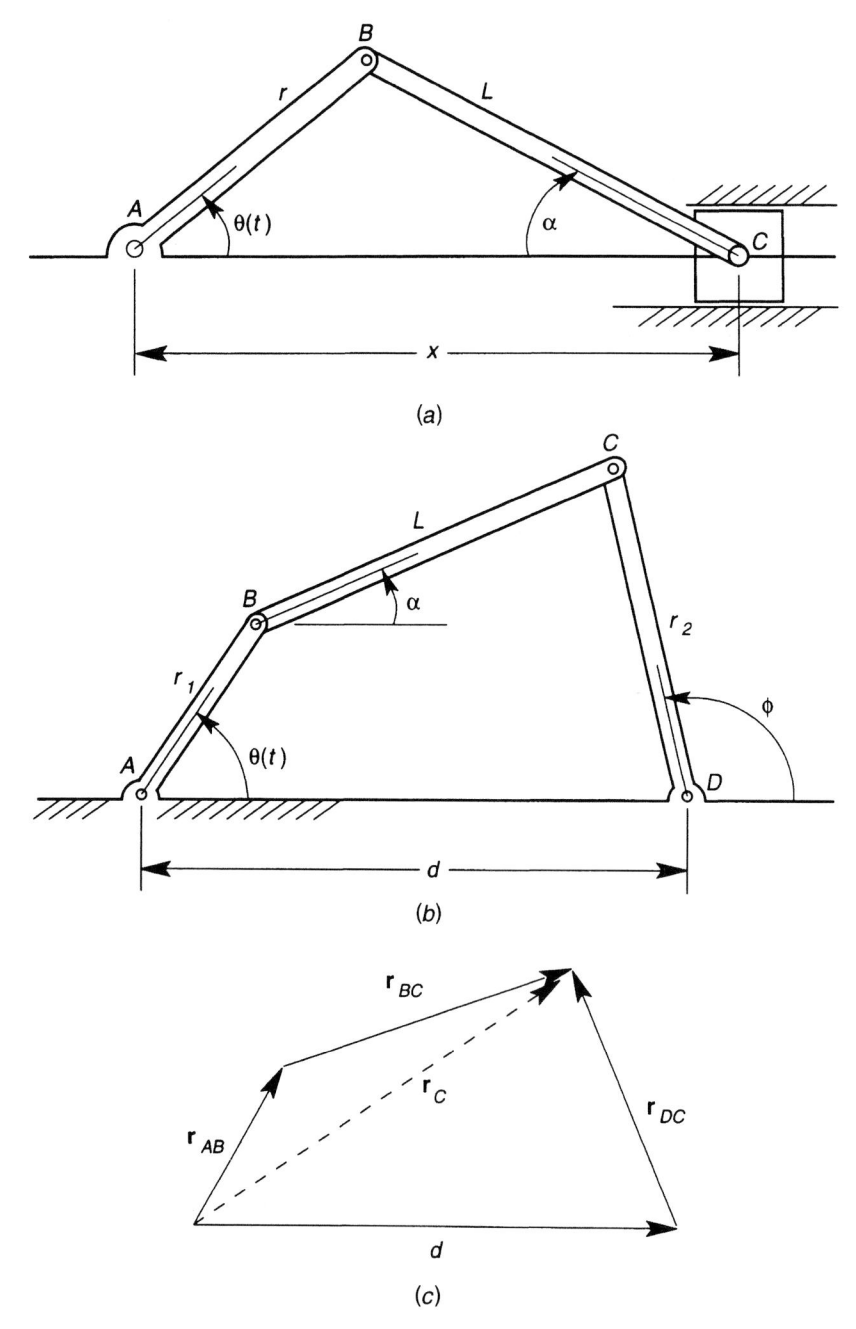

Fig. 3.16 (a) Slider crank mechanism. (b) and (c) Four-bar planar linkage.

by a transformation matrix, as in Section 3.7. Thus the revolute joint in Fig. 3.15, has a single degree of freedom in rotation. However the cylindrical joint in Fig. 3.15 has both rotary and translation degrees of freedom. In robotics, joints with pure rotation are called *revolute*, whereas those with translation are called *prismatic*.

Analysis of Motion in a Closed Kinematic Chain

Two classic planar mechanisms are the 4-bar linkage and the slider crank shown in Fig. 3.16a and b. One link in each case is ground. The latter is, of course, employed in millions of reciprocating engines and may be the world's most ubiquitous mechanism. A classic problem is to find the motion of the remaining links given the motion of link 1, usually thought of as having steady circular motion. There are two basic methods to solve such problems:

1. Geometric and trigonometric method
2. Vector calculus

Method 1 is illustrated in the following example.

Example 3.6 Slider Crank. From the geometry in Fig. 3.16a, the following scalar equations result

$$r \cos\theta + L \cos\alpha = x$$
$$r \sin\theta = L \sin\alpha \tag{3.68}$$

or

$$L^2 = x^2 + r^2 - 2rx \cos\theta$$
$$\alpha = \sin^{-1}\left(\frac{r}{L} \sin\theta\right) \tag{3.69}$$

These two equations determine $x(\theta)$ and $\alpha(\theta)$. To find the velocity relations we differentiate (3.68) and obtain a set of linear equations for $\dot{x}$ and $\dot{\alpha}$ in terms of $\dot{\theta}$, i.e.,

$$\dot{\alpha} = \frac{r}{L}\dot{\theta}\frac{\cos\theta}{\cos\alpha}$$
$$\dot{x} = -r\dot{\theta}(\sin\theta + \tan\alpha \cos\theta) \tag{3.70}$$

This simple example can be solved using a symbolic computer code. The following text in bold is a program written in MATHEMATICA (see Appendix B). The operation **D[%, t]** performs symbolic differentiation on the expressions preceding it indicated by the symbol %. The operation Solve is an algebraic equation solver for the generalized velocities **q1'[t]** and **q3'[t]**, in terms of the velocity **q2'[t]**. The user inputs the statements in bold and the computer program returns the solution that follows. Similar solutions can be obtained using MAPLE and MACSYMA.

Input Statements:

```
x=q1[t]
theta=q2[t]
alpha=q3[t]
r Cos[theta] + L Cos[alpha] - x == 0
r Sin[theta] - L Sin[alpha] == 0
D[%%,t]
D[%%,t]
Solve [{%, %%},{q1'[t],q3'[t]}]
```

Output Statement:

```
{{q1'[t] -> -((Sec[q3[t]])
        (L r Cos[q3[t]]    Sin[q2[t]] q2'[t] +
           L r Cos [q2[t]]    Sin[q3[t]] q2'[t] )) / L ),
  q3'[t] -> r Cos [q2[t]]    Sec[q3[t]] q2'[t]}}
                              L
```

In method 2 we use the closed chain property and write the position vector to point C in the 4-bar example in two forms. The equality of the velocity calculated from each vector relation provides a vector equation that yields two scalar relations between the circular motion and the unknown motion of the center link.

Example 3.7 4-Bar Closed Chain. In Fig. 3.16b, point C can be described by two different sums of vectors:

$$\mathbf{r}_{AC} = \mathbf{r}_{AB} + \mathbf{r}_{BC} = +\mathbf{r}_{DC} \tag{3.71}$$

The velocity $\dot{\mathbf{r}}_C$ is obtained by recognizing that each of the vectors is of constant length so that the angular-velocity cross-product theorem can be used.
For example:

$$\dot{\mathbf{r}}_{AB} = \boldsymbol{\omega}_1 \times \mathbf{r}_{AB} = \dot{\theta} r_1 (\cos\theta \hat{\mathbf{j}} - \sin\theta \hat{\mathbf{i}}) \tag{3.72}$$

Differentiating the vector identity [Eq. (3.71)] then yields two equations linear in $\dot{\phi}$, $\dot{\alpha}$, $\dot{\theta}$, or

$$\begin{bmatrix} r_2 \sin\phi & -L \sin\alpha \\ r_2 \cos\phi & -L \cos\alpha \end{bmatrix} \begin{bmatrix} \dot{\phi} \\ \dot{\alpha} \end{bmatrix} = r_1 \dot{\theta} \begin{bmatrix} \sin\theta \\ \cos\theta \end{bmatrix} \tag{3.73}$$

One can show that

$$\dot{\alpha} = \frac{r_1}{L} \dot{\theta} \frac{\sin(\phi - \theta)}{\sin(\alpha - \phi)} \tag{3.74}$$

Note that although this relation is independent of r_2, $\phi(\theta)$ and $\alpha(\theta)$ depend on r_2.

The determinant of the 2 × 2 matrix in Eq. (3.73) is $-r_2 L \sin(\phi - \alpha)$. We can see that when $\sin(\phi - \alpha) = 0$, the rates $\{\dot{\phi}, \dot{\alpha}\}$ cannot be determined, given $\dot{\theta}$. These points are known as *singular points* of the mechanism and are usually avoided in practical design.

Grübler's Mobility Criterion: Topological Invariants in Linkages

Aside from the basic laws of physics, assemblies of rigid bodies are subject to mathematical laws of geometry and topology. One of the basic ideas in kinematics of mechanisms is the notion of degrees of freedom. Thus if one is given a set of n rigid bodies or *links* and r geometric constraints between each of the bodies and the other links, it is important to be able to calculate the number of degrees of freedom in the device. A large number of rigid bodies can be connected in a way such that there are no degrees of freedom, except elastic deformation, as in a grounded three-link planar truss. However with the addition of one more link and pin joints between each of the links the linkage has one degree of freedom when one link is grounded.

The idea of a mechanism as a chain of links with different geometric constraints was formulated by the German engineer Franz Reuleaux in 1875. Each of the constraints may diminish the relative degrees of freedom in different ways. For example the revolute joint, between two links removes five degrees of freedom, allowing only rotary motion between the parts. A ball joint removes three translational degrees of freedom but allows three rotational degrees of freedom. These ideas lead to the algebraic relation below called *Grübler's criterion*.

The four-bar mechanism shown in Fig. 3.17, is the quintessential example of Reuleaux's kinematic chain of constrained links. In analogy to electrical circuits, the closed mechanical circuit can be generalized into multiple circuits or a kinematic network, called compound linkages (Fig. 3.17).

Such mechanisms were used by James Watt in the design of his steam engines, as well as by Stephenson in his steam locomotive engines. One of the properties of compound linkages is the existence of topological invariants. Topological relationships describe general properties of geometric objects independent of their specific dimensions and shapes. For example, if n denotes the number of links in a kinematic network, r the number of revolute or cylindrical joints, and F the number of degrees of freedom in the mechanism, sometimes called the *mobility*, then the following relationship can be established:

$$2r - 3n + (3 + F) = 0 \quad \text{or} \quad F = 3(n - 1) - 2r \tag{3.75}$$

For $F = 1$,

$$r = \frac{3}{2}n - 2 \tag{3.76}$$

The conventional derivation of the expression (3.75) begins with a set of n links in the plane with $3n$ degrees of freedom (two translations and one rotation) with

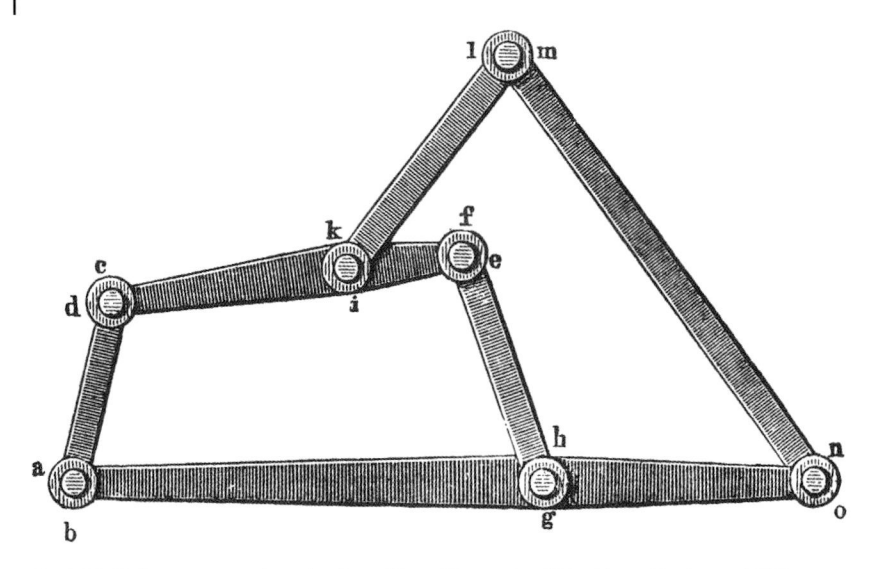

Fig. 3.17 Six-bar compound mechanism. (From *Kinematics of Machinery*, Reuleaux, 1876.)

one grounded link leaving $3(n-1)$ degrees of freedom. A set of r lower pair joints such as turning pairs or revolute joints removes $2r$ degrees of freedom hence the expression above. The interpretation of F is that $F = 1$ indicates a perfect mechanism where the movement of one link determines the movement of all the rest of the links. If $F = 0$, the arrangement of links forms a statically determinate rigid truss or structure and if $F = -1$, the structure is statically indeterminate and the internal forces are dependent on the elastic properties of the links.

The relation (3.76) for $F = 1$ was posited by Martin Grübler in an 1883 paper in *Der Civilingenieur* and later in his 1917 book on kinematics, *Getriebelehre* (Berlin). Grübler (1851–1935) was a professor at the Technische Hochschule Dresden and was influenced by the work of Franz Reuleaux. These relations hold for planar mechanisms. Similar equations can be written for spatial linkages. Grübler credited two mathematicians for this criteria, Sylvester (1874) and the Russian Chebyshev (1869). However this so-called *mobility criterion* has many notable exceptions. There are special geometric arrangements of links that allow small motions of a set of links with $F = 0, -1$. An example is the weighing mechanism, the Reuleaux–Voigt Model T-6, shown in Fig. 3.18.

This parallel linkage has $n = 8$ links and $r = 11$ joints, counting the double joints on the upper balance arm. Using (3.75) we find that $F = -1$, which might led one to conclude that the device should be rigid. However it is clear that the linkage has limited motion about the horizontal position of the balance arm. The relations (3.75), (3.76) can also be used when some of the joints are prismatic or sliding joints. However there are problems when all the joints are sliding, again a value of $F = 0$ or -1 can imply rigidity when in fact the mechanism has some lim-

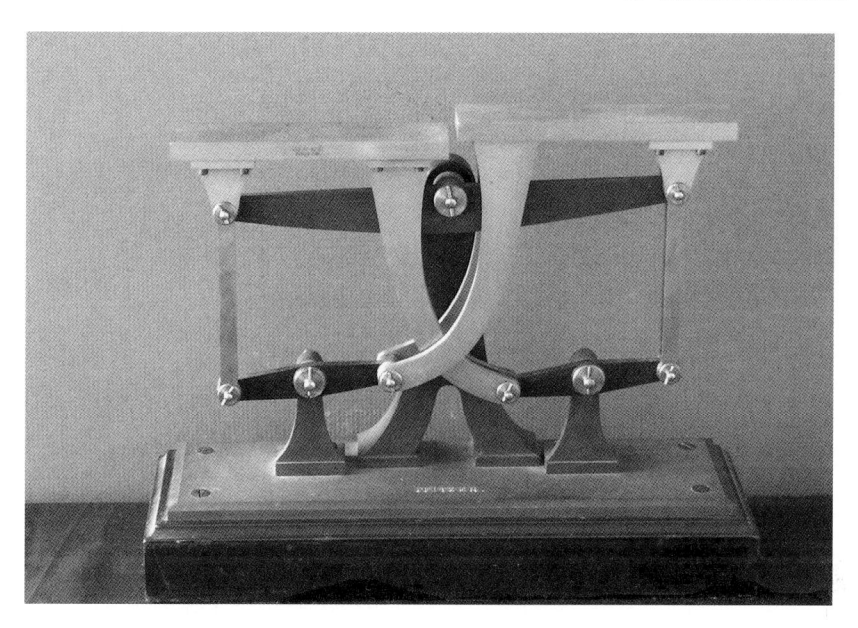

Fig. 3.18 Over-constrained compound mechanism with 8 links and 11 joints that admits small motions ($F = -1$). Used for a weighing mechanism. (Cornell Collection of Reuleaux Kinematic Mechanisms, see KMODDL Model T06.)

ited motions. A modern discussion of the use and limitations of mobility criteria may be found in the English texts of Burton Paul (1979), D. C. Tao (1967), Joseph Shigley (1969), Richard Hartenberg and Jacques Denavit (1964), and Arthur Erdman, George Sandor, and S. Kota (2001).

In relation (3.76), r, n, F are integers and the number of degrees of freedom assumes that one link of the network is grounded. The minimum number of links for $F = 1$, is $n = 4$, which gives $r = 4$. The integer requirement implies that the number of links n must be even, which leads to the sequence of possible single degree of freedom compound mechanisms:

$$\{(n, r) = (4, 4), (6, 7), (8, 10), (10, 13), \ldots\} \tag{3.77}$$

A generalization of the closed 4-bar linkage is to include rigid links with more that two revolute joints. Thus if n_i denotes the number of links with i joints, then the equations relating the number of sub-links to the total number of links and joints are (Grübler, 1917)

$$n = \sum_{i=2} n_i \tag{3.78}$$

$$r = \frac{1}{2} \sum_{i=2} i n_i \quad [F = 1] \tag{3.79}$$

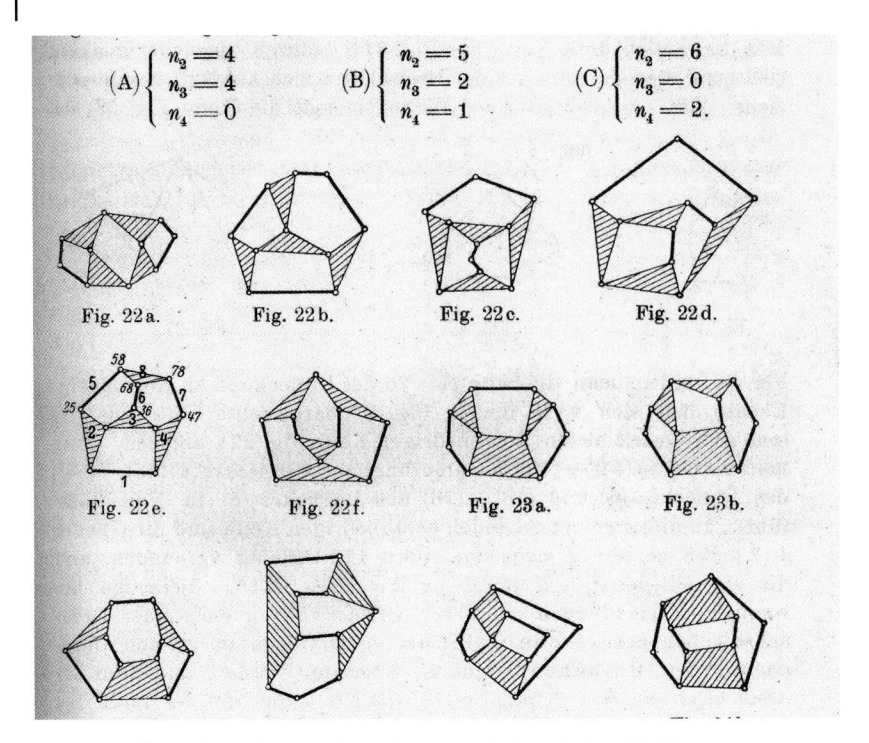

Fig. 3.19 Table of 8-bar linkages with mobility $F = +1$, from Grübler (1917).

Note that for closed chains the minimum number of joints per link is $i = 2$.

In the case of 8-link mechanisms, one can have 4-joint links. It is easy to show using the above equations that there are three classes of 8-link mechanisms with $F = 1$:

$$\{n_2 = 4,\ n_3 = 4,\ n_4 = 0\}, \qquad \{n_2 = 5,\ n_3 = 2,\ n_4 = 1\}$$
$$\{n_2 = 6,\ n_3 = 0,\ n_4 = 2\} \tag{3.80}$$

Within a single class of 8-link mechanisms there can be multiple distinct topologies as shown in Fig. 3.19 from Grübler (1917).

In typical sketches of compound linkages, 2-joint links are drawn as straight lines, 3-joint links as triangles, and 4-joint links as trapezoids. The revolute joints are drawn as open circles. In actual compound mechanisms however, the links can have any shape.

An example of a compound mechanism is the straight-line mechanism of Peaucellier shown in Fig. 3.20. In the classic Peaucellier linkage, lengths $a = b = c = d$ and lengths e and f are equal. As the links AE and FD rotate, the point C traces an exact straight line, provided links AF, FD have the same lengths, i.e., $h = g$. When these links are not equal, the point C traces an exact arc of a circle.

This was first recognized as an exact straight-line mechanism by Peaucellier in 1867 and a few years later by Lipkin a student of Chebyshev. This mechanism has

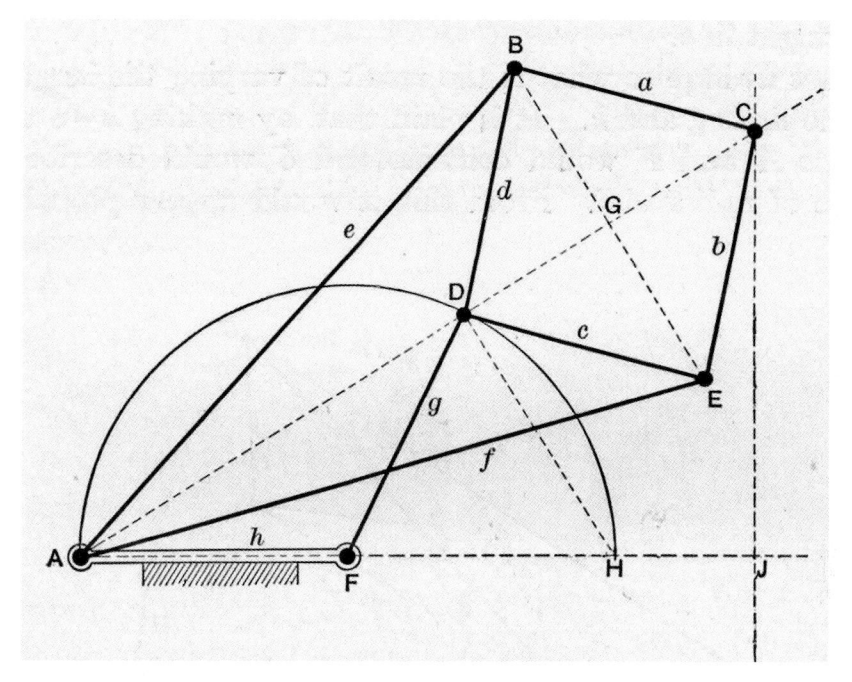

Fig. 3.20 Peaucellier exact straight-line mechanism. This compound mechanism has 8 essential links and 10 revolute kinematic pairs and mobility $F = +1$.

$n = 8$ links, and $r = 10$ turning joints, counting the four double pin joints at A, D, E, B. For this arrangement, the number of degrees of freedom, $F = 1$ when one link is grounded. The outer pin can trace either an exact straight line or an exact arc of a circle of any radius.

Mobility for Spatial Mechanisms: Kutzbach Criterion

The mobility equation (3.75) can be generalized for linkages that move in a three-dimensional space. Some references credit Grübler with this generalization in 1917 while others cite a paper by K. Kutzbach in 1929. (See Phillips (1984) for discussion of mobility in special mechanisms.) The central idea is that a collection of n free rigid links has $6n$ degrees of freedom and if one is grounded, there will be $6(n-1)$ degrees of freedom left. If one adds constraints between kinematic pairs this will reduce further the possible motions. Each kinematic pair can be assigned an integer measure of freedom f_i for the ith joint. For example a revolute joint has one degree of freedom, $f = 1$, while a cylindrical joint in 3D has two degrees of freedom, rotation and sliding or $f = 2$. A spherical joint has $f = 3$ and so on. Thus each joint reduces the mobility by $(6 - f_i)$ so that the combined mobility of the mechanism becomes

$$F = 6(n-1) - \sum(6 - f_i)$$

If there are g joints, then this relation becomes

$$F = 6(n - g - 1) + \sum f_i \tag{3.81}$$

This relation is called in some kinematics texts, the *Kutzbach criterion* for kinematic mobility in a spatial mechanism. If $F = 1$, the mechanism has complete movement for any general geometric parameters of the linkage. If $F = 0$, the linkage is technically a structure without any possibility of motion, however there are special geometries for which such "structural" mechanisms can have finite motions. The case of $F = -1$ represents an indeterminate structure. Again there are special cases where such over-constrained mechanisms can have both small and large finite motions.

In the case of a simple 3D spatial chain with one loop or circuit, there would be $n = g$ links and joints, or

$$F = + \sum f_i - 6 \tag{3.82}$$

Thus for a perfect movable spatial mechanism, $\sum f_i = 7$. For all revolute joints, (i.e., $f = 1$), one needs *seven links for a single closed kinematic chain*. For three cylindrical joints ($3f = 6$) and one revolute joint, we can have a four link spatial mechanism with complete freedom of motion ($F = +1$) and one grounded link.

Centrodes: Mechanisms as Rolling Motions

A little known idea, found in older texts on kinematics of mechanisms is the fact that all planar motions are equivalent to pure rolling. In Euler's theorem for finite motions of rigid bodies, he stated that any finite motion can be separated into a translation of a given point in the body and a rotation about an axis through this point. In the case of planar motions of two bodies one can further state that the relative motion between two rigid bodies can be synthesized by the rolling of one body on another. By rolling we mean that there is a line of contact between the two bodies and that the relative velocity of the contact points in each body is zero.

Consider the cylinder in Fig. 3.21, rolling on the plane surface of the fixed or grounded body. The only degree of freedom allowable is an instantaneous rotation of one body relative to the other about the line of contact. For finite motions however the line of contact translates along the plane surface. This line is called an *instantaneous center of rotation*. The *instant center*, sometimes called a *centro*, can have a velocity even though the relative velocity of the particles in either body at the contact point is zero. When the contact plane is fixed, points on the rim of the circular cross-section trace out curves called *cycloids*, and interior points within the circular boundary trace out curves called *trochoids*.

Now suppose that we attach an arbitrary body B_1 to the end of the cylinder and another body B_2 to the edge of the grounded plane and imagine that the cylinder and the fixed plane become transparent so that all we can see is body B_1, moving

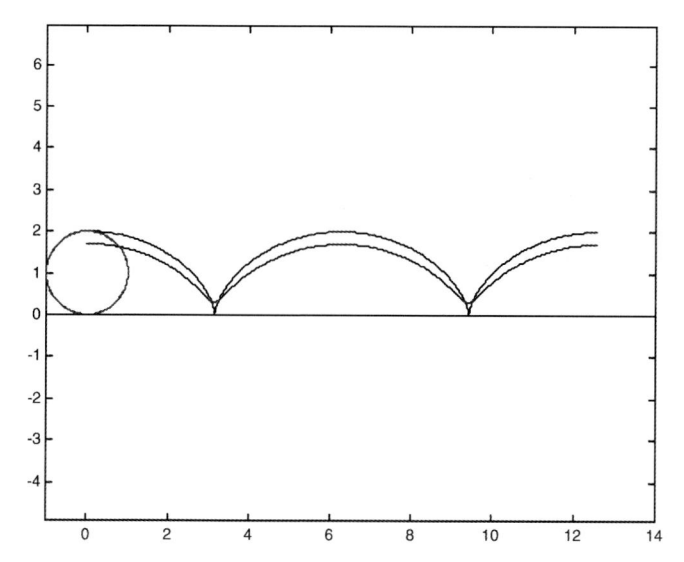

Fig. 3.21 Rolling of a circular body on a flat surface.

with respect to body B_2. However complex the motions of the two bodies appear, we know that the underlying motion is simple rolling of one body on another. In this problem, we say that the cylindrical surface and the plane are *centrodes* of the two moving bodies. This analogy can be generalized to all bodies in planar relative motion.

The German engineering professor Franz Reuleaux (1829–1905) stated this theorem in his 1875 book on theoretical kinematics of machinery, namely that: the general planar motion of two bodies is equivalent to the pure rolling of one surface on another. This pair of centrode curves is unique and that the relative motions of all points in the bodies is known once the centrodes are constructed.

The origins of this theorem go back to Euler, Chasles and Poinsot and examples may be found in the work of Robert Willis (1841/1870), but it was Reuleaux who explicitly described the theorem and the means by which one could apply it to the mechanisms of machines. In his German text, Reuleaux called this pair of rolling curves in the plane *Polbahnen*. In English they are called *centrodes*. The importance of this concept for machine design is that different design solutions may be found to produce a desired relative motion between two bodies as long as each mechanism has the same pair of centrodes. Two examples of multiple mechanisms with identical centrodes are described below: (i) the double slider and planetary gear mechanisms and (ii) the crossed four-bar linkage and elliptical gears.

(i) Double-slider Four-bar Linkage

Consider the four-link, double-slider mechanism shown in Fig. 3.22a. The coupler link is constrained between motion along the vertical and horizontal slots. Although the motions of the sliding links are rectilinear, the general motion of the coupler link involves both rotation and translation. However it is not difficult to

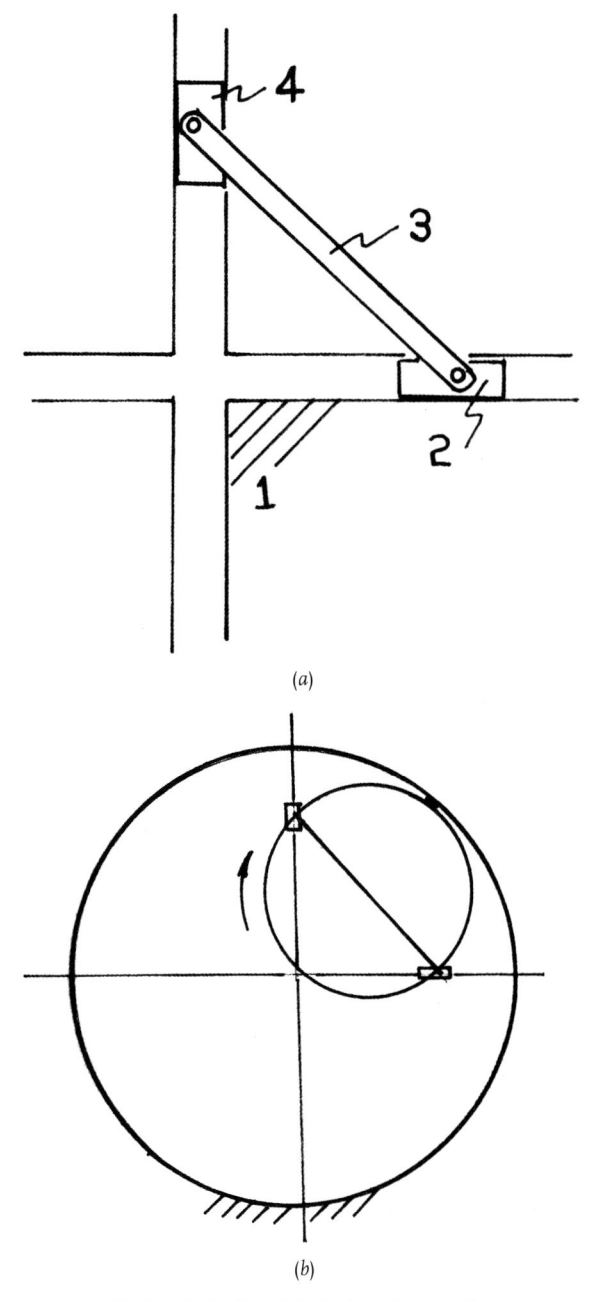

Fig. 3.22 (a) Sketch of a four-link, double-slider mechanism. (b) Sketch of a planetary gear mechanism with the ratio of diameters of planet and sun gears 1:2.

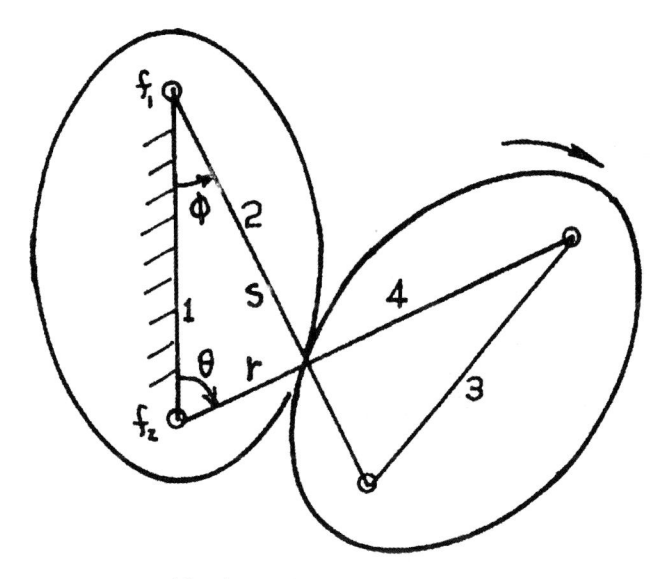

Fig. 3.23 Crossed four-bar mechanism with elliptical centrode curves.

show that the motion of the coupler link relative to the grounded link holding the slots is equivalent to a planetary gear or wheel rolling on the inside of a ring gear as shown in Fig. 3.22b. Here the diameter of the inner circle is half the diameter of the outer surface.

The two circles are said to be the *centrodes* of the relative motion of the double slider. The curves traced out by points on either the link of the double slider or the rolling circle are *hypocycloids*, for points on the rim of the circle, or *hypotrochoids* for interior points in the circle. One consequence of this fact is that a point on the rim of the planet gear will trace out a straight line.

(ii) Crossed Four-bar Linkage

Another example of centrodes is the motion of opposite links in a symmetric, crossed-link 4-bar mechanism shown in Fig. 3.23. Here one can show that the relative motion of the two links L_1, L_3 is equivalent to the rolling of one ellipse on another, where the four foci of the two ellipses lie at the positions of the four revolute joints of the crossed-link mechanism. One application of the equivalent rolling centrodes has been applied to bio mechanics of knee prostheses as shown in Fig. 3.24.

A simple proof of this fact can be understood by examining the geometry in Fig. 3.23. First we ground one of the links L_1. Next notice that the joints of the opposite link move in circles whose radii must meet at the instant center. Since the fixed centrode is the locus of instant centers, the crossing point of the two crank links must lie on the contact of the two centrode curves. We then observe, again because of the symmetry, that the lengths of the radii r, s from f_1, f_2 to the crossing point must sum to the length of the crank links L_2. Thus as the mechanism moves,

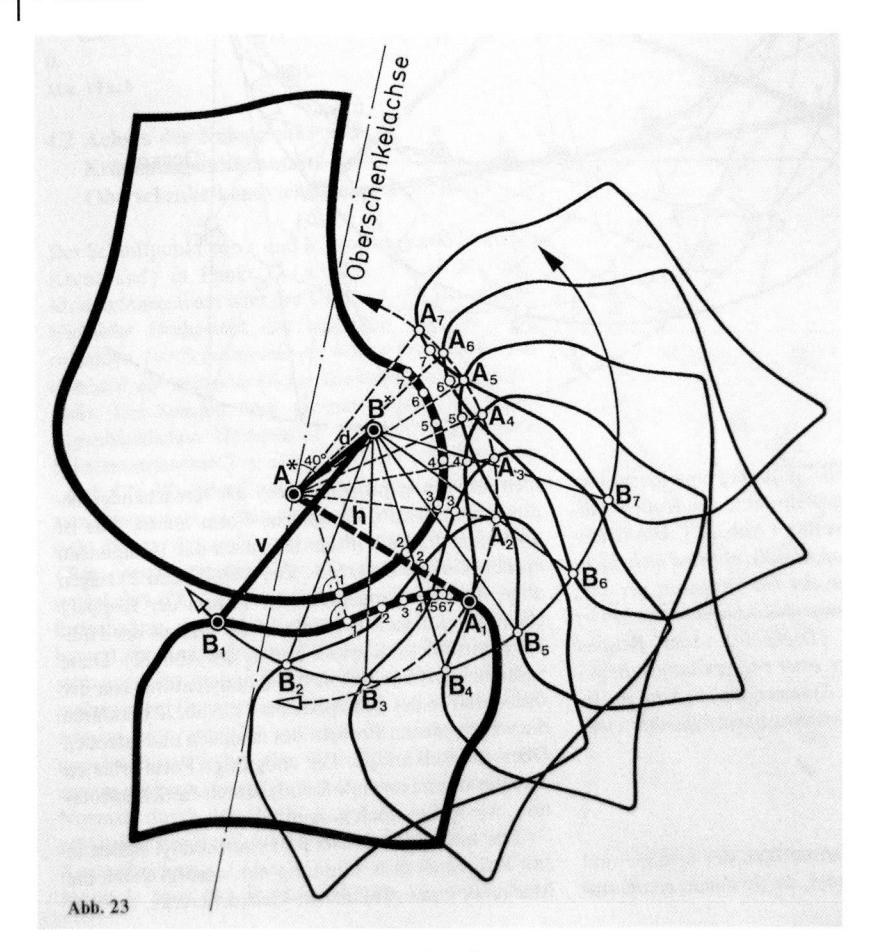

Fig. 3.24 Four-bar linkage model to generate the relative motion of lower leg and femur (Menschik, 1987).

the crossing point traces out the fixed centrode with a constraint that $r + s = L_2$. This is like fixing a string at two points and moving a pen such that the string remains taut. One can show that such a construction will generate an exact ellipse where the end points of the string lie at the foci of the ellipse. The mathematical expression of an ellipse in polar coordinates has the form

$$\frac{1}{r} = p - d \cos \theta \tag{3.83}$$

Using the symbols in Fig. 3.23, one can manipulate the equations for the geometry of the triangle formed by the fixed link and the lower portions of the crank links into the above form and prove that the curve traced by the moving string is an ellipse;

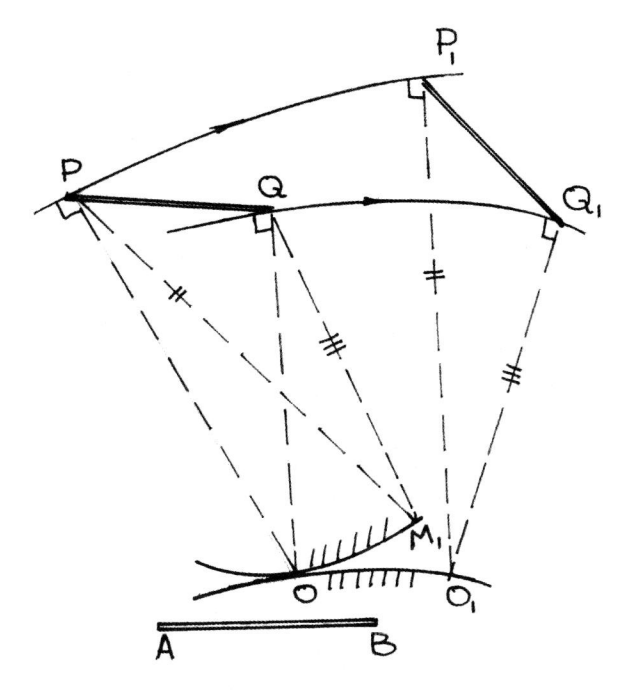

Fig. 3.25 Sketch of geometric construction of centrode curves for the motion of PQ relative to AB. (From Reuleaux, 1876, Fig. 19, p. 66, Dover Edition, 1963.)

$$L_1 = r \cos \theta + s \cos \phi$$

$$0 = r \sin \theta - s \sin \phi$$

or solving for r in terms of θ,

$$\frac{1}{r} = \frac{2}{L_2^2 - L_1^2}[L_2 - L_1 \cos \theta] \tag{3.84}$$

To find the moving centrode we invoke symmetry again and conclude that it too must be an ellipse identical to the fixed centrode and that the relative motion of the two opposite links in the crossed-link mechanism is equivalent to one ellipse rolling on another ellipse.

In general the two centrodes are not simple curves as described in the three examples. In fact some centrode curves are not closed. For example, if two links of a four bar linkage become parallel, the curvature at the contact point goes to zero, and the radius of curvature goes to infinity.

Application of Centrodes to Biomechanics

This idea of representing the relative motion of two bodies by rolling centrode curves has been exploited by German biomechanics engineers as illustrated in

Fig. 3.24. The figure shows the relative motion of the lower leg and the femur and an equivalent crossed 4-bar linkage ABA_1B_1 with the same pair of centrodes. This visualization model is useful to show that the knee is not equivalent to a fixed revolute joint. This figure is taken from the German text of Menschik (1987). It illustrates that a knee prosthesis could be constructed of a 4-bar linkage if the motion is restricted to the plane. In real knee joints there is also some out of plane relative motion.

Reuleaux's Method for Constructing Centrodes

A general graphical method for constructing centrodes based on Reuleaux's book will be described. This method can be adapted for computer solution using MAT-LAB. This description will refer to Fig. 3.25 where the fixed body is represented by the line AB, and the moving body by the line PQ. In order to draw the rolling centrode curves we assume that we know the motion of points P and Q relative to AB for a finite time period at the end of which the line ends up at position P_1Q_1.

First one draws the instant center (or centro in some books) by constructing normals to the two path curves of P and Q and extend these normals until they intersect at a point O, which is the instantaneous center of rotation. Performing the same construction for time t_1, at points P_1 and Q_1 the intersection of the normals we name as O_1. Thus the path of instant centers must go through points O and O_1, as shown in Fig. 3.25. This path is the fixed centrode attached to the body represented by AB.

Next we must find the second rolling centrode associated with the rolling of AB relative to PQ. This curve is shown in Fig. 3.25 as OM_1, using the same notation as Reuleaux. To find the point M_1, Reuleaux argues that body AB would roll such that points O_1 and M_1, would be in contact and that the triangle $P_1O_1Q_1$, would be congruent with the triangle formed by PQM_1, since the lengths of AB and PQ do not change in pure rotation. This identity, that results from the assumption of pure rolling, gives us the geometric relation to draw point M_1. If the time steps are close and the centrode curves are continuous with smooth derivatives in space, then we can approximate the lines OO_1 and OM_1, as straight lines and construct an approximation to the centrode curves. Again, because of the rolling constraint, the lengths of the curves OO_1 and OM_1 are equal.

To summarize the interpretation of the centrode curves OO_1 and OM_1: one imagines that OO_1 is attached to the body associated with the line AB and that curve OM_1 is attached to the body associated with PQ. Then the motion of PQ relative to AB is accomplished by the rolling of OM_1 on the curve OO_1.

Homework Problems

3.1 (*Angular Velocity*) Consider the swept wing aircraft shown in Fig. P3.1. Use the matrix representation of the angular velocity vector equation (3.8) to cal-

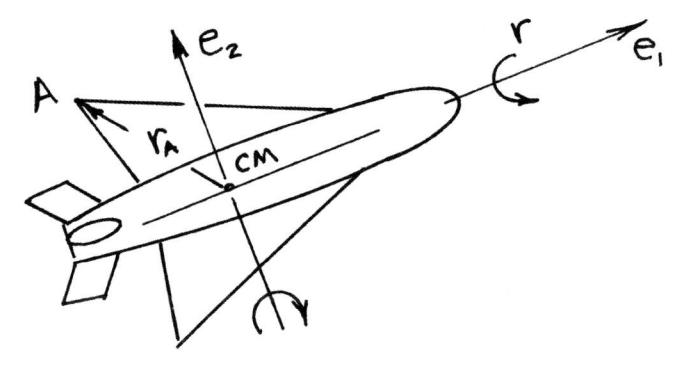

Fig. P3.1

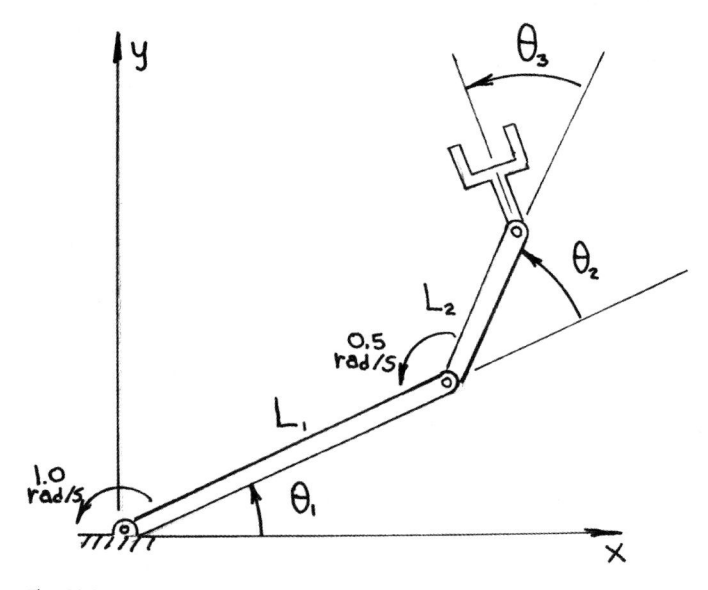

Fig. P3.2

culate the velocity of the wing tip point A when the aircraft in a roll with rate r and a pitch rate p.

3.2 *(Relative Motion)* A planar three-link robot manipulator arm is shown in Fig. P3.2. Consider only the wrist motion. Use the direct vector calculus method (3.10) and the relative velocity formula method to calculate the velocity of the wrist point when the shoulder motor rate is 1 radian/sec and the elbow motor rate 0.5 radian/sec for link parameters of $L_1 = 1$ meter and $L_2 = 0.5$ meter.

3.3 *(Acceleration; Relative Motion)* Refer to Problem 3.2 and calculate the acceleration of the wrist point when the shoulder motor rate is constant and the elbow motor rotation rate is constant. Use both the direct vector calculus method and the relative acceleration formula (3.18). Note that there may be several

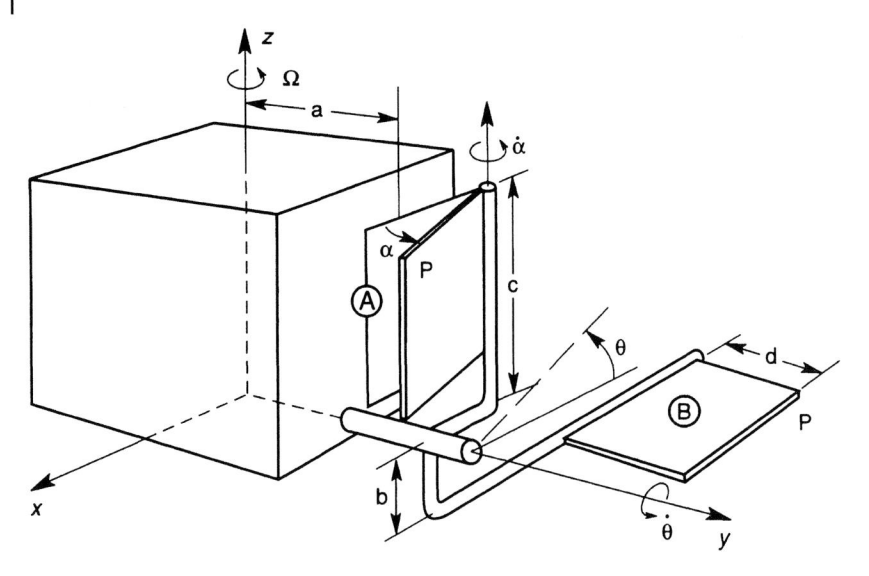

Fig. P3.4

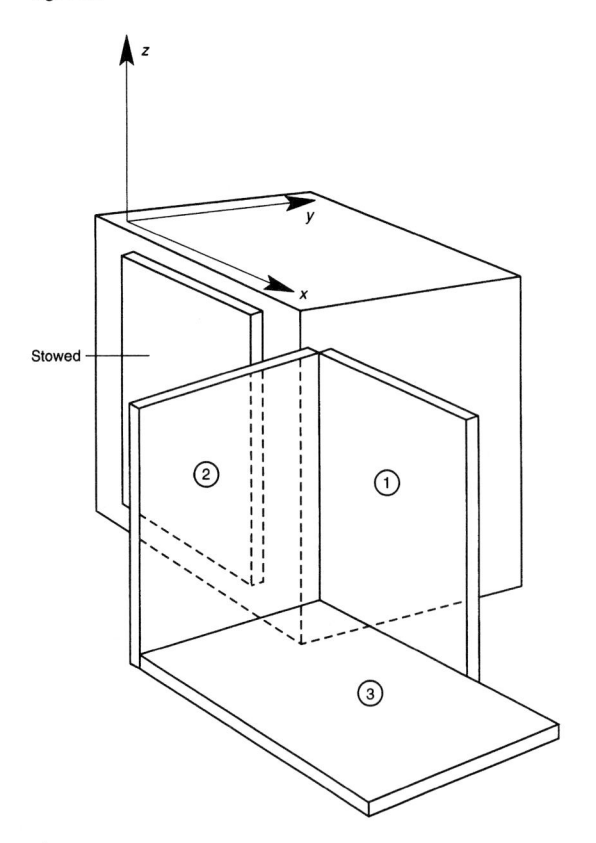

Fig. P3.5

methods in using (3.18) depending on the choice of the secondary reference frames.

3.4 A spacecraft spins about the vertical axis e_3 in Fig. P3.4 with constant rate Ω. A solar panel is deployed by rotation about two nonintersecting axes separated by a distance b.
   (a) If $\dot{\alpha}, \dot{\theta}$ are constant, what is the angular acceleration of the panel for an arbitrary angle θ?
   (b) Find the velocity $\mathbf{v}$, and acceleration, $\dot{\mathbf{v}}$, of point P on the panel for $\dot{\alpha}, \dot{\theta}, \Omega$ constant as a function of α, θ.
   (c) Find a 4×4 transformation matrix for finite motion from the stowed position A to the deployed position B.

$$\text{Answer:} \quad T = \begin{bmatrix} 0 & 0 & -1 & 0 \\ 1 & 0 & 0 & b \\ 0 & -1 & 0 & -b \\ 0 & 0 & 0 & 1 \end{bmatrix}$$

3.5 A stowed solar panel on the satellite shown in Fig. P3.5 undergoes one translation and two finite rotations in the sequence shown.
   (a) Write a rotation matrix for each of the three deployment motions of the panel.
   (b) Find a 4×4 transformation for the complete motion.

3.6 The two-link serial mechanism shown in Fig. P3.6 rotates about two axes as shown.
   (a) Find expressions for the velocity and acceleration of the end point B given $\{\theta_1, \theta_2, \dot{\theta}_1 = \omega_1, \dot{\theta}_2 = \omega_2\}$ and $\dot{\omega}_1 = \dot{\omega}_2 = 0$.
   (b) Calculate the Jacobian matrix that relates the velocity $\mathbf{v}_B$ to the generalized velocities $\{\dot{\theta}_1, \dot{\theta}_2\}$.
   (c) Write the velocity $\mathbf{v}_B$ in three different basis vectors: $\{\hat{\mathbf{i}}, \hat{\mathbf{j}}, \hat{\mathbf{k}}\}$, $\{\mathbf{e}_1, \mathbf{e}_2, \mathbf{e}_3\}$ on link #1, and $\{\mathbf{e}_1, \mathbf{e}_2, \mathbf{e}_3\}$ on link #2.

3.7 A rod-shaped lever rotates about the z-axis, and a plate is constrained to rotate about the x-axis as shown in Fig. P3.7. The end of the rod-shaped lever is constrained to remain in contact with the plate as it rotates.
   (a) Show that the path traced on the plate by the rod is a circle of radius $(a^2 + b^2)^{1/2}$.
   (b) Find the relation between the vertical rate $\dot{\phi}$ and the horizontal axis rate $\dot{\theta}$ as a function of ϕ.
   (c) If $\dot{\phi}$ is constant, find the angular acceleration of the plate. (*Hint*: Use the constraint $\mathbf{v}_{\text{plate}} \cdot \mathbf{n} = \mathbf{v}_{\text{lever}} \cdot \mathbf{n}$; $\mathbf{n}$ is normal to the plate.)
   (Answer: $\ddot{\theta} = \dot{\phi}ab \cos\phi / [b^2 + a^2 \sin^2\phi]$.)

3.8 Given two finite rotations of $\pi/2$ about two orthogonal axes that intersect at a point, find an equivalent Euler rotation, i.e., a single axis and single angle of rotation about that axis (see Beatty, 1986, p. 214). [*Hint*: Find T in Eq. (3.31) and use the relations in Eqs. (3.39).]

3.9 A general displacement of a rigid body consists of a translation and a rotation about an axis. Show that any general displacement of a rigid body is

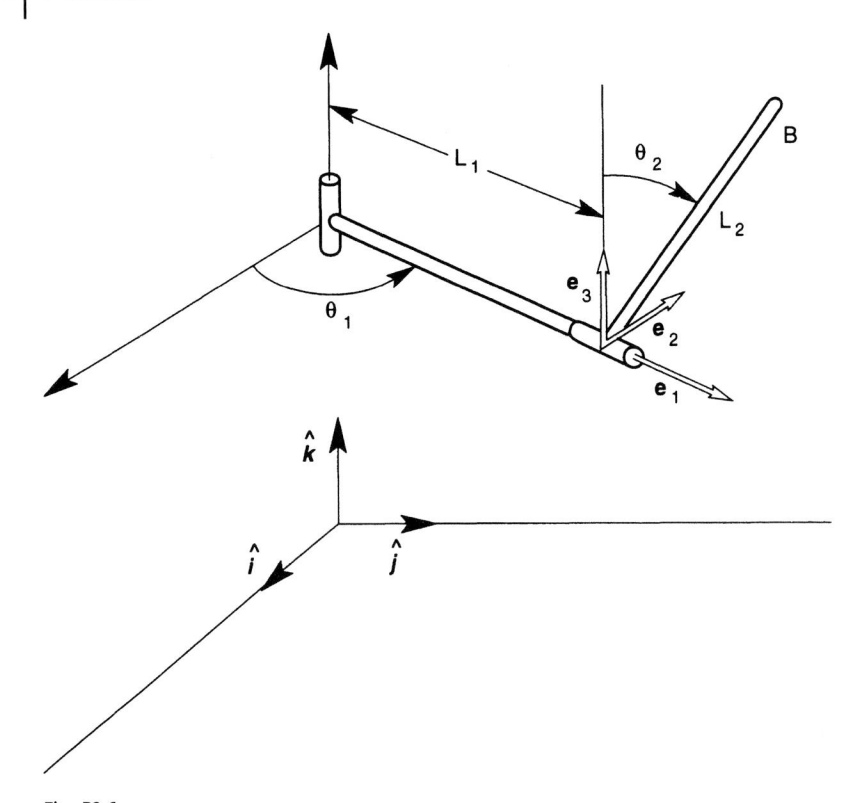

Fig. P3.6

equivalent to a rotation about a unique axis and a translation parallel to that axis. This is called a *screw displacement*. The theorem is attributed to Chasles (1843). (See, e.g., Beatty, 1986, p. 197.)

3.10 A compound solar panel on a satellite is shown in Fig. P3.10. The inner panel undergoes a rotation about the vertical axis, e_2, while the outer panel undergoes a rotation about the e_1 axis.
  (a) Find the velocity and acceleration of the point P on the outer edge, assuming that the scalar value of the rotation rates are constant. Express the answer in components of the local coordinate system $\{e_1, e_2, e_3\}$.
  (b) What is the angular acceleration of the outer panel?

3.11 Find a homogeneous (4 × 4) transformation that will move the object in Fig. P3.11 from position A to position B as shown. What is the inverse transformation?

3.12 Write the 4 × 4 transformation matrices for a kinematic pair that relate the position of points in one body relative to another connected by (a) a revolute joint, (b) a prismatic joint, (c) a screw joint (Fig. 3.15).

3.13 (*Screw Transformations*) In the description of the position of one link in a robot arm relative to a neighboring link, a 4 × 4 transformation with two screw transformations about two perpendicular axes is sometimes used (see,

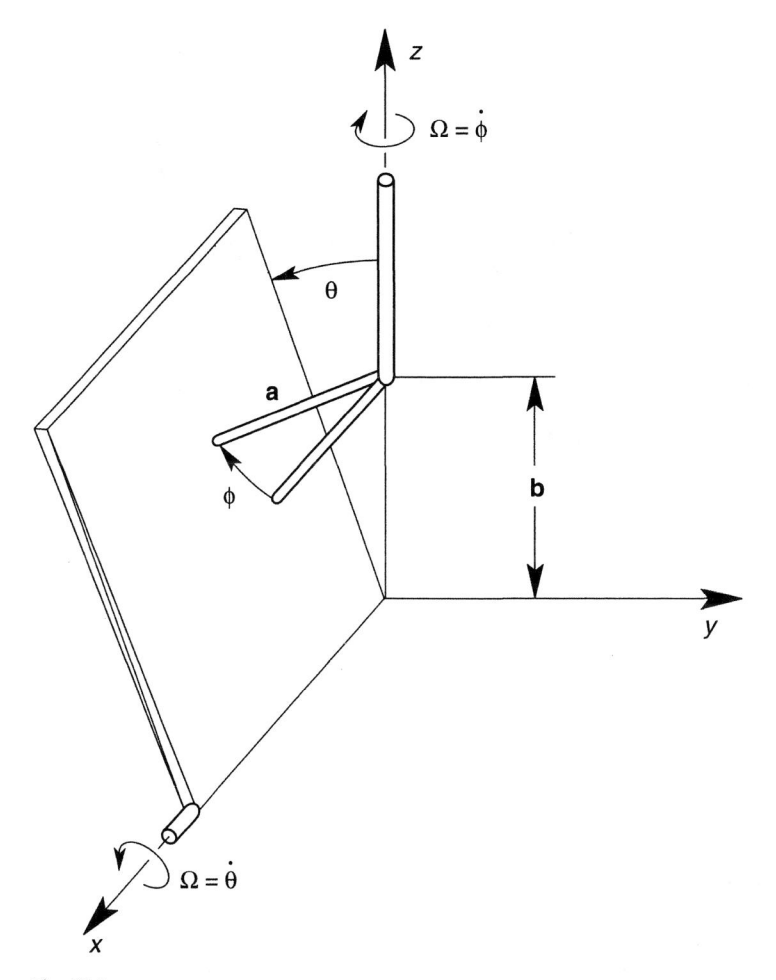

Fig. P3.7

e.g., Craig, 2005); $\mathbf{T} = \mathbf{S}_X \mathbf{S}_Z = \mathbf{R}_X(\phi)\mathbf{D}_X(d_x)\mathbf{R}_Z(\theta)\mathbf{D}_Z(d_z)$. Show that this matrix has the form

$$\mathbf{T} = \begin{bmatrix} c\theta & -s\theta & 0 & d_x \\ s\theta c\phi & c\theta c\phi & -s\phi & -s\phi d_z \\ s\theta s\phi & c\theta s\phi & c\phi & c\phi d_z \\ 0 & 0 & 0 & 1 \end{bmatrix}$$

where $s\phi = \sin(\phi)$, $c\theta = \cos(\theta)$, etc.

3.14 The end of a rotating link (Fig. P3.14) is connected by an inextensible cable to a pin constrained to move along the vertical axis.
   (a) Find the relations between the generalized velocities $\dot{\theta}, \dot{z}$.
   (b) Find an expression for the acceleration $\ddot{z}$ when the arm rotates with constant angular velocity.

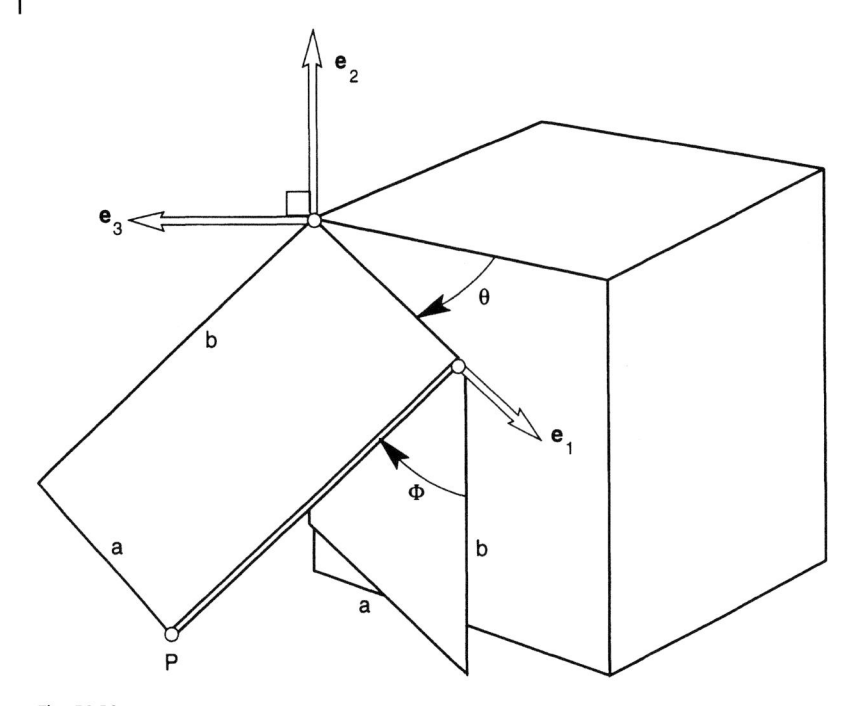

Fig. P3.10

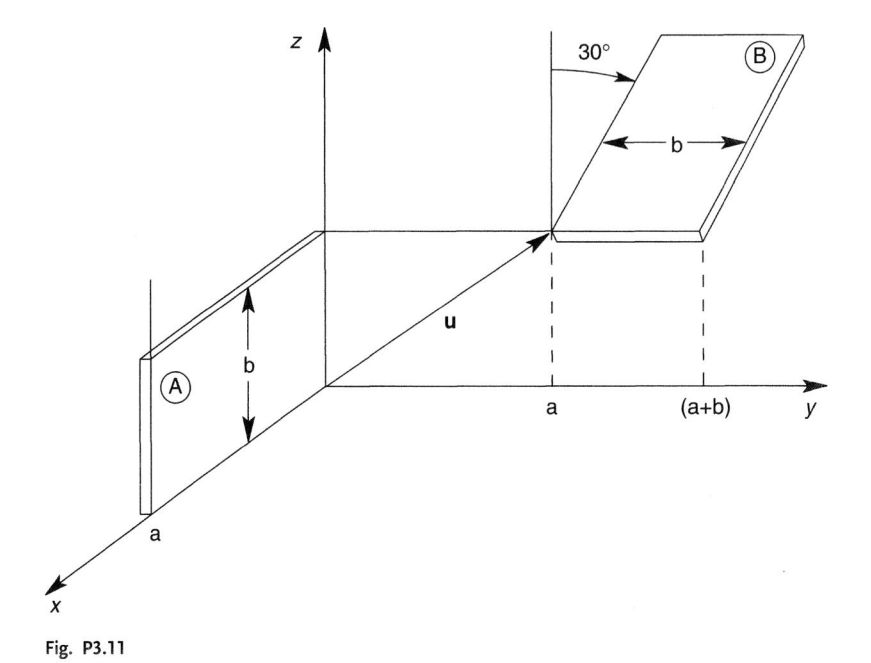

Fig. P3.11

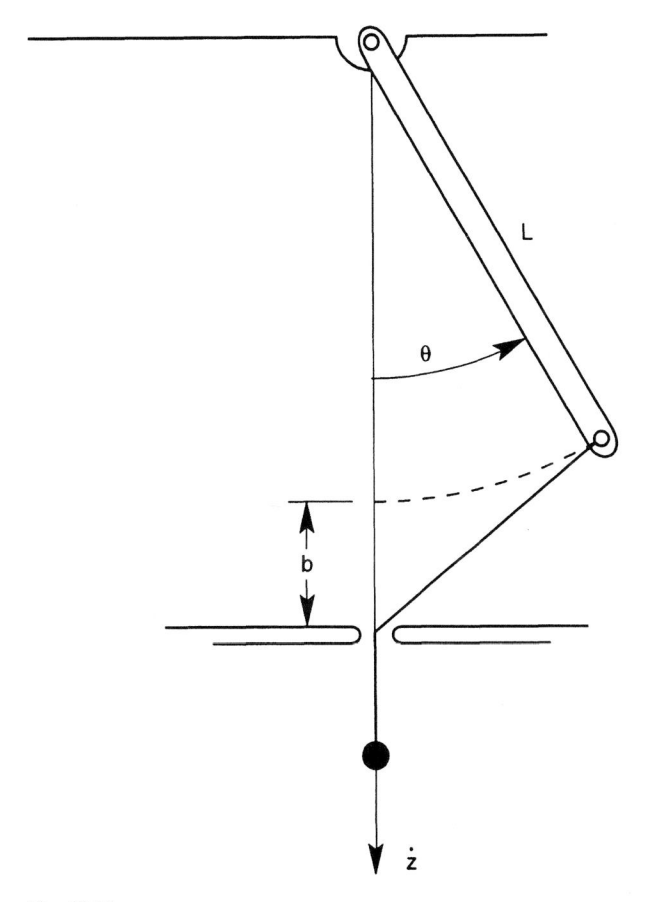

Fig. P3.14

3.15 A parallelepiped of dimensions $\{a, b, c\}$ is shown in Fig. P3.15. Suppose the object is subject to three finite motions: (1) translate along the x-axis by a; (2) translate along the y-direction by a; (3) rotate about the z axis by $\pi/4$.
   (a) Sketch the final position.
   (b) Find the 4×4 transformation matrix.

3.16 Two levers shown in Fig. P3.16 rotate about the x, y axes. The rod-shaped lever end is constrained to remain in contact with the flat-plate surface of the second lever.
   (a) If $0 \leqslant \phi \leqslant \phi^*$, find the curve traced by the contact point on the plate. (ϕ^* is the angle where contact is no longer possible.)
   (b) Given that ω_y is constant, find ω_x of the rod shaped lever.
   (Answer: $\omega_x = -\omega_y (b/a) \cos^2 \phi \cos \theta$.)

3.17 Using part of the MATLAB program in Example 3.4, animate the three link motion in Example 3.5. Move the box from the configuration $\theta_1 = \theta_2 = \theta_3 = 0$, to a position determined by $\theta_1 = \pi/4$, $\theta_2 = \pi/2$, $\theta_3 = -\pi/4$. Use the transformation matrices T_1, T_2, T_3 in the example. To complete the anima-

3 Kinematics

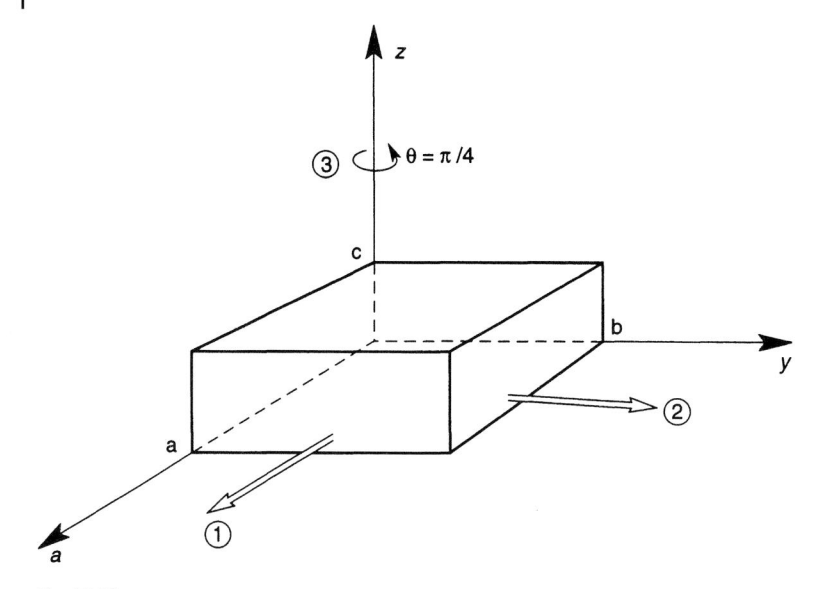

Fig. P3.15

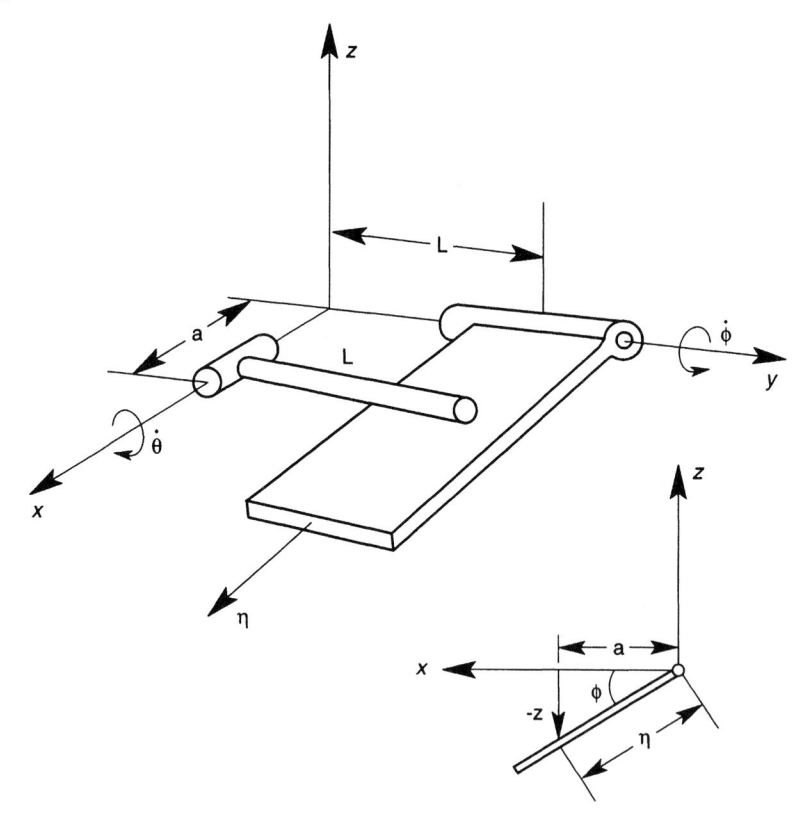

Fig. P3.16

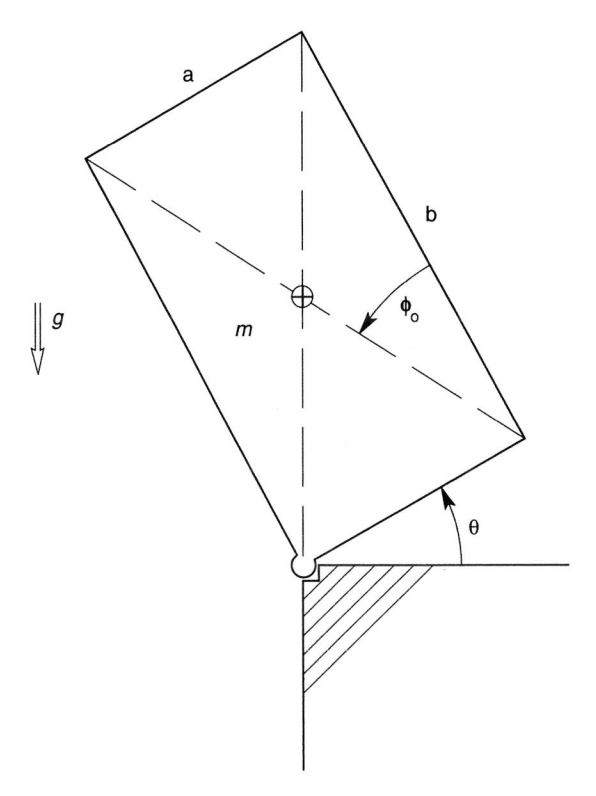

Fig. P3.18

tion, draw wire cartoons for the three links and several intermediate angles between the initial and final positions.

3.18 Using the MATLAB program in Example 3.4 as a model, write a program that will animate the falling of a box off the ledge shown in Fig. P3.18. Give the box an angular impulse at $\theta = 0$, and release the pin constraint when $\theta = \pi/2$.

3.19 Consider the three-dimensional 4-link mechanism $ABCD$ shown in Fig. P3.19. The link AB is constrained to move in the x–y plane about the z-axis with angular rate $\dot{\theta}$. The link CD is constrained to move in the yz plane about an axis parallel to the x-axis with angular rate $\dot{\phi}$.
  (a) Find a kinematic relation between $\dot{\theta}, \dot{\phi}$.
  (b) What is the vector expression for the angular velocity of link BD in terms of $\dot{\theta}$?

3.20 (*Planetary Gear Kinematics*) Planetary gear mechanisms are used in automotive, aircraft engine and robotic applications. The basic planetary gear geometry is shown in Fig. P3.20 and consists of a central sun gear S, a rolling planet gear P, a radial arm A to hold the central axis of the planet gear and often an outer ring gear R. Often there are three or four planet gears held in place with one arm called a spider arm. The planetary gear mechanism is often used as a speed reduction device. Suppose the ring gear R is fixed and the sun gear S

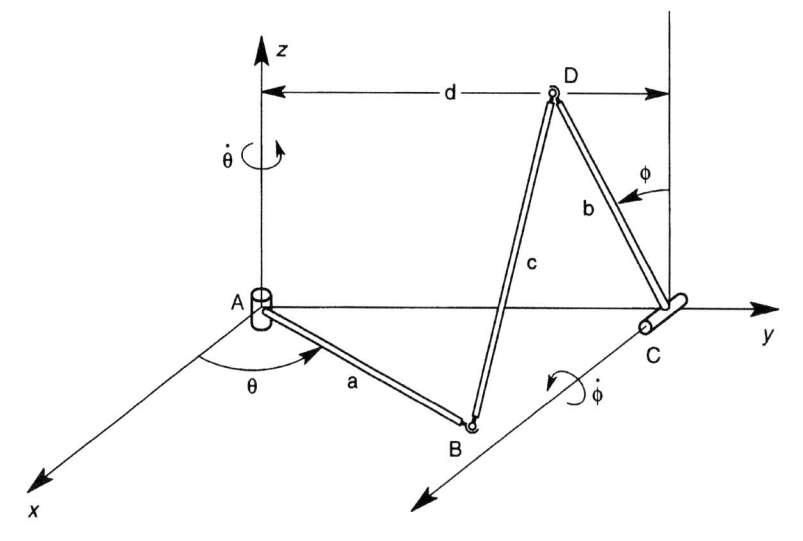

Fig. P3.19

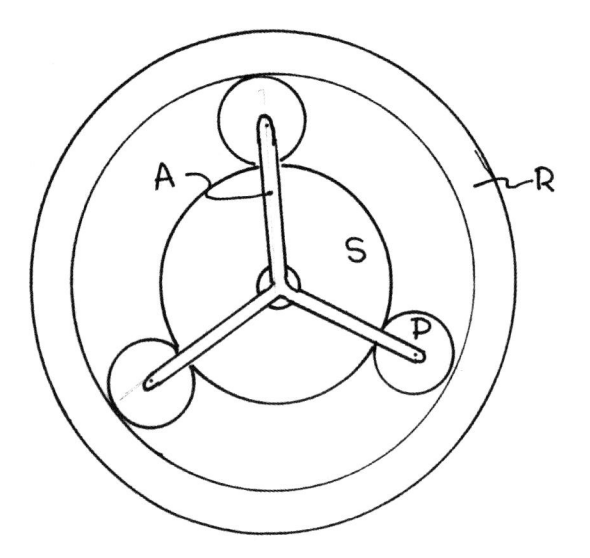

Fig. P3.20

rotates with Ω_S. Find the ratios of the angular velocities Ω_P/Ω_S, Ω_A/Ω_S in terms of the diameter ratios of the sun, planet and arm.

3.21 *(Planetary Gear Mechanism)* For the planetary gear mechanism in Fig. P3.20, assume that the sun gear S is fixed and the angular velocity of the ring gear R is given Ω_R. Find the ratios of the angular velocities Ω_A/Ω_R, Ω_P/Ω_R in terms of the ratios of the sun planet and arm radii.

3.22 *(Four-bar Linkage Jacobian)* Consider the special four-bar linkage with crossed links and equal opposite sides is used to position a platform shown in

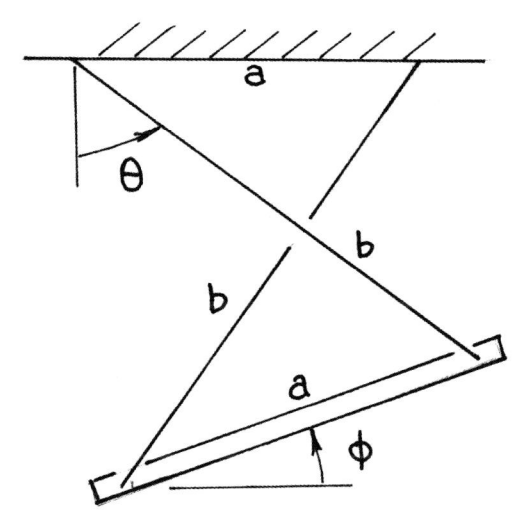

Fig. P3.22

Fig. P3.22. Find the relation between the joint angle velocity $\dot{\theta}$ and the angular velocity of the platform angle $\dot{\phi}$; i.e. find the Jacobian.

3.23 (*Mountain Bike Kinematics*) Modern mountain bikes now have added degrees of freedom in the frame in order to insert shock absorber springs and dampers. One such design is shown in Fig. P3.23 where the rear wheel rigid frame has been replaced by a four-bar linkage. (See the web for trademark bikes called "Banshee.") For the idealized geometry in the figure, find the Jacobian relation between the vertical velocity of the rear wheel and the rocker arm shock absorber pin velocity assuming that the seat frame is fixed.

3.24 (*Biomechanics Knee Kinematics*) The motion of a lower leg relative to the femur limb in humans can sometimes be replicated with a four-bar linkage shown in Fig. P3.24. Use the graphical method of Reuleaux to find the two centrodes for equivalent rolling of two opposite legs of the four-bar mechanism assuming that the femur link is grounded. Can you write a MATLAB routine to replace Reuleaux's graphical method to find the centrodes?

3.25 (*Rolling Centrodes*) Suppose a rod AB of length L is confined such that its end points A and B move in linear slots inclined at an angle ϕ (Fig. P3.25). Using the idea of instant centers and centrode surfaces, show that the equivalent motion of AB is the rolling of a wheel of radius $R = L/\sin\phi$ on another fixed circle of radius $2R$. [See Tao (1967, p. 71) for a solution.]

3.26 (*Sterling Engine Rhombus Mechanism*) A Sterling engine can run on a closed gas cycle but requires two moving pistons out of phase with each other. One mechanism to achieve two out of phase motions is a rhombus linkage mechanism shown in Fig. P3.26. The lower part of the rhombus drives the power piston while the upper part drives the displacement piston. Use the methods of this chapter to find the relative motion between the two pistons.

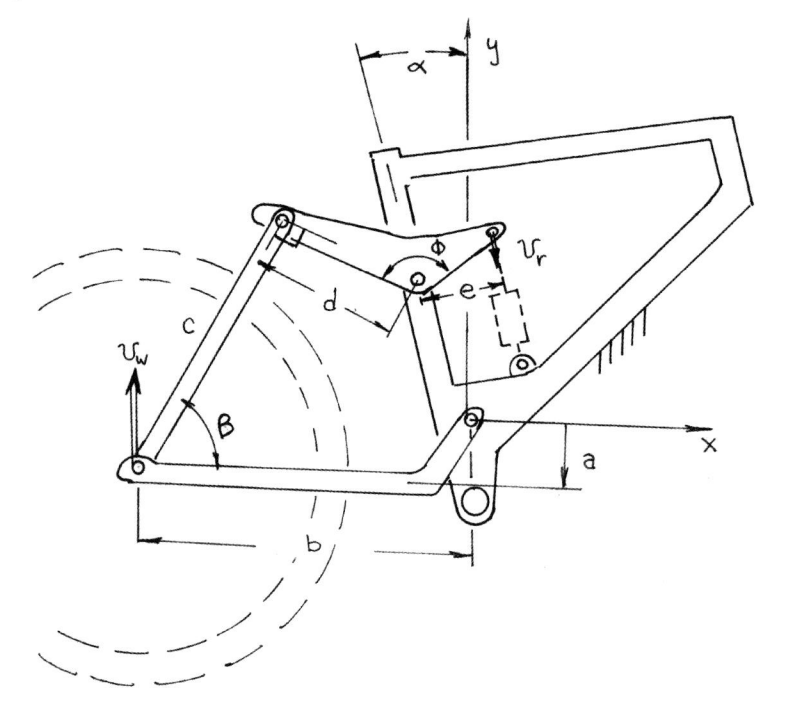

Fig. P3.23

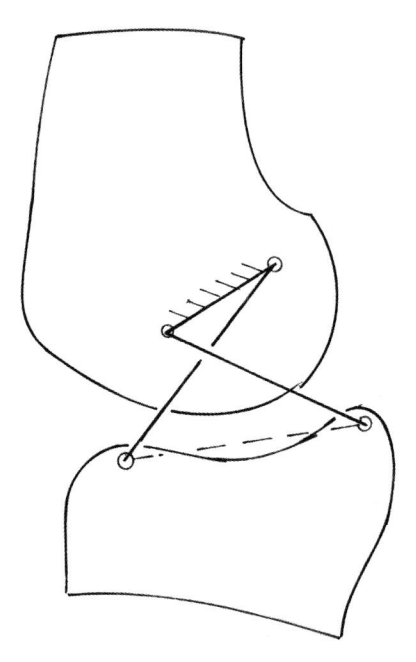

Fig. P3.24

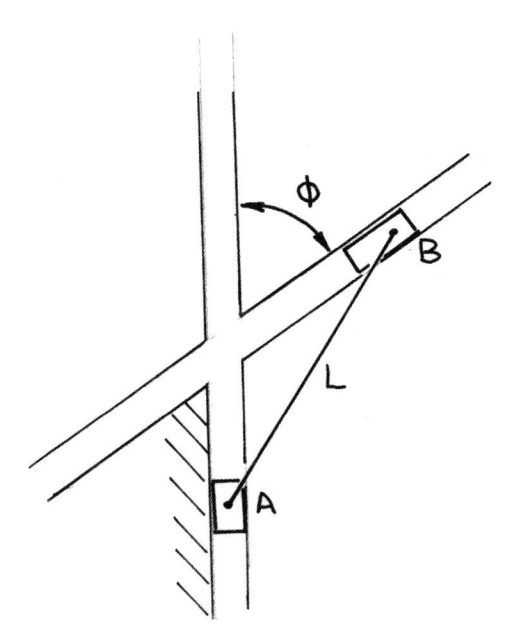

Fig. P3.25

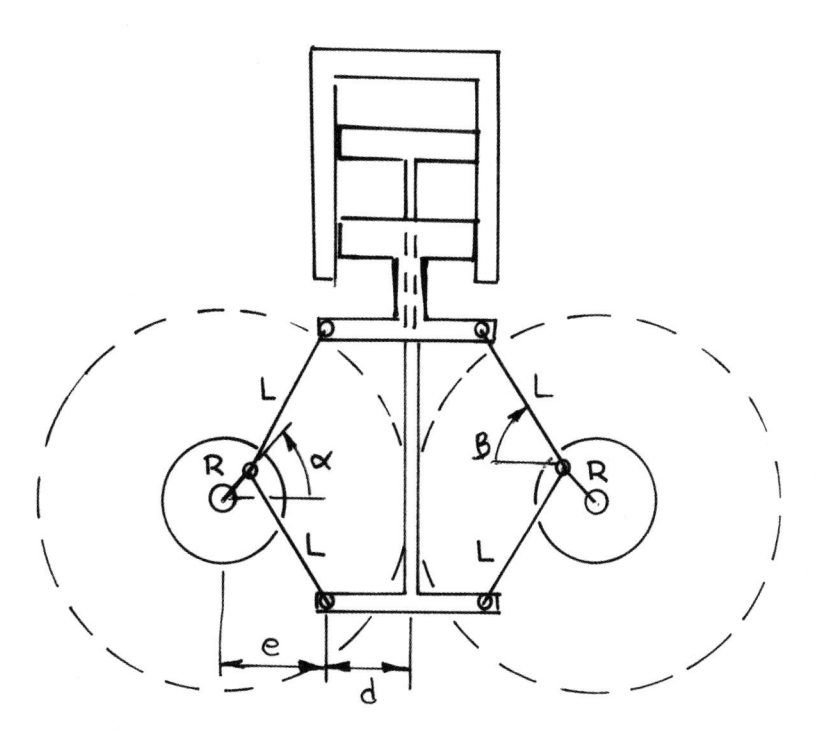

Fig. P3.26

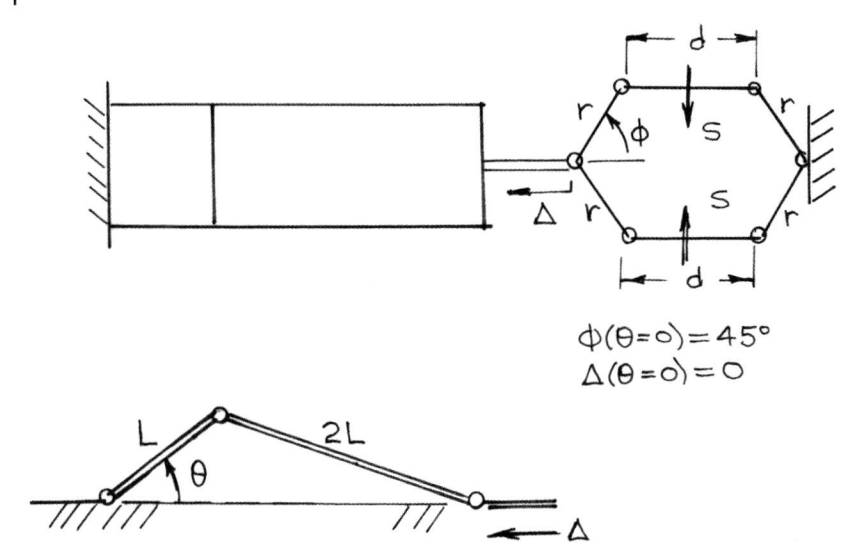

Fig. P3.27

3.27 (*MEMS Optical Mirror Rhombus Drive Mechanism*) A sketch of a MEMS slider crank mechanism in Fig. P3.27 shows a planar, rhombus linkage driving the out of plane motion of an optical mirror. Find the geometric relation between the lateral displacement of the rhombus linkage, d, and the angle of the mirror, θ. Also find the Jacobian relating the planar lateral velocity of the rhombus links and the angular velocity of the mirror.

4
Principles of D'Alembert, Lagrange's Equations, and Virtual Power

4.1
Introduction

In this chapter we present methods to formulate equations of motion for complex dynamical systems. Although a machine can be designed and created without direct knowledge of equations of motion, the complexity and cost of development puts a burden on the engineer to try and understand the behavior of these devices before they are built or put into service. Hence, dynamic models become the fundamental basis for constructing "virtual reality" simulations of dynamic machines and structures.

The method of Newton and its extension by Euler and others is essentially geometry- and vector-based. The methods presented in this chapter are more scalar-based and depend heavily on the differential calculus of several variables. These methods, which sometime involve kinetic and potential energy functions, are called *variational methods* or *energy methods*. They actually predate the time of Newton (i.e., late seventeenth century). Some historians of mechanics such as Dugas (1988) even suggest that the *method of virtual velocities* goes back to the time of Aristotle in Greece, around 300 B.C. Some of these methods are based on the notion of minimization of energy and are rooted in theological ideas of an efficient Creator, concepts that were popular in the Middle Ages. However, the modern development is linked with names such as Joseph-Louis Lagrange (1736–1813), Sir William Rowan Hamilton (1805–1865) among many others.

The basis of these methods for deriving equations of motion are the ideas of virtual work and the "force of inertia" (see Chapter 2). D'Alembert, in 1743, proposed to treat the acceleration times the mass as a negative "effective force." Lagrange then used this idea to formulate a dynamic principle of virtual work that today is known as D'Alembert's principle [see Section 2.6]. For a single particle under several forces, this principle takes the form [see Section 2.6, Eq. (2.67)]

$$\left(\sum \mathbf{F}_i - m\dot{\mathbf{v}}\right) \cdot \delta \mathbf{r} = 0 \tag{4.1}$$

Applied Dynamics. With Applications to Multibody and Mechatronic Systems. Second Edition. Francis C. Moon
Copyright © 2008 WILEY-VCH Verlag GmbH & Co. KGaA, Weinheim
ISBN: 978-3-527-40751-4

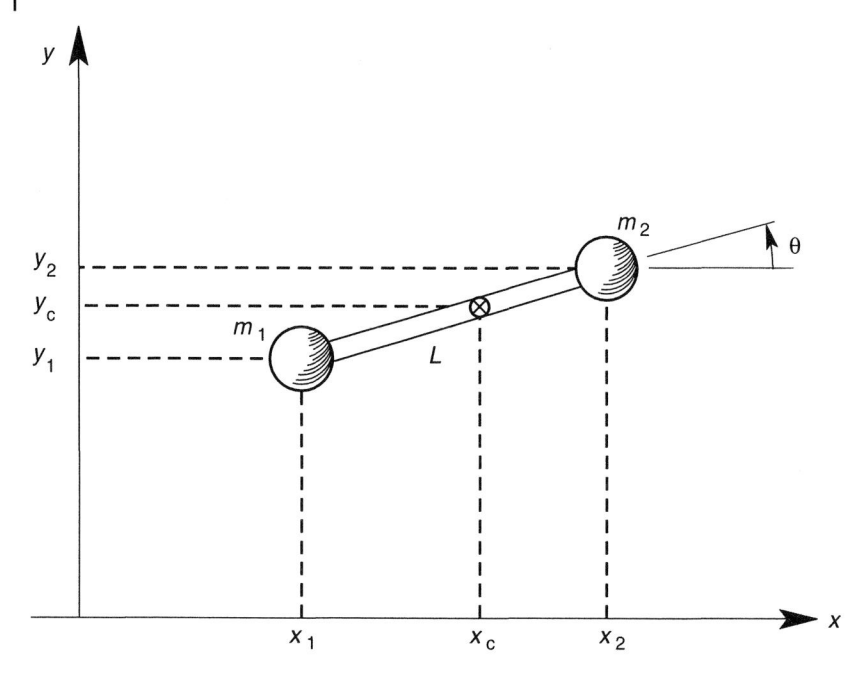

Fig. 4.1 Constrained motion of two masses in the plane.

The advantage of these methods is that they eliminate forces that do no work from the equations of motion. Such forces arise naturally when the system moves under certain types of constraints, such as linked bodies in a robotic device.

Constraints and Degrees of Freedom

Constraints in mechanical systems reduce the number of degrees of freedom. In the example below, we introduce the idea of generalized coordinates $\{q_j(t); j = 1, 2, \ldots, M\}$ associated with the M degrees of freedom of the system. We also introduce the concept of the Jacobian, defined in Chapter 3. The Jacobian can be a scalar, vector or matrix function that relates the physical velocities $\mathbf{v}_k(t)$ in the system to the generalized velocities $\{dq_j(t)/dt\}$. In general this relationship is a linear one of the form

$$\mathbf{v}_k = \sum J_{kj} \dot{q}_j \tag{4.2}$$

where $j = 1, 2, \ldots, M$.

Consider the planar motion of two masses m_1, m_2 shown in Fig. 4.1 constrained by a rigid link of length L. In general there are four degrees of freedom (x_1, x_2, y_1, y_2). Because of the rigid link, the coupled dynamic system has only three degrees of freedom (x_c, y_c, θ): the two scalar positions of the mass center, and the angular position of the connecting rod. Thus, in any variational method such as D'Alembert's principle, the variations $(\delta x_1, \delta x_2, \delta y_1, \delta y_2)$ are not all independent.

The geometric constraint equation is nonlinear, which makes direct elimination of one of the four variables difficult:

$$f(x_k, y_k) = (x_1 - x_2)^2 + (y_1 - y_2)^2 - L^2 = 0 \qquad (4.3)$$

However, a standard trick is to write the variational form of the constraint by assuming that each of the variables depends on time;

$$2(x_1 - x_2)(\dot{x}_1 - \dot{x}_2) + 2(y_1 - y_2)(\dot{y}_1 - \dot{y}_2) = 0$$

Next, if we assume that $(x_1 - x_2)$ and $(y_1 - y_2)$ remain approximately constant during some arbitrarily small time, the variational constraint equation becomes

$$2(x_1 - x_2)(\delta x_1 - \delta x_2) + 2(y_1 - y_2)(\delta y_1 - \delta y_2) = 0 \qquad (4.4)$$

This equation is linear in the virtual displacements, and we can choose to represent one in terms of the others.

In formulating the equations of motion with K constraints such as (4.4) we have two choices:

1. Choose a convenient set of generalized coordinates $\{q_i(t); \; i = 1, 2, \ldots, M\}$, which may not be independent, and solve the K constraint equations along with the equations of motion. This results in $M + K$ equations to solve.
2. Use the constraint equations to eliminate some of the dependent coordinates in terms of the other coordinates, or choose $M - K$ different but independent coordinates that are compatible with the constraints. This results in having to solve $M - K$ differential equations.

If we denote the position variables by the set of scalars $\{q_i(t)\}$, then Eq. (4.4) takes the form

$$\sum a_i(q_k)\delta q_i = 0 \qquad (4.5)$$

where

$$a_i(\delta_k) = \partial f(q_k)/\partial q_i$$

and $\delta q_1 = \delta x_1, \delta q_2 = \delta x_2, \delta q_3 = \delta y_1, \delta q_4 = \delta y_2$.

In order to generalize the preceding example to a system of N particles, we define position vectors $\mathbf{r}_k$ for each particle in Fig. 4.2. Without constraints the system will have $3N$ degrees of freedom. If we impose K geometric constraints, we will be left with $M = 3N - K$ degrees of freedom. To choose a set of M independent coordinates, we have the choice of picking M of the original coordinates as our independent variables or choosing M new coordinates. We denote the set of independent coordinates by $\{q_j\}$, where $j = 1, 2, \ldots, M$. If we use the notation $[\delta \mathbf{r}]^T$ to denote the virtual variation in the string of position vectors, i.e.,

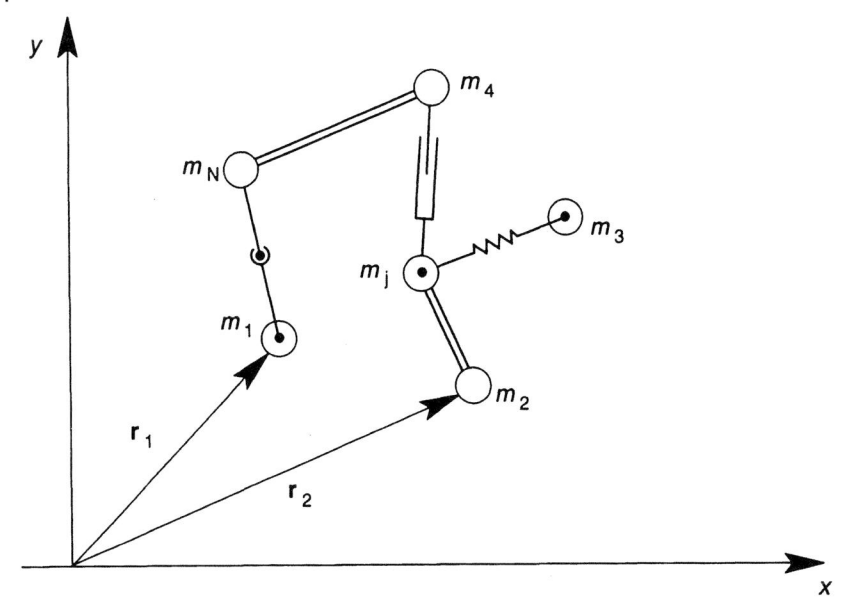

Fig. 4.2 System of N particles with constraints.

$$[\delta \mathbf{r}]^T = \left[[\delta \mathbf{r}_1]^T, [\delta \mathbf{r}_2]^T, \ldots, [\delta \mathbf{r}_N]^T\right]$$

$$= \left[\delta x_1, \delta y_1, \delta z_1, \ldots, \delta x_N, \delta y_N, \delta z_N\right] \tag{4.6}$$

then the constraints provide a relation between the old and new variables [see (2.12) and (2.13)]:

$$[\delta \mathbf{r}] = [J][\delta \mathbf{q}] \tag{4.7}$$

Here $[\delta \mathbf{q}]$ is a column vector of the scalar virtual displacements, i.e., $[\delta \mathbf{q}]^T = [\delta q_1, \delta q_2, \ldots, \delta q_M]$. The Jacobian matrix $[J]$ is $3N$ rows by M columns. In explicit notation

$$\delta \mathbf{r}_k = \sum_{j=1}^{m} \mathbf{J}_{kj} \delta q_j, \quad k = 1, 2, \ldots, N \tag{4.8}$$

where $\mathbf{J}_{kj} = \partial \mathbf{r}_k / \partial q_j$. To make the matrix notation more transparent, the $3N \times M$ matrix $[J]$ is composed of a vertical stack of the $\mathbf{J}_{kj}$ vectors:

$$[J] = \begin{bmatrix} [\ ]_{11} & [\ ]_{12} & \cdots & [\ ]_{1M} \\ \vdots & & & \\ [\ ]_{N1} & & \cdots & [\ ]_{NM} \end{bmatrix} \tag{4.9}$$

4.2
D'Alembert's Principle[1]

If the forces acting on the particle are separated into workless reaction forces of constraint, **R**, and active forces, $\mathbf{F}^a$, then Lagrange's generalization of D'Alembert's principle for a single particle with a single generalized position variable, $q(t)$, is (see Section 2.6)

$$\mathbf{R} \cdot \delta \mathbf{r} = 0$$

$$(\mathbf{F}^a - m\dot{\mathbf{v}}) \cdot \delta \mathbf{r} = 0 \qquad (4.10a)$$

$$\delta \mathbf{r} = \frac{\partial \mathbf{r}}{\partial q} \delta q$$

$$(\mathbf{F}^a - m\dot{\mathbf{v}}) \cdot \frac{\partial \mathbf{r}}{\partial q} = 0 \qquad (4.10b)$$

For several mass particles and one degree of freedom we shall show below that D'Alembert's principle takes the form

$$\sum_{i=1}^{N} \left(\mathbf{F}_i^a - m\dot{\mathbf{v}}_i \right) \cdot \frac{\partial r_i}{\partial q} = 0 \qquad (4.10c)$$

We illustrate the application of (4.10c) for a constrained system of two particles in Example 4.2.

The displacements $\delta \mathbf{r}$ are called "virtual," but the term is not a descriptive one. Instead, one should think of $\delta \mathbf{r}$ as arbitrarily small displacements that do not violate the constraints. Thus, if the particle moves on a curved surface, then $\delta \mathbf{r}$ is arbitrary, but tangent to the surface.

There are two advantages of this method. First, it eliminates constraint forces or internal forces. Second, it can be extended to many particle systems and to rigid bodies.

Example 4.1. A direct application of the principle of virtual work and D'Alembert's principle is illustrated in the motion of a cable-wrapped cylinder shown in Fig. 4.3. (This example is modified after one in Szabó, 1987.)

We wish to find the angular acceleration $\ddot{\phi}$ as a function of the two masses m_1, m_2 and the geometry. The position coordinates to each mass center are z_1, z_2, but because of the cable wrap-up, the change in z_1, z_2 is a function of the change of the cylinder angular position, i.e.,

$$\delta z_1 = -r \delta \phi$$

---

[1] Jean Le Rond D'Alembert (1717–1783) was born in Paris. He was trained in mathematics, mechanics and philosophy. He contributed to the development of the calculus. His work *Traite de Dynamique* (1743) reframed the problems of dynamics in the form of a statics problem. In 1747 he developed a theory of the vibration of strings using the solutions of the wave equation.

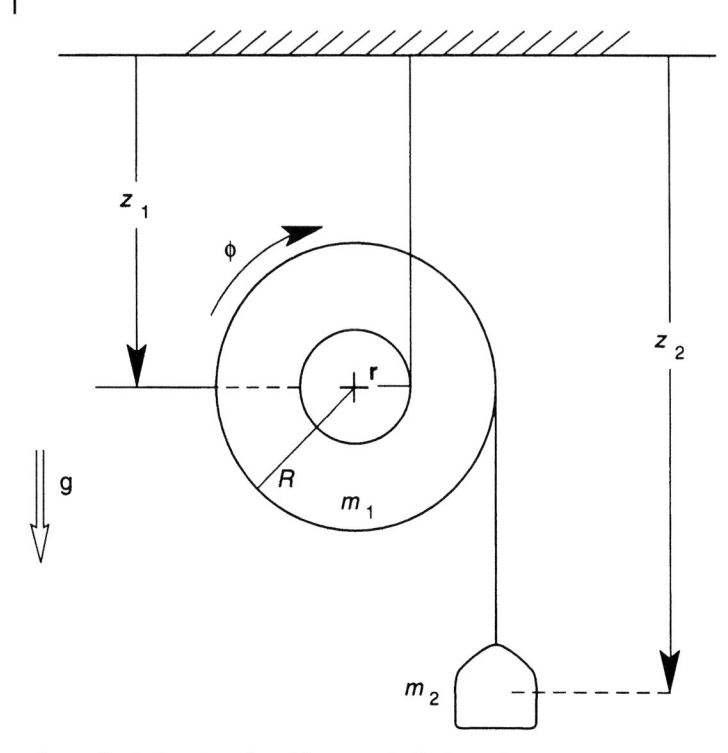

Fig. 4.3 Vertical motion of a cable-wrapped cylinder and mass.

$$\delta(z_2 - z_1) = R\delta\phi \tag{4.11}$$

These constraints also lead to the following kinematic relations

$$\ddot{z}_1 = -r\ddot{\phi}$$
$$\ddot{z}_2 = \ddot{z}_1 + R\ddot{\phi} \tag{4.12}$$

As a simplifying assumption, we assume that the mass m_1 is concentrated at the point 0 on the pulley. In this case, Lagrange's form of D'Alembert's principle (4.10) extended to two masses becomes

$$(m_1 g - m_1 \ddot{z}_1)\delta z_1 + (m_2 g - m_2 \ddot{z}_2)\delta z_2 = 0 \tag{4.13}$$

or using the constraint conditions,

$$\{[m_2 g - m_2(R-r)\ddot{\phi}](R-r) - [m_1 g + m_1 r\ddot{\phi}]r\}\delta\phi = 0 \tag{4.14}$$

Since $\delta\phi$ is assumed to be arbitrary, we can set the term in the bracket equal to zero and solve for $\ddot{\phi}$.

$$\ddot{\phi} = g\frac{m_2(R-r) - m_1 r}{m_2(R-r)^2 + m_1 r^2} \tag{4.15}$$

4.2 D'Alembert's Principle[3]

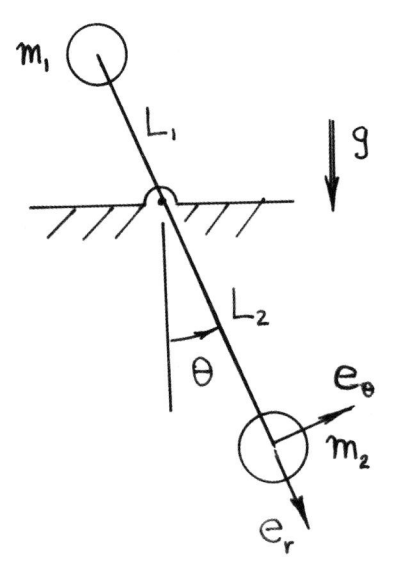

Fig. 4.4 Sketch of a two-mass model of a compound pendulum.

By cutting cables and explicitly incorporating the tensions T_1, T_2 in the analysis, one can also show that (neglecting rotary inertia)

$$T_1 = \frac{R}{r} T_2$$

$$T_2 = m_2 g \left(1 - \frac{R-r}{g} \ddot{\phi}\right) \qquad (4.16)$$

Example 4.2 Compound Pendulum. In the classic pendulum (1.7), the mass is concentrated at the end of the rotary link. (See Chapter 1.) In real systems the mass is distributed throughout the link and the revolute joint is at the interior of the body. A concentrated particle model for a more realistic pendulum is shown in Fig. 4.4 in which there are two masses: one at either end of the link and the joint located between the masses. There are two questions related to the dynamics; what is the natural frequency for small oscillations and how does it differ from the classic pendulum? What is the dynamic force at the bearing and how does it differ from the static force?

In the formulation for D'Alembert's Principle, we first identify a single degree of freedom and designate the angle theta, θ, as the generalized coordinate. Each of the masses has a circular motion so the use of the acceleration of each particle in polar coordinates is appropriate. Next one needs to calculate the projection vector or Jacobian vector:

$$\mathbf{r}_2 = L_2[\sin\theta \, \mathbf{e}_x - \cos\theta \, \mathbf{e}_z] = L_2 \mathbf{e}_r \qquad (4.17)$$

$$\frac{\partial \mathbf{r}_2}{\partial \theta} = L_2[\cos\theta \, \mathbf{e}_x + \sin\theta \, \mathbf{e}_z] = L_2 \mathbf{e}_\theta \qquad (4.18)$$

Likewise

$$\frac{\partial \mathbf{r}_1}{\partial \theta} = -L_1 \mathbf{e}_\theta \tag{4.19}$$

The active forces in this problem are the gravitational forces on the two particles. Since the projection vectors in D'Alembert's Principle (4.10c) for this problem are proportional to the circumferential vector the generalized forces are the projection or inner product of the vertical gravitational forces onto the unit vectors $\mathbf{e}_\theta$:

$$\dot{\mathbf{v}}_2 = L_2[\ddot{\theta}\mathbf{e}_\theta - \dot{\theta}^2 \mathbf{e}_r] \tag{4.20a}$$

$$\dot{\mathbf{v}}_1 = L_1[-\ddot{\theta}\mathbf{e}_\theta + \dot{\theta}^2 \mathbf{e}_r] \tag{4.20b}$$

$$\dot{\mathbf{v}}_1 \cdot \frac{\partial \mathbf{r}_1}{\partial \theta} = L_1^2 \ddot{\theta} \tag{4.21a}$$

$$\dot{\mathbf{v}}_2 \cdot \frac{\partial \mathbf{r}_2}{\partial \theta} = L_2^2 \ddot{\theta} \tag{4.21b}$$

$$\mathbf{f}_1^a \cdot \frac{\partial \mathbf{r}_1}{\partial \theta} = -m_1 g \mathbf{e}_z \cdot (-L_1 \mathbf{e}_\theta) = m_1 g L_1 \sin \theta \tag{4.22a}$$

$$\mathbf{f}_2^a \cdot \frac{\partial \mathbf{r}_2}{\partial \theta} = -m_2 g \mathbf{e}_z \cdot L_2 \mathbf{e}_\theta = -m_2 g L_2 \sin \theta \tag{4.22b}$$

Summing all the scalar projections of the two accelerations and the two active forces, as in (4.10c), one obtains the equation of motion in the form of a single, ordinary differential equation:

$$(L_1^2 m_1 + L_2^2 m)\ddot{\theta} + g(L_2 m_2 - L_1 m_1) \sin \theta = 0 \tag{4.23}$$

As reviewed in Chapters 1, 2, for small motions the approximate equation of motion takes the form of that for a harmonic oscillator with sinusoidal motions of the form

$$\ddot{\theta} + \Omega^2 \theta = 0 \tag{4.24}$$

where

$$\Omega^2 = g(L_2 m_2 - L_1 m_1)/(L_1^2 m_1 + L_2^2 m_2)$$

$$\theta(t) = A \cos \Omega t + B \sin \Omega t \tag{4.25}$$

These solutions have a time period $\tau = 1/f = 1/2\pi\Omega$ where the natural frequency Ω is given as a function of the masses and their distribution. The frequency of these solutions differs from the classical pendulum in that the period can be tuned depending on the size and position of the mass above the revolute joint. The period of the classical pendulum (1.8), can only be increased by increasing the length of the link.

The standard use of D'Alembert's Principle eliminates the internal, constraint forces that do no work. However, for design purposes, one must often find the magnitude of the constraint forces in order to properly design the bearings. One method to determine the bearing forces is to draw a free body diagram and uncover these forces and apply the direct form of Newton's law of motion to each of the particles.

$$\mathbf{f}_1^c = m_1 g \mathbf{e}_z + m_1 \dot{\mathbf{v}}_1 \tag{4.26a}$$

$$\mathbf{f}_2^c = m_2 g \mathbf{e}_z + m_2 \dot{\mathbf{v}}_2 \tag{4.26b}$$

$$\mathbf{F}_B = \mathbf{f}_1^c + \mathbf{f}_2^c \tag{4.27}$$

From these calculations, one can see that the bearing forces to support harmonic motion of the compound pendulum consist of three parts: a static part, a sinusoidal function with frequency omega and a sinusoidal function with frequency twice omega due to the square of $\dot{\theta}$ in the acceleration (4.20).

D'Alembert's Principle for a System of Particles

Consider a system of N particles of mass $\{m_i\}$ where the total force on each particle is $\mathbf{F}_i$. Then Lagrange's generalization of D'Alembert's principle becomes

$$\sum (\mathbf{F}_i - m_i \ddot{\mathbf{r}}_i) \cdot \delta \mathbf{r}_i = 0 \tag{4.28}$$

When there are either *external geometric constraints*, such as motion confined to a surface or *internal constraints*, such as rigid links between the particles, then elements of the set $\{\delta \mathbf{r}_i\}$ are not independent and the set must be reduced to, or replaced by, a set of independent, generalized coordinates, as in the examples given earlier.

Also the forces on each particle may be separated into three categories (Fig. 4.5)

$$\mathbf{F}_i = \mathbf{F}_i^a + \sum_{i \neq j} \mathbf{f}_{ij} + \mathbf{R}_i \tag{4.29}$$

The forces noted by $\mathbf{F}_i^a$ are the *external* or *active forces*, the set $\{\mathbf{f}_{ij}\}$ are the *internal forces* between the particles. The forces $\{\mathbf{R}_i\}$ are the reaction or constraint forces. In the case where there is no friction or plastic deformation between the particles and constraint surfaces, we assume that

$$\sum \mathbf{R}_i \cdot \delta \mathbf{r}_i = 0 \tag{4.30}$$

when the virtual displacements are chosen to be compatible with the constraints. Thus, the statement of the generalization of the principle of virtual work for a system of particles becomes

$$\sum (\mathbf{F}_i^a - m_i \ddot{\mathbf{r}}_i) \cdot \delta \mathbf{r}_i + \sum_{i \neq j} \mathbf{f}_{ij} \cdot \delta \mathbf{r}_i = 0 \tag{4.31}$$

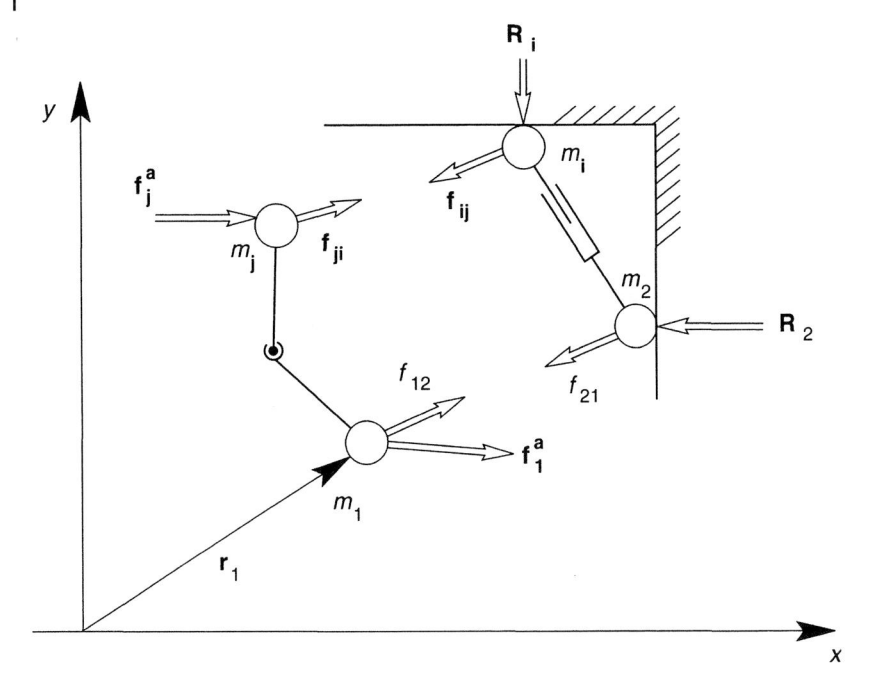

Fig. 4.5 System of masses with internal forces $\mathbf{f}_{ij}$ and constraint forces $\mathbf{R}_i$.

The second term may be thought of as the internal energy storage or dissipation. In the case of a collection of particles with rigid internal constraints, this term is zero. In the case of a set of particles with internal elastic links, we can replace the right-hand term with a potential energy function (see Section 4.3).

When the set of independent generalized coordinates has been chosen, $\{q_k(t)\}$ ($k = 1, 2, \ldots, M$), the statement of virtual work must be transformed into these variables. In applications, this is usually the hard part. However, if we can obtain explicit relations between the position vectors and the new variables $\{\mathbf{r}_i(q_k(t), t)\}$, this transformation is straightforward:

$$\delta \mathbf{r}_i = \sum_k \frac{\partial \mathbf{r}_i}{\partial q_k} \delta q_k \tag{4.32}$$

Note here that in a virtual displacement, the explicit dependence on time is suppressed, i.e., $\delta t = 0$. Only the dependence of $\mathbf{r}_i$ on the q_k variables is needed. [See Goldstein (1980) for a more detailed discussion of the theory.]

To simplify the notation, the idea of a generalized force is introduced, defined by the equation

$$\sum \mathbf{F}_i \cdot \delta \mathbf{r}_i = \sum Q_k \delta q_k \tag{4.33}$$

or using (4.32)

$$Q_k = \sum \mathbf{F}_i \cdot \frac{\partial \mathbf{r}_i}{\delta q_k} \tag{4.34}$$

Let us assume that the internal forces do not store any elastic energy or dissipate energy. Then D'Alembert's principle takes the form

$$\sum_{k=1}^{M} \left[\sum m_i \ddot{\mathbf{r}}_i \cdot \frac{\partial \mathbf{r}_i}{\partial q_k} - Q_k \right] \delta q_k = 0 \tag{4.35}$$

If we choose the new set of variables to be independent, each of the bracketed terms can be set to zero, or

$$\sum m_i \ddot{\mathbf{r}}_i \cdot \frac{\partial \mathbf{r}_i}{\partial q_k} = Q_k \tag{4.36}$$

where $i = 1, 2, \ldots, N$ (sum over all particles) and $k = 1, 2, \ldots, M < 3N$.

This is the form of D'Alembert's principle that we use to derive Lagrange's equations. However, it should be emphasized that we can use the equations directly to solve problems in dynamics. Until recently this fact has not been emphasized in dynamics textbooks. The form (4.36) is very useful for computer derivation and solutions of equations of motion.

To solve problems directly using D'Alembert's principle we can view the Jacobian column vectors as projection or tangent vectors,

$$\boldsymbol{\beta}_{ik} = \frac{\partial \mathbf{r}_i}{\partial q_k} \tag{4.37}$$

Then the equations of motion are simply Newton's law projected onto these projection directions

$$\sum_{i=1}^{N} (m_i \dot{\mathbf{v}}_i - \mathbf{F}_i) \cdot \boldsymbol{\beta}_{ik} = 0 \tag{4.38}$$

An example of the direct application of D'Alembert's principle of Lagrange is given in the following example. [See also Lesser (1995, p. 96) for a discussion of D'Alembert's principle and tangent vectors.]

Example 4.3. As a direct application of D'Alembert's principle, consider the two masses constrained to move in a radial slot on a freely rotating platform of negligible inertia (Fig. 4.6). The masses are secured to the platform by linear springs whose forces go to zero when the masses are at a radius a. We assume that the forces of constraint in the slot are frictionless and do no work. The slot constraint on the masses reduces the 4-degree-of-freedom problem to three. We choose the three generalized variables as $\{q_k\} = \{\rho_1, \rho_2, \theta\}$. The position vectors of each of the masses written in terms of $\{q_k\}$ become

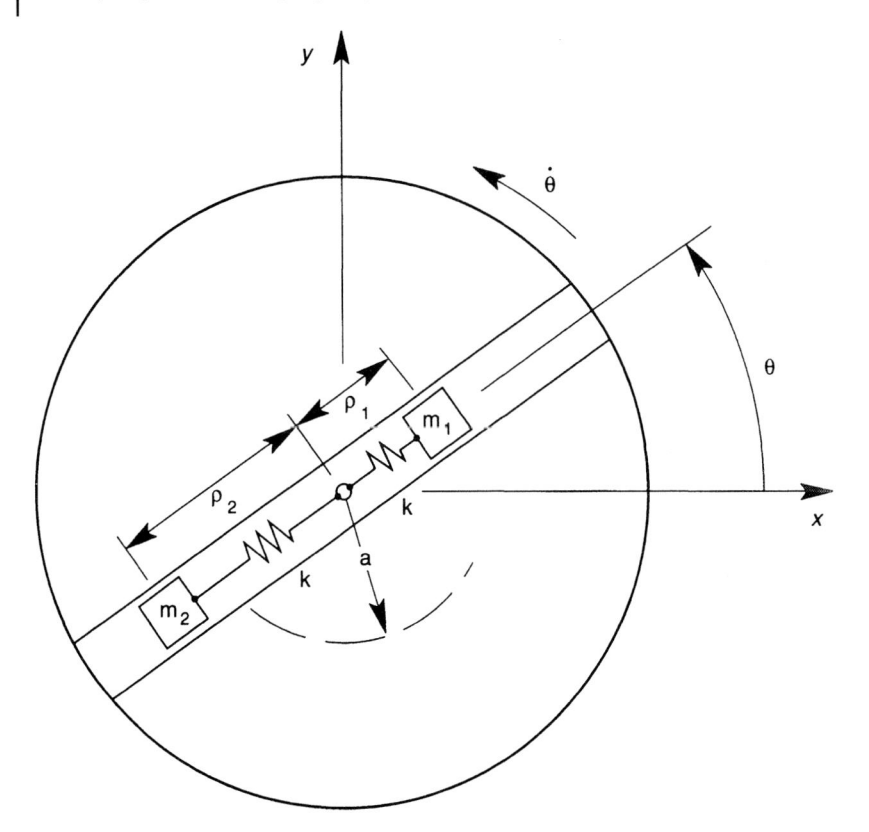

Fig. 4.6 Two masses constrained to a radial slot on a freely rotating platform.

$$\mathbf{r}_1 = \rho_1(\cos\theta\,\hat{\mathbf{i}} + \sin\theta\,\hat{\mathbf{j}}) = \rho_1 \mathbf{e}_r$$
$$\mathbf{r}_2 = -\rho_2(\cos\theta\,\hat{\mathbf{i}} + \sin\theta\,\hat{\mathbf{j}}) = -\rho_2 \mathbf{e}_r \qquad (4.39)$$

The only active forces are those of the springs that we write in the form

$$\mathbf{F}_1^a = -k(\rho_1 - a)\mathbf{e}_r$$
$$\mathbf{F}_2^a = k(\rho_2 - a)\mathbf{e}_r \qquad (4.40)$$

The projection or tangent vectors $\boldsymbol{\beta}_{ik}$ are then given by

$$\boldsymbol{\beta}_{11} = \frac{\partial \mathbf{r}_1}{\partial \rho_1} = \mathbf{e}_r, \quad \boldsymbol{\beta}_{12} = \frac{\partial \mathbf{r}_1}{\partial \rho_2} = 0, \quad \boldsymbol{\beta}_{13} = \frac{\partial \mathbf{r}_1}{\partial \theta} = \rho_1 \mathbf{e}_\theta$$

$$\boldsymbol{\beta}_{21} = \frac{\partial \mathbf{r}_2}{\partial \rho_1} = 0, \quad \boldsymbol{\beta}_{22} = \frac{\partial \mathbf{r}_2}{\partial \rho_2} = -\mathbf{e}_r, \quad \boldsymbol{\beta}_{23} = \frac{\partial \mathbf{r}_2}{\partial \theta} = -\rho_2 \mathbf{e}_\theta \qquad (4.41)$$

Note that $\partial \mathbf{e}_r / \partial \theta = \mathbf{e}_\theta$.

The acceleration vectors are best written in polar coordinates, e.g.,

$$\dot{\mathbf{v}}_1 = (\ddot{\rho}_1 - \dot{\theta}^2 \rho_1)\mathbf{e}_r + (\rho_1 \ddot{\theta} + 2\dot{\rho}_1 \dot{\theta})\mathbf{e}_\theta \qquad (4.42)$$

The first equation of motion using (4.38) is then

$$(m_1 \dot{\mathbf{v}}_1 - \mathbf{F}_1^a) \cdot \boldsymbol{\beta}_{11} = 0$$

or

$$m_1(\ddot{\rho}_1 - \dot{\theta}^2 \rho_1) + k(\rho_1 - a) = 0 \qquad (4.43)$$

The second equation is similar to the first

$$m_2(\ddot{\rho}_2 - \dot{\theta}^2 \rho_2) + k(\rho_2 - a) = 0 \qquad (4.44)$$

The third equation of motion becomes

$$(m_1 \dot{\mathbf{v}}_1 - \mathbf{F}_1^a) \cdot \boldsymbol{\beta}_{13} + (m_2 \dot{\mathbf{v}}_2 - \mathbf{F}_2^a) \cdot \boldsymbol{\beta}_{23} = 0 \qquad (4.45)$$

It is straightforward to show that this is equivalent to conservation of angular momentum, or,

$$\frac{d}{dt}(m_1 \rho_1^2 + m_2 \rho_2^2)\dot{\theta} = 0 \qquad (4.46)$$

If H_0 is the initial angular momentum of the masses, we can eliminate $\dot{\theta}$ in the first two equations using

$$\dot{\theta} = \frac{H_0}{m_1 \rho_1^2 + m_2 \rho_2^2} \qquad (4.47)$$

4.3
Lagrange's Equations[2]

Single Particle: One Degree of Freedom

Before discussing the general case, let us look at a single particle whose position vector depends on one generalized coordinate $q(t)$ as well as time:

$$\mathbf{r}(t) = \mathbf{r}(q(t), t) \qquad (4.48)$$

---

[2] Joseph Louis Lagrange (1736–1813) was a French-Italian mathematician and dynamicist. His treatise *Méchanique Analytique* was first published in 1788 in which his famous equations appear. He was proud that his proofs did not require geometric proofs.

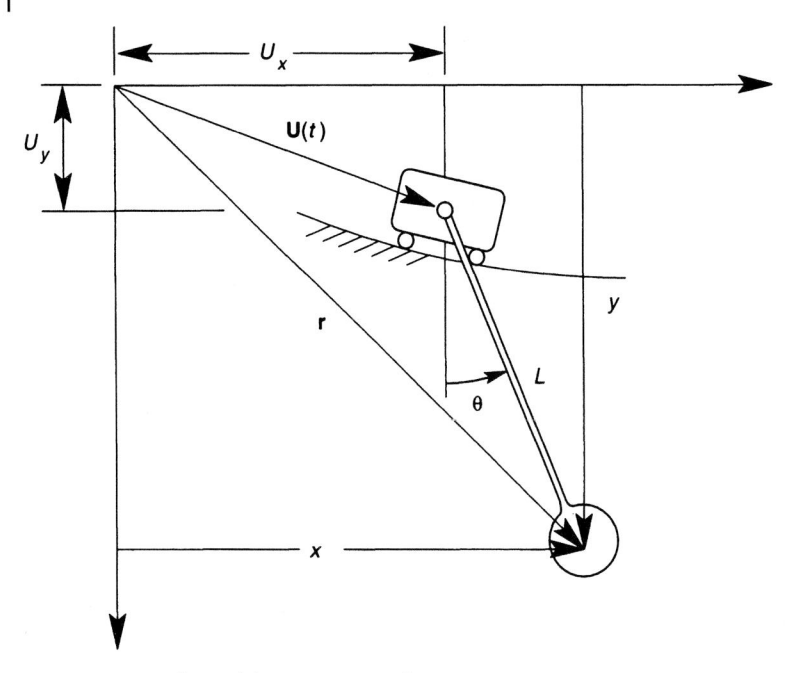

Fig. 4.7 Motion of a pendulum on a moving base.

An example is shown in Fig. 4.7 for a pendulum attached to a massless, moving base. If $r = [x, y]^T$, then

$$x(t) = U_x(t) + L \sin \theta(t)$$
$$y(t) = U_y(t) + L \cos \theta(t) \tag{4.49}$$

Here $q(t) = \theta(t)$.

In general, the velocity of a particle with a single degree of freedom is given by

$$\mathbf{v} = \dot{\mathbf{r}} = \frac{\partial \mathbf{r}}{\partial q} \dot{q} + \frac{\partial \mathbf{r}}{\partial t} \tag{4.50}$$

The second term on the right-hand side does not depend on $\dot{q}$, so that we have the identity:

$$\frac{\partial \mathbf{v}}{\partial \dot{q}} = \frac{\partial \mathbf{r}}{\partial q} \tag{4.51}$$

The virtual displacement is given by

$$\delta \mathbf{r} = \frac{\partial \mathbf{r}}{\partial q} \delta q \tag{4.52}$$

As in Section 4.2, let us assume that the forces on the particle involve an active force $\mathbf{F}^a$ which does work and a workless constraint force $\mathbf{R}$. Using (4.52), D'Alembert's principle (4.10) becomes

$$(m\ddot{\mathbf{r}} - \mathbf{F}^a) \cdot \frac{\partial \mathbf{r}}{\partial q} = 0 \tag{4.53}$$

Following the previous section, we define the projection of the active force $\mathbf{F}^a$ onto the direction $\boldsymbol{\beta} = (\partial \mathbf{r}/\partial q)$ as the generalized force Q so that (4.53) takes the form

$$m\dot{\mathbf{v}} \cdot \frac{\partial \mathbf{r}}{\partial q} = Q \tag{4.54}$$

Or using the identity (4.51), we have

$$m\dot{\mathbf{v}} \cdot \frac{\partial \mathbf{v}}{\partial \dot{q}} = Q \tag{4.55}$$

Note that the product $Q\dot{q}$ has units of power. This equation can be used directly to solve problems. We shall see, in the next section, that this equation is essentially the *principle of virtual power*. But, for now, we wish to transform the equation further using the kinetic energy;

$$T = \frac{1}{2} m \mathbf{v} \cdot \mathbf{v} \tag{4.56}$$

In general T can depend on both $q(t)$ and $\dot{q}(t)$, so that we can write

$$\frac{\partial T}{\partial q} = m\mathbf{v} \cdot \frac{\partial \mathbf{v}}{\partial q}$$

$$\frac{\partial T}{\partial \dot{q}} = m\mathbf{v} \cdot \frac{\partial \mathbf{v}}{\partial \dot{q}} = m\mathbf{v} \cdot \frac{\partial \mathbf{r}}{\partial q} \tag{4.57}$$

This last term is a projection of the momentum onto the direction $(\partial \mathbf{r}/\partial q)$, hence, we define the *generalized momentum* by

$$p = \frac{\partial T}{\partial \dot{q}} \tag{4.58}$$

In Newton's formulation, the time derivative of the momentum $m\mathbf{v}$ is equal to the force. Let us see how the time derivative of the generalized momentum p is related to the generalized force Q:

$$\frac{dp}{dt} = \frac{d}{dt}\left(m\mathbf{v} \cdot \frac{\partial \mathbf{r}}{\partial q}\right)$$

$$= m\dot{\mathbf{v}} \cdot \frac{\partial \mathbf{r}}{\partial q} + m\mathbf{v} \cdot \frac{d}{dt}\frac{\partial \mathbf{r}}{\partial q} \tag{4.59}$$

In the last term we can interchange derivatives so that (4.59) becomes

$$\frac{d}{dt}\frac{\partial T}{\partial \dot{q}} = m\dot{\mathbf{v}} \cdot \frac{\partial \mathbf{r}}{\partial q} + m\mathbf{v} \cdot \frac{\partial \dot{\mathbf{r}}}{\partial q} \qquad (4.60)$$

The second term on the right is just $\partial T/\partial q$. Using Eqs. (4.60), (4.55) and (4.51), we can show that

$$m\dot{\mathbf{v}} \cdot \frac{\partial \mathbf{v}}{\partial \dot{q}} = \frac{d}{dt}\frac{\partial T}{\partial \dot{q}} - \frac{\partial T}{\partial q}$$

or

$$\frac{dp}{dt} = Q + \frac{\partial T}{\partial q} \qquad (4.61)$$

Thus, when the particle is constrained to one degree of freedom, the projection of the equation of motion onto $\boldsymbol{\beta}$, (4.54) is analogous to Newton's form, with the added term reflecting the dependence of the kinetic energy on the generalized displacement. The standard form for *Lagrange's equation of motion* in terms of the kinetic energy $T(q,\dot{q})$ is then

$$\frac{d}{dt}\frac{\partial T}{\partial \dot{q}} - \frac{\partial T}{\partial q} = Q \qquad (4.62)$$

where

$$Q = \mathbf{F}^a \cdot \boldsymbol{\beta}$$

Example 4.4 Oscillation of a Double-slider Mechanism. As a simple example of the application of Lagrange's form of Newton's Law, consider the double-slider crank mechanism shown in Fig. 4.8. Two masses are constrained to move along horizontal and vertical tracks. One can show that the distance to the center point is constant and equal to the half length of the connecting link. Thus this radius can serve as a crank to turn the mechanism. If one extends the link to the point P, one can show that P will describe an ellipse given by

$$\frac{x_P^2}{a^2} + \frac{y_P^2}{(a+2L)^2} = 1 \qquad (4.63)$$

where a is the length of the link extension. The great kinematician Franz Reuleaux attributed the invention of the double-slider ellisograph to Leonardo da Vinci (see Moon, 2007).

In this example however we wish to find the dynamics of the double slider mechanism under the force of gravity. We begin with equations of constraint, using ϕ as a generalized position, namely

$$x = 2L\sin\phi; \qquad y = 2L\cos\phi \qquad (4.64)$$
$$\dot{x} = 2L\dot{\phi}\cos\phi; \qquad \dot{y} = -2L\dot{\phi}\sin\phi \qquad (4.65)$$

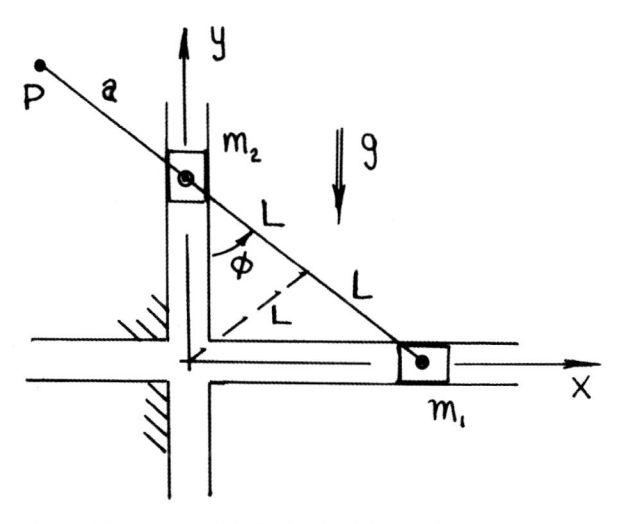

Fig. 4.8 Two-mass model of a double-slider mechanism under gravity.

As we shall see in the derivation below, when the system has several masses, the kinetic energy is the sum of all the kinetic energy of the masses. When the masses are equal $m_1 = m_2 = m$, the kinetic energy function can be shown to be

$$T = \frac{1}{2}m_1 \dot{x}^2 + \frac{1}{2}m_2 \dot{y}^2$$

and

$$T = \frac{1}{2}m 4L^2 \dot{\phi}^2 \qquad (4.66)$$

Assuming frictionless constraints, gravity only does work on the mass in the vertical track. The generalized force Q in (4.62) can be replaced with a potential function $V(\phi)$, i.e., $Q = -\partial V/\partial \phi$:

$$V = -mgy = -mg2L \cos \phi \qquad (4.67)$$

Evaluating the two partial derivatives in Lagrange's equation (4.62) and the total derivative we arrive at the ordinary differential equation of motion of the constrained link under gravity:

$$\frac{\partial T}{\partial \phi} = 0; \qquad \frac{\partial T}{\partial \dot{\phi}} = 4mL^2 \dot{\phi}; \qquad -Q = \frac{\partial V}{\partial \phi} = 2mgL \sin \phi$$

$$\frac{d}{dt}\frac{\partial T}{\partial \dot{\phi}} - \frac{\partial T}{\partial \phi} + \frac{\partial V}{\partial \phi} = 4mL^2 \ddot{\phi} + 2mgL \sin \phi = 0 \qquad (4.68)$$

$$\ddot{\phi} + \frac{g}{2L} \sin \phi = 0 \qquad (4.69)$$

This equation is independent of the mass, depending only on the length of the connecting link. In fact this equation is identical to that of a simple pendulum of length $2L$ as described in Chapter 1. For small oscillations the natural frequency in radians per second is given by

$$\Omega = \sqrt{g/2L}$$

or the time period of the oscillation is given by

$$\tau = 2\pi\sqrt{2L/g} \tag{4.70}$$

Example 4.5 Multiwell Part Holder. Consider the manufacturing fixture shown in Fig. 4.9, in which parts are positioned for some operation. The base is moved in a prescribed manner so as to shift a part from one well to the next. We wish to find the equation of motion of a part as it goes from one well to another, so that a base motion can be designed to produce the jump. As an idealization, we neglect friction and assume that during the motion the mass follows the sinusoidal surface given by the constraint equation:

$$x(t) = s(t) + q(t)$$
$$y(q) = A_0(1 - \cos kq) \tag{4.71}$$

The wavelength between wells is $\Lambda = 2\pi/k$. We choose the relative position q as the generalized coordinate. The mass position vector is $\mathbf{r} = x\mathbf{e}_x + y\mathbf{e}_y$. The constrained velocity is then

$$\dot{\mathbf{r}} = (\dot{s} + \dot{q})\mathbf{e}_x + A_0 k\dot{q} \sin kq \, \mathbf{e}_y \tag{4.72}$$

The D'Alembert projection vector is then given by

$$\frac{\partial \mathbf{r}}{\partial q} = \mathbf{e}_x + A_0 k \sin kq \, \mathbf{e}_y = \frac{\partial \dot{\mathbf{r}}}{\partial \dot{q}} \tag{4.73}$$

Note the identity (4.73). The only active force on the mass is gravity, so that the generalized force is the projection of gravity onto the D'Alembert direction

$$Q = -mg\mathbf{e}_y \cdot \frac{\partial \mathbf{r}}{\partial q}$$
$$Q = -mg A_0 k \sin kq \tag{4.74}$$

As a special case, consider a zero base motion or $s(t) = 0$. To obtain Lagrange's equations we must calculate the kinetic energy T:

$$T(q, \dot{q}) = \frac{1}{2}m(\dot{x} + \dot{y}^2) = \frac{1}{2}m\left(1 + A_0^2 k^2 \sin^2 kq\right)\dot{q}^2 \tag{4.75}$$

Lagrange's equation is then

$$\frac{d}{dt}\frac{\partial T}{\partial \dot{q}} - \frac{\partial T}{\partial q} = Q \tag{4.76}$$

4.3 Lagrange's Equations

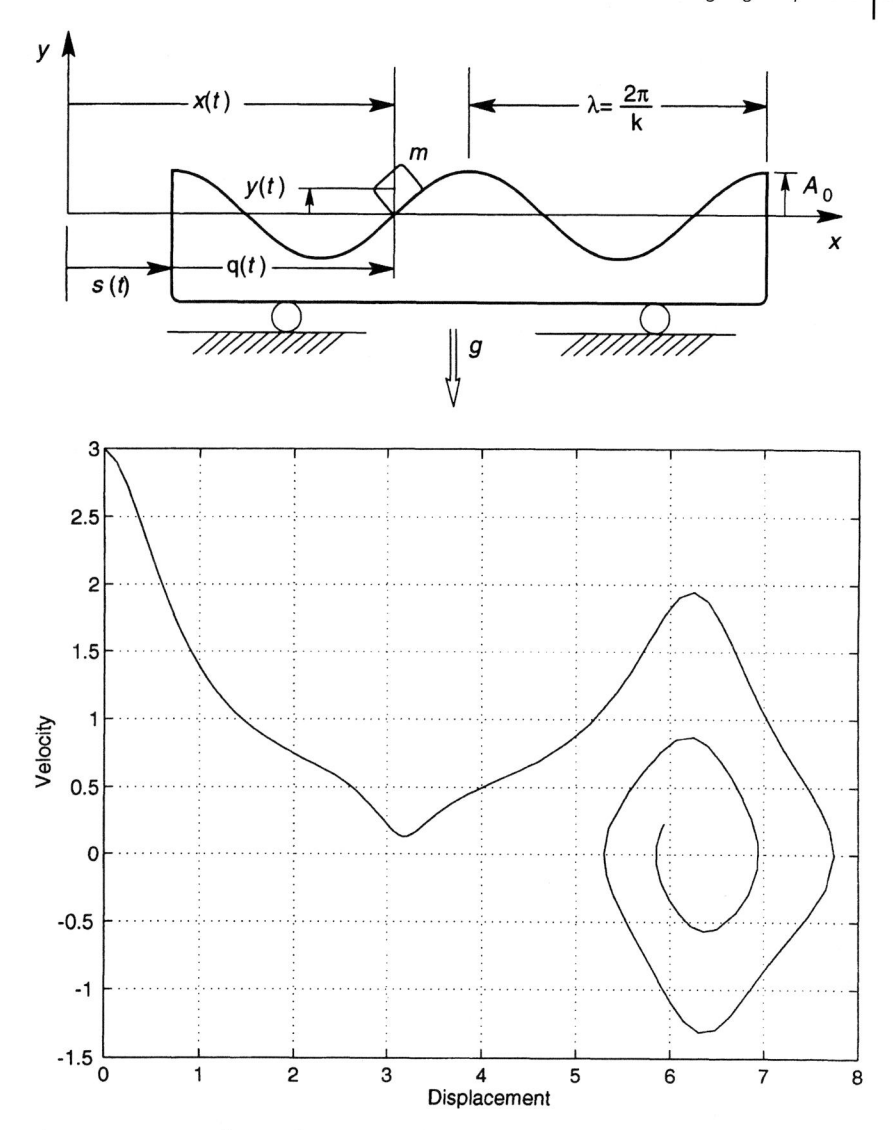

Fig. 4.9 *Top*: Motion of a particle in a sinusoidal potential well. *Bottom*: MATLAB integration of (4.76) Initial velocity $\dot{q}(0) = 3$, moves the particle from the first well at $q = 0$, to the second well at $q = 2\pi$. Damping added.

Carrying out the details of the two terms on the left-hand side, one can see common terms that partially cancel. The resulting equation of motion is then

$$m\left(1 + A_0^2 k^2 \sin^2 kq\right)\ddot{q} + m A_0^2 k^3 \dot{q}^2 \sin kq \cos kq + mgk A_0 \sin kq = 0 \quad (4.77)$$

This equation can be obtained directly from D'Alembert's principle (4.36), by calculating the acceleration and projecting onto the D'Alembert vector $(\partial \mathbf{r}/\partial q)$ using

$$\ddot{\mathbf{r}} = \ddot{q}\mathbf{e}_x + \left(A_0 k \ddot{q} \sin kq + A_0 k^2 \dot{q}^2 \cos kq\right)\mathbf{e}_y \qquad (4.78)$$

The equation of motion, (4.77), is nonlinear and it is difficult to obtain a solution without recourse to numerical integration. However, for small motions in one well we can linearize the terms, assuming that $\sin kq \sim kq$, and $\cos kq \sim 1$. In this limit the equation becomes

$$\ddot{q} + gk^2 A_0 q = 0 \qquad (4.79)$$

This is the equation for a harmonic oscillator with a natural frequency in radius per time

$$\omega = (gk^2 A_0)^{1/2} \qquad (4.80)$$

Using the wavelength Λ ($k = 2\pi/\Lambda$), the frequency is the same as a pendulum with length $L = \Lambda^2/4\pi^2 A_0$. Thus to move the base so that the mass jumps from one well to another, the base motion $s(t)$ should have a frequency component near this frequency.

The nonlinear differential equation (4.77) can be numerically integrated. The following is a simple MATLAB program (version 4.2) which uses a fourth-order Runge–Kutta algorithm (see Pratap, 2004).

The program requires that the second-order differential equation be written as two first-order differential equations. We have nondimensionalized the equation using $x = kq$ and a time scale based on the natural frequency $(gk^2 A_0)^{-1/2}$. (This program can be used for other second-order equations in this text.) Also some linear damping has been added to the equation.

$$\dot{x} = v$$
$$\dot{v} = \frac{[-\gamma v - \sin x - a^2 v^2 \sin x \cos x]}{[1 + a^2 \sin^2 x]}$$

where $a = A_0 k$.

The values used in the numerical integration are $\gamma = 0.3$, $a = \pi/2$, $x(0) = 0$, $v(0) = 3$. The time interval is $0 \leqslant t \leqslant 25$ nondimensional time units.

A phase plane plot is shown in Fig. 4.9b. It shows that the initial velocity has moved the mass over the first well peak and shows damped oscillations in the next well valley.

The MATLAB integration scheme requires the definition of a function which here is called *multiwell*

```
function dx = multiwell (t,x)
a = 3.14 / 2 ;
dx = zeros (2,1);
dx(1)= x(2);
```

```
dx(2)=(-0.3 .* x(2) - sin(x(1)) - (a ^2) .* (x(2) .^2)...
    .* sin(x(1)) .* cos(x(1))) ./ (1 + a .* sin(x(1))^2) ;

t0 = 0; tf = 25;
x0 = [0 ; 3] ;
[t.x] = ode45('multiwell',t0,tf,x0);
plot(x(:,1),x(:,2))
xlabel ('x Displacement'), ylabel('Velocity')
grid on
```

Potential Energy and Conservative Forces

In elementary physics one learns that the work done by some forces in nature can be represented by the change in a scalar function of the position called *potential energy* $V(\mathbf{r})$. For such forces the work done in moving from position $\mathbf{r}_1$ to $\mathbf{r}_2$ is independent of the path:

$$\int_1^2 \mathbf{F} \cdot d\mathbf{r} = V(\mathbf{r}_1) - V(\mathbf{r}_2) \tag{4.81}$$

If the motion is constrained to move along some path with a generalized coordinate q, then

$$\mathbf{F} \cdot d\mathbf{r} = -\frac{dV}{dq} dq = -dV \tag{4.82}$$

In general, where $\mathbf{r} = [x, y, z]^T$ in a three-dimensional space, $\mathbf{F}$ can be represented by the gradient operation;

$$\mathbf{F}(\mathbf{r}) = -\nabla V(\mathbf{r}) \tag{4.83}$$

Forces that can be expressed in this way are called *conservative forces*. In Cartesian coordinates the operator becomes

$$\nabla = \mathbf{e}_x \frac{\partial}{\partial x} + \mathbf{e}_y \frac{\partial}{\partial y} + \mathbf{e}_z \frac{\partial}{\partial z} \tag{4.84}$$

Examples of common conservative forces include (Fig. 4.10):

Gravity near the Earth's Surface

$$\mathbf{F} = -mg\mathbf{e}_z, \quad V = mgz \tag{4.85}$$

Gravitation Force

$$\mathbf{F} = -\frac{GMm}{r^2}\mathbf{e}_r, \quad V = -\frac{GMm}{r} \tag{4.86}$$

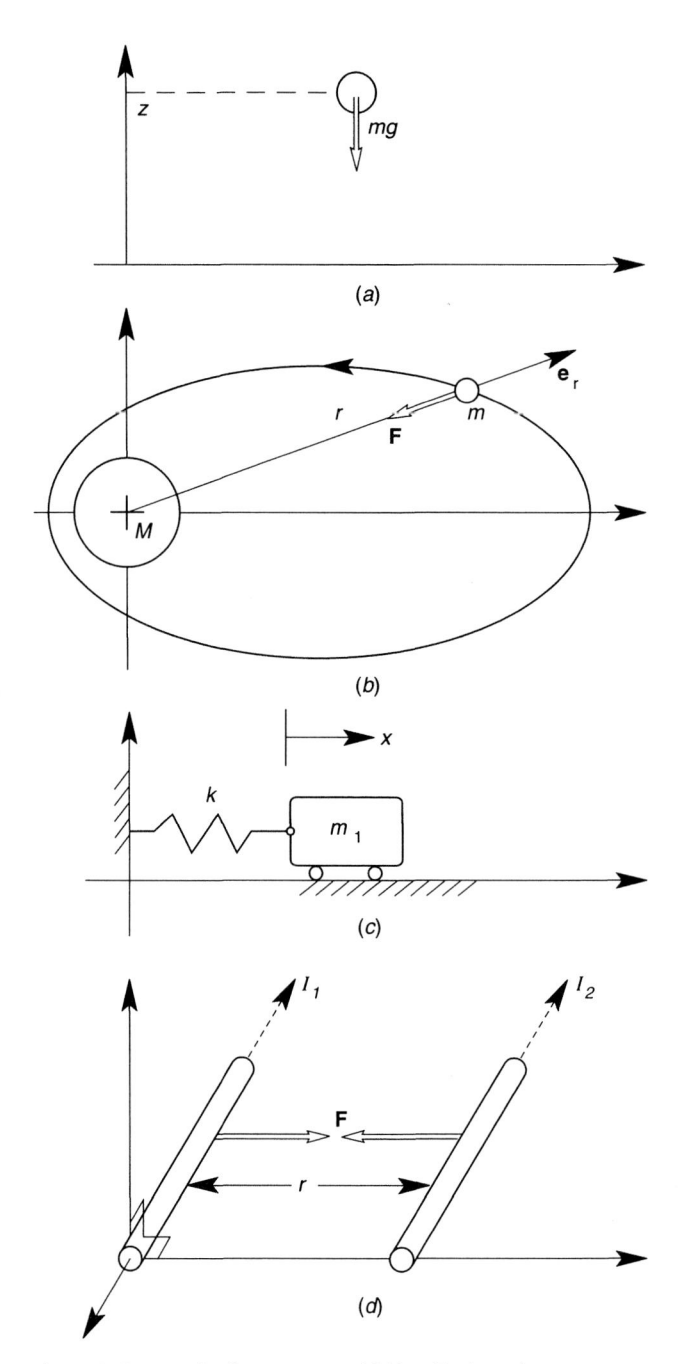

Fig. 4.10 Conservative force systems. (*a*) Near Earth gravity. (*b*) Gravity force. (*c*) Elastic spring. (*d*) Magnetic force.

4.3 Lagrange's Equations

Elastic Spring Forces

$$F = -kx, \quad V = \frac{1}{2}kx^2 \qquad (4.87)$$

Magnetic Force Between Two Wires

$$F = \frac{\mu_0}{2\pi}\frac{I_1 I_2}{r}, \quad V = -\frac{\mu_0}{2\pi}I_1 I_2 \log r \qquad (4.88a)$$

($\mu_0 = 4\pi \cdot 10^{-7}$ in mks units).

Electric Force Between Two Charges

$$F = \frac{Q_1 Q_2}{4\pi\varepsilon_0 r^2}, \quad V = \frac{Q_1 Q_2}{4\pi\varepsilon_0 r} \qquad (4.88b)$$

($1/4\pi\varepsilon_0 = 8.99 \cdot 10^9$, in mks units).

Since the force potential is defined in terms of work, it is easy to show that for a single-degree-of-freedom system, the generalized force is given by

$$Q = \mathbf{F} \cdot \frac{\partial \mathbf{r}}{\partial q} = -\frac{\partial V(\mathbf{r}(q))}{\partial q} \qquad (4.89)$$

Thus, for a single particle constrained to move in a conservative force field, Lagrange's equation of motion becomes

$$\frac{d}{dt}\frac{\partial T}{\partial \dot{q}} - \frac{\partial T}{\partial q} + \frac{\partial V}{\partial q} = 0 \qquad (4.90)$$

It is traditional to define a function $\mathcal{L}$ called the Lagrangian

$$\mathcal{L}(q, \dot{q}) = T(q, \dot{q}) - V(q) \qquad (4.91)$$

In terms of this function, Lagrange's equation becomes

$$\frac{d}{dt}\frac{\partial \mathcal{L}}{\partial \dot{q}} - \frac{\partial \mathcal{L}}{\partial q} = 0 \qquad (4.92)$$

Conservation of Energy

Educated intuition would lead one to conclude that if the generalized forces are related to a potential energy function and the constraints are fixed, then the total mechanical energy, kinetic plus potential energies, is conserved in any dynamic motion of the particle. To see this we return to the D'Alembert form of the equation of motion (4.54), multiplied by the generalized velocity $\dot{q}$ and use (4.89):

$$m\dot{\mathbf{v}} \cdot \frac{\partial \mathbf{r}}{\partial q}\dot{q} = -\frac{\partial V}{\partial q}\dot{q} = -\frac{dV}{dt} \qquad (4.93)$$

Now consider the total time derivative of the kinetic energy

$$\frac{dT}{dt} = m\dot{\mathbf{v}} \cdot \mathbf{v} \tag{4.94}$$

If the constraints are fixed, i.e., $\mathbf{r} = \mathbf{r}(q(t))$, then the kinematic relation between the velocity $\mathbf{v}$ and the generalized velocity $\dot{q}$ is

$$\mathbf{v} = \frac{\partial \mathbf{r}(q)}{\partial q} \dot{q} \tag{4.95}$$

(*Note*: This is a special case of a more general constraint, Eq. (4.48), $\mathbf{r} = \mathbf{r}(q, t)$.) Substituting this expression into Eq. (4.93) and using Eq. (4.94), we can see that

$$\frac{dT}{dt} = -\frac{dV}{dt} \quad \text{or} \quad T + V = \text{constant} \tag{4.96}$$

Rayleigh Dissipation Function

Many mechanical systems contain linear viscous damping forces. Such forces can be introduced into Lagrange's equation of motion by choosing a generalized force proportional to the generalized velocity $Q = -b\dot{q}(t)$. However, another method is to define a quadratic *dissipation function*, $\mathcal{R}$,

$$\mathcal{R} = \frac{1}{2} b \dot{q}^2$$

and modifying Lagrange's equation (4.92)

$$\frac{d}{dt}\frac{\partial \mathcal{L}}{\partial \dot{q}} - \frac{\partial \mathcal{L}}{\partial q} + \frac{\partial \mathcal{R}}{\partial \dot{q}} = 0 \tag{4.97}$$

The *Rayleigh dissipation function* is also useful in the application of Lagrange's equation to electric circuits (see Chapter 8). For example, the generalized voltage across a resistor is derived from $\mathcal{R} = \frac{1}{2} R I^2$, where R is the electrical resistance and $\dot{q} = I$ is the electric current.

Example 4.6 Vibration of an Electromechanical Oscillator. The expression for the electric field energy in (4.11) assumes localized charges. However in practical devices such as micro-electromechanical systems (MEMS) the charges are likely to be distributed. Looking ahead to Chapter 8, the electric voltage E between two charged bodies can be expressed in terms of a parameter called the *capacitance*, C, where $E = Q/C$. The electrical energy stored in the space between two charged bodies is then expressed as $V_e = Q^2/2C$. As we shall show in Chapter 8, when Q is fixed and the capacitance is assumed to be a function of a generalized coordinate, $C(x)$, then we can use the expression for $V_e(x)$ in Lagrange's equation for the dynamics of $x(t)$, provided the charge is held fixed. Consider then the geometry of two charged bodies in Fig. 4.11 with positive and negative charges between a hollow cylinder

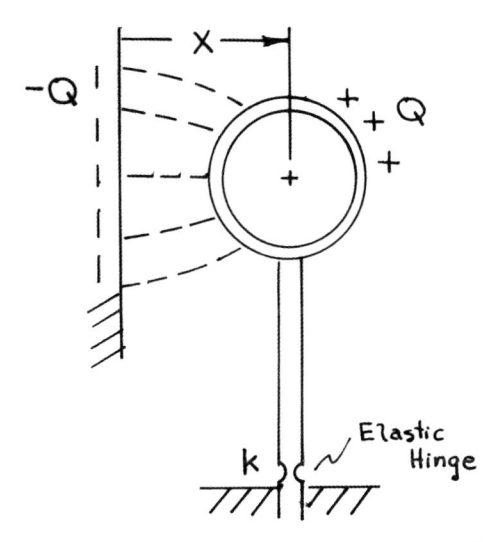

Fig. 4.11 MEMS scale compliant lever mechanism of a conducting shell with electric charge near a conducting wall.

of radius R, and a plane, where the separation distance between the center of the cylinder and the plane is x. The capacitance for this configuration is given by

$$C(x) = \pi \varepsilon \left[\cosh^{-1}(x/R) \right]^{-1} \tag{4.98}$$

(Note that $x/R > 1$.) Suppose the cylinder is fixed to a flexible beam with elastic constant k then the equation of motion can be derived from Lagrange's equation (4.90)

$$\frac{d}{dt}\frac{\partial T}{\partial \dot{x}} - \frac{\partial T}{\partial x} + \frac{\partial (V_b + V_e)}{\partial x} = 0$$

$$T = \frac{1}{2}m\dot{x}^2, \quad V_b = \frac{1}{2}kx^2, \quad V_e = \frac{1}{2}Q^2/C(x) \tag{4.99}$$

Carrying out the derivatives, the equation of motion becomes

$$m\ddot{x} + kx = \frac{1}{2}\frac{Q^2}{C^2}\frac{dC(x)}{dx} \tag{4.100}$$

where

$$\frac{dC}{dx} = \frac{1}{R\sqrt{(x/R)^2 - 1}}$$

One can see from this example, that although the mechanical problem is linear in the bending displacement, $x(t)$, the coupled electromechanical problem is nonlinear in both Q and x.

Lagrange's Equations for a System of Particles

In this section, we generalize Lagrange's equations to many interacting particles, and in Chapter 5 we extend the principle to rigid bodies. The application of Lagrange's equations to continuous deformable systems is not treated in this text. One should consult more advanced texts such as the second edition of Goldstein (1980).

We begin with a set of N interacting particles of masses $\{m_i\}$, which are acted upon by mutual forces between them $\{\mathbf{f}_{ij}\}$, external active forces $\mathbf{F}_i^a$ that do work, and reaction forces of constraint $\mathbf{R}_i$ that, under virtual displacements, do no work. Following the procedure in Section 4.2, we write a generalization of the principle of virtual work:

$$\sum \left(m_i \ddot{\mathbf{r}}_i - \mathbf{F}_i^a - \mathbf{R}_i - \sum \mathbf{f}_{ij} \right) \cdot \delta \mathbf{r}_i = 0 \tag{4.101a}$$

and assume that

$$\sum \mathbf{R}_i \cdot \delta \mathbf{r}_i = 0 \tag{4.101b}$$

We next have to identify the number of constraints and assume a set of M generalized position coordinates $\{q_k\}$ that can be displacements, angles, or other measures of the configuration of the system. We also assume that the constraints are geometric and of the form;

$$\mathbf{r}_i = \mathbf{r}_i(q_k, t) \tag{4.102}$$

The velocity of each mass then depends on all the M generalized velocities:

$$\mathbf{v}_i = \dot{\mathbf{r}}_i = \sum \frac{\partial \mathbf{r}_i}{\partial q_k} \dot{q}_k + \frac{\partial \mathbf{r}_i}{\partial t} \tag{4.103}$$

From these kinematic relations we can again obtain the identity

$$\frac{\partial \mathbf{v}_i}{\partial \dot{q}_k} = \frac{\partial \mathbf{r}_i}{\partial q_k}, \quad k = 1, 2, \ldots, M \tag{4.104}$$

This is true because we assumed that the constraint relations $\mathbf{r}_i(q_k, t)$ do not depend on the generalized velocities $\{\dot{q}_k\}$. Such constraints are called *holonomic*.

Next we write the virtual displacements of each mass, $\delta \mathbf{r}_i$ in terms of the generalized virtual displacements δq_k, i.e.,

$$\delta \mathbf{r}_i = \sum \frac{\partial \mathbf{r}_i}{\partial q_k} \delta q_k \tag{4.105}$$

Note the contrast of this equation with the expressions for the velocities, Eq. (4.103). In the concept of virtual work, we freeze time, i.e., $\delta t = 0$; only geometric variations come into play.

To define the generalized forces Q_k we consider the work of active forces in (4.101a)

$$\sum \mathbf{F}_i^a \cdot \delta \mathbf{r}_i = \sum \sum \mathbf{F}_i^a \cdot \frac{\partial \mathbf{r}_i}{\partial q_k} \delta q_k \qquad (4.106)$$

Exchanging the order of the summation operators, we define

$$Q_k = \sum \mathbf{F}_i \cdot \frac{\partial \mathbf{r}_i}{\partial q_k} = \sum \mathbf{F}_i \cdot \boldsymbol{\beta}_{ik} \qquad (4.107)$$

Here the vectors $\boldsymbol{\beta}_{ik} = (\partial \mathbf{r}_i / \partial q_k)$ represent projection directions on which we take the component of the corresponding force.

Using all these definitions, the principle of virtual work takes the form of the following scalar equation:

$$\left(\sum m_i \dot{\mathbf{v}}_i \cdot \frac{\partial \mathbf{r}_i}{\partial q_k} - Q_k - \sum \sum \mathbf{f}_{ij} \cdot \frac{\partial \mathbf{r}_i}{\partial q_k} \right) \delta q_k = 0 \qquad (4.108)$$

Since we have assumed all the $\{q_k\}$ and the variations $\{\delta q_k\}$ are independent, we can set each of the bracketed terms equal to zero, resulting in M equations of motion.

$$\sum m_i \dot{\mathbf{v}}_i \cdot \frac{\partial \mathbf{r}_i}{\partial q_k} = Q_k + \sum \sum \mathbf{f}_{ij} \cdot \frac{\partial \mathbf{r}_i}{\partial q_k} \qquad (4.109)$$

As discussed in Section 4.2, these equations can be used to directly solve problems without using Lagrange's form of these equations.

To derive Lagrange's equations of motion, we must relate the kinetic energy T to the left-hand side of (4.109).

The kinetic energy of the ensemble is the sum of the kinetic energies of each of the particles, i.e.,

$$T = \sum T_i$$

$$T_i = \frac{1}{2} m_i \mathbf{v}_i \cdot \mathbf{v}_i \qquad (4.110)$$

In Section 4.2 for a single particle we showed that [see Eq. (4.61)]

$$m_i \dot{\mathbf{v}}_i \cdot \frac{\partial \mathbf{r}_i}{\partial q_k} = \frac{d}{dt} \frac{\partial T_i}{\partial \dot{q}_k} - \frac{\partial T_i}{\partial q_k} \qquad (4.111)$$

Substituting this expression into Eq. (4.109), using (4.110), we obtain

$$\frac{d}{dt} \frac{\partial T}{\partial \dot{q}_k} - \frac{\partial T}{\partial q_k} = Q_k + \sum \sum \mathbf{f}_{ij} \cdot \frac{\partial \mathbf{r}_i}{\partial q_k} \qquad (4.112)$$

where $k = 1, 2, \ldots, M$. Thus there are as many equations as there are independent degrees of freedom. These equations are generally not used in this form. We usually try to separate out active forces and internal forces that can be represented by a potential energy function $V(q_k)$. Such forces are sometimes called *conserva-*

tive forces. Thus we assume that the generalized force Q_k can be separated into conservative and nonconservative parts, i.e.,

$$Q_k = -\frac{\partial V_1}{\partial q_k} + Q_k^{(nc)} \tag{4.113}$$

In the case of the internal forces we have $\mathbf{f}_{ij} = -\mathbf{f}_{ji}$ which expresses Newton's law of action and reaction. Thus the double sum over internal forces in (4.112) can be written in terms of pairs of particles;

$$\mathbf{f}_{ij} \cdot \left(\frac{\partial \mathbf{r}_i}{\partial q_k} - \frac{\partial \mathbf{r}_j}{\partial q_k} \right) \tag{4.114}$$

There are three possibilities here. First, in the case of rigid constraints between all the particles it can be shown that the double sum over $\mathbf{f}_{ij}$ is zero (see, e.g., Goldstein, 1980; Goldstein et al., 2002). In the second case, the pairwise terms represent elastic or conservative forces that can be represented by a potential energy function, i.e.,

$$\sum \sum \mathbf{f}_{ij} \cdot \frac{\partial \mathbf{r}_i}{\partial q_k} = -\frac{\partial V_2}{\partial q_k} \tag{4.115}$$

In the third case some of the pairwise forces may contain nonconservative or dissipative elements, such as friction or viscous dampers. In this case we can sometimes use a Rayleigh dissipation function to represent these forces, as in (4.97).

In the case of only conservative forces, we combine the potential energy functions of the active and internal forces, i.e., $V(q_k) = V_1 + V_2$.

Thus for the case of conservative forces, Lagrange's equations take the form

$$\frac{d}{dt}\frac{\partial T}{\partial \dot{q}_k} - \frac{\partial T}{\partial q_k} + \frac{\partial V}{\partial q_k} = 0 \tag{4.116}$$

for $k = 1, 2, \ldots, M$. We note again that the kinetic energy function may depend on both generalized position and velocity, $q_k, \dot{q}_k$. But the potential energy can only depend on the generalized position. With this understanding, a Lagrangian function $\mathcal{L}(q_k, \dot{q}_k)$ is defined;

$$\mathcal{L} = T(q_k, \dot{q}_k) - V(q_k) \tag{4.117}$$

and Lagrange's equations take a more compact form

$$\frac{d}{dt}\frac{\partial \mathcal{L}}{\partial \dot{q}_k} - \frac{\partial \mathcal{L}}{\partial q_k} = 0, \quad k = 1, 2, \ldots, M \tag{4.118}$$

The *generalized momentum* is defined as in Section 4.2:

$$p_k = \frac{\partial \mathcal{L}}{\partial \dot{q}_k} = \frac{\partial T}{\partial \dot{q}_k} \tag{4.119}$$

Thus if the Lagrangian function $\mathcal{L}$ is independent of the generalized position q_k, the corresponding momentum p_k is conserved [see (4.116)], i.e., if

$$\frac{\partial \mathcal{L}}{\partial q_k} = 0$$

then

$$p_k = \frac{\partial T}{\partial \dot{q}_k} = \text{constant} \qquad (4.120)$$

Thus, one can sometimes obtain a constant of the motion merely by inspecting the kinetic energy function. A generalized coordinate q_k that does not appear in the Lagrangian $\mathcal{L}$, is called a *ignorable* or *cyclic* coordinate.

Example 4.7. Consider the system shown in Fig. 4.12 in which a large elastic structure sits on an oscillating base. Engineering examples include a building or water tower under earthquake excitation. One of the methods to minimize the vibration of the large mass is to attach a smaller mass with a restoring force as a *vibration absorber*. In this example we have attached a pendulum of mass m_2. We wish to derive equations of motion to determine under what mass and stiffness ratios will the small mass "quench" the motion of mass m_1.

The idealized model chosen is shown in Fig. 4.12b. Here the structural stiffness is replaced by a single spring of stiffness k. We assume the mass m_1 undergoes pure translation. Obviously, in a real structure there are many vibration modes. We have chosen to model only the lowest one with frequency given by $(k/m_1)^{1/2}$. To use Lagrange's equations we chose the two generalized coordinates as

$$q_1 = x, \qquad q_2 = \theta \qquad (4.121)$$

In the first analysis, we examine the zero-base motion case (see the Homework Problem 4.3 for the case of forced, sinusoidal-base motion). The kinetic energy function is given by

$$T = \frac{1}{2}m_1\dot{x}^2 + \frac{1}{2}m_2(\dot{x}\mathbf{e}_x + L\dot{\theta}\mathbf{e}_\theta)^2 \qquad (4.122)$$

where the second term represents an inner product. The vertical motion of the structural mass is neglected here. The resulting scalar function becomes

$$T = \frac{1}{2}(m_1 + m_2)\dot{x}^2 + \frac{1}{2}m_2 L^2 \dot{\theta}^2 + m_2 \dot{x} L \dot{\theta} \cos\theta \qquad (4.123)$$

It is clear that T depends on the angle θ as well as $\dot{\theta}$.

The potential energy function is the sum of the energy stored in the spring and the gravitational potential of the pendulum mass:

$$V = \frac{1}{2}kx^2 + m_2 g L(1 - \cos\theta) \qquad (4.124)$$

We then use Lagrange's equations in the form

$$\frac{d}{dt}\frac{\partial T}{\partial \dot{q}_i} - \frac{\partial T}{\partial q_i} + \frac{\partial V}{\partial q_i} = 0 \qquad (4.125)$$

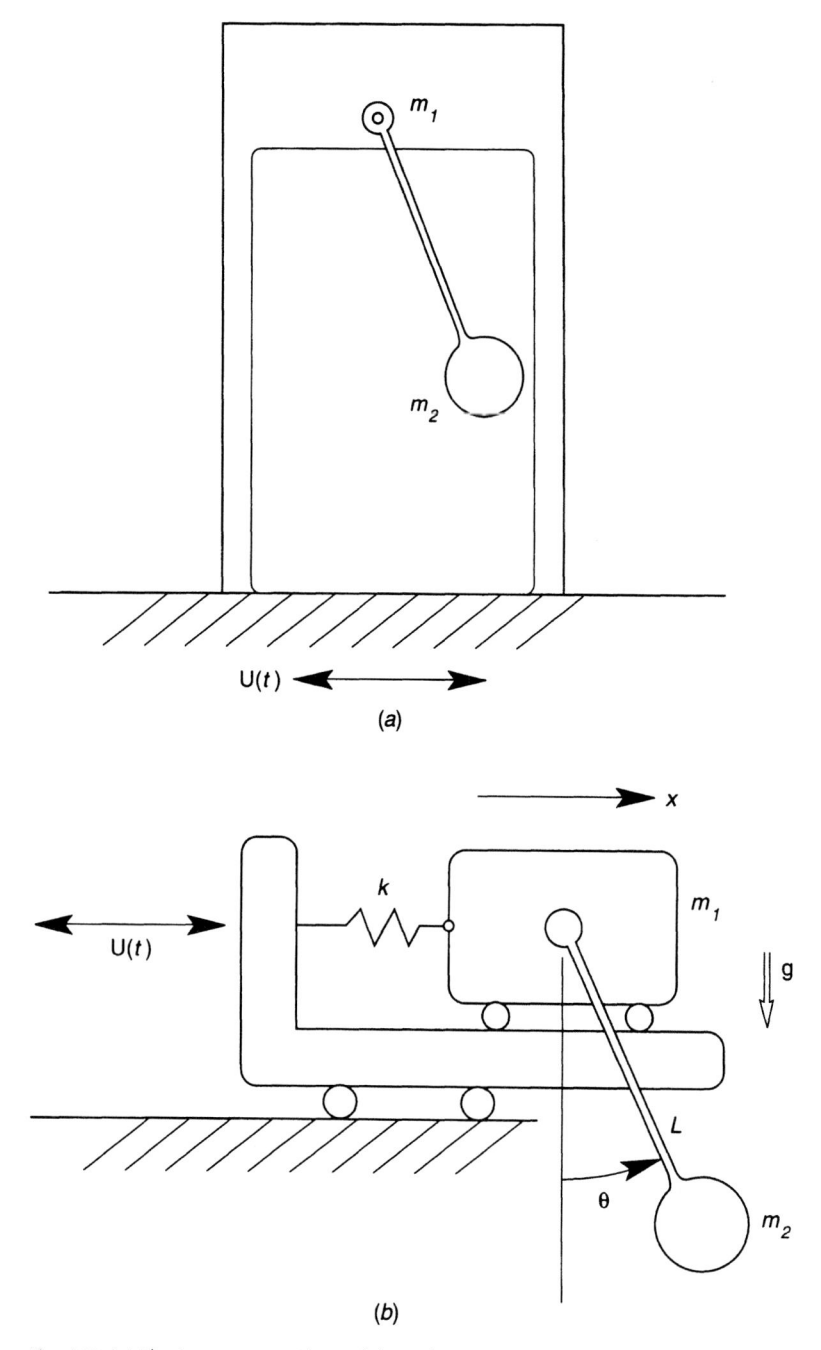

Fig. 4.12 (a) Elastic structure with pendulum vibration absorber. (b) Simple model of pendulum attached to an elastically constrained base.

where for the unforced case the generalized force $Q_i = 0$. Carrying out the derivatives, we obtain two ordinary, nonlinear, differential equations of motion:

$$(m_1 + m_2)\ddot{x} + m_2 L \ddot{\theta} \cos\theta - m_2 L \dot{\theta}^2 \sin\theta + kx = 0 \tag{4.126}$$

$$m_2 L^2 \ddot{\theta} + m_2 L \ddot{x} \cos\theta + m_2 g L \sin\theta = 0 \tag{4.127}$$

(*Note*: In the second equation two terms equal to $m_2 \dot{x} \dot{\theta} L \sin\theta$ canceled out. This can be a problem in symbolic math programs that may keep both terms and may result in calculation inefficiency if the equations are then integrated numerically.)

In the case of small motions of the pendulum, these equations can be linearized and put into the following matrix form

$$\begin{bmatrix} (m_1 + m_2) & m_2 L \\ m_2 L & m_2 L^2 \end{bmatrix} \begin{bmatrix} \ddot{x} \\ \ddot{\theta} \end{bmatrix} + \begin{bmatrix} k & 0 \\ 0 & m_2 g L \end{bmatrix} \begin{bmatrix} x \\ \theta \end{bmatrix} = 0 \tag{4.128}$$

where we have set $\cos\theta \sim 1$ and dropped $\dot{\theta}^2 \sin\theta$ in the x equation. We can see here that the two motions are coupled through the off-diagonal terms of the mass matrix.

To find a solution we assume a sinusoidal motion

$$\begin{bmatrix} x \\ \theta \end{bmatrix} = \begin{bmatrix} a \\ b \end{bmatrix} \cos(\omega t + \phi_0) \tag{4.129}$$

It is straightforward to show that the two vibration frequencies are determined by setting the following determinant equal to zero:

$$\det \begin{bmatrix} k - \omega^2(m_1 + m_2) & -m_2 L \omega^2 \\ -m_2 L \omega^2 & m_2 g L - m_2 L^2 \omega^2 \end{bmatrix} = 0 \tag{4.130}$$

or

$$\omega^4 (1 - \mu) - (\omega_1^2 + \omega_2^2)\omega^2 + \omega_1^2 \omega_2^2 = 0 \tag{4.131}$$

where

$$\mu = m_2/(m_1 + m_2), \quad \omega_1^2 = k/(m_1 + m_2), \quad \omega_2^2 = g/L \tag{4.132}$$

To design a vibration absorber under forced sinusoidal base motion, the frequency ω_2 is tuned to the frequency of the driven base motion.

4.4
The Method of Virtual Power

In Section 4.1, we saw how Lagrange's form of D'Alembert's principle of virtual work could be directly used to find equations of motion. In recent years, several authors in the United States and Europe (Kane and Levinson, 1985; Pfeiffer, 1989;

Schiehlen, 1986; Lesser, 1995) have taught a new method based on an old idea, namely, the principle of virtual power or virtual velocities. Why do we need another method to derive equations of motion?

We have seen that the Newtonian formulation employs all the forces whether active or produced from constraints. Both D'Alembert's method and Lagrange's equations allow us to formulate equations of motion when there are geometric constraints (so called *holonomic constraints*). However, there are important constraints such as rolling of one body on another or of feedback control that are often expressed in the form of velocity constraints or *nonholonomic constraints* of the form

$$\sum a_{ij} \dot{q}_j + b_i(t) = 0, \quad i = 1, 2, \ldots, k \tag{4.133}$$

In some cases, these equations can be integrated to obtain equivalent geometric constraints, and the direct form of Lagrange's equations is valid. In three-dimensional rolling problems, Lagrange's equations must be modified to handle such nonholonomic constraints (see Section 4.5).

However, the principle of virtual power as stated in Chapter 2, can be directly used to solve both geometric- and velocity dependent constraints. In this section we show how one can apply the method to problems with geometric or holonomic constraints.

The principle of virtual power can be considered as an extension of D'Alembert's principle[3] or as a separate principle in its own right (see Jourdain, 1909). We will not enter the philosophical debate on this point. Also, there are several variations of the method, such as Kane's equations (Kane, 1961). Jourdain's work of 1909 was a theoretical paper. In 1961, Thomas Kane of Stanford University formulated a variation of the virtual power principle explicitly for rigid body systems. The historian Dugas (1988) suggests that the essence of the idea may go back to Aristotle. The important point here is that for certain classes of problems expressing the constraints in terms of velocities and kinematics may provide an easier way to find projection vectors with which to derive the equations of motion.

As a simple example, we treat the case of a single particle constrained to move along a curve described by the variables (Fig. 4.13). We define a unit vector, $\mathbf{e}_s$, tangent to the path and a normal unit vector, $\mathbf{e}_n$, pointing toward the instantaneous center of curvature. When the curve is known, it is convenient to use path coordinates, i.e., the velocity is expressed as a vector tangent to the curve:

$$\mathbf{v} = \dot{s}\mathbf{e}_s \tag{4.134}$$

where $s(t)$ is the distance along the path. The acceleration is then given by

$$\mathbf{a} = \ddot{s}\mathbf{e}_s + \frac{\dot{s}^2}{\rho}\mathbf{e}_n \tag{4.135}$$

---

[3] Neimark and Fufaev (1972) discuss similar methods by V. Volterra and G. A. Massi published a century ago.

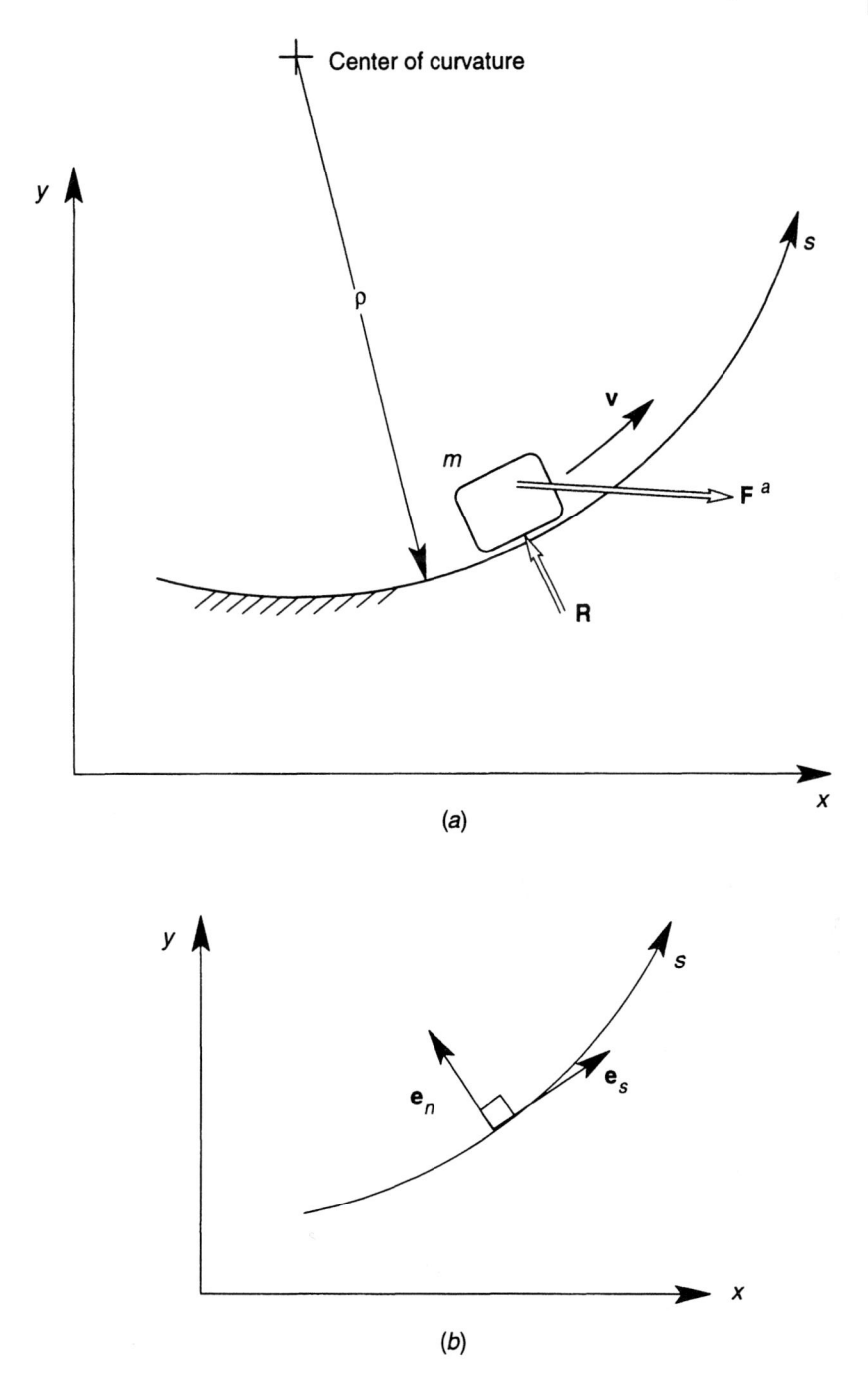

Fig. 4.13 (a) Motion of a constrained mass under an active force $\mathbf{F}^a$, and a workless constraint force $\mathbf{R}_0$. (b) Path coordinates.

If we assume that the path constraint has no friction and $\mathbf{F}^a$ is an applied force, Newton's law for the particle is given by

$$m\ddot{s}\mathbf{e}_s + \frac{m\dot{s}^2}{\rho}\mathbf{e}_n = \mathbf{F}^a + \mathbf{R} \tag{4.136}$$

where we have, by assumption of a frictionless constraint,

$$\mathbf{R} \cdot \mathbf{e}_s = 0 \tag{4.137}$$

Then, taking the projection of the vector Eq. (4.136) onto the direction $\mathbf{e}_s$, we have

$$m\ddot{s} = \mathbf{F}^a \cdot \mathbf{e}_s \tag{4.138}$$

However, we note that the tangent vector $\mathbf{e}_s$ is given by [see (4.134)]

$$\mathbf{e}_s = \frac{\partial \mathbf{v}}{\partial \dot{s}} \tag{4.139}$$

Therefore, we can write the projected equation of motion (4.138) in a form that is vector-basis independent:

$$(m\dot{\mathbf{v}} - \mathbf{F}^a) \cdot \frac{\partial \mathbf{v}}{\partial \dot{s}} = 0 \tag{4.140}$$

This equation expresses the fact that the power of the active force under unit velocity must balance the power of the D'Alembert inertial "force," $-m\mathbf{a}$. Another statement of this principle is that *the power of the reaction force R under frictionless constraints is zero.*

For this example, it is easy to show that Eq. (4.140) is equivalent to D'Alembert's principle of virtual work. When the position vector is constrained by a relation $\mathbf{r} = \mathbf{r}(q(t))$, where q is a generalized coordinate, then the equation of motion takes the form

$$(m\mathbf{a} - \mathbf{F}^a) \cdot \frac{\partial \mathbf{r}}{\partial q} = 0 \tag{4.141}$$

However, in Chapter 2 and from Eq. (4.104) we can easily establish the identity

$$\frac{\partial \mathbf{r}}{\partial q} = \frac{\partial \mathbf{v}}{\partial \dot{q}} \tag{4.142}$$

Thus, in the case of geometric or holonomic constraints, the projection vector can be determined either from the geometry $\mathbf{r}(q, t)$ or from the kinematic relations $\mathbf{v}(\dot{q}, q)$, which are in some cases easier to establish.

Example 4.8. To show how we can derive the equation of motion from the principle of virtual power, consider the motion of a particle on a helical path wrapped around a cylinder with nondimensional pitch κ (Fig. 4.14). In this problem we

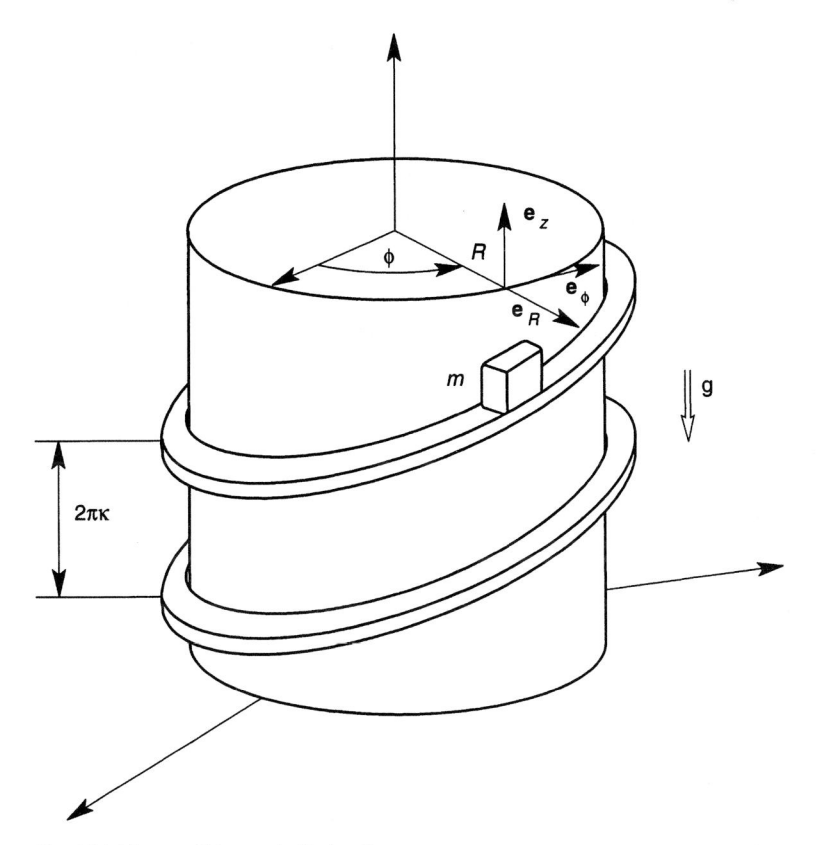

Fig. 4.14 Mass particle on a helical path.

choose cylindrical coordinates (R, ϕ). Here we have one degree of freedom $\phi(t)$, and we choose this as our generalized coordinate. The geometric and kinematic equations are easily established:

$$\mathbf{r} = R\mathbf{e}_R + \kappa R\phi\mathbf{e}_z$$
$$\mathbf{v} = R\dot{\phi}\mathbf{e}_\phi + \kappa R\dot{\phi}\mathbf{e}_z$$
$$\mathbf{a} = -R\dot{\phi}^2\mathbf{e}_R + R\ddot{\phi}\mathbf{e}_\phi + \kappa R\ddot{\phi}\mathbf{e}_z \tag{4.143}$$

We wish to determine the velocity as a function of time under a gravity force,

$$\mathbf{F}^a = -mg\mathbf{e}_z \tag{4.144}$$

To apply this method we use the velocity relation to project the force and acceleration onto the vector:

$$\frac{\partial \mathbf{v}}{\partial \dot{\phi}} = R\mathbf{e}_\phi + \kappa R\mathbf{e}_z \tag{4.145}$$

(*Note*: In this case the projection vector is not a unit vector and it has the dimension of length.) The advantage of this method (4.140) over using $\mathbf{r}(\phi)$ in (4.141) is that $\mathbf{e}_\phi$ is independent of $\dot{\phi}$, but in differentiating the position vector, we must find the dependence of $\mathbf{e}_R$ on the ϕ variable, i.e.,

$$\frac{\partial \mathbf{r}}{\partial \phi} = \kappa R \mathbf{e}_z + R \frac{\partial \mathbf{e}_R}{\partial \phi}, \qquad \frac{\partial \mathbf{e}_R}{\partial \phi} = \mathbf{e}_\phi \tag{4.146}$$

Using the velocity expression $\mathbf{v}(\dot{\phi})$ avoids this extra step. Projecting the gravity force and the acceleration onto the vector $\partial \mathbf{v}/\partial \dot{\phi}$, we obtain

$$mR^2(1+\kappa^2)\ddot{\phi} = -mgR\kappa \tag{4.147}$$

Integrating with respect to time

$$\dot{\phi} = -\frac{g\kappa}{R(1+\kappa^2)}t + \Omega_0 \tag{4.148}$$

The scalar velocity is related to $\dot{\phi}$ by

$$v = \dot{s} = R\dot{\phi}(1+\kappa^2)^{1/2} \tag{4.149}$$

You should check this solution by using the conservation of energy principle (2.46) or (2.49) to find $v(z)$.

Principle of Virtual Power for a System of Particles

Following the pattern in earlier sections, we assume a set of N particles with masses $\{m_i\}$, each with a position vector $\mathbf{r}_i(t)$. We assume that the force on each particle can be separated into three parts; active or external forces $\mathbf{F}_i^a$, internal forces between the ith and jth particles $\mathbf{f}_{ij} = -\mathbf{f}_{ji}$, and workless forces of constraint $\mathbf{R}_i$, i.e.,

$$\mathbf{F}_i = \mathbf{F}_i^a + \sum_{j=1}^{N} \mathbf{f}_{ij} + \mathbf{R}_i, \quad i \neq j \tag{4.150}$$

Then the general statement of the principle of virtual power is similar to the principle of virtual work for dynamic systems (see Jourdain, 1909 or Papastravidis, 1992)

$$\sum_i \left(m_i \dot{\mathbf{v}}_i - \mathbf{F}_i - \sum_j \mathbf{f}_{ij} \right) \cdot \delta \mathbf{v}_i = 0 \tag{4.151}$$

Like virtual displacements, the virtual velocities of each of the particles $\delta \mathbf{v}_i$ represent a set of arbitrary vectors that are compatible with the constraints. In order to solve a holonomic constraint problem, we have to choose a set of independent coordinates $\{q_i(t)\}$, say $i = 1, 2, \ldots, M$, where $M \leqslant 3N$. Associated with each coordinate is a generalized velocity $\dot{q}_i(t)$. To obtain a set of M independent equations of motion we assume

$$\mathbf{v}_i = \mathbf{v}_i(q_i, \dot{q}_i, t) \tag{4.152}$$

In a virtual velocity we assume that the configuration and time are frozen (see, e.g., Jourdain, 1909), i.e.,

$$\delta q_i = \delta t = 0 \tag{4.153}$$

Then the variation of v is just due to small variations in the generalized velocities, i.e.,

$$\delta \mathbf{v}_i = \sum_k \frac{\partial \mathbf{v}_i}{\partial \dot{q}_k} \delta \dot{q}_k \tag{4.154}$$

Substituting this expression into Eq. (4.151) and changing the order of summation, we obtain

$$\sum_{k=1}^{M} \sum_{i=1}^{N} \left(m_i \dot{\mathbf{v}}_i - \mathbf{F}_i^a - \sum_{j=1}^{N} \mathbf{f}_{ij} \right) \cdot \frac{\partial \mathbf{v}_i}{\partial \dot{q}_k} \delta \dot{q}_k = 0 \tag{4.155}$$

This equation has the form

$$\sum b_k x_k = 0 \tag{4.156}$$

If the set $\{x_k\}$ is linearly independent, then the only solution for arbitrary choices of the x_k is $b_k = 0$. Since we have assumed a set of M independent variations $\delta \dot{q}_k$, we can set the term multiplying each variation equal to zero.

$$\sum_{i=1}^{N} \left(m_i \dot{\mathbf{v}}_i - \mathbf{F}_i^a - \sum_{j=1}^{N} \mathbf{f}_{ij} \right) \cdot \frac{\partial \mathbf{v}_i}{\partial \dot{q}_k} = 0 \tag{4.157}$$

There is one scalar equation for each independent generalized velocity ($k = 1, \ldots, M$). This is the form we will use to solve problems. If the constraints are velocity dependent or nonholonomic, the same equations hold, except the choice of independent velocities $\{\dot{q}_k\}$ may not be easily related to a set of generalized coordinates. This is discussed further in Section 4.5.

To further simplify these equations, we make the assumption that the internal forces do not store energy or dissipate power, i.e.,

$$\sum \sum \mathbf{f}_{ij} \cdot \delta \mathbf{v}_i = 0 \tag{4.158}$$

This statement is close to assuming that the collection of particles is either rigidly connected or some particles are free of mutual interaction forces.

Professor Kane of Stanford University calls the projection vectors *partial velocities*. Other authors call them *tangent vectors* (see, e.g., Lesser, 1995). We shall use the latter terminology and define the tangent vectors $\{\boldsymbol{\beta}_{ik}\}$

$$\boldsymbol{\beta}_{ik} = \frac{\partial \mathbf{v}_i}{\partial \dot{q}_k} \tag{4.159}$$

The equations of motion then take the form

$$\sum_{i=1}^{N} (m_i \dot{\mathbf{v}}_i - \mathbf{F}_i^a) \cdot \boldsymbol{\beta}_{ik} = 0, \quad k = 1, 2, \ldots, M \tag{4.160}$$

The relation (4.159) assumes that each velocity can be expressed in a linear sum of the tangent vectors, i.e.,

$$\mathbf{v}_i = \sum_{k=1}^{M} \dot{q}_k \boldsymbol{\beta}_{ik} = \mathbf{b}_i(t) \tag{4.161}$$

The term *tangent vector* is motivated by (4.139) where $\boldsymbol{\beta} = \mathbf{e}_s$ is a vector tangent to the trajectory.

Equations (4.160) are similar to what some practitioners call *Kane's equations*, named after Professor Thomas Kane of Stanford University. Using the notation in this book, we can write (4.160) as,

$$Q_k + Q_k^* = 0 \tag{4.162}$$

where Q_k is called the *generalized active force* and Q_k^* is called the *generalized inertia force* defined by

$$Q_k = \sum_{i=1}^{N} \mathbf{F}_i^a \cdot \boldsymbol{\beta}_{ik} \tag{4.163a}$$

$$Q_k^* = -\sum_{i=1}^{N} m_i \dot{\mathbf{v}}_i \cdot \boldsymbol{\beta}_{ik} \tag{4.163b}$$

In the text of Kane and Levinson (1985), they define a set of *generalized speeds* $\{u_r\}$ as a linear combination of the generalized velocities $\{\dot{q}_s\}$;

$$u_r = \sum_{i=1}^{M} Y_{rs} \dot{q}_s + Z_r \tag{4.164}$$

where Y_{rs} and Z_r may depend on the $\{q_s\}$ and time. The velocities of the particles are given by

$$\mathbf{v}_i = \sum_{i=1}^{M} \mathbf{v}_r^i u_r + \mathbf{v}_t$$

where $\mathbf{v}_r^i$ are called *partial velocities*. Kane's equations for holonomic systems becomes

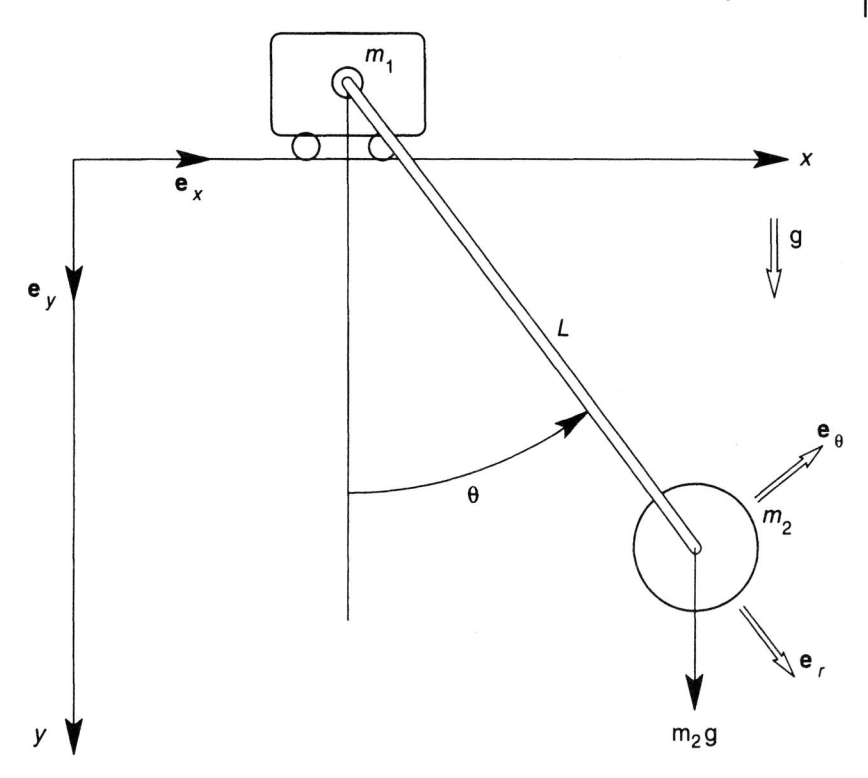

Fig. 4.15 Two-degrees-of-freedom, two-mass pendulum.

$$\mathcal{F}_n + \mathcal{F}_n^* \tag{4.165}$$

where

$$\mathcal{F}_n = \sum_{i=1}^{N} \mathbf{F}_i^a \cdot \mathbf{v}_k^i$$

$$\mathcal{F}_n^* = \sum_{i=1}^{N} m_i \dot{\mathbf{v}}_i \cdot \mathbf{v}_k^i$$

Thus when the generalized speeds $\{u_r\}$ are simply chosen as the set $\{\dot{q}_s\}$, the partial velocities $\mathbf{v}_k^i$ are identical to the Jacobian vector $\boldsymbol{\beta}_{ik}$ in (4.159), derived from the principle of virtual power.

Example 4.9. To illustrate the method, consider the two-particle pendulum under gravity and masses m_1, m_2 constrained to remain on $y = 0$ and a rigid link between m_1 and m_2 of length L (Fig. 4.15). The elements of the method are as follows:

1. Identify the numbers of degrees of freedom: $M = 2$.
2. Choose independent generalized velocities: $\dot{q}_1 = \dot{x}, \dot{q}_2 = \dot{\theta}$,

$$\mathbf{v}_1 = \dot{x}\mathbf{e}_x$$
$$\mathbf{v}_2 = \dot{x}\mathbf{e}_x + L\dot{\theta}\mathbf{e}_\theta \tag{4.166}$$

3. Calculate the tangent vectors or partial velocities

$$\boldsymbol{\beta}_{11} = \frac{\partial \mathbf{v}_1}{\partial \dot{x}} = \mathbf{e}_x, \quad \boldsymbol{\beta}_{12} = 0$$
$$\boldsymbol{\beta}_{21} = \frac{\partial \mathbf{v}_2}{\partial \dot{x}} = \mathbf{e}_x, \quad \boldsymbol{\beta}_{22} = \frac{\partial \mathbf{v}_2}{\partial \dot{\theta}} = L\mathbf{e}_\theta \tag{4.167}$$

4. Calculate generalized active forces (project active forces onto the tangent vector)

$$\mathbf{F}_1^a = 0, \quad \mathbf{F}_2^a = -mg\mathbf{e}_y$$
$$Q_1 = \mathbf{F}_2^a \cdot \boldsymbol{\beta}_{21} = -mg\mathbf{e}_y \cdot \mathbf{e}_x = 0$$
$$Q_2 = \mathbf{F}_2^a \cdot \boldsymbol{\beta}_{22} = -mgL\mathbf{e}_y \cdot \mathbf{e}_\theta = -mgL\sin\theta \tag{4.168}$$

5. Calculate accelerations

$$\dot{\mathbf{v}}_1 = \ddot{x}\mathbf{e}_x$$
$$\dot{\mathbf{v}}_2 = \ddot{x}\mathbf{e}_x + L\ddot{\theta}\mathbf{e}_\theta - L\dot{\theta}^2\mathbf{e}_r \tag{4.169}$$

6. Calculate generalized inertia (project accelerations onto the tangent vectors)

$$\sum m_i \dot{\mathbf{v}}_i \cdot \boldsymbol{\beta}_{i1} = m_1\dot{\mathbf{v}}_1 \cdot \mathbf{e}_x + m_2\dot{\mathbf{v}}_2 \cdot \mathbf{e}_x$$
$$= (m_1 + m_2)\ddot{x} + m_2(L\ddot{\theta}\cos\theta - L\dot{\theta}^2\sin\theta)$$
$$\sum m_i \dot{\mathbf{v}}_i \cdot \boldsymbol{\beta}_{i2} = m_1\dot{\mathbf{v}}_1 \cdot 0 + m_2\dot{\mathbf{v}}_2 \cdot L\mathbf{e}_\theta$$
$$= m_2L(\ddot{x}\cos\theta + L\ddot{\theta}) \tag{4.170}$$

Thus the resulting equations of motion (4.160) become

$$(m_1 + m_2)\ddot{x} + m_2L(\ddot{\theta}\cos\theta - \dot{\theta}^2\sin\theta) = 0$$
$$m_2L\ddot{x}\cos\theta + m_2L^2\ddot{\theta} + m_2gL\sin\theta = 0 \tag{4.171}$$

These equations are nonlinear ordinary differential equations and, as such, an analytic solution cannot be found. One can try to integrate these equations numerically, or one can linearize the equations about an equilibrium point and study the local motions near this point. The equilibrium points are $\theta = 0, \pi$, with $\theta = 0$ obviously a stable configuration. For motions close to $\theta = 0$, we assume $\cos\theta \approx 1$, $\sin\theta \approx \theta$ to obtain

$$\ddot{x} + \mu L\ddot{\theta} = 0$$
$$\ddot{x} + L\ddot{\theta} + g\theta = 0 \tag{4.172}$$

or

$$(1-\mu)L\ddot{\theta} + g\theta = 0 \quad (\mu = m_2/(m_1+m_2)) \tag{4.173}$$

This equation describes an oscillating solution

$$\theta = A\cos(\omega t + \phi_0) \tag{4.174}$$

where

$$\omega^2 = \frac{g(m_1+m_2)}{Lm_1} \tag{4.175}$$

When $m_1/m_2 \gg 1$, the system behaves as a classic pendulum. When $m_1 \ll m_2$, the frequency increases as m_1 decreases.

As an exercise, the student can examine the case where m_1 is attached to a linear elastic spring of constant k.

Example 4.10 Virtual Power Method Using Symbolic Mathematics Software: MATHEMATICA. The development of computer codes to manipulate mathematical symbols and perform algebraic and calculus operations without numbers has been a quiet revolution in computer software. Popular codes on the market today (2008) include *MACSYMA*, *MATHEMATICA*, and *MAPLE*. In this example we illustrate the use of the principle of virtual power in a *MATHEMATICA* format to derive equations of motion for the double pendulum shown in Fig. 4.16. In the following, comment statements are shown. *MATHEMATICA* input statements are shown in bold, and *MATHEMATICA* output statements are unitalicized. Students should consult a *MATHEMATICA* handbook for precise instructions. In this example we assume lossless pinned joints and that the only active force on the two masses is gravity.

Define Position Vectors

r1 = {L1 Sin[q1[t]], L1 Cos[q1[t]]}
r2 = r1 + {L2 Sin[q1[t]+q2[t]], L2 Cos[q1[t] + q2[t]]}

Define Velocity Vectors

v1 = D[r1, t]
{L1 Cos[q1[t]] q1'[t], - (L1 Sin[q1[t]] q1'[t])}
v2 = D[r2, t]
{L1 Cos[q1[t]] q1'[t] + L2 Cos[q1[t] + q2[t]]
    (q1'[t] + q2'[t]), - (L1 Sin[q1[t]] q1'[t]) -
  L2 Sin[q1[t] + q2[t]] (q1'[t] + q2'[t])}

Define Acceleration Vectors

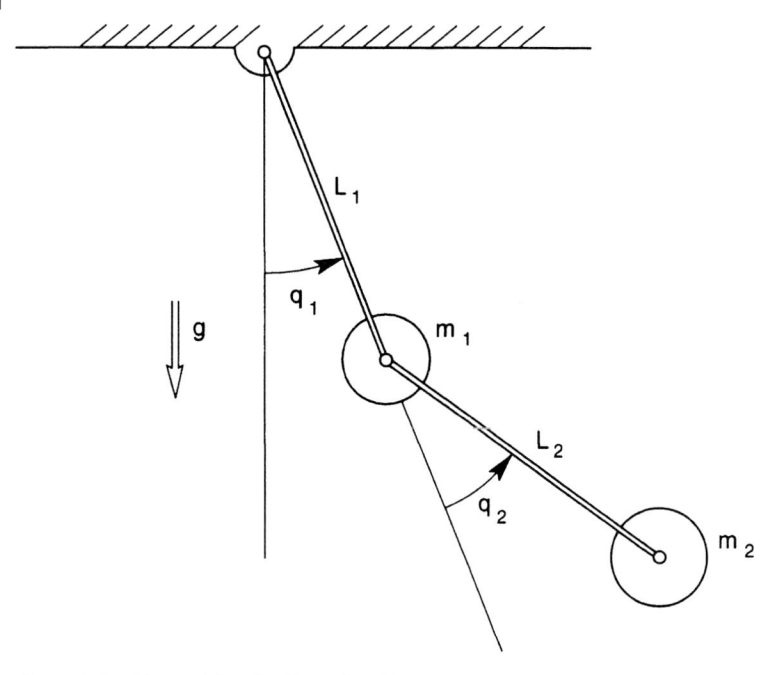

Fig. 4.16 Double pendulum for Example 4.10.

```
a1 = D[v1, t]
{-(L1 Sin[q1[t]] q1'[t]^2) + L1 Cos[q1[t]] q1''[t],
   -(L1 Cos[q1[t]] q1'[t]^2) - L1 Sin[q1[t]] q1''[t]}
a2 = D[v2, t]
{-(L1 Sin[q1[t]] q1'[t]^2) -
   L2 Sin[q1[t] + q2[t]] (q1'[t] + q2'[t])^2 +
   L1 Cos[q1[t]] q1''[t] +
   L2 Cos[q1[t] + q2[t]] (q1''[t] + q2''[t]),
   -(L1 Cos[q1[t]] q1'[t]^2) -
   L2 Cos[q1[t] + q2[t]] (q1'[t] + q2'[t])^2 -
   L1 Sin[q1[t]] q1''[t] -
   L2 Sin[q1[t] + q2[t]] (q''[t] + q2''[t])}
```

Define Active Gravity Forces

```
Fa1 = {0, m1 1/2 g}
Fa2 = {0, m2 1/2 g}
```

Define Tangent Vectors or Jacobian

```
Jcb11 = D[v1, q1'[t]]
{L1 Cos[q1[t]], - (L1 Sin[q1[t]])}
```

4.4 The Method of Virtual Power

```
Jcb21 = D[v2, q1'[t]]
{L1 Cos[q1[t]] + L2 Cos[q1[t] + q2[t]],
    -(L1 Sin[q1[t]]) - L2 Sin[q1[t] + q2[t]]}
Jcb12 = D[v1, q2'[t]]
{0, 0}
Jcb22 = D[v2, q2'[t]]
{L2 Cos[q1[t] + q2[t]], - (L2 Sin[q1[t] + q2[t]])}
```

Virtual Power Equations for Generalized Velocity q1'[t]

```
eqn1 = (Fa1 -m1 a1) . Jcb11 + (Fa2 - m2 a2) . Jcb21;
```

Simplify Expressions by Combining Trig Terms

```
Expand [%, Trig ⟶ True]
-(g L1 m1 Sin[q1[t]]) - g L1 m2 Sin[q1[t]] -
    g L2 m2 Sin[q1[t] + q2[t]] +
    2 L1 L2 m2 Sin[q2[t]] q1'[t] q2'[t] +
    L1 L2 m2 Sin[q2[t]] q2'[t]² - L1² m1 q1''[t] -
    L1² m2 q1''[t] - L2² m2 q1''[t] -
    2 L1 L2 m2 Cos[q2[t]] q1''[t] - L2² m2 q2''[t] -
    L1 L2 m2 Cos[q2[t]] q2''[t]
eqn4 = %
```

Virtual Power Equation for Generalized Velocity q2'[t]

```
eqn2 = (Fa1 - m1 a1) . Jcb12 + (Fa2 - m2 a2) . Jcb22;
Expand [%, Trig ⟶ True]
-(g L2 m2 Sin[q1[t] + q2[t]]) -
    L1 L2 m2 Sin[q2[t]] q1'[t]² - L2² m2 q1''[t] -
    L1 L2 m2 Cos[q2[t]] q1''[t] - L2² m2 q2''[t]
eqn3 = %
```

Evaluate Constants in Equations of Motion

```
eqn4 /. L1 ⟶ 1; L2 ⟶ 1; m1 ⟶ 1; m2 ⟶ 1; g ⟶ 10
-20 Sin[q1[t]] -10 Sin[q1[t] + q2[t]] +
    2 Sin [q2[t]] q1'[t] q2'[t] + Sin[q2[t]] q2'[t]² -
    3 q1''[t] - 2 Cos[q2[t]] q1''[t] - q2''[t] -
    Cos[q2[t]] q2''[t]
eqn5 = %
eqn3 /. L1 ⟶ 1; L2 ⟶ 1; m1 ⟶ 1; m2 ⟶ 1; g ⟶ 10
-10 Sin[q1[t] + q2[t]] - Sin[q2[t]] q1'[t]² - q1''[t]-
    Cos[q2[t]] q1''[t] - q2''[t]
eqn6 = %
```

Integrate the Equations of Motion for Initial Conditions

```
Sol = NDSolve [{eqn5 == 0, eqn6 == 0,
    q1[0] == q2[0] == 0, q1'[0] == 0,
    q2'[0] == 1}, {q1, q2, {t, 10}]
```

Plot Time History From Time $= 0$ to 10 Seconds

```
Plot[Evaluate [q2[t] /. Sol], {t, 0, 10}]
```

The results are shown in Fig. 4.17a, and b is for another set of initial conditions.

4.5
Nonholonomic Constraints: Lagrange Multipliers

In most dynamics problems, one tries to formulate the model using the least number of variables. In this sense the use of constraints is always implicit in formulating a problem, e.g. assuming that the motion is planar or that the motion is confined to a circle. But there are some constraints that require special treatment, especially those that depend on velocities and so-called *nonholonomic constraints* (see, e.g., Neimark and Fufaev, 1972).

Example 4.11. An example of a kinematic constraint is shown in Fig. 4.18 where the velocity of one particle is constrained to be proportional to the velocity of another particle through the use of electrical feedback. In this case the horizontal velocity of mass m_2 is constrained to be proportional to the velocity of mass m_1;

$$\mathbf{v}_2 \cdot \mathbf{e}_x = -\Gamma \dot{x} \tag{4.176}$$

The velocity of mass m_2 is given by

$$\mathbf{v}_2 = \dot{x}\mathbf{e}_x + L\dot{\theta}\mathbf{e}_\theta \tag{4.177}$$

Thus the constraint becomes

$$\dot{x} + L\dot{\theta}\cos\theta + \Gamma\dot{x} = 0 \tag{4.178}$$

or using the notation

$$a_{11}\dot{x} + a_{12}\dot{\theta} = 0 \tag{4.179}$$

where

$$a_{11} = (1 + \Gamma), \qquad a_{12} = L\cos\theta \tag{4.180}$$

We note in this example that the constraint (4.178) can be integrated to obtain a geometric constraint in terms of $\{x, \theta\}$.

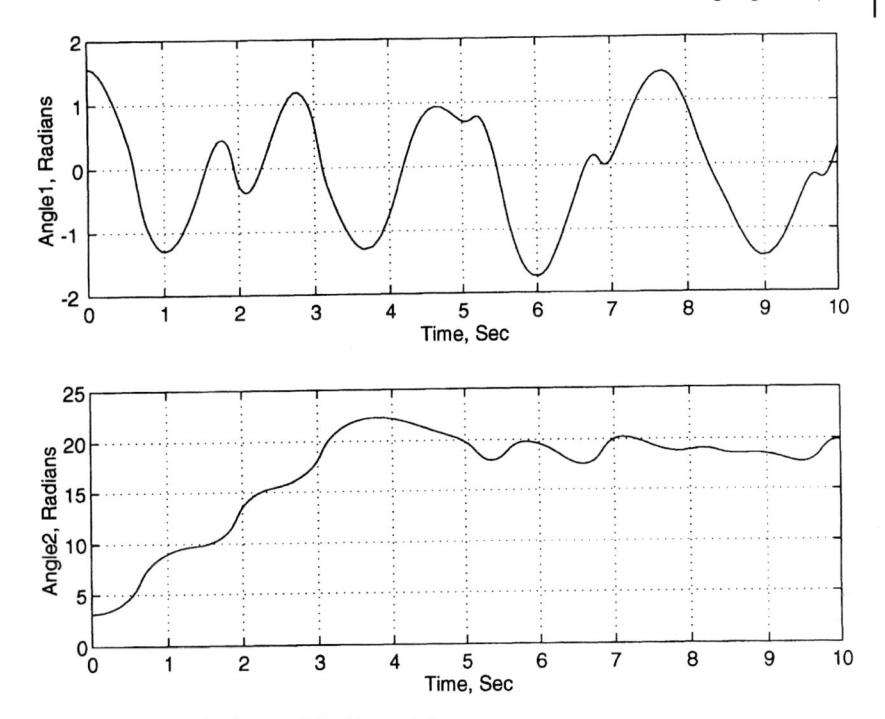

Fig. 4.17 Numerical solution of double pendulum in Example 4.10, Fig. 4.16. *Top:* Time history of $q_1(t)$ from numerical simulation showing oscillatory motion. *Bottom:* Time history of $q_2(t)$ from numerical simulation showing rotary motion. Initial conditions $q_1(0) = \pi/2$, $q_2(0) = \pi$, zero initial velocities.

Nonholonomic Constraints and Lagrange's Equations

Let us assume that the problem has been reduced to N independent degrees of freedom with N generalized coordinates $\{q_k\}$. Also, we assume that we have conservative forces with a potential energy function $V(q_k)$ and kinetic energy function $T(\dot{q}_k, q_k)$ with no dissipation. Now let us introduce $M < N$ constraint equations. These can be of two kinds: first relations between the N coordinates *(holonomic constraints)*,

$$f_i(q_1, q_2, \ldots, q_N, t) = 0 \qquad (4.181)$$

and second constraint equations, linear in the generalized velocities

$$\sum a_{ij}\dot{q}_k + b_i = 0 \qquad (4.182)$$

where the a_{ik} are not related to a set of functions f_i. In other words, we assume that, in terms of the velocities, these new constraints cannot be integrated into a relationship between the coordinates only. Such constraints are called *nonholo-*

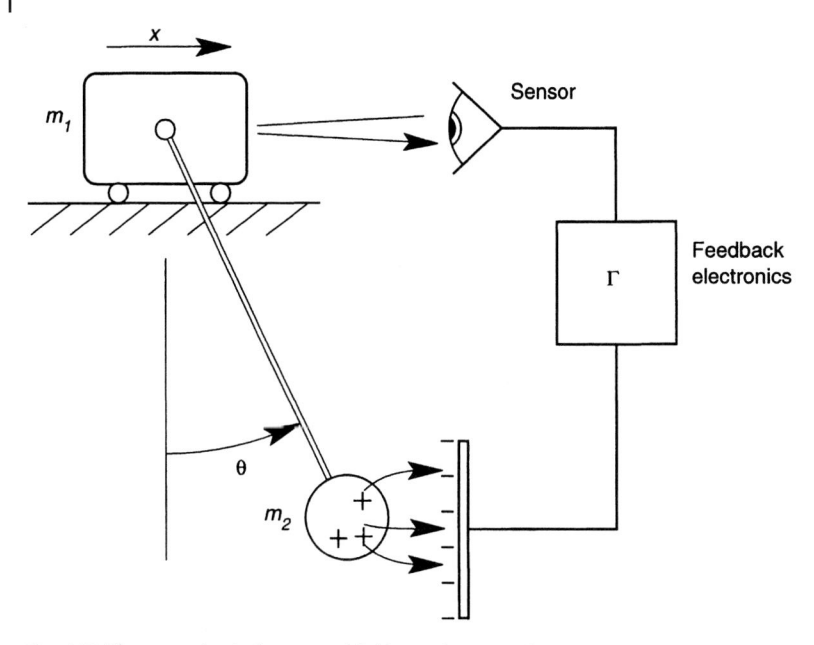

4 Principles of D'Alembert, Lagrange's Equations, and Virtual Power

Fig. 4.18 Electromechanical system with kinematic constraints.

nomic. By differentiating equation (4.181) with respect to time, we also get a linear relationship between the velocities

$$\sum \frac{\partial f_i}{\partial q_k}\dot{q}_k + \frac{\partial f_i}{\partial t} \equiv \sum a_{ij}(q_j, t)\dot{q}_k + b_i(q_j, t) = 0 \quad (4.183)$$

Note, however, that the a_{ik} are all related to the M functions f_i.

Example 4.12. An example of a nonholonomic constraint is rolling of a disc on a plane, where the disc is always vertical (Fig. 4.19). Without rolling, the degrees of freedom involve four variables, x_c, y_c, θ, ϕ.

The rolling constraint is of the form

$$dx_c = Rd\phi \cos\theta$$
$$dy_c = Rd\phi \sin\theta \quad (4.184)$$

In terms of velocities

$$\dot{x}_c - R\dot{\phi}\cos\theta = 0, \qquad \dot{y}_c - R\dot{\phi}\sin\theta = 0 \quad (4.185)$$

In the notation of nonholonomic constraints (4.182), $\sum a_{ik}\dot{q}_k = 0$, we have

$$a_{11} = 1, \quad a_{12} = 0, \quad a_{13} = 0, \quad a_{14} = -R\cos\theta$$
$$a_{21} = 0, \quad a_{22} = 1, \quad a_{23} = 0, \quad a_{24} = -R\sin\theta \quad (4.186)$$

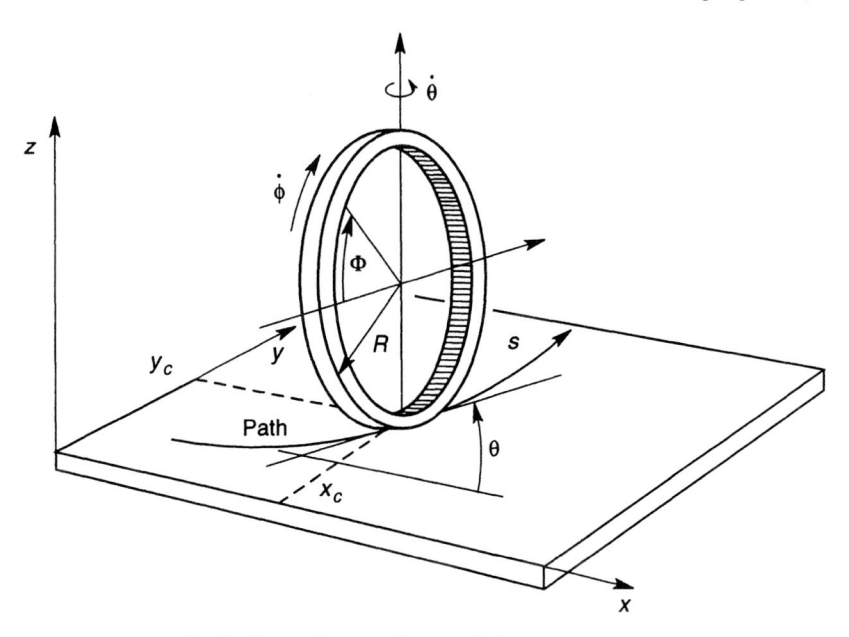

Fig. 4.19 Rolling of a wheel on a planar surface. Nonholonomic constraints.

In order to modify Lagrange's equations with the constraints of the form of (4.182) or (4.183) we must introduce constraint forces Q_k. (Note that we cannot derive a set of Lagrange's equations in the original N variables $\{q_k\}$ because they are not independent under the added constraints.) However, we can think of the N variables $\{q_k\}$ as independent if we enforce the constraints by the generalized forces Q_k. This allows us to write a set of N Lagrange's equations,

$$\frac{d}{dt}\frac{\partial \mathcal{L}}{\partial \dot{q}_k} - \frac{\partial \mathcal{L}}{\partial q_k} = Q_k, \quad k = 1, 2, \ldots, N \tag{4.187}$$

Next we assume that these constraint forces do zero virtual work, i.e.,

$$\delta W = \sum_k Q_k \delta q_k = 0 \tag{4.188}$$

From the constraint equation (4.182) or (4.183), we also have

$$\sum_k a_{\ell k} \delta q_k = 0 \tag{4.189}$$

One can show that (4.188) can be satisfied if we assume that Q_k is a linear combination of the $a_{\ell k}$, i.e.,

$$Q_k = \sum_\ell \lambda_\ell a_{\ell k} \tag{4.190}$$

where the λ_ℓ must be determined. Substituting this expression into (4.188) for δW, we get

$$\delta W = \sum_k \sum_\ell \lambda_\ell a_{\ell k} \delta q_k \qquad (4.191)$$

If the order of summation is changed and the constraint conditions (4.189), are involved, then $\delta W = 0$. (This argument may be found in Goldstein, 1980, pp. 48–49.)

Substituting (4.190) into (4.187), Lagrange's equations take the form

$$\frac{d}{dt}\frac{\partial T}{\partial \dot{q}_k} - \frac{\partial T}{\partial q_k} + \frac{\partial V}{\partial q_k} = \sum_{\ell=1}^{M} \lambda_\ell a_{\ell k}, \quad k = 1, 2, \ldots, N \qquad (4.192)$$

Now, however, both the $\{q_k\}$ and $\{\lambda_\ell\}$ as unknowns, i.e., we have $N+M$ unknowns. To the N Lagrange's equations, we must add the M constraint equations (4.177)

$$\sum a_{ik}\dot{q}_k + b_i(t) = 0 \qquad (4.193)$$

The added variables $\{\lambda_\ell\}$ are called Lagrange multipliers and can be used to calculate the forces of constraint.

It should be noted that for some problems in which the Lagrangian does not depend on some of the coordinates q_k, one can use the relations (4.193) to eliminate some of the variables directly and reduce the degrees of freedom to $N - M$, and avoid the use of the Lagrange multipliers.

Example 4.13 Rolling of a Wide Tread Tire on a Planar Incline. Consider the inclined plane with a wide tread tire on it as shown in Fig. 4.19. Assume that the plane is inclined to gravity by an angle ψ. The potential energy of gravity is given by

$$V = mg(\sin\psi) y_c \qquad (4.194)$$

To calculate the kinetic energy we assume a distribution of particles around the rim of mass $\gamma R\, d\Phi$. The position vector components of this differential mass element is given by

$$\mathbf{r} = \mathbf{r}_c + R\mathbf{e}_R$$

The velocity of a particle at position Φ on the ring circumference is given by

$$\mathbf{v}(\Phi) = \mathbf{v}_c + R\dot{\phi}\mathbf{e}_\phi + R\dot{\theta}\cos\Phi\,\mathbf{e}_\theta$$

Note that $\mathbf{v}_c$ is in the plane of the ring and $\mathbf{e}_\theta$ is normal to this plane. Thus we have

$$v^2(\Phi) = v_c^2 + R^2\dot{\phi}^2 + R^2\dot{\theta}^2\cos^2\Phi + 2v_c R\dot{\phi}\sin\Phi$$

The kinetic energy of the ring can be found by integrating with respect to Φ, holding $\theta, \dot{\phi}, \dot{\theta}$ fixed:

$$T = \frac{1}{2} \int_0^{2\pi} v^2(\Phi) \gamma R \, d\Phi$$

The resulting expression is given by (see also Chapter 5)

$$T = \frac{1}{2} m (\dot{x}_c^2 + \dot{y}_c^2) + \frac{1}{2} I_1 \dot{\phi}^2 + \frac{1}{2} I_2 \dot{\theta}^2 \tag{4.195}$$

where $I_1 = mR^2$ is the moment of inertia about the axel, and $I_2 = mR^2/2$ is the moment of inertia about a diameter. The wide-tread assumption ensures that the plane of the tire remains perpendicular to the inclined plane. Using the constraint equations (4.185), (4.186) in Example 4.12, the equations of motion (4.192) become:

$$m\ddot{x}_c = \lambda_1 a_{11} + \lambda_2 a_{21} = \lambda_1$$
$$m\ddot{y}_c = -mg \sin\psi + \lambda_1 a_{12} + \lambda_2 a_{22} = -mg \sin\psi + \lambda_2$$
$$I_2 \ddot{\theta} = \lambda_1 a_{13} + \lambda_2 a_{23} = 0$$
$$I_1 \ddot{\phi} = \lambda_1 a_{14} + \lambda_2 a_{24} = -\lambda_1 R \cos\theta + -\lambda_2 R \sin\theta$$
$$\dot{x}_c - R\dot{\phi} \cos\theta = 0, \qquad \dot{y}_c - R\dot{\phi} \sin\theta = 0 \tag{4.196}$$

The third equation says that the angular momentum about the axis normal to the plane is conserved, i.e.,

$$I_2 \dot{\theta} = p_\theta = \text{constant} \tag{4.197}$$

or $\theta = a + bt$, where a, b depend on initial conditions.

Next, using the first two equations we eliminate λ_1, λ_2 from the fourth equation (also using the constraint conditions)

$$I_1 \ddot{\phi} = -m\ddot{x}_c R \cos\theta - (m\ddot{y}_c + mg \sin\psi) R \sin\theta \tag{4.198}$$

Note that by taking the time derivative of the constraint equations (4.196), we can show that

$$\cos\theta \ddot{x}_c + \sin\theta \ddot{y}_c = R\ddot{\phi} \tag{4.199}$$

Thus (4.198) becomes

$$(I_1 + R^2 m) \ddot{\phi} = -mg (\sin\psi) r \sin(a + bt) \tag{4.200}$$

This last equation can be integrated directly to get $\dot{\phi}(t)$ and $\phi(t)$ as functions of time.

The Lagrange multiplier method can be compared with the direct Newton–Euler method for the special case of a cylinder rolling down an incline. In Fig. 4.19, set

$\theta = \pi/2$, $\dot\theta = 0$, $\dot x_c = x_c = 0$. One must include a tangential friction force N_2. The Newton–Euler equations (2.31), (2.64) become

$$m\ddot y_c = -mg\sin\psi - N_2$$
$$I_2\ddot\phi = N_2 r$$
$$\dot y_c = R\dot\phi \tag{4.201}$$

This yields the same equation as Lagrange's equation, with $a = \pi/2$ and $b = 0$. Also we see that the Lagrange multiplier $\lambda_2 = -N_2$ represents a constraint force.

4.6
Variational Principles in Dynamics: Hamilton's Principle

The modern student of dynamics is usually taught the Newtonian force vector approach in introductory dynamics. However, historically dynamics principles were approached from the concepts of virtual work, energy, and virtual power. This approach goes back to Greek science in the fourth century B.C. (see Dugas, 1988). These concepts evolved into variational or optimization mathematics. Such problems involve minimizing drag, time, thrust forces, or cost under a set of constraints. In fact, Lagrange's equations can be derived from a minimization or extremum problem called *Hamilton's principle*. In words this principle states:

> *The motion of a conservative system under a set of geometric (holonomic) constraints is such that the integral of the Lagrangian $\mathcal{L} = T - V$ between any two times t_1, t_2 is an extremum.*

In mathematical terms, this principle is written in the form

$$\delta\int_{t_1}^{t_2}\mathcal{L}(q_i(t),\dot q_i(t),t)\,dt = 0 \tag{4.202}$$

The symbol $\delta[\ ]$ means that the generalized coordinates are varied about the correct or natural time history $q_i^*(t)$, i.e.,

$$q_i(t) = q_i^*(t) + \alpha\eta_i(t) \tag{4.203}$$

Then Hamilton's principle takes the form

$$\frac{d}{d\alpha}\int_{t_1}^{t_2}\mathcal{L}(t;\alpha)\,dt = 0 \tag{4.204}$$

As outlined below, this so-called extremum problem leads to the differential equation form of Lagrange's equation:

$$\frac{d}{dt}\frac{\partial L}{\partial\dot q} - \frac{\partial L}{\partial q} = 0 \tag{4.205}$$

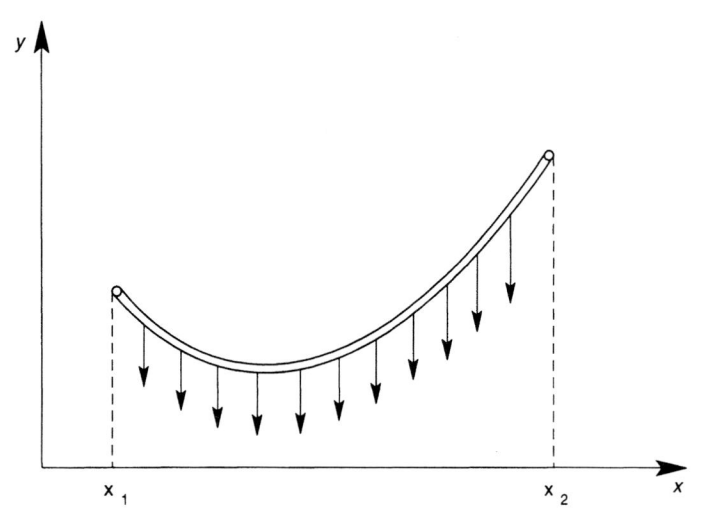

Fig. 4.20 Minimum potential energy problem of the hanging chain or catenary.

under a suitable set of arbitrary test functions $\eta_i(t)$ and for $\alpha \longrightarrow 0$ (see, e.g., Goldstein, 1980; Goldstein et al., 2002).

The origins of variational methods predate Newton's *Principia* (1687). For example, Simon Stevin (1548–1620) analyzed the pulley using the idea of virtual power. Galileo (1564–1642) used the concept of virtual work to solve the problem of the static inclined plane, and extended this principle to a system of connected bodies under gravity. He deduced that the common center of gravity was as near as possible to the center of the earth. This is essentially the principle of minimum potential energy, which follows from the idea of virtual work. Jean (or Johann) Bernoulli[4] (1667–1748) in 1717 called this the principle of *vitesse virtuelle* or virtual velocities. Later he and his brother, James (Jacob or Jacques) (1654–1705) posited the variational mechanics problem of a hanging chain or catenary (Fig. 4.20). They reasoned that the chain would take the configuration for which the center of gravity would be lowest. Mathematically they sought to minimize the potential energy

$$V = \int_{x_1}^{x_2} \rho g y(x)\, ds \tag{4.206}$$

The solution to this problem is the *catenary*:

$$y = A \cosh kx + B \tag{4.207}$$

Around the time of Newton, Jean Bernoulli, a student of James, posited the famous brachistochrone problem in 1696. He sought to find a curve $y(x)$ along which a

---

[4] The Bernoulli family produced three generations of Swiss mathematicians and scientists in the Seventeenth and Eighteenth centuries.

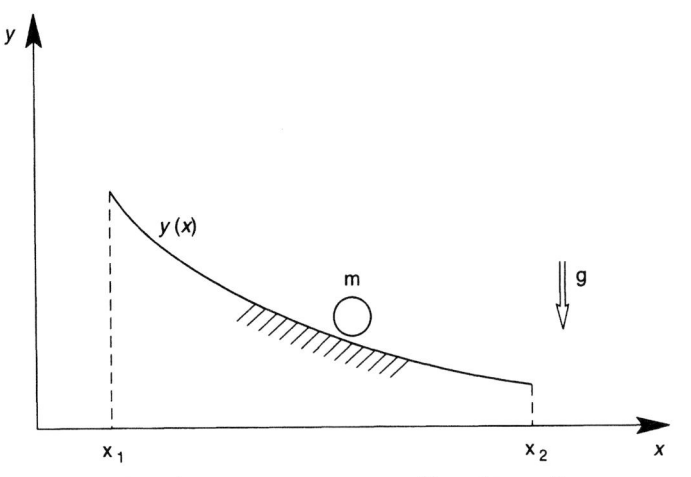

Fig. 4.21 Brachistochrone or minimum-time problem of Bernoulli.

particle will travel from x_1 to x_2 under gravity in the shortest time (Fig. 4.21). Thus he sought to minimize the integral:

$$T = \int_{x_1}^{x_2} \frac{ds}{v} = \int_{1}^{2} \frac{ds}{[2g(y-a)]^{1/2}} \tag{4.208}$$

This problem is similar to one in optics posed by Pierre de Fermat (1601–1665) in 1662. He sought the path of least time that a light ray would travel between two points in an optical medium.

The solution to the brachistochrone problem is the *cycloid* curve.

Finally Euler in 1744 and later Lagrange in 1762–1770 presented an analytical method to solve more general extremum problems. They sought to find solutions $y(x)$ that would extremize integrals of the form

$$I = \int_{x_1}^{x_2} f\left(y(x), \frac{dy}{dx}\right) dx \tag{4.209}$$

The result is the famous Euler–Lagrange equation: i.e., $f(y(x), y'(x))$ must satisfy

$$\frac{d}{dx}\frac{\partial f}{\partial y'} - \frac{\partial f}{\partial y} = 0 \tag{4.210}$$

The Euler–Lagrange equation is derived by varying the parameter functions of the integrand of the integral

$$\delta \int_{x_1}^{x_2} f(y(x), y'(x), x) \, dx = 0 \tag{4.211}$$

This is accomplished by varying path function $y(x)$, namely adding an arbitrary function, $\eta(x)$, to the true function designated by $y^*(x)$:

$$y(x) = y^*(x) + \alpha\eta(x) \tag{4.212}$$

such that

$$\eta(x_1) = \eta(x_2) = 0$$

Here $y^*(x)$ is the function that either minimizes or maximizes the integral. The perturbation function $\eta(x)$ vanishes at the end points and is arbitrary. The variation of the integral is then equivalent to setting

$$\int_{x_1}^{x_2} \frac{\partial}{\partial\alpha} f(\ldots) \, dx = 0 \tag{4.213}$$

The integrand expanded becomes

$$\frac{\partial f}{\partial y}\eta(x) + \frac{\partial}{\partial y'}\eta'(x)$$

The second term can be rewritten by performing integration by parts:

$$\int_{x_1}^{x_2} u(x)\frac{dv}{dx}\, dx = u(x)v(x)\Big|_{x_1}^{x_2} - \int_{x_1}^{x_2} v(x)\frac{du}{dx}\, dx$$

or

$$\int_{x_1}^{x_2} \frac{\partial f}{\partial y'}\eta'(x)\, dx = \frac{\partial f}{\partial y'}\eta(x)\Big|_{x_1}^{x} - \int_{x_1}^{x_2} \eta(x)\frac{d}{dx}\frac{\partial f}{\partial y'}\, dx \tag{4.214}$$

The first term on the right-hand side is zero because by assumption $\eta(x) = 0$ at the end points. Thus the variation of the integral leads to the condition for the max or min as

$$\int \left[\frac{\partial f}{\partial y} - \frac{d}{dx}\frac{\partial f}{\partial y'}\right]\eta(x)\, dx = 0 \tag{4.215}$$

The mathematical logic follows that since $\eta(x)$ is an arbitrary function, then the integrand must be zero and

$$\frac{\partial f}{\partial y} - \frac{d}{dx}\frac{\partial f}{\partial y'} = 0 \tag{4.216}$$

As an example of the application of this equation to the hanging chain problem, if $f = V(x)$ is the potential energy of all the particles in a chain, then the shape or deflection of the chain that minimizes the potential energy must satisfy the equation

$$\frac{d}{dx}\frac{\partial V}{\partial y'} - \frac{\partial V}{\partial y} = 0 \tag{4.217}$$

It was Hamilton who saw the connection between Lagrange's equations of motion and the Euler–Lagrange equation. He defined the Lagrangian $\mathcal{L}[q(t), \dot{q}(t)]$ and used the extremum of the integral of $\mathcal{L}$ and the Euler–Lagrange condition to arrive at Lagrange's equation i.e., let $f = \mathcal{L}$, $y = q$, $y' = \dot{q}$, $x \longrightarrow t$ in Eq. (4.210) [compare with (4.92)]:

$$\frac{d}{dt}\frac{\partial \mathcal{L}}{\partial \dot{q}} - \frac{\partial \mathcal{L}}{\partial q} = 0 \qquad (4.218)$$

Another variational principle that embodies Newton's laws of motion is the *principle of least action* proposed by Pierre-Louis de Maupertuis in 1744. Inspired by Fermat's principle of least time in optics, he reasoned by analogy that during a motion of a particle under given forces, nature would choose the path that would minimize the "action" integral

$$A = \int_{x_1}^{x_2} mv\,ds \qquad (4.219)$$

> *The general principle..., is that the quantity of action necessary to produce some change in Nature is the smallest that is possible.*

(See Dugas, 1988, pp. 260–269.) Maupertuis applied this principle to solve the problem of the impact of two bodies. However, the variation in the principle of least action is different from that in Hamilton's principle. A theoretical discussion of these variational principles can be found in Goldstein (1980), Goldstein et al. (2002), and Meirovitch (1970). A historical discussion may be found in Szabó (1987).

Homework Problems

4.1 Consider the motion of a particle constrained to follow a sinusoidal track under gravity as shown in Fig. P4.1. The constraint is given by the equation $x = A\sin(2\pi y/L)$. Use Lagrange's equation for one degree of freedom to find the equation of motion using $y(t)$ as the generalized coordinate.

4.2 A MEMS device is modeled as a cantilevered beam with a concentrated tip mass m and a spring constant k characterizing the resistance to lateral deformation of the tip $x(t)$ (Fig. P4.2). Suppose there is a charge $+Q$ on the end of the beam and a charge $-Q$ on a nearby stationary node whose undeformed beam distance is x_0. Use the potential energy representation for the force between the two charges along with Lagrange's equation to derive an equation of motion for the beam tip motion, $x(t)$, neglecting gravity. Show that the effective electric force is given by

$$F_e = -\frac{\partial V_e}{\partial x}$$

where

$$V_e = -\frac{Q^2}{4\pi\varepsilon_0(x_0 - x)}$$

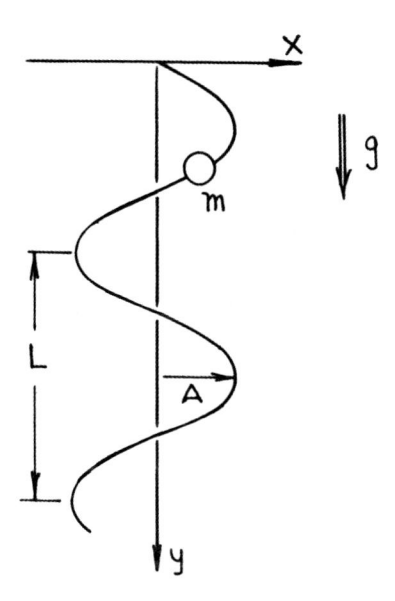

Fig. P4.1

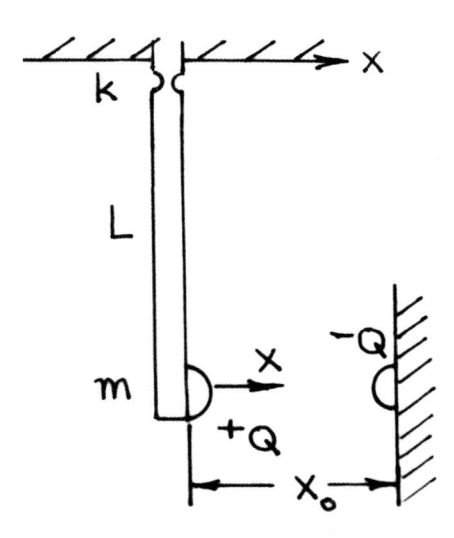

Fig. P4.2

4.3 (*Vibration Absorber*) In Example 4.7, let the base motion be equal to $A \cos \Omega t$. Find the value of L, m_2 such that the steady-state motion of m_1 is identically zero.

4.4 The massless rod in Fig. P4.4 has two masses on it, one mass m_1 is fixed at the end, while the other m_2, is constrained to move along the radius by a linear spring k. Use Lagrange's equations to find the equations of motion for a constant torque. Assume the spring is unstretched when $r = L/2$.

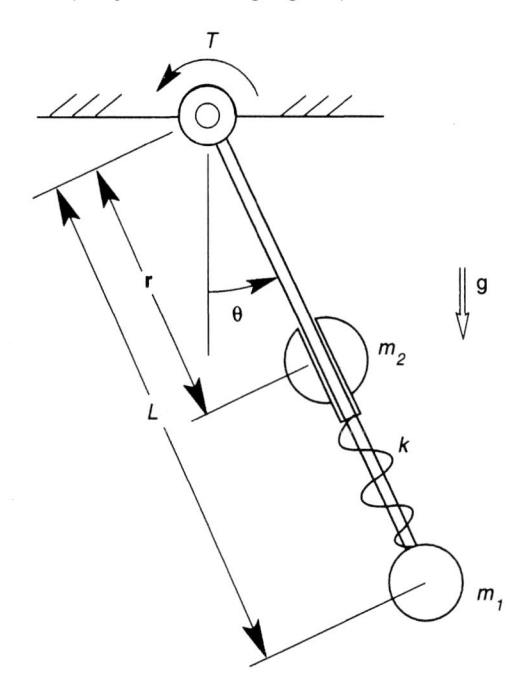

Fig. P4.4

[Answer:

$$m_2(\ddot{r} - r\dot{\theta}^2) + k(r - L/2) - m_2 g \cos\theta = 0$$

$$m_1 L^2 \ddot{\theta} + m_2(2r\dot{r}\dot{\theta} + r^2\ddot{\theta}) + (m_1 L + m_2 r)g \sin\theta = T]$$

4.5 Two masses are constrained by the three-bar linkages shown in Fig. P4.5. Mass m_1 is constrained to move in the vertical direction, while mass m_2 moves in a circular motion. Use θ as a generalized coordinate and find the Lagrangian. Show that the equation of motion is given by

$$\ddot{\theta} L^2 (m_2 + 4m_1 \sin^2\theta) + \dot{\theta}^2 4L^2 m_1 \sin\theta \cos\theta - (2m_1 + m_2)gL \sin\theta = 0$$

4.6 (a) In Problem 4.5, derive the equation of motion using the principle of virtual power or D'Alembert's method of virtual work. (*Hint:* show that the projection vectors are given by $\boldsymbol{\beta}_1 = -2L \sin\theta \, \mathbf{e}_z$, $\boldsymbol{\beta}_2 = L\mathbf{e}_\theta$.)
(b) Reverse the direction of gravity and show that the frequency of small vibration is given by $[(1 + 2m_1/m_2)g/L]^{1/2}$.

4.7 A particle m_1 is constrained to move on a conical surface shown in Fig. P4.7, while a second mass m_2 is constrained to the vertical direction. The two masses are connected by an inextensible string of length L. Choose r, θ as generalized coordinates and find the equations of motion using Lagrange's equations.

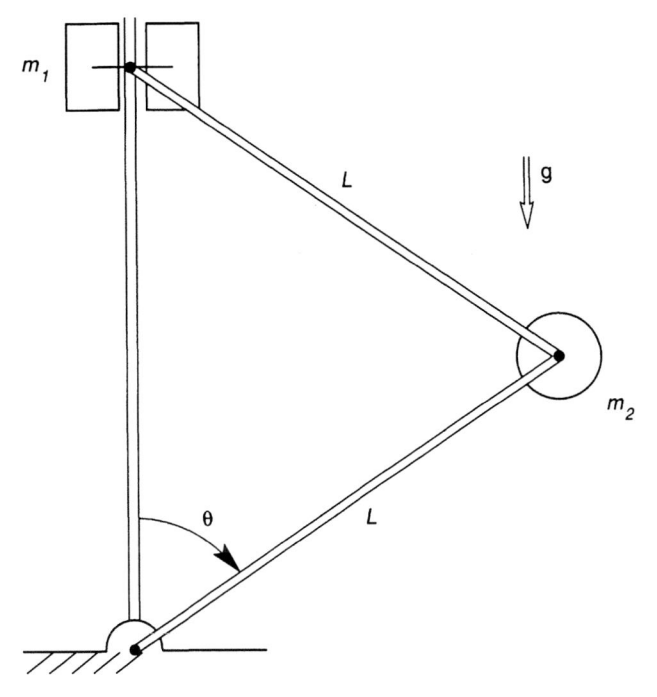

Fig. P4.5

[Answer:

$$(m_1 + m_2)\ddot{r} - m_1 r\dot{\theta}^2 \sin^2\alpha + m_1 g \cos\alpha + m_2 g = 0$$
$$r\ddot{\theta} + 2\dot{r}\dot{\theta} = 0]$$

Use a constant of the motion to reduce the problem to a single differential equation for $r(t)$.

4.8 A particle m is constrained to move on a frictionless circular path. A linear spring force acts on the particle proportional to $k(d - d_0)$ and always in the horizontal direction. Choose $\dot{\theta}$ as a generalized velocity and find the projection vector $\boldsymbol{\beta}$. Use the principle of virtual power to derive the equation of motion for $\theta(t)$.

[Answer: $mR^2\ddot{\theta} + kR^2(1 - \cos\theta)\sin\theta = 0$]

4.9 A block shown in Fig. P4.9 is suspended by four linear springs of stiffness k each. Assume that the unstretched state is when $y = 0$. Derive the equation of motion for vertical motion. Show that the equilibrium position under gravity is approximately given by $y_e^3 = mgL^2/2k$.

4.10 Two particles are constrained to motion in the plane by two inextensible cables of lengths R, $L = \rho + r$ (Fig. P4.10). Use the principle of virtual power to derive the equations of motion with θ_1, θ_2 as generalized coordinates.

4.11 Four equal masses carry positive and negative charges, as shown in Fig.

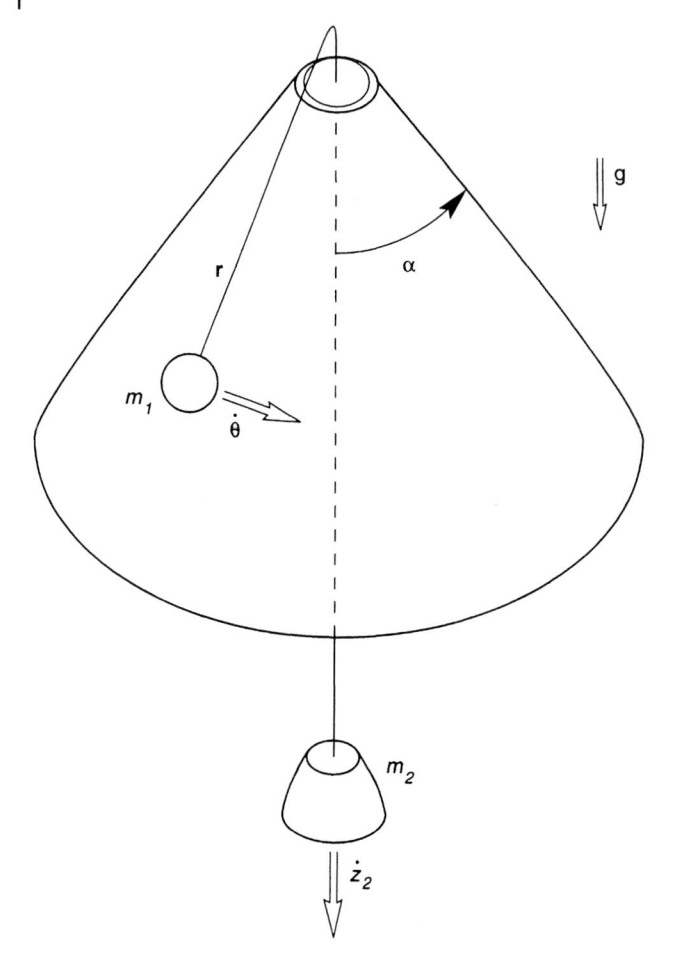

Fig. P4.7

P4.11. The electric force between any two pairs of charges is given by

$$F_{12} = \gamma Q_1 Q_2 / r_{12}^2$$

with the rule that like charges repel and unlike charges attract one another. Assume that the two middle charges can move in the horizontal direction.
(a) Find the equations of motion.
(b) What is the potential energy for this problem?
(c) Derive the equations of motion using Lagrange's equations and compare with Newton's force method.

4.12 A massless rod rolls on a half cylinder with masses m_1, m_2 at the ends (Fig. P4.12). Assume that when $\theta = 0$ the lengths are $x_1 = x_2 = L$.
(a) Use the method of virtual power to derive the equation of motion.
(b) What is the equilibrium position?

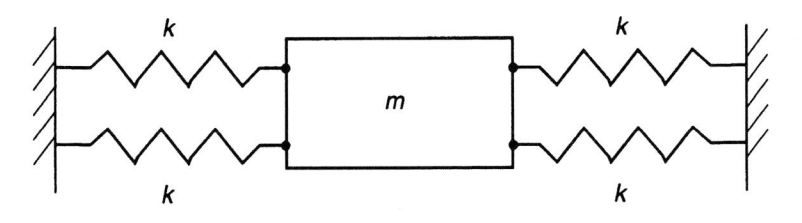

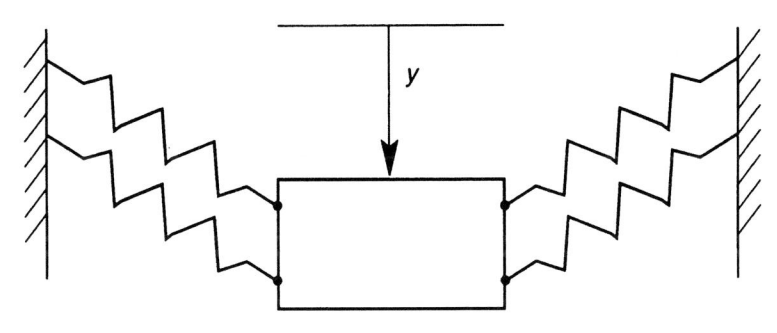

Fig. P4.9

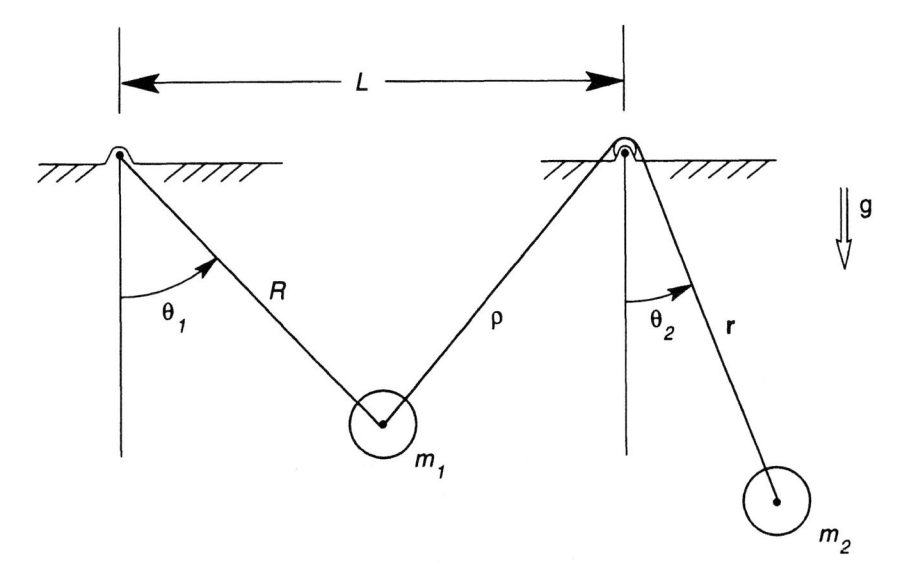

Fig. P4.10

(c) For the case of equal masses $m_1 = m_2$ find the natural frequency for small motion.

(*Hint*: Show that the rolling constraint yields $x_1 = L + a\theta$, $x_2 = L - a\theta$.)

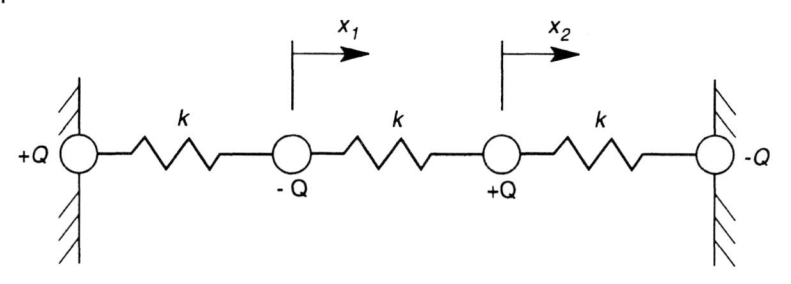

Fig. P4.11

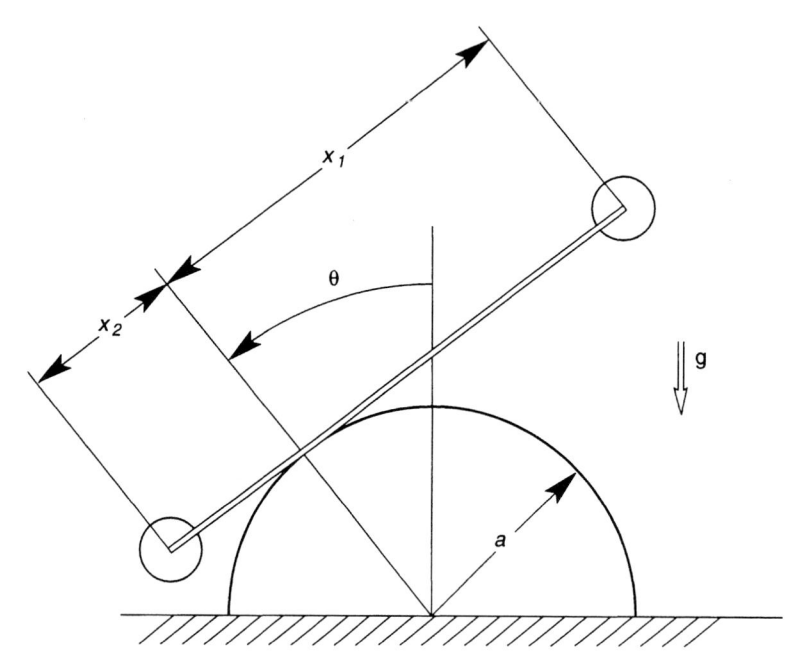

Fig. P4.12

[Answer: (a) $m_1(L + a\theta)\{(L + a\theta)\ddot{\theta} + a\dot{\theta}^2\}$
$+ m_2(L - a\theta)\{(L - a\theta)\ddot{\theta} - a\dot{\theta}^2\}$
$+ m_1 g \cos\theta(L + a\theta) - m_2 g \cos\theta(L - a\theta) = 0$
(c) $\omega^2 = ga/L^2$]

4.13 A massless cylinder of radius R rolls on a horizontal surface (Fig. P4.13). A concentrated mass acts at a radius $\rho < R$.
(a) Use the method of virtual power to find the equation of motion.
(b) Show that the natural frequency for small angular motions is given by $[g\rho/(R - \rho)^2]^{1/2}$.

4.14 A particle of mass m is constrained to move on a circular track, as shown in Fig. P4.14. A linear spring, one end secured to the mass is anchored at a

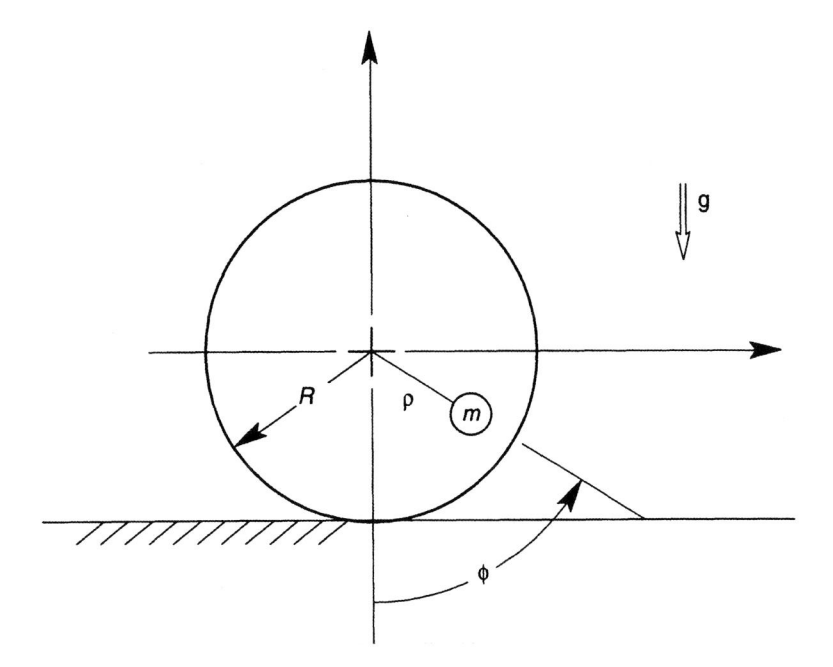

Fig. P4.13

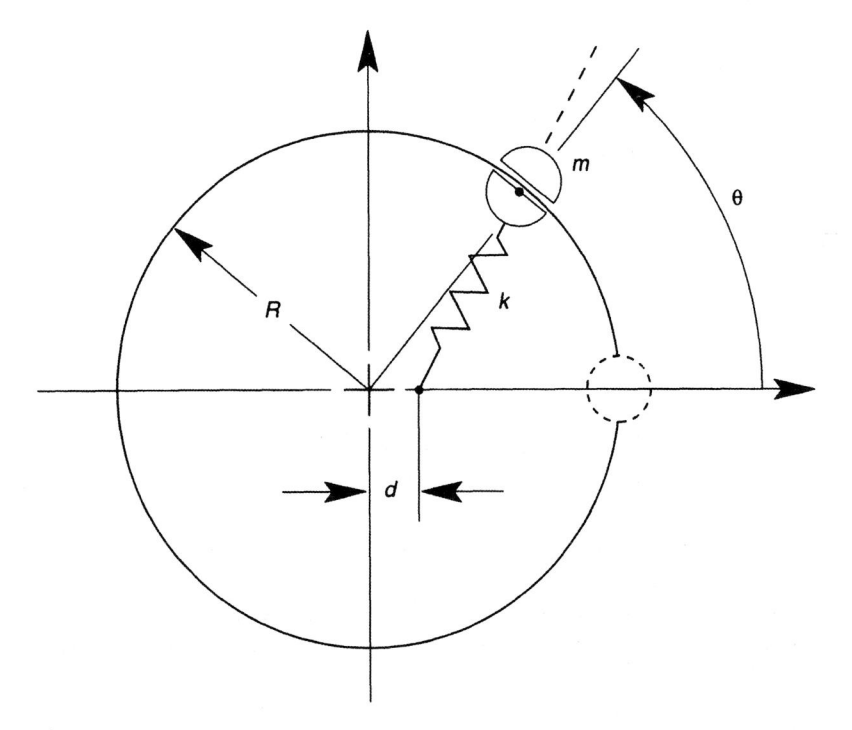

Fig. P4.14

![Fig. P4.15](fig-p4.15)

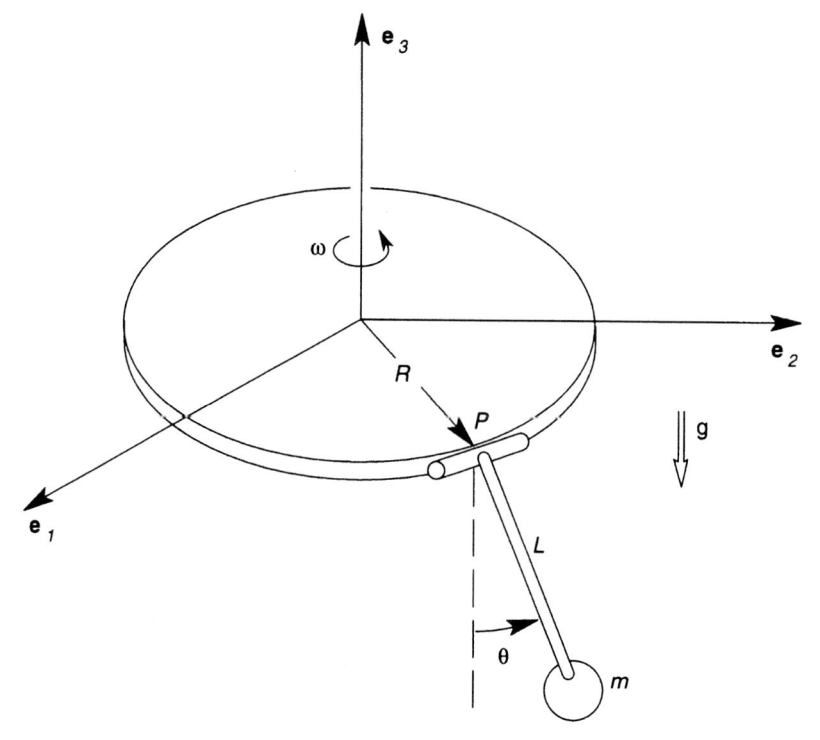

Fig. P4.15

radial offset d and exerts a prestress F_0 when $\theta = 0$. Use the principle of virtual work (D'Alembert's method) to find the equation of motion (neglect friction and gravity). Find the natural frequency of oscillation when θ is small.

4.15 A pendulum axis is attached to a spinning disc, as shown in Fig. P4.15. The disc rotates about the vertical $\mathbf{e}_3$ axis with constant speed ω. The pendulum is constrained to rotate in the plane formed by the $\mathbf{e}_3$ axis and the attachment point P. Use the angle θ as a generalized coordinate and derive the equation of motion using Lagrange's equation.

$$\left[\text{Answer:}\quad \ddot{\theta} + \frac{g}{L}\sin\theta = \omega^2\left(\frac{R}{L} + \sin\theta\right)\cos\theta\right]$$

4.16 Derive the equation of motion for the rotating pendulum in Problem 4.15 using the principle of virtual power. For very large ω^2 compare this problem with Example 5.10.

4.17 Two masses m_1, m_2 rotate about the vertical axis and are attached to the same inextensible string (Fig. P4.17). Let $\omega_1 = \dot{\theta}_1$, $\omega_2 = \dot{\theta}_2$ and assume that both ω_1, ω_2 are perturbed from an initially large angular rotation Ω_0, i.e., $\dot{\phi} = \Omega_0$,

$$\theta_1 = \phi + \beta_1, \qquad \theta_2 = \phi + \beta_2$$

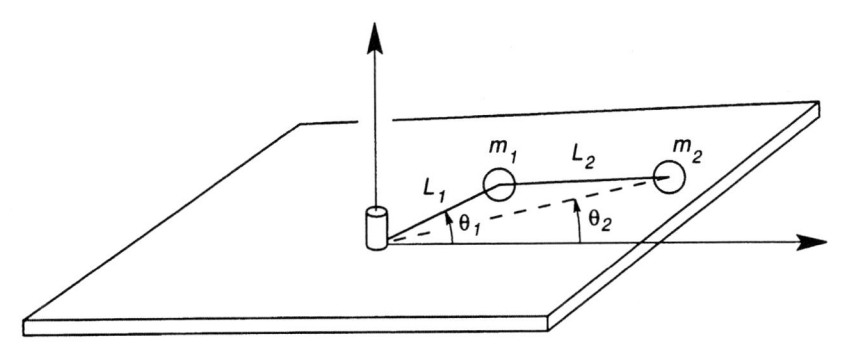

Fig. P4.17

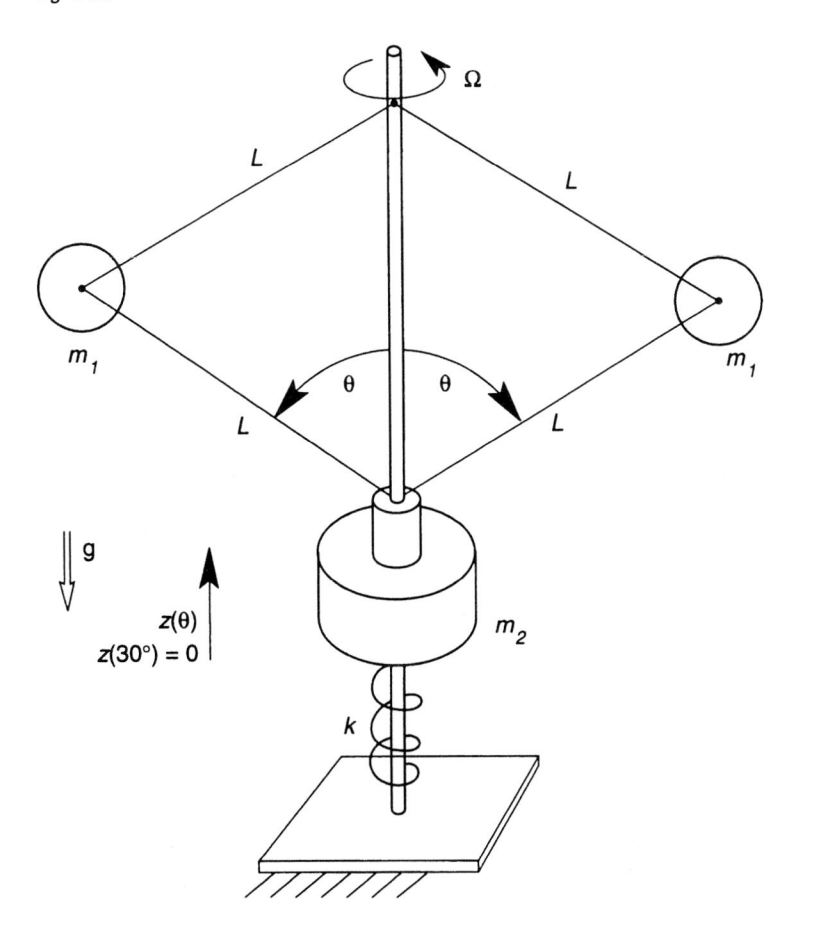

Fig. P4.18

where both β_1, β_2 are small. Derive the equations of motion in the horizontal plane using Lagrange's equations. Linearize and find the natural frequency. Compare with the double pendulum problem, Fig. 4.16, Example 4.10.

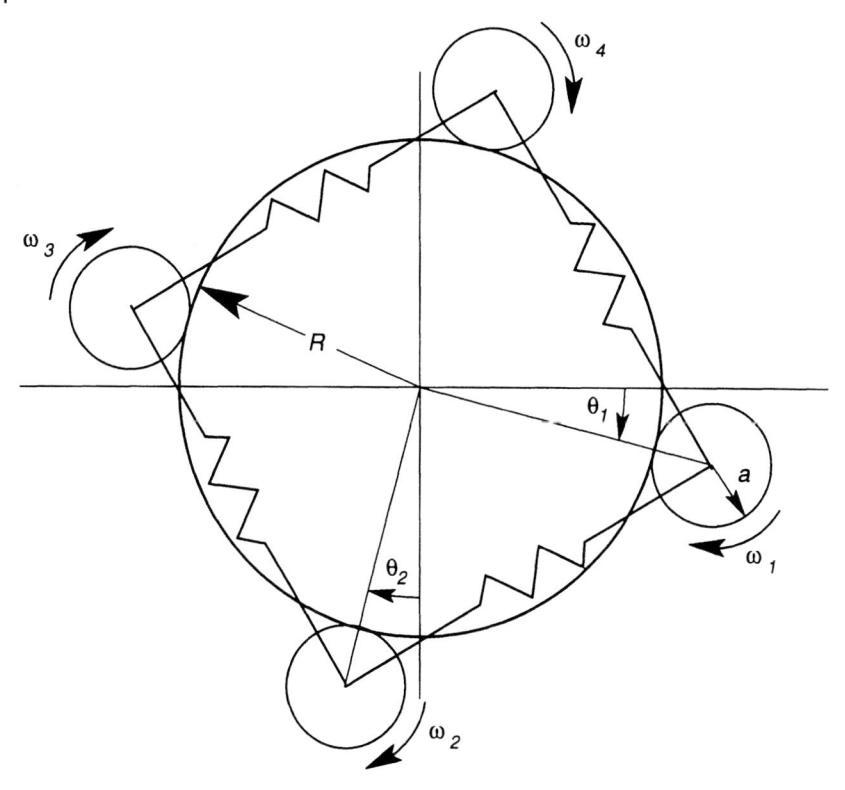

Fig. P4.19

4.18 Consider the three mass mechanism shown in Fig. P4.18. Suppose some applied torque rotates the masses about the vertical axis with constant speed Ω. Use the principle of virtual power to find the equation of motion.

4.19 Four identical planetary gears roll without slip on a sun gear of radius R (Fig. P4.19). Four springs of equal length $L_0 = \sqrt{2}R$ are connected between the gears.
   (a) Neglect the rotary inertia of the planet gears (see Chapter 5 for the correction terms) and derive the equations of motion using Lagrange's equations. Neglect gravity. Assume that the perturbed angles $\{\theta_i\}$ are small.
   (b) What are the modes of vibration and natural frequencies? (*Hint:* Use the relation for the new spring length, $L = \sqrt{2}R(1 + \sin(\theta_i - \theta_{i-1})^{1/2})$.)

4.20 A two-link mechanism has a pin at the center and four equal masses at the ends, as shown in Fig. P4.20. This problem is similar to one in Greenwood (1988) who asks the student to derive the equation of motion using Lagrange's equations. In this problem derive the equation of motion using D'Alembert's method of virtual work. (*Hint:* Show that the projection vector for the two top masses is $\boldsymbol{\beta} = (L\cos\theta \mathbf{e}_x - 2L\sin\theta \mathbf{e}_y)/2$.)

[Answer: $\ddot{\theta}(1 + \sin^2\theta) + \dot{\theta}^2 \sin\theta \cos\theta = 2(g/L)\sin\theta$]

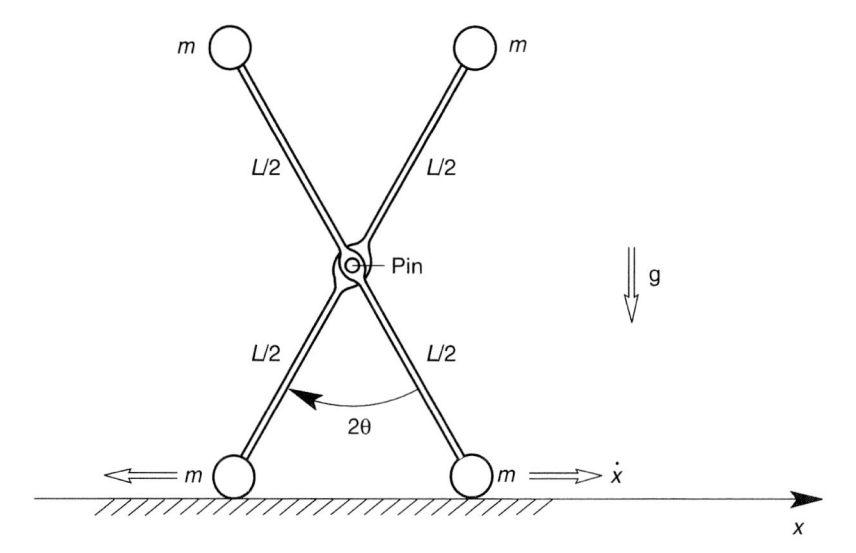

Fig. P4.20

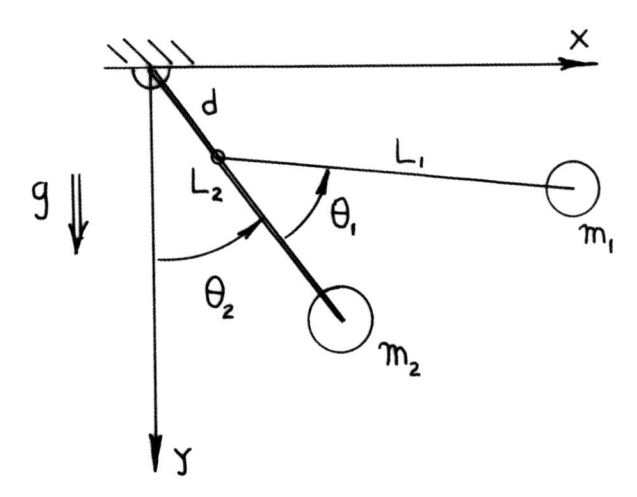

Fig. P4.21

4.21 (*Double Pendulum*) The double pendulum is a classic paradigm of nonlinear dynamics and also chaos theory as well as a model for a planar revolute robot manipulator arm. The classic problem with concentrated masses at the ends of each link is discussed in Chapters 1, 2. Shown in Fig. P4.21 is a variation of the double pendulum in which the revolute joint of the lower link is between the second mass and the grounded revolute joint. Use Lagrange's equations to obtain the equations of motion.

4.22 (*Double Pendulum: MATLAB*) In Problem 4.21 write the second-order equations of motion as a set of first-order differential equations using the angles and their time derivatives as state variables and find numerical solutions us-

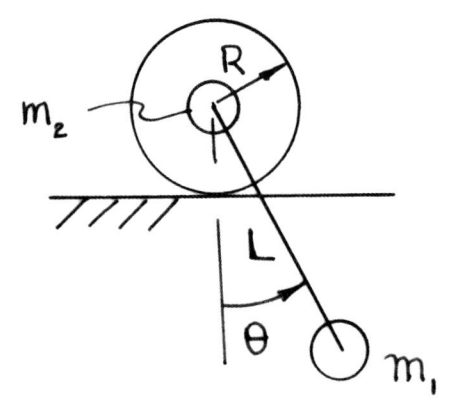

Fig. P4.24

ing a integration code such as MATLAB. First find solutions for small amplitudes and then explore solutions for large initial angles. Use values of $m_1 = 1$, $m_2 = 1$, $L_1 = L_2 = 1$, $d = 0.5$.

4.23 (*Double Pendulum: Linear Modes*) In the equations of motion in Problem 4.21 that were derived using Lagrange's equations, identify the linear and nonlinear terms. Linearize the equations of motion and write in the form of 2×2 matrices, and find the eigenfrequencies and eigenmodes. Use the values of mass and link geometry in Problem 4.22. Compare the eigenmodes with the classical case when $d = L_2$.

4.24 (*Rolling Pendulum*) In the rolling pendulum shown in Fig. P4.24 the fixed revolute joint of the classical pendulum is replaced with a rolling circular arc. Use the virtual power or Kane's method to derive an equation of motion for the rolling pendulum. Linearize the equation of motion about the lower equilibrium point and find the natural frequency. Compare this frequency with that of the classical pendulum as discussed in Chapter 2.

4.25 (*Rolling Pendulum: MATLAB*) Write the equation of motion derived in Problem 4.24 for the rolling pendulum as a set of two first-order differential equations and integrate with MATLAB. First choose initial condition and observe small oscillations as predicted in Problem 4.24. Imagine the circular arc in Fig. P4.24 as a closed curve and choose initial conditions to achieve rollover. Can you use this idea to design a safe but exciting amusement ride?

4.26 (*Electric Pendulum*) Consider the cantilevered MEMS rod in Fig. P4.26 with a large mass on the free end. Imagine that a charge Q_1 is placed on this mass as well as charges of Q_2 on nearby fixed nodes. Assume that the MEMS rod is notched at the base so that the rod behaves as a hinged rigid rod. Use the potential energy for the electric charges (4.88b) in Lagrange's equations and derive an equation of motion for small motions $U(t)$, and $Q_2 = -Q_1$. Find the linearized natural frequency.

4.27 (*Hooke's Joint: Spherical Pendulum*) In Fig. P4.27 is shown a universal or Hooke's joint with a single mass suspended under gravity. The mechanism

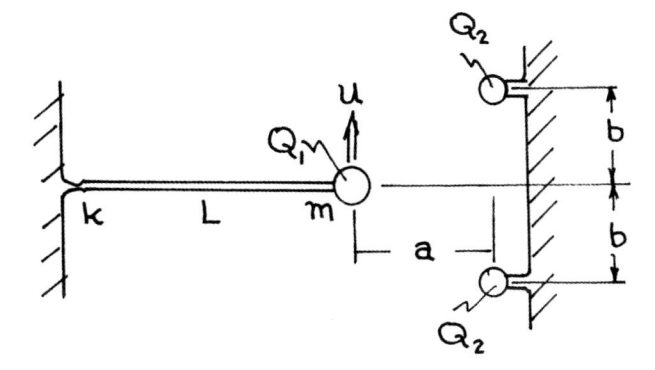

Fig. P4.26

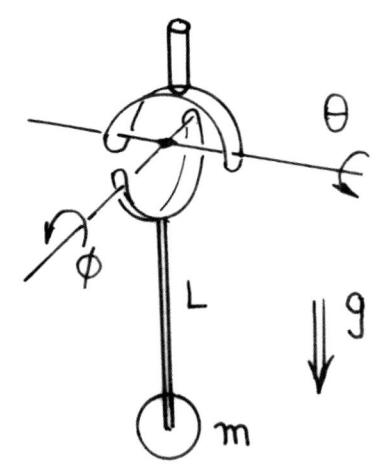

Fig. P4.27

has two degrees of freedom representing the two angles of rotation about the U-joint axes. Use Lagrange's equations to derive equations of motion. Linearize and determine the natural frequencies of small vibrations.

4.28 (*Spherical Pendulum with Dumbbell*) Consider the Hooke's joint in Problem 4.27 with two masses at the ends of a dumbbell as shown in Fig. P4.28. Use Lagrange's equations to derive equations of motion. Determine the natural frequencies for small motions and compare to those in Problem 4.27.

4.29 (*Slider Crank Mechanism: MATLAB*) The slider crank mechanism is an integral part of internal combustion engines. In Fig. P4.29 is shown a lumped mass model. In this problem we examine the forced motion of the piston with the enclosed gas represented as a linear spring where the spring force is zero when the crank angle $\theta = \pi$. Use the principle of virtual power to derive the equation of motion for a constant torque applied to the crank. As an extra credit choose parameters and integrate with MATLAB.

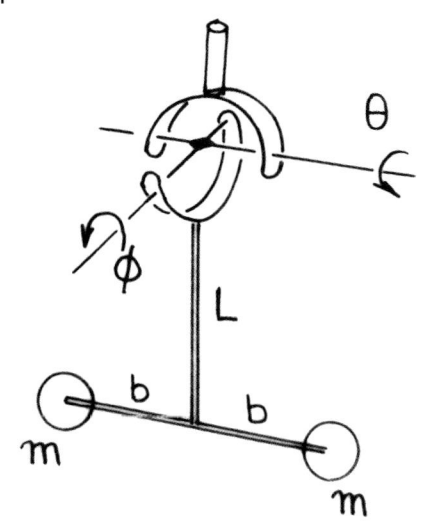

Fig. P4.28

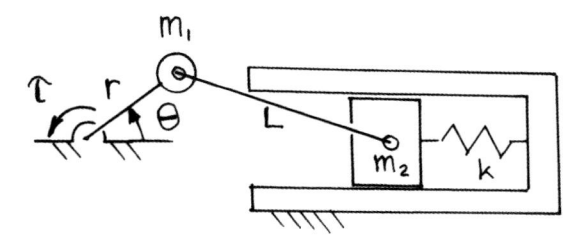

Fig. P4.29

4.30 (*Lagrange Multipliers: Rolling Wheel*) In Example 4.13 (Fig. 4.19), add a mass M on a radial line between the axel and the outer rim and determine the new equations of motion. Note that the non-holonomic constraints remain as in Example 4.13.

4.31 (*Vibration of Two Particles: Lagrange's Equations*) Consider a circular ring along which two particles are constrained to move as shown in Fig. P4.31. Assume the particles are connected by a linear spring with an initial undeformed angle of 45 degrees. Assume that the ring lies in a plane normal to the force of gravity. Use Lagrange's equations to derive the two equations of motion. Show that there exists one ignorable coordinate. Find the natural frequencies. Show that one corresponds to steady circular motion.

4.32 (*Four-particle Periodic Structure Vibrations*) Extend Problem 4.31 to four particles on a circular ring (Fig. P4.32) with four equal masses and springs. Find the equations of motion. Find four eigenmodes of small vibrations and their associated natural frequencies.

4.33 (*MEMS Slider Crank Micro-mirror*) Optical switching devices as well as digital mirror projectors use small micro-mirrors. One such design is shown

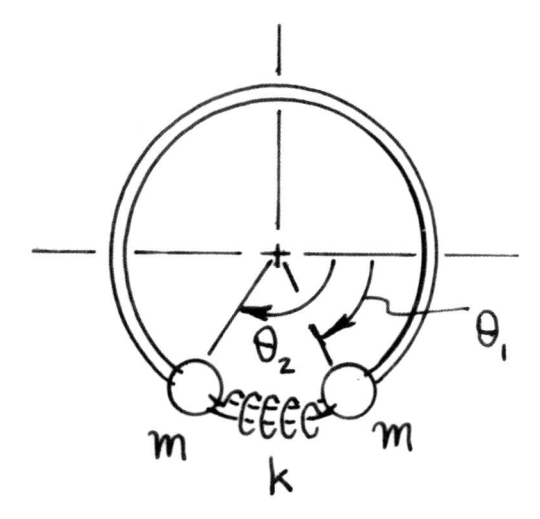

Fig. P4.31

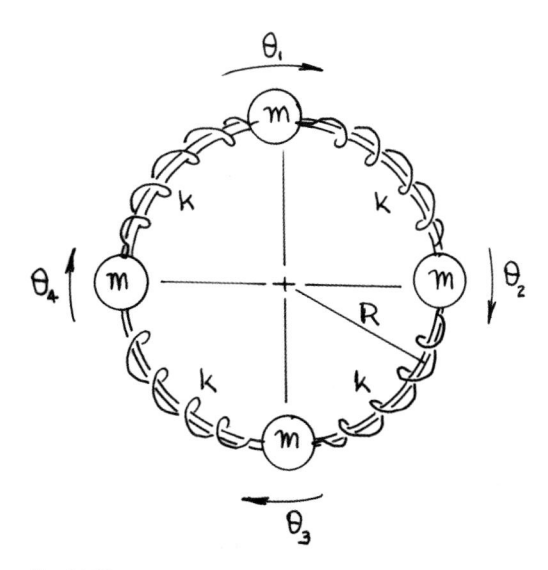

Fig. P4.32

in Fig. P4.33 and consists of a slider crank type of compliant mechanism. In compliant MEMS mechanisms the joint is etched out to form a weak bending center that can act as a rotary spring. In this problem we use an approximate model where the mass of the slider and the crank link is replaced by concentrated masses m_1, m_2. Assume that the slider is driven by a constant force F_0 from an electric comb drive actuator and that the joint can be characterized by a torsional spring of strength k. Find an equation of motion for small angle α, and estimate the time to lift the mirror as a function of the applied force F_0.

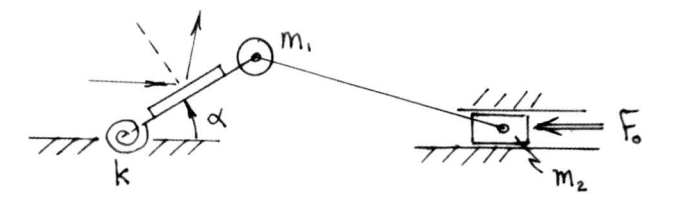

Fig. P4.33

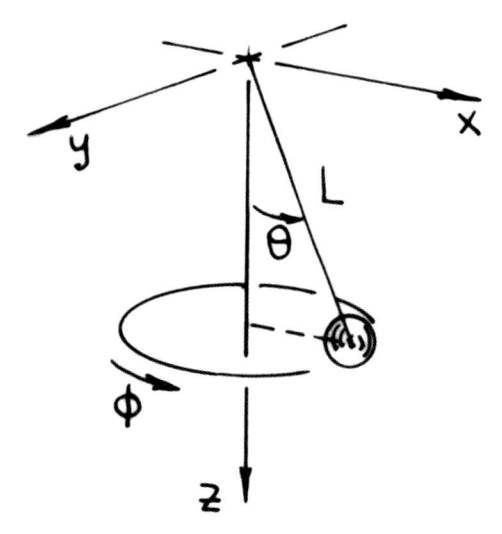

Fig. P4.34

4.34 (*Spherical Pendulum*) Consider a mass on a string that is constrained to move on a spherical surface (Fig. P4.34). Derive the equations of motion using Lagrange's equation. Show that there is an ignorable coordinate. Identify the conserved generalized momentum. Compare this problem with the universal joint pendulums in Problems 4.27 and 4.28.

4.35 (*Rolling Rings: Virtual Power Method*) In a problem similar to Problem 4.12, consider a circular ring of radius R rolling on another smaller cylinder of radius a. The larger ring has two equal masses at the diameter positions. Assume that the two rings roll with no slip, use the principle of virtual power to derive the equation of motion. For small motions under gravity, find the natural frequency.

4.36 (*Parallelogram Robot Arm*) Shown in Fig. P4.36 is a lumped mass model of a planar, two degree of freedom parallelogram robotic manipulator arm. In this model a payload mass M acts at the wrist point and a mass m acts at the end of the parallel link as shown in the figure. The two joint angles are assumed to driven by two torque actuators τ_1, τ_2, where the input power is given by

$$P = \tau_1 \dot{\theta}_1 + \tau_2 \dot{\theta}_2$$

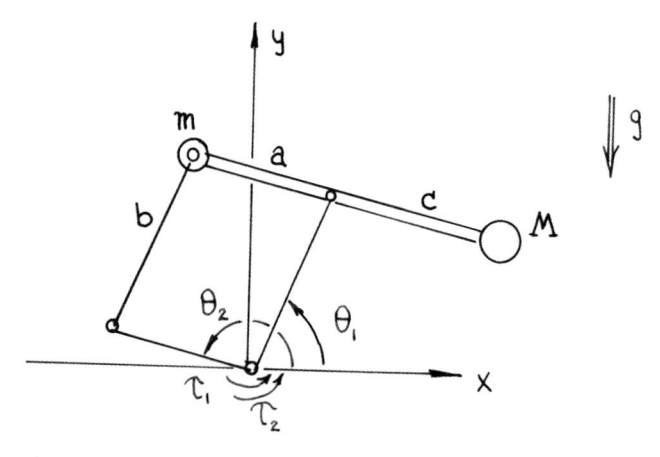

Fig. P4.36

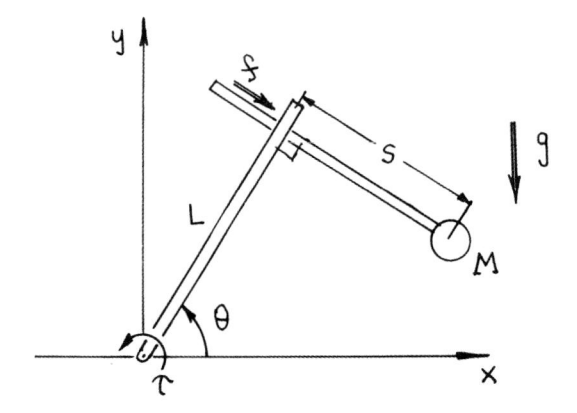

Fig. P4.37

Use the principle of virtual power to derive the two coupled equations of motion.

4.37 (*Robot Manipulator Arm Equations*) A planar manipulator arm is controlled by a prismatic and a revolute joint actuators as shown in Fig. P4.37, where the prismatic joint motion is perpendicular to the link arm of the revolute joint. Use Lagrange's equations to derive the equations of motion where the input power is related to the actuator force and torque by $P = f\dot{s} + \tau\dot{\theta}$.

5
Rigid Body Dynamics

5.1
Introduction

The applications of rigid-body dynamics include aircraft, vehicle and satellite dynamics, robotic devices, gyroscopic instruments, biomechanical prostheses, to name a few. In fact, much of our common experience of dynamical phenomena is governed by the principles of rigid-body kinematics and dynamics. This includes the motions of humans and animals as a connected set of rigid bodies. These principles also govern the behavior of sport objects such as footballs, baseballs, and bowling balls and pins. In power-producing machines, rigid-body mechanics is used to calculate the forces between components in internal combustion machines, power transmission devices, and gearing systems.

In order to appreciate the ubiquitous nature of rigid-body mechanics, the student should examine a common consumer item such as a bicycle or even a VCR and dissect it into rigid-body and nonrigid components. Then the student should identify which of these components requires knowledge of rotational motion, force moments, and torques as well as accelerations and forces in understanding or designing the system. Two examples are the reciprocating engine shown in Fig. 5.1 and the Hubble space telescope shown in Fig. 5.2.

Example 5.1. In the exploded view of a reciprocating engine (Fig. 5.1), combustion pressure on the top face of the piston is transmitted through the connecting rods to produce torque on the crankshaft assembly. The rigid-body dynamics of the connecting rod determine the forces on the crank pin. Also the distribution of mass in the crank shaft determines the dynamic forces on the bearings. This mechanism is a three-dimensional version of the slider-crank mechanism discussed at the end of Chapter 3, Example 3.6.

Applying the Newton–Euler equations of motion (2.30) and (2.34) to each of the principal components, there are three types of dynamical equations. The piston has

Applied Dynamics. With Applications to Multibody and Mechatronic Systems. Second Edition. Francis C. Moon
Copyright © 2008 WILEY-VCH Verlag GmbH & Co. KGaA, Weinheim
ISBN: 978-3-527-40751-4

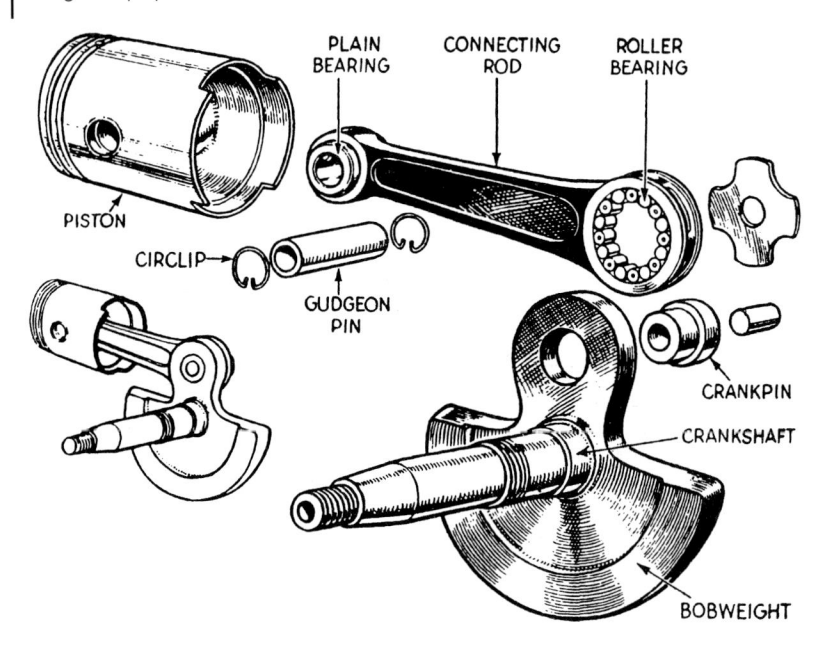

Fig. 5.1 Exploded view of internal combustion engine crankshaft and pistons. (From *Restoring Motorcycles, Two-Stroke Engines*, R. Bacon, Osprey Publ. Ltd., London.)

a prismatic or linear motion that is governed by a single component of Newton's equation:

$$m_1 \dot{v}_x = f_x \quad \text{(Piston)}$$

Here m_1 is the mass of the piston and f_x is the sum of all the components of the forces on the piston in the direction of motion x. The equation of motion for the crankshaft involves pure rotational motion about a fixed axis and the connecting rod has both coupled translational and rotational dynamics governed by equations of the form

$$I_3 \ddot{\phi} = T \quad \text{(Crankshaft-flywheel assembly)}$$
$$m_2 \dot{V}_{cx} = F_x; \quad m_2 \dot{V}_{cy} = F_y \quad \text{(Connecting rod)}$$
$$I_2 \ddot{\theta} = M_z$$

Here T is the net torque on the crankshaft from the connecting rod and the load and M_z is the moment of all the forces on the connecting rod. $\mathbf{V}_c$ is the vector velocity of the center of mass of the connecting rod. All these forces and motions are related through the kinematic constraints of the slider crank mechanism (see Example 3.6). There are five coupled differential equations. In Chapters 5 and 6, we present alternative methods of Lagrange and Virtual Power that will simplify the analysis of such complex systems without having to disassemble the machine. Such

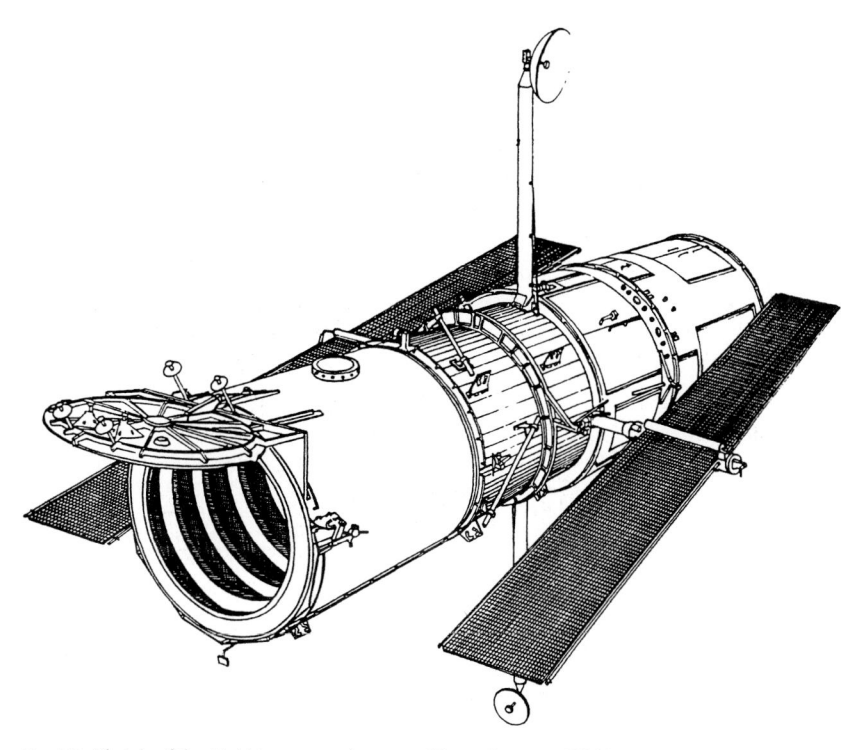

Fig. 5.2 Sketch of the Hubble space telescope. (From Caprara, 1986.)

variational methods of dynamics will reduce the above five equations of motion to a single differential equation.

Methods of Formulation

As in the case of dynamics of particles, there are a number of different formulations for rigid body problems:

1. Newton–Euler formulation
2. D'Alembert's method and principle of virtual work
3. Lagrange's equations
4. Principle of virtual power; methods of Jourdain and Kane

For simple planar problems with one degree of freedom, we can also use energy and momentum methods as described in elementary texts.

Problem Solving in Rigid Body Dynamics

The basic method to solve problems in rigid-body systems can be broken down into five steps:

- Develop a geometric and physical model
- Apply kinematics and constraints
- Derive equations of motion using dynamical principles
- Obtain analytical or numerical solutions
- Use simulation for design and optimization

Dynamics texts often place more emphasis on the second and third topics. However, in many problems modeling is a key element in obtaining a correct dynamic analysis. Modeling not only involves assigning geometric measures such as moments of inertia and center of mass, but often requires material properties such as friction between bodies or making idealized assumptions about forces and moments. Modeling is a skill not obtained with mathematical acumen alone, but with a strong grounding in physics and mechanics. It is learned through solving many different kinds of problems related to the technical world.

The student is often confused by the many variations of Newton's laws; *energy methods, force-acceleration, moment-angular momentum methods, D'Alembert's method, Lagrange's equation*, and the *principle of virtual power*. In recent years the choice of method has been guided by the need to obtain numerical solutions. Modern codes and computers are sometimes configured to treat problems efficiently in a matrix formulation. This has tended to favor methods such as D'Alembert's method or the virtual power methods over the more traditional Lagrange's equations. Also, new computer algorithms such as symbolic calculus codes, *MACSYMA, MATHEMATICA*, or *MAPLE* can now be used to handle pages of algebraic terms that sometimes result from the simplest formulation of a rigid body, e.g., the rolling of one body on a rigid surface.

Thus, while the principles of kinematics and dynamics are embodied in mathematical rigor, the route from model to solution, from a design goal to a controlled multibody machine often has several paths, some more efficient than others, but often chosen as a matter of personal style and experience.

5.2
Kinematics of Rigid Bodies

In Chapter 3 we reviewed the concept of a rotation-rate vector $\boldsymbol{\omega}$ and finite-rotation transformations. Both of these concepts are extremely important in the dynamics of rigid bodies. It is the author's experience that of the many difficulties students have in advanced dynamics, most are traceable to a lack of clear understanding of kinematics.

Instantaneous Motion of a Rigid Body

The fixed-distance constraint between any two points in a rigid body leads to the basic relation between the velocity of two points, A, P in the body: Let $\boldsymbol{\rho}_{AP}$ be a fixed length vector from point A to point P and let $\boldsymbol{\omega}$ be the rotation rate vector of the body relative a fixed reference. Then

$$\mathbf{v}_P = \mathbf{v}_A + \boldsymbol{\omega} \times \boldsymbol{\rho}_{AP} \tag{5.1}$$

From this expression, we can derive the relation between the acceleration between the two points:

$$\mathbf{a}_P = \mathbf{a}_A + \dot{\boldsymbol{\omega}} \times \boldsymbol{\rho}_{AP} + \boldsymbol{\omega} \times (\boldsymbol{\omega} \times \boldsymbol{\rho}_{AP}) \tag{5.2}$$

Here we have used the fact that

$$\dot{\boldsymbol{\rho}}_{AB} = \boldsymbol{\omega} \times \boldsymbol{\rho}_{AB} \tag{5.3}$$

since $\boldsymbol{\rho}_{AB}$ is a constant length vector [see (3.3)]. It should also be noted that the angular acceleration vector $\dot{\boldsymbol{\omega}}$ can be nonzero even though scalar components of $\boldsymbol{\omega}$ are constant.

Example 5.2 Angular Acceleration. A classic example of angular acceleration is the *precession* of a rigid body shown in Fig. 5.3. We assume that the body spins about a moving axis with constant rate $\dot{\phi}$ at the same time the spin axis rotates about the vertical axis with constant rate $\dot{\psi}$, i.e.,

$$\boldsymbol{\omega} = \dot{\phi}\mathbf{e}_a + \dot{\psi}\mathbf{e}_z \tag{5.4}$$

Then

$$\dot{\boldsymbol{\omega}} = \dot{\phi}\dot{\mathbf{e}}_a$$

and

$$\dot{\boldsymbol{\omega}} = \dot{\phi}\dot{\psi}\mathbf{e}_z \times \mathbf{e}_a \tag{5.5}$$

where $\mathbf{e}_z \times \mathbf{e}_a = \mathbf{e}_n$ is a unit vector normal to the plane formed by $\mathbf{e}_z$ and $\mathbf{e}_a$.

This term is often called a *gyroscopic effect* and refers to the coupling between rotations about two different axes.

For computer-based numerical calculation, it is sometimes more convenient to work with a matrix representation of the kinematics. In Chapter 3 we introduced a rotation-rate matrix

$$[\tilde{\omega}] = \begin{bmatrix} 0 & -\omega_3 & \omega_2 \\ \omega_3 & 0 & -\omega_1 \\ -\omega_2 & \omega_1 & 0 \end{bmatrix} \tag{5.6}$$

Using this notation, the formula (5.3) becomes

$$[\dot{\rho}_{AB}] = [\tilde{\omega}][\rho_{AB}] \tag{5.7}$$

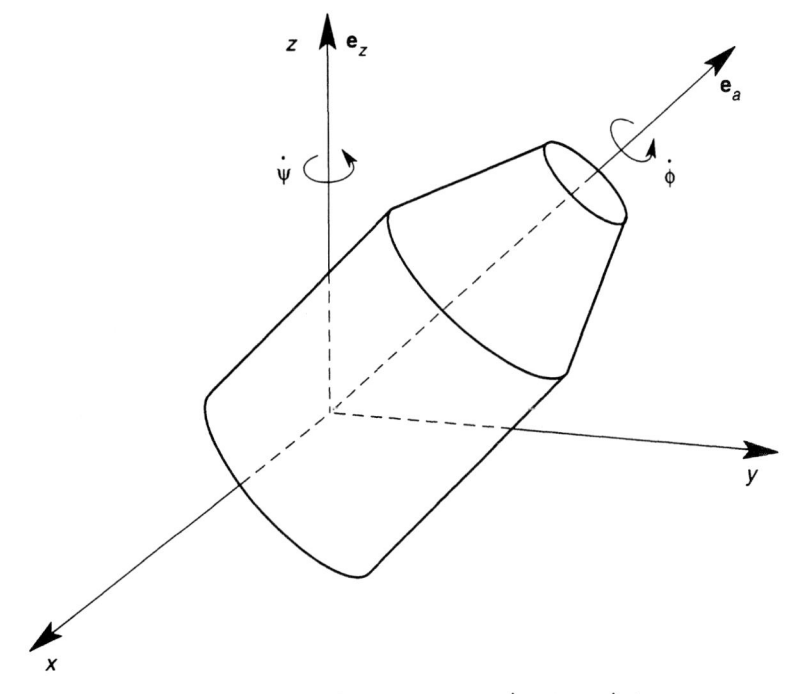

Fig. 5.3 Sketch of a rigid symmetric body in precession $\dot{\psi}$ and spin $\dot{\phi}$ about its symmetric axis.

In using this notation it is necessary that all three terms in (5.3) be written in the same coordinate system. The disadvantage of this notation is that the reference basis vectors are implicit.

Finite Motions of a Rigid Body

Finite motions require a definition of angular position. Traditional angle systems have been defined in the theory of gyroscopes, mechanisms and aircraft. One set is called *Euler angles*, although different books often use slightly different notations.

In Sections 3.6 and 3.7 we introduced 3×3 finite-rotation matrices and 4×4 transformation matrices, sometimes called homogeneous transformations. If the body is in pure rotation about a point in a fixed reference (Fig. 5.4), a point in the rigid body denoted by the column position vector ρ before the rotation, is transformed into the column vector $\mathbf{r}$ by the formula

$$\mathbf{r} = \mathsf{A}\rho \tag{5.8}$$

where A is an orthogonal matrix with the property

$$\mathsf{A}^T \mathsf{A} = \mathsf{I} \tag{5.9}$$

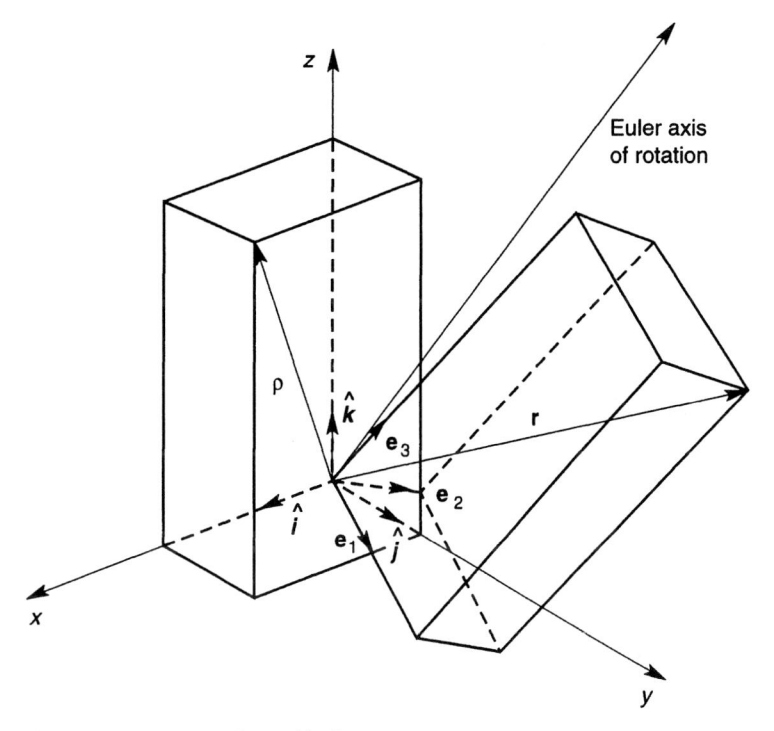

Fig. 5.4 Finite rotation of a rigid body.

If the reference basis vectors are $\{\hat{\mathbf{i}}, \hat{\mathbf{j}}, \hat{\mathbf{k}}\}$ and the body fixed basis vectors are $\{\mathbf{e}_1, \mathbf{e}_2, \mathbf{e}_3\}$, then $\boldsymbol{\rho} = \rho_1\hat{\mathbf{i}} + \rho_2\hat{\mathbf{j}} + \rho_3\hat{\mathbf{k}}$, and

$$\mathbf{r} = \rho_1\mathbf{e}_1 + \rho_2\mathbf{e}_2 + \rho_3\mathbf{e}_3 = \rho_1\mathsf{A}\hat{\mathbf{i}} + \rho_2\mathsf{A}\hat{\mathbf{j}} + \rho_3\mathsf{A}\hat{\mathbf{k}} \tag{5.10}$$

We can interpret the rotation matrix A as operating on the basis vectors, i.e.,

$$\mathbf{e}_1 = \mathsf{A}[1, 0, 0]^T$$
$$\mathbf{e}_1 = A_{11}\hat{\mathbf{i}} + A_{12}\hat{\mathbf{j}} + A_{13}\hat{\mathbf{k}} \tag{5.11}$$

Thus A_{ij} are the direction cosines of the basis vector $\{\mathbf{e}_i\}$ with respect to the fixed reference. The matrix operator A is sometimes called the *direction cosines matrix*. Note that in this interpretation $\mathbf{e}_1 = [1, 0, 0]^T$ in the body reference and is given by (5.11) in the fixed base reference. Another way to write this is to think of $\boldsymbol{\rho} = \mathbf{r}(t = 0)$, so that (5.8) becomes

$$\mathbf{r}(t) = \mathsf{A}(t)\mathbf{r}(0)$$

or

$$\mathbf{r}(0) = \mathsf{A}^T\mathbf{r}(t) \tag{5.12}$$

5 Rigid Body Dynamics

To find the relation between the angular velocity vector $\boldsymbol{\omega}$ and $A(t)$, we differentiate (5.12a) with respect to time.

$$\dot{\mathbf{r}}(t) = \dot{A}(t)\mathbf{r}(0) = \dot{A}A^T \mathbf{r}(t) \tag{5.13}$$

But in Eqs. (5.6) and (5.7) we saw that we could represent the angular velocity as a matrix $\tilde{\boldsymbol{\omega}}$. So we are led to the relation

$$\tilde{\boldsymbol{\omega}} = \dot{A}A^T \tag{5.14}$$

Example 5.3. Suppose A represents a rotation about the x-axis where $\theta = \omega_x t$,

$$A = \begin{bmatrix} 1 & 0 & 0 \\ 0 & \cos\theta & -\sin\theta \\ 0 & \sin\theta & \cos\theta \end{bmatrix} \tag{5.15}$$

To find the angular velocity matrix, $\tilde{\boldsymbol{\omega}}$, we perform a term-by-term derivative of A with respect to time,

$$\dot{A} = \omega_x \begin{bmatrix} 0 & 0 & 0 \\ 0 & -\sin\theta & -\cos\theta \\ 0 & \cos\theta & -\sin\theta \end{bmatrix}$$

Performing the matrix multiplication $\dot{A}A^T$, we obtain

$$\tilde{\boldsymbol{\omega}} = \dot{A}A^T = \omega_x \begin{bmatrix} 0 & 0 & 0 \\ 0 & 0 & -1 \\ 0 & 1 & 0 \end{bmatrix} \tag{5.16}$$

Comparing this matrix with Eq. (5.6), we see that

$$\boldsymbol{\omega} = \omega_x \hat{\mathbf{i}} \tag{5.17}$$

For the general case of translation and rotation, we learned in Chapter 3 that we could artificially extend the vector dimension to 4 and incorporate a rotation A and a translation $\mathbf{R}$ in a single 4×4 transformation T. Writing $\mathbf{r} = [r_x, r_y, r_z, 1]^T$, $\boldsymbol{\rho} = [\rho_1, \rho_2, \rho_3, 1]^T$. The new position $\mathbf{r}$, is related to the original position by the equation

$$\mathbf{r} = T\boldsymbol{\rho} \tag{5.18}$$

where

$$T = \begin{bmatrix} & & & | & R_x \\ & A & & | & R_y \\ & & & | & R_z \\ \hline 0 & 0 & 0 & | & 1 \end{bmatrix} \tag{5.19}$$

Here as in (5.8), A is an orthogonal rotation matrix. At any time, we can interpret this transformation as a finite rotation of the body, followed by a translation **R**. We recall, however, that a translation by **R**, followed by a rotation A, is not represented by (5.19) [see (3.56b)].

Kinematic Relations: Euler Angles

The integration of the dynamic equations of motion for a rigid body, Eqs. (2.30) and (2.37), yields the velocity of the center of mass $\mathbf{v}_c(t)$ and the rotation rate vector $\boldsymbol{\omega}(t)$. The translation position $[x_c(t), y_c(t), z_c(t)]^T$ can be found by integrating the following set of first-order differential equations

$$\dot{x}_c = v_{cx}(t)$$

$$\dot{y}_c = v_{cy}(t)$$

$$\dot{z}_c = v_{cz}(t) \tag{5.20}$$

These equations are known as *kinematic equations* of motion. To find the analogous set or relations for the angular orientation of the body, we must define a set of three independent angular transformations. There are many possibilities. One class of such transformations is called *Euler angles*, and this contains two popular sets, one used to describe *gyroscopic* problems (called Set A in this book), such as spinning satellites, and the other set used to describe *aircraft*, *ship*, or *vehicle* dynamics. (Set B is sometimes called Bryan angles. See, e.g., Goldstein, 1980.)

The Set A Euler angles are defined in Fig. 5.5. The base reference has a frame labeled $\{X, Y, Z\}$, and the final reference is labeled $\{x, y, z\}$, with two intermediate references $\{x', y', z\}$, $\{x', y'', z\}$. The sequence of rotations are defined by

Rotate about Z-axis by $\phi(t)$;   Precession
Rotate about x'-axis by $\theta(t)$;   Nutation
Rotate about z-axis by $\psi(t)$;   Spin

The angular-velocity vector can be written as the sum of three rate vectors:

$$\boldsymbol{\omega} = \dot{\phi}\hat{\mathbf{k}} + \dot{\theta}\mathbf{e}_{x'} + \dot{\psi}\mathbf{e}_z \tag{5.21}$$

The vector $\boldsymbol{\omega}$ can be written in either the fixed-base coordinates or the body based coordinates. For many rigid-body problems, it is more convenient to write $\boldsymbol{\omega}$ in terms of the body basis vectors:

$$\boldsymbol{\omega} = \omega_x \mathbf{e}_x + \omega_y \mathbf{e}_y + \omega_z \mathbf{e}_z$$

Thus we must write $\hat{\mathbf{k}}$ and $\mathbf{e}_{x'}$ in terms of the body reference, e.g.,

$$\mathbf{e}_{x'} = \cos\psi\, \mathbf{e}_x - \sin\psi\, \mathbf{e}_y \tag{5.22}$$

220 | 5 Rigid Body Dynamics

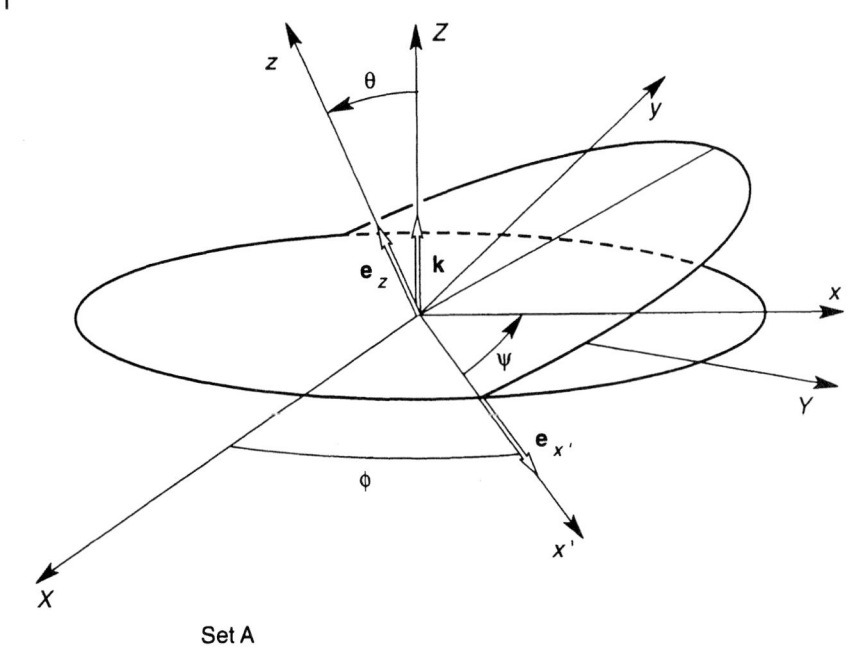

Set A

Fig. 5.5 The $Z - x' - z$ set of Euler angles (Set A) to describe the finite rotation of a rigid body.

The resulting scalar equations relating the components of $\boldsymbol{\omega}$ to the generalized velocities $\dot{\phi}, \dot{\theta}, \dot{\psi}$, can be written in the form of a matrix equation

$$\begin{bmatrix} \sin\theta \sin\psi & \cos\psi & 0 \\ \sin\theta \cos\psi & -\sin\psi & 0 \\ \cos\theta & 0 & 1 \end{bmatrix} \begin{bmatrix} \dot{\phi} \\ \dot{\theta} \\ \dot{\psi} \end{bmatrix} = \begin{bmatrix} \omega_x(t) \\ \omega_y(t) \\ \omega_z(t) \end{bmatrix} \quad (5.23)$$

This set of *kinematic relations* has a solution, provided the determinant is not zero. We can show, however, that the determinate $= -\sin\theta$. Thus, we cannot obtain a unique solution relating $\boldsymbol{\omega}$ and $\{\dot{\phi}, \dot{\theta}, \dot{\psi}\}$ near $\theta = 0, n\pi$.

Set B Euler angles are designed to avoid the singular nature of Set A near $\theta = 0$. In aircraft or vehicle problems, we want to operate near $\theta = 0$. For these problems, the Euler angles are defined by the following transformation shown in Fig. 5.6.

Rotate about Z-axis by $\psi(t)$;   Heading (yaw)
Rotate about y'-axis by $\theta(t)$;   Attitude (pitch)
Rotate about z-axis by $\phi(t)$;   Bank (roll)

Goldstein (1980) calls these the Tait–Bryan angles. The kinematic relations can be shown to be

$$\begin{bmatrix} 1 & 0 & -\sin\theta \\ 0 & \cos\phi & \cos\theta \sin\phi \\ 0 & -\sin\phi & \cos\theta \cos\phi \end{bmatrix} \begin{bmatrix} \dot{\phi} \\ \dot{\theta} \\ \dot{\psi} \end{bmatrix} = \begin{bmatrix} \omega_x \\ \omega_y \\ \omega_z \end{bmatrix} \quad (5.24)$$

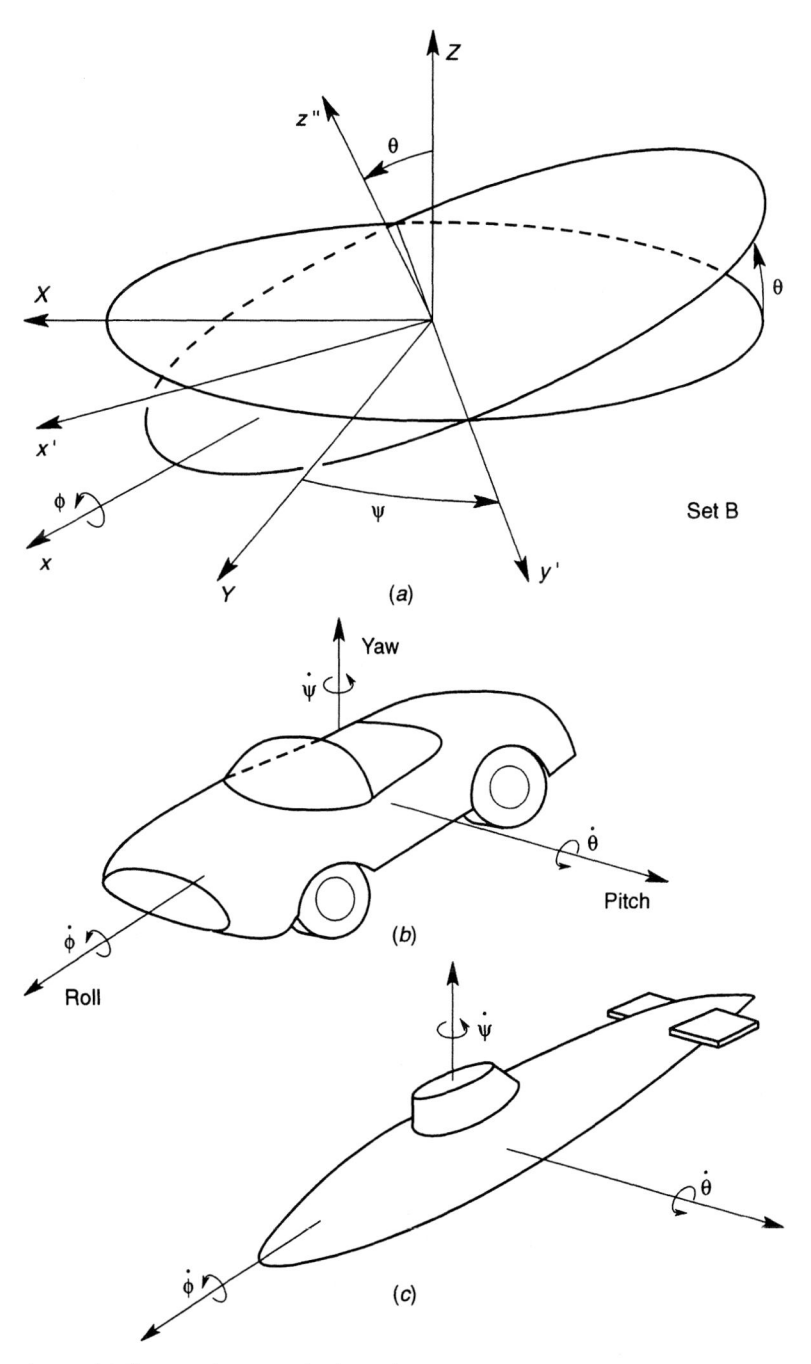

Fig. 5.6 (a) The $Z - y' - x$ set of Euler angles (Set B).
(b), (c) Yaw, pitch, and roll axes of vehicle dynamics.

Two of the Euler angles are defined as transformations relative to the intermediate reference systems not with respect to the $\{\hat{\mathbf{i}}, \hat{\mathbf{j}}, \hat{\mathbf{k}}\}$ system. Suppose R_ψ, R_θ, R_ϕ are the three Euler transformation matrices. If $\boldsymbol{\rho}^0$ is a vector in the untransformed body, what are the coordinates of $\boldsymbol{\rho}^0$ after the three Euler transformations? In other words, can we find a transformation A such that the new coordinates in $\{\hat{\mathbf{i}}, \hat{\mathbf{j}}, \hat{\mathbf{k}}\}$ are given by

$$\boldsymbol{\rho} = A\boldsymbol{\rho}^0 \tag{5.25}$$

It is straight forward to show that the transformation matrix A is given by the reverse sequence of rotations, i.e.,

$$A = R_\psi R_\theta R_\phi \tag{5.26}$$

$$R_\psi = \begin{bmatrix} \cos\psi & -\sin\psi & 0 \\ \sin\psi & \cos\psi & 0 \\ 0 & 0 & 1 \end{bmatrix} \quad \text{(relative to the } \{\mathbf{e}_\ell^0\} \text{ or } \{\hat{\mathbf{i}}, \hat{\mathbf{j}}, \hat{\mathbf{k}}\} \text{ basis)} \tag{5.27}$$

$$R_\theta = \begin{bmatrix} \cos\theta & 0 & \sin\theta \\ 0 & 1 & 0 \\ -\sin\theta & 0 & \cos\theta \end{bmatrix} \quad \text{(relative to the } \{\mathbf{e}_\ell'\} \text{ basis)} \tag{5.28}$$

$$R_\phi = \begin{bmatrix} 1 & 0 & 0 \\ 0 & \cos\phi & -\sin\phi \\ 0 & \sin\phi & \cos\phi \end{bmatrix} \quad \text{(relative to the } \{\mathbf{e}_\ell''\} \text{ basis)} \tag{5.29}$$

To see this we define four sets of basis vectors, $\{\mathbf{e}_\ell^0\}$, $\{\mathbf{e}_\ell'\}$, $\{\mathbf{e}_\ell''\}$, $\{\mathbf{e}_\ell\}$ ($\ell = 1, 2, 3$) such that $\{\mathbf{e}_\ell\}$ are the body fixed basis vectors and $\{\mathbf{e}_\ell^0\}$ are identical with $\{\hat{\mathbf{i}}, \hat{\mathbf{j}}, \hat{\mathbf{k}}\}$.

These vectors are related by the transformations

$$\mathbf{e}_\ell' = R_\psi \mathbf{e}_\ell^0, \qquad \mathbf{e}_\ell'' = R_\theta \mathbf{e}_\ell', \qquad \mathbf{e}_\ell = R_\phi \mathbf{e}_\ell'' \tag{5.30}$$

where the components of $\mathbf{e}_\ell'$ in $\{\mathbf{e}_\ell^0\}$ are the components of the ℓth column vector of R_ψ. The transformation sequence takes $\boldsymbol{\rho}^0 \to \boldsymbol{\rho}' \to \boldsymbol{\rho}'' \to \boldsymbol{\rho}$. Since these are rigid-body transformations, the components of $\boldsymbol{\rho}'$, $\boldsymbol{\rho}''$, $\boldsymbol{\rho}$ in the local coordinate bases remain fixed, i.e.,

$$\boldsymbol{\rho}^0 = \sum x_\ell \mathbf{e}_\ell^0, \qquad \boldsymbol{\rho}' = \sum x_\ell \mathbf{e}_\ell', \qquad \boldsymbol{\rho}'' = \sum x_\ell \mathbf{e}_\ell''$$

$$\boldsymbol{\rho} = \sum x_\ell \mathbf{e}_\ell \tag{5.31}$$

Combining (5.30) and (5.31), we find that the new components of $\boldsymbol{\rho}''$ in the original basis after two transformations R_ψ, R_θ are given by

$$\boldsymbol{\rho}'' = \sum_\ell x_\ell'' \mathbf{e}_\ell^0 = \sum_\ell \sum_m R(\psi)_{\ell m} \sum_k R(\theta)_{mk} x_k \mathbf{e}_\ell^0 \tag{5.32}$$

i.e. the components of $\boldsymbol{\rho}''$ in the $\{\mathbf{e}_\ell^0\}$ (or $\{\hat{\mathbf{i}}, \hat{\mathbf{j}}, \hat{\mathbf{k}}\}$) basis are

$$x_\ell'' = \sum_k \left[\sum_m R(\psi)_{\ell m} R(\theta)_{mk} \right] x_k \tag{5.33}$$

or $\boldsymbol{\rho}'' = \mathsf{R}_\psi \mathsf{R}_\theta \boldsymbol{\rho}^0$.

It follows that after the three Euler transformations, the components of the original vector $\boldsymbol{\rho}^0$ in the basis $\{\mathbf{e}_\ell^0\}$ or $\{\hat{\mathbf{i}}, \hat{\mathbf{j}}, \hat{\mathbf{k}}\}$ are given by (5.25), (5.26), or

$$\boldsymbol{\rho} = \mathsf{R}_\psi \mathsf{R}_\theta \mathsf{R}_\phi \boldsymbol{\rho}^0$$

Example 5.4 Euler Angles. Suppose a slender body like a submarine (Fig. 5.6c) undergoes a yaw rate $\dot{\psi} = A \cos \Omega t$, and a pitch rate $\dot{\theta} = B \sin \Omega t$. If the local x-axis is in the long-body direction, describe the motion of the bow of the submarine relative to its center of mass. Use the Set B Euler angles.

To solve this we define a position vector $\boldsymbol{\rho} = L \mathbf{e}_x$, from the mass center to the bow of the body. The velocity of this point is given by

$$\mathbf{v} = \boldsymbol{\omega} \times \boldsymbol{\rho} = \tilde{\boldsymbol{\omega}} \boldsymbol{\rho} \tag{5.34}$$

where

$$\boldsymbol{\omega} = \dot{\psi} \hat{\mathbf{k}} + \dot{\theta} \mathbf{e}_y + \dot{\phi} \mathbf{e}_x$$

In this example $\dot{\phi} = 0$.

In the body references, the components of $\boldsymbol{\omega}$ depend not only on $\dot{\psi}, \dot{\theta}$, but also on the angles (ψ, θ, ϕ). Choosing zero initial values, we can integrate the angle rate equations to find

$$\psi = \frac{A}{\Omega} \sin \Omega t$$
$$\theta = -\frac{B}{\Omega} \cos \Omega t$$
$$\phi = 0 \tag{5.35}$$

Using the kinematic equations (5.24), we find that

$$\boldsymbol{\omega} = -\dot{\psi} \sin \theta \mathbf{e}_x + \dot{\theta} \mathbf{e}_y + (\dot{\psi} \cos \theta) \mathbf{e}_z$$

where we have used $\dot{\phi} = 0$, $\phi = 0$ in (5.24). Then the velocity is given by

$$\mathbf{v} = -\dot{\theta} L \mathbf{e}_z + L(\dot{\psi} \cos \theta) \mathbf{e}_y$$

or

$$\mathbf{v} = -BL \sin \Omega t \mathbf{e}_z + AL \cos \Omega t \cos \left(\frac{B}{\Omega} \cos \Omega t \right) \mathbf{e}_y \tag{5.36}$$

For small values of B/Ω, the velocity vector performs circular motion and the vector ρ describes an elliptical motion of the end of the body.

5.3
Newton–Euler Equations of Motion

In Chapter 2, we reviewed the basic principles of dynamics presented by Newton and Euler for a system of particles

$$m\frac{d\mathbf{v}_c}{dt} = \mathbf{F} \tag{2.30}$$

$$\frac{d\mathbf{H}}{dt} = \mathbf{M} \tag{2.34}$$

In this section we must interpret these equations as applied to rigid body systems.

In the direct application of Newton's laws to a particle, one works with forces, velocity, and rate of change of momentum. In rigid-body dynamics one deals with moments of force and the moment of momentum called angular momentum, $\mathbf{H}$. Defined as an integral over a continuous mass density, $\mathbf{H}$ is given by (see Fig. 5.7)

$$\mathbf{H} = \int \mathbf{r} \times \mathbf{v}\, dm \tag{5.37}$$

where $\mathbf{r}$ is a position vector from some reference point to a differential mass element dm.

The expression (5.37) implies that $\mathbf{H}$ is relative to a specific reference point. In practical terms this reference point is usually the center of mass of the body, or a fixed point of pure rotation. The definition [Eq. (5.37)] is not specific to rigid bodies, however. The rigid body is defined by the statement (2.46). If $\mathbf{r}_A(t)$ is a point that moves with the body and ρ is a vector from point A to another point in the body, then $\mathbf{r}$, $\mathbf{v}$ become

$$\mathbf{r} = \mathbf{r}_A + \rho$$
$$\mathbf{v} = \mathbf{v}_A + \boldsymbol{\omega} \times \rho \tag{5.38}$$

This expression embodies the assumption that the distance between any two points in a body, e.g., ρ, is a constant length vector [see Eq. (3.3)]. The angular velocity vector is a property of the entire body and thus can be removed from the integral in (5.37). This leads to the concept of second moments of mass or the moment of inertia matrix. [See (2.52), (2.53).]

Mass Moments for a Rigid Body

When the velocity is written in terms of the angular velocity $\boldsymbol{\omega}$ for a rigid body (5.38), and substituted into (5.37), three mass integrals emerge in the calculation of $\mathbf{H}$, the mass, vector moment of mass, and second moment of mass matrix:

5.3 Newton–Euler Equations of Motion

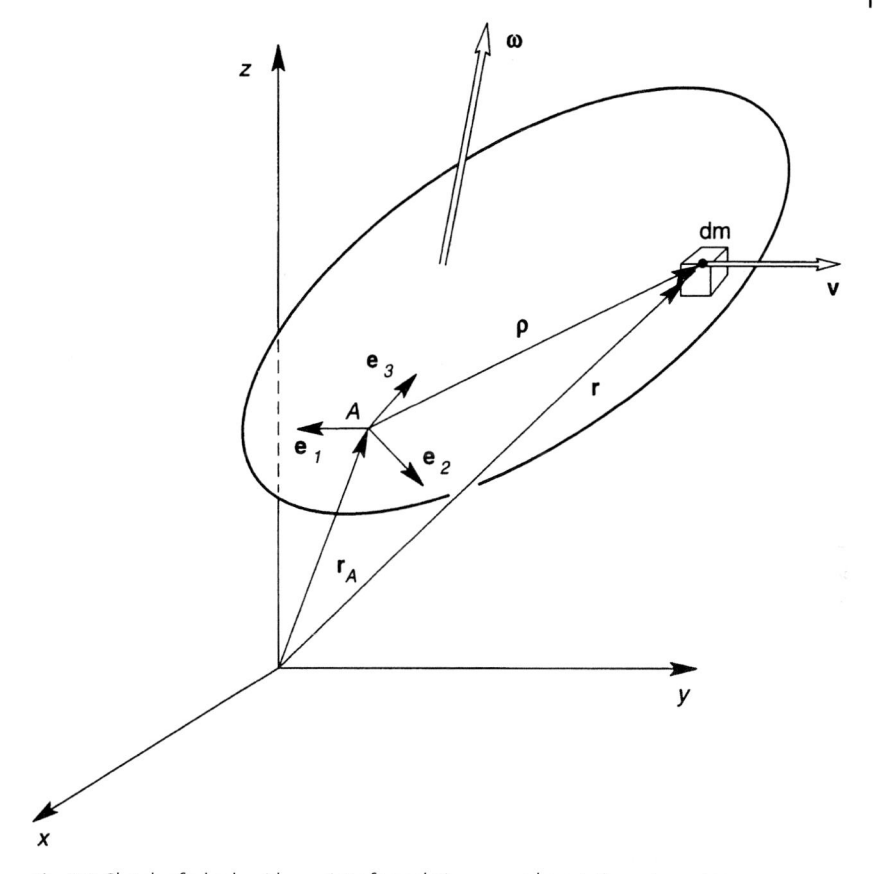

Fig. 5.7 Sketch of a body with a point of translation $\mathbf{r}_A$, and a rotation rate vector $\boldsymbol{\omega}$.

$$\mathbf{H} = \mathbf{r}_A \times \mathbf{v}_A \int dm + \left[\int \boldsymbol{\rho}\, dm\right] \times \mathbf{v}_A + \mathbf{r}_A \times \left[\boldsymbol{\omega} \times \int \boldsymbol{\rho}\, dm\right]$$
$$+ \int \boldsymbol{\rho} \times (\boldsymbol{\omega} \times \boldsymbol{\rho})\, dm \tag{5.39}$$

In the first two integrals we have the total mass and definition of center of mass relative to the point A:

$$m = \int dm$$
$$m\boldsymbol{\rho}_c = \int \boldsymbol{\rho}\, dm \tag{5.40}$$

If the point A is the center of mass, then $\boldsymbol{\rho}_c = 0$. With these definitions we can rewrite the angular momentum about the origin of reference

$$\mathbf{H} = \mathbf{H}_A + \mathbf{r}_A \times m\mathbf{v}_A + \boldsymbol{\rho}_c \times m\mathbf{v}_A + \mathbf{r}_A \times (\boldsymbol{\omega} \times \boldsymbol{\rho}_c)m \tag{5.41}$$

where

$$\mathbf{H}_A = \int \boldsymbol{\rho} \times (\boldsymbol{\omega} \times \boldsymbol{\rho}) \, dm$$

In this last integral, it is practical to bring the angular velocity vector $\boldsymbol{\omega}$ outside the integral. Then we can perform an integral over the mass distribution that is independent of the motion. To do this, we use an expression for the triple-vector product:

$$\mathbf{A} \times (\mathbf{B} \times \mathbf{C}) = (\mathbf{A} \cdot \mathbf{C})\mathbf{B} - (\mathbf{A} \cdot \mathbf{B})\mathbf{C}$$

or

$$\boldsymbol{\rho} \times (\boldsymbol{\omega} \times \boldsymbol{\rho}) = \boldsymbol{\rho} \cdot \boldsymbol{\rho} \frac{1}{2} \boldsymbol{\omega} - \boldsymbol{\rho}\boldsymbol{\rho} \cdot \boldsymbol{\omega} \tag{5.42}$$

It is standard practice to write both terms as a dot or scalar product, with $\boldsymbol{\omega}$ using either a matrix or dyadic notation. When $\boldsymbol{\rho}, \boldsymbol{\omega}$ are expressed in an orthonormal vector basis, the first term is easily written in matrix notation.

$$\boldsymbol{\rho} \cdot \boldsymbol{\rho} \frac{1}{2} \boldsymbol{\omega} = \begin{bmatrix} \rho^2 & 0 & 0 \\ 0 & \rho^2 & 0 \\ 0 & 0 & \rho^2 \end{bmatrix} \begin{bmatrix} \omega_1 \\ \omega_2 \\ \omega_3 \end{bmatrix}$$

or using compact notation

$$\boldsymbol{\rho} \cdot \boldsymbol{\rho} \frac{1}{2} \boldsymbol{\omega} = \rho^2 \boldsymbol{\delta} \cdot \boldsymbol{\omega} \tag{5.43}$$

and the matrix $\boldsymbol{\delta}$ is the identity matrix

$$[\boldsymbol{\delta}] = \begin{bmatrix} 1 & 0 & 0 \\ 0 & 1 & 0 \\ 0 & 0 & 1 \end{bmatrix}$$

Note that the matrix product

$$\boldsymbol{\delta} \cdot \boldsymbol{\omega} = \boldsymbol{\omega}$$

The second term of (5.42) is more difficult. Consider the expression

$$\mathbf{D} = \mathbf{AB} \cdot \mathbf{C}$$

where all vectors are planar and an orthonormal basis set is $\{\mathbf{e}_1, \mathbf{e}_2\}$ so that $\mathbf{A} = A_1 \mathbf{e}_1 + A_2 \mathbf{e}_2$, etc. Then we know that

$$D_1 = A_1(B_1 C_1 + B_2 C_2)$$

$$D_2 = A_2(B_1 C_1 + B_2 C_2)$$

5.3 Newton–Euler Equations of Motion

Expressed as a matrix operation on **C**, we can show that

$$\mathbf{D} = \begin{bmatrix} D_1 \\ D_2 \end{bmatrix} = \begin{bmatrix} A_1 B_1 & A_1 B_2 \\ A_2 B_1 & A_2 B_2 \end{bmatrix} \begin{bmatrix} C_1 \\ C_2 \end{bmatrix} \tag{5.44}$$

In the case of the angular momentum $\mathbf{H}_A$ ($\mathbf{A} = \boldsymbol{\rho}$, $\mathbf{B} = \boldsymbol{\rho}$, $\mathbf{C} = \boldsymbol{\omega}$), we have

$$\boldsymbol{\rho}\boldsymbol{\rho} \cdot \boldsymbol{\omega} = \begin{bmatrix} x^2 & xy & xz \\ \cdot & y^2 & yz \\ \cdot & \cdot & z^2 \end{bmatrix} \begin{bmatrix} \omega_1 \\ \omega_2 \\ \omega_3 \end{bmatrix} \tag{5.45}$$

where $\boldsymbol{\rho} = x\mathbf{e}_1 + y\mathbf{e}_2 + z\mathbf{e}_3$ and $\rho^2 = x^2 + y^2 + z^2$. The matrices in Eqs. (5.43) and (5.45) must be integrated over the body.

The final expression of $\mathbf{H}_A$ is given in the form

$$\mathbf{H}_A = \begin{bmatrix} H_{1A} \\ H_{2A} \\ H_{3A} \end{bmatrix} = \begin{bmatrix} I_{11} & I_{12} & I_{13} \\ \cdot & I_{22} & I_{23} \\ \cdot & \cdot & I_{33} \end{bmatrix} \begin{bmatrix} \omega_1 \\ \omega_2 \\ \omega_3 \end{bmatrix} \tag{5.46}$$

$$\mathbf{H}_A = \mathbf{I}\boldsymbol{\omega}$$

The matrix **I** is called the *moment of inertia* or simply *inertia matrix*. (Some texts use the term "inertia tensor.") However, the name is a misnomer since we have taken the kinematics outside of the integral (5.41). It should properly be called the *second moment of mass matrix*, as is clear from the expressions for I_{ij}:

$$I_{11} = \int (y^2 + z^2)\, dm$$

$$I_{12} = -\int xy\, dm \tag{5.47}$$

The first integral represents an integral of mass elements times the square of the perpendicular distance from the $\mathbf{e}_1$ axis through the point A. The off-diagonal terms are called *cross products of inertia*.

From the symmetry of I_{ij}, it follows that one can find a set of three orthogonal axes for which I_{ij} is diagonal. These axes are called *principal axes* and the diagonal elements $\{I_1, I_2, I_3\}$ are called *principal inertias*. They can be shown to be the eigenvalues of the matrix I_{ij}. [See also the discussion before (2.54).]

Properties of the Inertia Matrix

The following properties of I_{ij} are listed without proof. The reader should consult a theoretical text or try the proofs as an exercise. (The proofs are not difficult):

1. I_{ij} is a symmetric matrix.
2. I_{ij} has positive eigenvalues and three orthogonal eigenvectors.
3. The off-diagonal terms are zero in a basis with symmetry planes.

4. The inertia-matrix calculation is an additive operator. The inertia matrix of a body with volumes $V_1 + V_2$ is the sum of the inertia matrix of V_1 and the inertia matrix of V_2.
5. *The parallel axes theorem.* The inertia matrix about a set of axes at $\mathbf{r}_A$ can be written in terms of inertias about a set of parallel axes at the center of mass $\mathbf{r}_c$ using the following formula:

$$I_{kk} = I_{kk}^c + m\Delta_k^2$$
$$I_{xy} = I_{xy}^c - mx_c y_c, \quad \text{etc.} \tag{5.48}$$

where Δ_k is the perpendicular distance between the two parallel axes. Also x_c, y_c are the components of the vector $\mathbf{r}_c - \mathbf{r}_A$.

Practical Methods to Determine the Inertia Matrix

In practice one rarely uses the expressions, Eq. (5.47), to calculate I_{ij} or $\mathbf{H}_A$ except in elementary physics and calculus exercises. The engineer usually resorts to one of three methods:

1. Look-up tables (see Appendix A)
2. Computer calculation; digital or symbolic
3. Experiments

Example 5.5 Center of Mass of an Incomplete Ring. As an application of the definition of the center of mass, (5.40), consider the incomplete ring defined by a radius R, mass density per unit length μ, total mass m, and angle α, as shown in Fig. 5.8. By symmetry, the center of mass must lie on the y axis. The position of the center of mass on the y axis is found by the following equations:

$$\boldsymbol{\rho}_c m = \int \mathbf{r}\, dm, \qquad y_c = \boldsymbol{\rho}_c \cdot \mathbf{e}_y$$
$$y_c m = \int_{-\pi-\alpha}^{\alpha} \mu R^2 \sin\theta\, d\theta = -\mu R^2 \cos\theta \Big|_{\pi-\alpha}^{\alpha}$$
$$y_c = -\frac{R}{\pi}\cos\alpha$$

The center of mass is below the x axis. The calculation of the moments of inertias about the x and y axes follows a similar path and is given as a homework exercise.

Example 5.6 Inertia Matrix for a Satellite Antenna. A small satellite antenna dish is modeled approximately by a thin circular disc of radius R and thickness Δ and is attached to an axis of rotation x_1 by three thin-walled hollow rods of thickness δ, as shown in Fig. 5.9. We are asked to find the inertia matrix relative to the reference axes $\{\mathbf{e}_1, \mathbf{e}_2, \mathbf{e}_3\}$ shown in the figure.

To solve this problem we note that the four sub-bodies have their symmetry axes aligned with the $\{x_1, x_2, x_3\}$ axes so that the off-diagonal products of inertia will

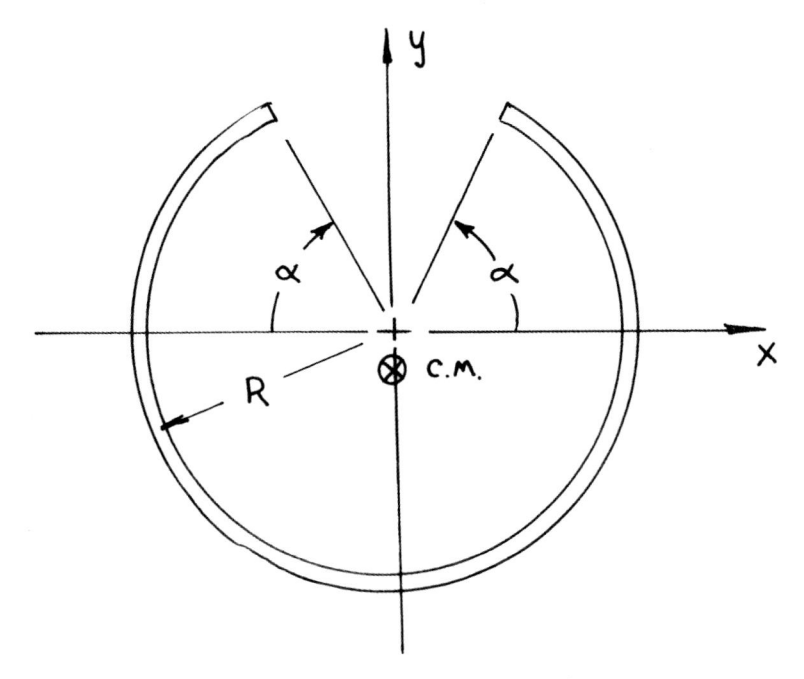

Fig. 5.8 Sketch of thin incomplete circular ring with distributed mass.

be zero. To solve the problem we make use of the formula in Appendix A for the principal inertias for a cylinder and a thin cylindrical shell about their centers of mass. We also make use of the parallel-axis theorem, relating the inertia about a principal axis through the center of mass I_c to the inertia about a parallel axis I separated by a perpendicular distance D:

$$I = I_c + mD^2 \tag{5.49}$$

As a simplification, we assume that the following ratios are small; Δ/R, δ/d, d/a, d/b, d/c. In the following we neglect the squares of these ratios compared to unity. With these approximations and assumptions, we can easily show that the three principal inertias are given by

$$I_1 = m_A \left[\frac{R^2}{4} + (R+b)^2 \right] + m_B \left[\frac{a^2}{12} + b^2 \right] + \frac{m_C b^2}{3}$$

$$I_2 = m_A \left[\frac{R^2}{2} + (R+b)^2 \right] + m_B b^2 + \frac{m_C b^2}{3} + \frac{m_D c^2}{12}$$

$$I_3 = m_A \left[\frac{R^2}{4} + a^2 \right] + \frac{m_B a^2}{3} + \frac{m_D c^2}{12} \tag{5.50}$$

If all the parts are made of the same material, with density, ρ, then the masses are given by

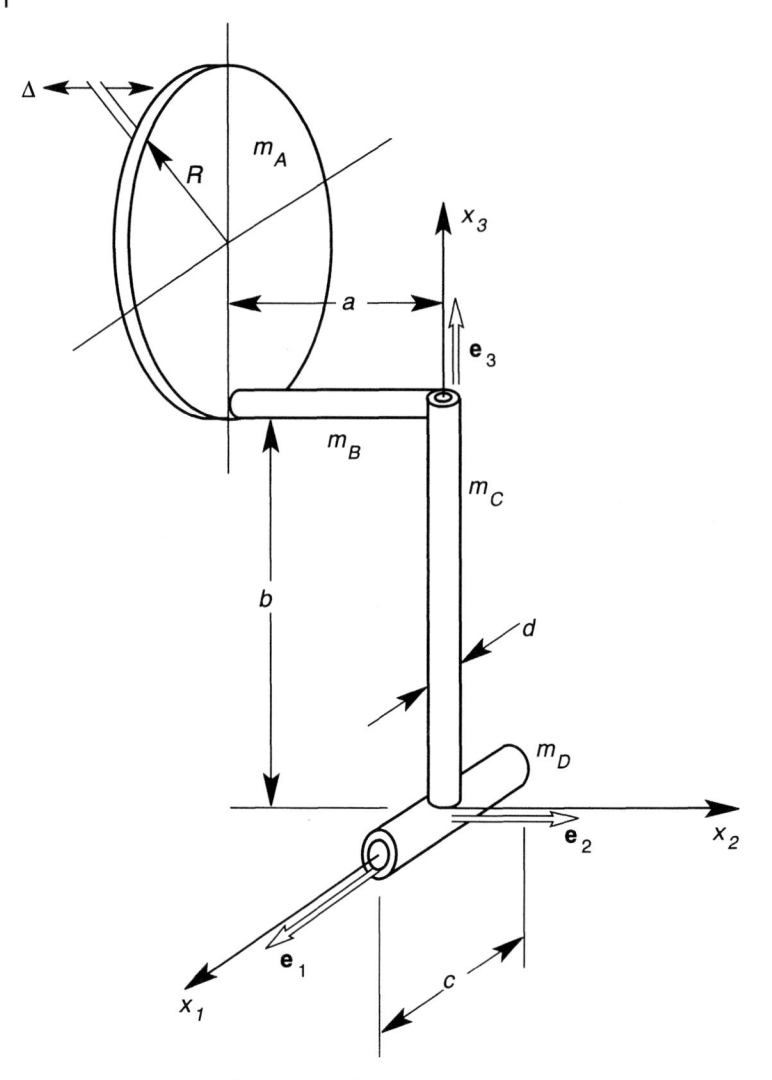

Fig. 5.9 Geometric model to calculate the moments of inertia of a deployable satellite dish (Example 5.6).

$$m_A = \pi R^2 \Delta \rho$$

$$m_B = \pi d \delta a \rho; \qquad m_C = \pi d \delta b \rho; \qquad m_D = \pi d \delta c \rho \tag{5.51}$$

Angular Momentum and Dynamics

As reviewed in Chapter 2, the laws governing the motion of a rigid body are the application of Newton's law to the motion of the center of mass and the application of the law of angular momentum to the rotational motion of the body. (In the latter, the moments in **M** and **H** can be calculated either about a fixed point or about the center of mass.)

$$\mathbf{F} = \frac{d}{dt}m\mathbf{v}_c$$
$$\mathbf{M} = \frac{d}{dt}\mathbf{H} \tag{5.52}$$

In the previous section we learned that the angular momentum **H** is linearly proportional to the angular velocity components [see (5.46) or (2.54)], through the principal second moments of mass $\{I_1, I_2, I_3\}$ and the orthonormal basis of principal axes $\{\mathbf{e}_1, \mathbf{e}_2, \mathbf{e}_3\}$:

$$\mathbf{H} = \mathbf{I}\boldsymbol{\omega} \tag{5.53}$$

or

$$\mathbf{H} = I_1\omega_1\mathbf{e}_1 + I_2\omega_2\mathbf{e}_2 + I_3\omega_3\mathbf{e}_3 \tag{5.54}$$

This relationship makes the law or angular momentum different from the law of linear momentum. In the latter, the force is directly proportional to the acceleration vector, $\dot{\mathbf{v}}_c$, but the force moment, **M**, is not always proportional to the angular acceleration vector $\dot{\boldsymbol{\omega}}$. Even in the case of motion about a fixed axis, there can arise a moment transverse to the axis of rotation if the rotation axis is not a principal axis. This difference makes the use of educated intuition in problems of angular motion more difficult than the application of Newton's law to particles where one learns about Coriolis and centripetal acceleration and can directly relate the accelerations to associated forces. Another difficulty in rigid-body dynamics is the problem of moment-free dynamics. When $\mathbf{F} = 0$, the velocity is a constant vector. When the moment is zero, e.g., $\mathbf{M} = 0$, the angular momentum vector, **H**, is constant, but the angular-velocity vector $\boldsymbol{\omega}$ may not be constant.

Example 5.7 Translating Pendulum. A long, thin solid rod is suspended at one end and attached with a low friction bearing to a massless moving base (Fig. 5.10). Starting from rest, the rod is to be moved a distance s_0 from point A to point B. An engineer is asked to design a time history $u(t)$ such that the rod comes to rest (i.e., $\theta = \dot{\theta} = 0$) at point B without the need of braking or friction.

To solve this problem we first write the equation of motion for the rod using a polar coordinate system attached to the rod. The velocity and acceleration of the center of mass of the rod is given by

$$\mathbf{v}_c = \dot{u}\mathbf{e}_x + \dot{\theta}\frac{L}{2}\mathbf{e}_\theta$$
$$\mathbf{a}_c = \ddot{u}\mathbf{e}_x + \ddot{\theta}\frac{L}{2}\mathbf{e}_\theta - \dot{\theta}^2\frac{L}{2}\mathbf{e}_r \tag{5.55}$$

The Newton–Euler equations of motion become

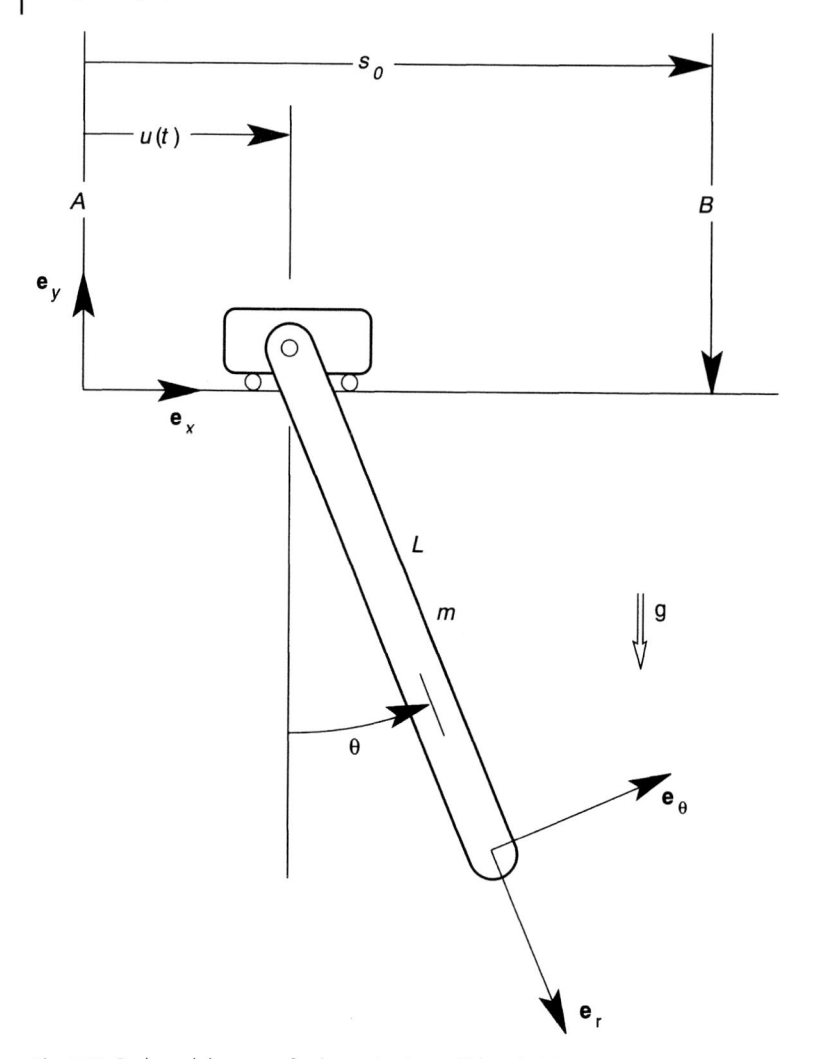

Fig. 5.10 Rod pendulum on a freely moving base (Example 5.7).

$$m\mathbf{a}_c = F_r \mathbf{e}_r + F_\theta \mathbf{e}_\theta - mg\mathbf{e}_y$$

$$I_c \ddot{\theta} = -F_\theta \frac{L}{2} \qquad (5.56)$$

This set involves three scalar equations in the three scalar unknowns, $\ddot{\theta}$, and the two components of the pin force F_r, F_θ. Writing $\mathbf{a}_c$ and $\mathbf{e}_y$ in polar coordinates, we can eliminate F_θ from the angular momentum equation (note $\mathbf{e}_y = -\cos\theta\,\mathbf{e}_r + \sin\theta\,\mathbf{e}_\theta$, $\mathbf{e}_x = \cos\theta\,\mathbf{e}_\theta + \sin\theta\,\mathbf{e}_r$):

$$I_c \ddot{\theta} = -\frac{L}{2}[ma_{c\theta} + mg\sin\theta]$$

or

$$I_0\ddot{\theta} + \frac{mgL}{2}\sin\theta = -\frac{mL}{2}\ddot{u}\cos\theta \qquad (5.57)$$

where $I_0 = I_c = mL^2/4$ is the moment of inertia about the axis of rotation.

As with many engineering problems, there is more than one solution. Here we choose one such that θ is small and $\sin\theta \sim \theta$ and $\cos\theta \simeq 1$. We also divide the time interval into two short-time, impulsive intervals near A, B, and an interval in between where $\dot{u} = v_0$ is constant, so that the elapsed time between A, B is $t_0 = s_0/v_0$.

During the impulsive interval near A, $\sin\theta \simeq 0$, and we can integrate the equation to obtain

$$I_0\dot{\theta}(t = 0^+) = -\frac{mL}{2}v_0 \qquad (5.58)$$

Since θ is small in the steady motion stage, we have

$$I_0\ddot{\theta} + \frac{mgL}{2}\theta = 0 \qquad (5.59)$$

or

$$\theta = C\sin\omega t$$

where $\omega^2 = mgL/2I_0$, and from the initial impulse, $\omega C = -mLv_0/2I_0$.

Near point B at $t = t_0$, we require that, $\theta = 0$. Integration of the equation of motion around t_0 yields

$$I_0\left[\dot{\theta}(t_0^+) - \dot{\theta}(t_0^-)\right] = -\frac{mL}{2}\left[\dot{u}(t_0^+) - \dot{u}(t_0^-)\right] \qquad (5.60)$$

where $t_0^+ - t_0^-$ is very small compared with t_0. We know that $\dot{u}(t_0^-) = v_0$ and require that $\dot{\theta}(t_0^+) = \dot{u}(t_0^+) = 0$. Thus, before the impulsive braking of the pivot, we must have

$$-I_0\dot{\theta}(t_0^-) = \frac{mL}{2}v_0; \qquad \theta(t_0^-) = 0 \qquad (5.61)$$

From the steady-motion solution we have

$$\dot{\theta}(t_0^-) = -\frac{mL}{2I_0}v_0\cos\omega t_0 \qquad (5.62)$$

If we choose $\omega t_0 = 2n\pi$ or

$$v_0 = \frac{s_0}{2n\pi}\sqrt{\frac{mgL}{2I_0}} \qquad (n \text{ is an integer}) \qquad (5.63)$$

we can bring the oscillation to a halt at $\theta = 0$ and satisfy (5.61).

This analysis can also be applied to a construction crane moving a load hanging from a cable. The student can verify the solution by moving a heavy hanging object from a string with his or her hand.

Euler's Equations for Rigid-body Dynamics

There are three levels of expressing the law of angular momentum developed in Chapter 2. The first level is simply a statement that the rate of change of angular momentum is equal to the applied moment vector:

$$\mathbf{M} = \dot{\mathbf{H}} \tag{5.64}$$

In this section we assume that either the body is in pure rotation about a fixed point 0, and $\mathbf{M}$, $\mathbf{H}$ are calculated with respect to this point, or in the case of general motion, $\mathbf{M}$, $\mathbf{H}$ involve moments about the center of mass.

The next level of writing the law of angular momentum is to express the equations of motion in terms of the angular velocity as in (5.46)

$$\mathbf{H} = \mathbf{I}\boldsymbol{\omega} \tag{5.65}$$

The third level is to add the kinematic equations relating $\boldsymbol{\omega}$ to the angular variables, e.g., (5.23) or (5.24) or to incorporate the angles or generalized position variables directly into $\mathbf{H}$.

Euler's equations express the law (5.64) at the second level, i.e., in terms of $\boldsymbol{\omega}$. To make this explicit we write $\boldsymbol{\omega}$ in a coordinate system with *principal* axes

$$\boldsymbol{\omega} = \omega_1 \mathbf{e}_1 + \omega_2 \mathbf{e}_2 + \omega_3 \mathbf{e}_3$$

There are two forms of these equations. In the first, $\{\mathbf{e}_1, \mathbf{e}_2, \mathbf{e}_3\}$ rotate with the body so that

$$\dot{\mathbf{e}}_1 = \boldsymbol{\omega} \times \mathbf{e}_1, \quad \text{etc.}$$

When the principal axes are fixed to the body, the calculation of the inertia matrix I is independent of time since $\{\mathbf{e}_1, \mathbf{e}_2, \mathbf{e}_3\}$ are fixed to the body. Therefore, the change in $\mathbf{H}$ comes about due to a change in $\boldsymbol{\omega}$ and more specifically, due to changes in both the components $\{\omega_1(t), \omega_2(t), \omega_3(t)\}$, as well as the change in the directions of $\{\mathbf{e}_i\}$. Carrying out the details we write

$$\dot{\mathbf{H}} = I_1 \dot{\omega}_1 \mathbf{e}_1 + I_1 \omega_1 \dot{\mathbf{e}}_1 + \text{etc.}$$

or

$$\dot{\mathbf{H}} = I_1 \dot{\omega}_1 \mathbf{e}_1 + I_1 \omega_1 \boldsymbol{\omega} \times \mathbf{e}_1, \quad \text{etc.} \tag{5.66}$$

Collecting all the terms for each direction, we obtain *Euler's equations of motion*

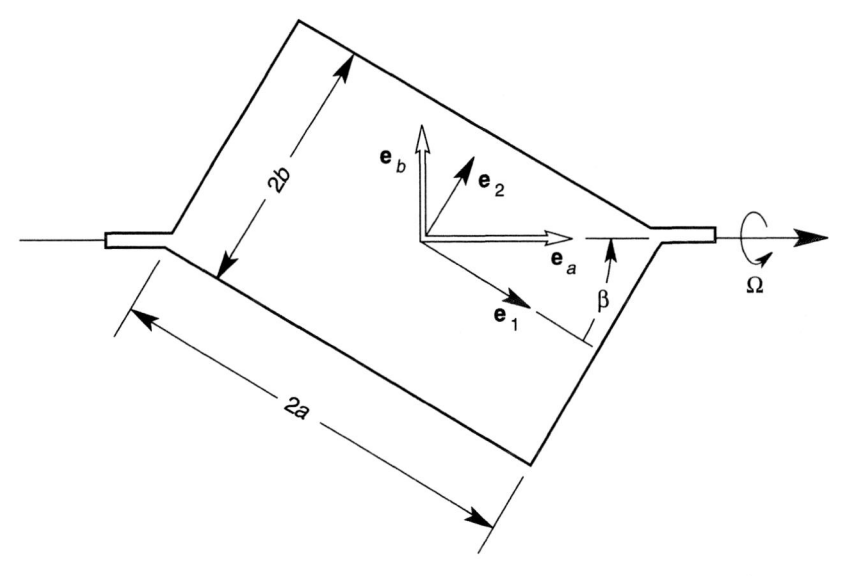

Fig. 5.11 Rotation of a rigid, thin, rectangular plate about a nonprincipal axis (Example 5.8).

$$M_1 = I_1\dot{\omega}_1 + (I_3 - I_2)\omega_2\omega_3$$

$$M_2 = I_2\dot{\omega}_2 + (I_1 - I_3)\omega_3\omega_1$$

$$M_3 = I_3\dot{\omega}_3 + (I_2 - I_1)\omega_1\omega_2 \tag{5.67}$$

This is a set of three coupled, first-order, nonlinear differential equations. To solve for $\omega_i(t)$, we require knowledge of the moments that may depend explicitly on time, on the angular velocity, and even on the angular variables.

In the second case or *modified Euler's equations*, one of the axes is an axis symmetry, say $\mathbf{e}_3$, with $I_1 = I_2$. The angular velocity is written in the form

$$\boldsymbol{\omega} = \boldsymbol{\Omega} + \omega_0 \mathbf{e}_3$$

where $\boldsymbol{\Omega}$ is the angular velocity of the basis vectors $\{\mathbf{e}_1, \mathbf{e}_2, \mathbf{e}_3\}$, i.e., $\dot{\mathbf{e}}_1 = \boldsymbol{\Omega} \times \mathbf{e}_1$, etc.

Thus the body spins about the $\mathbf{e}_3$ axis. When $\dot{\omega}_0 = 0$, the modified Euler equations become

$$M_1 = I_1\dot{\Omega}_1 + (I_3 - I_1)\Omega_2\Omega_3 + I_3\omega_0\Omega_2$$

$$M_2 = I_1\dot{\Omega}_2 + (I_1 - I_3)\Omega_3\Omega_1 - I_3\omega_0\Omega_1$$

$$M_3 = I_3\dot{\Omega}_3 \tag{5.68}$$

There are several important cases where Euler's equations can be solved explicitly. These are discussed below.

Example 5.8 Motion About a Fixed Axis. Consider the example of a plate in rotation about a fixed axis, as shown in Fig. 5.11. We write the angular velocity in the form

$$\boldsymbol{\omega} = \Omega \mathbf{e}_a = \omega_1 \mathbf{e}_1 + \omega_2 \mathbf{e}_2 + \omega_3 \mathbf{e}_3$$

or

$$\omega_1 = \Omega \mathbf{e}_a \cdot \mathbf{e}_1 = \Omega d_1, \quad \text{etc.}$$

where $\{d_i\}$ are direction cosines and

$$d_1^2 + d_2^2 + d_3^2 = 1$$

We also define

$$M_a = \mathbf{M} \cdot \mathbf{e}_a$$

In this case, we can easily show that the first two Euler's equations (5.67) reduce to a scalar equation

$$M_a = I_a \dot{\Omega} \tag{5.69}$$

where since $d_3 = 0$

$$I_a = I_1 d_1^2 + I_2 d_2^2$$

$$M_a = d_1 M_1 + d_2 M_2$$

and $d_1 = \cos\beta$, $d_2 = \sin\beta$. In this example $\omega_1 = \Omega \cos\beta$, $\omega_2 = \Omega \sin\beta$, $\omega_3 = 0$. Using the third equation in Euler's equations (5.67), we obtain

$$M_3 = (I_2 - I_1)\Omega^2 \cos\beta \sin\beta \tag{5.70}$$

This moment must be created by bearing forces normal to the plate surface

$$F = M_3/2(a^2 + b^2)^{1/2}$$

If Ω is related to an angular variable θ,

$$\Omega = \frac{d\theta}{dt}$$

then a class of problems where the applied moment is linearly proportional to θ, ω can be solved using standard techniques of linear ordinary differential equations, i.e., if M_a is the sum of applied and reaction torques, then (5.69) becomes

$$M_a = M_0(t) - \gamma\Omega - \kappa\theta$$

5.3 Newton–Euler Equations of Motion

$$I_a \ddot{\theta} + \gamma \dot{\theta} + \kappa \theta = M_0(t) \tag{5.71}$$

The third term on the left-hand side of (5.71) is a torsional stiffness. For most motor-generator, pump, and engine applications, $\kappa = 0$, and (5.71) takes the simple form of a first-order differential equation:

$$I_a \dot{\Omega} + \gamma \Omega = M_0(t) \tag{5.72}$$

or

$$\Omega(t) = C_1 e^{-(\gamma/I_a)t} + \frac{1}{I_a} \int e^{(\gamma/I_a)(t-\tau)} M_0(\tau)\, d\tau$$

The first term represents the damped decay of the angular velocity, and the second term represents a particular solution depending on the time history of $M_0(t)$.

Gyroscopic Dynamics

The most tangible effect of gyroscopic motion occurs when the axis of rotation of a body is itself made to rotate about a transverse axis, thereby generating a moment, couple, or torque about an axis orthogonal to the two rotation axes. The reader can experience this troika or three-axis effect by demounting a front wheel from a mountain bike, spinning the wheel while holding the axle, and sitting in a rotary office chair. The effect, illustrated in Fig. 5.12a, will be the feeling of a torque about the axis parallel to the arms.

This effect can be best understood by using the law of angular momentum (Chapter 1) and the rule for calculating the derivative of a rotating vector of constant length. Assume that a body spins with rate $\dot{\phi}$ about one of its principal axes, say $\mathbf{e}_1$, which is made to rotate about the $\mathbf{e}_3$ axis with rate $\dot{\psi}$, shown in Fig. 5.12b. The angular momentum is given by

$$\mathbf{H} = I_1 \dot{\phi} \mathbf{e}_1 + I_3 \dot{\psi} \mathbf{e}_3 \tag{5.73}$$

with $\dot{\phi}, \dot{\psi}$ held constant. The rate of change of $\mathbf{H}$ becomes

$$\dot{\mathbf{H}} = I_1 \dot{\phi} \dot{\mathbf{e}}_1 = I_1 \dot{\phi} \dot{\psi} \mathbf{e}_3 \times \mathbf{e}_1 \tag{5.74}$$

Thus the applied moment requiring this to happen is given by a vector orthogonal to the $\{\mathbf{e}_1, \mathbf{e}_3\}$ plane:

$$\mathbf{M} = I_1 \dot{\phi} \dot{\psi} \mathbf{e}_2 \tag{5.75}$$

From the figure, it is clear that as the tip of the angular momentum vector rotates with $\dot{\psi}$ in a circle, the change in $\mathbf{H}$ must occur in the direction of the circle or perpendicular to the axes of $\dot{\phi}$ and $\dot{\psi}$. Another way to express this is to write

$$\mathbf{M} = \dot{\boldsymbol{\psi}} \times \mathbf{H} \tag{5.76}$$

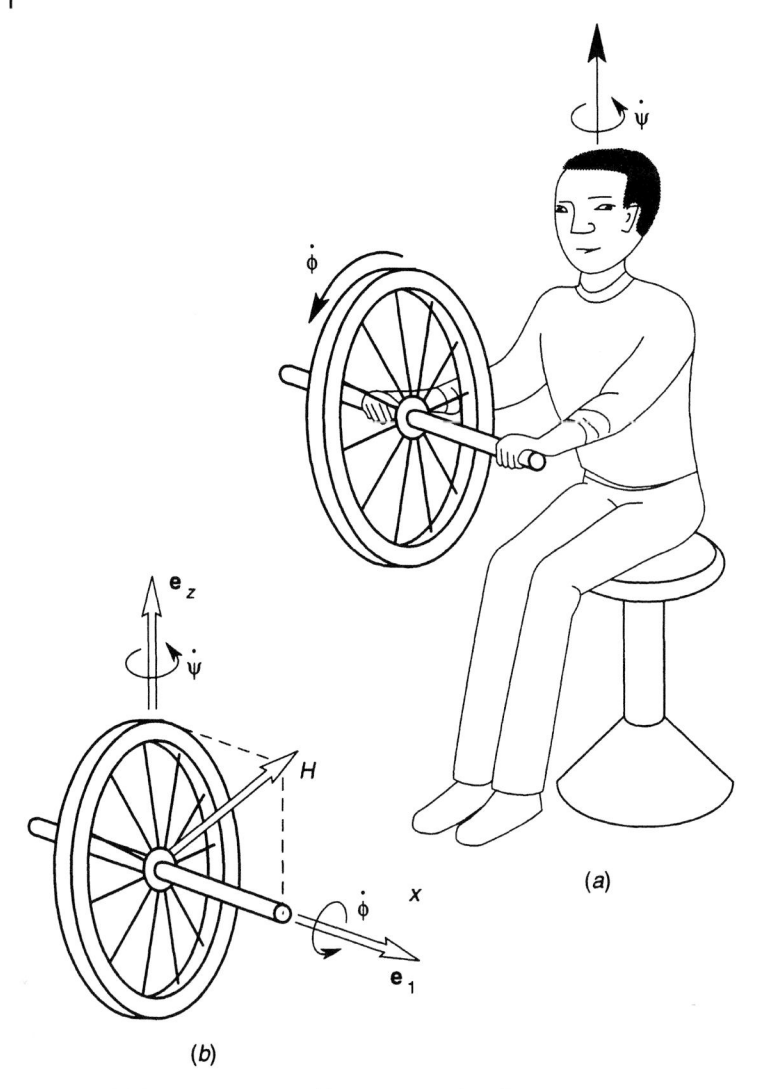

Fig. 5.12 (a) Gyro experiment: Holding a spinning bicycle wheel while sitting on a swivel chair, try to turn the wheel about an axis perpendicular to the spin axis. (b) Sketch of the angular-momentum components of the wheel.

Steady Precession of a Gyro Top

The gyro effect can also be seen in the steady precession of a spinning symmetric body about a point (Fig. 5.13). Here we take moments about the fixed point, 0. Choosing $\mathbf{e}_3$ for the symmetry axis and assuming that $I_1 = I_2$, the angular-momentum vector and the rotation vector can be expressed as

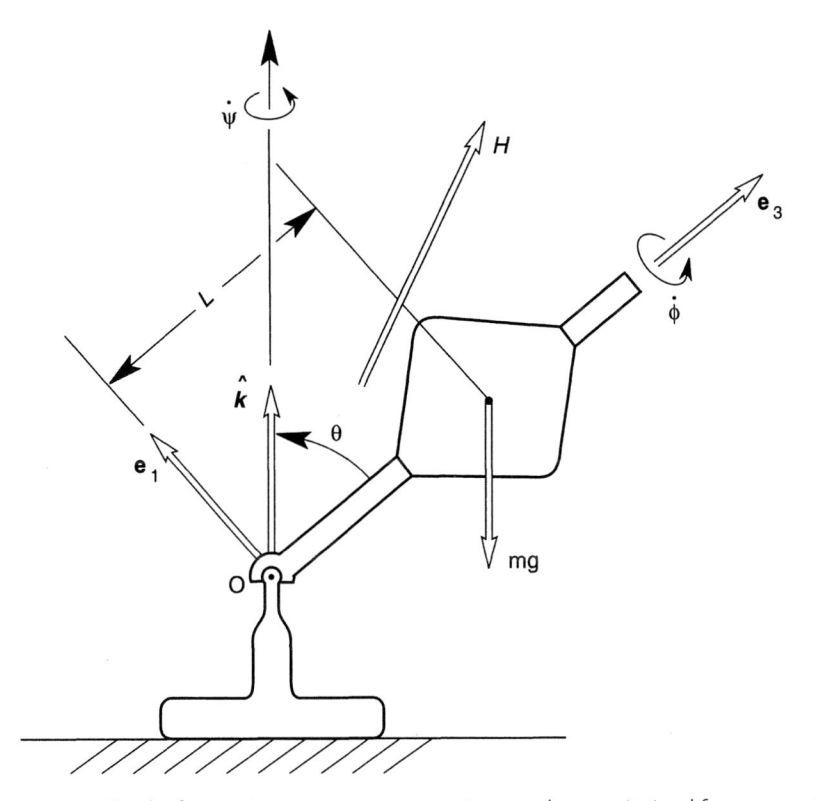

Fig. 5.13 Sketch of a spinning, precessing, symmetric top under a gravitational-force moment.

$$\mathbf{H}_0 = I_1\omega_1\mathbf{e}_1 + I_1\omega_2\mathbf{e}_2 + I_3\omega_3\mathbf{e}_3$$

$$\boldsymbol{\omega} = \dot{\phi}\mathbf{e}_3 + \dot{\psi}\cos\theta\mathbf{e}_3 + \dot{\psi}\sin\theta\mathbf{e}_1 \tag{5.77}$$

where the precession $\dot{\psi}$ occurs about the vertical or $\hat{\mathbf{k}}$ axis.

Thus $\omega_1 = \dot{\psi}\sin\theta$, $\omega_2 = 0$, $\omega_3 = \dot{\phi} + \dot{\psi}\cos\theta$. For steady motion we assume that $\dot{\theta} = 0$, and $\dot{\omega}_1 = \dot{\omega}_2 = \dot{\omega}_3 = 0$. (Note that the vector rate $\dot{\boldsymbol{\omega}}$ is not zero.) As in the previous example, the change in $\mathbf{H}_0$ is given by

$$\dot{\mathbf{H}}_0 = \dot{\psi}\hat{\mathbf{k}} \times \mathbf{H}_0 = (I_1\omega_1\cos\theta - I_3\omega_3\sin\theta)\dot{\psi}\mathbf{e}_2 \tag{5.78}$$

This change in $\mathbf{H}_0$ must equal the applied-force moment, $\mathbf{M}_0$. In this problem the gravitational force mg produces a moment about the point 0 given by

$$\mathbf{M}_0 = -mgL\mathbf{e}_3 \times \hat{\mathbf{k}} = -mgL\sin\theta\mathbf{e}_2 \tag{5.79}$$

Thus, for steady rotation about the point 0 to occur, we must satisfy one of two conditions

$$\sin\theta = 0$$

or

$$mgL = I_3\dot{\psi}\dot{\phi} + (I_3 - I_1)\dot{\psi}^2 \cos\theta \tag{5.80}$$

We also note that since the gravity force acts through the $\mathbf{e}_3$ axis, the H_3 component is conserved, i.e.,

$$\mathbf{H}_3 = I_3\omega_3 = I_3(\dot{\phi} + \dot{\psi}\cos\theta) = \text{constant} \tag{5.81}$$

Thus if we choose the angle θ, Eqs. (5.80) and (5.81) determine the values of $\dot{\phi}$ and $\dot{\psi}$ for steady rotation under the gravity moment.

Special Case $\theta = \pi/2$
For this case we see that $\dot{\phi}$ = constant, and

$$mgL = I_3\dot{\psi}\dot{\phi}_0$$

Thus the precession speed $\dot{\psi}$ is determined by

$$\dot{\psi} = \frac{mgL}{I_3\dot{\phi}_0} \tag{5.82}$$

Note also that since I_3 is proportional to the mass times the radius of gyration squared, i.e., $I_3 = mr_G^2$, $\dot{\psi}$ is independent of the mass

$$\dot{\psi} = \frac{gL}{r_G^2\dot{\phi}_0} \tag{5.83}$$

Increasing the spin $\dot{\phi}_0$ decreases the precession and increasing the moment arm increases $\dot{\psi}$. For a more advanced analysis of the spinning top and related gyroscopic dynamics, look at Meirovitch (1970) or Greenwood (1988).

Example 5.9 Energy-storage Flywheel on a Magnetic Thrust Bearing. Conventional energy-storage flywheels suffer energy losses from windage and mechanical bearings. In recent designs the rotor is operated in a vacuum and levitated on either active electromagnetic or passive superconducting magnetic bearings. One such design is shown in Fig. 5.14 using cryogenically cooled yttrium–barium–copper–oxide (YBCO) super-conducting material as a thrust bearing (see Moon, 1994). The magnetic-field source is a rare-earth permanent magnet (see Chapter 8 for a discussion of magnetic forces). In this example we wish to determine the natural frequencies of the rotor as a function of the rotation speed. For this problem it is easier to use the so-called *modified Euler's equations* (5.68). This formulation is useful for axially symmetric bodies for which the inertia matrix does not change with motion around the spin axes. The rotation of the symmetric body is separated into a spin component and a gimbal component $\mathbf{\Omega}$, i.e.,

$$\boldsymbol{\omega} = \omega_s \mathbf{e}_3 + \mathbf{\Omega} \tag{5.84}$$

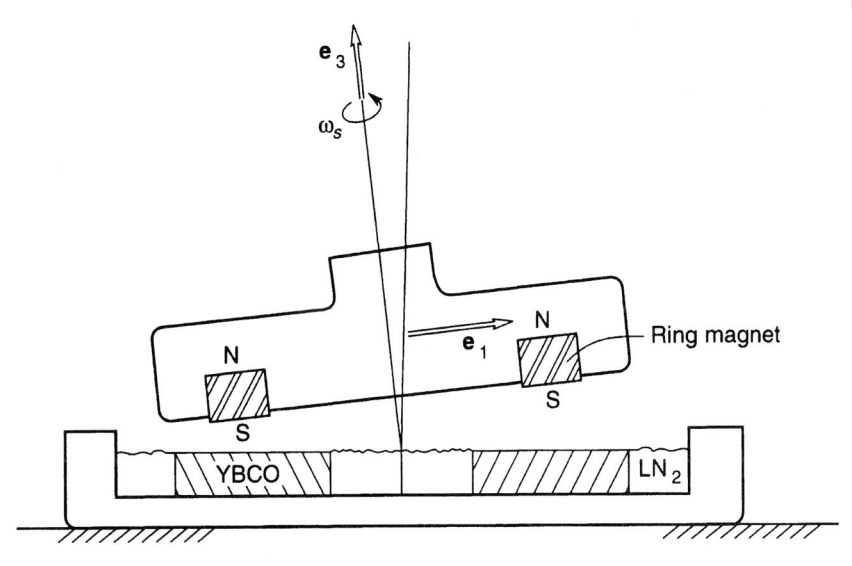

Fig. 5.14 Spinning, magnetically levitated flywheel suspended above a ceramic superconductor (YBCO) (Example 5.9).

The gimbal frame unit vectors are denoted by ($\mathbf{e}_x, \mathbf{e}_y, \mathbf{e}_z = \mathbf{e}_3$). The angular momentum of the flywheel is given by

$$\mathbf{H} = I_1 \Omega_x \mathbf{e}_x + I_1 \Omega_y \mathbf{e}_y + I_3 (\Omega_z + \omega_s) \mathbf{e}_z \tag{5.85}$$

The resulting modified Euler's equations (5.68) become

$$I_1 \dot{\Omega}_x + I_3 (\Omega_z + \omega_s) \Omega_y - I_1 \Omega_y \Omega_z = M_x$$
$$I_1 \dot{\Omega}_y + I_1 \Omega_x \Omega_z - I_3 (\Omega_z + \omega_s) \Omega_x = M_y$$
$$I_3 (\dot{\Omega}_z + \dot{\omega}_s) = M_z \tag{5.86}$$

To investigate the small vibrations of the spinning flywheel, we assume that ω_s is large compared to the components of the gimbal angular velocity Ω. Keeping only linear terms in Ω, the previous equations become

$$I_1 \dot{\Omega}_x + I_3 \Omega_y \omega_s = M_x$$
$$I_1 \dot{\Omega}_y - I_3 \Omega_x \omega_s = M_y$$
$$I_3 (\dot{\Omega}_z + \dot{\omega}_s) = M_z \tag{5.87}$$

We next choose a set of kinematic relations relating $\mathbf{\Omega}$ to a set of Euler angles. To first order we have

$$\Omega_x = \dot{\phi}, \qquad \Omega_y = \dot{\theta}$$

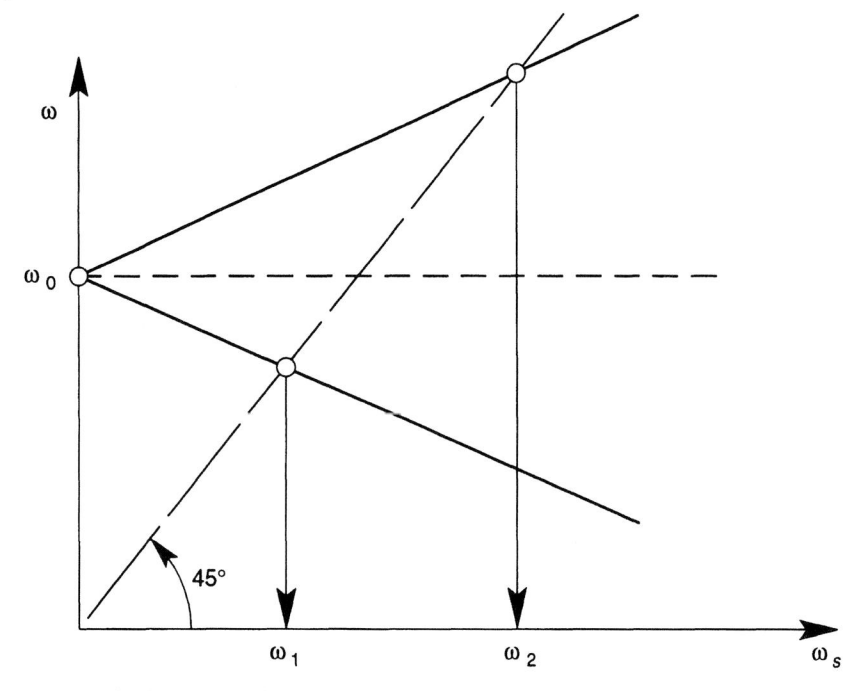

Fig. 5.15 The dependence of the natural frequencies of a spinning levitated flywheel as a function of the spin rate ω_s (Example 5.9).

As a further simplification we assume that the angular motion is decoupled from the center of mass motion. This is tantamount to assuming restoring magnetic torques in the form

$$M_x = -\kappa\phi, \qquad M_y = -\kappa\theta \qquad (5.88)$$

These assumptions result in the following coupled second-order differential equations

$$I_1\ddot{\phi} + I_3\omega_s\dot{\theta} = -\kappa\phi$$
$$I_1\ddot{\theta} - I_3\omega_s\dot{\phi} = -\kappa\theta \qquad (5.89)$$

The second terms on the left-hand side are called gyroscopic coupling. When $\omega_s = 0$, there is a double natural frequency for pitch and roll; $\omega_0 = (\kappa/I_1)^{1/2}$. Gyroscopic coupling induces coupled pitch and roll 90° out of phase, i.e.,

$$\phi = A\cos\omega t, \qquad \theta = B\sin\omega t$$

The characteristic equation for the natural frequencies is given by

$$(\omega^2 - \omega_0^2)^2 - \left(\frac{I_3 \omega_s}{I_1}\right)^2 \omega^2 = 0 \tag{5.90}$$

The two branches as a function of spin rate ω_s are shown in Fig. 5.15. If there is a slight imbalance in the flywheel rotor, there will be a driving frequency ω shown as the 45° line in Fig. 5.15. The splitting of the frequencies means that there can be two resonant conditions where the imbalance frequency can excite large oscillations in the flywheel. This problem is similar to a spinning rotor on air bearings or other soft suspension.

Moment-free Dynamics

Spinning satellites, gymnasts, the free flight of sports balls, such as footballs, are all governed to some extent by moment-free dynamics, neglecting aerodynamic forces, of course. It was noted that Euler's equations of motion (5.67) for the angular motion of a rigid body are incomplete. We must define the applied moments as well as append kinematic relations between the angular variables and the components of angular velocity. However, in the moment-free case an important problem can be addressed with Euler's equations, namely, the stability of rotation of a body about one of its principal axes. By stability we ask the question: If the initial rotation vector is ever so close to a principal axis, will it remain close, or will the body begin to see increasing rotation about one of the other axes?

A complete theoretical analysis of moment-free motion can be found in classic texts such as Goldstein. Here we present a more limited analysis and give the important results of the theory.

The main results of this theory can be observed easily with an experiment. The reader should take a book, like this one, and secure the covers with a strong rubber band, as illustrated in Fig. 5.16. The experimenter throws the book up into the air in such a way that the book initially rotates about one of the three axes. What are the observations?

1. Rotary motion about either of two principal axes persists
2. Rotation about one of the three principal axes begins to wobble and complex motions of the book are observed

The question then arises: What distinguishes the unstable principal axis from the two stable axes?

We begin by writing Euler's equations (5.68) under the assumption that the net force moment about the center of mass is zero:

$$I_1 \dot{\omega}_1 + (I_3 - I_2)\omega_3 \omega_2 = 0$$
$$I_2 \dot{\omega}_2 + (I_1 - I_3)\omega_3 \omega_1 = 0$$
$$I_3 \dot{\omega}_3 + (I_2 - I_1)\omega_1 \omega_2 = 0 \tag{5.91}$$

Now assume that there is an initial motion about one of the axes, ω_0, and that a small perturbation vector $[\eta_1(t), \eta_2(t), \eta_3(t)]^T$ is added to the angular velocity

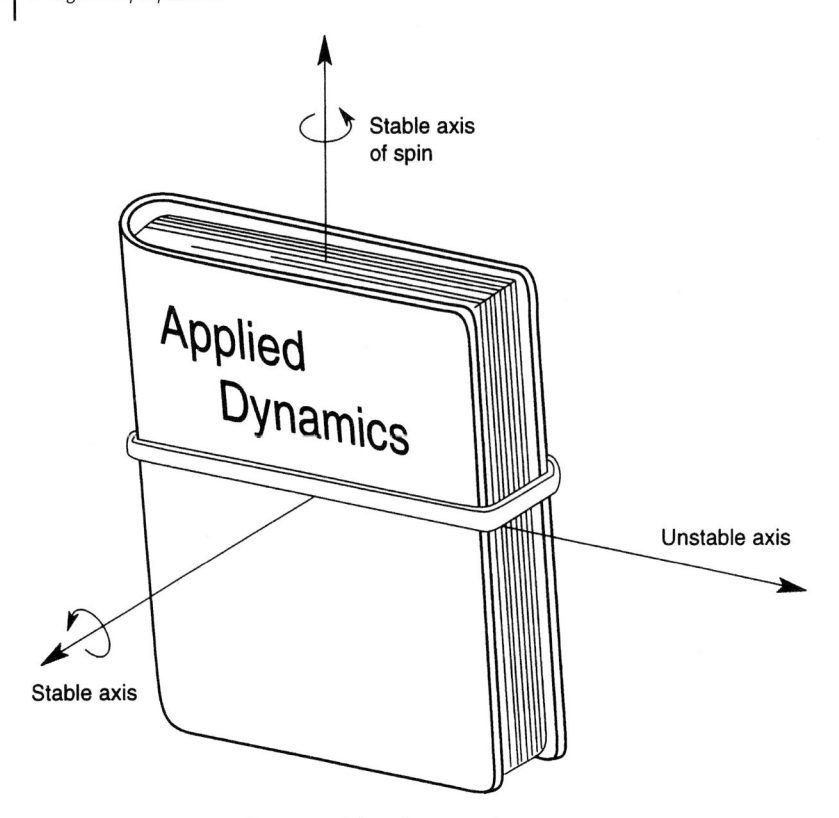

Fig. 5.16 Experiment on the spin stability of moment-free rotation of a rigid body. (Use a rubber band to keep the book covers closed while spinning.)

vector

$$\omega_1 = \omega_0 + \eta_1(t)$$
$$\omega_2 = \eta_2(t)$$
$$\omega_3 = \eta_3(t) \tag{5.92}$$

These expressions are put into Euler's equations (5.91) and all nonlinear terms of the type, $\eta_i \eta_j$ are dropped.

$$I_1 \dot{\eta}_1 = 0$$
$$I_2 \dot{\eta}_2 + (I_1 - I_3)\omega_0 \eta_3 = 0$$
$$I_3 \dot{\eta}_3 + (I_2 - I_1)\omega_0 \eta_2 = 0 \tag{5.93}$$

This method of analysis is called *perturbation theory*. The theoretical basis for dropping the nonlinear terms can be found in books by Nayfeh and Mook (1979) or Guckenheimer and Holmes (1983).

A solution to the second and third first-order linear differential equations can be found by writing

$$\begin{bmatrix} \eta_2 \\ \eta_3 \end{bmatrix} = \begin{bmatrix} a \\ b \end{bmatrix} e^{\lambda t} \tag{5.94}$$

It is easy to see that for a solution to exist

$$\lambda^2 = (I_1 - I_3)(I_2 - I_1)\omega_0^2 / I_3 I_2 \tag{5.95}$$

Thus there are two solutions of the form (5.94) corresponding to the two roots of (5.95). If $\lambda > 0$, one solution is unstable since perturbations about the off-axis spin will grow and the body will experience severe wobbling motions. However, if $\lambda^2 < 0$, oscillating solutions will exist and the spin will remain close to the original axis. For stable spin we require

$$I_1 > I_2 > I_3$$

or

$$I_1 < I_2 < I_3$$

One concludes that the $\mathbf{e}_1$ or spin axis is stable if I_1 is either the *maximum* or *minimum* principal inertia. On the other hand, the spin is unstable if the $\mathbf{e}_1$ axis is the intermediate principal inertia axis. This can be confirmed with the rotating book experiment of Fig. 5.16.

The general theory of moment-free dynamics of a rigid body can be understood by reference to the *Poinsott ellipsoid* shown in Fig. 5.17 (see Goldstein, 1980). Since $\mathbf{M} = 0$, $\dot{\mathbf{H}} = 0$, the angular-momentum vector $\mathbf{H}$ is of constant length and fixed in space. What is counter-intuitive, is that the angular velocity vector $\boldsymbol{\omega}$ is not fixed in general, but can vary in time. In addition to the conservation of angular momentum, we also have conservation of kinetic energy (see Section 5.4 below), i.e.,

$$\frac{1}{2}\boldsymbol{\omega} \cdot \mathbf{H} = \frac{1}{2}(I_1\omega_1^2 + I_2\omega_2^2 + I_3\omega_3^2) = \text{constant} \tag{5.96}$$

This fact allows us to construct an ellipsoid in the $(\omega_1, \omega_2, \omega_3)$ space. The possible paths of motion of $\boldsymbol{\omega}$ are then drawn on this ellipsoid.

In this figure, $\mathbf{e}_1, \mathbf{e}_3$ represent maximum and minimum principal axes, while $\mathbf{e}_2$ is an intermediate principal axis. It can be seen that for initial motions near $\mathbf{e}_1$ or $\mathbf{e}_3$ the angular velocity vector will move in a small orbit or precession about the principal axis. But at the intermediate inertia axis, a saddle point exists. Motions near the saddle are unstable (see, e.g., Chapter 9), i.e., the angular-velocity components about $\mathbf{e}_1$ and $\mathbf{e}_3$ will become very large (not infinite) and the body will experience large wobble, precession, and large nutation.

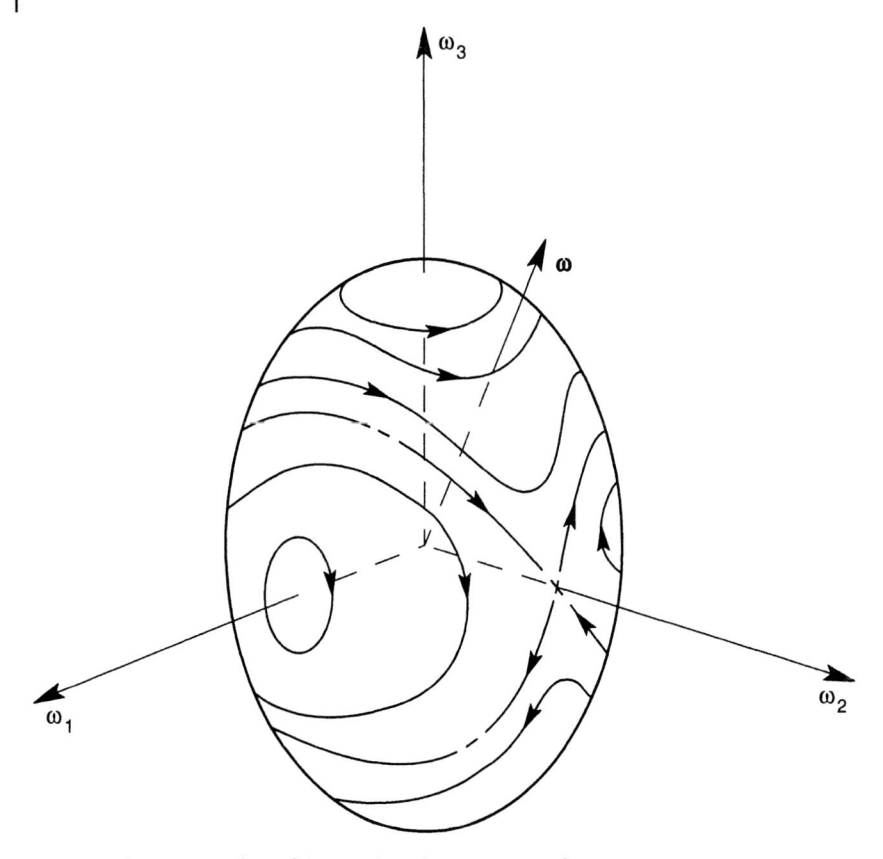

Fig. 5.17 Phase space orbits of the angular velocity vector $\boldsymbol{\omega}$ of moment-free rotation of a rigid body. Initial motions near the small ellipses about axes 1 and 3 are *stable*. Initial motions near the saddle point on axis 2 are *unstable*.

Impulse–Momentum Relations

In many problems the applied forces have a very short time history as in space satellite thruster maneuvers. For these cases one can integrate the Newton–Euler differential equations of motion to obtain the following algebraic equations. The integral of the force time history is called the *impulse* $\mathscr{I}$, and the integral of the force moments and torques is called the *angular impulse*, $\mathscr{A}$. These terms are defined as follows:

$$\mathscr{I} = \int_{t_1}^{t_2} \mathbf{F}(t)\,dt; \qquad \mathscr{A} = \int_{t_1}^{t_2} \mathbf{M}(t)\,dt \tag{5.97}$$

The integration of the Newton–Euler equations of motion (5.52) then takes the form

$$\mathscr{I} = m\Delta\mathbf{v}_c \tag{5.98a}$$

$$\mathscr{A}_c = \Delta\mathbf{H}_c \tag{5.98b}$$

If one defines a set of basis unit vectors along the principal axes of the inertia matrix $\{\mathbf{e}_1, \mathbf{e}_2, \mathbf{e}_3\}$, and projects the angular velocity vector along these directions, $\omega = \omega_1\mathbf{e}_1 + \omega_2\mathbf{e}_2 + \omega_3\mathbf{e}_3$, the angular momentum vector can easily be expressed by the vector sum of the angular momentum components about each of the principal axes: $\mathbf{H} = I_1\omega_1\mathbf{e}_1 + I_2\omega_2\mathbf{e}_2 + I_3\omega_3\mathbf{e}_3$. If we assume that the geometric orientation remains fixed during the short impulse period, then $\Delta\mathbf{e}_i = 0$. Then the impulse–momentum equations for an unconstrained body take the form

$$\mathscr{I} = \mathbf{e}_1 m\Delta v_{c1} + \mathbf{e}_2 m\Delta v_{c2} + \mathbf{e}_3 m\Delta v_{c3} \tag{5.99a}$$

$$\mathscr{A}_c = \mathbf{e}_1 I_1 \Delta\omega_1 + \mathbf{e}_2 I_2 \Delta\omega_2 + \mathbf{e}_3 I_3 \Delta\omega_3 \tag{5.99b}$$

Thus the vector components of the impulses are proportional to their respective changes in linear and angular momentum. These equations must be modified when the body is constrained. In this case the impulses of the constraint forces must enter into the analysis as well as the kinematic constraints.

Rigid Body Rotation Dynamics: General Observations

Inspired by J. P. Den Hartog's 1948 text *Mechanics*, it is useful to summarize some of the general properties of rotating rigid bodies. These principles are based on understanding the vector relationships of forces, force-moments, acceleration and angular momentum.

(i) In general, the angular momentum vector and the angular velocity vector are not co-linear. The exception is where the axis of rotation is about one of the principal axes of inertia of the body.
(ii) For every rigid body there exists three mutually perpendicular axes for which the moment of inertia is diagonal. This is true for any point fixed relative to the body.
(iii) The total angular momentum vector is the vector sum of the three angular momentum vectors about each of the principal axes.
(iv) The vector sum of the forces on a body is equal to the change of linear momentum of the center of mass.
(v) When a body is in pure rotation about a fixed point, the dynamical equations are found by equating the moment of all forces about the fixed point to the change of angular momentum about the same point.
(vi) When a rigid body has geometric or kinematic constraints, one must determine a set of kinematic equations in addition to the dynamical equations of motion.

(vii) For general rigid body motions, the angular momentum equations of Euler can be found by evaluating the force-moments and the angular momentum about the center of mass.

One can add other general rules for rigid body dynamics problems, but the above statements provide a good guide to dynamical modeling of such problems. Variational methods such as Lagrange's equations and others based on scalar mathematics sometimes mask some of these principles but the student is always advised to understand the vector principles of Newton and Euler.

5.4
Lagrange's Equations for a Rigid Body

The general method of Lagrange's equations for a system of particles was described in Chapter 4. To solve problems using Lagrange's equations, one must identify the generalized position variables $q_k(t)$ and the generalized velocities $\dot{q}_k(t)$. Then the kinetic energy, T, must be determined. Next the forces that do reversible work are expressed in terms of a potential energy function, V, or generalized forces, Q_k; then the equivalent form of Newton's laws of motion are obtained by Lagrange's equations (Eq. (4.105)), one for each of the generalized variables q_k:

$$\frac{d}{dt}\frac{\partial T}{\partial \dot{q}_k} - \frac{\partial T}{\partial q_k} + \frac{\partial V}{\partial q_k} = Q_k \tag{5.100}$$

where Q_k are generalized forces not accounted for in the potential energy function V.

Extension of Lagrange's equations to rigid bodies is straightforward, as shown below.

Kinetic Energy

There are two steps to formulate the kinetic energy function for use in Lagrange's equations of motion. First, the kinetic energy is expressed in terms of the angular velocity vector, and then the angular velocity must be expressed in terms of the generalized velocities. In the general case, the center of mass is in motion and the instantaneous axis of rotation itself is in rotation. This is the case for an aircraft, ship, or ground vehicles (Fig. 5.18). In such problems it is usually convenient to choose a set of axes that is centered at the center of mass, and whose axes are aligned with the principal axes. For these conditions we will show below that the kinetic energy associated with the translation of the center of mass can be decoupled from the kinetic energy of rotation about the center of mass:

$$T = \frac{1}{2}m\mathbf{v}_c \cdot \mathbf{v}_c + \frac{1}{2}\mathbf{H} \cdot \boldsymbol{\omega} \tag{5.101}$$

where the angular momentum $\mathbf{H}$ is calculated about the center of mass. Using (5.54), the second term can be written in an explicit form

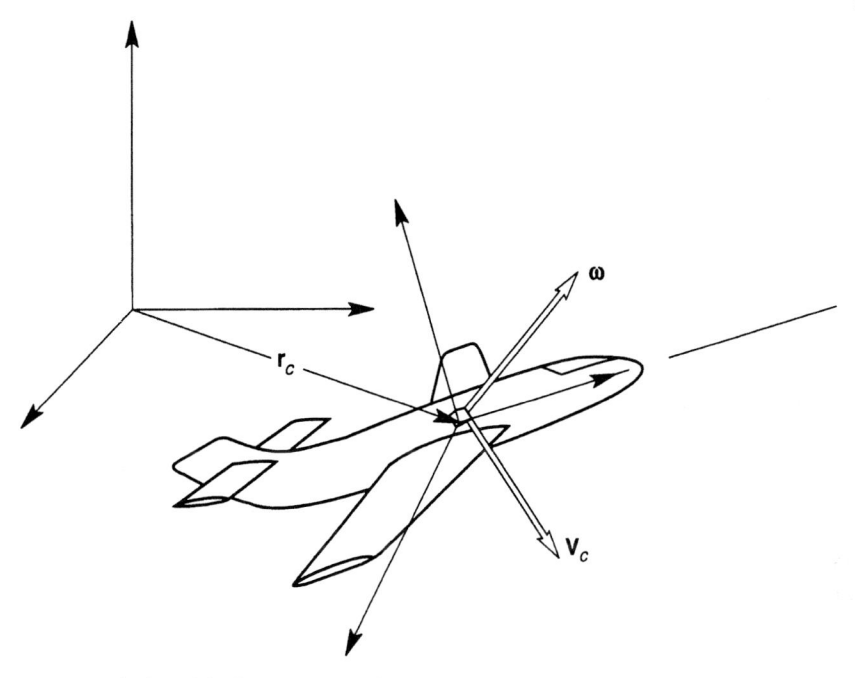

Fig. 5.18 Vehicle with both translation and rotational kinetic energy.

$$T_R = \frac{1}{2}\mathbf{H} \cdot \boldsymbol{\omega} = \frac{1}{2}[I_1\omega_1^2 + I_2\omega_2^2 + I_3\omega_3^2] \tag{5.102}$$

where $\{I_1, I_2, I_3\}$ are the principal inertias and

$$\boldsymbol{\omega} = \omega_1\mathbf{e}_1 + \omega_2\mathbf{e}_2 + \omega_3\mathbf{e}_3$$

In matrix notation the expression (5.102), is written in the form

$$T_R = \frac{1}{2}[\tilde{\omega}]^T[I][\tilde{\omega}] \tag{5.103}$$

The next step is to write the vector $\boldsymbol{\omega}$, or the column matrix $[\omega]$ in terms of generalized velocities. When a body is in simple rotation about a fixed axis $\mathbf{e}_0$ and $\boldsymbol{\omega} = \Omega\mathbf{e}_0$, then the variable $\Omega(t)$ can be directly related to the time derivative of an angular variable, say $\phi(t)$:

$$\Omega = \frac{d\phi}{dt}$$

Therefore, $\{\phi, \Omega\}$ represent a generalized position-velocity pair. In general, the angular velocity components $\{\omega_1, \omega_2, \omega_3\}$ are not related simply to the time derivative of three angles.

In Section 5.2 we explained the use of Euler angles to orient a body in space. We also derived the *kinematic relations* (5.23) or (5.24) between the angular velocity and the time derivative of the angular variables:

$$\begin{bmatrix} \omega_1 \\ \omega_2 \\ \omega_3 \end{bmatrix} = [J] \begin{bmatrix} \dot{\phi} \\ \dot{\theta} \\ \dot{\psi} \end{bmatrix}$$

Thus to complete the expression for the kinetic energy, this relation is substituted into (5.103) to obtain an explicit formula for $T_R(\phi, \theta, \psi, \dot{\phi}, \dot{\theta}, \dot{\psi})$. Examples 5.10 and 5.11 below illustrate the method.

The expression for the kinetic energy for a rigid body (5.101), can be derived from the integral

$$T = \frac{1}{2} \int \mathbf{v} \cdot \mathbf{v} \, dm \tag{5.104}$$

where

$$\mathbf{v} = \mathbf{v}_c + \boldsymbol{\omega} \times \boldsymbol{\rho}$$

Here $\mathbf{v}_c$ is the velocity of the center of mass and $\boldsymbol{\rho}$ is a vector from the center of mass to the differential mass dm. The vectors $\mathbf{v}_c, \boldsymbol{\omega}$, can be brought outside the integral. Also the first moment of mass is zero, i.e.,

$$\int \boldsymbol{\rho} \, dm = 0$$

This leads to the expression

$$T = \frac{1}{2} m \mathbf{v}_c \cdot \mathbf{v}_c + \frac{1}{2} \int (\boldsymbol{\omega} \times \boldsymbol{\rho}) \cdot (\boldsymbol{\omega} \times \boldsymbol{\rho}) \, dm$$

or

$$T = \frac{1}{2} m \mathbf{v}_c \cdot \mathbf{v}_c + \frac{1}{2} \boldsymbol{\omega} \cdot \int \boldsymbol{\rho} \times (\boldsymbol{\omega} \times \boldsymbol{\rho}) \, dm \tag{5.105}$$

The second term uses the following property of a scalar triple product,

$$(\mathbf{A} \times \mathbf{B}) \cdot \mathbf{C} = \mathbf{A} \cdot (\mathbf{B} \times \mathbf{C})$$

The integral in (5.105) can be seen to be the definition of the angular momentum of a rigid body about the center of mass $\mathbf{H}_c$, as in (5.41). The expression for T, (5.105), is identical to (5.101).

Generalized Forces for Rigid Bodies

In the application of Lagrange's equations to rigid bodies the active forces that do work must be incorporated into generalized forces $\{Q_k\}$ or a potential energy

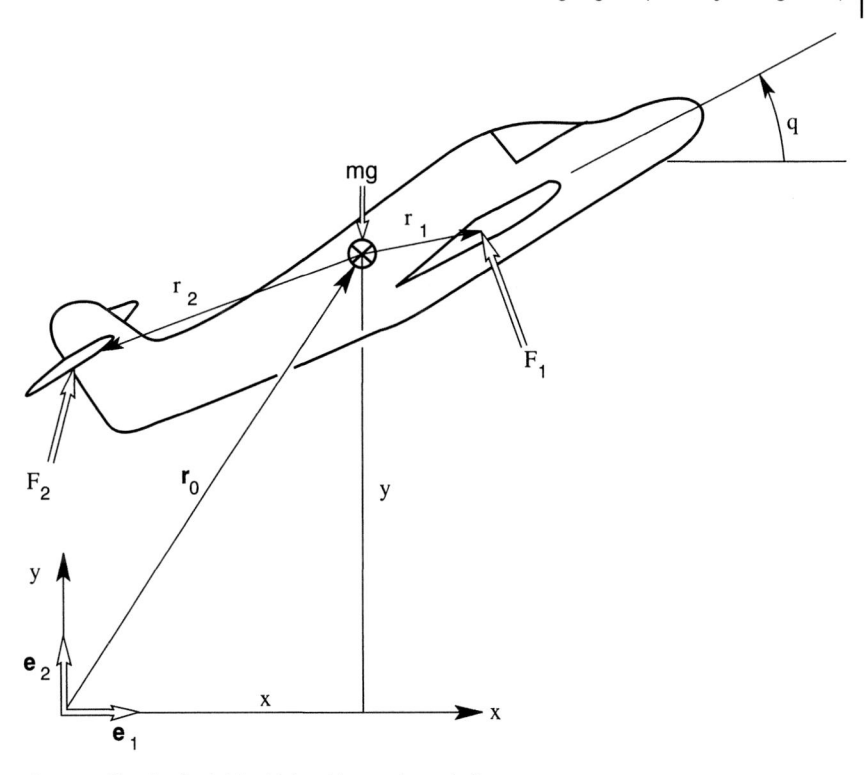

Fig. 5.19 Sketch of a rigid vehicle with aerodynamic force resultants $\mathbf{F}_1, \mathbf{F}_1$ and their respective moment arm vectors ρ_1, ρ_2 about the center of mass.

function $V(q_k)$. The idea of generalized forces originated with the foundation of Lagrange's equations, namely D'Alembert's principle of virtual work (Chapter 4). Here the virtual work of all the active forces is equal to the work of the generalized forces, i.e.,

$$\sum \mathbf{F}_i^a \cdot \delta \mathbf{r}_i = \sum Q_i \delta q_i \tag{5.106}$$

Consider the example shown in Fig. 5.19 of the planar motion of an aircraft with integrated lift and control forces $\mathbf{F}_1, \mathbf{F}_2$. Then according to the ideas of virtual work we imagine a small change in the position and orientation of the body so that

$$\mathbf{r}'_1 = \mathbf{r}_1 + \delta \mathbf{r}_1, \qquad \mathbf{r}'_2 = \mathbf{r}_2 + \delta \mathbf{r}_2$$

For our generalized position variables, we choose

$$q_1 = x, \qquad q_2 = y, \qquad q_3 = \theta$$

where

$$\mathbf{r}_0 = x\mathbf{e}_1 + y\mathbf{e}_2$$
$$\mathbf{r}_1 = \mathbf{r}_0 + \boldsymbol{\rho}_1, \qquad \mathbf{r}_2 = \mathbf{r}_0 + \boldsymbol{\rho}_2 \tag{5.107}$$

Under the variation $\{\delta x, \delta y, \delta\theta\}$, $\mathbf{r}_0 \to \mathbf{r}_0 + [\delta x, \delta y, 0]^T$. But the lengths of $\boldsymbol{\rho}_1, \boldsymbol{\rho}_2$ do not change since this is a rigid body. To find the change in these vectors we can use

$$\delta\mathbf{r}_i = \delta\mathbf{r}_0 + \delta\theta\mathbf{e}_3 \times \boldsymbol{\rho}_i \tag{5.108}$$
$$\delta\mathbf{r}_0 = \delta x\mathbf{e}_1 + \delta y\mathbf{e}_2 \tag{5.109}$$

Using the triple scalar product theorem,

$$\mathbf{A} \cdot (\mathbf{B} \times \mathbf{C}) = \mathbf{B} \cdot (\mathbf{C} \times \mathbf{A})$$

the expression for the virtual work (5.106) becomes

$$\begin{aligned}\delta W &= \mathbf{F}_1 \cdot \delta\mathbf{r}_1 + \mathbf{F}_2 \cdot \delta\mathbf{r}_2 \\ &= (\mathbf{F}_1 + \mathbf{F}_2) \cdot \delta\mathbf{r}_0 + (\boldsymbol{\rho}_1 \times \mathbf{F}_1 + \boldsymbol{\rho}_2 \times \mathbf{F}_2) \cdot \mathbf{e}_3\delta\theta \\ &= F_x\delta x + F_y\delta y + M_3\delta\theta\end{aligned} \tag{5.110}$$

The three generalized forces then become by inspection

$$Q_1 = (\mathbf{F}_1 + \mathbf{F}_2) \cdot \mathbf{e}_1, \qquad Q_2 = (\mathbf{F}_1 + \mathbf{F}_2) \cdot \mathbf{e}_2$$
$$Q_3 = M_3 = (\boldsymbol{\rho}_1 \times \mathbf{F}_1 + \boldsymbol{\rho}_2 \times \mathbf{F}_2) \cdot \mathbf{e}_3 \tag{5.111}$$

The last term is the total moment of the forces about the point 0 projected onto the axis $\mathbf{e}_3$.

Example 5.10. Consider the circular cam mechanism shown in Fig. 5.20. The off-axis pin is constrained to slide in a frictionless horizontal slot, while the circular edge is constrained by the plane at $x = 0$. An elastic spring is used to pull the pin toward the contact plane. When the cam rotates we assume that a constant dynamic friction force, f_0 exists between the wall and the cam. In this example, the rotational and translational motion are coupled. The constraint equation can be written as

$$\mathbf{r}_c = x_c\hat{\mathbf{i}} + y_c\hat{\mathbf{j}} = R\hat{\mathbf{i}} + a\cos\theta\hat{\mathbf{j}}$$

or

$$\dot{x}_c = 0, \qquad \dot{y}_c = -a\dot\theta\sin\theta$$

In order to apply Lagrange's equation, we need to find the kinetic energy function $T(\theta, \dot\theta)$. For this planar problem T takes the form

$$T = \frac{1}{2}mv_c^2 + \frac{1}{2}I_c\omega^2$$

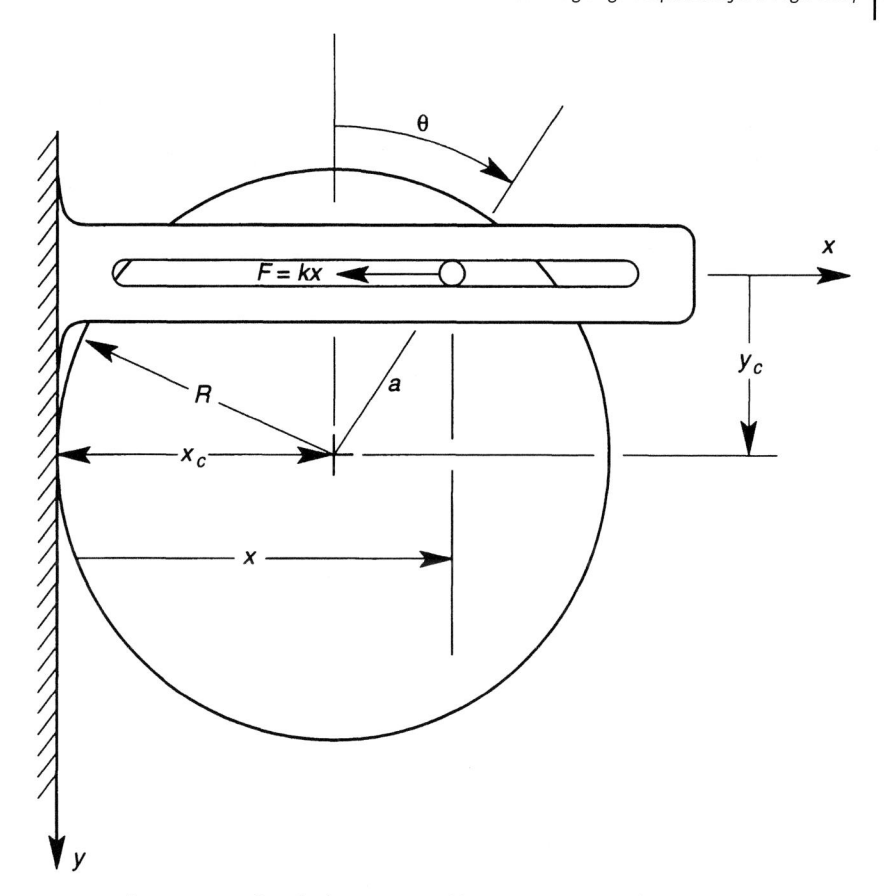

Fig. 5.20 Planar motion of a cylinder constrained by a frictionless pin in a horizontal slot and by a frictionless fixed wall (Example 5.10).

or

$$T = \frac{1}{2}m\dot{\theta}^2 a^2 \left(\frac{R^2}{2a^2} + \sin^2 \theta \right) \tag{5.112}$$

where $I_c = mR^2/2$. We assume that the spring force is proportional to the x position of the pin or that the potential energy is given by

$$V = \frac{1}{2}kx^2 = \frac{1}{2}k(R + a\sin\theta)^2 \tag{5.113}$$

To calculate the generalized force due to friction between the wall and the cam, we write the energy dissipation rate and integrate to get the virtual work.

$$P = -f_0|v_w| = -f_0|\dot{y}_c - R\dot{\theta}|$$

or

$$\delta W = -f_0(R + a\sin\theta)\,\text{sign}(\dot\theta)\delta\theta$$

and

$$Q = -f_0(R + a\sin\theta)\,\text{sign}(\dot\theta)$$

Lagrange's equation, with θ as a generalized coordinate is then

$$\frac{d}{dt}\frac{\partial T}{\partial \dot\theta} - \frac{\partial T}{\partial \theta} + \frac{\partial V}{\partial \theta} = Q$$

$$\left(\frac{R^2}{2a^2} + \sin^2\theta\right)\ddot\theta + \dot\theta^2 \sin\theta\cos\theta$$

$$+ \frac{k}{m}\left(\frac{R}{a} + \sin\theta\right)\cos\theta = -\frac{f_0}{ma^2}(R + a\sin\theta)\,\text{sign}(\dot\theta) \tag{5.114}$$

This is a nonlinear, second-order differential equation. When $f_0 = 0$, and both $\dot\theta$ and $\ddot\theta$ are zero, there are two equilibrium positions for the cam, $\theta = \pi/2$ and $-\pi/2$.

By intuition, the position $\theta = \pi/2$ is unstable. Near the stable position $\theta = -\pi/2$, we can write $\theta = -\pi/2 + \phi$, where ϕ is a small angle. The linearized equation near $\theta = -\pi/2$ becomes (for zero friction)

$$\left(1 + \frac{R^2}{2a^2}\right)\ddot\phi + \frac{k}{m}\left(\frac{R}{a} - 1\right)\phi = 0 \tag{5.115}$$

which is the equation of a harmonic oscillator with a natural frequency f (Hz) given by

$$2\pi f = \left(\frac{k}{m}\right)^{\frac{1}{2}}\left(\frac{R}{a} - 1\right)^{\frac{1}{2}}\left(1 + \frac{R^2}{2a^2}\right)^{-\frac{1}{2}} \tag{5.116}$$

Example 5.11 Rotation Induced Stiffness. Consider the rotating cylinder with a movable flap shown in Fig. 5.21. Assume that the motion of the flap does not significantly change the rotation of the large rotor, Ω. We also neglect gravity, so there are no forces that do work in the system. With these assumptions the problem has one degree of freedom, which is the angular motion of the hinged flap. In order to derive Lagrange's equation, we need to calculate the kinetic energy as a function of the flap angle $\theta(t)$. To this end we assign a local coordinate system $\{\mathbf{e}_x, \mathbf{e}_y, \mathbf{e}_z\}$ at the center of mass and in the principal moments of inertia directions of the flap. Then we project the angular velocity and center-of-mass velocity onto these coordinates.

$$\boldsymbol\omega = \Omega(\cos\theta\,\mathbf{e}_z + \sin\theta\,\mathbf{e}_x) - \dot\theta\,\mathbf{e}_y$$

$$\mathbf{v}_c = (R + b\cos\theta)\Omega\,\mathbf{e}_y + b\dot\theta\,\mathbf{e}_z \tag{5.117}$$

5.4 Lagrange's Equations for a Rigid Body

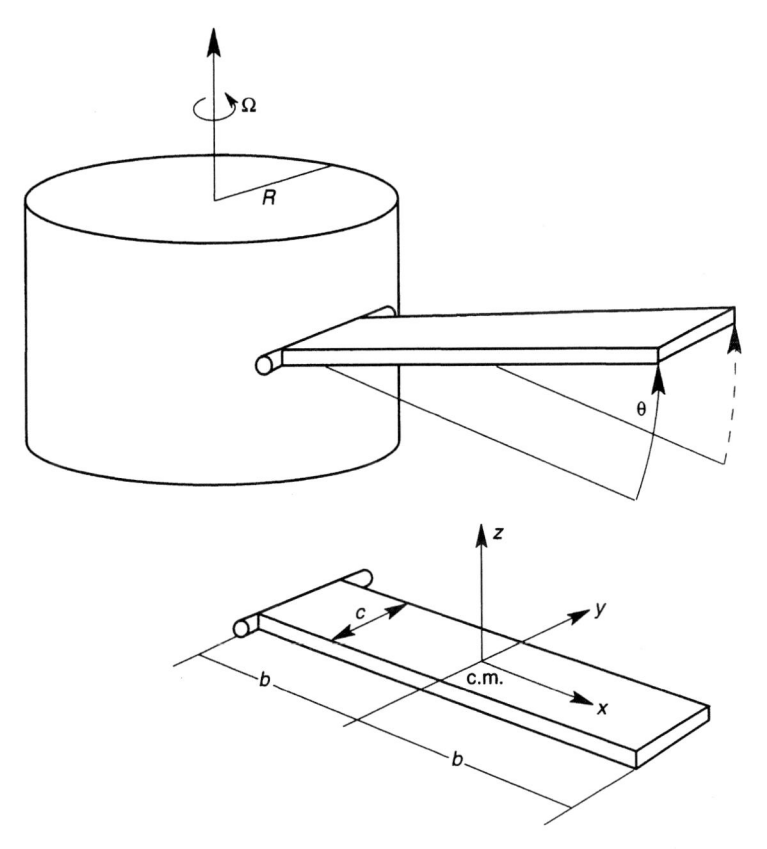

Fig. 5.21 Rotating cylinder with a hinged plate flap (Example 5.11). The rotation Ω induces a restoring moment or stiffness about the θ axis.

The kinetic energy function, (5.101), is given by

$$T = \frac{1}{2}mv_c^2 + \frac{1}{2}\left(I_{cx}\omega_x^2 + I_{cy}\omega_y^2 + I_{cz}\omega_z^2\right) \quad (5.118)$$

From a table of moments of inertia, we can find

$$I_{cx} = \frac{mc^2}{12}, \quad I_{cy} = \frac{mb^2}{3}, \quad I_{cz} = \frac{m}{12}(4b^2 + c^2) \quad (5.119)$$

Under all the assumptions the kinetic energy function can be found to be

$$T(\theta, \dot{\theta}) = \frac{1}{2}m\left((R + b\cos\theta)^2\Omega^2 + b^2\dot{\theta}^2\right)$$

$$+ \frac{1}{2}m\left(\frac{c^2}{12}\Omega^2 + \frac{b^2}{3}\dot{\theta}^2 + \frac{b^2}{3}\Omega^2\cos^2\theta\right) \quad (5.120)$$

With no active forces in the problem, Lagrange's equation of motion is

$$\frac{d}{dt}\frac{\partial T}{\partial \dot\theta} - \frac{\partial T}{\partial \theta} = 0$$

Carrying out the derivatives we obtain

$$\ddot\theta + \frac{3}{4}\Omega^2 \frac{R}{b}\left(1 + \frac{4b}{3R}\cos\theta\right)\sin\theta = 0 \qquad (5.121)$$

For small motions of the flap, $\cos\theta \approx 1$, $\sin\theta \approx \theta$, and the equation of motion resembles a harmonic oscillator with natural frequency f (Hz)

$$2\pi f = \frac{\sqrt{3}}{2}\Omega\left(\frac{R}{b} + \frac{4}{3}\right)^{1/2} \qquad (5.122)$$

Thus we see that a large rotation can induce an effective kinematic stiffness. This occurs in helicopters and turbine components.

Example 5.12 Double-slider Dynamics: Comparison of Lagrange and Newton–Euler Methods. In Example 4.4 of Chapter 4 we examined the dynamics of a double-slider pendulum consisting of two particles using the energy method of Lagrange. In this example we derive the equation of motion for a uniform rod constrained to move in two perpendicular slots (i.e., a double-slider mechanism) using both Lagrange's equation and the Newton–Euler equations of dynamics. We shall see that although the Lagrange method is more efficient in arriving at the equation of motion, the Newton–Euler method gives us explicit equations for the internal constraint forces.

In Fig. 5.22a, a rod is constrained in the horizontal plane by two frictionless slots and a linear spring along the x direction. In Fig. 5.22b, a free body diagram of the rod shows the all the forces in the plane of motion. (We neglect forces due to gravity normal to the plane of motion.) Our basic method is to first write down the equations of geometric and kinematic constraints. Then we derive the equation of motion using the two methods choosing the angle $\theta(t)$, as our generalized position variable.

Geometric and Kinematic Constraints

$$x = 2x_c = a\cos\theta; \qquad y = 2y_c = a\sin\theta$$

$$\mathbf{v}_c = \dot x_c \mathbf{e}_x + \dot y_c \mathbf{e}_y = \frac{a}{2}\dot\theta(-\sin\theta\,\mathbf{e}_x + \cos\theta\,\mathbf{e}_y)$$

$$\dot{\mathbf{v}}_c = \frac{a}{2}\ddot\theta(-\sin\theta\,\mathbf{e}_x + \cos\theta\,\mathbf{e}_y) - \frac{a}{2}\dot\theta^2(\cos\theta\,\mathbf{e}_x + \sin\theta\,\mathbf{e}_y) \qquad (5.123)$$

In these equations one can recognize that the center of mass point moves in a circular motion with radius $a/2$. Also the angular velocity of the rod is given by $\omega = \dot\theta$.

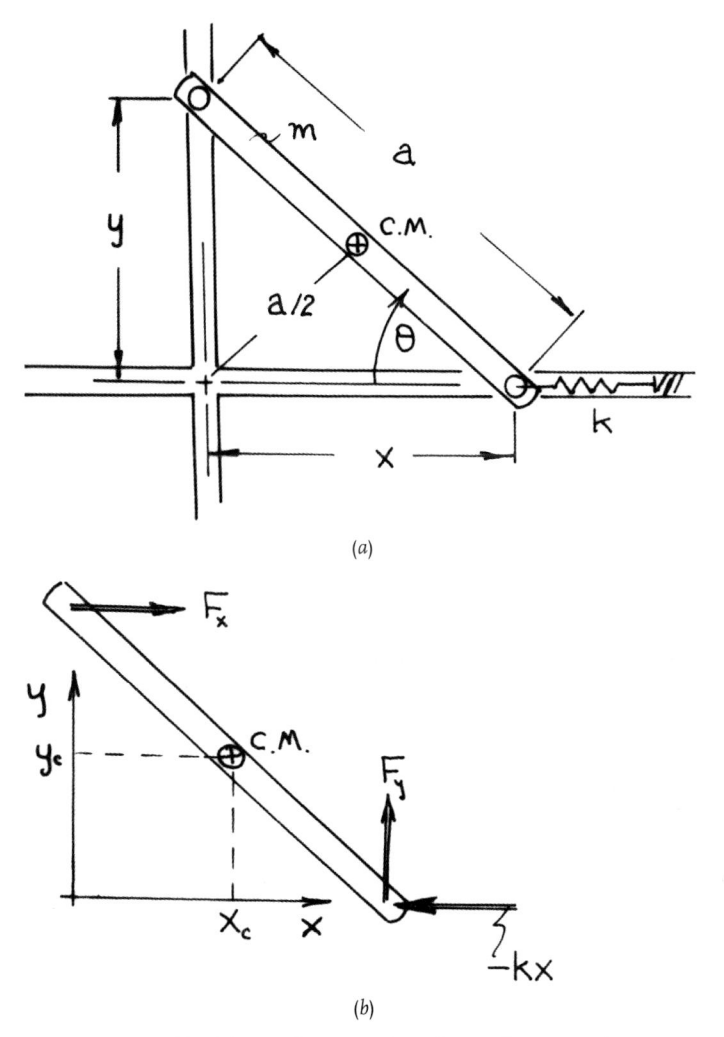

Fig. 5.22 (a) Double-slider mechanism with uniform rod constrained by an elastic spring on the x axis. (b) Free-body diagram of the double-slider rod in (a).

Dynamics

(i) Method of Lagrange's Equation

In this method we derive the kinetic energy expression $T(\theta, d\theta/dt)$ as a function of the generalized coordinate $\theta(t)$. We also classify the forces that do work and those that do not and represent the conservative forces in terms of potential energy functions of the generalized coordinate. In this problem we neglect friction and assume that the constraint forces don't do any work and that the force of the spring can be represented by a potential energy function $V(\theta)$.

$$T = \frac{1}{2}m\mathbf{v}_c \cdot \mathbf{v}_c + \frac{1}{2}\mathbf{H}_c \cdot \boldsymbol{\omega}$$

$$T = m\frac{a^2}{8}\dot{\theta}^2 + \frac{1}{2}I_c\dot{\theta}^2 = \frac{1}{2}I_0\dot{\theta}^2 \tag{5.124}$$

where

$$I_0 = I_c + m\frac{a^2}{4}$$

The resulting equation of motion becomes

$$\frac{d}{dt}\frac{\partial T}{\partial \dot{\theta}} - \frac{\partial T}{\partial \theta} + \frac{\partial V}{\partial \theta} = 0$$

$$I_0\ddot{\theta} - ka^2 \cos\theta \sin\theta = 0 \tag{5.125}$$

(ii) Newton–Euler Method

In the standard method of Newton and Euler, it is advisable to draw a free body diagram with all the forces (Fig. 5.22b). Then one identifies the vector components of all the forces and moments and equates the force components to the corresponding accelerations of the center of mass. Finally the force-moment components are equated to the change in angular momentum as illustrated in the following equations. We note first that because the forces are planar, the force moments about the center of mass are calculated about an axis normal to the plane. Therefore only the scalar component of angular momentum $\mathbf{H}_c$ is calculated.

$$F_x = m\dot{v}_{cx} + kx$$

$$F_y = m\dot{v}_{cy} \tag{5.126}$$

$$M_{cz} = \dot{H}_{cz}$$

where $H_{cz} = I_c\dot{\theta}$, or

$$y_c F_x + y_c kx - x_c F_y = I_c\ddot{\theta} \tag{5.127}$$

Substituting the expressions for F_x, F_y into the force-moment equation, one obtains the identical equation of motion for $\theta(t)$ that was found using Lagrange's equation (5.125):

$$\left(I_c + \frac{ma^2}{4}\right)\ddot{\theta} - a^2 k \sin\theta \cos\theta = 0$$

It is interesting to note that the centripetal acceleration terms $\dot{\theta}^2$ that appear in the acceleration components of the center of mass, drop out when combined in the

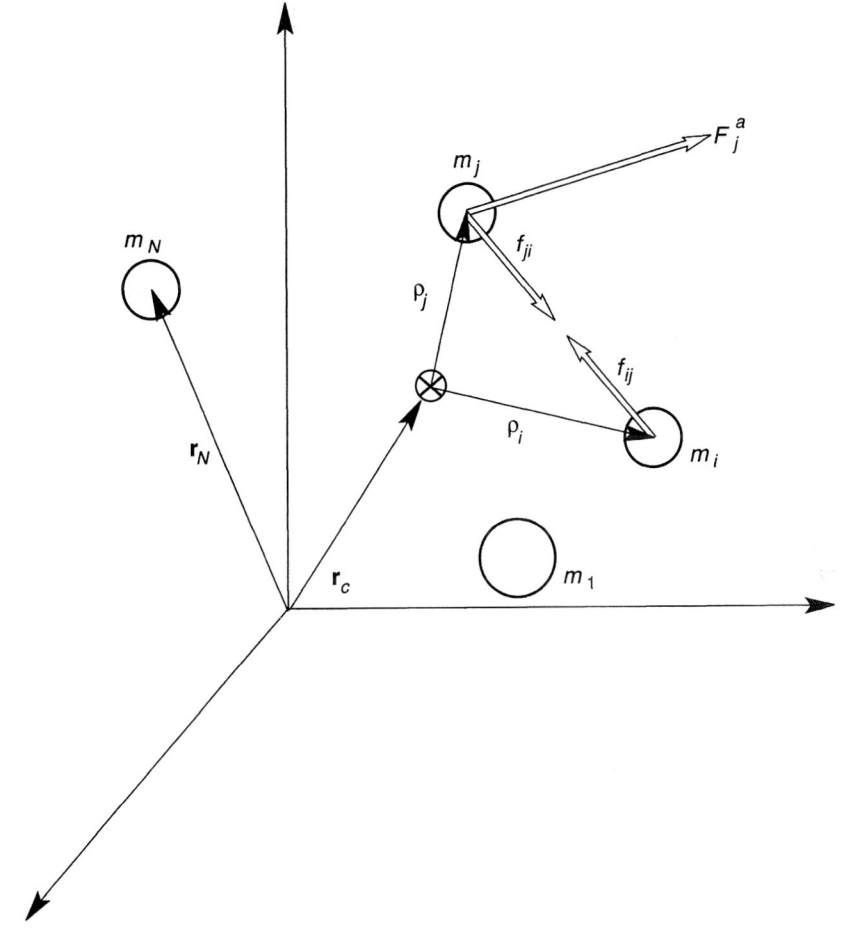

Fig. 5.23 A collection of rigidly constrained particles showing internal forces f_{ij}, and applied external forces $\mathbf{F}_j^a$.

final equation of motion. An interesting exercise is to write the equation of motion for small motions $x(t)$. In this case we use the angle of the rod off the vertical.

$$\theta + \phi = \pi/2$$

$$I_0\ddot{\phi} + a^2 k \sin\phi \cos\phi = 0$$

For small ϕ

$$I_0\ddot{\phi} + a^2 k \phi = 0 \tag{5.128}$$

The natural frequency in radians per second is then $f = 2\pi\sqrt{(a^2 k/I_0)}$.

5.5
Principle of Virtual Power for a Rigid Body

The concept of virtual power applied to a system of particles was discussed in Section 4.4. The extension of this principle to a rigid body can be found in an early paper by Kane (1961) and is derived in several textbooks such as Kane and Levinson (1985) and Pfeiffer (1989). We begin with a statement of the principle [see (4.146)] for N particles with M degrees of freedom and a corresponding set of M generalized position variables $\{q_k(t)\}$ (Fig. 5.23)

$$\sum_i \left(m_i \dot{\mathbf{v}}_i - \mathbf{F}_i^a - \sum_j \mathbf{f}_{ij} \right) \cdot \frac{\partial \mathbf{v}_i}{\partial \dot{q}_k} = 0, \quad k = 1, 2, \ldots, M \tag{5.129}$$

We recall that $\{\mathbf{F}_i^a\}$ are the forces that do work and that constraint forces that do not produce power have been eliminated. The internal forces between particles of mass m_i and m_j are $\{\mathbf{f}_{ij}\}$.

The projection vectors $\partial \mathbf{v}/\partial \dot{q}_k$ are obtained from the kinematic constraint relations.

For a rigid body the distances between any two particles is constant in time and the maximum number of degrees of freedom is six, i.e., $M \leq 6$. Also for a rigid body it can be shown as an exercise that the internal forces produce no net power, i.e.,

$$\sum_i \sum_j \mathbf{f}_{ij} \cdot \frac{\partial \mathbf{v}_i}{\partial \dot{q}_n} = 0 \tag{5.130}$$

To extend the principle of virtual power to a rigid body we must express the velocities of all the particles in terms of a velocity of one particle, say $\mathbf{v}_0$, and the rotation velocity, $\boldsymbol{\omega}$, i.e.,

$$\mathbf{v}_i = \mathbf{v}_0 + \boldsymbol{\omega} \times \boldsymbol{\rho}_i \tag{5.131}$$

where $\boldsymbol{\rho}_i$ is a position vector from point 0 to the mass m_i.

In order to see the basic form that this principle takes for a rigid body, we first consider the restricted problem of motion about a fixed point.

Special Case: Motion About a Fixed Point
In this case we set $\mathbf{v}_0 = 0$. Thus there are at most three degrees of freedom, $M \leq 3$. The generalized velocity variables $\{\dot{q}_i\}$ will, in general, represent the time rate of change of three angular positions of the body. We note that the local position vectors $\boldsymbol{\rho}_i$ are not functions of the three generalized velocities $\{\dot{q}_i\}$. Therefore, the kinematic constraints and the Jacobians become

$$\mathbf{v}_i = \boldsymbol{\omega} \times \boldsymbol{\rho}_i$$
$$\frac{\partial \mathbf{v}_i}{\partial \dot{q}_k} = \frac{\partial \boldsymbol{\omega}}{\partial \dot{q}_k} \times \boldsymbol{\rho}_i \tag{5.132}$$

5.5 Principle of Virtual Power for a Rigid Body

With all the assumptions thus far, (5.129) now takes the form

$$\sum (m_i \dot{\mathbf{v}}_i - \mathbf{F}_i^a) \cdot \frac{\partial \boldsymbol{\omega}}{\partial \dot{q}_k} \times \boldsymbol{\rho}_i = 0, \quad k = 1, \ldots, M \tag{5.133}$$

Since $\boldsymbol{\omega}$ and its derivative are independent of the mass points, we manipulate this expression so that $\boldsymbol{\omega}$ is outside the summation. To do this we use a vector identity for a triple-scalar product; $\mathbf{A} \cdot (\mathbf{B} \times \mathbf{C}) = \mathbf{B} \cdot (\mathbf{C} \times \mathbf{A})$. Consider the force terms first

$$\sum \mathbf{F}_i^a \cdot \frac{\partial \mathbf{v}_i}{\partial \dot{q}_k} = \left(\sum \boldsymbol{\rho}_i \times \mathbf{F}_i^a \right) \cdot \frac{\partial \boldsymbol{\omega}}{\partial \dot{q}_k}$$

$$= \mathbf{M}^a \cdot \frac{\partial \boldsymbol{\omega}}{\partial \dot{q}_k} \tag{5.134}$$

The term in brackets is simply the total moment of all the active forces about the point 0. The inertia terms are a little bit more complicated, but follow in the same fashion as the force summation:

$$\sum m_i \dot{\mathbf{v}}_i \cdot \left(\frac{\partial \boldsymbol{\omega}}{\partial \dot{q}_k} \times \boldsymbol{\rho}_i \right) = \frac{\partial \boldsymbol{\omega}}{\partial \dot{q}_k} \cdot \sum \boldsymbol{\rho}_i \times m_i \dot{\mathbf{v}}_i$$

$$= \frac{\partial \boldsymbol{\omega}}{\partial \dot{q}_k} \cdot \dot{\mathbf{H}} \tag{5.135}$$

The second equality follows, since

$$\boldsymbol{\rho}_i \times m_i \dot{\mathbf{v}}_i = \frac{d}{dt}(\boldsymbol{\rho}_i \times m_i \mathbf{v}_i) - \dot{\boldsymbol{\rho}}_i \times m_i \mathbf{v}_i \tag{5.136}$$

and $\dot{\boldsymbol{\rho}}_i \times \mathbf{v}_i = 0$. Also the angular momentum about the point 0 is easily recognized.

$$\mathbf{H} = \sum \boldsymbol{\rho}_i \times m_i (\boldsymbol{\omega} \times \boldsymbol{\rho}_i) = I \cdot \boldsymbol{\omega} \tag{5.137}$$

Thus, in its simplest form, the *principle of virtual power* for a rigid body takes the form

$$(\dot{\mathbf{H}} - \mathbf{M}^a) \cdot \frac{\partial \boldsymbol{\omega}}{\partial \dot{q}_k} = 0, \quad k = 1, \ldots, M \tag{5.138}$$

Both $\mathbf{H}$ and $\mathbf{M}^a$ are calculated about the fixed point of rotation. The equation looks like the projection of Euler's equation onto the direction $\partial \boldsymbol{\omega}/\partial \dot{q}_k$. However, the $\mathbf{M}^a$ only includes the moment of the forces that do work. The moments due to zero-work constraint forces do not appear.

Example 5.13 Stability of Spinning Rod: MATLAB Example. In the classic pendulum a body rotates about a fixed point under gravity and has a vibratory solution about the vertical equilibrium position. In this example we demonstrate the instability of the pendulum when it is subject to centripetal effects of a constant rotation

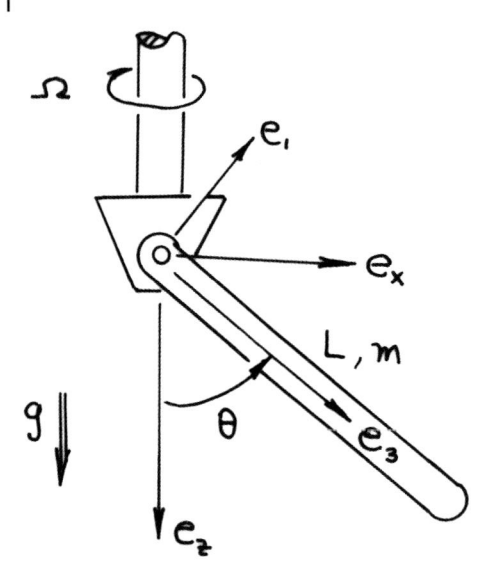

Fig. 5.24 Uniform rod on a hinged rotary joint, spinning with constant angular velocity Ω about the vertical z axis under gravity.

of the plane of oscillation. Consider a thin rod of length L constrained to a revolute joint whose support is spun about the vertical axis with constant rate Ω as shown in Fig. 5.24. To derive the equation of motion we will use the principle of virtual power. Since the rigid body is in pure rotation we can use the restricted form of the principle (5.138)

$$\left(\dot{\mathbf{H}}_0 - \mathbf{M}_0^a\right) \cdot \frac{\partial \boldsymbol{\omega}}{\partial \dot{\theta}} = 0$$

In this one degree of freedom problem we identify $\dot{\theta}$ as the generalized velocity. We also choose a coordinate system with basis vectors $\{\mathbf{e}_x, \mathbf{e}_y, \mathbf{e}_z\}$ that rotates with the plane of the rod with the vector $\mathbf{e}_z$ aligned with the vertical axis of rotation and pointing in the direction of gravity. The angular velocity vector for the rod is then written as $\boldsymbol{\omega} = \Omega \mathbf{e}_z + \dot{\theta} \mathbf{e}_y$. We introduce another coordinate basis set fixed to the rod aligned with the principle axes $\{\mathbf{e}_1, \mathbf{e}_2, \mathbf{e}_3\}$ where $\mathbf{e}_2 = \mathbf{e}_y$. In this basis vector system the angular momentum vector, $\mathbf{H}$, can be found to be

$$\mathbf{H}_0 = -I_0 \Omega \sin\theta \, \mathbf{e}_1 + I_0 \dot{\theta} \mathbf{e}_2 \tag{5.139}$$

where

$$I_1 = I_2 = \frac{1}{3} m L^2 \quad \text{and} \quad I_3 = 0$$

To calculate the change of angular momentum vector we must consider not only the change in the scalar components of **H**, but also the change in the vectors $\{\mathbf{e}_1, \mathbf{e}_2, \mathbf{e}_3\}$.

$$\dot{\mathbf{H}} = -I_0 \Omega \dot{\theta} \cos\theta\, \mathbf{e}_1 - I_0 \Omega \theta \sin\theta\, \dot{\mathbf{e}}_1 + I_0 \ddot{\theta}\mathbf{e}_2 + I_0 \dot{\theta} \dot{\mathbf{e}}_2 \tag{5.140}$$

where

$$\dot{\mathbf{e}}_i = \Omega \mathbf{e}_z \times \mathbf{e}_i \quad (i = 1, 2)$$

The projection vector is proportional to $\mathbf{e}_2$, so that only this component of the change of angular momentum vector (5.140) enters the equation of motion:

$$\dot{\mathbf{H}} \cdot \frac{\partial \omega}{\partial \dot{\theta}} = I_0 \ddot{\theta} - \Omega^2 I_0 \sin\theta \cos\theta$$

Neglecting friction in the joint, the only active force is gravity and the projection of its moment onto the $\mathbf{e}_2$ axis is easily found as well as the resulting equation of motion:

$$\mathbf{M}_0^a \cdot \frac{\partial \omega}{\partial \dot{\theta}} = \mathbf{M}_0^a \cdot \mathbf{e}_2 = -mg\frac{L}{2}\sin\theta \tag{5.141}$$

and

$$I_0 \ddot{\theta} - \Omega^2 I_0 \sin\theta \cos\theta + mg\frac{L}{2}\sin\theta = 0 \tag{5.142}$$

For small angles from the vertical axis, we can see that the rotation of the plane of the pendulum introduces an unstable term in the differential equation of motion:

$$\ddot{\theta} + \left[\frac{mgL}{2I_0} - \Omega^2\right]\theta = 0 \tag{5.143}$$

At a critical value of the rotational speed, the centrifugal acceleration moment will overcome the restoring moment of gravity and the rod will move away from the vertical. Thus for stability we arrive at the following criterion:

$$\Omega^2 < \frac{mgL}{2I_0} = \frac{3g}{2L} \tag{5.144}$$

MATLAB Solution

To obtain numerical solutions to the nonlinear differential equation (5.142) using the MATLAB code, one must write the second-order differential equation as a set of first-order equations. To review the essentials of MATLAB the student is referred to the short guide *Getting Started With MATLAB7* by Rudra Pratap.

$$x_1 = \theta; \qquad x_2 = \dot{\theta}$$

$$\dot{x}_1 = x_2$$

$$\dot{x}_2 = \Omega^2 \sin x_1 \cos x_1 - \frac{3g}{2L} \sin x_1 - \gamma x_2 \qquad (5.145)$$

We have added a small amount of viscous damping in the last term of (5.145).

MATLAB has an integration subprogram ode45('Function', 'Time' 'IC') that requires one to define three parameters: the first-order ODE function as in (5.145), the time integration range, 'Time' and initial conditions on the vector $[x_1, x_2]$.

In MATLAB format the function is defined by:

```
function dx = spinningrod (t,x)
       Omega = 3; g = 10; L = 1; %Define constants
dx = zeros(2,1);
dx(1) = x(2);
dx(2) = Omega^2*sin(x(1))*cos(x(1)) - (3*m/2*L) sin(x(1)) -
       0.1*x(2);

%Next one defines the MATLAB program that performs the
integration and plots the numerical results;

%Set the time parameters
t0 = 0;
tf = 30;
%Set the initial conditions vector
x0 = [0.1; 0];
%Call the ODE integration subroutine
[t,x] = ode45(@spinningrod, [t0,tf], x0);

%Plot the results in the x vector file as a phase plane:
angle vs angular velocity
figure(1);
plot(x(:,1),x(:,2));
grid on;
```

Two integration plots are shown in Fig. 5.25. The first, Fig. 5.25a shows the case when the spin rate is less that critical, i.e. the rod behaves like a classic pendulum; under a little damping the motion decays to the vertical position. In Fig. 5.25b however the spin exceeds the critical value and the rod moves away from the vertical and settles into a new equilibrium position when a little friction is added.

Example 5.14 Gyropendulum. Consider the disc of radius R and mass m rolling on an inclined plane shown in Fig. 5.26. The effective active gravity force in the plane is $mg \sin \phi$. Assume that the disc rolls without slipping and find the natural frequency for small motions of the angle θ.

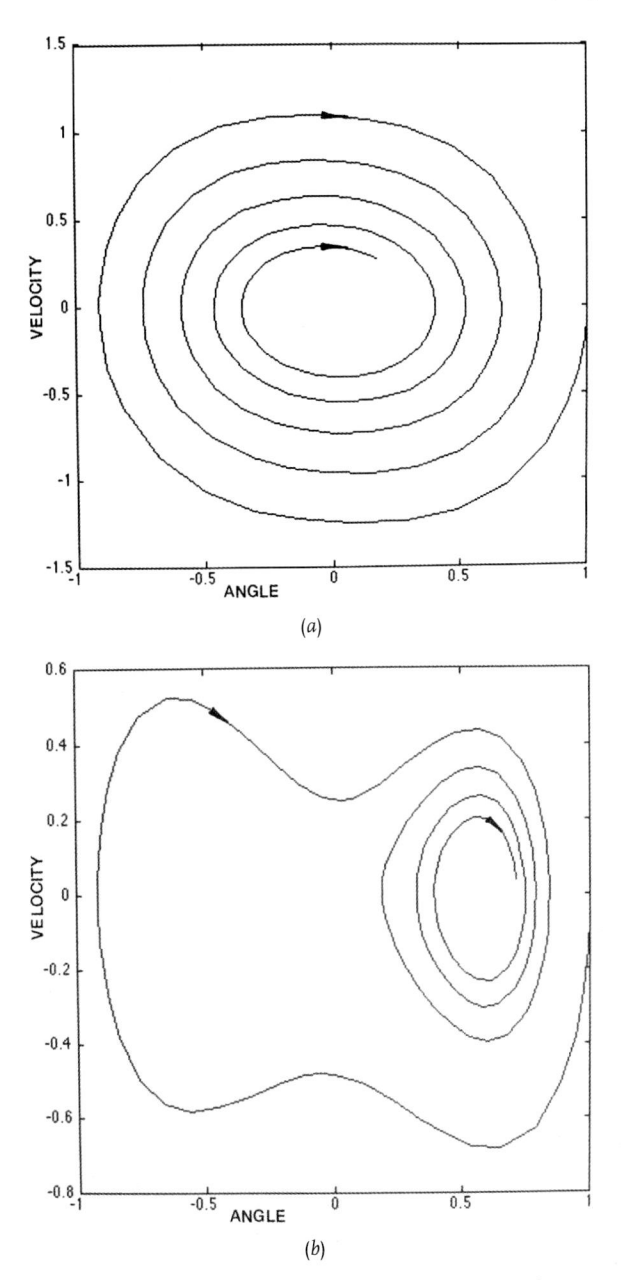

Fig. 5.25 (a) MATLAB integration of differential equation of motion (5.142) with damping or (5.145) for the spinning hinged rod: Phase plane orbit (angle versus angular velocity) for spin speed Ω below the critical velocity. (b) MATLAB integration of the equation of motion (5.145) for the spinning hinged rod for the angular spin speed Ω greater that the critical speed showing motion away from the axis of rotation.

5 Rigid Body Dynamics

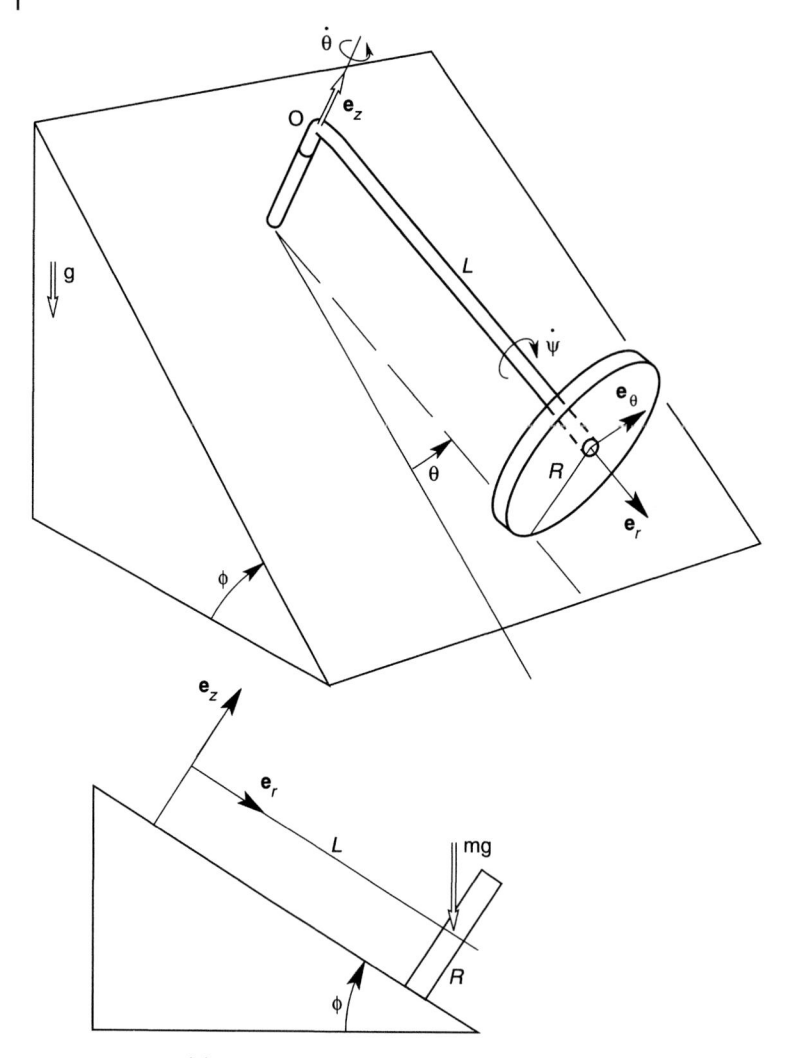

Fig. 5.26 *Gyro pendulum.* A thin disc of radius R rotating on an inclined surface (Example 5.14).

The number of degrees of freedom is one and we choose $\dot{\theta}$ as the generalized velocity. The total angular velocity of the disc is given by

$$\boldsymbol{\omega} = \dot{\theta}\mathbf{e}_z - \dot{\psi}\mathbf{e}_r$$

The rolling constraint is given by calculating the velocity of the center of the disc from two references

$$\dot{\theta}L = R\dot{\psi} \tag{5.146}$$

Thus

$$\frac{\partial \boldsymbol{\omega}}{\partial \dot{\theta}} = \mathbf{e}_z - \frac{L}{R}\mathbf{e}_r$$

Since the disc is in pure rotation about the hinge point, we calculate the angular momentum about this point.

$$\mathbf{H} = I_z \dot{\theta}\mathbf{e}_z - I_0 \dot{\psi}\mathbf{e}_r$$

$$\dot{\mathbf{H}} = I_z \ddot{\theta}\mathbf{e}_z - I_0 \ddot{\psi}\mathbf{e}_r + I_0 \dot{\psi}\dot{\theta}\mathbf{e}_\theta \tag{5.147}$$

The moment of the gravity force about the hinge point is given by

$$\mathbf{M} = -mgL \sin\phi \sin\theta \mathbf{e}_z$$

The generalized force is given by

$$\mathbf{M} \cdot \frac{\partial \boldsymbol{\omega}}{\partial \dot{\theta}} = -mgL \sin\phi \sin\theta \tag{5.148}$$

Also

$$\dot{\mathbf{H}} \cdot \frac{\partial \boldsymbol{\omega}}{\partial \dot{\theta}} = I_z \ddot{\theta} + I_0 \frac{L}{R}\ddot{\psi} \tag{5.149}$$

The disc is in pure rotation about the point 0 so that $\mathbf{H}, \mathbf{M}, I_z, I_0$ are all calculated with respect to this point. Since the body is in pure rotation, we can use the principle of virtual power in the form of (5.138):

$$(\dot{\mathbf{H}} - \mathbf{M}) \cdot \frac{\partial \boldsymbol{\omega}}{\partial \dot{\theta}} = \left(I_z + I_0 \frac{L^2}{R^2}\right)\ddot{\theta} + mgL \sin\phi \sin\theta = 0 \tag{5.150}$$

The effective inertia term is simplified using the parallel axis theorem:

$$\left(I_z + I_0 \frac{L^2}{R^2}\right) = \left(\frac{1}{4}mR^2 + mL^2 + \frac{1}{2}mR^2 \frac{L^2}{R^2}\right)$$

$$= \frac{3}{2}mL^2 + \frac{1}{4}mR^2$$

For small angles, $\theta \ll 1$, the equation takes the form

$$\ddot{\theta} + \alpha^2 \theta = 0$$

Thus an oscillating solution has a frequency given by

$$\alpha^2 = \frac{2gL \sin\phi}{3L^2 + R^2/2} \tag{5.151}$$

Virtual Power: General Motion of a Rigid Body

In the general case of both translation and rotation we assume that the point 0 in (5.131) is the center of mass of the rigid body. This will allow us to use the identity

$$\sum m_i \boldsymbol{\rho}_i = 0 \tag{5.152}$$

when $\{\boldsymbol{\rho}_i\}$ are measured from the center of mass. We also note that the number of degrees of freedom may be as high as six. The Jacobian matrix for each mass m_i for the general case takes the form

$$\frac{\partial \mathbf{v}_i}{\partial \dot{q}_k} = \frac{\partial \mathbf{v}_c}{\partial \dot{q}_k} + \frac{\partial \boldsymbol{\omega}}{\partial \dot{q}_k} \times \boldsymbol{\rho}_i \tag{5.153}$$

We will also need an expression for the acceleration of each particle

$$\dot{\mathbf{v}}_i = \dot{\mathbf{v}}_c + \dot{\boldsymbol{\omega}} \times \boldsymbol{\rho}_i + \boldsymbol{\omega} \times (\boldsymbol{\omega} \times \boldsymbol{\rho}_i) \tag{5.154}$$

where the last term on the right-hand side expresses the fact that each $\boldsymbol{\rho}_i$ is a constant-length vector and can only change by rotation (see Chapter 3). When these expressions are put into (5.129), there will be six groups of terms

$$\sum m_i \dot{\mathbf{v}}_c \cdot \frac{\partial \mathbf{v}_c}{\partial \dot{q}_k} + \sum m_i \dot{\mathbf{v}}_c \cdot \frac{\partial \boldsymbol{\omega}}{\partial \dot{q}_k} \times \boldsymbol{\rho}_i$$

$$- \sum \mathbf{F}_i^a \cdot \frac{\partial \mathbf{v}_c}{\partial \dot{q}_k} - \sum \mathbf{F}_i^a \cdot \frac{\partial \boldsymbol{\omega}}{\partial \dot{q}_k} \times \boldsymbol{\rho}_i$$

$$\sum m_i [\dot{\boldsymbol{\omega}} \times \boldsymbol{\rho}_i + \boldsymbol{\omega} \times (\boldsymbol{\omega} \times \boldsymbol{\rho}_i)] \cdot \frac{\partial \mathbf{v}_c}{\partial \dot{q}_k}$$

$$\sum m_i [\dot{\boldsymbol{\omega}} \times \boldsymbol{\rho}_i + \boldsymbol{\omega} \times (\boldsymbol{\omega} \times \boldsymbol{\rho}_i)] \cdot \frac{\partial \boldsymbol{\omega}}{\partial \dot{q}_k} \times \boldsymbol{\rho}_i = 0 \tag{5.155}$$

Because of the assumption on the center of mass (5.152), the second and fifth summations are zero. Using the definition of total mass, force, and moment about the center of mass, the remaining terms can be rewritten employing the scalar triple product theorem:

$$(m\dot{\mathbf{v}}_c - \mathbf{F}^a) \cdot \frac{\partial \mathbf{v}_c}{\partial \dot{q}_k}$$

$$+ \left(\sum m_i \boldsymbol{\rho}_i \times (\dot{\boldsymbol{\omega}} \times \boldsymbol{\rho}_i + \boldsymbol{\omega} \times (\boldsymbol{\omega} \times \boldsymbol{\rho}_i)) - \mathbf{M}_c^a \right) \cdot \frac{\partial \boldsymbol{\omega}}{\partial \dot{q}_k} = 0 \tag{5.156}$$

where

$$m = \sum m_i$$
$$\mathbf{M}_c^a = \sum \boldsymbol{\rho}_i \times \mathbf{F}_i^a$$
$$\mathbf{F}^a = \sum \mathbf{F}_i^a$$

The last term can be shown to be related to the change in angular-momentum vector calculations with reference to the center of mass. Thus, we define

$$\mathbf{H}_c = \sum \boldsymbol{\rho}_i \times m_i \mathbf{v}_i = \sum m_i \boldsymbol{\rho}_i \times (\boldsymbol{\omega} \times \boldsymbol{\rho}_i) \tag{5.157}$$

where we have used

$$\sum \boldsymbol{\rho}_i \times m_i \mathbf{v}_c = 0$$

Then it is easy to see that the second bracketed term in (5.156) is $\dot{\mathbf{H}}_c$:

$$\dot{\mathbf{H}}_c = \sum m_i \boldsymbol{\rho}_i \times [\dot{\boldsymbol{\omega}} \times \boldsymbol{\rho}_i + \boldsymbol{\omega} \times (\boldsymbol{\omega} \times \boldsymbol{\rho}_i)] \tag{5.158}$$

Thus the final form of the principle of virtual power for general motion of a rigid body is

$$(m\dot{\mathbf{v}}_c - \mathbf{F}^a) \cdot \frac{\partial \mathbf{v}_c}{\partial \dot{q}_k} + (\dot{\mathbf{H}}_c - \mathbf{M}_c^a) \cdot \frac{\partial \boldsymbol{\omega}}{\partial \dot{q}_k} = 0 \tag{5.159}$$

or

$$(m\dot{\mathbf{v}}_c - \mathbf{F}^a) \cdot \boldsymbol{\beta}_k + (\dot{\mathbf{H}}_c - \mathbf{M}_c^a) \cdot \boldsymbol{\gamma}_k = 0$$

where $\boldsymbol{\beta}_k, \boldsymbol{\gamma}_k$ are *projection vectors*, defined by,

$$\boldsymbol{\beta}_k = \frac{\partial \mathbf{v}_c}{\partial \dot{q}_k}, \qquad \boldsymbol{\gamma}_k = \frac{\partial \boldsymbol{\omega}}{\partial \dot{q}_k}$$

Comparison of Virtual Power and Newton–Euler Methods

It is easy to write down general equations like Lagrange's equations (5.100) or the principle of virtual power (5.159). Since the "devil is in the details," we examine a popular example from many textbooks; the rolling of a disc on a flat surface as shown in Fig. 5.27 (see, e.g., Greenwood, 1988, pp. 459–462, pp. 434–437). We derive equations of motion suitable for numerical simulation, using both the Newton–Euler method and principle of virtual power (Jourdain–Kane Method).

This rolling problem is an example of a nonholonomic constraint (see Section 4.5). Such constraints involve a relation between the generalized velocities of the form (4.167)

$$\sum a_{ij} \dot{q}_j(t) + b_i(t) = 0$$

In Section 5.6 we discuss methods to solve such problems. One common method is to choose a set of generalized coordinates and velocities that automatically satisfy the rolling constraint. This method is used in this example.

In both the Newton–Euler and virtual power methods we need to specify the kinematics and determine the acceleration of the center of mass and the time rate of change of angular momentum about the center of mass.

We first assign reference axes in which the x-axis is normal to the disc, the y-axis is parallel to the plane, and the z-axis lies in the plane of the disc and goes through the contact point, 0.

The disc is constrained to the plane by means of a planar vector $\mathbf{r}_0$ to the contact point 0, and three angular coordinates $\{\theta, \phi, \psi\}$ to specify the orientation. As shown in Fig. 5.27, ψ specifies the precession angle in the plane, θ specifies the tilt or nutation of the disc, and 4 is similar to a spin angle in gyro dynamics [see also Euler angles Set A, (5.23)].

The rolling constraint couples the angular rates and the velocity of the center of mass through the relation

$$\mathbf{v}_c = \boldsymbol{\omega} \times \boldsymbol{\rho}$$

where

$$\boldsymbol{\rho} = -r\mathbf{e}_z$$

and

$$\boldsymbol{\omega} = (\dot{\phi} - \dot{\psi}\sin\theta)\mathbf{e}_x + \dot{\theta}\mathbf{e}_y + \dot{\psi}\cos\theta\,\mathbf{e}_z \tag{5.160}$$

The acceleration of the center of mass is then

$$\mathbf{a}_c = \dot{\mathbf{v}}_c = \dot{\boldsymbol{\omega}} \times \boldsymbol{\rho} + \boldsymbol{\omega} \times \dot{\boldsymbol{\rho}} \tag{5.161}$$

The only active force in the problem is the gravitational force that we assume is concentrated at the origin:

$$\mathbf{F}^a = mg(-\sin\theta\,\mathbf{e}_x + \cos\theta\,\mathbf{e}_z) = mg\mathbf{n} \tag{5.162}$$

where $\mathbf{n}$ is normal to the plane. The active moment is zero since $\mathbf{F}^a$ acts through the center of mass. The angular momentum vector is given by

$$\mathbf{H}_c = I_a\omega_x\mathbf{e}_x + I_t\omega_y\mathbf{e}_y + I_t\omega_z\mathbf{e}_z \tag{5.163}$$

The calculation of $\dot{\mathbf{H}}_c$ requires not only $\dot{\omega}_x$, etc., but also $\dot{\mathbf{e}}_x$, etc. Since these are constant length vectors, we must use

$$\dot{\mathbf{e}}_x = \boldsymbol{\omega}_1 \times \mathbf{e}_x, \quad \text{etc.}$$

where $\boldsymbol{\omega}_1$ is the rotation vector of the reference system, which differs from $\boldsymbol{\omega}$ by the spin $\dot{\phi}$,

$$\boldsymbol{\omega}_1 = (-\dot{\psi}\sin\theta)\mathbf{e}_x + \dot{\theta}\mathbf{e}_y + \dot{\psi}\cos\theta\,\mathbf{e}_z \tag{5.164}$$

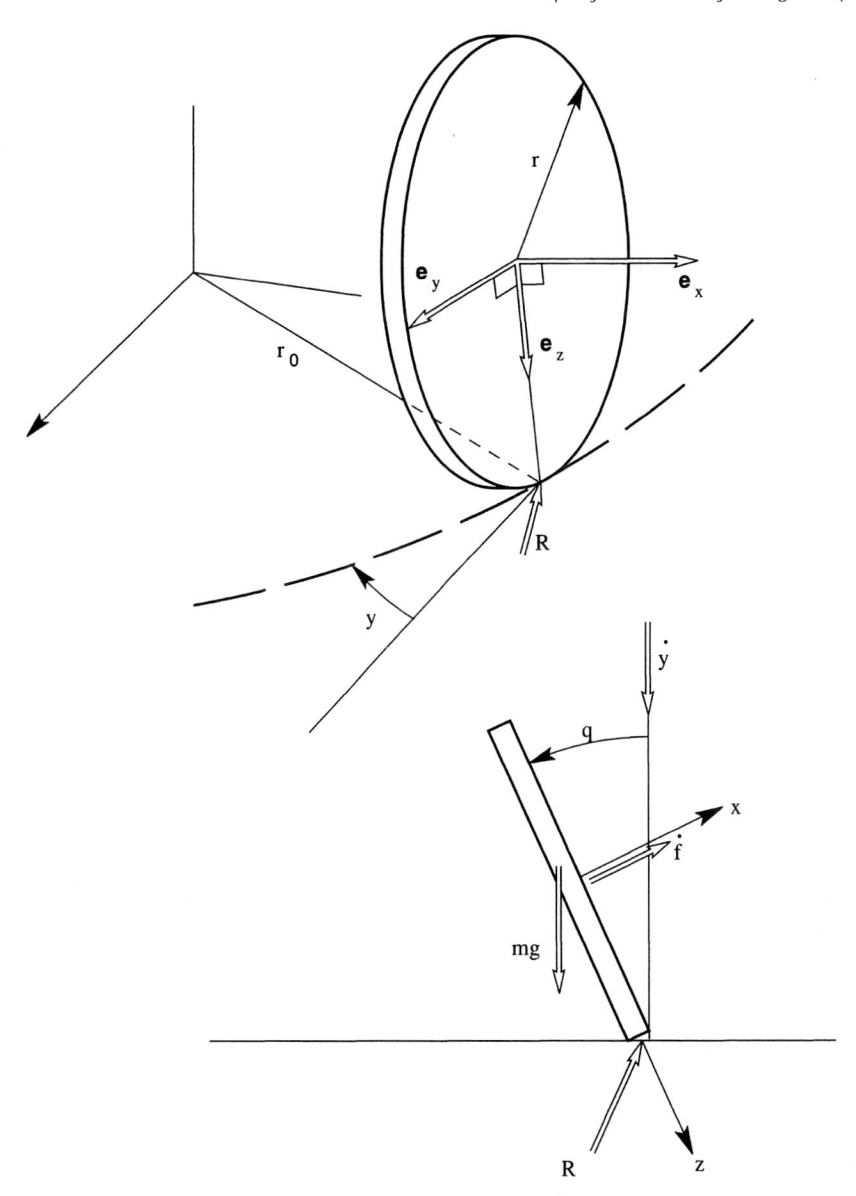

Fig. 5.27 Sketch of the geometry of a rolling disc on a horizontal surface.

From both $\mathbf{a}_c$ and $\dot{\mathbf{H}}_c$ we can obtain components of each in the coordinate system $\{\mathbf{e}_x, \mathbf{e}_y, \mathbf{e}_z\}$, i.e.,

$$\mathbf{a}_c = a_x \mathbf{e}_x + a_y \mathbf{e}_y + a_z \mathbf{e}_z \tag{5.165}$$

Note that $\mathbf{a}_c$ is the absolute acceleration, but we have written its components in a moving reference.

Newton–Euler Method

Here we must write the total force in the body, which includes both gravity and the reaction force:

$$\mathbf{F} = \mathbf{R} + mg(\cos\theta\, \mathbf{e}_z - \sin\theta\, \mathbf{e}_x) \tag{5.166}$$

The force moment about the center of mass is then

$$\mathbf{M} = r\mathbf{e}_z \times \mathbf{R} \tag{5.167}$$

The equations of motion then take the form

$$ma_x = R_x - mg\sin\theta$$
$$ma_y = R_y$$
$$ma_z = R_z + mg\cos\theta$$
$$\dot{H}_x = -rR_y$$
$$\dot{H}_y = -rR_x$$
$$\dot{H}_z = 0 \tag{5.168}$$

Eliminating R_x, R_y from the angular momentum equation, we get

$$\dot{H}_x = -rma_y$$
$$\dot{H}_y = rma_x + rmg\sin\theta$$
$$\dot{H}_z = 0 \tag{5.169}$$

Now, let us compare this with the virtual power method.

Principle of Virtual Power Method

In this method, the constraint force $\mathbf{R}$ does not enter the equation since it does no work. However, we must calculate Jacobian matrices or partial velocities in the language of Kane's method. Following the steps in Example 4.7, we first choose generalized velocities:

$$\dot{q}_1 = \dot{\psi}, \qquad \dot{q}_2 = \dot{\theta}, \qquad \dot{q}_3 = \dot{\phi} \tag{5.170}$$

Then we write the velocity constraint,

$$\mathbf{v}_c = -r\dot{\theta}\mathbf{e}_x + r(\dot{\phi} - \dot{\psi}\sin\theta)\mathbf{e}_y \tag{5.171}$$

Next calculate the tangent vectors (partial velocities) or Jacobians:

$$\frac{\partial \mathbf{v}_c}{\partial \dot{q}_1} = -r\sin\theta\, \mathbf{e}_y, \qquad \frac{\partial \mathbf{v}_c}{\partial \dot{q}_2} = -r\mathbf{e}_x, \qquad \frac{\partial \mathbf{v}_c}{\partial \dot{q}_3} = -r\mathbf{e}_y$$

$$\frac{\partial \boldsymbol{\omega}}{\partial \dot{q}_1} = -\sin\theta\, \mathbf{e}_x + \cos\theta\, \mathbf{e}_z \equiv \mathbf{n}$$

$$\frac{\partial \boldsymbol{\omega}}{\partial \dot{q}_2} = \mathbf{e}_y, \qquad \frac{\partial \boldsymbol{\omega}}{\partial \dot{q}_3} = \mathbf{e}_x \qquad (5.172)$$

These terms are then used in (5.159)

$$(m\dot{\mathbf{v}}_c - \mathbf{F}^a) \cdot \frac{\partial \mathbf{v}_c}{\partial \dot{q}_i} + (\dot{\mathbf{H}}_c - \mathbf{M}_c^a) \cdot \frac{\partial \boldsymbol{\omega}}{\partial \dot{q}_i} = 0 \qquad (5.173)$$

where

$$\mathbf{M}^a = 0$$

Carrying out the inner products or projections we obtain

$$\dot{H}_x(-\sin\theta) + \dot{H}_z \cos\theta - m a_y r \sin\theta = 0 \qquad (5.174a)$$

$$\dot{H}_y - m a_x r - mgr \sin\theta = 0 \qquad (5.174b)$$

$$\dot{H}_x + m a_y r = 0 \qquad (5.174c)$$

The first and third equations yield

$$\dot{H}_z = 0 \qquad (5.175)$$

which agrees with the Newton–Euler equations (5.169). The second and third equations are identical to those in the Newton–Euler method (5.169).

For completeness, we write out the acceleration terms

$$a_x = r[-\ddot{\theta} - \dot{\psi}\cos\theta(\dot{\phi} - \dot{\psi}\sin\theta)]$$

$$a_y = r[\ddot{\phi} - \ddot{\psi}\sin\theta - 2\dot{\psi}\dot{\theta}\cos\theta]$$

$$\dot{H}_x = I_a[\ddot{\phi} - \ddot{\psi}\sin\theta - \dot{\psi}\dot{\theta}\cos\theta]$$

$$\dot{H}_y = I_t[\ddot{\theta} + \dot{\psi}^2 \sin\theta\cos\theta] + I_a\dot{\psi}\cos\theta(\dot{\phi} - \dot{\psi}\sin\theta)$$

$$\dot{H}_z = I_t[\ddot{\psi}\cos\theta - 2\dot{\psi}\dot{\theta}\sin\theta] - I_a\dot{\theta}(\dot{\phi} - \dot{\psi}\sin\theta) \qquad (5.176)$$

The equations of motion, (5.174), are three coupled second-order nonlinear differential equations in the unknown functions $\{\psi(t), \theta(t), \phi(t)\}$. In numerical integration schemes (5.174) are written as a set of six first-order differential equations, and an integration algorithm such as the Runge–Kutta method is used. This solution will determine the set

$$\{\psi(t), \dot{\psi}(t), \theta(t), \dot{\theta}(t), \phi(t), \dot{\phi}(t)\}$$

To determine the position of the contact point we must go back and integrate the constraint equation (5.171).

$$\dot{\mathbf{r}}_c = -r\dot{\theta}\mathbf{e}_x + r(\dot{\phi} - \dot{\psi}\sin\theta)\mathbf{e}_y \tag{5.177}$$

Stability of a Rolling Disc

General analytic solutions to the rolling disc equations derived above have not been found due to the nonlinear nature of the equations. However, a standard technique in nonlinear systems is to seek a solution close to a steady motion. Such techniques are called *perturbation methods* (see, e.g., Nayfeh and Mook, 1979).

One can easily conduct a stability experiment by rolling a coin on a rough surface. The coin will roll on a straight path if the speed is greater than some critical value. At slower speeds than the critical value, the coin rolls over on its side.

We begin the analysis by looking for a solution to (5.174) close to $\theta = 0$. Then we determine the critical value at which the motion loses stability. In the case of near-vertical motion we write the generalized velocity variables in the form

$$\dot{\phi} = \omega_0 + \eta_1(t)$$

$$\dot{\psi} = \eta_2(t)$$

$$\dot{\theta} = \eta_3(t) \tag{5.178}$$

where we assume η_1, η_2, η_3 are small compared with the spin ω_0.

In the perturbation method we linearize the equations of motion. Thus, we assume

$$\sin\theta \sim \theta$$

$$\cos\theta \sim 1$$

and we drop all nonlinear terms in (5.174) such as $\dot{\phi}\dot{\psi}$, $\dot{\psi}\theta$, $\dot{\psi}^2$, etc. When this is done the equations for $\{\eta_1, \eta_2, \eta_3\}$ take the form

$$\begin{cases} \dot{\eta}_1 = 0 \\ I_t\dot{\eta}_2 - I_a\omega_0\eta_3 = 0 \\ [mr^2 + I_t]\dot{\eta}_3 + [mr^2 + I_a]\omega_0\eta_2 = mgr\theta \end{cases} \tag{5.179}$$

These equations correspond to (5.174c), (5.174a), (5.174b) in that order. Taking the derivative of the last equation and using $\dot{\theta} = \eta_3$, we obtain a single second-order differential equation:

$$I'_t\ddot{\eta}_3 + \left[-mgr + \frac{I_a I'_a}{I_t}\omega_0^2\right]\eta_3 = 0 \tag{5.180}$$

where

$$I'_t = I_t + mr^2, \qquad I'_a = I_a + mr^2 \tag{5.181}$$

This second-order equation has oscillatory solutions of the form $\eta_3 = C_1 \cos(\Omega t + C_2)$, provided that

$$\left[-mgr + \frac{I_a I'_a}{I_t} \omega_0^2 \right] > 0$$

or

$$\omega_0^2 > \frac{I_t mgr}{I_a [I_a + mr^2]} \equiv \omega_c^2 \tag{5.182}$$

When $\omega_0 < \omega_c$, the solution grows exponentially with time, $\eta_3(t)$ becomes large, and the linear approximation breaks down. Thus ω_c is a critical rolling speed to avoid growth in the roll angle $\theta(t)$.

For a solid disc of the size of a U.S. quarter coin ($0.25), $2r = 2.4$ cm and $I_a = 2I_t = mr^2/2$. For this case the spin rate $\omega_c/2\pi$ is equal to 2.65 cycles per second (cps). Note that ω_c is independent of the mass:

$$\omega_c = (g/3r)^{1/2}$$

This is also the natural frequency of a particle pendulum on a string of length $3r$.

In order to ensure the no-slip rolling constraint in testing out the theory with a coin, you may have to carefully wrap a small rubber band around the coin edge.

This example, like the one in Example 5.11, shows how a rotary motion can induce a kinematic stiffness.

Rolling Dynamics: Simulation and Animation

Until recently dynamicists could do little to make use of the highly nonlinear equations of rigid-body dynamics, like those for the rolling coin (5.174) beyond linear stability analysis or numerical integration. However, advances in both computer software and hardware permit more direct application of nonlinear equations of motion. The rolling disc problem serves as a model for these applications. Our goal here is to integrate the equations of motion to obtain the state vector as a function of time $[\theta, \dot{\theta}, \psi, \dot{\psi}, \phi, \dot{\phi}_c]^T$.

It is important for the dynamicist to define the goals for numerical analysis. For example if we wanted to know the contact force as a function of time or to check if the rolling friction force exceeded the material properties, a more complicated analysis would be needed. In this section our goal is to provide a numerical algorithm to obtain a three-dimensional animation of the rolling as a function of time. This entails collecting a sequence of time-frozen images of the disc to enable a graphics software package to create a movie of the rolling dynamics. We outline the steps to create an animation movie:

1. Set up the equations of motion in first-order format for numerical integration.
2. Choose a numerical integration algorithm.
3. Define an object set of vectors in three space $\{\boldsymbol{\rho}_k\}$ that will embody the approximate geometry of the disc in a reference state: $t = t_0$.
4. Establish a transformation matrix that will use the output of the integration subprogram to create a new set of object vectors at each of the movie frame times $\{t_k\}$, $t_k > t_0$.
5. Input the sets of object vectors into a graphics plotting subroutine to obtain a set of time-sequential images.
6. Input the time-series images in a movie subroutine at a certain graphics refresh rate to produce an illusion of a moving object.

Each of these steps will be explained below.

1. *First-order Format of Equations of Motion* Most numerical integration methods are based on the dynamic equations in the form

$$\dot{\mathbf{x}} = \mathbf{f}(\mathbf{x}, t)$$

where $\mathbf{x}$ is called a state vector. In the case of the rolling disc it is convenient to use a set of generalized coordinates and velocities

$$\mathbf{x} = [\theta, \dot{\theta}, \psi, \dot{\psi}, \Phi, \dot{\Phi}]$$

where $\dot{\Phi}$ is the relative spin rate defined by

$$\dot{\Phi} = \dot{\phi} - \dot{\psi} \sin\theta \tag{5.183}$$

To obtain a set of first-order equations from (5.174) we define the angular rates

$$\dot{\theta} = \tilde{\omega}_\theta$$

$$\dot{\psi} = \tilde{\omega}_\psi$$

$$\dot{\Phi} = \tilde{\omega}_\Phi$$

Then we solve (5.174) for the second derivative of $\{\dot{\theta}, \dot{\psi}, \dot{\Phi}\}$

$$\dot{\tilde{\omega}}_\theta = \frac{mgr}{I'_t} \sin\theta - \tilde{\omega}_\psi^2 \sin\theta \cos\theta - \frac{I'_a}{I'_t} \tilde{\omega}_\psi \tilde{\omega}_\Phi$$

$$\dot{\tilde{\omega}}_\psi = 2\tilde{\omega}_\psi \tilde{\omega}_\theta \frac{\sin\theta}{\cos\theta} + \frac{I_a}{I_t} \tilde{\omega}_\theta \tilde{\omega}_\Phi \frac{1}{\cos\theta}$$

$$\dot{\tilde{\omega}}_\Phi = \frac{mr^2}{I'_a} \tilde{\omega}_\psi \tilde{\omega}_\theta \cos\theta \tag{5.184}$$

The equation for $\dot{\tilde{\omega}}_\theta$ is from (5.174b), that for $\dot{\tilde{\omega}}_\psi$ is from $\dot{H}_z = 0$ [use (5.174a) and (5.174c)] and $\dot{\tilde{\omega}}_\Phi$ is from (5.174c). The two sets of equations form a coupled set. In addition to these equations, we must also integrate the equations for the center-of-mass position: $\mathbf{r}_c = X_c\mathbf{i} + Y_c\mathbf{j} + Z_c\mathbf{k}$

$$\dot{\mathbf{r}}_c = -r\dot{\theta}\mathbf{e}_x + r\dot{\Phi}\mathbf{e}_y$$

Note that the vertical position off the plane Z_c is equal to $r\cos\theta$.

2. *Numerical Integration Scheme* There are many numerical integration algorithms, including Euler, Runge–Kutta, and Adams. Many of these methods contain variable time-step sizes. If one wishes to avoid interpolating, a fixed time-step method should be chosen in order to have a correct time flow in the movie or animation.

3. *Define a Set of Object Vectors* Typical computer graphics packages draw lines between points defined by position vectors. In the case of the circular disc, we define a reference position in the vertical plane with $X_c = Y_c = \theta = \psi = 0$. The circumference is divided into N segments, and a set of position vectors corresponding to these N points must be defined.

4. *Finite Transformation Matrix* At each time, the integration subroutine will return a set of generalized positions $\{\theta_k, \psi_k, \Phi_k, X_{ck}, Y_{ck}, Z_{ck}\}$. Here we can use the finite-rotation transformation matrices in Chapter 3. Using a 4×4 matrix format, the new set of object vectors $\boldsymbol{\rho}_k$ can be determined by the matrix operation

$$\boldsymbol{\rho}_k = \boldsymbol{\rho}(t = t_k) = \mathsf{T}_k \boldsymbol{\rho}_0 \tag{5.185}$$

where

$$\mathsf{T}_k = \begin{bmatrix} & & & X_{ck} \\ & \mathcal{R}_k & & Y_{ck} \\ & & & Z_{ck} \\ 0 & 0 & 0 & 1 \end{bmatrix}$$

where $Z_{ck} = r\cos\theta_k$, and $\mathcal{R}_k$ is the product of the three Euler angle rotation matrices (5.27), (5.28), (5.29) discussed in Section 5.2.

5, 6. *Graphics Plotting and Movie Subroutines* MATLAB is one of the more widely available multipurpose codes. In Fig. 5.28, a MATLAB subroutine called *surf* is used to draw a reference surface with a grid to simulate the rolling constraint surface. The input involves initial conditions, the time step size, and the number of frames N. The output matrix M is a set of "frames" i.e., position of the vector $\{\boldsymbol{\rho}_k\}$ at different times. MATLAB then displays each frame sequentially with a subroutine called *movie* (M). Fig. 5.28 shows a sequence of frames showing the rolling disc. (This program was written by a former Cornell University graduate student Dr. E. Catto.)

Caution

As in all numerical computer programs, one must always be skeptical of the output until one have verified the code. One source of error is the numerical integration scheme, which can sometimes add a negative damping (energy input) to the system. In the coin problem, there is no energy dissipation. Therefore one should

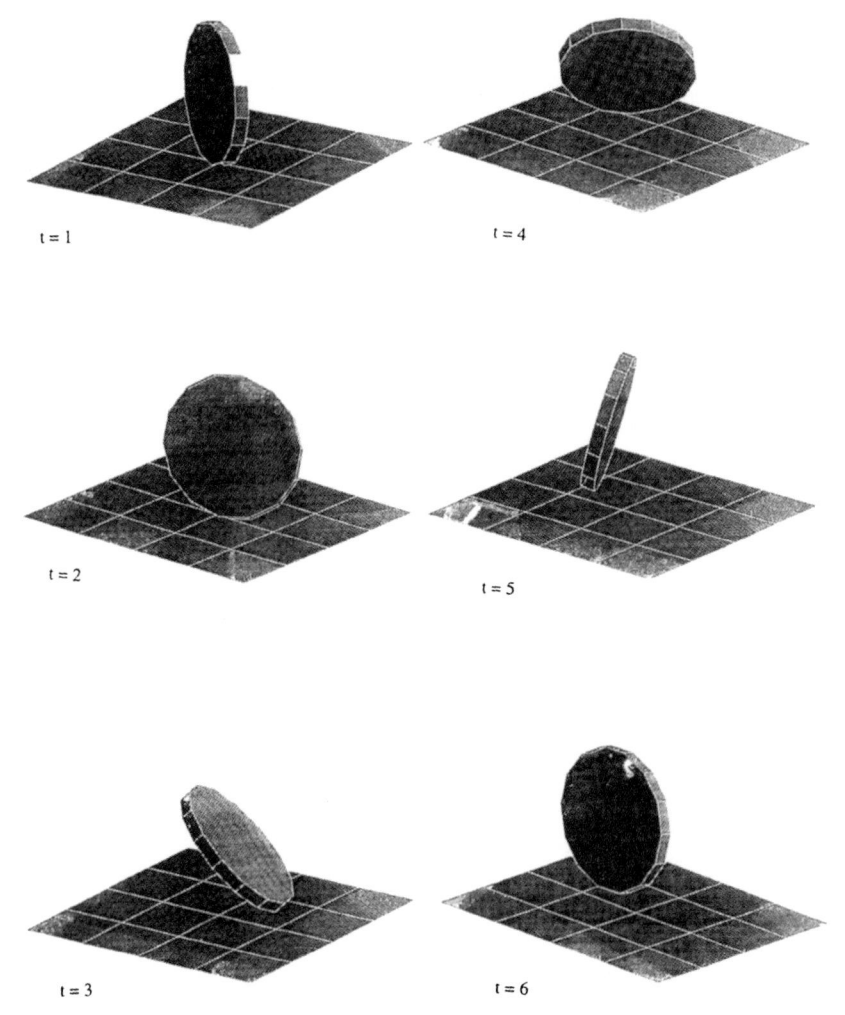

Fig. 5.28 Time sequence of numerically integrated motions of a rolling disc. Graphical output using MATLAB software program.

check that the total kinetic plus potential energy is conserved. Make the time step smaller if one sees serious deviations from conservation of energy.

5.6
Nonholonomic Rigid Body Problems

Nonholonomic problems involve kinematic constraints between the generalized velocities that cannot be integrated to give a pure geometric constraint between the generalized coordinates. Examples include rolling, some types of feedback control forces, and certain types of voltage–current constraints involving sliding contacts in electric machines. In nonholonomic problems the N unconstrained degrees of freedom or generalized coordinates $\{q_k(t)\}$ are reduced in number by the M nonholonomic constraint equations that usually take the form

$$\sum_{j=1}^{N} a_{ij}\dot{q}_j(t) + b_i(t) = 0 \quad (i = 1, 2, \ldots, M) \tag{5.186}$$

There are three commonly used methods to solve such problems:

1. Choose another set of $N - M$ generalized coordinates that identically satisfy the kinematic constraints.
2. Use Lagrange multipliers. This technique was introduced in Section 4.5.
3. Use the principle of virtual power.

Technically, rigid-body problems with nonholonomic constraints involve the stability of rolling, such as shimmy and vehicle skid stability.

A general discussion of the dynamics of nonholonomic constraints can be found in the excellent monograph by Neimark and Fufaev (1972). They discuss many rolling problems as well as nonholonomic problems in electro-mechanical systems such as arise in sliding contacts. They also treat practical problems of the stability of aircraft landing gear, automobile steering and shimmy, and the stability of railroad wheels.

In Chapter 4 we described the principle of virtual power as a method that naturally incorporated nonholonomic constraints and attributed its theory and application to Jourdain (1909) and Kane (1961). However, Neimark and Fufaev describe several other formulations and modifications of Lagrange's equations for incorporating nonholonomic constraints. [See references to Volterra (1898), Appel (1899) and Voronec (1901) in Neimark and Fufaev (1972).] Anyone with a more theoretical interest in these problems is encouraged to read Neimark and Fufaev, who give several hundred references on the subject.

In this section, we present a classic skidding problem of rigid bodies that helps illustrate the methods discussed in the previous sections.

Vehicle Stability in a Skid

The problem of vehicle stability in a skid is shown in Fig. 5.29, in which a two-axel vehicle skids and we are asked to determine the stability of motion of the body rolling on one axle. This problem may be found in Greenwood (1988) and also in Neimark and Fufaev (1972). The Author also attended a discussion of the problem by Professor Thomas Kane of Stanford University at a guest lecture at Cornell University in the early 1980s.

In all three treatments of the problem the rolling constraint is simplified to what some call a skating or *sliding knife-edge constraint*. Mathematically the velocity par-

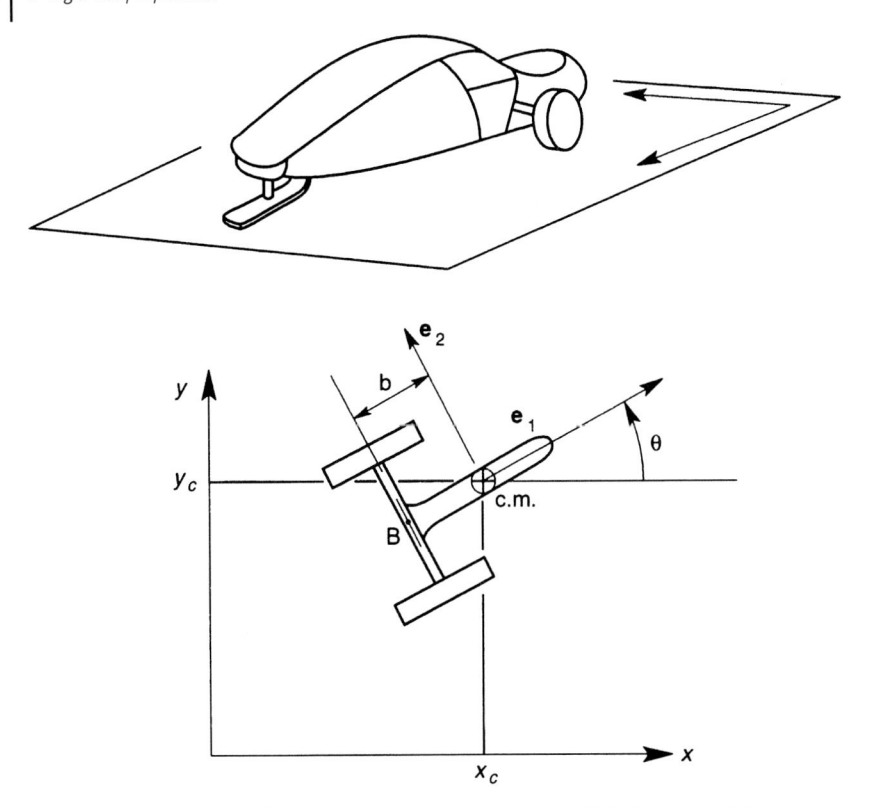

Fig. 5.29 Nonholonomic rolling or skating dynamics of a rigid vehicle (Example 5.14).

allel to the nonskidding axle is assumed to be zero. Thus if B is a point on the axle and the axes $\{\mathbf{e}_1, \mathbf{e}_2\}$ are fixed to the vehicle, then we require that

$$\mathbf{v}_B \cdot \mathbf{e}_2 = 0 \tag{5.187}$$

Denoting the velocity of the center of mass by $\mathbf{v}_c = [v_1, v_2]^T$, and θ, the angular position of the $\mathbf{e}_1$ axis relative to the horizontal x-axis, the constraint (5.187), takes the form of a scalar equation

$$v_2 - b\dot{\theta} = 0 \tag{5.188}$$

Thus of the three possible generalized velocities $\{v_1, v_2, \omega = \dot{\theta}\}$, only two are independent. We present two methods of analysis for comparison: (1) the principle of *virtual* power, and (2) Lagrange's equations with Lagrange multipliers.

The virtual power method has the advantage of being closer to the physics of Newton's law and involves calculating accelerations and forces and moments. Lagrange's method of multipliers is more abstract, but has less dependence on vector calculus. The student should judge for him or herself which method is easier.

5.6 Nonholonomic Rigid Body Problems

Example 5.15 Principle of Virtual Power. In this method we choose $\{v_1, v_2\}$ for the two independent generalized velocities. In this analysis, we neglect the rotary inertia of the rolling wheels. Following (5.159), the equations of motion take the form

$$(m\dot{\mathbf{v}}_c - \mathbf{F}^a) \cdot \frac{\partial \mathbf{v}_c}{\partial \dot{q}_j} + (I\dot{\boldsymbol{\omega}} - \mathbf{M}^a) \cdot \frac{\partial \boldsymbol{\omega}}{\partial \dot{q}_j} = 0 \tag{5.189}$$

where from the constraint, Eq. (5.188),

$$\boldsymbol{\omega} = \dot{\theta}\mathbf{e}_3 = \frac{v_2}{b}\mathbf{e}_3 \tag{5.190}$$

The velocity of the center of mass is, $\mathbf{v}_c = v_1\mathbf{e}_1 + v_2\mathbf{e}_2$. The projection vectors are then given by

$$\frac{\partial \mathbf{v}_c}{\partial \dot{q}_1} = \mathbf{e}_1, \qquad \frac{\partial \mathbf{v}_c}{\partial \dot{q}_2} = \mathbf{e}_2$$

$$\frac{\partial \boldsymbol{\omega}}{\partial \dot{q}_1} = 0, \qquad \frac{\partial \boldsymbol{\omega}}{\partial \dot{q}_2} = \frac{1}{b}\mathbf{e}_3 \tag{5.191}$$

We assume for the moment that the vehicle is on an inclined plane with a grade ϕ, and that the gravity-force component in the plane is in the negative y-direction, or

$$\mathbf{F}^a = -mg\sin\phi\,\hat{\mathbf{j}} \tag{5.192}$$

With no power on the wheeled axle, we also have $\mathbf{M}^a = 0$. The two equations of motion are given by

$$ma_1 - F_1^a = 0$$

$$ma_2 - F_2^a + I\dot{\omega}\left(\frac{1}{b}\right) = 0$$

where

$$F_1^a = -mg\sin\theta\sin\phi$$

$$F_2^a = -mg\cos\theta\sin\phi \tag{5.193}$$

The acceleration of the center of mass $\dot{\mathbf{v}}_c$ is written in components relative to the body axes,

$$\dot{\mathbf{v}}_c = \dot{v}_1\mathbf{e}_1 + \dot{v}_2\mathbf{e}_2 + v_1\dot{\mathbf{e}}_1 + v_2\dot{\mathbf{e}}_2$$

$$= a_1\mathbf{e}_1 + a_2\mathbf{e}_2$$

where

$$a_1 = \dot{v}_1 - \dot{\theta}v_2 = \dot{v}_1 - v_2^2/b$$
$$a_2 = \dot{v}_2 + \dot{\theta}v_1 = \dot{v}_2 + v_1 v_2/b \tag{5.194}$$

For zero grade, $\phi = 0$, $\mathbf{F}^a = 0$, the equations of motion become

$$\dot{v}_1 = v_2^2/b$$
$$\dot{v}_2 = -\frac{v_1 v_2}{b(1 + r_g^2/b^2)} \tag{5.195}$$

where we have replaced the moment of inertia by its radius of gyration about the center of mass; $I = mr_g^2$. We shall solve these coupled first-order nonlinear differential equations after the next example.

Example 5.16 Lagrange Multiplier. The kinetic energy of the vehicle is given by

$$T = \frac{1}{2}m v_c^2 + \frac{1}{2}I\omega^2 \tag{5.196}$$

In this method we must choose generalized coordinates $\{q_i(t)\}$. Thus we are not free to choose the path velocities as in the virtual power method, since we cannot integrate them directly to obtain a set of $\{q_i(t)\}$. For the Lagrange method we choose Cartesian components of the position vector as well as the angle, θ; i.e., $\{q_i\} = \{x, y, \theta\}$, so that

$$T = \frac{1}{2}m(\dot{x}^2 + \dot{y}^2) + \frac{1}{2}I\dot{\theta}^2 \tag{5.197}$$

In these coordinates the constraint (5.187) becomes

$$-\dot{x}\sin\theta + \dot{y}\cos\theta - b\dot{\theta} = 0 \tag{5.198}$$

which has the form

$$\sum \alpha_i \dot{q}_i = 0$$

where

$$\alpha_1 = -\sin\theta, \qquad \alpha_2 = +\cos\theta, \qquad \alpha_3 = -b$$

Lagrange's equations with one Lagrange multiplier have the form (see Section 4.5)

$$\frac{d}{dt}\frac{\partial T}{\partial \dot{q}_i} - \frac{\partial T}{\partial q_i} - Q_i = \lambda \alpha_i \tag{5.199}$$

which in this problem becomes

$$m\ddot{x} = -\lambda \sin\theta$$
$$m\ddot{y} = \lambda \cos\theta - mg\sin\phi$$
$$I\ddot{\theta} = -b\lambda \tag{5.200}$$

Thus in the Lagrange formulation we end up with three differential equations of motion (5.200) plus the constraint equation (5.198) in four unknown functions of time $\{x, y, \theta, \lambda\}$. We can show that $\lambda(t)$ is proportional to the lateral constraint force on the fixed, nonskid axle. Knowledge of this force may be important for a designer. But if we want a minimum set of equations, we have to eliminate λ, θ. To show that these equations are equivalent to those found from the principle of virtual power we transform the velocity to body fixed axes using a rotation matrix:

$$\begin{bmatrix} \dot{x} \\ \dot{y} \end{bmatrix} = \begin{bmatrix} \cos\theta & -\sin\theta \\ \sin\theta & \cos\theta \end{bmatrix} \begin{bmatrix} v_1 \\ v_2 \end{bmatrix} \tag{5.201}$$

Eliminating λ from the first two equations of (5.200), one obtains for $\phi = 0$,

$$\ddot{x}\cos\theta + \ddot{y}\sin\theta = 0 \tag{5.202}$$

Using the transformation equation, one can then obtain

$$\dot{v}_1 - v_2\dot{\theta} = 0 \tag{5.203}$$

or using the constraint (5.187),

$$\dot{v}_1 = v_2^2/b \tag{5.204}$$

which is identical to that obtained by the principle of virtual power (5.195). The second equation for $\dot{v}_2$ in (5.195) can also be obtained from (5.200) and (5.198).

Stability of a Vehicle Skid

The virtual power and the Lagrange multiplier methods applied to the planar dynamics of a vehicle under the nonholonomic skating or skid constraints result in equations of motion involving the velocity components measured relative to the principal axes (5.195) (Fig. 5.29):

$$\begin{aligned} \dot{v}_1 &= v_2^2/b \\ \dot{v}_2 &= -v_1 v_2/b\left(1 + r_g^2/b^2\right) \end{aligned} \tag{5.205}$$

In this problem r_g is the radius of gyration about the center of mass and b is the distance between the center of mass and the skating or non-skid axle as shown in Fig. 5.29.

This set of first-order differential equations is amenable to the general methods of nonlinear dynamical systems (see Chapter 9). In these methods, we describe the dynamics using trajectories in a phase plane with a vector $[v_1, v_2]^T$, as shown in Fig. 5.30. Equilibrium points in the phase plane are given by $\dot{v}_1 = 0, \dot{v}_2 = 0$, or

$$v_2 = 0, \quad v_1 = \pm V_0 \quad \text{(constant velocity)}$$

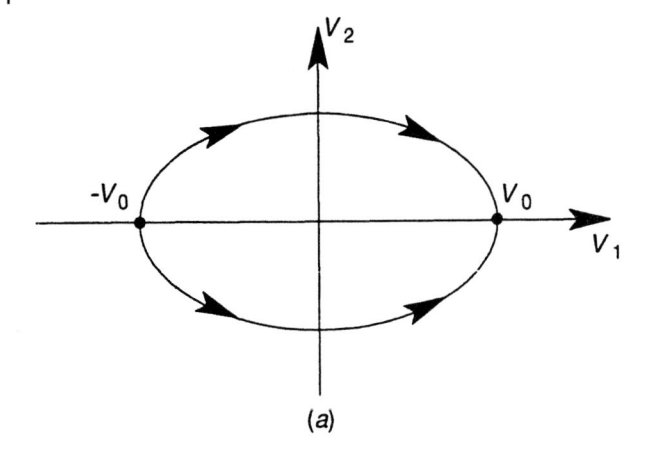

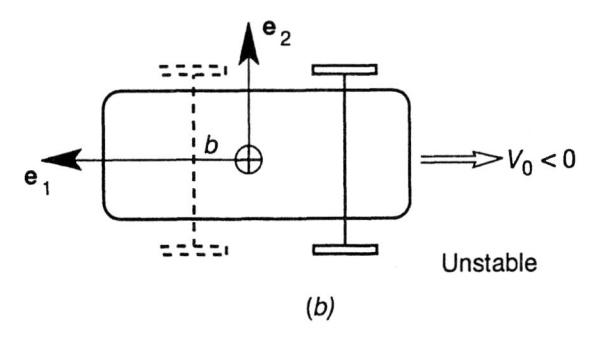

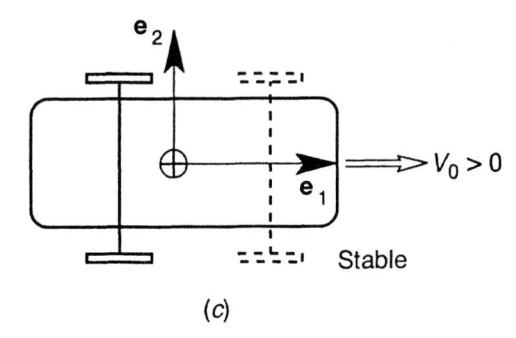

Fig. 5.30 (a) Phase plane diagram showing trajectories of the unstable motion with negative initial velocity V_0. (b) Unstable and (c) stable skid configurations of a vehicle. The dotted axel shows the skidding wheels.

Dividing $\dot{v}_1$ by $\dot{v}_2$, we can show that the following relation is valid:

$$\frac{dv_1}{dv_2} = -\left(1 + \frac{r_g^2}{b^2}\right)\frac{v_2}{v_1} \tag{5.206}$$

This equation can be integrated to obtain a conservation-of-energy relation,

$$\frac{1}{2}v_1^2 + \frac{1}{2}(1 + r_g^2/b^2)v_2^2 = \frac{1}{2}V_0^2 \tag{5.207}$$

This trajectory is drawn in the (v_1, v_2) plane through the two equilibrium points on the $v_2 = 0$ axis. One can also show that time flows away from the negative steady velocity $-V_0$, as shown in Fig. 5.30a since $\dot{v}_1$ is always positive (5.200). Thus a small departure from a negative velocity (i.e., rear wheel skid) leads to an increase in v_2 or a rotation of the body (Fig. 5.30b). On the other hand, if the front wheels skid and the rear do not (Fig. 5.30c), the initial velocity in the body axes is positive and small departures tend to reorient the body away from rotation.

Homework Problems

5.1 (*Moment of Inertia*) Consider a thin semi-circular ring of mass, m. Use the definition of the center of mass to find the location of the center of mass. Denoting the distance of the center mass from the circle center as d_c, use the parallel axis theorem to find the moment of inertia about an axis through the center of mass and normal to the plane of the ring.

5.2 (*Moment of Inertia, Incomplete Circular Ring*) Following the calculations in Example 5.5, calculate the moments of inertia about the x, y and z axes where the z axis is normal to the plane of the ring. Use the parallel axis theorem to calculate the complete moment of inertia matrix about the center of mass.

5.3 (*Robot Arm Kinematics*) The robotic manipulator arm shown in Fig. P5.3 consists of three serial links with three revolute joints. Suppose the waist motor, shoulder motor and elbow motor rates are respectively, Ω_W, Ω_S, Ω_E, and are constant in magnitude. If the origin of the upper arm link is coincident with the waist motor axis, find an expression for the vector velocity and acceleration of the wrist point P. Choose a cylindrical coordinate system that rotates with the waist motor.

5.4 (*Reuleaux Triangle*) The Reuleaux triangle is a closed curve of constant width formed with three circular arcs drawn from the vertices of an equilateral triangle (Fig. P5.4). It is named after the German kinematics theorist Franz Reuleaux (1839–1905) who studied curves of constant width. The so-called curved triangle was used in positive return cams in mechanical control systems and a similar shape is used in the piston of the Wankel rotary engine. The Reuleaux triangle can rotate in a square hole as shown in Fig. P5.4. Find an expression for the velocity and acceleration of the center point of the curved triangle as it is constrained in the square hole.

5.5 (*Rolling and Cycloid Curves*) Consider the rolling of a circular cylinder on a flat plane (Fig. P5.5). The path of a point on the circumference of the wheel traces out a curve called a cycloid. (i) Derive an analytical expression for cycloid curve. (ii) Write vector expressions for the velocity and acceleration of a rim point when the wheel rolls with constant angular velocity and without slip.

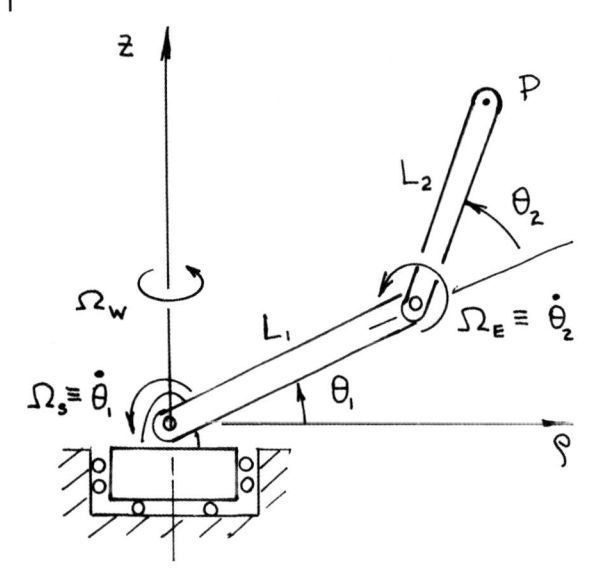

Fig. P5.3

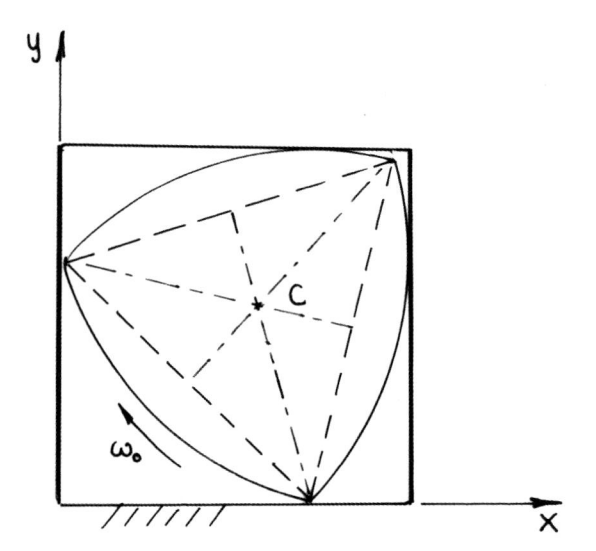

Fig. P5.4

(iii) Write a MATLAB program to draw the cycloid curve as well as plot the acceleration of a rim point as a function of the angle of rolling.

5.6 A spacecraft similar to the Hubble telescope rotates about an axis parallel to its diameter with constant rate Ω, while at the same time the solar panels rotate relative to the main structure with angular velocity and acceleration $\dot{\theta}, \ddot{\theta}$ (Fig. P5.6).

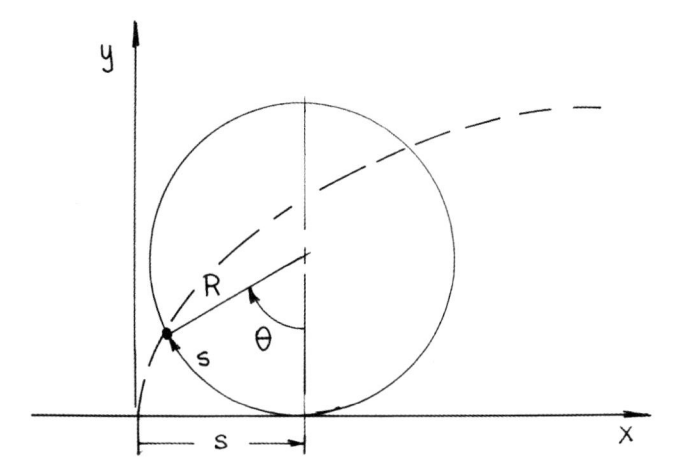

Fig. P5.5

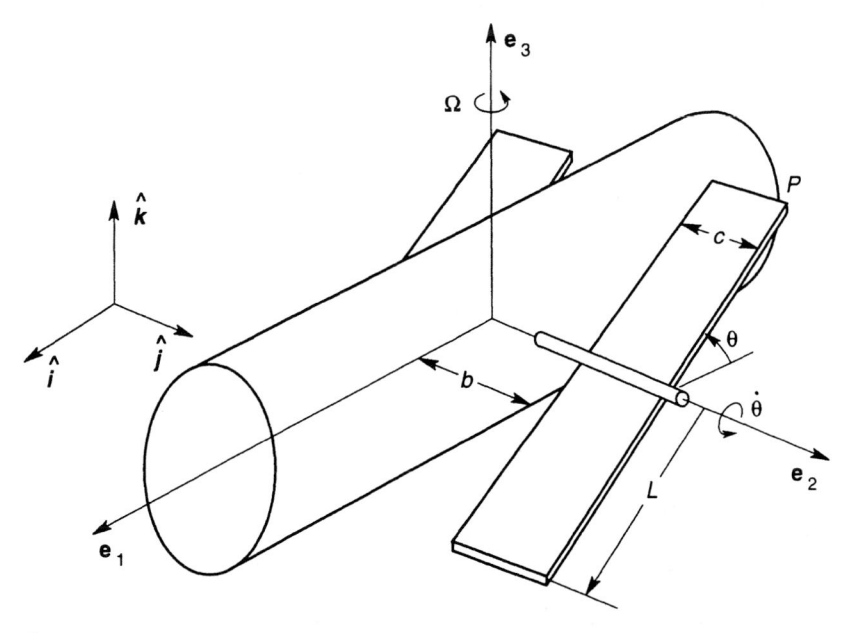

Fig. P5.6

(a) Write an expression for the velocity and acceleration of the point P on the end of the solar panel. (Assume $\mathbf{e}_3$ fixed in space.)
(b) What is the angular acceleration vector for the cylindrical body?
(c) What is $\dot{\boldsymbol{\omega}}$ for the solar panel?

5.7 In Problem 5.6, suppose the spacecraft first performs a finite rotation about the $\mathbf{e}_3$ or $\hat{k}$-axis of $\pi/2$ radian and then the solar panel rotates relative to the spacecraft by an angle of $\theta = 60$ degrees. Find a 3×3 rotation matrix that gives the new position vector for the point P relative to an unrotated reference

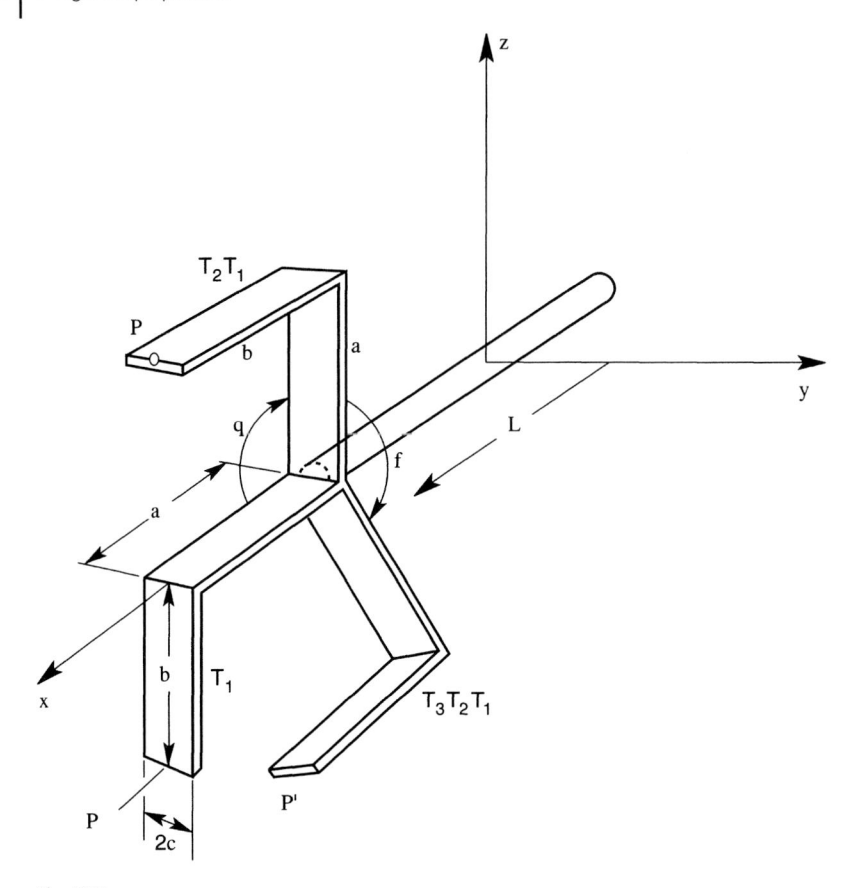

Fig. P5.9

frame $\{\hat{\mathbf{i}}, \hat{\mathbf{j}}, \hat{\mathbf{k}}\}$ at the origin. (*Note*: The original position vector is $\mathbf{r} = (b+c)\hat{\mathbf{j}} - L\hat{\mathbf{i}}$.) [*Hint*: Use the relations in (3.39).]

5.8 In the previous problem, use the vector representation in Chapter 3, (3.38), to find a single axis of rotation $\mathbf{n}$ and angle ϕ that describes the same finite rotation of the solar panel. (*Note*: $\mathbf{r}' = \mathbf{n}(\mathbf{n}\cdot\mathbf{r}) + [\mathbf{r} - \mathbf{n}(\mathbf{n}\cdot\mathbf{r})]\cos\phi + (\mathbf{n}\times\mathbf{r})\sin\phi$. See Problem 3.5 and use the relations (3.39).)

5.9 A two-link mechanism undergoes three sequential finite motions (Fig. P5.9). With the hinge originally at the origin, the initial position vector of the point P is given by $\mathbf{r} = a\hat{\mathbf{i}} - b\hat{\mathbf{k}}$. Find a 4×4 transformation that describes the new position vector after three subtransformations:

(a) T_1: Translation along x-axis, distance L.
(b) T_2: Rotation about the y-axis, $\theta = -\pi/2$.
(c) T_3: Rotation about the x-axis, $\phi = -2\pi/3$.

5.10 Write a MATLAB program (or use equivalent software) to simulate the finite rotation of the second (bent) link in Problem 5.9. (*Hint*: Use a *wire model* for the bent link that has initial vertices given by $[0, c, 0]$, $[a, c, 0]$, $[a, c, -b]$,

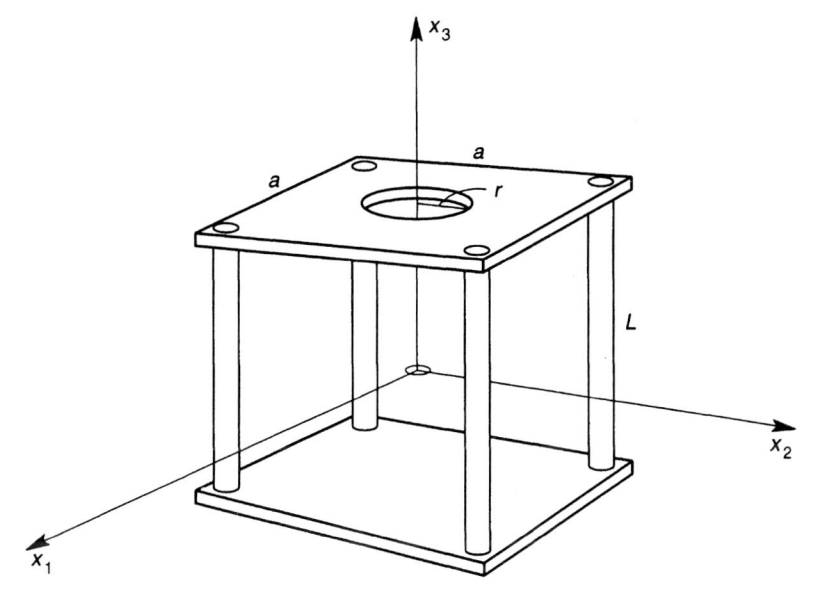

Fig. P5.13

[a, −c, −b], [a, −c, 0], [0, −c, 0]. Use the transformation matrices defined in Problem 5.9 to draw the wire model after each of the three transformations.)

5.11 An axisymmetric body similar to that shown in Fig. 5.3 undergoes precession, nutation, and spin $\{\phi, \theta, \psi\}$ using the Set A Euler angles. Assuming that the spin is about the axisymmetric z-axis in Fig. 5.5, derive kinematic relations similar to (5.23), but where the components of $\boldsymbol{\omega}$ are in the nonspin intermediate axes or $\boldsymbol{\omega} = [\omega_{x'}, \omega_{y''}, \omega_z]^T$.

5.12 A vehicle-like rigid body similar to that in Fig. 5.6b undergoes a rotation rate projected in the principal axes coordinates given by $\omega_x = 0.1$, $\omega_y = 0.3$, $\omega_z = -0.5$ rad per min. Find the yaw, pitch, and roll rates when $\theta = 0$, $\phi = 10°$.

5.13 The frame structure in Fig. P5.13 is made out of two square aluminum plates and four thin aluminum rods. The top plate has a circular opening of radius r. Find the center of mass as well as the principal inertias about the center of mass.

5.14 A satellite-type structure shown in Fig. P5.14 consists of a thin-walled cylinder of mass m_1 and two rectangular solar panels, each of mass m_2 oriented at an angle θ to the long axis of the cylinder. If both cylinder and panels have the same angular velocity vector $\boldsymbol{\omega}$, find an expression for the angular momentum in terms of the cylinder-based reference $\{x_1, x_2, x_3\}$. (*Hint*: Use the principal axes formula for each sub-body $\mathbf{H} = \sum I_i \omega_i \mathbf{e}_i$ and transform the panel basis vectors in terms of the cylinder principal axis vectors.)

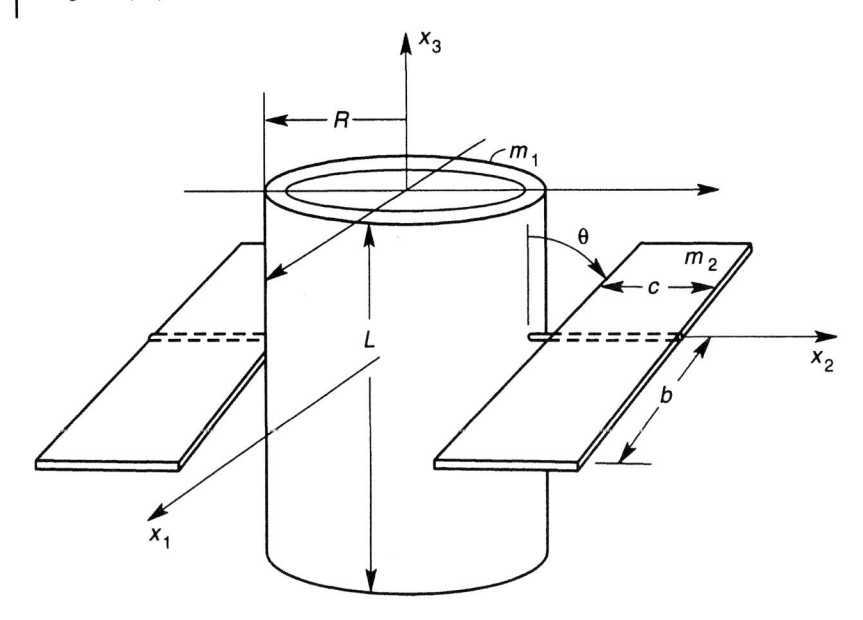

Fig. P5.14

5.15 Consider the pendulum shown in Fig. P5.15. In this problem we assume that the connecting pin block sits on a rough surface with mass m_1 and friction coefficient μ, i.e., the horizontal friction force $F_x = \mu N$ if the mass does not move. Suppose the rod oscillates with small angles $\theta(t)$. How large must the pin block be for a given maximum angle θ_{max} so that the block does not slip?

5.16 Derive the equations of motion for Problem 5.15 when the block slips under Coulomb friction. (Assume static and dynamic coefficients of friction are equal.) Write a MATLAB code to integrate the equations of motion or initial conditions $\dot{x}(0) = 0$, $x(0) = 0$, $\theta(0) = 0$, $\dot{\theta}(0) = \omega_0$, where ω_0 is large enough so that the block slides sometime after $t > 0$ and $|\theta_{max}| < \pi/2$. Show that eventually the slipping will stop and that the rod oscillates with periodic motion.

5.17 A wheel of mass m and radius R shown in Fig. P5.17 rotates about two axes with constant rotation rates $\dot{\theta}$, $\dot{\psi}$. (a) Calculate the angular momentum vector $\mathbf{H}_0$ and the applied moment $\mathbf{M}_0$ necessary to maintain this motion. (b) Suppose the vertical strut is a hollow, thin-walled tube of diameter d and thickness Δ. Find the average torsional stress τ required to produce this motion. Assume the mass is concentrated at the outer rim of the wheel. [Answer: $\tau = mR^2 \dot{\psi} \dot{\theta} / \pi d \Delta$.]

5.18 A thin rod of mass m and length L is constrained to move in the plane by two massless cables of lengths a, b with fixed points separated by a distance c (Fig. P5.18). Use Lagrange's equation to find the equation of motion. When the cable lengths are equal, $a = b$, can you show that the rotation rate $\boldsymbol{\omega}$, is always zero? (*Hint*: see Example 3.7.)

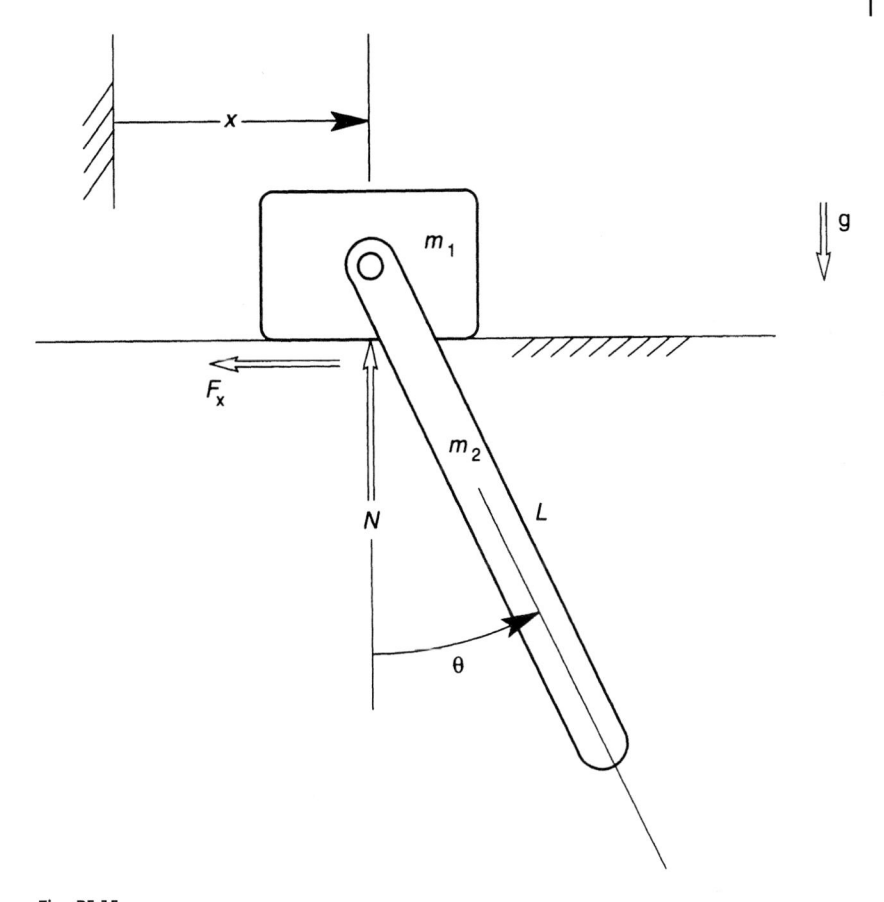

Fig. P5.15

5.19 A particle m_1 is attached to a rolling cylinder by an inextensible cable, as shown in Fig. P5.19. The roll angle $\theta = 0$ when the entire cable length hangs below the rolling surface. Use the Newton–Euler equations of motion to derive the equations of motion. For small angular motions, show that the cable tension is approximately equal to $m_1 g$ and that the roll oscillations mimic that of a pendulum with a natural frequency, $\Omega = [m_1 g R/(I_0 + m_2 R^2)]^{1/2}$.

5.20 A half cylinder of radius R rolls without slipping on a horizontal surface (Fig. P5.20). Use Lagrange's equation to derive the equation of motion. Assume a small angle of rotation and find the natural frequency of oscillation.

5.21 A cylinder of elliptic cross section rolls on a horizontal surface without slipping (Fig. P5.21). Assume that the semimajor and minor axes are a, b, respectively. Use Lagrange's equation to find the equation of motion. For small oscillations find the natural frequency.

5.22 A three-body satellite structure spins about the e_3 axis with initial angular velocity Ω_0 (Fig. P5.22). A pantograph mechanism moves the outer masses m_2 away from the central structure.

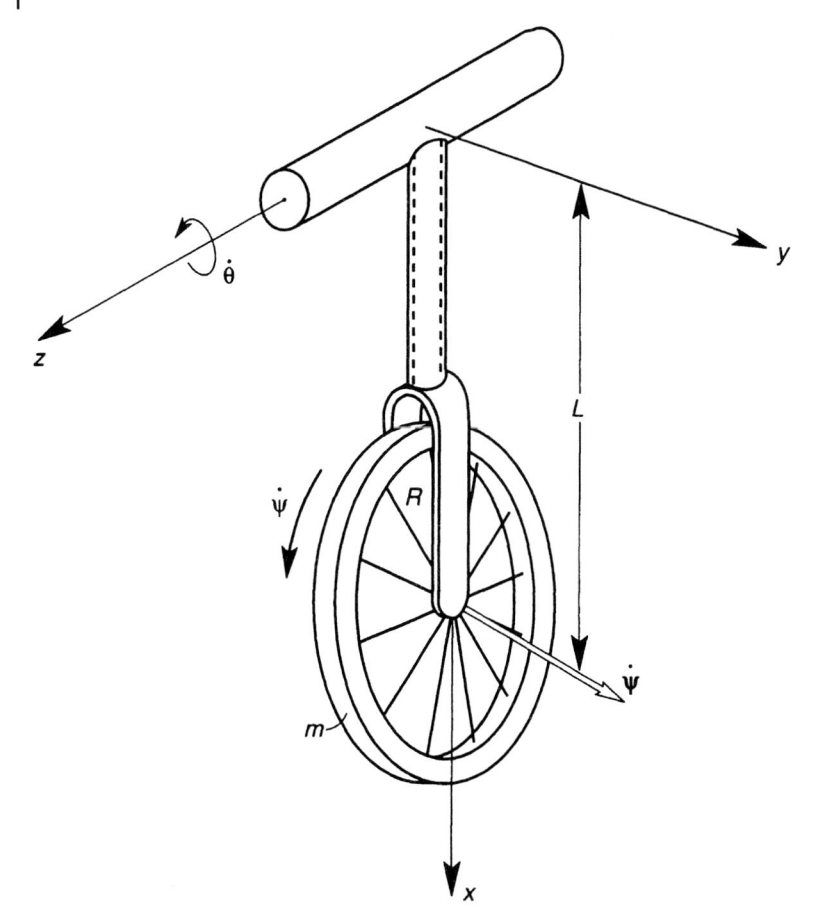

Fig. P5.17

(a) At what value of d will the spin become unstable?
(b) At the critical value of d, what is the value of the sign?

5.23 A thin rectangular plate of mass m_1 rotates about an axis along the diagonal, as shown in Fig. P5.23. Four small masses m_2 are screwed onto the plate along the minor plate axis front and back for symmetry.
  (a) Find the lateral reaction forces on the bearings due to rotation when $m_2 = 0$ (neglect gravity).
  (b) Can you find an offset ε and mass m_2 that makes the reaction force become zero?

5.24 A thin rod with a pin joint at its end is spun about the vertical axis with constant angular velocity Ω (Fig. P5.24).
  (a) Use the *principle of virtual power* to derive the equations of motion for the generalized coordinate $\theta(t)$. (*Hint*: Show that the Jacobian vector is given by $\partial \boldsymbol{\omega}/\partial \dot{\theta} = \mathbf{e}_2$, where $\mathbf{e}_2$ is a vector normal to the plane of the rod and vertical axis of rotation for Ω.)

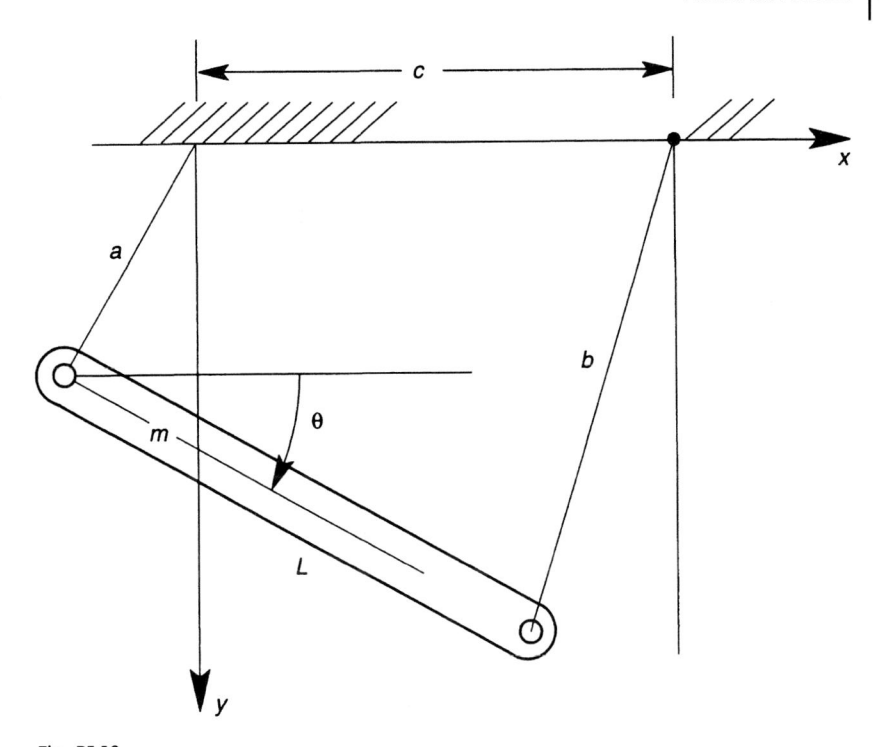

Fig. P5.18

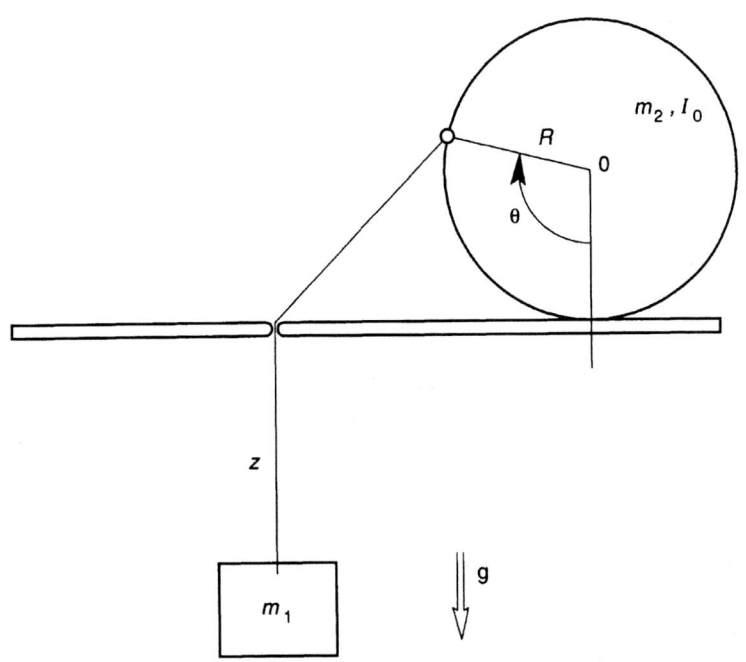

Fig. P5.19

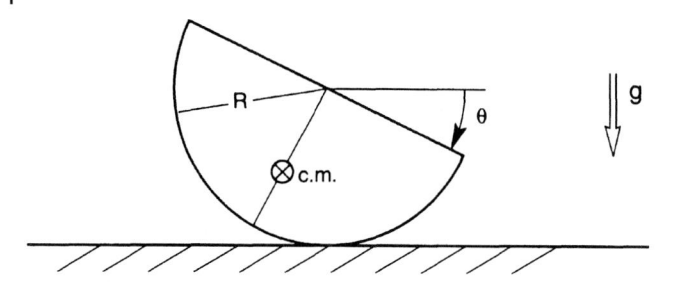

Fig. P5.20

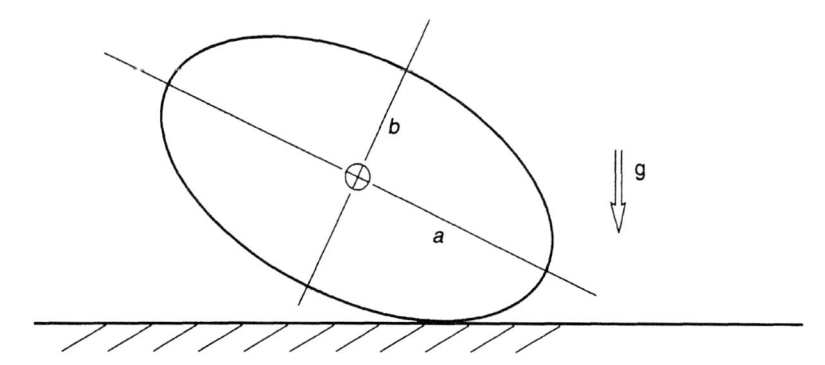

Fig. P5.21

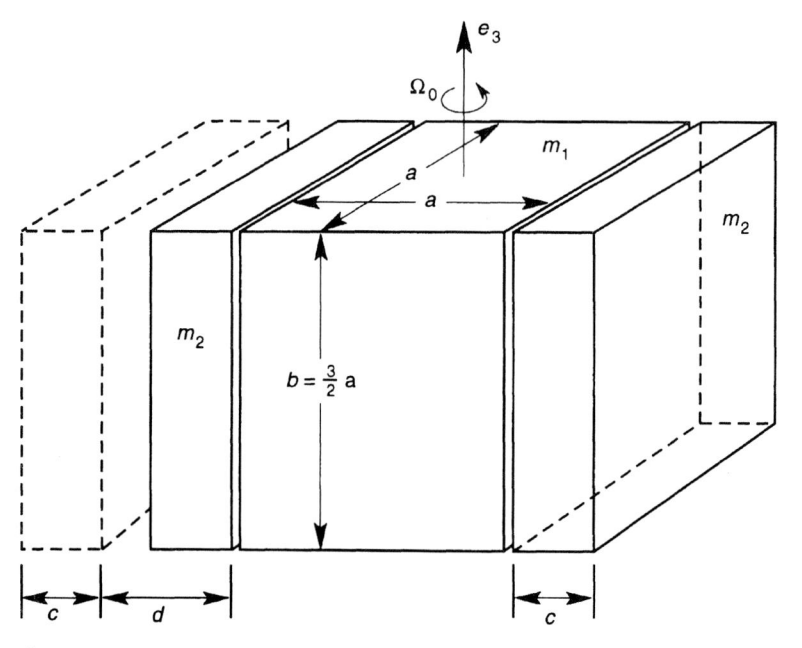

Fig. P5.22

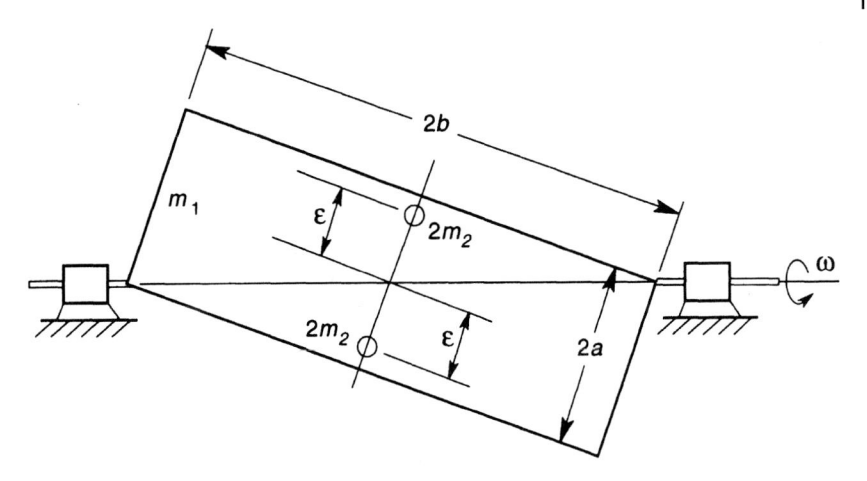

Fig. P5.23

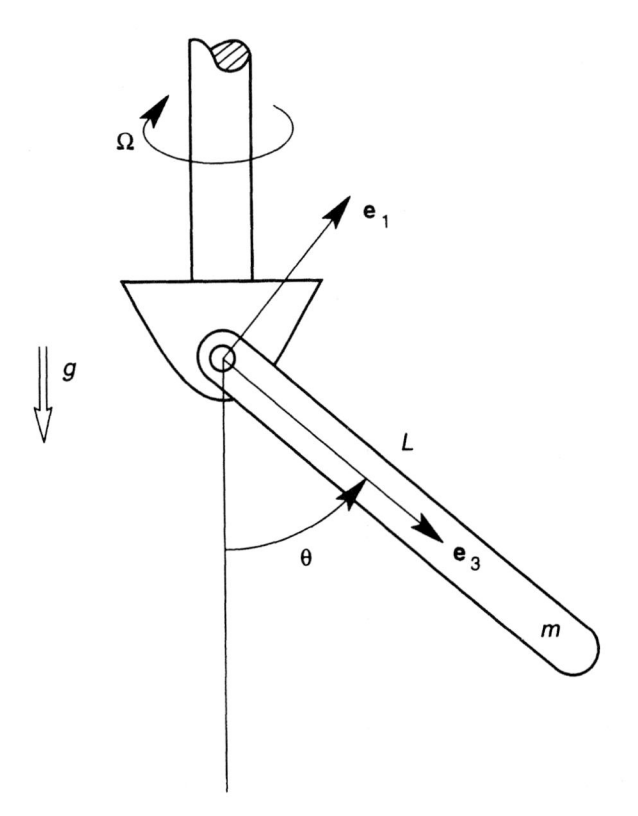

Fig. P5.24

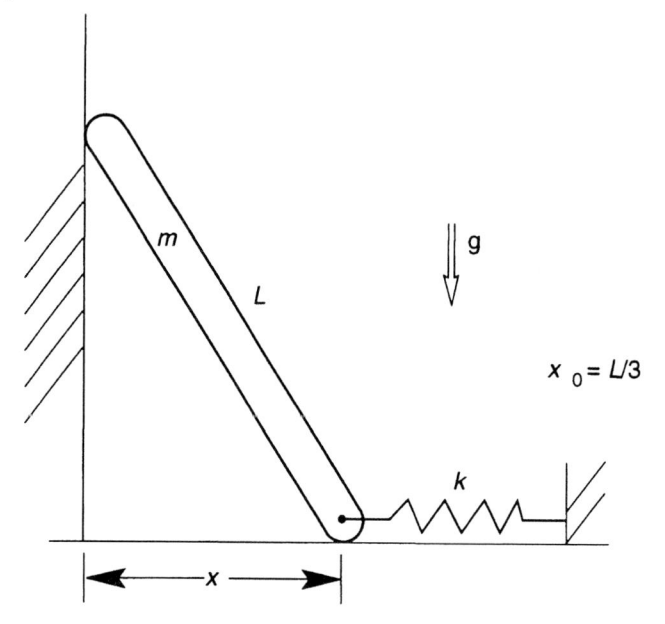

Fig. P5.25

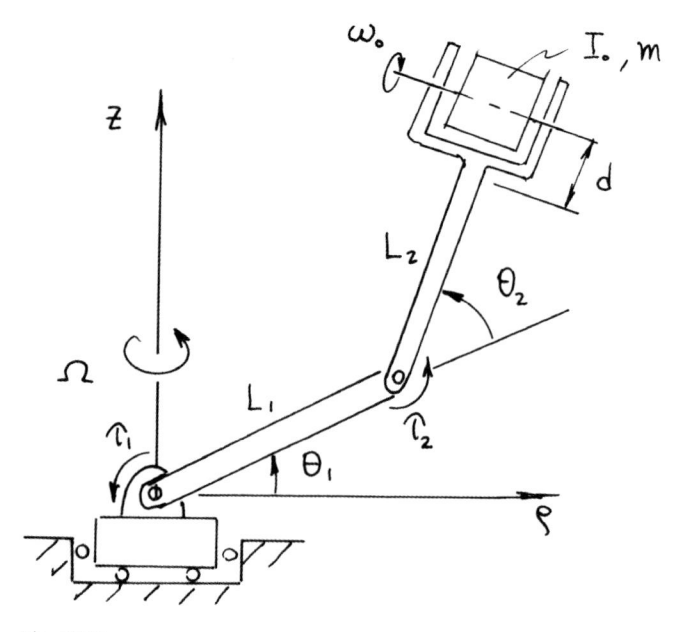

Fig. P5.26

(b) Show that $\theta = 0$ is a stable equilibrium point for $\Omega < \Omega_c$, where $\Omega_c^2 = mgL/2I_0$.

[Answer: $I_0 \ddot{\theta} - \Omega^2 \cdot I_0 \sin\theta \cos\theta + \dfrac{mgL}{2} \sin\theta = 0$]

5.25 The ends of a thin rod are constrained to follow the rigid vertical and horizontal surfaces as shown in Fig. P5.25. The bottom end is attached to a linear spring whose uncompressed state is at $x = x_0 = L/3$.
  (a) Use the *principle of virtual work* to find the equilibrium position.
  (b) Use the *principle of virtual power* to find the equation of motion for the generalized coordinate $x(t)$.

5.26 A heavy spinning rotor is attached to a lightweight robot manipulator arm as shown in Fig. P5.26. Suppose the waist motor is turning about the vertical axis with angular velocity Ω, with the shoulder and elbow joint angles fixed. What torques must be applied to the waist motor, shoulder motor and elbow motor to keep the axis of the spinning rotor horizontal to the base plane. Neglect gravity and the mass of the manipulator arm.

5.27 (Rod Pendulum) In Chapter 1, Eq. (1.7), we calculated the natural frequency of a point mass pendulum for small amplitude vibrations. Consider a thin rod constrained to rotate in the plane about a fixed axis at one of its ends. Use the Newton–Euler method and show that the natural frequency is given by $f = (1/2\pi)\sqrt{3g/2L}$. Compare this frequency with that of the point mass pendulum. Find the natural frequency when the rod pivots about a point between the center of mass and one of its ends.

5.28 (Rolling Rod Pendulum) A heavy rod of length L is welded to a thin circular shell of radius R as shown in Fig. P5.28. The circular shell is constrained to roll on a flat surface. Use Lagrange's equation to derive an equation of motion for the rolling pendulum.

5.29 (Rolling Pendulum) In Problem 5.28 assume small vibrations and the special case of $R = L$. Also assume a periodic motion of the form $\theta(t) = A \sin\omega t$. Derive the natural frequency using the equation relating the maximum kinetic energy of motion at time $t = 0$ to the maximum potential energy of the rod under gravity. Neglect the mass of the ring. Show that the natural frequency for this special case is that same as that for the small vibrations of the fixed pivot rod under gravity in Problem 5.27.

5.30 A thin rod is constrained to lie in contact with a circular arc of radius R as shown in Fig. P5.30. The plane of the rod and circular constraint is then rotated with angular velocity Ω. Use Lagrange's equation to derive the equation of motion of the rod. Is there a critical velocity Ω for which the motion about the vertical axis becomes unstable? Neglect the friction of constraints.

5.31 A *rate gyro* is a spinning device that produces a measurable nutation angle θ proportional to the precessional or yaw angular velocity. A sketch of such a sensor is shown in Fig. P5.31. Assume that the precession angular velocity is constant and is directed along the vertical axis. Also assume that the linear spring restricts the angle to small values. Use the direct form of the law

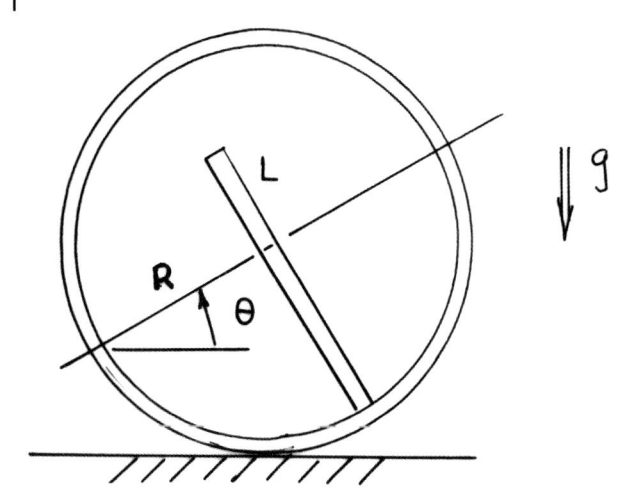

Fig. P5.28

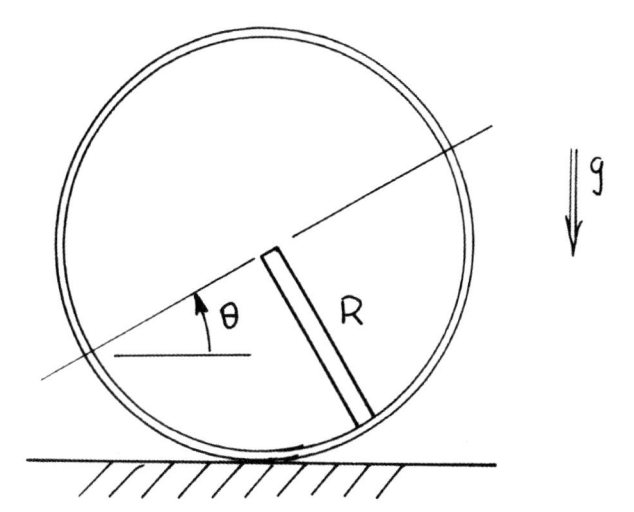

Fig. P5.29

of angular momentum, $\mathbf{M} = \dot{\mathbf{H}}$, to derive the equation of motion. Assume that the spin rate $\dot{\phi}_0$ is constant. Show that when $\ddot{\theta} = 0$ and $\dot{\theta}_0 \gg \dot{\psi}_0$, θ is proportional to $\dot{\psi}_0$.

[Answer: $I_1\ddot{\theta} + b\dot{\theta} - \dot{\psi}_1(\dot{\phi}_0 + \dot{\psi}_0\theta)I_3 + \dot{\psi}_0^2\theta I_1 + L^2k\theta = 0$, where I_1, I_3 are the principal moments of inertia.]

5.32 One end of a thin rod is constrained to follow a path $y = \alpha x^2/2$ (Fig. P5.32).
  (a) Use *Lagrange's equations* to derive equations of motion for the generalized coordinates $q_1 = x$, $q_2 = \theta$.
  (b) Linearize the equations of motion for small x, θ and find the natural frequencies of the system.

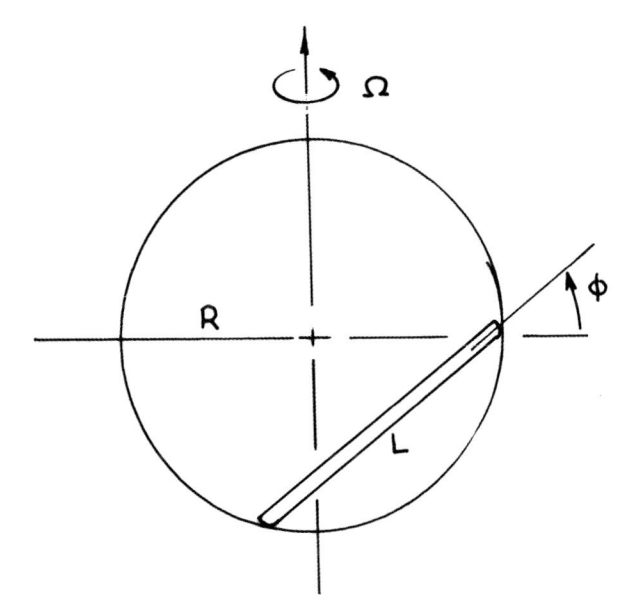

Fig. P5.30

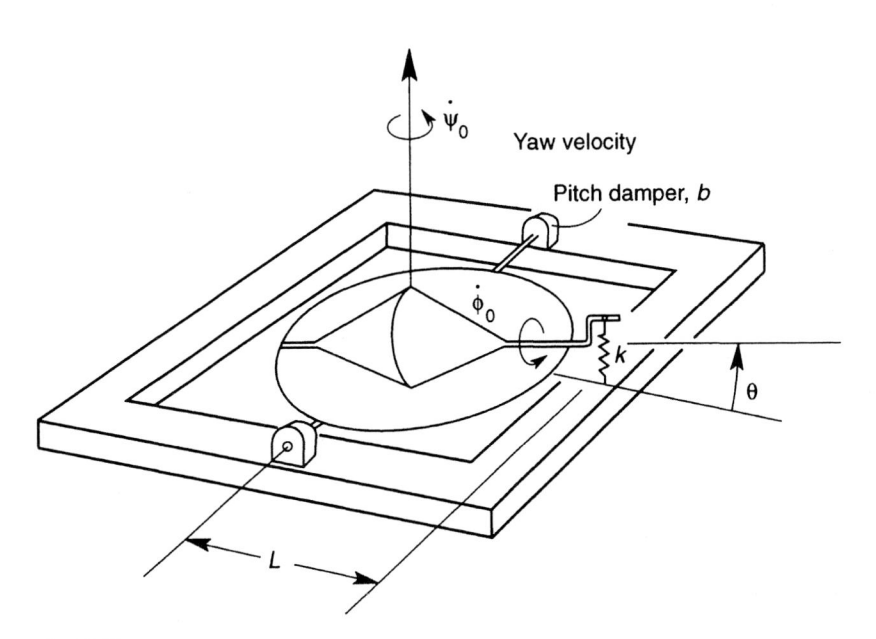

Fig. P5.31

5.33 (*Spinner Toy*) A handheld children's toy is shown in Fig. P5.33. A rotor is free to turn about a pin joint. The hand moves in an oscillatory manner. By changing the oscillation frequency and amplitude, one can obtain rotary, oscillatory,

5 Rigid Body Dynamics

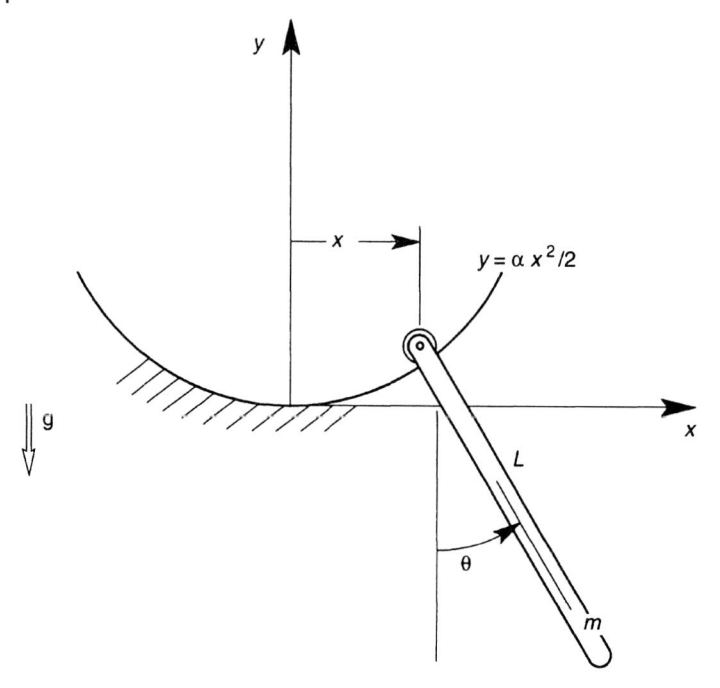

Fig. P5.32

or chaotic motion of the spinner. A dynamics model is shown in Fig. P5.33b, in which the base motion is prescribed: $x(t) = A \sin \omega_0 t$. Use the principle of virtual power to derive the equation of motion. As a special project numerically integrate the equation and look for the three different types of motion. This problem is similar to a magnetic dipole (e.g., compass needles) in an oscillatory magnetic field (Fig. P5.33c), which has been shown to exhibit chaotic motions.

[Answer: In nondimensional form, the equation of motion becomes:

$$\ddot{\theta} + \alpha \sin \tau \sin \phi = 0$$

For numerical integration add a small damping term and define

$$\dot{x} = y, \qquad \dot{y} = -\gamma y - \alpha \sin x = 0]$$

5.34 In Section 5.6 we examined a nonholonomic problem of a vehicle skid (Examples 5.15 and 5.16). In the examples, friction was neglected. Suppose we assume that there is a skid pad at point A shown in Fig. P5.34. One of the models for friction is the Coulomb model where the friction force is a constant magnitude μ and directed opposite to the velocity, i.e.,

$$\mathbf{F} = -\mu (\text{siqn } \mathbf{v}_A) \frac{\mathbf{v}_A}{|\mathbf{v}_A|}$$

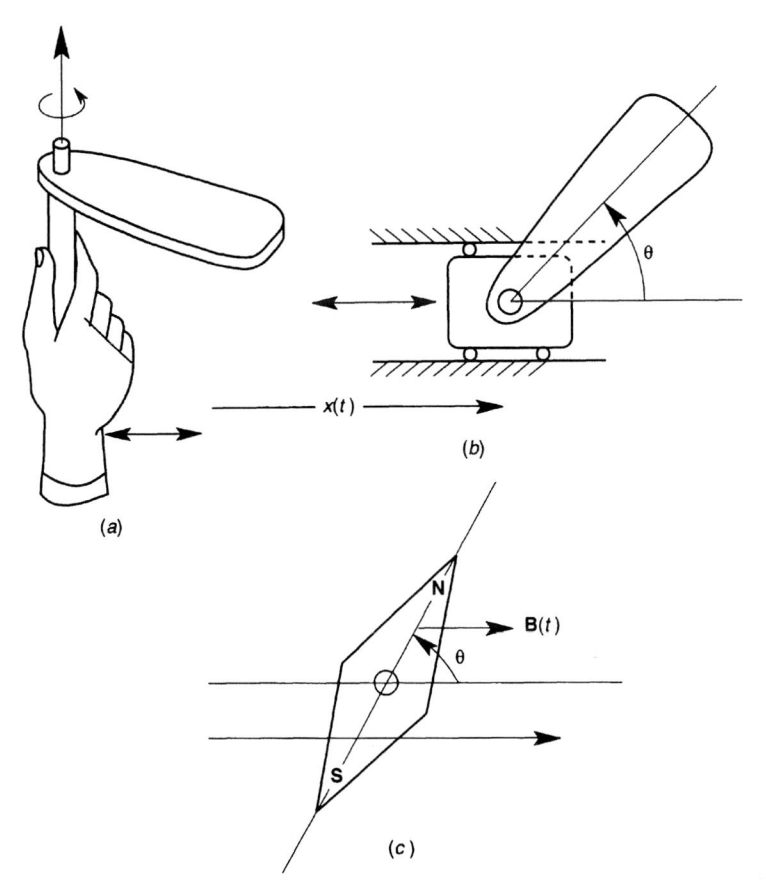

Fig. P5.33

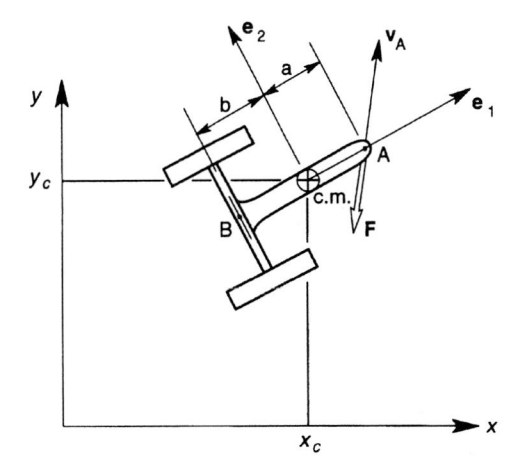

Fig. P5.34

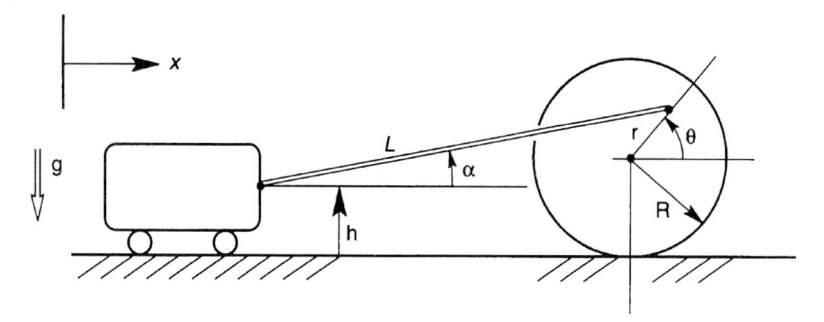

Fig. P5.35

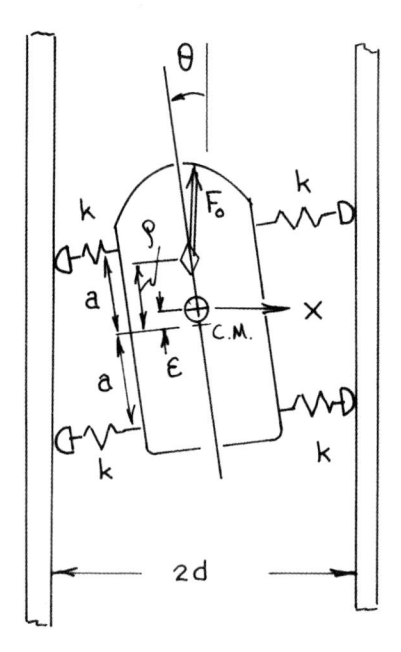

Fig. P5.37

Derive the equations of motion in the vehicle basis vectors $\{\mathbf{e}_1, \mathbf{e}_2\}$ when friction is present.

5.35 Two bodies are connected by a rigid massless link of length L, as shown in Fig. P5.35. The cylindrical body rolls without slipping. Use the method of Lagrange multipliers to determine the force in the link. (*Hint*: Use x, θ as initially independent generalized coordinates and introduce the link as a constraint.)

5.36 In the gyropendulum problem of Example 5.14, Fig. 5.26, derive the equations of motion using the Newton–Euler equations. For small oscillations, find the dynamic component of the contact force on the disc.

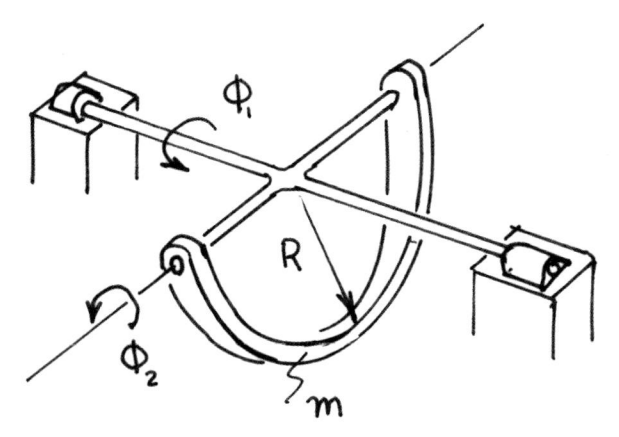

Fig. P5.38

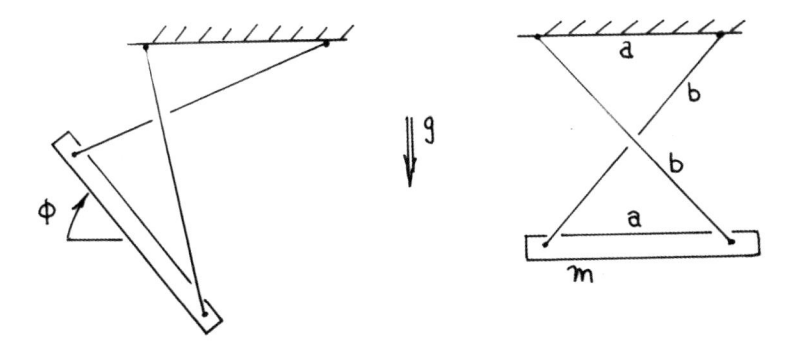

Fig. P5.39

5.37 (*Bobsled Dynamics*) Imagine the motion of a rectangular shaped rigid body confined to slide between two planes by four spring loaded contact slides (Fig. P5.37). Assume that the classical Coulomb dry friction law holds relating the friction force and the normal force. Each of the springs with constant k is assumed to be initially compressed by x_0. Also assume that the motion takes place such that a component of gravity acts with a constant force in the y direction at the center of mass point. Use the Newton–Euler method to derive the equations of motion for the linearized case for two degrees of freedom; lateral motion $x(t)$, and yaw motion $\theta(t)$. Simulate the dynamics in MATLAB to find out if the yaw motion becomes unstable.

5.38 (*Vibration of a Circular Ring on Gimbals*) A semi-circular ring of radius R is pivoted on the rotating axis of crossed gimbals as shown in Fig. P5.38. Write an expression for the angular velocity vector in terms of components relative to the gimbal axes. Use either Lagrange's equations or Virtual Power principle to derive the two equations of motion of the ring under gravity. Note that the body is in pure rotation about the fixed point O.

5.39 (*Vibration of Crossed Four-bar Mechanism*) The crossed four-bar mechanism is sometimes used in knee replacement prostheses. The moving link *AB* exhibits translation and rotation similar the human lower leg. Use either the Virtual Power principle or Lagrange's equation to derive an equation of motion for small motions for the special geometric case shown in Fig. P5.39.

6
Introduction to Robotics and Multibody Dynamics

6.1
Introduction

In this chapter we present methods to formulate dynamics problems involving connected rigid bodies. The animal and human skeletal system is the most ubiquitous multibody mechanism. However, in the technical world almost all machines are multibody systems. Robotic manipulator devices are the most anthropomorphic class of problems. Other problems including vehicle-suspension systems, truck tandem trailers, trains, geared power transmissions, construction vehicles such as front-end loaders, all involve the coupled dynamics between two or more rigid bodies (Fig. 6.1).

The coupling between different rigid bodies is of three types:

1. Kinematic
2. Force elements
3. Control elements

Kinematic coupling usually occurs at a local region called a *joint*, which constrains the motion to limited actions such as rotation about an axis, rotation about a point, translation along a line or in a plane, or a combined motion such as the helical motion in a screw joint (Fig. 6.2).

A force element connection includes springs and dampers, but could also include fluid or aerodynamic forces and magnetic or electric actuators.

In mechatronic or controlled machines, forces, torques or motions are sometimes applied through actuators which depend on the feedback of some or all of the state variables. These devices include rotary and linear motors such as servomotors, hydraulic actuators and piezoelectric and shape memory material actuators.

Another force coupling involves so-called *unimodal constraints*. These include impact and friction, stiction, or cold welding between bodies in contact. In these problems the constraints are sometimes short-lived, as in impact, or discontinuous, as in friction or the breaking of a stiction or cold-weld junction. These problems

Applied Dynamics. With Applications to Multibody and Mechatronic Systems. Second Edition. Francis C. Moon
Copyright © 2008 WILEY-VCH Verlag GmbH & Co. KGaA, Weinheim
ISBN: 978-3-527-40751-4

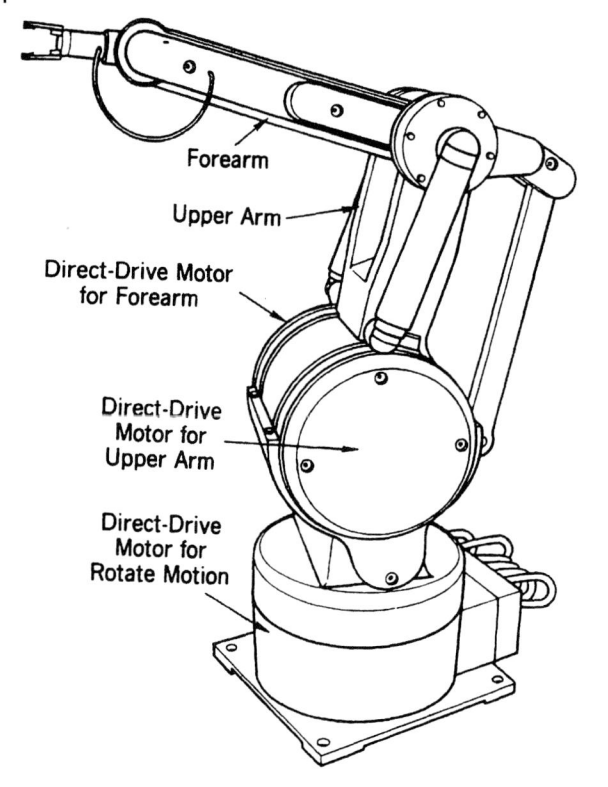

Fig. 6.1 Sketch of a commercial robot manipulator arm.
Reprinted with permission from Rosheim (1994).

present special difficulties. A discussion of impact problems for a rigid body is given at the end of this chapter (see also Brach, 1991 or Pfeiffer and Glocker, 1996).

A series of connected rigid bodies can be further classified according to the topology of the connections (Fig. 6.3). Thus, connected rigid bodies can have an open- or closed-chain structure, or have several branches as in a tree. In computer codes that simulate the dynamics of multibody systems one must adopt a labeling model to denote which body is connected to which. One such system is based on the *theory of graphs* (see, e.g., Wittenburg, 1977). A brief description of this convention is given in Section 6.2.

The solution of the dynamics of multibody systems has a number of steps:

- Modeling; limiting assumptions, approximations
- Geometric description
- Interconnection convention
- Formulation of equations of motion
- Analytical and/or numerical integration of equations of motion
- Graphical display, e.g., phase plane, animation

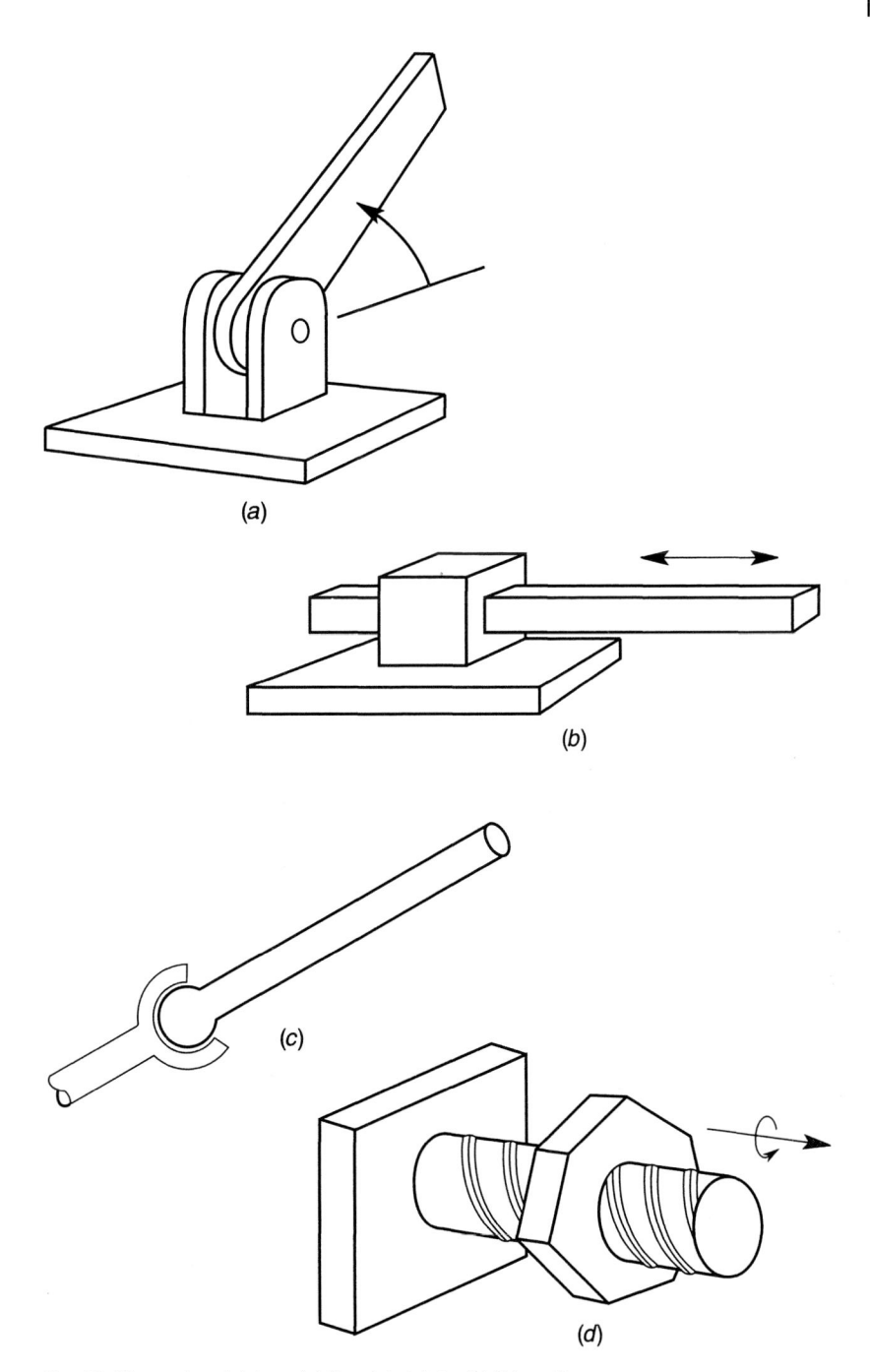

Fig. 6.2 Kinematic pair joints. (*a*) Revolute joints. (*b*) Prismatic joint. (*c*) Ball or spherical joint. (*d*) Helical or screw joint.

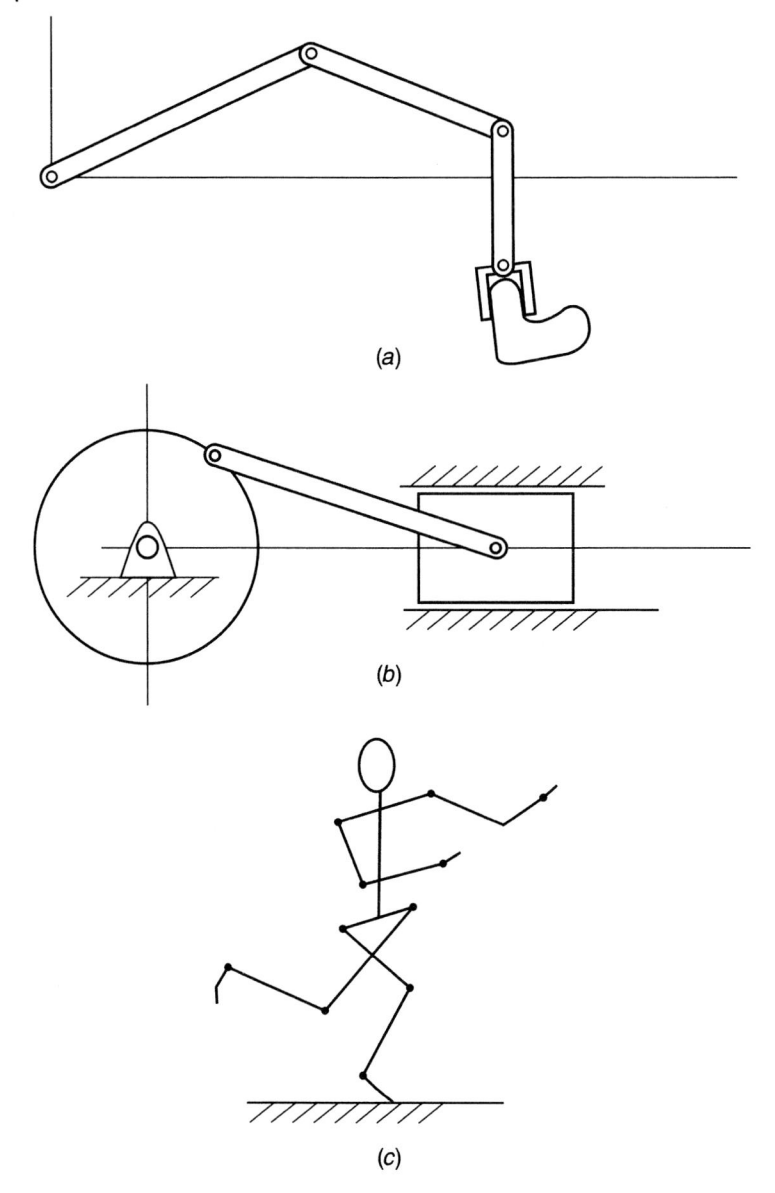

Fig. 6.3 (*a*) Open chain of connected bodies. (*b*) Closed chain. (*c*) Branched system of connected bodies.

- Dynamic data analysis; Fourier transform, modal decomposition, fractal dimension
- Design implications

The methods of formulating equations of motion are based on all the principles discussed in Chapters 2, 3, 4, and 5. The principal methods used today are the

Newton–Euler, Lagrange, and D'Alembert's virtual work and virtual power methods. In modern codes this is often done using symbolic computer software such as *MACSYMA, MATHEMATICA,* and *MAPLE.*

Multibody Codes

Often the engineer or dynamic analyst will use a packaged software system that combines all the elements in the list just described. Such codes include *Working Model, Adams,* or *DADS.* A description of some of these codes can be found in Appendix B and in the reviews by Schiehlen (1990) and Erdman (1993). For two or three connected bodies however, the student can often accomplish a good deal by combining a symbolic code such as *MATHEMATICA* with numerical and graphical packages such as *MATLAB.*

One word of caution: In many codes, the methods used to derive and integrate the equations of motion are proprietary and not available to the user, so one may not know the assumptions in a simulation displayed on the computer screen. The phrase "garbage in–garbage out" is still valid here. When using numerical codes you should always be cautious and skeptical. Always try the code out on several problems with known analytical solutions. Another test of packaged codes is to check conservation of energy and momentum in the output of a simulation.

6.2
Direct Newton–Euler Method Using Graph Theory[1]

Anyone who has taken apart a machine or looked at an exploded view graphic of one is struck with the enormous number of parts (see, e.g., Figs. 5.1, 5.2). Some clusters of parts move as one, as in a control circuit card, but other components, such as ball bearings, have many moving parts. In order to analyze the dynamics of these machines, we must assemble equations of motion for each component. Modern multibody codes require some method to link the motion of one body with another. There are several schemes to do this. In this text we provide a brief introduction to the application of graph theory to this task. The treatment follows the conventions in the advanced text by Wittenburg (1977).

We consider a multibody system as a set of linked rigid bodies in an open or closed chain or tree structure. To each body we assign an index S_i, and to each link or connection between bodies we assign an index u_j. In graph theory the set $\{S_i\}$ are called the *vertices* and the set $\{u_a\}$ are called the *edges* of the graph. In this abstract picture of a machine or mechanism, the geometric model is replaced by a set of points representing the different bodies, connected with arrows or arcs corresponding to the links as shown in Fig. 6.4. The direction of the arrows is

---

1) Section 6.2 may be skipped for those interested in variational methods such as Lagrange's equations or virtual power.

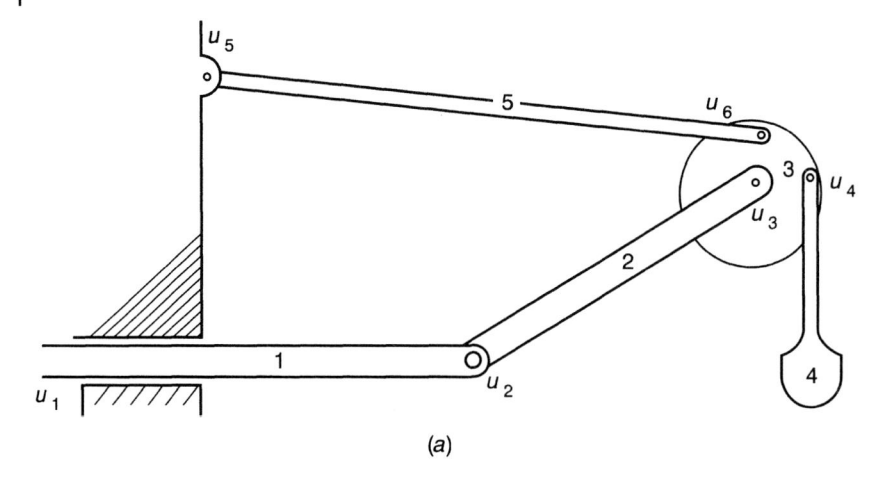

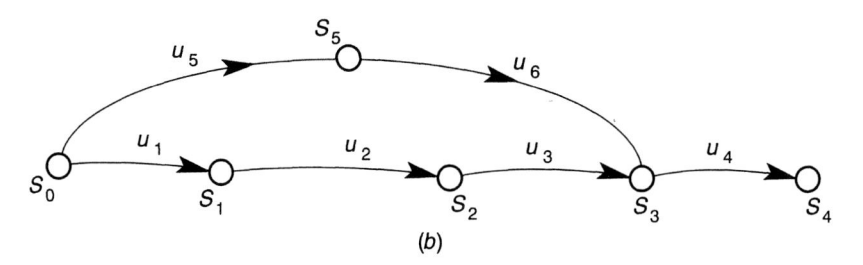

Fig. 6.4 (a) Closed-link and branched system of rigid bodies. (b) Graph convention of the system in part (a).

arbitrary. The points and the arcs are numbered according to the indices $\{S_i\}$ and $\{u_a\}$.

Next, we speak of paths along the graph, say from body S_0 to body S_K denoted by the sequence of hinge numbers along the path $(u_1 \cdots u_a \cdots u_b \cdots)$. A hinge or arc u_a is said to be *incident* with (S_i, S_j) if it connects the two bodies. This leads to one of the important constructs in the theory, the *incidence matrix* $[S_{ia}]$. The incidence matrix elements take only three values: $+1, -1, 0$. These numbers essentially indicate to the computer algorithm the sense of the hinge force acting on a body. The elements of $[S_{ia}]$ are defined by

$$S_{ia} = \begin{cases} +1; & \text{if the arc } u_a \text{ is directed away from } S_i \\ -1; & \text{if the arc } u_a \text{ is directed toward } S_i \\ 0; & \text{if } u_a \text{ is not incident on } S_i \end{cases} \quad (6.1)$$

It is useful to separate $[S_{ia}]$ into a row vector and an $n \times n$ matrix

$$[S_{ia}] = \begin{bmatrix} [S_{0a}] \\ [S] \end{bmatrix} \tag{6.2}$$

The $n \times n$ matrix S can be shown to have an inverse T

$$T = S^{-1} \tag{6.3}$$

The matrix T has the property

$$T^T S_0^T = -1 \tag{6.4}$$

which is a $n \times 1$ column matrix of -1 entries. It can be shown that the elements of T have the following interpretation:

$$T_{ai} = \begin{cases} +1; & \text{if } u_a \text{ is on the path } [S_0, S_i] \text{ and points toward body } S_0 \\ -1; & \text{if } u_a \text{ is on the path } [S_0, S_i] \text{ and points away from } S_0 \\ 0; & \text{if } u_a \text{ is not on the path between body } S_0 \text{ and } S_a \end{cases} \tag{6.5}$$

Example 6.1 Two Link Serial Mechanism. Consider the two-link armlike mechanism shown in Fig. 6.5. By convention the base is labeled S_0, and the system graph drawn in Fig. 6.5b. Thus we assume we have three bodies connected by two hinges or arcs. The S_{ia} matrix is 3 rows by 2 columns as shown below

$$[S_{ia}] = \begin{bmatrix} 1 & 0 \\ -1 & 1 \\ 0 & -1 \end{bmatrix} \tag{6.6}$$

By convention, we define the hinge connected to S_0, to be directed away from S_0, so that $S_{10} = 1$. If this is done right, the columns should sum to zero. We also separate $[S_{ia}]$ into a 1×2 and 2×3 matrices

$$[S_0] = \begin{bmatrix} 1 & 0 \end{bmatrix}$$

$$[S] = \begin{bmatrix} -1 & 1 \\ 0 & -1 \end{bmatrix} \tag{6.7}$$

Using (6.5) we can find the **T** matrix and establish (6.3), (6.4).

$$[T] = \begin{bmatrix} -1 & -1 \\ 0 & -1 \end{bmatrix}$$

$$[T]^T [S_0]^T = \begin{bmatrix} -1 \\ -1 \end{bmatrix}$$

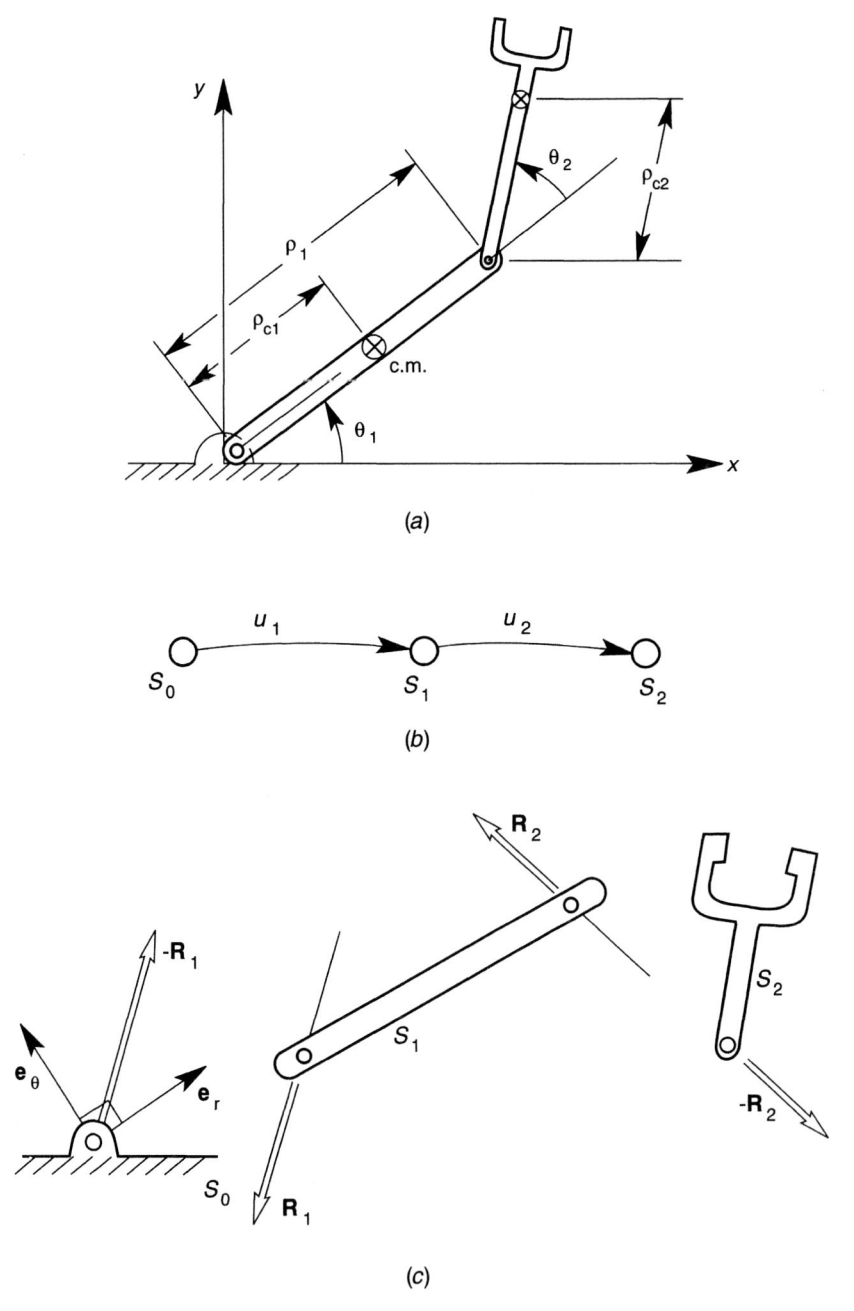

Fig. 6.5 (a) Two-link serial or open-chain system (Examples 6.1, 6.2). (b) Graph of system in part (a). (c) Internal constraint forces in the system in part (a).

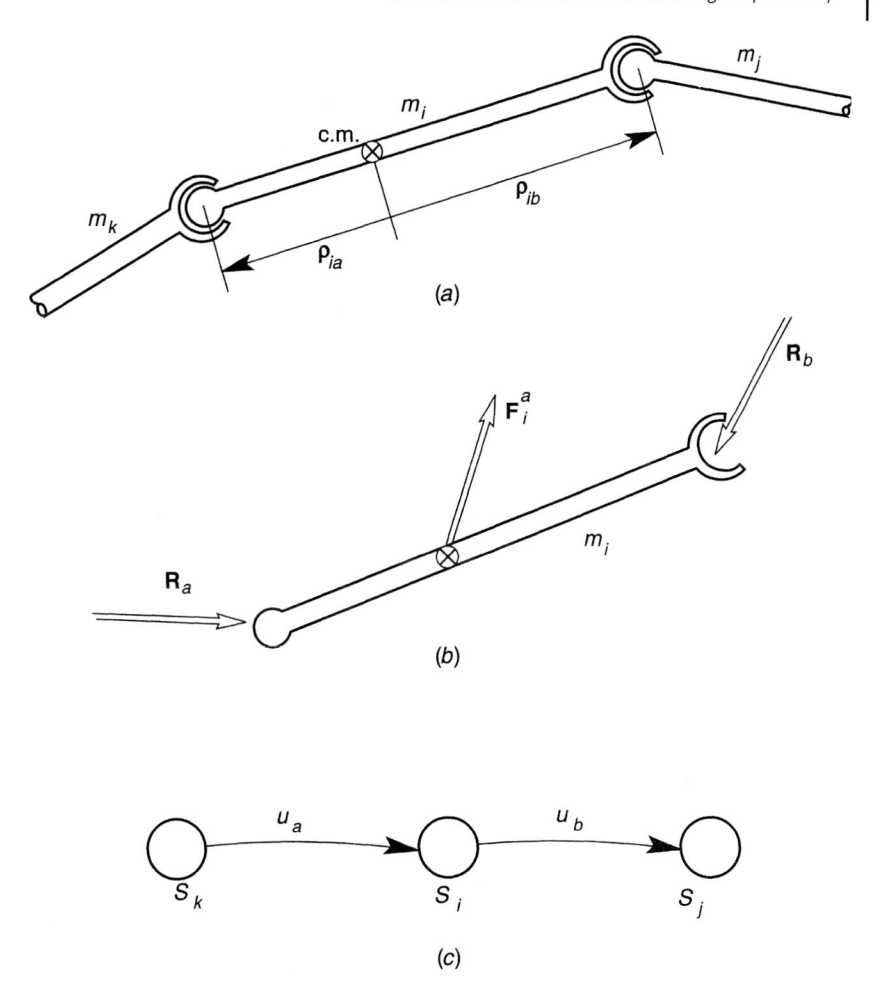

Fig. 6.6 Two joints and three links in a multibody system.

Incidence Matrices and Equations of Motion

To see how the incidence matrix for a system of linked bodies is applied, we consider the special case of a serial link system in which all bodies are connected with ball-and-socket joints, as shown in Fig. 6.6. We focus on body S_i, which has hinges u_a, u_b, etc., that connect the body S_i to other bodies. Acting at each of the hinges are constraint forces $\mathbf{R}_a, \mathbf{R}_b$, and hinge moments or couples $\mathbf{C}_a, \mathbf{C}_b$. These moments may result from friction, internal springs, or dampers. We choose a sign convention such that $\mathbf{R}_a$ is positive when it acts in the direction assigned to u_a. Thus if the arc u_a is directed toward S_i and away from S_k, $\mathbf{R}_a$ is directed away from body S_k and points toward body S_i. Newton's law for body i can be written as

$$m_i \ddot{\mathbf{r}}_i = \mathbf{F}_i^A + S_{ia}\mathbf{R}_a + S_{ib}\mathbf{R}_b$$

or in general

$$m_i \ddot{\mathbf{r}}_i = \mathbf{F}_i^A + \sum_{a=1}^{n} S_{ia} \mathbf{R}_a \tag{6.8}$$

To obtain the law of angular momentum, we define local position vectors $\boldsymbol{\rho}_{ia}$ in body S_i from the center of mass to the hinge point $\mathbf{r}_{ia}$. It can be shown (see Wittenburg, 1977) that

$$\dot{\mathbf{H}}_i = \mathbf{M}_i^A + \sum_{a=1}^{n} S_{ia} (\boldsymbol{\rho}_{ia} \times \mathbf{R}_a + \mathbf{C}_a) \tag{6.9}$$

In this equation, we assume that the moments of the active forces $\mathbf{M}_i^A$ are taken about the center of mass of body S_i.

In both the methods of Lagrange and virtual power, workless constraint forces such as $\{\mathbf{R}_a\}$ do not appear in the equations of motion. However, for design purposes, knowledge of hinge forces may be important. If the motions of all the mass centers in the multibody system are determined, i.e., we know $\{\ddot{\mathbf{r}}_i\}$, then the constraint forces can be found using the inverse of S_{ia} or T_{ai}

$$\mathbf{R}_a = \sum_{i=1}^{n} T_{ai} (m_i \ddot{\mathbf{r}}_i - \mathbf{F}_i^A) \tag{6.10}$$

Example 6.2. We wish to write the equations of planar motion for the system of two bodies in Fig. 6.5, where we assume that the internal hinge moments, $\mathbf{C}_a$ in (6.9), are zero and that gravity is the only applied force on each body.

For these assumptions the equations of motion (6.8), (6.9) become

$$m_1 \ddot{\mathbf{r}}_1 = -m_1 g \hat{\mathbf{j}} + \sum_{a=1}^{2} S_{ia} \mathbf{R}_a$$

$$m_2 \ddot{\mathbf{r}}_2 = -m_2 g \hat{\mathbf{j}} + \sum_{a=2}^{2} S_{ia} \mathbf{R}_a \tag{6.11}$$

$$I_1 \dot{\omega}_1 = \hat{\mathbf{k}} \cdot \sum_{a=1}^{2} S_{ia} (\boldsymbol{\rho}_{1a} \times \mathbf{R}_a)$$

$$I_2 \dot{\omega}_2 = \hat{\mathbf{k}} \cdot \sum_{a=1}^{2} S_{2a} (\boldsymbol{\rho}_{2a} \times \mathbf{R}_a) \tag{6.12}$$

where

$$S_{ia} = \begin{bmatrix} -1 & 1 \\ 0 & -1 \end{bmatrix} \tag{6.13}$$

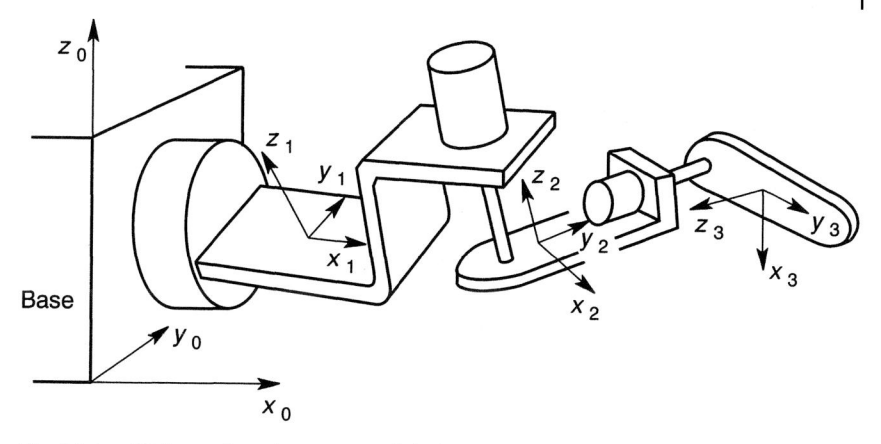

Fig. 6.7 Local link coordinate bases in a multibody system.

and

$$\sum S_{ia} \mathbf{R}_a = -\mathbf{R}_1 + \mathbf{R}_2$$

$$\sum S_{2a} \mathbf{R}_a = -\mathbf{R}_2$$

The use of the machine matrix S_{ia} in this example is trivial. Its usefulness may become more important in more complex multibody systems.

6.3
Kinematics

One of the first steps in formulating equations of motion for a multibody system is the choice of a kinematic formalism. This choice has two parts: a choice of reference frame and the type of mathematical description, e.g., whether vector or matrix representation. General theories of multibody dynamics can be found in more advanced books. In this introduction we discuss the principles by examples.

Consider the open chain, serial link mechanism shown in Fig. 6.7. This topology is characteristic of serial-link robot manipulator arms. Shown in the figure are four reference frames: one is called the *base coordinate* frame, in which Newton's laws are valid, and three other reference frames are attached to the links. These local reference frames move and rotate with each link. In deriving the equation of motion for a single rigid body we found that using a local frame simplified the calculation of the angular momentum (Chapter 5). We can choose to represent all dynamic vectors in the base coordinates. However, in robotics, actuation of each link is relative to the neighboring links and the use of local frames is often more useful.

In Chapter 3 we saw that the location and motion of a rigid body could be represented by three conventions:

1. Position vectors and rotation rate vectors $\boldsymbol{\omega}$
2. Position vectors and 3×3 finite rotation matrices
3. 4×4 transformation matrices representing rotations and translation

We illustrate the kinematics description with the first and second methods. See Paul (1981) for a treatment using 4×4 matrices.

Degrees of Freedom: Generalized Coordinates

A system of N rigid bodies has at most $6N$ degrees of freedom. The use of hinges and connections restricts the degrees of freedom. When we use relative frames of reference, it is natural to choose the generalized coordinates $\{q_1(t), q_2(t), \ldots, q_m(t)\}$, which describe the motion of one body relative to one of its neighboring links. Thus in the example in Fig. 6.8, where revolute joints are used, there is one angular degree of freedom for each link. It is natural to choose $q_i = \theta_i$ where, θ_i is the relative motion of link i with respect to link $i - 1$. In vector notation, the angular velocity of each link in the chain depends on all the relative angular velocities, i.e.,

$$\boldsymbol{\omega}_i = \sum_{j=1}^{i} \dot{\theta}_j \mathbf{b}_j \tag{6.14}$$

where $\mathbf{b}_i$ is a unit vector along the joint axis of rotation connecting the ith and $(i-1)$th links. This convention is used in the robotics book by Asada and Slotine (1986). The preceding formula is useful in calculating the Jacobian matrix used in the virtual power method of deriving equations of motion [Chapter 5, (5.143)].

Jacobians

In deriving equations of motion for connected rigid bodies, we need to know the relation between the velocities $\{\dot{\mathbf{r}}_{ci}\}$ and the generalized velocities $\{\dot{\theta}_i\}$.

To write a position vector to the center of mass of the ith link, $\mathbf{r}_{ci}$, we must define vectors that give the position of hinge $(i - 1)$ relative to hinge i, denoted by $\boldsymbol{\rho}_{(i-1)i}$, and vectors that give the relative position of the center of mass of each link $\boldsymbol{\rho}_{ci}$, as shown in Fig. 6.8.

In vector notation we can then write

$$\mathbf{r}_{ci} = \sum_{j=1}^{i} \boldsymbol{\rho}_{(j-1)j} + \boldsymbol{\rho}_{ci}$$

or

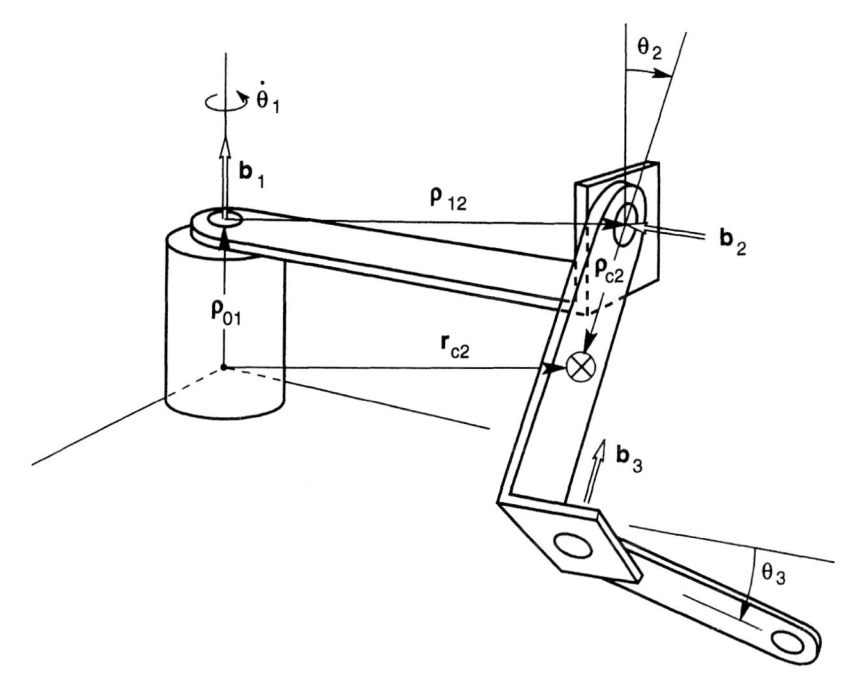

Fig. 6.8 Local axes of rotation $\{\mathbf{b}_i\}$ in a serial, revolute-link multibody system.

$$\mathbf{r}_{ci} = \mathbf{R}_i + \boldsymbol{\rho}_{ci} \tag{6.15}$$

where

$$\mathbf{R}_i = \sum_{j=1}^{i} \boldsymbol{\rho}_{(i-1)j}$$

The velocity is then given by

$$\mathbf{v}_{ci} = \dot{\mathbf{R}}_i + \boldsymbol{\omega}_i \times \boldsymbol{\rho}_{ci} \tag{6.16}$$

Thus

$$\frac{\partial \mathbf{v}_{ci}}{\partial \dot{\theta}_i} = \frac{\partial \boldsymbol{\omega}_i}{\partial \dot{\theta}_i} \times \boldsymbol{\rho}_{ci}$$

An alternative representation is to use relative rotation matrices $A_i(\theta_i)$ that describe the position of the local vectors $\boldsymbol{\rho}_{(i-1)i}$ and $\boldsymbol{\rho}_{ci}$ relative to the local frame of the next lower link $(i-1)$ [see (3.35) etc.]. In this notation we define a vector $\hat{\boldsymbol{\rho}}_{ci}$ to be the position of $\boldsymbol{\rho}_{ci}$ relative to the $(i-1)$ frame when $\theta_i = 0$. For this representation, we have

$$\mathbf{r}_{ci} = \mathbf{R}_i + A_i \hat{\boldsymbol{\rho}}_{ci} \tag{6.17}$$

and

$$\mathbf{v}_{ci} = \dot{\mathbf{R}}_i + \dot{\mathbf{A}}_i(\theta_i(t))\hat{\boldsymbol{\rho}}_{ci}$$

or

$$\mathbf{v}_{ci} = \dot{\mathbf{R}}_i + \mathbf{A}'_i(\theta_i)\hat{\boldsymbol{\rho}}_{ci}\dot{\theta}_i \qquad (6.18)$$

where the prime on A_i indicates a derivative with respect to θ_i. Note that $\hat{\boldsymbol{\rho}}_{ci}$ does not depend on time. Again this is useful in calculating the Jacobian matrix or the projection vectors. For example

$$\frac{\partial \mathbf{v}_{ci}}{\partial \dot{\theta}_i} = \mathsf{A}'_i \hat{\boldsymbol{\rho}}_{ci} \qquad (6.19)$$

To calculate the Jacobian for the rotation projection vectors γ_i, in the Virtual Power method (5.143) we use (6.14),

$$\frac{\partial \boldsymbol{\omega}_i}{\partial \dot{\theta}_j} = \mathbf{b}_j \quad \text{for } j \leqslant i$$

$$= 0 \quad \text{for } j > i \qquad (6.20)$$

Example 6.3. Consider the planar motion of a two-link serial mechanism shown in Fig. 6.5. We define a system velocity vector

$$\mathbf{v} = [\dot{x}_{c1}, \dot{y}_{c1}, \omega_1, \dot{x}_{c2}, \dot{y}_{c2}, \omega_2]^T \qquad (6.21)$$

and a generalized configuration vector

$$\mathbf{q} = [\theta_1, \theta_2]^T \qquad (6.22)$$

We want to show that there exists a linear relation between $\mathbf{v}$ and $\dot{\mathbf{q}}$ of the form

$$\mathbf{v} = \mathsf{J}\dot{\mathbf{q}} \qquad (6.23)$$

where J is a 6×2 system Jacobian matrix. This global Jacobian can be broken into smaller matrices, i.e.,

$$\mathsf{J} = [\mathsf{J}_{T1}, \mathsf{J}_{R1}, \mathsf{J}_{T2}, \mathsf{J}_{R2}]^T \qquad (6.24)$$

where

$$\mathsf{J}_{T1} = \frac{\partial \mathbf{v}_{c1}}{\partial \dot{\mathbf{q}}} = \begin{bmatrix} -\rho_{c1} \sin \theta_1 & 0 \\ \rho_{c1} \cos \theta_1 & 0 \end{bmatrix}^T$$

$$J_{T2} = \frac{\partial \mathbf{v}_{c2}}{\partial \dot{\mathbf{q}}} = \begin{bmatrix} -\rho_1 S\theta_1 - \rho_{c2} S(\theta_1 + \theta_2) & -\rho_{c2} S(\theta_1 + \theta_2) \\ \rho_1 C\theta_1 + \rho_{c2} C(\theta_1 + \theta_2) & \rho_{c2} C(\theta_1 + \theta_2) \end{bmatrix}^T$$

(The notation here is $S\theta = \sin\theta$, $C\theta = \cos\theta$.)

The other two sub-matrices are given by

$$J_{R1} = \frac{\partial \omega_1}{\partial \dot{\mathbf{q}}} = [1, 0]^T$$

$$J_{R2} = \frac{\partial \omega_2}{\partial \dot{\mathbf{q}}} = [1, 1]^T \tag{6.25}$$

6.4
Equations of Motion: Lagrange's Equations and Virtual Power Method

As illustrated in Chapter 5, the application of the Newton–Euler equations requires that all the forces and moments on each of the bodies be made explicit. In the case of rigid bodies connected by kinematic constraints, this means that the forces of constraint must be included in the analysis using the Newton–Euler methodology. When one assumes that the constraint forces do not produce any work, as in frictionless kinematic constraints, it is often easier to derive equations of motion for connected bodies using a variational method such as Lagrange's equations or virtual power method.

Lagrange's Equations for Multibody Systems

In this section we illustrate the procedure for using Lagrange's equations for a planar, serial link robot manipulator arm. As the reader may have already discovered, the algebra in carrying out dynamics calculations for more than one body can be very daunting and in practice one would use a symbolic math code such as *MAPLE* or *MATHEMATICA* to keep track of the algebraic equations. Therefore for pedagogical reasons, we will illustrate Lagrange's method with a specific problem.

Referring to Fig. 6.9, we use the joint variables $\{\theta_1, \theta_2, \theta_3\}$ as our generalized coordinates for the three-link arm. Since the angles are measured relative to the neighboring link, the angular velocities (whose axes are normal to the plane) become

$$\omega_1 = \dot{\theta}_1, \qquad \omega_2 = \dot{\theta}_1 + \dot{\theta}_2, \qquad \omega_3 = \dot{\theta}_1 + \dot{\theta}_2 + \dot{\theta}_3 \tag{6.26}$$

In order to derive an explicit equation for the kinetic energy, we also need to calculate the squares of the velocities of the center of masses of each of the three bodies in terms of the generalized coordinates and generalized velocities. The lengths of each of the links are designated as $\{L_1, L_2, L_3\}$ and the distance from the joint to the center of each mass is denoted by $\{\rho_{1c}, \rho_{2c}, \rho_{3c}\}$. Using this notation the velocities of each of the center of masses are given by

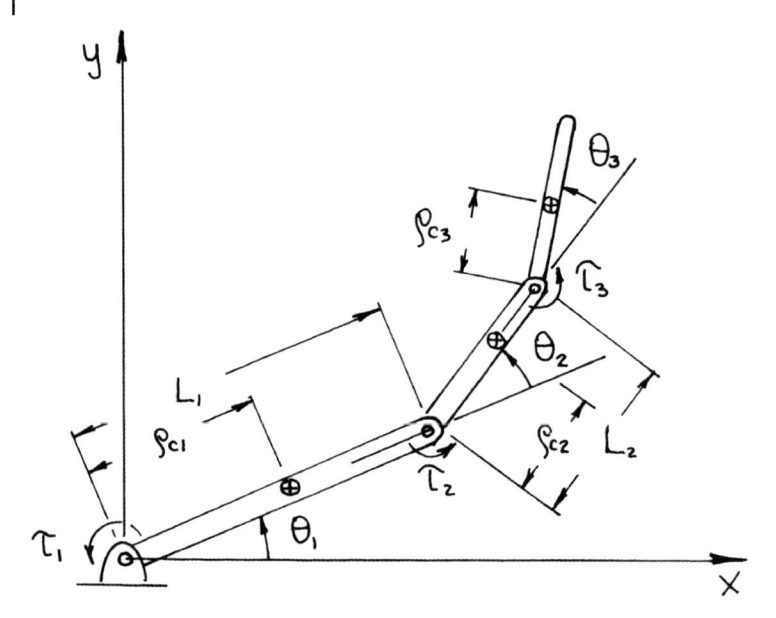

Fig. 6.9 Sketch of a three degree-of-freedom, planar, serial-link manipulator arm with control torques at the hinges.

$$\mathbf{v}_{1c} = \rho_{1c}\dot{\theta}_1 \mathbf{e}_{\theta 1}$$

$$\mathbf{v}_{2c} = L_1\dot{\theta}_1 \mathbf{e}_{\theta 1} + \rho_{2c}(\dot{\theta}_1 + \dot{\theta}_2)\mathbf{e}_{\theta 2}$$

$$\mathbf{v}_{3c} = L_1\dot{\theta}_1 \mathbf{e}_{\theta 1} + L_2(\dot{\theta}_1 + \dot{\theta}_2)\mathbf{e}_{\theta 2} + \rho_{3c}(\dot{\theta}_1 + \dot{\theta}_2 + \dot{\theta}_3)\mathbf{e}_{\theta 3} \qquad (6.27)$$

where $\mathbf{e}_{\theta i}$ is normal to the ith link.

The kinetic energy of the system is the sum of the translational kinetic energy of each of the bodies and the rotational kinetic energy of the links in the form

$$T = \frac{1}{2}\sum m_i \mathbf{v}_{ic} \cdot \mathbf{v}_{ic} + \frac{1}{2}\sum I_i \omega_i^2 \qquad (6.28)$$

The second term represents the planar form of the inner product of the angular momentum and the angular velocity [see (5.101)]: $\mathbf{H}_i \cdot \omega_i$.

In this problem we assume that the arm operates in a horizontal plane so that the gravity forces will do no work and the potential energy function in Lagrange's equation, V, will be assumed to be zero. The form of the equations of motion is then

$$\frac{d}{dt}\frac{\partial T}{\partial \dot{\theta}_i} - \frac{\partial T}{\partial \theta_i} = Q_i \quad \{i = 1, 2, 3\} \qquad (6.29)$$

For a robot manipulator arm, we assume that there are three applied control torques on each of the three joints, $\{\tau_1, \tau_2, \tau_3\}$. To determine the generalized forces Q_i, in (6.29) we need to make some assumptions as to how the control torques are applied. In some robot designs the torques are applied through cables and pulleys to ground. However for this problem, we assume that each joint has a direct drive motor actuator connected to ground for driving joint #1, connected to link #1 for driving joint #2, and connected to link #2 for driving joint #3. In this case it is easy to calculate the applied input power and the generalized forces:

$$P = \sum Q_i \dot{\theta}_i = \tau_1 \dot{\theta}_1 + \tau_2 \dot{\theta}_2 + \tau_3 \dot{\theta}_3 \tag{6.30}$$

or

$$Q_1 = \tau_1, \qquad Q_2 = \tau_2, \qquad Q_3 = \tau_3 \tag{6.31}$$

Using the definitions of T and the associated velocities, the equations of motion can be represented in the following form

$$\begin{bmatrix} m_{11} & m_{12} & m_{13} \\ m_{12} & m_{22} & m_{23} \\ m_{13} & m_{23} & m_{33} \end{bmatrix} \begin{bmatrix} \ddot{\theta}_1 \\ \ddot{\theta}_2 \\ \ddot{\theta}_3 \end{bmatrix} + \begin{bmatrix} c_1 \\ c_2 \\ c_3 \end{bmatrix} = \begin{bmatrix} \tau_1 \\ \tau_2 \\ \tau_3 \end{bmatrix} \tag{6.32}$$

where a few of the mass matrix and the centripetal and Coriolis terms in (6.32) are given by

$$c_1 = \mu_{122} \dot{\theta}_2^2 + \mu_{133} \dot{\theta}_3^2 + \mu_{112} \dot{\theta}_1 \dot{\theta}_2 + \mu_{113} \dot{\theta}_1 \dot{\theta}_3 + \mu_{123} \dot{\theta}_2 \dot{\theta}_3 \tag{6.33}$$

$$m_{11} = I_1 + I_2 + I_3 + \rho_{1c}^2 m_1 + m_2 \left(L_1^2 + \rho_{2c}^2 + 2 L_1 \rho_{2c} \cos \theta_2 \right)$$

$$+ m_3 \left(L_1^2 + L_2^2 + \rho_{3c}^2 + 2 L_1 L_2 \cos \theta_2 \right) \tag{6.34}$$

$$\mu_{122} = -m_2 L_1 \rho_{2c} \sin \theta_2 - m_3 L_1 \left(L_2 \sin \theta_2 + \rho_{3c} \sin(\theta_2 + \theta_3) \right) \tag{6.35}$$

There is not much pedagogical value in the specific details of these algebraic expressions except to note that the mass matrix is symmetric, and the c_i terms (6.33) are quadratic in the generalized velocities. Also the coefficients (6.34), (6.35) depend on trigonometric functions of the joint angles. If these equations are used in an on-line calculation as the robot is performing operations, then it may be of some advantage to study the efficiency of making these calculations as well as making approximations such as neglecting terms that remain small when the arm is in the operating envelope of the machine.

The equations of motion can be derived using several methods related to Newton's laws and the principles of virtual work and power. In the following we outline the method of virtual power (Jourdain's principle), which was formulated for rigid bodies by Kane (1961). In all methods we must define a vector of generalized coordinates

$$\mathbf{q} = \left[q_1(t), q_2(t), \ldots, q_M(t)\right]^T$$

where $M \leqslant 6N$. With complete generality, the resulting equations of motion for a multibody system will have the following form (see, e.g., Asada and Slotine, 1986 or Wittenburg, 1977):

$$\sum m_{ij}\ddot{q}_j + \sum\sum \mu_{ijk}\dot{q}_j\dot{q}_k = Q_i \qquad (6.36)$$

In general, m_{ij} and μ_{ijk} will be nonlinear functions of $\mathbf{q}$. It can be shown that the *mass matrix* $[m_{ij}]$ is symmetric, i.e., $m_{ij} = m_{ji}$. Also there is symmetry in the coefficients in the second term, i.e., $\mu_{ijk} = \mu_{ikj}$.

The first expression on the left represents the linear acceleration terms, while the second term represents nonlinear accelerations that are similar to centripetal and Coriolis accelerations that appear in cylindrical or polar coordinates (see, e.g., Chapter 2). The term on the right is the generalized force. Note that the product $Q_i\dot{q}_i$ has units of power. The generalized force Q_i contains gravity and load forces as well as feedback forces in the case of controlled machines.

Principle of Virtual Power for Connected Rigid Bodies

The principle of virtual power formulation for one rigid body (5.143) can be generalized to N connected bodies under the assumption that the constraint forces between them do no work: for each generalized velocity $\dot{q}_k$, we have

$$\sum_{i=1}^{N}\left(m_i\dot{\mathbf{v}}_{ci} - \mathbf{F}_i^a\right)\cdot\boldsymbol{\beta}_{ik} + \sum_{i=1}^{N}\left(\dot{\mathbf{H}}_i - \mathbf{M}_i^a\right)\cdot\boldsymbol{\gamma}_{ik} = 0 \qquad (6.37)$$

where the projection vectors are given by

$$\boldsymbol{\beta}_{ik} = \frac{\partial\mathbf{v}_{ci}}{\partial\dot{q}_k}, \qquad \boldsymbol{\gamma}_{ik} = \frac{\partial\boldsymbol{\omega}_i}{\partial\dot{q}_k}$$

In the preceding equation $\mathbf{v}_{ci}$ is the velocity of the center of mass of the ith body, and $\mathbf{H}_i$ is the angular momentum about the center of mass of the ith body. As in the single-body case, $\mathbf{F}_i^a$, $\mathbf{M}_i^a$ represent the active forces and moments that do work on the ith body. It is clear that the forces of constraint between the bodies do not enter the equations of motion.

We present a limited proof for the case of two bodies connected by a ball-and-socket-joint, shown in Fig. 6.10. The pin forces acting on each body $\mathbf{F}_1^p$, $\mathbf{F}_2^p$ are assumed to obey Newton's third law, $\mathbf{F}_1^p = -\mathbf{F}_2^p$. We neglect moments in the joint. To obtain the two-body form of (6.37) we first write the principle of virtual power for each link.

$$\left(m_1\dot{\mathbf{v}}_{c1} - \mathbf{F}_1^a - \mathbf{F}_1^p\right)\cdot\boldsymbol{\beta}_{1k} + \left(\dot{\mathbf{H}}_1 - \mathbf{M}_1^a - (\boldsymbol{\rho}_1\times\mathbf{F}_1^p)\right)\cdot\boldsymbol{\gamma}_{1k} = 0$$

$$\left(m_2\dot{\mathbf{v}}_{c2} - \mathbf{F}_2^a - \mathbf{F}_2^p\right)\cdot\boldsymbol{\beta}_{2k} + \left(\dot{\mathbf{H}}_2 - \mathbf{M}_2^a - (\boldsymbol{\rho}_2\times\mathbf{F}_2^p)\right)\cdot\boldsymbol{\gamma}_{2k} = 0 \qquad (6.38)$$

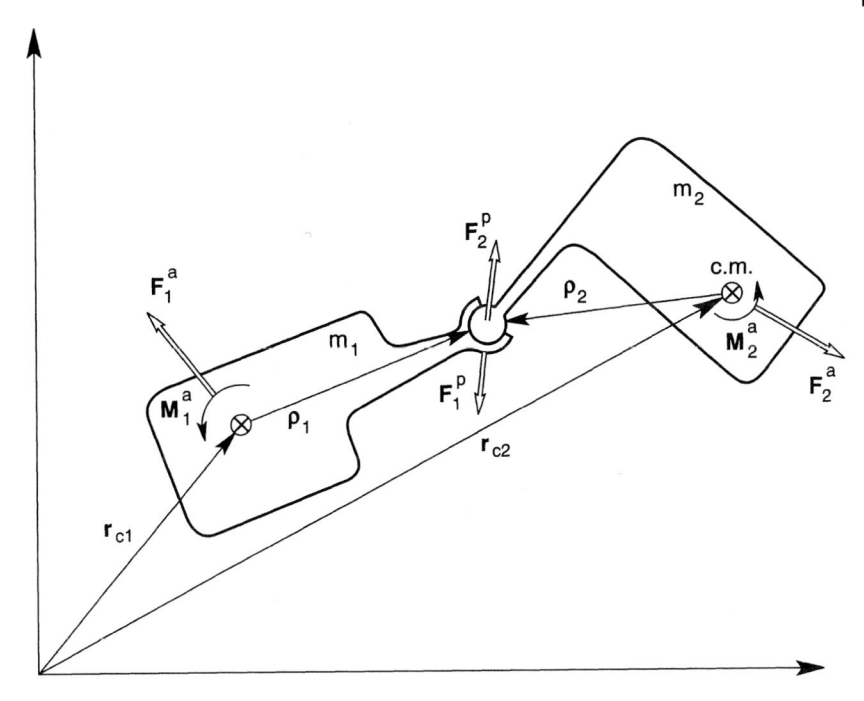

Fig. 6.10 Geometry of two connected bodies with a ball joint.

The sum of these two equations will yield the two body version of (6.37), provided we can show that

$$\mathbf{F}_1^p \cdot (\boldsymbol{\beta}_{2k} - \boldsymbol{\beta}_{1k}) + \boldsymbol{\rho}_2 \times \mathbf{F}_1^p \cdot \boldsymbol{\gamma}_{2k} - \boldsymbol{\rho}_1 \times \mathbf{F}_1^p \cdot \boldsymbol{\gamma}_{1k} = 0 \tag{6.39}$$

To relate the Jacobian or projection vectors we use the kinematic constraints;

$$\mathbf{r}_{c2} = \mathbf{r}_{c1} + \boldsymbol{\rho}_1 - \boldsymbol{\rho}_2$$

$$\mathbf{v}_{c2} = \mathbf{v}_{c1} + \boldsymbol{\omega}_1 \times \boldsymbol{\rho}_1 - \boldsymbol{\omega}_2 \times \boldsymbol{\rho}_2$$

$$\boldsymbol{\beta}_{2k} = \frac{\partial \mathbf{v}_{c2}}{\partial \dot{q}_k} = \boldsymbol{\beta}_{1k} + \boldsymbol{\gamma}_{1k} \times \boldsymbol{\rho}_1 - \boldsymbol{\gamma}_{2k} \times \boldsymbol{\rho}_2 \tag{6.40}$$

Substituting this expression into (6.39), we obtain the identity.

An alternative statement of the principle of virtual power for multibody systems is the statement that the inner product of a global constraint force vector $\mathbf{F}^c$ with a global Jacobian J is zero (see, e.g., Schiehlen, 1986). In the special case of a system of constrained particles (neglecting angular momentum), this statement takes the form

$$\mathbf{J}^T \mathbf{F}^c = 0 \tag{6.41}$$

where $\mathbf{F}^c = [\mathbf{F}_1^{cT}, \mathbf{F}_2^{cT}, \ldots, \mathbf{F}_N^{cT}]^T$ and J is defined by

$$\dot{\mathbf{r}} = \mathsf{J}\dot{\mathbf{q}}$$

and $\dot{\mathbf{r}} = [\dot{\mathbf{r}}_1^T, \dot{\mathbf{r}}_2^T, \ldots, \dot{\mathbf{r}}_N^T]^T$, $\dot{\mathbf{q}} = [\dot{q}_1, \dot{q}_2, \ldots, \dot{q}_\mu]^T$.

This abstract statement is made more explicit in the example below.

Example 6.4. In order to clarify the preceding Eqs. (6.36), (6.41), consider the simple example of the two-particle system shown in Fig. 6.11.

In the example shown in Fig. 6.11 we restrict the motion of mass m_1 to the plane, and the motion of mass m_2 to a linear direction normal to the plane. Thus without constraints there are *three* degrees of freedom $\{x_1, y_1, z_2\}$. However, if a cable of fixed length is connected between the two particles, we are left with only *two* degrees of freedom, which we elect to choose as $\mathbf{q} = [r, \theta]^T$. Thus our global configuration vector is $\mathbf{r} = [x_1, y_1, z_2]^T$ and it is straightforward to find a relation

$$\dot{\mathbf{r}} = \mathsf{J}\dot{\mathbf{q}} \tag{6.42}$$

where J is a 3×2 Jacobian matrix. It can be shown that the constraint condition

$$r + z = \text{Constant}$$

or

$$\dot{r} + \dot{z} = 0$$

leads to

$$\mathsf{J} = \begin{bmatrix} \cos\theta & -r\sin\theta \\ \sin\theta & r\cos\theta \\ -1 & 0 \end{bmatrix} \tag{6.43}$$

The equations of motion can be written in the form of matrix equations

$$\begin{bmatrix} m_1 & 0 & 0 \\ & m_1 & 0 \\ & & m_2 \end{bmatrix} \begin{bmatrix} \ddot{x}_1 \\ \ddot{y}_1 \\ \ddot{z}_2 \end{bmatrix} = \begin{bmatrix} -T\cos\theta \\ -T\sin\theta \\ -T \end{bmatrix} + \begin{bmatrix} 0 \\ 0 \\ m_2 g \end{bmatrix} \tag{6.44}$$

where we assume frictionless sliding in the plane. These equations can be written in compact matrix notation

$$\widehat{\mathsf{M}}\ddot{\mathbf{r}} = \mathbf{F}^c + \mathbf{F}^a \tag{6.45}$$

where the mass matrix $\widehat{\mathsf{M}}$ is given by

$$\widehat{\mathsf{M}} = \begin{bmatrix} m_1 & 0 & 0 \\ 0 & m_1 & 0 \\ 0 & 0 & m_2 \end{bmatrix}$$

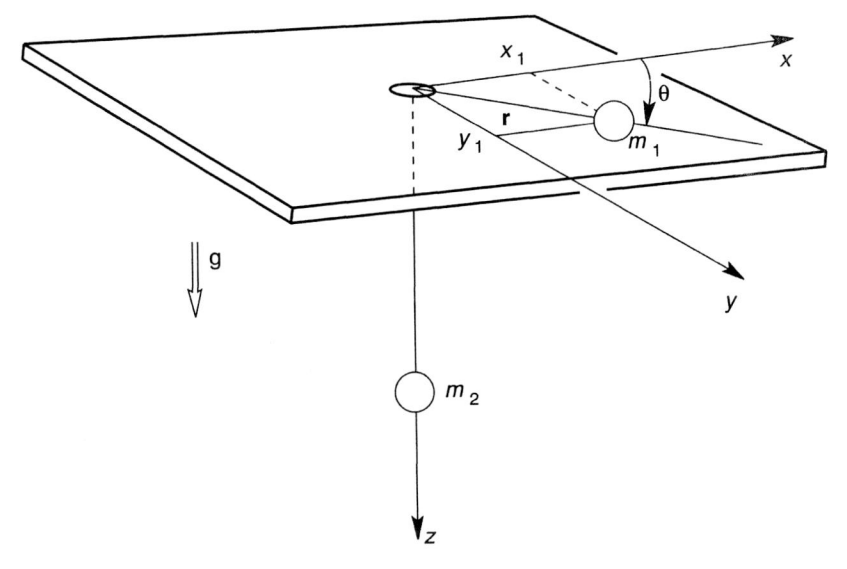

6.4 Equations of Motion: Lagrange's Equations and Virtual Power Method | 325

Fig. 6.11 Cable constrained motion of two connected masses.

the constraint force $\mathbf{F}^c = [-TC\theta, -TS\theta, -T]^T$ and the active or work-producing force $\mathbf{F}^a = [0, 0, m_2 g]^T$.

When the kinematic Eqs. (6.42), are substituted into (6.44), we find

$$\widehat{M} J \ddot{q} + \widehat{M} \dot{j} \dot{q} = \mathbf{F}^c + \mathbf{F}^a \tag{6.46}$$

According to the principle of virtual power, the power of the constraint force is zero, (6.41),

$$J^T \mathbf{F}^c = 0$$

Thus multiplying the equation of motion (6.46), by J^T, we obtain

$$J^T \widehat{M} J \ddot{q} + J^T \widehat{M} \dot{j} \dot{q} = J^T \mathbf{F}^a \tag{6.47}$$

It remains to show that this has the form of (6.36). In particular we note that $\dot{j}$ is linear in $\dot{q}$:

$$\dot{j} = \dot{\theta} \begin{bmatrix} -S\theta & -rC\theta \\ C\theta & -rS\theta \\ 0 & 0 \end{bmatrix} + \dot{r} \begin{bmatrix} 0 & -S\theta \\ 0 & C\theta \\ 0 & 0 \end{bmatrix} \tag{6.48}$$

This leads to

$$J^T \widehat{M} \dot{j} \dot{q} = m_1 r \dot{\theta}^2 \begin{bmatrix} -1 \\ 0 \end{bmatrix} + m_1 2 \dot{r} \dot{\theta} \begin{bmatrix} 0 \\ r \end{bmatrix}$$

Also we can define a new 2×2 mass matrix

$$\mathsf{M} = \mathsf{J}^T \widehat{\mathsf{M}} \mathsf{J} = \begin{bmatrix} (m_1 + m_2) & 0 \\ 0 & r^2 m_1 \end{bmatrix} \qquad (6.49)$$

The generalized force is easily seen to be

$$\mathsf{J}^T \mathbf{F}^a = \begin{bmatrix} C\theta & S\theta & -1 \\ -rS\theta & rC\theta & 0 \end{bmatrix} \begin{bmatrix} 0 \\ 0 \\ m_2 g \end{bmatrix} = \begin{bmatrix} -m_2 g \\ 0 \end{bmatrix}$$

The resulting equations of motion in the reduced coordinates become

$$\begin{bmatrix} (m_1 + m_2) & 0 \\ 0 & rm_1 \end{bmatrix} \begin{bmatrix} \ddot{r} \\ \ddot{\theta} \end{bmatrix} + m_1 r \dot{\theta}^2 \begin{bmatrix} -1 \\ 0 \end{bmatrix} + m_1 2\dot{r}\dot{\theta} \begin{bmatrix} 0 \\ r \end{bmatrix} = \begin{bmatrix} -m_2 g \\ 0 \end{bmatrix} \qquad (6.50)$$

These equations could have been derived more simply by using polar coordinates and using Newton's direct method. Lagrange's equations would lead to the same equations. However, the formulation shows how the virtual power method can be used in multibody problems as is illustrated below.

The principle of virtual power in the form of (6.41) can be used in a symbolic code, such as *MAPLE* or *MATHEMATICA*, to derive equations of motion. The idea is to write Newton–Euler equations of motion in the form (6.44), and premultiply both sides of (6.44) by J^T to obtain equations of the form (6.36) or (6.50). Note, one does not need to know the constraint forces $\mathbf{F}^c$ since they dropout via (6.41).

Serial-link Robotic Manipulators

The most common robotic mechanism is the serial-link manipulator arm shown in Fig. 6.12. Modern realization of these robotic devices had its origins in the 1950s and 1960s, and began to be deployed in industry in the 1970s. The classic device is broken down into the arm and the end effector or gripper. In many commercial robot arms, different types of end effectors can be used. Almost all of these devices are constructed by a series of rigid bodies connected at a joint. Typical joints are the revolute joint (an axis of rotation) and a prismatic joint (an axis of translation) (see Fig. 6.2). Each joint typically has one degree of freedom. Thus to move and position a workpiece with a serial link arm requires at least six rigid links.

There are many excellent books on the dynamics of robotic devices. A classic treatment is by Paul (1981), in which he derives the basic equation set using Lagrange's equations and uses the 4×4 transformation kinematics. Other excellent texts are by Asada and Slotine (1986) and Craig (1986, 2005). Asada and Slotine use both the Newton–Euler approach and Lagrange's equations.

To illustrate the methodology, we consider a system of planar links with pin-joint connections, as shown in Fig. 6.12. The equations of motion for each link are given

6.4 Equations of Motion: Lagrange's Equations and Virtual Power Method

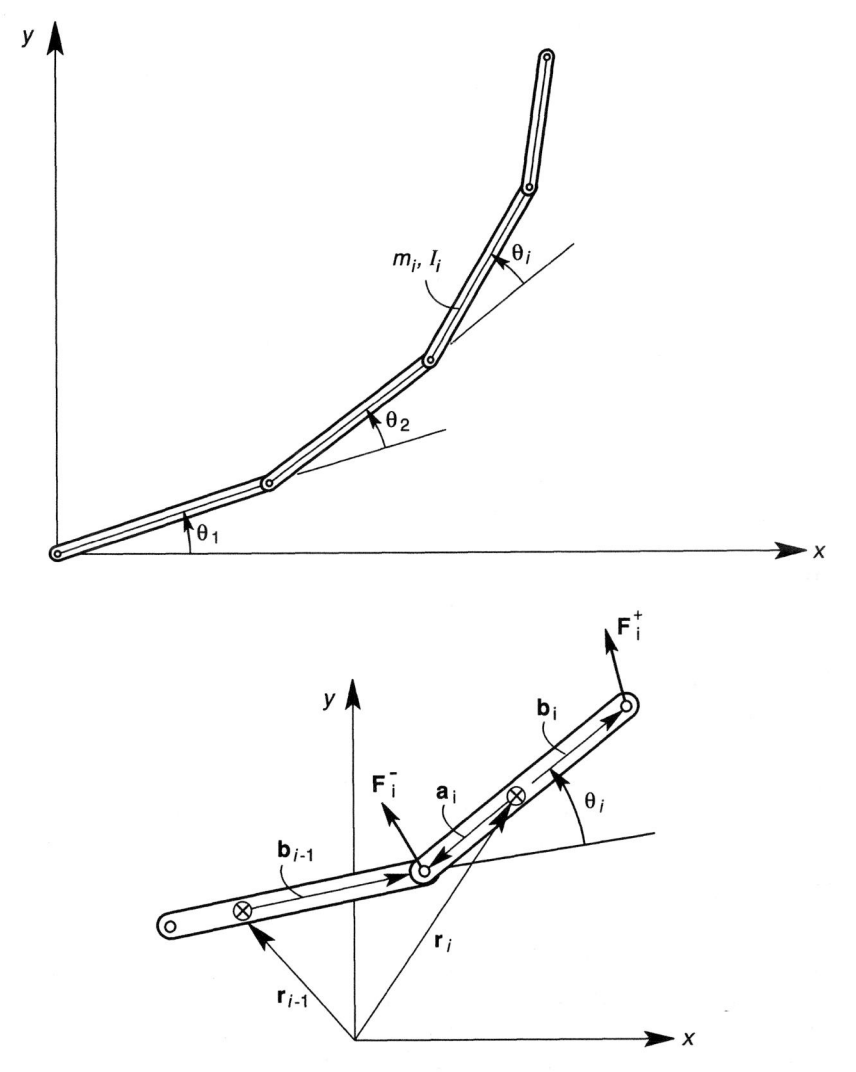

Fig. 6.12 Geometry of a planar system of revolute, serially connected bodies.

by

$$m_i \dot{\mathbf{v}}_{ci} = \mathbf{F}_i^a + \mathbf{F}_i^+ + \mathbf{F}_i^-$$
$$I_i \dot{\omega}_i = M_i^a + \mathbf{n} \cdot \left(\mathbf{a}_i \times \mathbf{F}_i^- + \mathbf{b}_i \times \mathbf{F}_i^+ \right) \qquad (6.51)$$

where $\mathbf{n}$ is a unit vector out of the plane, $\mathbf{F}_i^+$, $\mathbf{F}_i^-$ are pin joint forces, and $\mathbf{a}_i$, and $\mathbf{b}_i$, are vectors from the center of mass to the pin joints. In robotic applications $\mathbf{F}_i^a$ is typically a gravity or load force and M_i^a are applied torques at the joints. A series of N links in an open chain with one link pinned to a base has N degrees of freedom.

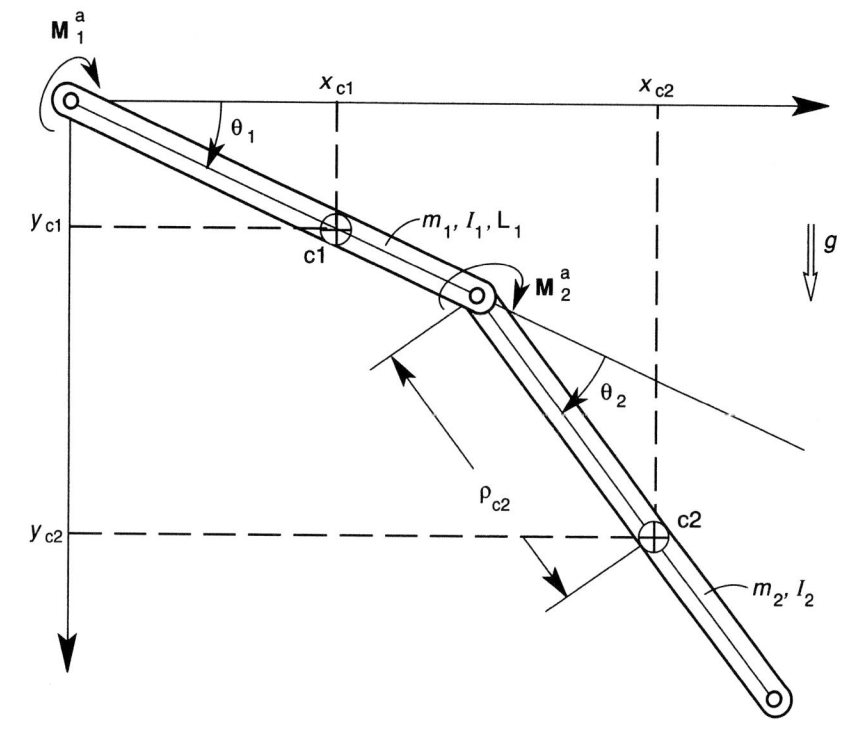

Fig. 6.13 Geometry of the two-link system in Example 6.5.

We choose the relative angles θ_i shown in Fig. 6.12 as generalized variables. Thus the angular velocities ω_i are given by

$$\omega_i = \sum_{k=1}^{i} \dot{\theta}_k \tag{6.52}$$

By using either D'Alembert's direct method of virtual work or virtual power we can eliminate the constraint forces $\mathbf{F}_i^+, \mathbf{F}_i^-$ at the frictionless pins. This principle has the form

$$\sum (m_i \dot{\mathbf{v}}_{ci} - \mathbf{F}_i^a) \cdot \frac{\partial \mathbf{v}_{ci}}{\partial \dot{\theta}_j} + \sum (I_i \dot{\omega}_i - M_i^a) \frac{\partial \omega_i}{\partial \dot{\theta}_j} = 0 \tag{6.53}$$

for $j = 1, 2, \ldots, M$.

The velocity constraints that define the Jacobian, $\partial \mathbf{v}_{ci}/\partial \dot{\theta}_j$ are derived from the geometric conditions

$$\mathbf{r}_i + \mathbf{a}_i = \mathbf{r}_{i-1} + \mathbf{b}_{i-1} \tag{6.54}$$

To see the details of this method we look at the two-link arm in the example below.

6.4 Equations of Motion: Lagrange's Equations and Virtual Power Method

Example 6.5 Two-Link Manipulator Arm. The geometry is defined in Fig. 6.13. For convenience we use the notation $\rho_{c1} = a_1$, $L_1 = a_1 + b_1$, $\rho_{c2} = a_2$, where $\{a_i, b_i\}$ are lengths in Eq. (6.54) and Fig. 6.12. The kinematics of this problem were discussed in the previous section. The global configuration vector is $\mathbf{r} = [x_{c1}, y_{c1}, \theta_1, x_{c2}, y_{c2}, \theta_1 + \theta_2]$. The global mass matrix $\widehat{M} = \text{diag}\{m_1, m_1, I_1, m_2, m_2, I_2\}$. The Newton–Euler equations in Cartesian coordinates takes the form

$$\widehat{M}[\dot{v}_{1x}, \dot{v}_{1y}, \dot{\omega}_1, \dot{v}_{2x}, \dot{v}_{2y}, \dot{\omega}_2]^T = [0, m_1 g, M_1^a, 0, m_2 g, M_2^a]^T + F^c$$

The principle of virtual power (6.53) can be applied by premultiplying this equation by J^T and using (6.41) or $J^T F^c$. The global matrix equation of motion has the standard form

$$M\ddot{q} + J^T \widehat{M} \dot{J} \dot{q} = Q \tag{6.55}$$

where $M = J^T \widehat{M} J$. The 2 row by 6 column transpose of the Jacobian matrix J^T is found from $\dot{\mathbf{r}} = J\dot{q}$,

$$J^T = \begin{bmatrix} -\rho_{c1} S\theta_1 & \rho_{c1} C\theta_1 & 1 & J_{41} & J_{51} & 1 \\ 0 & 0 & 0 & -\rho_{c2} S\theta_{12} & \rho_{c2} C\theta_{12} & 1 \end{bmatrix}$$

and

$$J_{41} = -L_1 S\theta_1 - \rho_{c2} S\theta_{12}, \qquad J_{51} = L_1 C\theta_1 + \rho_{c2} C\theta_{12}$$

(Note, $S\theta_2 = \sin\theta_1$, $S\theta_{12} = \sin(\theta_1 + \theta_2)$ etc.)

The generalized applied force/torque is defined by

$$Q = J^T F^a \tag{6.56}$$

where

$$F^a = [0, m_1 g, M_1^a, 0, m_2 g, M_2^a]^T \tag{6.57}$$

The equations of motion for this problem are derived in Asada and Slotine (1986) using the direct Newton–Euler method. The principle of virtual power yields exactly the same equations which take the form

$$\begin{bmatrix} m_{11} & m_{12} \\ m_{12} & m_{22} \end{bmatrix} \begin{bmatrix} \ddot{\theta}_1 \\ \ddot{\theta}_2 \end{bmatrix} + \begin{bmatrix} -\mu\dot{\theta}_2^2 - 2\mu\dot{\theta}_1\dot{\theta}_2 \\ +\mu\dot{\theta}_1^2 \end{bmatrix} - \begin{bmatrix} G_1 \\ G_2 \end{bmatrix} = \begin{bmatrix} \tau_1 \\ \tau_2 \end{bmatrix} \tag{6.58}$$

where $\mu = m_2 L_1 \rho_{c2} \sin\theta_2$. The mass matrix terms can be shown to be

$$m_{11} = I_1 + I_2 + m_1\rho_{c1}^2 + m_2(L_1^2 + \rho_{c1}^2 + 2L_1\rho_{c2}\cos\theta_2)$$

$$m_{12} = I_2 + m_2\rho_{c2}^2 + m_2 L_1 \rho_{c2} \cos\theta_2$$

$$m_{22} = I_2 + m_2\rho_{c2}^2 \tag{6.59}$$

The generalized gravity terms are

$$G_1 = m_1 \rho_{c1} g \cos\theta_1 + m_2 g\left[\rho_{c2} \cos(\theta_1 + \theta_2) + L_1 \cos\theta_1\right]$$

$$G_2 = m_2 \rho_{c2} g \cos(\theta_1 + \theta_2) \tag{6.60}$$

The generalized applied torques $\tau_1 = M_1^a + M_2^a$, $\tau_2 = M_2^a$ have either specified time histories or are tied into a feedback control scheme.

Note that the centripetal and Coriolis acceleration terms in the second bracketed term in (6.58) are similar to the general form of (6.37). Compare this example with Example 4.10.

Numerical Integration of Equations of Motion

Multibody equations of the form (6.36), (6.50), or (6.58), are coupled second-order nonlinear differential equations. Many standard numerical integration algorithms, such as Runge–Kutta and some variable time step methods, are formulated for coupled first-order differential equations. In the case of the multilink manipulator with equations of motion of the form (6.55), one can put them into a first-order form by defining $\mathbf{q} = [\theta_1, \theta_2, \ldots, \theta_n]$ and

$$\dot{\mathbf{q}} = \boldsymbol{\omega}$$

$$\dot{\boldsymbol{\omega}} = \mathsf{M}^{-1}\left[\mathbf{Q} - \mathsf{J}^T \widehat{\mathsf{M}} \mathsf{J} \boldsymbol{\omega}\right]$$

provided M^{-1} exists, i.e., M is not singular or $\det \mathsf{M} \neq 0$ for any configuration $\mathbf{q}$.

Problem Formulation in Three-dimensional Linked Bodies

The preceding examples illustrate the basic method for deriving equations of motion for linked multibody problems. This method can be summarized as follows:

1. Choose a fixed reference and local coordinate frames in each body. Usually these are located at the centers of mass or at the joints.
2. Choose N independent generalized coordinates corresponding to the N degrees of freedom.
3. Write expressions for the angular velocity $\boldsymbol{\omega}_k$ of each body in terms of the generalized velocities $\{\dot{q}_k(t)\}$.
4. Use the angular velocities $\boldsymbol{\omega}_k$ to obtain expressions for the velocities of the centers of mass of all the bodies $\mathbf{v}_{ck}$ in terms of the generalized coordinates $\{q_k(t)\}$ and generalized velocities $\{\dot{q}_k(t)\}$. These expressions allow us to calculate the Jacobian J or projection vectors $\partial \mathbf{v}_{ci}/\partial \dot{q}_k$ and $\partial \boldsymbol{\omega}_i/\partial \dot{q}_k$.
5. Calculate the acceleration of each body $\mathbf{a}_{ci}$ as well as the rate of change of angular momentum of each body $\dot{\mathbf{H}}_{ci}$.
6. Identify active forces and moments (torques) on each of the bodies.

7. Apply the principle of virtual power:

$$\sum_i (m_i \mathbf{a}_{ci} - \mathbf{F}_i^a) \cdot \frac{\partial \mathbf{v}_{ci}}{\partial \dot{q}_k} + \sum_i (\dot{\mathbf{H}}_{ci} - \mathbf{M}_i^a) \cdot \frac{\partial \boldsymbol{\omega}_i}{\partial \dot{q}_k} = 0 \quad (6.61)$$

(Note that for planar problems $\dot{\mathbf{H}}_{ci}$ is proportional to the angular acceleration of each link $\ddot{\theta}_i$ as in (6.51) and (6.53).)

This method is illustrated below for a nonplanar, three-degree-of-freedom serial-link manipulator. This problem is adapted from the German text by Professor F. Pfeiffer of the Technical University of Munich (1992). Pfeiffer also derives the equations using Lagrange's equations and, of course, arrives at the same result.

These steps can be performed by a symbolic code such as *MAPLE* or *MATHEMATICA*. The example below provides motivation for using such codes for equation formulation.

The geometry of a typical serial-link robot arm with revolute joints is sketched in Fig. 6.14. In this example we look at the three-link, three-degree-of-freedom mechanism to illustrate the steps in deriving the equations of motion. Note that Link #1 is simply a rotor with motion ϕ_1 about the vertical axis.

Step 1

Assign four-coordinate systems. The fixed-base triad unit vectors are $\{\hat{\mathbf{i}}, \hat{\mathbf{j}}, \hat{\mathbf{k}}\}$, as shown in Fig. 6.14. Local coordinate systems are assigned to each of the three links $\{\mathbf{e}_{ix}, \mathbf{e}_{iy}, \mathbf{e}_{iz}\}$, $i = 1, 2, 3$. The y-axes are chosen as the axes of rotation, and the x-axes for links 2 and 3 are parallel to the joint location vectors $\mathbf{r}_2, \mathbf{r}_3$. These local coordinate references are each centered at the center of mass of the respective links. As we shall see in Steps 5 and 7 we need to express the basis vectors of one link in terms of basis vectors in another link. To this end we can use either vector expressions such as

$$\mathbf{e}_{1y} = S\phi_2 \mathbf{e}_{2x} - C\phi_2 \mathbf{e}_{2z}$$
$$\mathbf{e}_{1y} = S\phi_{23} \mathbf{e}_{3x} - C\phi_{23} \mathbf{e}_{3z} \quad (6.62)$$

where $S\phi = \sin\phi$, $C\phi = \cos\phi$ and $\phi_{23} = \phi_2 + \phi_3$. One can also use matrix transformation rules as illustrated for relative motion between links 2 and 3:

$$\begin{bmatrix} \mathbf{e}_{2x} \\ \mathbf{e}_{2y} \\ \mathbf{e}_{2z} \end{bmatrix} = \begin{bmatrix} C\phi_3 & 0 & S\phi_3 \\ 0 & 1 & 0 \\ -S\phi_3 & 0 & C\phi_3 \end{bmatrix} \begin{bmatrix} \mathbf{e}_{3x} \\ \mathbf{e}_{3y} \\ \mathbf{e}_{3z} \end{bmatrix} \quad (6.63)$$

Step 2

Choose generalized coordinates. The three independent joint-rotation angles are the natural choice for generalized coordinates $\{\phi_1, \phi_2, \phi_3\}$. They each measure the rotation of the ith link relative to the $(i-1)$th link.

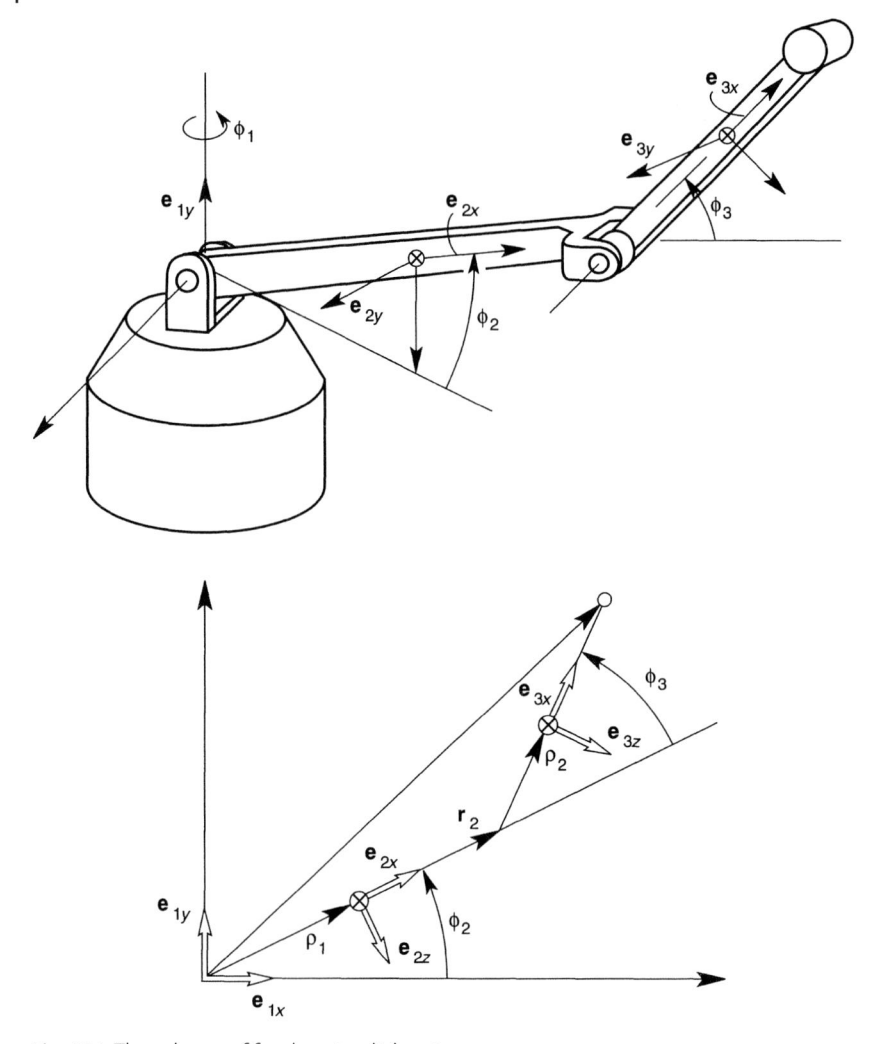

Fig. 6.14 Three-degree-of-freedom, two-link system.

Step 3

Determine the angular velocities. The angular-velocity vectors of each of the rigid bodies are written in terms of the generalized velocities $\{\dot{\phi}_1, \dot{\phi}_2, \dot{\phi}_3\}$ as in (6.14).

$$\boldsymbol{\omega}_1 = \dot{\phi}_1 \mathbf{e}_{1y}$$
$$\boldsymbol{\omega}_2 = \boldsymbol{\omega}_1 + \dot{\phi}_2 \mathbf{e}_{2y}$$
$$\boldsymbol{\omega}_3 = \boldsymbol{\omega}_2 + \dot{\phi}_3 \mathbf{e}_{3y} \qquad (6.64)$$

Step 4

Calculate the projection vectors or Jacobian. First calculate the velocity of the center of mass of the three links. For the system in Fig. 6.14, these become [see (6.16)]

$$\mathbf{v}_1 = 0$$
$$\mathbf{v}_2 = \boldsymbol{\omega}_2 \times \boldsymbol{\rho}_1$$
$$\mathbf{v}_3 = \boldsymbol{\omega}_2 \times \mathbf{r}_2 + \boldsymbol{\omega}_3 \times \boldsymbol{\rho}_2 \tag{6.65}$$

where the vectors $\{\boldsymbol{\rho}_i\}$ specify the location of the center of mass of the ith link relative to the $(i-1)$th joint, as shown in Fig. 6.14. From the expressions (6.64) and (6.65) we can derive the Jacobian vectors or projection vectors for the principle of virtual power.

$$\frac{\partial \mathbf{v}_2}{\partial \dot{\phi}_i} = \frac{\partial \boldsymbol{\omega}_2}{\partial \dot{\phi}_i} \times \boldsymbol{\rho}_1$$
$$\frac{\partial \mathbf{v}_3}{\partial \dot{\phi}_i} = \frac{\partial \boldsymbol{\omega}_2}{\partial \dot{\phi}_i} \times \mathbf{r}_2 + \frac{\partial \boldsymbol{\omega}_3}{\partial \dot{\phi}_i} \times \boldsymbol{\rho}_2 \tag{6.66}$$

Also

$$\frac{\partial \boldsymbol{\omega}_1}{\partial \dot{\phi}_1} = \mathbf{e}_{1y}, \qquad \frac{\partial \boldsymbol{\omega}_1}{\partial \dot{\phi}_2} = \frac{\partial \boldsymbol{\omega}_1}{\partial \dot{\phi}_3} = 0$$
$$\frac{\partial \boldsymbol{\omega}_2}{\partial \dot{\phi}_1} = \mathbf{e}_{1y}, \qquad \frac{\partial \boldsymbol{\omega}_2}{\partial \dot{\phi}_2} = \mathbf{e}_{2y}, \qquad \frac{\partial \boldsymbol{\omega}_3}{\partial \dot{\phi}_3} = 0$$
$$\frac{\partial \boldsymbol{\omega}_3}{\partial \dot{\phi}_1} = \mathbf{e}_{1y}, \qquad \frac{\partial \boldsymbol{\omega}_3}{\partial \dot{\phi}_2} = \mathbf{e}_{2y}, \qquad \frac{\partial \boldsymbol{\omega}_3}{\partial \dot{\phi}_3} = \mathbf{e}_{3y} \tag{6.67}$$

Step 5

Calculate the linear and angular momentum rates. It is important to decide in which set of basis vectors to write these expressions. It is a matter of taste and convenience as to which basis to choose. Since terms in the virtual power equations such as $\dot{\mathbf{v}}_i \cdot \partial \mathbf{v}_i / \partial \dot{\phi}_k$ are scalars, we choose to write the acceleration $\dot{\mathbf{v}}_i$ and the Jacobian vector $\partial \mathbf{v}_i / \partial \dot{\phi}_k$ in the *local* link basis vectors $\{\mathbf{e}_{ix}, \mathbf{e}_{iy}, \mathbf{e}_{iz}\}$. The general expressions for the accelerations follow from Chapters 3 and 5 [e.g., (5.1), (5.2)]:

$$\dot{\mathbf{v}}_1 = 0$$
$$\dot{\mathbf{v}}_2 = \dot{\boldsymbol{\omega}}_2 \times \boldsymbol{\rho}_1 + \boldsymbol{\omega}_2 \times \dot{\boldsymbol{\rho}}_1$$
$$= \dot{\boldsymbol{\omega}}_2 \times \boldsymbol{\rho}_1 + \boldsymbol{\omega}_2 \times (\boldsymbol{\omega}_2 \times \boldsymbol{\rho}_1)$$
$$\dot{\mathbf{v}}_3 = \dot{\boldsymbol{\omega}}_2 \times \mathbf{r}_2 + \boldsymbol{\omega}_2 \times (\boldsymbol{\omega}_2 \times \mathbf{r}_2)$$
$$+ \dot{\boldsymbol{\omega}}_3 \times \boldsymbol{\rho}_2 + \boldsymbol{\omega}_3 \times (\boldsymbol{\omega}_3 \times \boldsymbol{\rho}_2) \tag{6.68}$$

Here we note that in calculating $\dot{\boldsymbol{\omega}}_2, \dot{\boldsymbol{\omega}}_3$ the angular change in the unit vectors must be considered; i.e., $\dot{\mathbf{e}}_{ix} = \boldsymbol{\omega}_i \times \mathbf{e}_{ix}$ etc.

Step 6

Identify the active forces. In this example the active forces are gravity and torque axis moments. We write the moment vectors in the form

$$\mathbf{M}_1^a = M_1 \mathbf{e}_{1y}$$
$$\mathbf{M}_2^a = M_2 \mathbf{e}_{2y}$$
$$\mathbf{M}_3^a = M_3 \mathbf{e}_{3y} \qquad (6.69)$$

In practice these moments will often be effected by servomotors connected to the axes of rotation by either cables, gearing, or belts. Sometimes the motors are directly connected, but then their masses must be considered in the dynamics. The torques on the servomotors are created by magnetic fields and electric currents whose dynamics must often be considered along with control electronics (see Chapter 8). Of course, mechanical actuation such as pneumatic or hydraulic actuators are also used in such devices.

The gravity forces on the second and third links are aligned with the base vector $\hat{\mathbf{j}} = \mathbf{e}_{1y}$, i.e.,

$$\mathbf{F}_2^a = -m_2 g \mathbf{e}_{1y}$$
$$\mathbf{F}_3^a = -m_3 g \mathbf{e}_{1y} \qquad (6.70)$$

These forces can be written in terms of the local unit vectors by using the transformation matrices such as (6.63).

Step 7

Derive the equations of motion. One of the four methods discussed in this book can be used:

Newton–Euler
Lagrange's equations
D'Alembert's method (virtual work)
Principle of virtual power (methods of Jourdain, Kane)

In the example below we give some of the details of this derivation using the virtual power method. Pfeiffer (1989) derives a simplified version of this problem ($\varepsilon_2 = \varepsilon_3 = 0$) in Example 6.6 using both Jourdain's principle (virtual power) and Lagrange's equations. In this example either method results in a system of three coupled, nonlinear ordinary equations of motion of the form similar to (6.36).

6.4 Equations of Motion: Lagrange's Equations and Virtual Power Method

$$\begin{bmatrix} m_{11} & 0 & 0 \\ 0 & m_{22} & m_{23} \\ 0 & m_{23} & m_{33} \end{bmatrix} \begin{bmatrix} \ddot{\phi}_1 \\ \ddot{\phi}_2 \\ \ddot{\phi}_3 \end{bmatrix}$$
$$+ \begin{bmatrix} \mu_{112}\dot{\phi}_1\dot{\phi}_2 + \mu_{113}\dot{\phi}_1\dot{\phi}_3 \\ \mu_{211}\dot{\phi}_1^2 + \mu_{233}\dot{\phi}_3^2 + \mu_{223}\dot{\phi}_2\dot{\phi}_3 \\ \mu_{311}\dot{\phi}_1^2 + \mu_{322}\dot{\phi}_2^2 \end{bmatrix}$$
$$+ g \begin{bmatrix} 0 \\ \gamma_1 \\ \gamma_2 \end{bmatrix} = \begin{bmatrix} M_1 \\ M_2 \\ M_3 \end{bmatrix} \qquad (6.71)$$

where the terms $\{m_{ij}\}, \{\mu_{ijk}\}, \{\gamma_i\}$ all depend on the generalized coordinates $\{\phi_2, \phi_2, \phi_3\}$.

Interpretation of Equations of Motion

The appearance of some Coriolis/centripetal terms and not others in the equations of motions can be understood by analogy to the acceleration of a particle in cylindrical coordinates (2.8):

$$\mathbf{a} = (\ddot{r} - r\dot{\theta}^2)\mathbf{e}_r + (r\ddot{\theta} + 2\dot{r}\dot{\theta})\mathbf{e}_\theta$$

We observe that the centripetal acceleration must be balanced by a radial force that cannot produce a torque about the axis of rotation. The Coriolis acceleration, on the other hand, results from both radial and circumferential motions and requires a torque or moment about the axis of rotation.

Consider first the virtual power equations for $\dot{\phi}_1$. Only the Coriolis terms $\dot{\phi}_1\dot{\phi}_2, \dot{\phi}_1\dot{\phi}_3$ appear. This is because they induce a torque about the $\mathbf{e}_{y1}$ axis, whereas the centripetal acceleration vectors, proportional to $\dot{\phi}_1^2, \dot{\phi}_2^2, \dot{\phi}_3^2$, pass through the axis $\mathbf{e}_{y1}$, and hence create no need for respective torques (see Fig. 6.14).

On the other hand, the second virtual power equation in (6.71), derives from the power about the link-2 axis of rotation. Here we see the terms $\dot{\phi}_1^2, \dot{\phi}_3^2, \dot{\phi}_2\dot{\phi}_3$ in the equation of motion, but not $\dot{\phi}_2^2$. The reason for the latter follows from the previous remarks, i.e., the partial acceleration vector proportional to $\dot{\phi}_2^2$ goes through the link-2 joint axis. However, the partial acceleration vectors associated with $\dot{\phi}_1^2, \dot{\phi}_3^2$ produce a moment about the joint axis that must be balanced by M_2.

The term $\dot{\phi}_2\dot{\phi}_3$ is a Coriolis term where $\dot{\phi}_2$ is analogous to $\dot{\theta}$ in Eq. (2.8) and $\dot{\phi}_3$ produces a radial velocity $\dot{r}$ relative to the link-2 axis on which M_2 is applied (see Fig. 6.13). Similar remarks follow for the terms in the third virtual power equation of motion (6.71).

Example 6.6 Three Link Robot Arm. In this example (Fig. 6.15) we show some of the details involved in deriving the equations of motion (6.71). In this problem the local y axis and unit vector $\mathbf{e}_{iy}$ were chosen to be parallel to the three axes of rotation. With the angular rates chosen as generalized velocities $\{\dot{\phi}_i\}$, the Jacobian or projection vectors for the principle of virtual power (6.67) can be derived. It is convenient in this problem to calculate $\dot{\mathbf{v}}_i, \dot{\mathbf{H}}_i, \partial\boldsymbol{\omega}/\partial\dot{\phi}_i, \partial\mathbf{v}_i/\partial\dot{\phi}_i$ in their respective

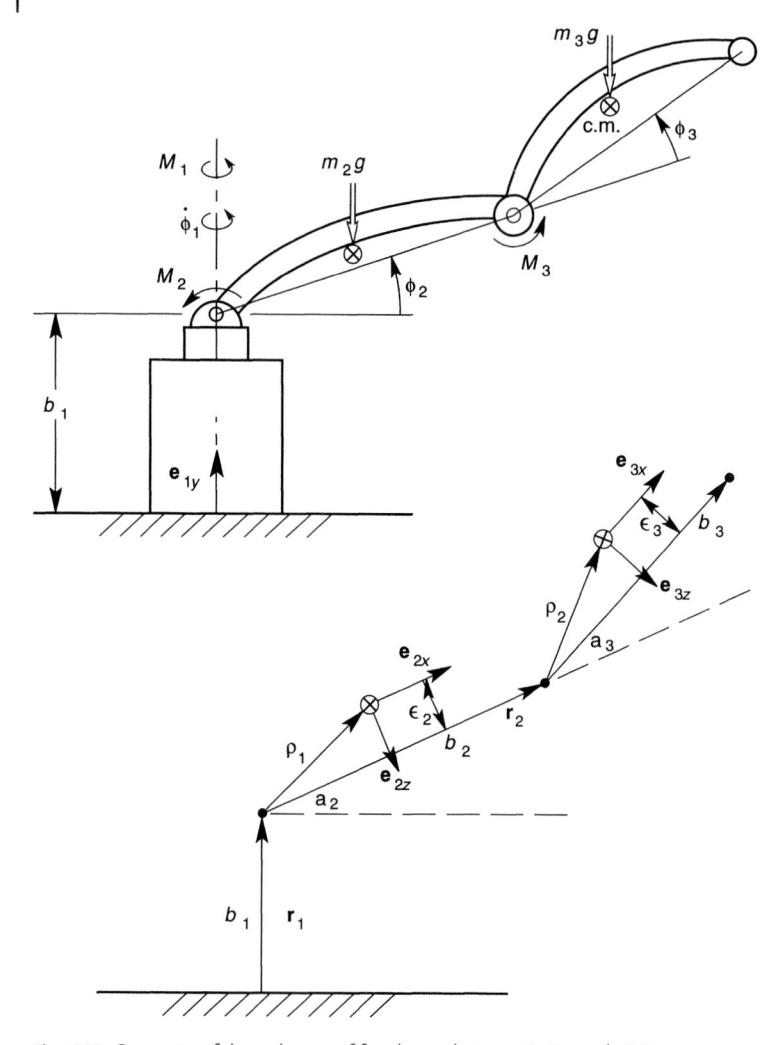

Fig. 6.15 Geometry of three-degree-of-freedom robot arm in Example 6.6.

local coordinates. The notation here and in Fig. 6.15 is that $\mathbf{r}_1$, $\mathbf{r}_2$ are joint-to-joint vectors and $\boldsymbol{\rho}_1$, $\boldsymbol{\rho}_2$ are joint-to-center of mass locator vectors.

As an example, consider the second link with mass m_2 and principal inertias $\{I_{1x}, I_{2y}, I_{3z}\}$. We assume that the basis vectors $\{\mathbf{e}_{ix}, \mathbf{e}_{iy}, \mathbf{e}_{iz}\}$ are principal axes. The velocity of the center of mass of link 2 can be shown to be using (6.65):

$$\mathbf{r}_1 = b_1 \mathbf{e}_{1y}, \qquad \mathbf{r}_2 = (a_2 + b_2)\mathbf{e}_{2x}$$

$$\boldsymbol{\rho}_1 = a_2 \mathbf{e}_{2x} - \varepsilon_2 \mathbf{e}_{2z}, \qquad \boldsymbol{\rho}_2 = a_3 \mathbf{e}_{3x} - \varepsilon_3 \mathbf{e}_{3z}$$

$$\mathbf{v}_2 = -\varepsilon_2 \dot{\phi}_2 \mathbf{e}_{2x} + \dot{\phi}_1(-a_2 C\phi_2 + \varepsilon_2 S\phi_2)\mathbf{e}_{2y} - a_2 \dot{\phi}_2 \mathbf{e}_{2z} \qquad (6.72)$$

To calculate the acceleration we must consider the rate of change of the local link basis vectors. Again these are expressed in components relative to the local link coordinates.

$$\dot{\mathbf{e}}_{2x} = -\dot{\phi}_2 \mathbf{e}_{2z} - \dot{\phi}_1 C\phi_2 \mathbf{e}_{2y}$$
$$\dot{\mathbf{e}}_{2x} = \dot{\phi}_1 C\phi_2 \mathbf{e}_{2x} + \dot{\phi}_1 S\phi_2 \mathbf{e}_{2z}$$
$$\dot{\mathbf{e}}_{2z} = \dot{\phi}_2 \mathbf{e}_{2x} - \dot{\phi}_1 S\phi_2 \mathbf{e}_{2y} \tag{6.73}$$

Using these expressions and (6.72), we can show that the acceleration vector for the center of mass of link 2 has components in the local basis given by

$$\dot{\mathbf{v}}_2 = A_{2x} \mathbf{e}_{2x} + A_{2y} \mathbf{e}_{2y} + A_{2z} \mathbf{e}_{2z}$$

where

$$A_{2x} = -\varepsilon_2 \ddot{\phi}_2 + \dot{\phi}_1^2 (-a_2 C\phi_2 + \varepsilon_2 S\phi_2) C\phi_2 - a_2 \dot{\phi}_2^2$$
$$A_{2y} = \ddot{\phi}_1 (-a_2 C\phi_2 + \varepsilon_2 S\phi_2) + 2\dot{\phi}_1 \dot{\phi}_2 (a_2 S\phi_2 + \varepsilon_2 C\phi_2)$$
$$A_{2z} = -a_2 \ddot{\phi}_2 + \varepsilon_2 \dot{\phi}_2^2 + \dot{\phi}_1^2 (-a_2 C\phi_2 + \varepsilon_2 S\phi_2) S\phi_2 \tag{6.74}$$

It is clear that these terms are quite complicated, with just three degrees of freedom.

Finally, in order to derive the equations of motion we must calculate the Jacobian vectors

$$\frac{\partial \mathbf{v}_2}{\partial \dot{\phi}_1} = (-a_2 C\phi_2 + \varepsilon_2 S\phi_2) \mathbf{e}_{2y}$$

$$\frac{\partial \mathbf{v}_2}{\partial \dot{\phi}_2} = -\varepsilon_2 \mathbf{e}_{2x} - a_2 \mathbf{e}_{2z}$$

$$\frac{\partial \mathbf{v}_2}{\partial \dot{\phi}_3} = 0 \tag{6.75}$$

We can then take the inner product or projection term in the virtual power equation, (6.61):

$$m_2 \dot{\mathbf{v}}_2 \cdot \frac{\partial \mathbf{v}_2}{\partial \dot{\phi}_i} = \begin{bmatrix} A_{2y}(-a_2 C\phi_2 + \varepsilon_2 S\phi_2) \\ -A_{2x}\varepsilon_2 - A_{2z}a_2 \\ 0 \end{bmatrix} \tag{6.76}$$

This calculation can also be carried out in matrix notation. For example, in the local coordinate system of link 2:

$$\mathbf{v}_2 = \begin{bmatrix} -\varepsilon_2 \dot{\phi}_2 \\ (-a_2 C\phi_2 + \varepsilon_2 S\phi_2)\dot{\phi}_1 \\ -a_2 \dot{\phi}_2 \end{bmatrix}$$

$$\left[\frac{\partial \mathbf{v}_2}{\partial \dot{\phi}_i}\right] = \begin{bmatrix} 0 & -\varepsilon_2 & 0 \\ (-a_2 C\phi_2 + \varepsilon_2 S\phi_2) & 0 & 0 \\ 0 & -a_2 & 0 \end{bmatrix}$$

or

$$\left[\frac{\partial \mathbf{v}_2}{\partial \dot{\phi}_i}\right]^T = \begin{bmatrix} 0 & (-a_2 C\phi_2 + \varepsilon_2 S\phi_2) & 0 \\ -\varepsilon_2 & 0 & -a_2 \\ 0 & 0 & 0 \end{bmatrix}$$

$$\left[m_2 \dot{\mathbf{v}}_2 \cdot \frac{\partial \mathbf{v}_2}{\partial \dot{\phi}_i}\right] = \left[\frac{\partial \mathbf{v}_2}{\partial \dot{\phi}_i}\right]^T [m_2 \dot{\mathbf{v}}_2] \tag{6.77}$$

To see how the angular acceleration projections are obtained, consider the second link. We calculate $\mathbf{H}_2$ about the center of mass of link 2. For the angular momentum terms we write

$$\mathbf{H}_2 = I_{2x}\omega_{2x}\mathbf{e}_{2x} + I_{2y}\omega_{2y}\mathbf{e}_{2y} + I_{2z}\omega_{2z}\mathbf{e}_{2z} \tag{6.78}$$

Again we calculate $\mathbf{H}_2$ and $\boldsymbol{\omega}_2$ in the local link coordinates.

$$\boldsymbol{\omega}_2 = \dot{\phi}_1 S\phi_2 \mathbf{e}_{2x} + \dot{\phi}_2 \mathbf{e}_{2y} - \dot{\phi}_1 C\phi_2 \mathbf{e}_{2z} \tag{6.79}$$

The first and third terms come from the relation

$$\mathbf{e}_{1y} = S\phi_2 \mathbf{e}_{2x} - C\phi_2 \mathbf{e}_{2z} \tag{6.80}$$

As in the case of the linear acceleration, when we calculate $\dot{\mathbf{H}}_2$, the time derivatives of the basis vectors must be considered. This results in the following expressions for the angular acceleration $\dot{\mathbf{H}}_2$ expressed in the local coordinates.

$$\begin{aligned} (\dot{\mathbf{H}}_2)_x &= I_{2x}\ddot{\phi}_1 S\phi_2 + \dot{\phi}_1 \dot{\phi}_2 [I_{2x} + (I_{2y} - I_{2z})]C\phi_2 \\ (\dot{\mathbf{H}}_2)_y &= I_{2y}\ddot{\phi}_2 + \dot{\phi}_1^2 (I_{2z} - I_{2x}) S\phi_2 C\phi_2 \\ (\dot{\mathbf{H}}_2)_z &= -I_{2z}\ddot{\phi}_1 C\phi_2 + \dot{\phi}_1 \dot{\phi}_2 [I_{2z} + (I_{2y} - I_{2x})]S\phi_2 \end{aligned} \tag{6.81}$$

The Jacobian vectors for rotation are then calculated using (6.79)

$$\frac{\partial \boldsymbol{\omega}_2}{\partial \dot{\phi}_1} = \mathbf{e}_{1y} = S\phi_2 \mathbf{e}_{2x} - C\phi_2 \mathbf{e}_{2z}$$

$$\frac{\partial \boldsymbol{\omega}_2}{\partial \dot{\phi}_2} = \mathbf{e}_{2y}, \qquad \frac{\partial \boldsymbol{\omega}_2}{\partial \dot{\phi}_3} = 0 \tag{6.82}$$

The contribution of the projection terms for the angular acceleration in the virtual power equations becomes

$$\dot{\mathbf{H}}_2 \cdot \frac{\partial \boldsymbol{\omega}_2}{\partial \dot{\phi}_i} = \begin{bmatrix} (\dot{\mathbf{H}}_2)_x S\phi_2 - (\dot{\mathbf{H}}_2)_z C\phi_2 \\ (\dot{\mathbf{H}}_2)_y \\ 0 \end{bmatrix} \tag{6.83}$$

These extensive calculations lead to the following expressions:

Mass matrix terms (Note: $r_2 = a_2 + b_2$)

$$m_{11} = I_{1y} + I_{2x}S\phi_2^2 + I_{2z}C\phi_2^2 + I_{3x}S\phi_{23}^2 + I_{3z}C\phi_{23}^2$$
$$+ m_2(-a_2C\phi_2 + \varepsilon_2 S\phi_2)^2 + m_3(-r_2C\phi_2 - a_3C\phi_{23} + \varepsilon_3 S\phi_{23})^2$$
$$m_{22} = I_{2y} + I_{3y} + m_2(a_2^2 + \varepsilon_2^2) + m_3[(r_2 S\phi_3 - \varepsilon_3)^2 + (a_3 + r_2 C\phi_3)^2]$$
$$m_{33} = I_{3y} + m_3(a_3^2 + \varepsilon_3^2)$$
$$m_{32} = m_{23} = I_{3y} + m_3(a_3^2 + \varepsilon_3^2) + m_3 r_2(a_3 C\phi_3 - \varepsilon_3 S\phi_3) \tag{6.84}$$

Selected Centripetal Acceleration Term

$$\mu_{322} = m_3 r_2(a_3 S\phi_3 + \varepsilon_3 C\phi_3) \tag{6.85}$$

Selected Coriolis Acceleration Term

$$\mu_{223} = -2m_2 r_2(a_3 S\phi_3 + \varepsilon_3 C\phi_3)$$

6.5
Inverse Problems

A multibody machine, such as a robot, is usually designed to fulfill a set of kinematic or dynamic tasks. For example, the end effector of a manipulator arm may be required to traverse a given path or to move an object from one location and orientation to another, as shown in Fig. 6.16. The *inverse problem* is defined as the determination of the time histories of the forces, torques, joint angles, and velocities necessary to produce a given kinematic output of one of the links of a multibody device. In contrast, the *direct problem* involves solution of the link dynamics output [as in (6.71)] in terms of given torques and forces. The inverse problem is closer to engineering design. The direct problem involves analytical or numerical solution of a set of differential equations with unique solutions. The inverse problem involves solution of nonlinear algebraic equations that often have multiple solutions. In this section we outline the inverse problem for a serial-link mechanism often found in robotic devices. We begin the discussion with the simple two-link, two- and three-degree-of-freedom problem and then give a brief discussion of the six-degree-of-freedom robot arm. The reader is referred to more advanced texts on robotics such as Craig (1986, 2005) or Asada and Slotine (1986) for more details on the inverse problem.

Example 6.7 Two-link Planar Arm. The two-link mechanism has been studied elsewhere in this text from the viewpoint of a direct problem (see, e.g., Chapters 4 and 6). The mechanism is shown in Fig. 6.17 with control torques $T_1(t)$, $T_2(t)$ as well as gravity. In the simplest problem we try to find the joint angles and torques

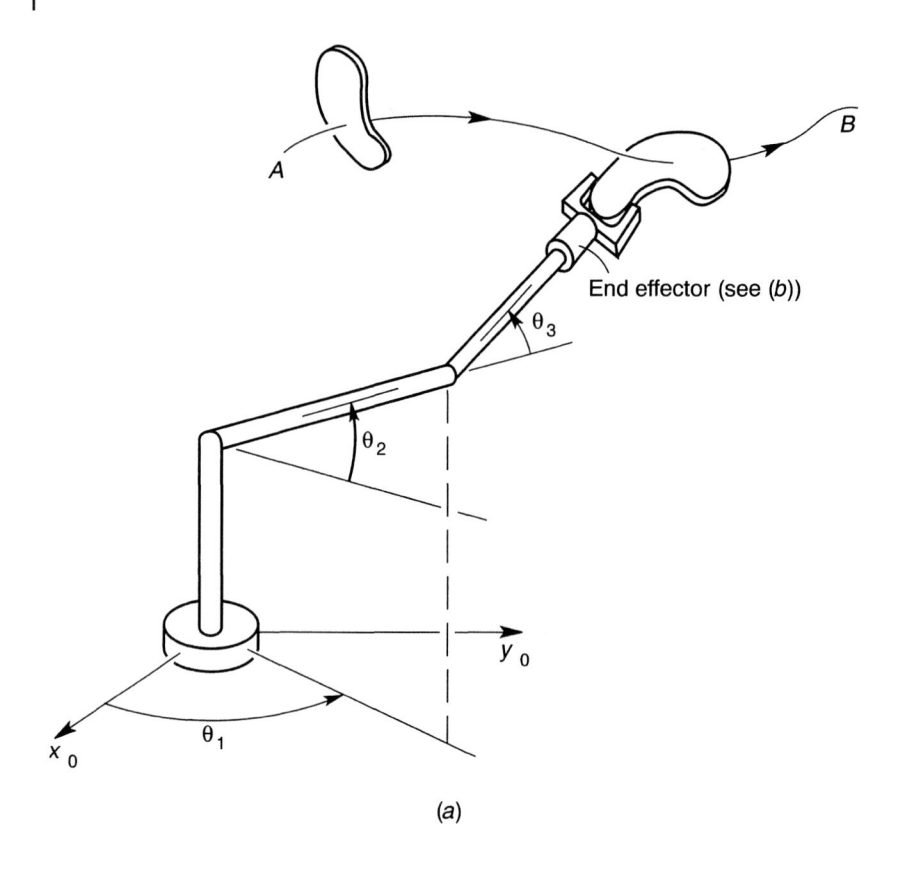

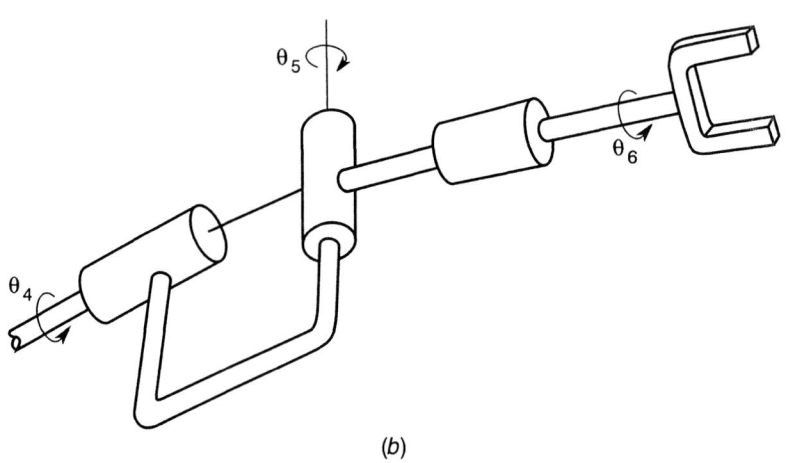

Fig. 6.16 (a) Sketch of three-link robot arm and end effector. (b) Three-degree-of-freedom end-effector wrist (after Asada and Slotine, 1986).

$(\theta_1, \theta_2; T_1, T_2)$ required to move the arm from position $\mathbf{r}_A$ to $\mathbf{r}_B$. Since neither the path nor the velocity along the path is specified, we write an expression for $\mathbf{r}(\theta_1, \theta_2)$, and then try to find the inverse expression for $\{\theta_1(\mathbf{r}), \theta_2(\mathbf{r})\}$.

The solution is determined from the geometry

$$\mathbf{r} = (L_1 \cos\theta_1 + L_2 \cos(\theta_2 + \theta_2))\mathbf{e}_x + (L_1 \sin\theta_1 + L_2 \sin(\theta_2 + \theta_2))\mathbf{e}_y$$

If we write $\mathbf{r} = [x, y]^T$, then we must solve the algebraic equation for θ_1, θ_2 given x, y:

$$L_1 C\theta_1 + L_2 C(\theta_1 + \theta_2) = x$$
$$L_1 S\theta_1 + L_2 S(\theta_1 + \theta_2) = y \tag{6.86}$$

A simple graphical solution shown in Fig. 6.17 shows clearly that two solutions are possible:

$$\theta_1 = \tan^{-1}\frac{y}{x} - \alpha$$

$$\alpha = \cos^{-1}\left(\frac{r^2 + L_1^2 - L_2^2}{2L_1 r}\right)$$

$$\theta_2 = \cos^{-1}\left(\frac{r^2 - (L_1^2 + L_2^2)}{2L_1 L_2}\right)$$

where

$$r^2 = x^2 + y^2 \tag{6.87}$$

The solution of the six-degree-of-freedom inverse kinematic problem is a highly nonlinear, often transcendental mathematical problem. According to Craig (1986, 2005), the serial-link problem can be solved in principle, at least, numerically. However, for on-line or "live" computation in a working environment, numerical solutions for inverse robotic problems are not desirable. In fact, sometimes the robot geometry is designed with the analytical or closed-form inverse solution in mind. Special configurations of manipulator–end effect or devices can be solved in closed form. One such arrangement is a three-link serial wrist-positioning system with an end-effector or wrist device that has three intersecting axes of rotation (see Fig. 6.16b). Thus, we can first solve for the wrist position (i.e., three degrees of freedom), and then use Euler angle kinematics to invert the angular orientation of the wrist mechanism. In the following example we consider a three degree-of-freedom serial mechanism. The interested reader is encouraged to consult more advanced books on robotics, such as Craig (1986, 2005) or Asada and Slotine (1986), for the complete six-degree-of-freedom problem.

In the robotics literature, a conventional choice of assignment of separate link coordinate systems is used, called the Denavit–Hartenberg notation after the two engineers who invented it in 1951. We do not digress here to define this conven-

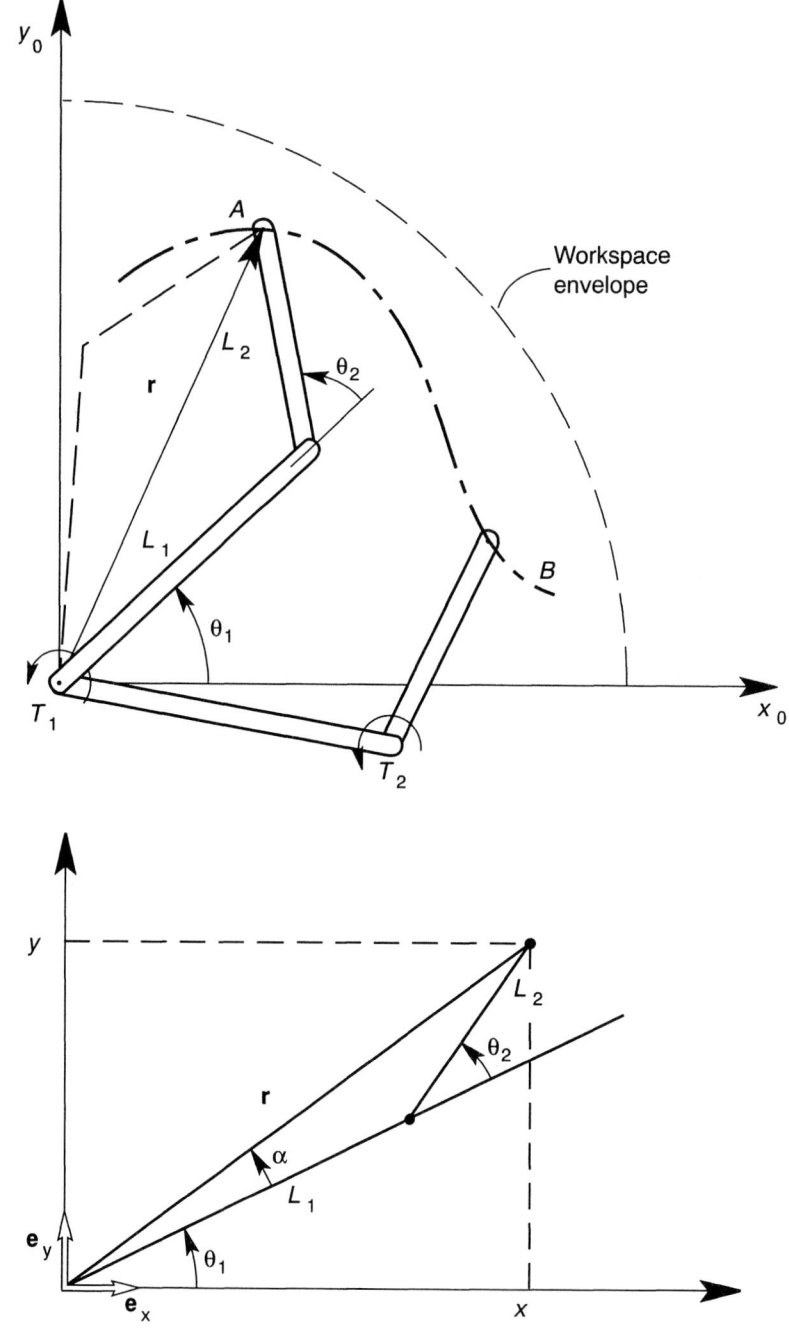

Fig. 6.17 Workspace envelope of a two-link arm and the two configurations possible in an inverse problem (Fig. 6.7).

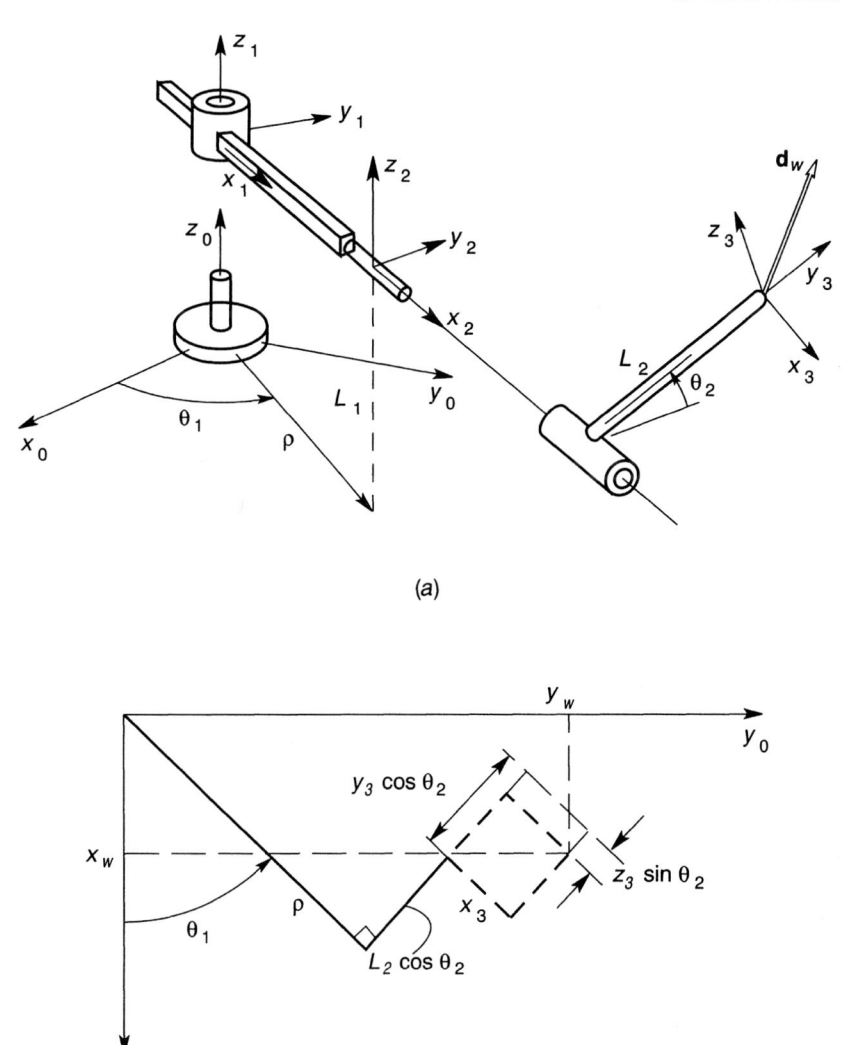

Fig. 6.18 Geometry for the inverse problem of a three-degree-of-freedom arm.

tion. Instead, we choose the most direct coordinate systems to illustrate the use of the 4×4 transformation matrices. We remind the reader of the nature of this transformation (see Chapter 3). For the example shown in Fig. 6.18, we have two revolute joints with an intermediate prismatic or translation joint. We suppose that a point on an object in the wrist coordinates is denoted by

$$\mathbf{d}_w = [x_3, y_3, z_3]^T \tag{6.88}$$

Then its position in the base coordinates is given by

$$\begin{bmatrix} x_{wd} \\ y_{wd} \\ z_{wd} \\ 1 \end{bmatrix} = T \begin{bmatrix} x_3 \\ y_3 \\ z_3 \\ 1 \end{bmatrix} \tag{6.89}$$

where the 4×4 transformation can be decomposed into three sequential transformations:

$$T = {}_1^0T\,{}_2^1T\,{}_3^2T \tag{6.90}$$

where we have used the notation of Craig. For example ${}_2^1T$ transforms components of a 4-vector written in the $\{x_2, y_2, z_2\}$ system into a vector in the $\{x_1, y_1, z_1\}$ system under the translation operation.

These "T" matrices can be shown to be given by

$${}_1^0T = \begin{bmatrix} C\theta_1 & -S\theta_1 & 0 & | & 0 \\ S\theta_1 & C\theta_1 & 0 & | & 0 \\ 0 & 0 & 1 & | & L_1 \\ \hline 0 & 0 & 0 & | & 1 \end{bmatrix}$$

$${}_2^1T = \begin{bmatrix} 1 & 0 & 0 & | & \rho \\ 0 & 1 & 0 & | & 0 \\ 0 & 0 & 1 & | & 0 \\ \hline 0 & 0 & 0 & | & 1 \end{bmatrix}$$

$${}_3^2T = \begin{bmatrix} 1 & 0 & 0 & | & 0 \\ 0 & C\theta_2 & -S\theta_2 & | & L_2 C\theta_2 \\ 0 & S\theta_2 & C\theta_2 & | & L_2 S\theta_2 \\ \hline 0 & 0 & 0 & | & 1 \end{bmatrix} \tag{6.91}$$

The last matrix is found by first translating a unit triad at the origin of $\{x_2, y_2, z_2\}$ along the y_2 direction an amount L_2 and then rotating the triad about the x_2 axis to place the unit triad at the wrist location and orientation of system $\{x_3, y_3, z_3\}$.

The inverse problem for the origin of the wrist is then solved by finding values of $\{\theta_1, \theta_2, \rho\}$, in terms of its position in the base coordinates, i.e.,

$$\begin{bmatrix} x_{w0} \\ y_{w0} \\ z_{w0} \\ 1 \end{bmatrix} = T(\theta_1, \theta_2, \rho) \begin{bmatrix} 0 \\ 0 \\ 0 \\ 1 \end{bmatrix} \tag{6.92}$$

where

$$T = \begin{bmatrix} C\theta_1 & -S\theta_1 C\theta_2 & S\theta_1 S\theta_2 & \rho C\theta_1 - L_2 C\theta_2 S\theta_1 \\ S\theta_1 & C\theta_1 C\theta_2 & -C\theta_1 S\theta_2 & \rho S\theta_1 + L_2 C\theta_2 C\theta_1 \\ 0 & S\theta_2 & C\theta_2 & L_1 + L_2 S\theta_2 \\ \hline 0 & 0 & 0 & 1 \end{bmatrix} \quad (6.93)$$

Thus the coordinates $[x_{w0}, y_{w0}, z_{w0}]^T$ are given by the rightmost column vector, i.e., we must solve the three nonlinear algebraic equations

$$x_{w0} = \rho \cos \theta_1 - L_2 \cos \theta_2 \sin \theta_1$$

$$y_{w0} = \rho \sin \theta_1 + L_2 \cos \theta_2 \cos \theta_1$$

$$z_{w0} = L_1 + L_2 \sin \theta_2 \quad (6.94)$$

The angle θ_2 is found easily:

$$\theta_2 = \sin^{-1} \frac{z_{w0} - L_1}{L_2} \quad (6.95)$$

To find θ_1, first multiply the first two equations of (6.94) by $\cos \theta_1$, and $\sin \theta_1$, respectively. Adding the resulting equations, we obtain

$$\rho = x_{w0} \cos \theta_1 + y_{w0} \sin \theta_1 \quad (6.96)$$

The value of θ_1 is found by defining

$$\tan \beta = x_{w0}/y_{w0}$$

$$A = \left(x_{w0}^2 + y_{w0}^2\right)^{\frac{1}{2}}$$

Then

$$\sin(\theta_1 + \beta) = \rho/A$$

(See also Craig, 1986, 2005 for another form of the solution.)

To move the wrist point along some path, we divide the path into path segments given by a set of position vectors $\{\mathbf{r}_{w0}^{(k)}\}$. The preceding algebraic solution is then used to generate a sequence of joint variables $\{\rho^{(k)}, \theta_1^{(k)}; \theta_2^{(k)}; k = 1, 2, \ldots\}$.

There are also problems when the dynamics along the path must be specified. Here we must then determine joint velocities as well, and knowledge of Jacobians, e.g., $\partial \mathbf{r}_{w0}/\partial \theta_i$ or $\partial \dot{\mathbf{r}}_{w0}/\partial \dot{\theta}$ must be used. But this topic is beyond the level of this book. One should consult more advanced texts on robotics, such as Asada and Slotine (1986) or Craig (1986, 2005).

6.6
PD Control of Robotic Machines

Robotic machines are distinguished by their ability to dramatically change their geometric configuration in response to some command or feedback from the environment or an intelligent operator. In certain manufacturing applications robotic machines are designed to move a part or a tool from one position and orientation to another in the workspace. The dynamic mathematical model of a robot manipulator arm expressed by Eqs. (6.32) or (6.36) is identical in form to any other multibody machine except for the nature of the applied torques which depend on the joint positions and velocities, i.e. $\{\tau_i = \tau_i(\theta_k(t), \dot{\theta}_k(t), f(t))\}$. There are many theories of control of mechanical systems embodying linear and nonlinear feedback. Although the dynamical models of robotic systems are strongly nonlinear, in many practical robotic machines, a simple *proportional-derivative* or PD control scheme is used. (See, e.g., Craig, 2005.) In such systems the control torques are assumed to take the form

$$\tau_i = -\sum \Gamma_{ik}\left(\theta_k - \tilde{\theta}_k(t)\right) - \sum \Lambda_{ik} \dot{\theta}_k \qquad (6.97)$$

In this form of feedback, torques are applied to the joints if any of the joint variables deviate from a reference time history $\tilde{\theta}_k(t)$. The functions $\tilde{\theta}_k(t)$, therefore, embody the desired path or end points in joint space. The gain matrices Γ and Λ are called the *control stiffness* and *damping matrices* since they mimic the addition of elastic springs and linear dampers to the joints.

In practice one first determines the desired path of the end-effector in the Cartesian space and then solves for the desired joint variable time history with an "inverse" kinematics program. If the machine is responding to inputs from its environment, the desired path may depend on the motion of the machine links, as in avoiding obstacles. Feedback laws of the form (6.97) effectively add control stiffness to the machine. To avoid overshooting the desired motion path, one also adds so-called *"damping" feedback* as expressed by the linear term in (6.97) proportional to the joint velocities. Of course the values of the matrices, Γ and Λ, are up to the choice of the machine designer. Often one of the *"Design Rules"* is to choose the control gains to make each of the joints appear as a damped linear oscillator as illustrated in the example below.

One Degree-of-Freedom Example
Suppose we freeze all the joints of the robot except one, whose joint variable is represented by $\theta(t)$. The single degree-of-freedom motion of the machine can then be represented by a single differential equation of the form where for the moment we neglect gravity forces:

$$I\ddot{\theta}(t) = T = -\Gamma\left(\theta - \tilde{\theta}(t)\right) - \Lambda \dot{\theta}$$

or

$$I\ddot{\theta}(t) + \Lambda\dot{\theta} + \Gamma\theta = \Gamma\tilde{\theta}(t) \tag{6.98}$$

The solution to Eq. (6.98) involves the sum of a particular solution dependent on the given reference function $\tilde{\theta}$, and a solution to the homogeneous equation with the right-hand side equal to zero. The so-called homogeneous solution reflects the transient behavior of the system. In our case one applies the design rule to make the transients behave as a critically damped oscillator. The characteristic equation for solutions of the form e^{st} is

$$Is^2 + \Lambda s + \Gamma = 0$$

$$s = -\frac{\Lambda}{2I} \pm \sqrt{\left(\frac{\Lambda}{2I}\right)^2 - \frac{\Gamma}{I}} \tag{6.99}$$

For critical damping, the square root term in (6.99) must vanish and the stiffness and damping gains must be related by the relation

$$\Lambda_{\text{crit}} = \sqrt{2\Gamma I} \tag{6.100}$$

In order to determine the value of the stiffness gain, one invokes a second *design rule* that seeks to avoid any resonance with the natural vibratory frequencies in the machine. For example, if the lowest resonant frequency is f_0 in Hertz of cycles per second, then the natural frequency of the controlled machine should be much lower than this value, or

$$\sqrt{\frac{\Gamma}{I}} < \frac{1}{n} 2\pi f_0 \tag{6.101}$$

where a typical value of n is two.

Lyapunov Stability for Robotic Systems

One of the chief concerns about using feedback control in mechanical systems is the potential for instability in the control system. One of the goals of feedback control in robotic machines is to introduce effective control stiffness and damping in the joints. However if the phase of the control becomes out of sync, positive stiffness can become negative stiffness and buckling or divergence instability can result. Also feedback control might introduce negative damping forces and a flutter, limit cycle or Hopf bifurcation instability might occur. However it is some comfort to know that some types of control have theoretical stability benefits provided certain properties are met by the control gains. This is especially important for multibody and robotic systems that are often inherently nonlinear.

As an illustration of the restrictions on control gains for stability, we examine the proportional-derivative or PD control of a robotic system of connected rigid bodies. (This example is taken from the robotics textbook of Craig, 2005, third edition.)

The general equations for a multibody system of rigid bodies are [see (6.36)]

$$M(\Theta)\ddot{\Theta} + V(\Theta, \dot{\Theta}) + G(\Theta) = \tau + f(\Theta, \dot{\Theta}) \tag{6.102}$$

where Θ is the vector of link position variables such as the relative angles between links for revolute joints or the relative translation displacement between links for a prismatic joint. For a six degree of freedom manipulator arm Θ is a six-dimensional column vector, i.e., $\Theta = [q_1(t), q_2(t), q_3(t), q_4(t), q_5(t), q_6(t)]^T$, where $q_k(t)$ are the joint position variables.

In Eq. (6.102), $M(\ )$ is an $n \times n$ symmetric mass matrix, $V(\ )$ is a column vector of centrifugal and centripetal acceleration terms quadratic in the generalized velocities, $G(\ )$ is a column vector of gravity related forces, τ is a column vector of control torques and control actuator forces. Finally, $f(\ )$ is a column vector of internal joint friction of either the dry friction or viscous damper type. In the following discussion we set the friction forces equal to zero, i.e., $f(\ ) = 0$.

In classic proportional-derivative (PD) control for point to point dynamics, the control torques $\tau(t)$ are comprised of two parts; one proportional to the static gravity loads G, and another proportional to the control error E, and the velocities, i.e.

$$\tau = \Gamma E - \Lambda \dot{\Theta} + G(\Theta) \tag{6.103}$$

where

$$E = \Theta_{\text{ref}} - \Theta$$

Here Γ and Λ are $n \times n$ control gain matrices. When Γ and Λ are diagonal with positive entries, this control is equivalent to inserting springs and dashpots into the mechanical system and one would expect the system if originally stable to remain so after adding the control. The Lyapunov stability theorem described below guarantees stability if the gain matrices Γ and Λ are positive definite or that the eigenvalues are positive. If Γ and Λ are diagonal matrices, this is satisfied if the diagonal control gains are positive.

Lyapunov Stability Theorem

This theorem is a way to establish stability in a linear or nonlinear system without having to find the explicit solution. This is important because in most nonlinear problems in dynamics, it is not possible to write down explicit mathematical functions to investigate whether the system state variable will remain bounded. In Lyapunov's method, one tries to find an energy-like function $\mathcal{F}$, and show that under the dynamical equations of motion, the function will decrease in time. A proof of Lyapunov's stability theorem is beyond the scope of this book. [See, e.g. Nayfeh and Balachandran (1995) or Guckenheimer and Holmes (1983).] However we state the theorem and see how it places restrictions on the control gains in order to maintain bounded motions of the joint variables.

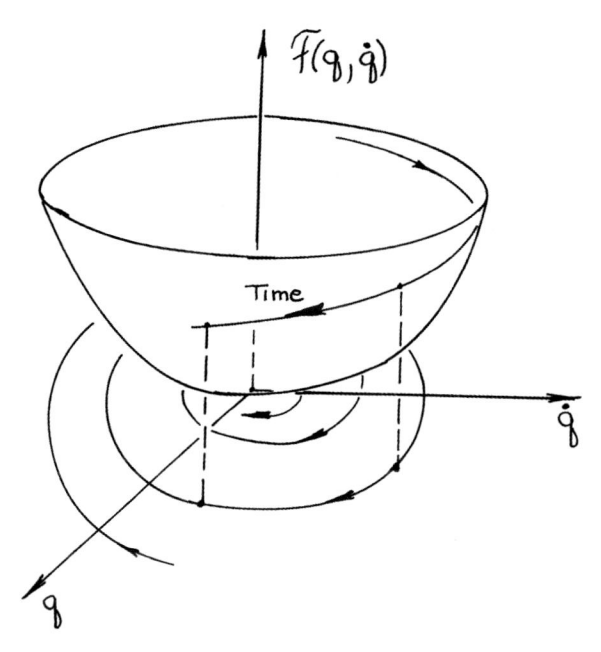

Fig. 6.19 Phase plane dynamics and Lyapunov function for stability of a single degree of freedom system.

First one must define stability. In this theorem one defines the concept of *asymptotic stability*: this means that near some reference point in state space, if the system is disturbed, the motion will approach the fixed point as time increases (Fig. 6.19).

Next we define a Lyapunov function $\mathscr{F}(q(t))$ for a single degree of freedom problem with the following properties:

(i) If $q = 0$, $dq/dt = 0$ is the fixed point in state space, then $F(0) = 0$.
(ii) $F(q(t))$ is a positive definite function in the neighborhood of a fixed point; $F(q(t)) > 0$. Also $F(q(t))$ is assumed to have continuous first derivatives. Then the stability theorem states:
(iii) Under the dynamics of the system, the system is stable if

$$\frac{d}{dt}\mathscr{F} \leqslant 0 \tag{6.104}$$

For example, consider a one degree of freedom nonlinear second-order system with linear PD control.

$$\ddot{q} + q^3 = -\Gamma q - \Lambda \dot{q} \tag{6.105}$$

The state space is described by the joint position and velocity $[q(t), dq/dt]$. An appropriate Lyapunov function would be of the form (see Fig. 6.19)

$$\mathscr{F} = \frac{1}{2}\left[\dot{q}^2 + \Gamma q^2 + \frac{1}{2}q^4\right]$$

$$\mathscr{F} > 0, \quad \text{near the origin if } \Gamma > 0$$

$$\dot{\mathscr{F}} = \dot{q}(\ddot{q} + \Gamma q + q^3) = -\Lambda \dot{q}^2 < 0, \quad \text{for } \Lambda > 0 \tag{6.106}$$

The theorem states that if stiffness and damping feedback gains are positive, then the Lyapunov function decreases with time and the system remains stable.

For a multi-degree-of freedom robotic system, Craig (2005) proposes a Lyapunov function of the form

$$\mathscr{F} = \frac{1}{2}\dot{\Theta}^T M(\Theta)\dot{\Theta} + \frac{1}{2}E^T \Gamma E \tag{6.107}$$

where the stiffness control matrix Γ is positive definite. E is the error function defined in (6.103).

Under the dynamics of the general robotic system with PD control (6.102), the Lyapunov function $\mathscr{F}$ will decrease in time if the control gain matrix is positive definite, i.e. one can show that for (6.107) under the dynamics (6.102) and (6.103),

$$\dot{\mathscr{F}} = -\dot{\Theta}^T \Lambda \dot{\Theta} < 0 \tag{6.108}$$

The details of this calculation may be found in Craig (2005). Our purpose in discussing this theorem has been to provide some mathematical underpinning for the use of proportional-derivative control of the form (6.103) in multibody systems. There are of course other control strategies for nonlinear dynamical systems for which the reader is referred to advanced books on nonlinear control theory.

Reflected Actuator Inertia in Robotic Machines

To move a multi-rigid body system through a desired task requires actuators to create forces and torques on the individual links. This is usually accomplished through electromagnetic motors or hydraulic piston actuators of both the rotary and linear type. In the case of DC electromagnetic servo motors, the torque is proportional to the electric current I in the motor armature winding. (See Chapter 8.) Often the motor is optimized to run at a speed that is one or several orders of magnitude speed above the robot link speed and a speed reduction mechanism is required. This can be accomplished by either belt and pulleys of different diameters, or by a gear transmission system.

Consider the robot arm in Fig. 6.20 with a single stage speed reducing gear pair. In general the motor moment of inertia I_m is much smaller than the link second moment of mass I_L. However because of the high speed of the motor, one cannot neglect the kinetic energy of the motor armature. This is often modeled by replacing the link inertia I_L, by an effective inertia $I_{\text{eff}} = I_L + n^2 I_m$, where n is the speed ratio of the armature and the output of the speed reduction shaft. This relation can easily be derived using Lagrange's equation for the system where the kinetic energy

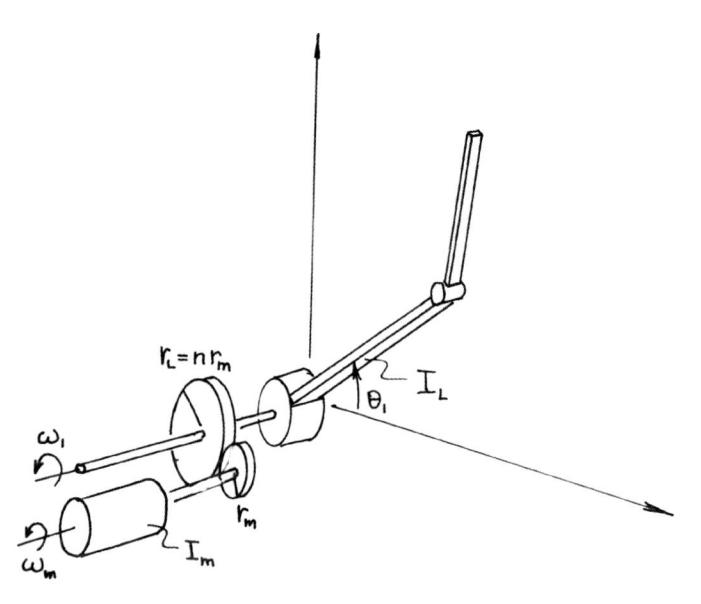

Fig. 6.20 Two-link planar manipulator arm driven by a motor with a one stage speed reducer.

is the sum of that of the link and the motor. Here the gear inertia is absorbed in the link and the pinion inertia included in the motor.

$$T = \frac{1}{2} I_m \omega_m^2 + \frac{1}{2} I_L \omega_1^2 \tag{6.109}$$

where

$$\omega_1 = \dot{\theta}_1, \qquad \omega_m = n\omega_1$$

$$T = \frac{1}{2}(I_L + n^2 I_m)\dot{\theta}_1^2 = \frac{1}{2} I_{\text{eff}} \dot{\theta}_1^2 \tag{6.110}$$

$$I_{\text{eff}} = I_L + n^2 I_m \tag{6.111}$$

It is easy to see that although I_L might be nonlinear in the other link variables, if the speed reduction factor n is large, the effective motor inertia can dominate the inertia terms and in some cases justify neglect of I_L.

6.7
Impact Problems

A foot hitting a soccer ball, backlash in gears, a meteorite hitting a satellite, these are common examples of transient contact of bodies that we call impact. Elementary treatment of impact problems usually concerns smooth spherical particles with no friction. In the simplest case, we have direct impact, that is, initial velocities directed along the line of the center of mass (Fig. 6.21). The problem in central impact

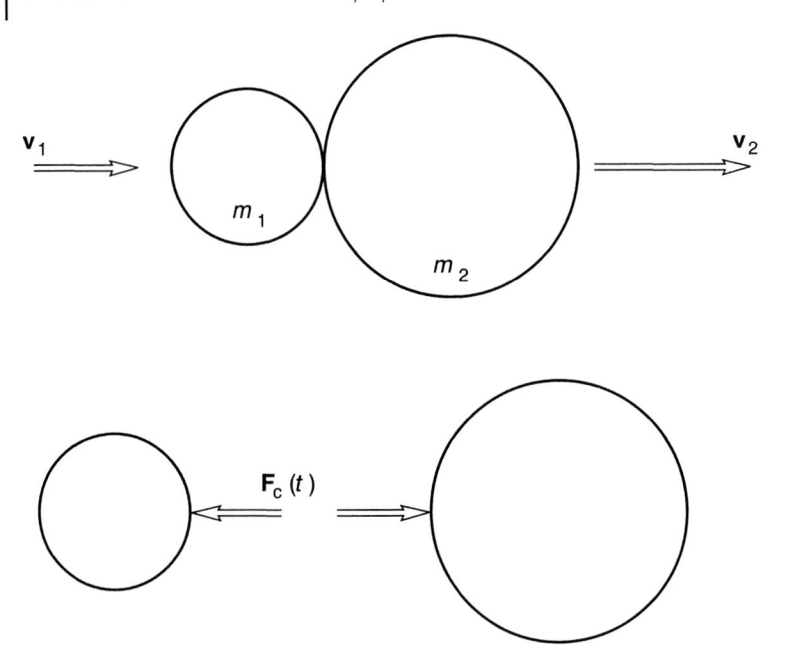

Fig. 6.21 Central impact of spherical bodies.

is to determine the relative separation velocity after impact as a function of the relative approach velocity before impact. The so-called *Newton law of impact* assumes a linear relationship between the final and initial relative velocities $\mathbf{v}_r^+$, $\mathbf{v}_r^-$, with a sign change, i.e.,

$$\mathbf{v}_r^+ \cdot \mathbf{n} = -\varepsilon \mathbf{v}_r^- \cdot \mathbf{n} \tag{6.112}$$

where ε is called the *coefficient of restitution* and $\mathbf{n}$ is a unit vector normal to the tangent plane of contact.

Contact problems between extended bodies is still an evolving field of study. As such, the use of numerical codes, either multibody or finite-element codes for complex impact problems must be used and interpreted with extreme caution. There are many impact problems where the deformation of the body is important, such as impact with cables or strings, fluid surfaces, extended vibratory structures, such as beams or plates, and highly deformable systems, such as sand or other granular media. In this section we restrict our attention to the impact between bodies where, except for the contact region, the overall motion can be described by rigid-body dynamics.

In the contact region, complex mechanical deformation processes such as elastic and inelastic deformation, sliding and static friction, stiction, cold welding, fracture, cratering, melting, and heating, can occur. There may be other physical processes, including light and electron emission, as well as chemical reactions. The traditional approach to solving the dynamics of impact problems has been to make as many simplifying assumptions as possible, as in the case of Newton's law of impact. In this section we show by example how this simple theory may be extended to planar rigid-body collisions. However, many practical impact problems require a more complex analysis. Two recent monographs on impact of rigid bodies include Brach (1991) and Pfeiffer and Glocker (1996).

The *Newton law of impact* (6.112) is called a kinematic law since it relates velocities and not forces. However, another method, called the *Poisson law*, relates the impulse during the approach or compression of the two bodies, to the impulse during the separation or expansion phase.

Consider again the central impact of spheres shown in Fig. 6.21. In the Newton theory, the impact time is infinitesimal, while in the Poisson theory we imagine a finite contact time. In the approach phase $t^- < t < t^*$, the compressive contact force $\mathbf{F}_c(t)$ between the bodies rises to a maximum value, and during the separation phase it drops to zero, $t^* < t < t^+$. The total impulse of the impact force is given by

$$\Lambda = \int_{t^-}^{t^+} \mathbf{F}_c(t)\, dt \tag{6.113}$$

This integral can be divided into integrals over the approach (compression) and separation (expansion) phases so that

$$\Lambda = \Lambda_C + \Lambda_E \tag{6.114}$$

where

$$\Lambda_C = \int_{t^-}^{t^*} \mathbf{F}_C\, dt, \qquad \Lambda_E = \int_{t^*}^{t^+} \mathbf{F}_C\, dt \tag{6.115}$$

The Poisson law of impact relates the two phases of impact, i.e.,

$$\Lambda_E = \bar{\varepsilon} \Lambda_C \tag{6.116}$$

When friction is not present, we can set $\varepsilon = \bar{\varepsilon}$ and the Newton and Poisson formulations yield the same result.

The basic procedure for solving rigid-body impact problems can be outlined as follows:

1. Integrate the differential equations of motion up to the time of impact $t = t^-$ (either numerically, or analytically).
2. Determine if impact is about to occur.
3. Solve the algebraic impact laws to find the separation velocities and angular velocities.
4. Continue integration of the equation of motion until the next contact event.

The preceding shows the importance of determining whether impact is about to occur. In some machine systems with loose-fitting parts, a whole sequence of im-

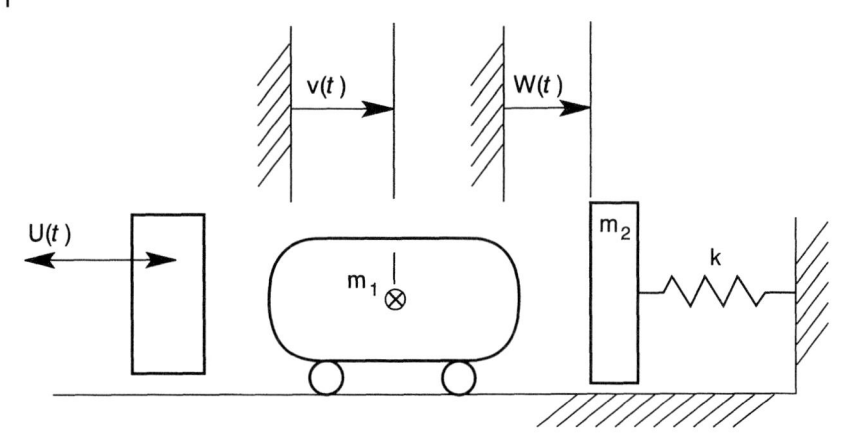

Fig. 6.22 Model for machine-generated noise. A free mass impacting two movable walls.

pacts can occur, so one must continually monitor at each time step of the numerical integration if impact will occur.

In the following examples we assume that there is just one contact event. For more complex problems, see Pfeiffer and Glocker (1996).

Example 6.8 Machine Rattling Noise. In many multiple-part machines, gaps and play between moving parts create impact that generates structure-borne and eventually air-borne noise. An idealized model problem is shown in Fig. 6.22. A moving mass suffers impact between two moving walls, one of which has a prescribed motion $U(t) = U_0 \sin \omega t$, while the right-hand wall motion $W(t)$ behaves like a linear oscillator with natural frequency Ω. Our goal here is to find the equations of motion.

To solve this problem we denote the free-mass velocity by $v(t)$, the sequence of impact times on the left wall by $\{t_n\}$, and those on the right wall by $\{t_\alpha\}$. Newton's law of impact applied to the impacts at the left and right walls are given by

$$V_n^+ - \dot{U}_n = \varepsilon(V_n^- + \dot{U}_n)$$

$$V_\alpha^+ + \dot{W}_\alpha^+ = \varepsilon(V_\alpha^- - \dot{W}_\alpha^-) \qquad (6.117)$$

where $\dot{U}_n = \omega U_0 \cos \omega t_n$, and

$$v(t_n^-) = -V_n^-, \qquad v(t_n^+) = V_n^+$$

$$v(t_\alpha^-) = V_\alpha^-, \qquad v(t_\alpha^+) = -V_\alpha^+$$

$$\dot{W}(t_\alpha^+) = \dot{W}_\alpha^+, \qquad \dot{W}(t_\alpha^-) = \dot{W}_\alpha^-$$

The + and − denote post- and pre-impact times. We assume that the left wall mass has sufficient mass to be unaffected by the impact. In the rightwall impact we assume conservation of linear momentum

$$m_1 V_\alpha^- + m_2 \dot{W}_\alpha^- = -m_1 V_\alpha^+ + m_2 \dot{W}_\alpha^+ \tag{6.118}$$

In some problems, we can have multiple impacts on one wall or the other. In this treatment we assume that an alternating impact sequence is possible, so that we have the ordered sequence $\{t_{\alpha-1}, t_n, t_\alpha, t_{n+1}\}$. Neglecting friction on mass m_1 as it moves from one wall to the other, we assume that

$$V_n^+ = V_\alpha^-, \qquad V_\alpha^+ = V_{n+1}^- \tag{6.119}$$

For alternating impact, the resulting impact times are determined by

$$t_\alpha = t_n + \frac{\Delta + W_\alpha - U_n}{|V_n^+|}$$

$$t_{n+1} = t_\alpha + \frac{\Delta + W_\alpha - U_{n+1}}{|V_\alpha^+|} \tag{6.120}$$

(If the impacts are indeed alternating, the absolute-value signs are superfluous.)

Finally, we need an equation for the motion of the right wall $W(t)$, which is governed by

$$m_2 \ddot{W} + kW = 0$$

where $\Omega^2 = k/m_2$. By assumption we assume that the displacement is continuous during the impact, i.e., $W(t_\alpha^-) = W(t_\alpha^+) = W_\alpha$, whereas the velocity suffers a discontinuity. If the state of the oscillator after an impact at $t = t_\alpha$ is $\{W_\alpha, \dot{W}_\alpha^+\}$, then we can show that the oscillatory solution yields a set of equations for the state at $t = t_{\alpha+1}$:

$$W_{\alpha+1} = W_\alpha \cos[\Omega(t_{\alpha+1} - t_\alpha)] + \frac{\dot{W}_\alpha^+}{\Omega} \sin[\Omega(t_{\alpha+1} - t_\alpha)]$$

$$\dot{W}_{\alpha+1}^- = -\Omega W_\alpha \sin[\Omega(t_{\alpha+1} - t)] + \dot{W}_\alpha^+ \cos[\Omega(t_{\alpha+1} - t_\alpha)] \tag{6.121}$$

The set of impact equations is a transcendental set of algebraic equations. A simplifying assumption is to assume that the gap Δ is much larger than the moving wall amplitudes, i.e., $\Delta \gg A_0$, $\Delta \gg |W_\alpha|$. In this case, the time of input is approximately given by

$$t_\alpha = t_n + \Delta/V_n^+$$

$$t_{n+1} = t_\alpha + \Delta/V_\alpha^+ \tag{6.122}$$

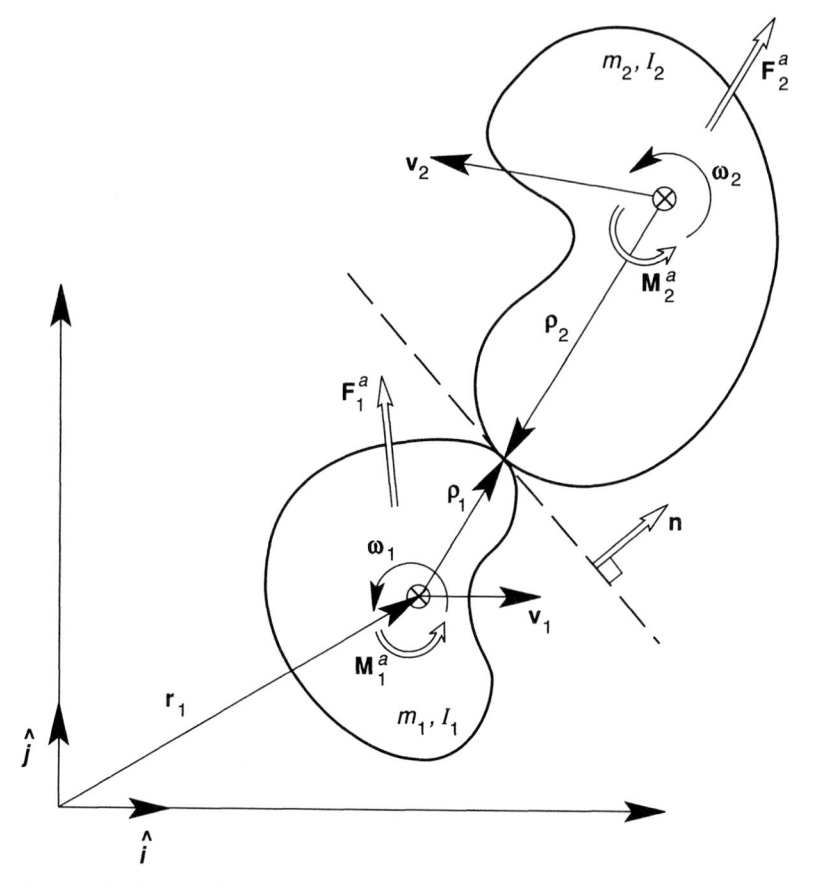

Fig. 6.23 Single-point planar impact of two smooth rigid bodies.

To solve this system for one cycle of alternating impact $\{t_\alpha, t_{n+1}, t_{\alpha+1}\}$, we assume that $\{t_\alpha, V_\alpha^+, \dot{W}_\alpha^+, W_\alpha\}$ are known, and we solve for the unknowns in the following order:

$$\{t_{n+1}, V_{n+1}^+, t_{\alpha+1}, \dot{W}_{\alpha+1}, V_{\alpha+1}^+, \dot{W}_{\alpha+1}^+, W_{\alpha+1}\}$$

For further details on this problem, see the paper by Moon and Broschart (1991).

Planar Impact Dynamics

The key difference between elementary particle impact dynamics and rigid body impact is that the velocities to be used in the impact laws are the *local contact-point velocities*, which must then be related to the rigid body center-of-mass velocity, and angular velocity. To outline the basic method consider the two bodies shown in Fig. 6.23, each with position vectors to their respective centers of mass, $\mathbf{r}_1$, $\mathbf{r}_2$

and corresponding center-of-mass velocities and angular velocities $\mathbf{v}_1$, $\mathbf{v}_2$, $\boldsymbol{\omega}_1$, $\boldsymbol{\omega}_2$. A fixed reference frame has planar orthogonal unit vectors $\hat{\mathbf{i}}$, $\hat{\mathbf{j}}$, and a unit vector normal to the plane $\hat{\mathbf{k}}$. We assume that the bodies are constrained to move in the plane so that the angular velocities take the form

$$\boldsymbol{\omega}_1 = \omega_1 \hat{\mathbf{k}} = \dot{\theta}_1 \hat{\mathbf{k}}$$
$$\boldsymbol{\omega}_2 = \omega_2 \hat{\mathbf{k}} = \dot{\theta}_2 \hat{\mathbf{k}} \tag{6.123}$$

We also denote the contact force on mass m_1 by $\mathbf{F}_c$ and applied forces and moments by $\mathbf{F}_1^a$, $\mathbf{F}_2^a$, $\mathbf{M}_1^a$, $\mathbf{M}_2^a$, where the applied moments contain the moments of the applied forces about the respective center of mass. The Newton–Euler equations of motion during impact then become

$$m_1 \dot{\mathbf{v}}_1 = \mathbf{F}_c + \mathbf{F}_1^a$$
$$m_2 \dot{\mathbf{v}}_2 = -\mathbf{F}_c + \mathbf{F}_2^a$$
$$I_1 \dot{\omega}_1 \hat{\mathbf{k}} = \boldsymbol{\rho}_1 \times \mathbf{F}_c + \mathbf{M}_1^a$$
$$I_2 \dot{\omega}_2 \hat{\mathbf{k}} = -\boldsymbol{\rho}_2 \times \mathbf{F}_c + \mathbf{M}_2^a \tag{6.124}$$

Here I_1, I_2 are the second moments of mass about the respective centers of mass $\mathbf{r}_1$, $\mathbf{r}_2$.

The total impulse of the impact force, $\boldsymbol{\Lambda}$, is defined in (6.113). It is implicitly assumed that at least one of the surfaces of either m_1 or m_2 has a definable normal at the contact point, $\mathbf{n}$. During impact we assume there is no stiction or welding so that, $\boldsymbol{\Lambda} \cdot \mathbf{n} < 0$.

Another key assumption is that the impact time is small enough so that $\mathbf{r}_1$, $\mathbf{r}_2$ and $\boldsymbol{\rho}_1$, $\boldsymbol{\rho}_2$ are unchanged during the impact time interval $t^- \leqslant t \leqslant t^+$. Integrating the equations of motion during this time interval we obtain the impact equation of motion

$$m_1(\mathbf{v}_1^+ - \mathbf{v}_1^-) = \boldsymbol{\Lambda}$$
$$m_2(\mathbf{v}_2^+ - \mathbf{v}_2^-) = -\boldsymbol{\Lambda} \tag{6.125}$$

This leads to a linear momentum conservation law

$$m_1 \mathbf{v}_1^+ + m_2 \mathbf{v}_2^+ = m_1 \mathbf{v}_1^- + m_2 \mathbf{v}_2^- \tag{6.126}$$

The angular momentum equations become

$$I_1(\omega_1^+ - \omega_1^-) = (\boldsymbol{\rho}_1 \times \boldsymbol{\Lambda}) \cdot \hat{\mathbf{k}}$$
$$I_2(\omega_2^+ - \omega_2^-) = (\boldsymbol{\rho}_2 \times \boldsymbol{\Lambda}) \cdot \hat{\mathbf{k}} \tag{6.127}$$

A quick tally shows that we have eight scalar unknowns $\{\mathbf{v}_1^+, \mathbf{v}_2^+, \omega_1^+, \omega_2^+, \boldsymbol{\Lambda}\}$ and only six equations of motion. Therefore we require two equations for the impact physics. In general this might involve either Newton's or Poisson's law of impact as well as a friction law (e.g., Coulomb's law) relating $\boldsymbol{\Lambda} \cdot \mathbf{n}$ and $\boldsymbol{\Lambda} \cdot \mathbf{t}$. In order to keep the problem simple, we assume smooth, *frictionless impact* and Newton's impact law. Thus our two additional equations become

$$\boldsymbol{\Lambda} \cdot \mathbf{t} = 0$$
$$(\mathbf{v}_{c1}^+ - \mathbf{v}_{c2}^+) \cdot \mathbf{n} = -\varepsilon (\mathbf{v}_{c1}^- - \mathbf{v}_{c2}^-) \cdot \mathbf{n} \tag{6.128}$$

where the velocities at the contact point $\mathbf{r}_c$ are given by

$$\mathbf{v}_{c1} = \mathbf{v}_1 + \boldsymbol{\omega}_1 \times \boldsymbol{\rho}_1$$
$$\mathbf{v}_{c2} = \mathbf{v}_2 + \boldsymbol{\omega}_2 \times \boldsymbol{\rho}_2$$

and

$$\mathbf{r}_c = \mathbf{r}_1 + \boldsymbol{\rho}_1 = \mathbf{r}_2 + \boldsymbol{\rho}_2 \tag{6.129}$$

(see Fig. 6.23).

The zero-friction assumption, (6.128a), leads to the solution of the center-of-mass velocities tangential to the plane of impact, i.e.,

$$(\mathbf{v}_1^+ - \mathbf{v}_1^-) \cdot \mathbf{t} = 0$$
$$(\mathbf{v}_2^+ - \mathbf{v}_2) \cdot \mathbf{t} = 0 \tag{6.130}$$

This reduces the number of unknowns after impact to five. Writing the set of unknowns as a column vector, $\boldsymbol{\eta} = [v_{1n}^+, v_{2n}^+, \omega_1^+, \omega_2^+, \Lambda_n]^T$, the impact equations take the form of a coupled set of linear algebraic equations:

$$A\boldsymbol{\eta} = \mathbf{B} \tag{6.131}$$

These equations can be solved and the solution written in the form using the notation of Brach (1991) [note, $\rho_{it} = \boldsymbol{\rho}_i \cdot \mathbf{t}, \mathbf{n} \times \mathbf{t} = \hat{\mathbf{k}}$]:

$$\omega_1^+ - \omega_1^- = v_{rn} \bar{m}(1+\varepsilon)\rho_{1t}q/I_1$$
$$\omega_2^+ - \omega_2^- = v_{rn} \bar{m}(1+\varepsilon)\rho_{2t}q/I_2$$
$$v_{1n}^+ - v_{1n}^- = v_{rn} \bar{m}(1+\varepsilon)q/m_1$$
$$v_{2n}^+ - v_{2n}^- = v_{rn} \bar{m}(1+\varepsilon)q/m_2 \tag{6.132}$$

Here v_{rn} is the initial relative velocity at the contact point:

$$v_{rn} = (v_{2n}^- - \rho_{2t}\omega_2^-) - (v_{1n}^- - \rho_{1t}\omega_1^-) \tag{6.133}$$

The other parameters in the preceding solution are defined by

$$\bar{m} = m_1 m_2 / (m_1 + m_2)$$

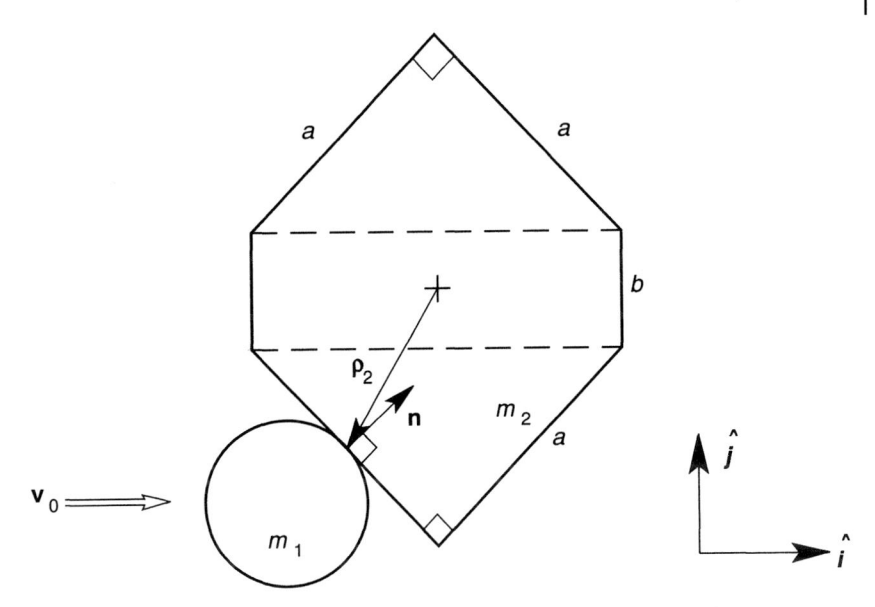

Fig. 6.24 Planar impact of a particle with a hexagonal-shaped rigid body (Example 6.9).

$$q = \left[1 + \frac{\bar{m}\rho_{2t}^2}{I_2} + \frac{\bar{m}\rho_{1t}^2}{I_1}\right]^{-1} \tag{6.134}$$

Example 6.9 Planar Impact of Two Discs. Consider the collision of two planar objects (Fig. 6.24), a circular disc and a six-sided polygonal disc. At the moment before impact, we assume that the circular disc has a horizontal velocity $v_0\hat{i}$, and that the polygon disc is stationary. The contact point is at the midpoint of one of the sides of length a. Under a frictionless impact assumption, we want to find the post impact velocities and angular velocity of the polygonal disc ω_2^+. We also want to verify that as the side $b \to 0$, the angular velocity ω_2^+ goes to zero. To solve this problem we use the impact Eqs. (6.125), (6.127), as well as the Newton impact law (6.112). The linear momentum Eqs. (6.125), become

$$m_1(\mathbf{v}_1^+ - v_0\hat{i}) = -\Lambda \frac{\sqrt{2}}{2}(\hat{i} + \hat{j})$$

$$m_2\mathbf{v}_2^+ = \Lambda \frac{\sqrt{2}}{2}(\hat{i} + \hat{j}) \tag{6.135}$$

The angular momentum Eqs. (6.127) become

$$I_1\omega_1^+ = 0$$

$$I_2\omega_2^+ = \Lambda d \tag{6.136}$$

where the moment arm $d = \sqrt{2}b/4$ for midface impact.

Newton's impact law (6.112), or (6.128), takes the form

$$(\mathbf{v}_1^+ - \mathbf{v}_2^+ - \omega_2^+ \hat{\mathbf{k}} \times \boldsymbol{\rho}_2) \cdot \mathbf{n} = -\varepsilon v_0 \hat{\mathbf{i}} \cdot \mathbf{n} \tag{6.137}$$

where

$$\mathbf{n} = \sqrt{2}(\hat{\mathbf{i}} + \hat{\mathbf{j}})/2$$
$$\boldsymbol{\rho}_2 = -\frac{b}{2}\hat{\mathbf{j}} - (\hat{\mathbf{i}} + \hat{\mathbf{j}})a\sqrt{2}/4 \tag{6.138}$$

In this last equation, we have used the fact that $\omega_1^+ = 0$. The number of scalar unknowns in this problem is seven, $\{v_{1x}^+, v_{1y}^+, v_{2x}^+, v_{2y}^+, \omega_1^+, \omega_2^+, \Lambda\}$. We have seven scalar equations, all linear in the unknowns.

From the linear-momentum equation (6.10), we deduce that the post impact velocity of the center of mass is directed in the 45° direction to the horizontal, i.e.,

$$\mathbf{v}_2^+ = v_2^+ (\hat{\mathbf{i}} + \hat{\mathbf{j}})/\sqrt{2}$$

and we also have that

$$\Lambda = m_2 v_2^+$$
$$\omega_2^+ = m_2 v_2^+ \sqrt{2} b / 4 I_2 \tag{6.139}$$

Adding the two linear-momentum equations, we have the result that

$$m_1 \mathbf{v}_1^+ = m_1 v_0 \hat{\mathbf{i}} - m_2 v_2^+ (\hat{\mathbf{i}} + \hat{\mathbf{j}})/\sqrt{2} \tag{6.140}$$

Thus we are left with one scalar unknown, v_2^+ which we solve for, using the scalar Newton's impact law (6.137).

$$v_2^+ = \frac{(1+\varepsilon)v_0}{\sqrt{2}} \left\{ 1 + \frac{m_2}{m_1} + \frac{m_2}{I_2} \frac{b^2}{4\sqrt{2}} \right\}^{-1} \tag{6.141}$$

In the case of $b = 0$, we can see from (6.139) that $\omega_2^+ = 0$, and the preceding result is identical to the central-impact problem.

Single-body Impact

In many problems one of the bodies in the impact is much more massive than the other, as shown in Fig. 6.25. In these problems we do not have momentum conservation laws as in the two-body case, but the solution follows using the impulse-momentum laws. Again we consider the planar-impact problem. Defining a local coordinate or normal and tangential unit vectors at the point of contact, we obtain the following equations:

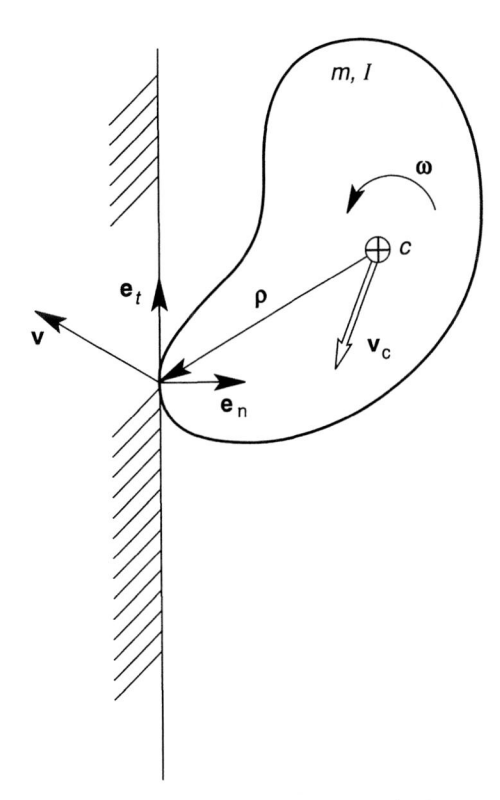

Fig. 6.25 Planar impact of a rigid body with a large immovable mass.

$$m(v_n^+ - v_n^-) = \Lambda_n$$
$$m(v_t^+ - v_t^-) = \Lambda_t$$
$$I(\omega^+ - \omega^-) = (\rho \times \Lambda) \cdot \hat{\mathbf{k}} \tag{6.142}$$

Following the analysis of the two-body problem, we use Newton's impact law and assume zero friction, i.e.,

$$\Lambda_t = 0 \quad \text{or} \quad v_t^+ = v_t^- \tag{6.143}$$
$$v_{cn}^- = -\varepsilon v_{cn}^- \tag{6.144}$$

The velocity at the point of contact is related to the center-of-mass velocity by the equation

$$\mathbf{v}_c = \mathbf{v} + \boldsymbol{\omega} \times \boldsymbol{\rho} \tag{6.145}$$

In the planar problem we have three unknowns $\{v_n^+, \omega^+, \Lambda_n\}$ and three algebraic equations linear in the unknowns. Eliminating the impulse, Λ_n, the solution for the kinematic variables $[v_n^+, \omega^+]$ is

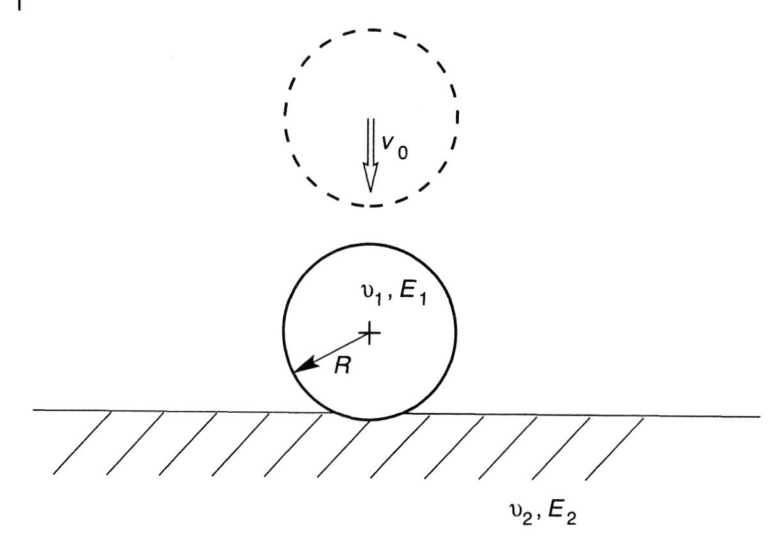

Fig. 6.26 Impact of an elastic sphere with an elastic half-space (Hertz contact) with respective values of Young's Modules E_i, and Poisson's ratio v_i.

$$v_n^+ - v_n^- = -(1+\varepsilon)(v_n^- + \rho_t\omega^-)\{1 + m\rho_t^2/I\}^{-1}$$
$$\omega_n^+ - \omega_n^- = -(v_n^+ - v_n^-)m\rho_t/I \qquad (6.146)$$

where

$$\boldsymbol{\rho} = -(\rho_n\mathbf{e}_n + \rho_t\mathbf{e}_t)$$

The solution can also be obtained from the two-body impact by letting $m_2 \to \infty$, $\bar{m} = 1$, $\mathbf{v}_2^- = 0$, $\omega_2^- = 0$.

Elastic-impact Theory

The study of the impact of elastic solids has a large literature. A classic study is the book by Goldsmith (1960). The elastic classic theory of impact by Hertz can be used to estimate the time of impact and the stresses involved in the contact zone. To summarize the results for a sphere hitting an elastic half-space consider the problem defined in Fig. 6.26. The impact force F is found to be a nonlinear function of the penetration displacement of the sphere into the half-space α:

$$F = \kappa\alpha^{3/2} \qquad (6.147)$$

where

$$\kappa = \frac{4}{3}[R^{1/2}/(\delta_1 + \delta_2)], \qquad \delta_i = (1 - v_i^2)/E_i$$

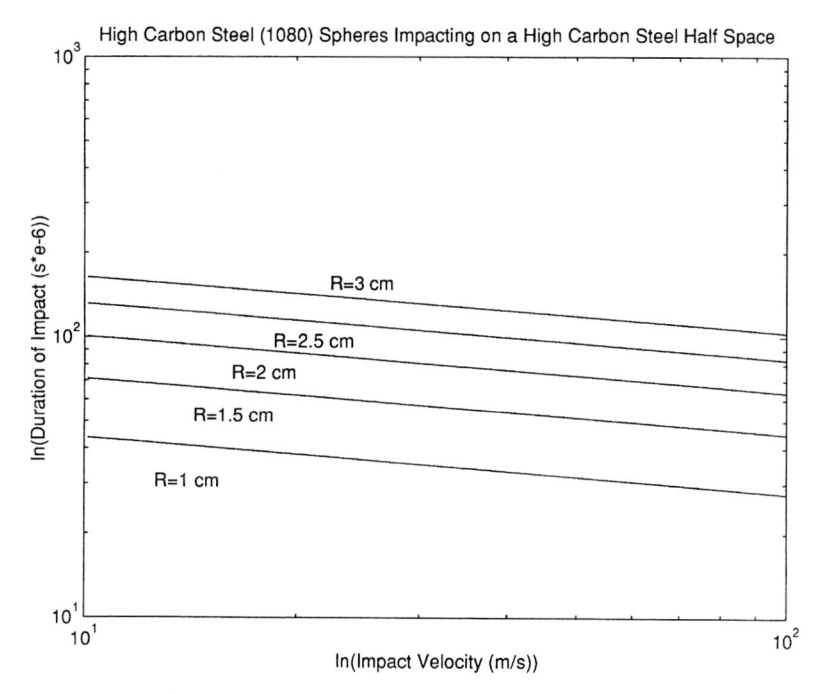

Fig. 6.27 Values of duration of impact times for Hertz contact of a sphere on a half space vs. impact velocity (6.148).

R is the radius of the sphere, v_i is the Poisson's ratio, and E_i is the elastic Young's modules for either the sphere ($i = 1$) or the half-space ($i = 2$). The impact time duration τ is found to depend on the initial normal impact velocity V_0,

$$\tau = 2.94 \alpha_m / V_0 \tag{6.148}$$

where

$$\alpha_m = \left[\frac{5}{4} (m V_0^2 / \kappa) \right]^{2/5}$$

The impact force in this theory is approximated by a half sine wave

$$F(t) = \begin{cases} \frac{1.14 m V_0^2}{\alpha_m} \sin\left(\frac{1.068 V_0 t}{\alpha_m}\right), & t \leqslant \tau \\ 0, & t > \tau \end{cases} \tag{6.149}$$

Here α_m is the maximum "rigid-body" penetration into the elastic half-space. These formulas reveal the following dependence of the contact time

$$\tau \simeq 1/V_0^{1/5} \tag{6.150}$$

For materials such as steel on steel, the contact times for small spheres ($R \sim 1$ cm) range from 10 μs to 100 μs for velocities in the range of 100 m/s (Fig. 6.27).

Table 6.1 Coefficient of restitution normal impact.

Hard sphere on mild steel target Normal velocity (m/s)	ε
50	0.26
100	0.17
200	0.10

Note: Adapted from Brach (1991), Fig. 6.22.
(Empirical relation, $\varepsilon = k/(v_n + k)$,
$k = 19.9$ m/s.)

Most impacts generate stresses that exceed the elastic limit, and elastic-plastic impact analysis must be used. However, the Hertz theory gives an estimate of the contact time that for many problems is very small. This justifies the use of the impulse-momentum theory in rigid-body impact dynamics.

Experimental Impact Data

Elementary physics and mechanics texts list coefficients of restitution for different materials for normal impact and moderate- to low-impact velocities, as illustrated in Table 6.1. However, when tangential relative velocities are involved, then the effects of friction (neglected in the previous analysis) and angle of incidence become important to the extent that the coefficient of restitution depends on the initial velocities and angle of incidence. We again raise the warning that complex impact problems may require a more detailed theory than the one based on simple impact contact assumptions of coefficients of restitution and friction.

Impact and Chaos

In the preceding discussion we considered only a single impact. However, in many machine problems, parts in relative motion to one another can undergo repeated impacts. Such examples include gear teeth rattling, ball bearings, rotating shafts in a loose bearing, and mating parts with play or gaps (see Example 6.8). In recent years it has been shown that such problems can exhibit *chaotic dynamics* (see, e.g., Pfeiffer, 1994 or Moon, 1992). In these problems the solution is very sensitive to initial conditions. Consider, for example, the problem shown in Fig. 6.28, where one of the rigid bodies is constrained to move in a sinusoidal motion. Then it has been observed in both experiments as well as numerical simulation, that the motion of the second body may not necessarily be sinusoidal. These problems are examples of nonlinear dynamics where the output dynamics in time is not similar in nature to the input kinematics (see also Chapter 9). When the motion is chaotic, the time of impact relative to the phase of the sinusoidal input cannot be determined when there is a small uncertainty in the initial conditions. In such problems, both the times of impact and the post impact velocities are best described by probability

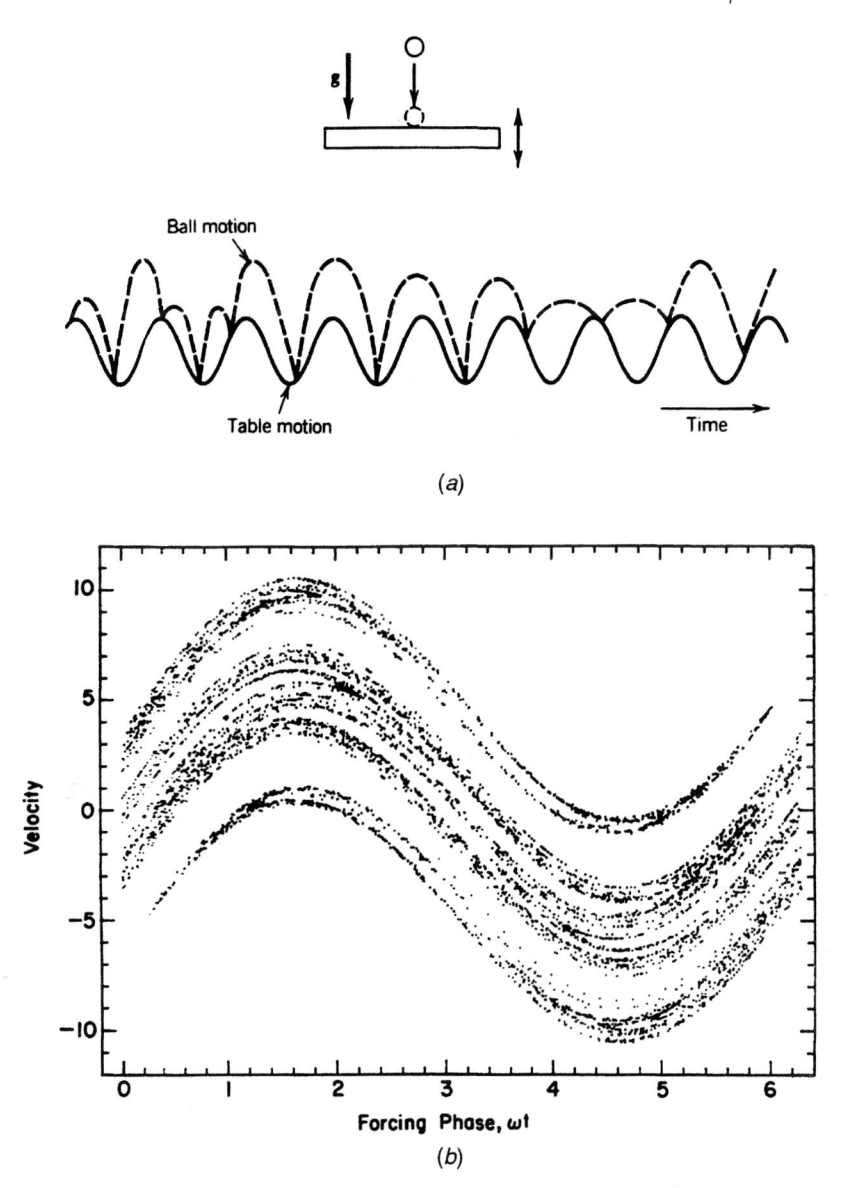

Fig. 6.28 (*a*) Chaotic bouncing of a sphere on an oscillating surface. (*b*) Fractal Poincare map of impact velocity vs. phase of the moving table at the time of contact (see Chapter 9, also Moon, 1992).

distributions even though the input motion is not random. These discoveries have shown how random-like noise can be generated in machines even when the input motions are deterministic or contain only a small amount of noise. There are many

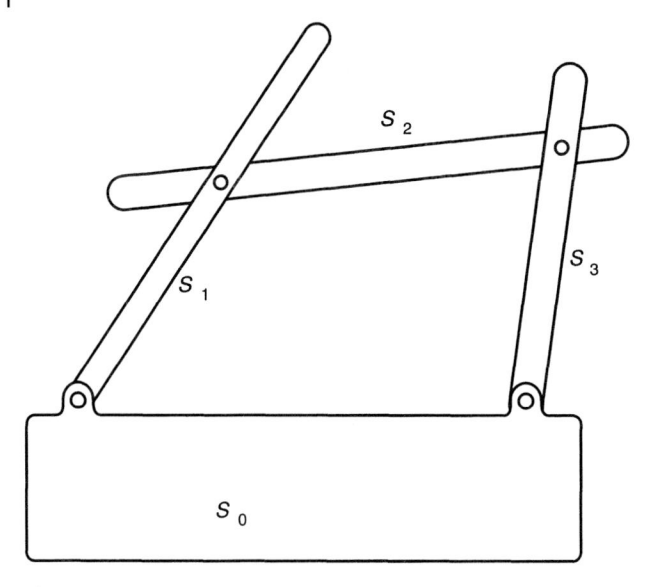

Fig. P6.1

introductory books on chaotic and nonlinear dynamics which discuss these ideas (see, e.g., Moon, 1992).

Homework Problems

6.1 The four bodies shown in Fig. P6.1 $\{S_0, S_1, S_3, S_4\}$ are connected by four hinges into a 4-bar linkage. (a) Use Grübler's theorem (Chapter 3) to show that linkage has one degree of freedom. (b) Find geometric conditions on the lengths of the links so that body S_1 is a crank and body S_3 has a rocking motion. (c) Using the graph theory notation of Section 6.2, find the incidence matrix $[S_{ia}]$. Also find the S and T matrices defined in (6.1) and (6.5).

6.2 In Fig. P6.2, two masses m_1 and m_2 are connected by a homogeneous rod of length L and mass m_3. Treating the constraints between body S_0 and S_1, S_2 as generalized hinges, find the incidence matrix $[S_{ia}]$. Assuming that the constraints are frictionless, use Eqs. (6.8), (6.9) to derive equations of motion for the four-body system in Cartesian coordinates (x, y).

6.3 Find the incidence matrix $[S_{ia}]$ for the four-body mechanism in Fig. P6.3.

6.4 In the four-body mechanism shown in Fig. P6.3, find the Jacobians $\partial x/\partial \theta$, $\partial x/\partial \phi$. For what values of ϕ does the mechanism have the greatest and least kinematic sensitivity, i.e., for what geometric parameter is $|\partial x/\partial \phi|$ or $|\partial \dot{x}/\partial \dot{\phi}|$ maximum or a minimum? You may use numerical simulation to obtain an answer. (Answer: $\partial x/\partial \theta = L \cos \theta$, $\partial x/\partial \phi = -(L/2) \cos \theta$, $2\theta = \pi - \phi$.)

6.5 Consider the four-body mechanism with three movable links (Fig. P6.5), where $0 \leqslant \phi_1 < \pi$, $0 \leqslant \phi_2 < 2\pi$. Find the workspace of the end point **r**.

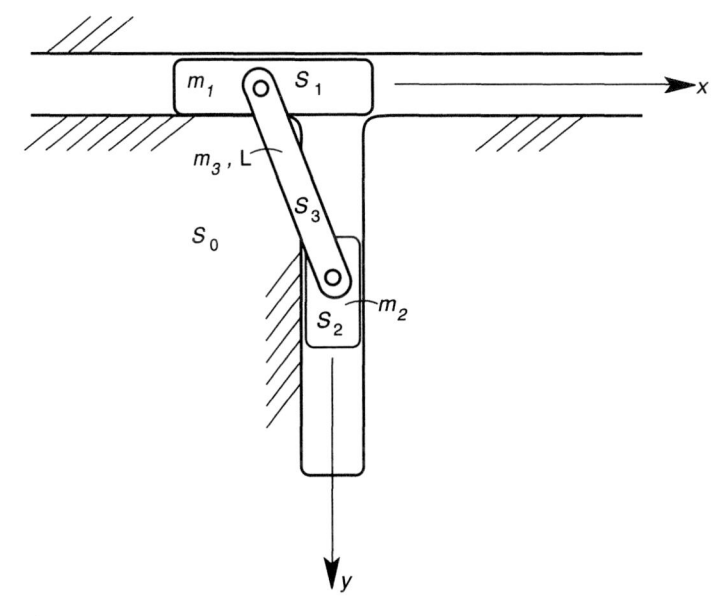

Fig. P6.2

Show that the Jacobian $\partial \mathbf{r}/\partial \phi_i$ is singular on the boundary of the workspace. Are there any singular points of the Jacobian inside the workspace? (See, e.g., Craig (1986, P146), or Asada and Slotine (1986, P65).)

6.6 (*Robot Arm Dynamics*) A two-link planar arm has two degrees of freedom with one revolute joint and one prismatic joint as shown in Fig. P6.6. (a) Sketch the workspace when the revolute joint angle ranges from zero to $\pi/2$ and the prismatic joint variable s spans zero to L. (b) Use Lagrange's equations for rigid bodies to derive the equations of motion for this arm assuming there is an actuator torque at the base and an actuator motor force at the sliding joint. Write the equations in the form (6.36) or in a matrix form similar to (6.58) or (6.71).

6.7 (*Robot Arm Dynamics*) A two-link planar arm has two degrees of freedom with one elbow revolute joint and the lower link is constrained to move in two prismatic joints as shown in Fig. P6.7. (a) Sketch the workspace when the revolute joint angle ranges from $-\pi/2$ to $\pi/2$ and the prismatic joint variable s spans zero to L. (b) Use the principle of virtual power for rigid bodies to derive the equations of motion for this arm assuming there is an actuator torque at the elbow and an actuator motor force at the horizontal sliding joint. Write the equations in the form (6.36) or in a matrix form similar to (6.58) or (6.71).

6.8 (*Robot Arm Dynamics with MATLAB*) Write the equations of motion for the two-link arm in Problem 6.7 in the form of first-order equations using matrix notation and write a MATLAB program to integrate the coupled first-order differential equations. Apply a prescribed actuator force on the sliding joint

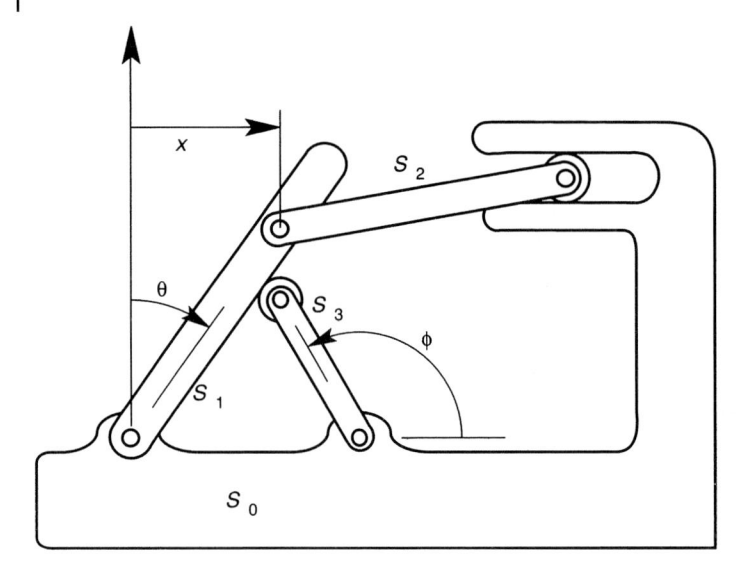

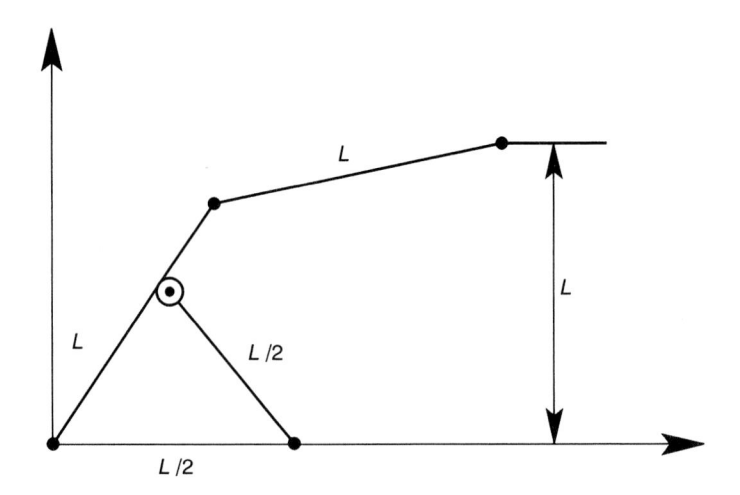

Fig. P6.3

and an elbow torque of the form $\sin(\pi t/2)$ for both actuators during the time interval $0 < t < 2$ seconds. If the link dimensions are each one meter in length and one kilogram in mass, determine the amplitude of the actuator force and torque to move the arm through its workspace. Plot the joint variables versus time as well as graph the position of the links for four intervals in the time span.

6.9 (*Robot Arm Kinematics*) The parallelogram mechanism shown in Fig. P6.9 is similar to that of the MIT planar robot manipulator arm. (a) Use Grübler's

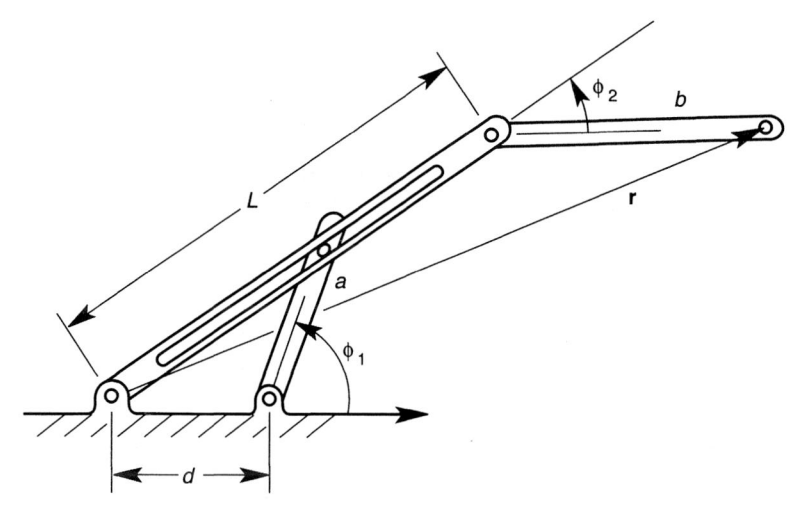

Fig. P6.5

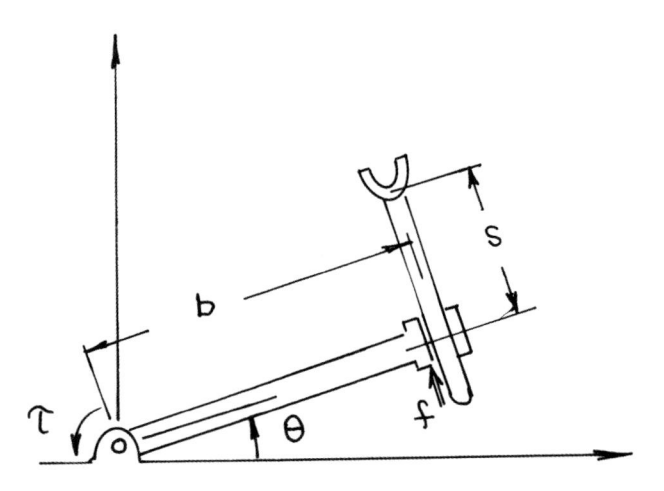

Fig. P6.6

criterion (Chapter 3) to show that the 5 link, 5 joint linkage has 2 degrees of freedom. (Note the ground is one link.) (b) Sketch the wrist workspace if the two base angles are controlled by independent motors and can range from 0 to $\pi/2$. (c) Find the Jacobian relating the velocity in Cartesian coordinates and the joint angle velocities. Determine if there is a singularity.

6.10 (*Robot Arm Dynamics*) For the MIT arm in Problem 6.9 (Fig. P6.9) assume that Link 2 and Link 4 have inertias I_{2c}, I_{4c}, with respect to their centers of mass and are much greater than the moments of inertias of Links 3, 5. Use the Principle of Virtual power to derive equations of motion for the arm assuming that there are two independent torque actuators at the base. Write the equations of motion in the form of a matrix equation.

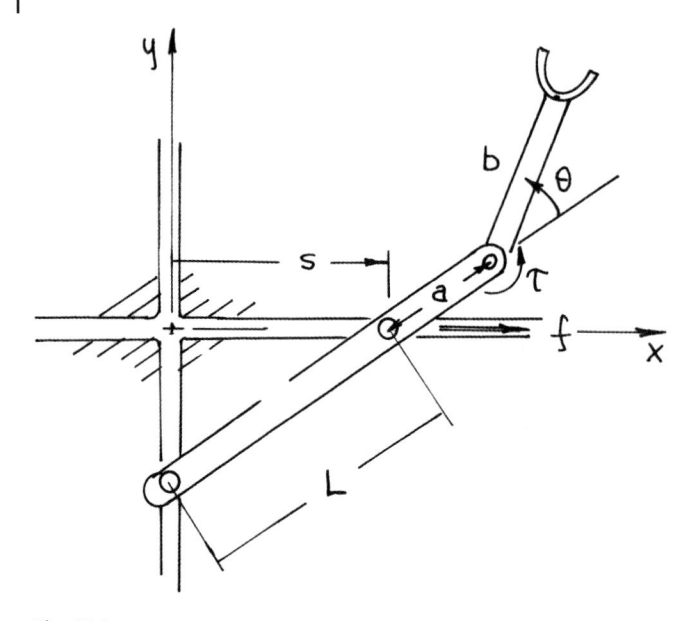

Fig. P6.7

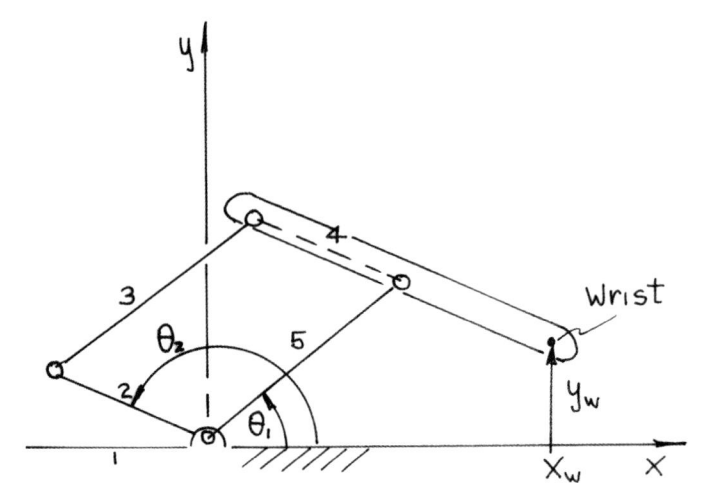

Fig. P6.9

6.11 Two bodies are connected by a rigid link, as shown in Fig. P6.11. Assume that m_1 is constrained to planar translation, while $m_2 = (m_0 + m_3)$ undergoes rolling in the plane without slip on the horizontal surface. Also assume that the connecting link is massless. Calculate the Jacobian $\partial\theta/\partial x$ for the one-degree-of-freedom motion, where x is the generalized variable. Determine the singularities of the Jacobian.

6.12 In Problem 6.11 assume that the mass m_2 consists of mass m_0 with a mass at the center of the disc and an unbalanced mass m_3 at a radius ε, as shown

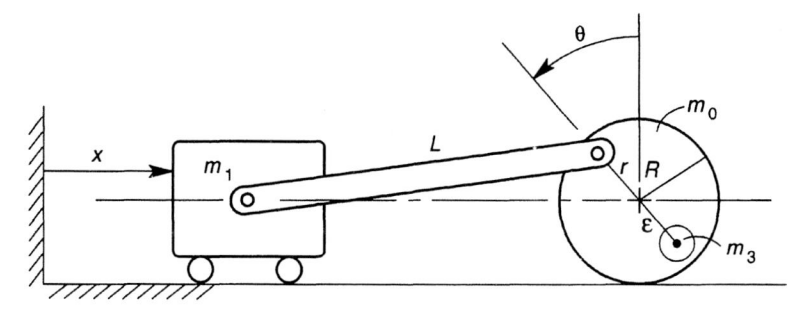

Fig. P6.11

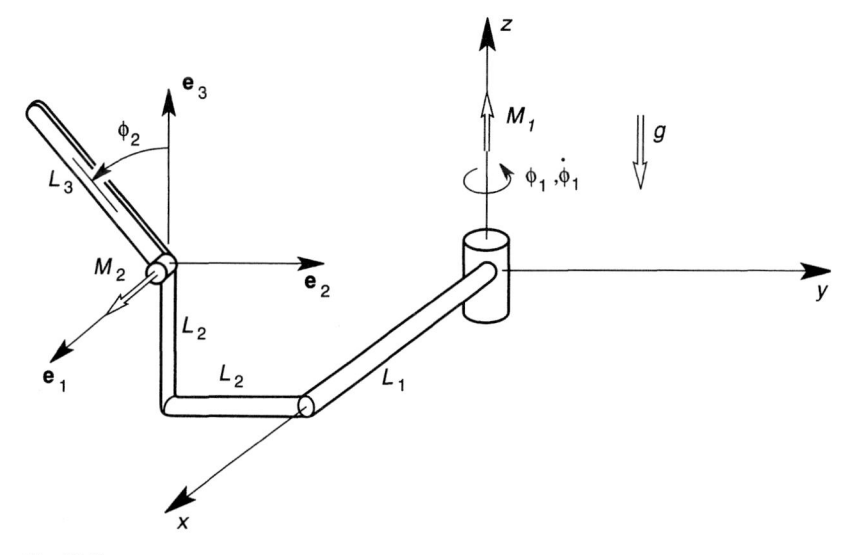

Fig. P6.13

in Fig. P6.11 ($m_2 = m_0 + m_3$). Use the principle of virtual power to determine the equation of motion in terms of the generalized variable θ. (Hint: In this three-body problem assume that the only active work-producing force is gravity force on m_3.)

6.13 Consider the two-link, 3-dimensional robot arm in Fig. P6.13. Use ϕ_1 and ϕ_2 as generalized coordinates and derive the equations for motion using the principle of virtual power. Find the mass matrix in (6.36) as well as the centripetal/Coriolis terms μ_{ijk} in (6.36). Show that the centripetal acceleration $\dot{\phi}_2^2$ produces a term in the equation for $\dot{\phi}_1$, i.e., $\mu_{122} \neq 0$. Why doesn't gravity produce a term in the $\dot{\phi}_1$ equation? Why is there a Coriolis term $\mu_{112}\dot{\phi}_2\dot{\phi}_1$? (Answer: $\mu_{122} = (m_2 L_1 L_3 S\phi_2)/2$; $\mu_{112} = m_2 L_3 [3 C\phi_2 (2L_2 + L_3 S\phi_2) + L_3 S\phi_2 C\phi_2]/6$.)

6.14 Two rods of equal mass, m, and length, L, are connected at the ends by a pin joint (Fig. P6.14). Choose the set of generalized coordinates $\{x_0, y_0, \phi, \theta\}$, and use the principle of virtual power to derive equations of motion for planar

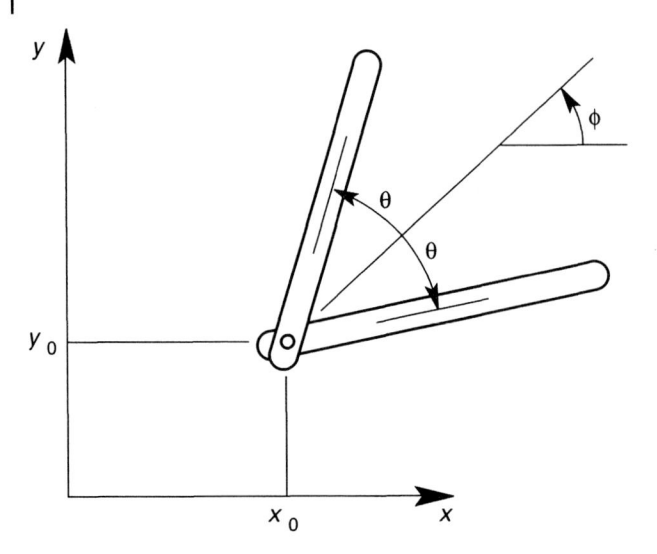

Fig. P6.14

motion of the connected rods. Assume that there are no active forces (e.g., gravity acts normal to this plane). Show that one equation is

$$\left(I_0 + \frac{mL^2}{4}\cos^2\theta\right)\ddot{\theta} - \frac{mL^2}{4}(\dot{\theta}^2 + \dot{\phi}^2)\sin\theta\cos\theta = 0$$

6.15 In the 3-dimensional robotic arm problem of Pfeiffer (1989) (Fig. 6.14), find the centripetal acceleration term μ_{322} in (6.71). Derive the rest of the μ_{ijk} terms in (6.71) using the method outlined in Example 6.6 using a symbolic programming code such as *MATHEMATICA* or *MAPLE*.

6.16 Another problem from Professor F. Pfeiffer of the Technical University of Munich is an analysis of an amusement ride called the "Wild Mouse," sometimes seen in the Oktoberfest in Munich (Fig. P6.16). A simple model consists of two connected bodies: the top carries the passengers, while the bottom follows a tortuous path under the force of gravity. In this problem we assume the path is sinusoidal and that both the tracked mass and the rotating car remain horizontal. Add a viscous drag term to the motion of the tracked mass and the rotating car, and use Lagrange's equations to derive equations of motion with generalized coordinates $\{\phi, \eta\}$. If η is the mean inclined path, then use the constraints $(z_0 - z) = \eta \sin\alpha$, $x = A\sin(2\pi\eta/\Lambda)$. Show that the Lagrangian is $\mathcal{L} = T - V$

$$T = \frac{1}{2}(m_1 + m_2)\dot{\eta}^2[1 + B^2\cos^2\kappa\eta] + \frac{1}{2}(I_c + m_2r_c^2)\dot{\phi}^2$$
$$+ m_2r_c\dot{\phi}\dot{\eta}[\cos\alpha\cos\phi - B\sin\phi\cos\kappa\eta]$$

$$V = (m_1 + m_2)gz$$

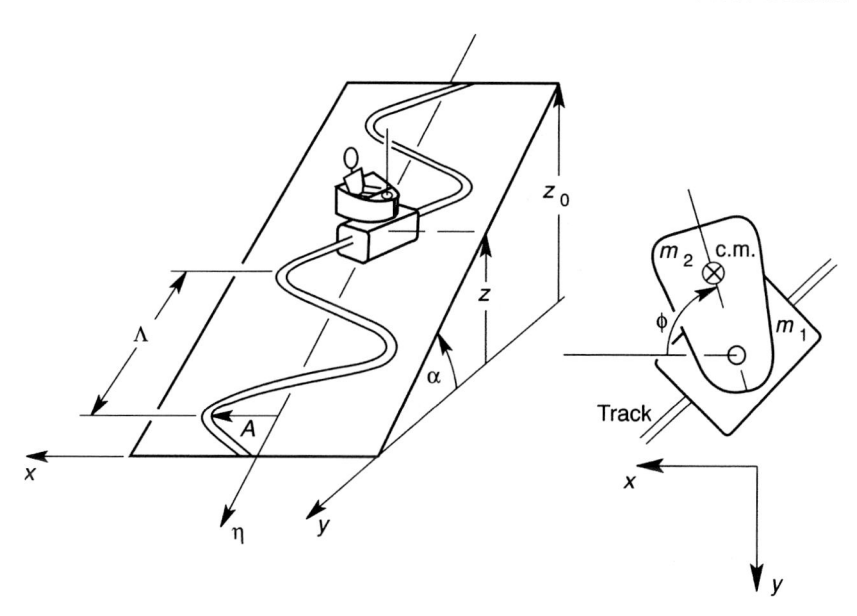

Fig. P6.16

where $B = 2\pi A/\Lambda$, $\kappa = 2\pi/\Lambda$. Add a Rayleigh dissipation function (see (4.89)), $\mathcal{R} = \frac{1}{2}c_1\dot{\phi}^2 + \frac{1}{2}c_2\dot{\eta}^2$ and derive the equation of motion.

6.17 In the Wild Mouse example of Problem 6.16, suppose the damping c_2 is large enough such that $\ddot{z} \simeq 0$, $\eta = v_0 t$. Then show that the limiting equation for the rotary motion of the car is given by

$$I_0\ddot{\phi} + c_1\dot{\phi} = -m_2 r_c B v_0^2 \kappa \sin\phi \sin\omega_0 t$$

where $\omega_0 = \kappa v_0$, $I_0 = I_c + m_2 r_c^2$. Show that this is similar to the "spinner toy" of Problem 5.23. Write a MATLAB program to integrate the equation $\ddot{\phi} + c_2\dot{\phi} + a\sin\omega_0 t \sin\phi = 0$, with $\phi(0) = 0.1$, $\dot{\phi}(0) = 0$, for $0 < t < 50$. Show that for $c_2 = 0.3$, $a = 10$, $\omega_0 = 1$, a chaotic-like bounded motion results, while $a = 1.0$, $\omega_0 = 1$, produces a modulated spin, i.e., $\dot{\phi} > 0$.

6.18 In Chapter 5, Example 5.15, we examined a single vehicle with one pair of wheels which exhibited a nonholonomic constraint. Consider the problem of two linked vehicles in Fig. P6.18, each with wheel constraints. Each wheel can rotate independently of the other, subject to the no-slip constraint parallel to the axels. Use the principle of virtual power to find the equations of motion. Neglect gravity. Use Example 5.15 as a guide.

6.19 A five-bar planar mechanism is shown in Fig. P6.19 with four movable links. What is the workspace for $0 \leqslant \phi_1 < \pi$, $0 \leqslant \phi_2 < \pi$? Find the Jacobian matrix $\partial \mathbf{r}_P/\partial \phi_i$, where $\mathbf{r}_P = [x, y]^T$ and find the singular points. Show that there are multiple solutions $\phi_i(x, y)$, given a position vector $\mathbf{r}_P$. Find the inverse relation $\phi_i(x, y)$.

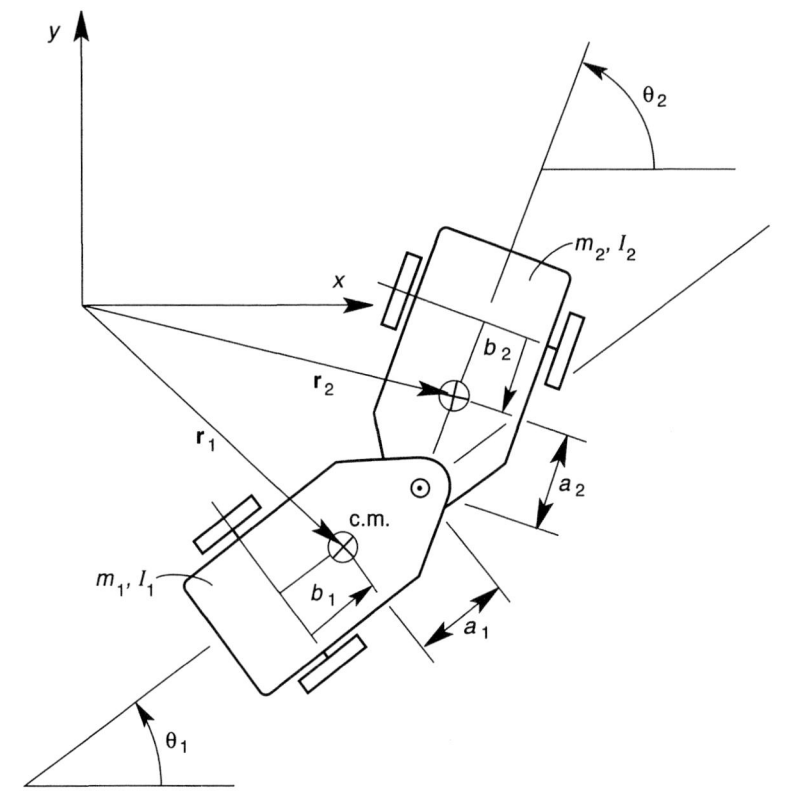

Fig. P6.18

6.20 Consider the three-link serial manipulator arm with two rotary and one prismatic joint (Fig. P6.20) (see also Asada and Slotine, 1986, p. 42). Find an expression for the Cartesian components of the position vector **r** to the point P in terms of the generalized joint coordinates $\{\phi_1, \phi_2, s\}$ and find the inverse functions $\phi_1(\mathbf{r}), \phi_2(\mathbf{r}), s(\mathbf{r})$. For what values of the generalized coordinates is the Jacobian singular?

6.21 For the normal impact of two spherical particles, solve for the post impact velocities using both Newton and Poisson impact laws (6.112) and (6.116). Show that in this special case, both laws give the same result.

6.22 In Fig. P6.22, two bodies are about to experience planar impact. The rigid body has half-cylinder ends of radius R. Use the method outlined in Example 6.9 to determine the post impact velocities of the center of masses as well as the angular velocity of the rigid body. Show that for the impact angle $\phi = 0, \omega^+ = 0$, the solution reduces to the normal impact solution between two spherical particles. [Partial answer:

$$\omega_2^+ = -\frac{V_0 L}{2} \sin\phi (1+\varepsilon) \left\{ \left(\frac{m_1 + m_2}{m_1 m_2} \right) I_2 + \frac{L^2}{4} \sin^2 \phi \right\}]$$

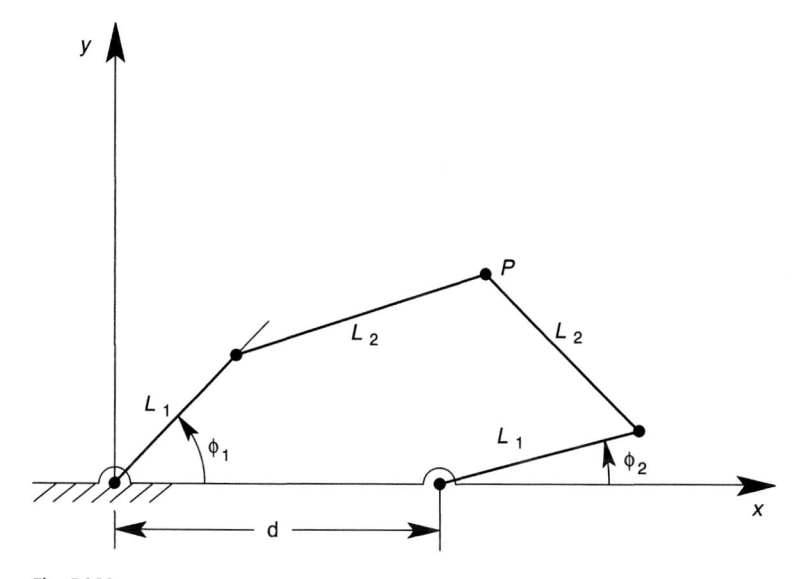

Fig. P6.19

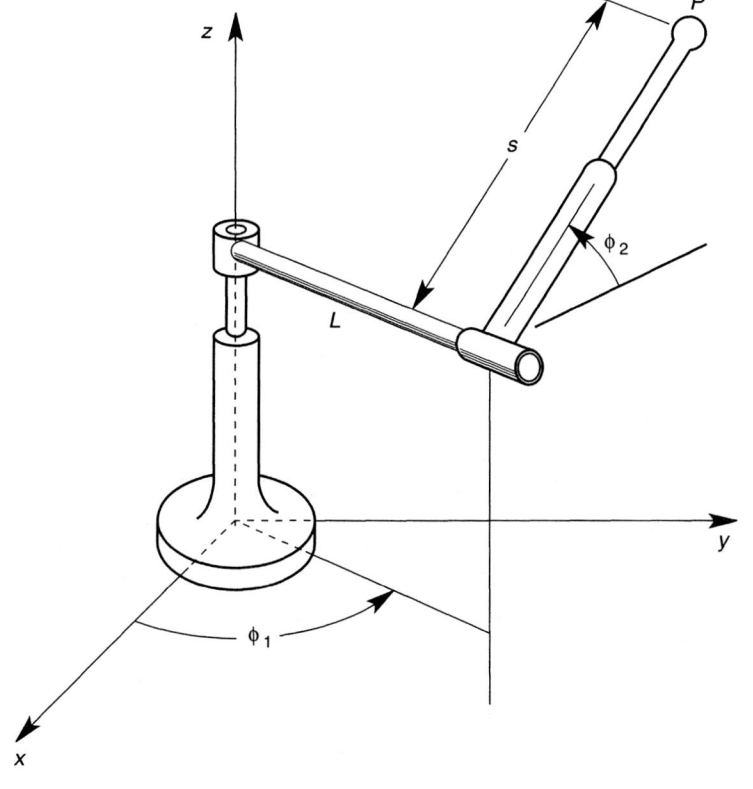

Fig. P6.20

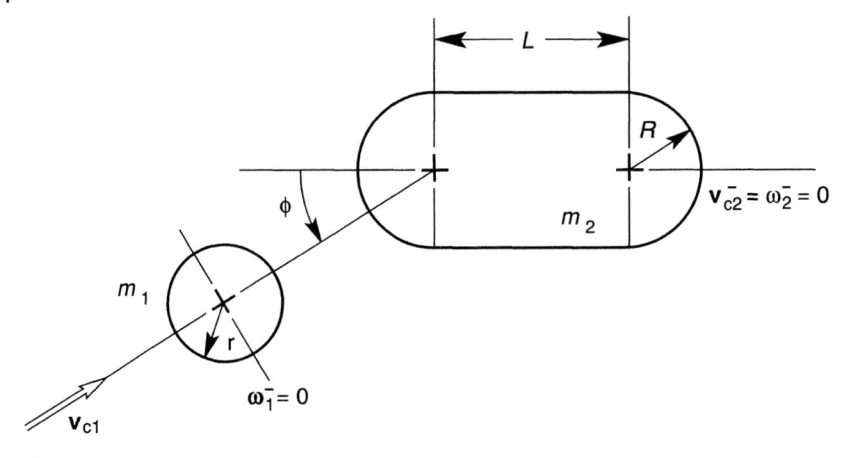

Fig. P6.22

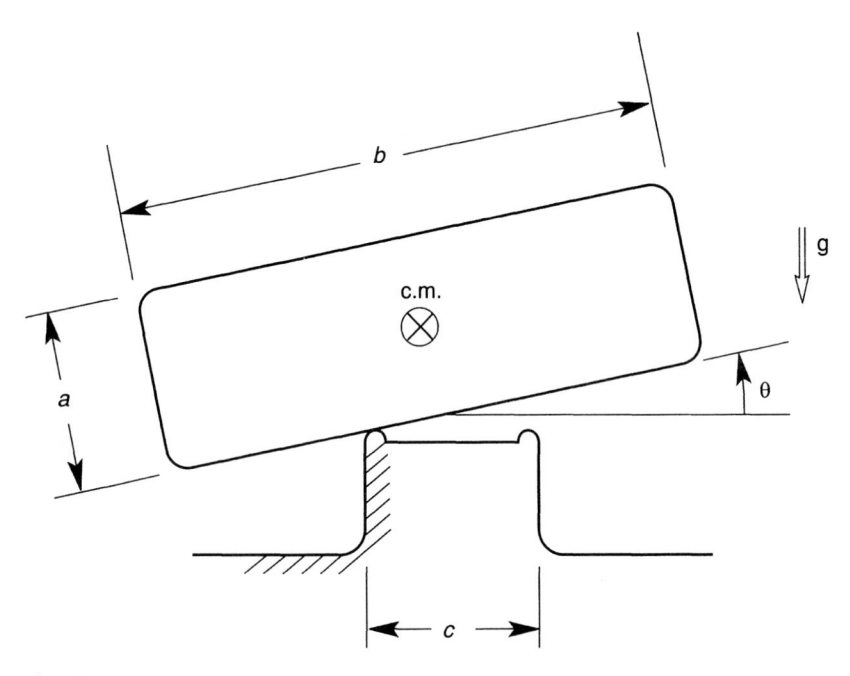

Fig. P6.23

6.23 Analyze the rocking oscillations of a rectangular block in planar impact on two stops (Fig. P6.23). Assume that at each impact the block-stop separation is zero or that the local impact is plastic. Find the time between impacts when the angle θ is small. Find the maximum angle after each impact for several cycles. Start with $\theta = \theta_0$, $\dot{\theta} = 0$ for the first cycle. Does the block come to rest in a finite number of cycles?

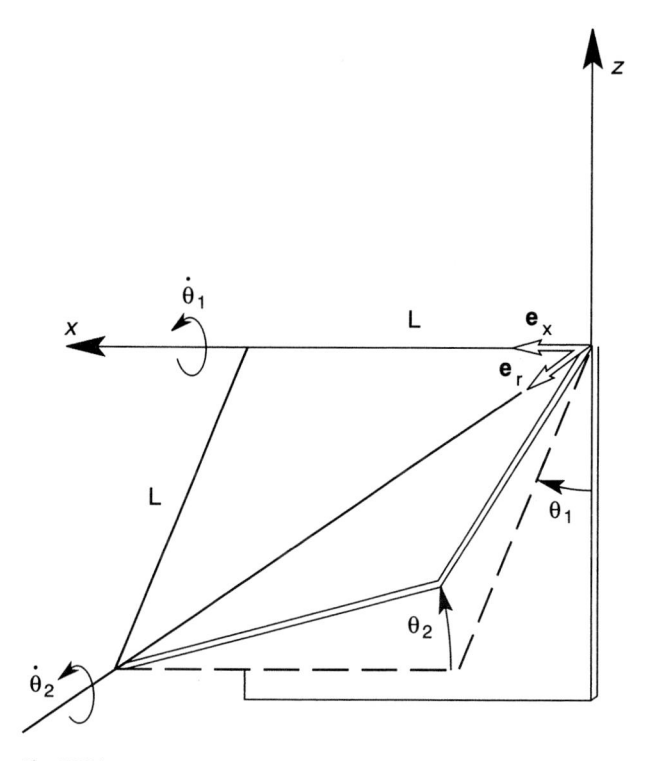

Fig. P6.24

6.24 Two triangular plates form a square of sides L when aligned (Fig. P6.24). The top plate is hinged on a revolute joint along the horizontal axis, while the bottom plate is hinged along the diagonal. Use either Lagrange's equation or the principle of virtual power to find the equations of motion under gravity. Use the two angles θ_1, θ_2 as shown as generalized coordinates. For $\theta_1 \theta_2$ small find the two linearized natural frequencies. [*Hint:* Use the fact that both plates are in pure rotation about the origin and $\boldsymbol{\omega}_1 = \dot{\theta}_1 \mathbf{e}_x, \boldsymbol{\omega}_2 = \boldsymbol{\omega}_1 + \dot{\theta}_1 \mathbf{e}_r.$]

6.25 A two link mechanism, similar to the double pendulum in Example 6.5, has an additional prismatic constraint at the end of the second link as shown in Fig. P6.25. Use the principle of virtual power to derive an equation of motion. Write expressions for the forces in the two revolute pin joints and for the contact constraint force in the slot. Write a *MATLAB* program to find the forces in the joints as functions of time when the mechanism is released from rest with initial condition $\theta = \pi/4$.

6.26 Consider the two link mechanism in Example 6.5 under gravity as shown in Fig. P6.26. Suppose a nonholonomic constraint is placed at the end of the second link in the form of a skate or rolling constraint such that the velocity at the end, $\mathbf{v}_2$, must be in the radial $\mathbf{e}_{r2}$ direction, i.e., $\mathbf{v}_2 \cdot \mathbf{e}_{\theta 2} = 0$. Show that the kinematic constraint is given by, $L_1 \dot{\theta}_1 \cos \theta_2 + L_2(\dot{\theta}_1 + \dot{\theta}_2) = 0$. Derive the equation of motion. [See also Lesser (1995) for a similar problem.]

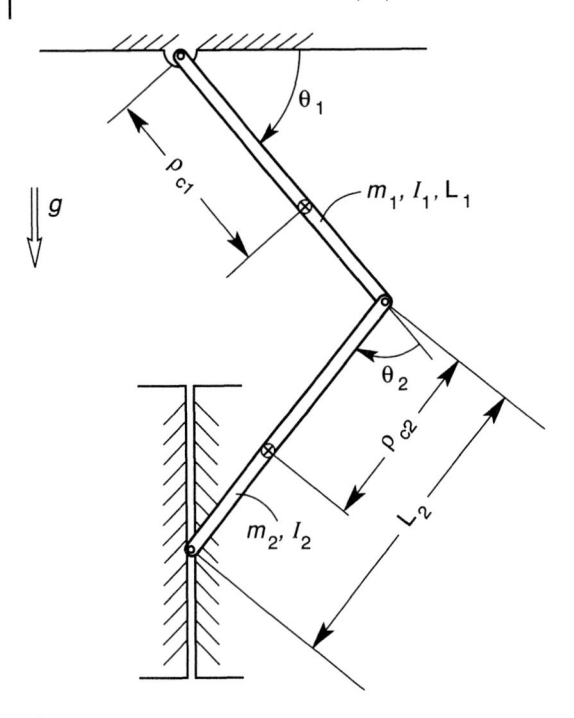

Fig. P6.25

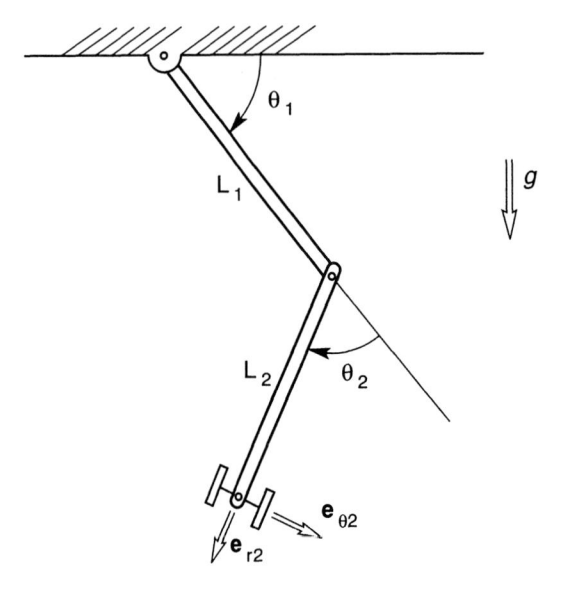

Fig. P6.26

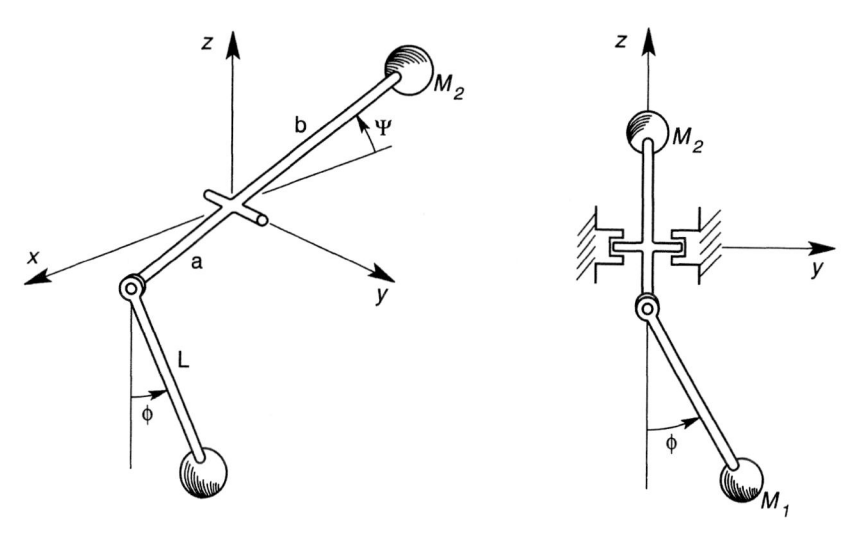

Fig. P6.28

6.27 Consider the equilibrium problem of the two-link arm in Example 6.5, $\tau_1 + G_1 = 0$, $\tau_2 + G_2 = 0$. (a) Given two fixed applied moments at the joints (usually through servo motors), find expressions for the equilibrium angles θ_1, θ_2. (b) Write a *MATLAB* program to look at the dynamics when constant moments M_1^a, M_2^a are suddenly applied. Add a torsional damper to each joint. Choose M_1^a, M_2^a such that the equilibrium or final state is $\theta_1(t \to \infty) = \pi$, $\theta_2(t \to \infty) = 0$, with initial conditions $\theta_1(0) = \pi/2, \theta_2(0) = 0$.

6.28 Leibniz (1646–1716) was a codiscoverer of the calculus along with Newton. His name is also associated with one of the first multibody dynamics problems shown in Fig. P6.28. According to Szabó (1987, Appendix), this problem was posed by Mariotte (1620–1684) and a solution was discussed by Leibniz. The lower pendulum is assumed to remain vertical but can rotate about point A normal to the plane. The equation of motion can be derived by Lagrange's equations with $\{\phi, \psi\}$ as generalized coordinates. Show that the equation takes the form

$$\begin{bmatrix} m_{11} & -(\beta \sin\phi \cos\psi) \\ -(\beta \sin\phi \cos\psi) & m_{22} \end{bmatrix} \begin{bmatrix} \ddot{\phi} \\ \ddot{\psi} \end{bmatrix}$$
$$+ \begin{bmatrix} \mu_{122}\dot{\psi}^2(\sin\phi \sin\psi) \\ \mu_{211}\dot{\phi}^2(\cos\phi \cos\psi) \end{bmatrix} + \begin{bmatrix} r_c g \sin\phi \\ 0 \end{bmatrix} = 0$$

Find $m_{11}, m_{22}, \beta, \mu_{122}, \mu_{211}, r_c$.

6.29 An elliptically shaped plate is dropped from a height h_0 at an angle α with zero angular momentum. Given the Newton impact coefficient of restitution of ε, with no friction, find the resulting velocity of the center of mass and post impact angular velocity.

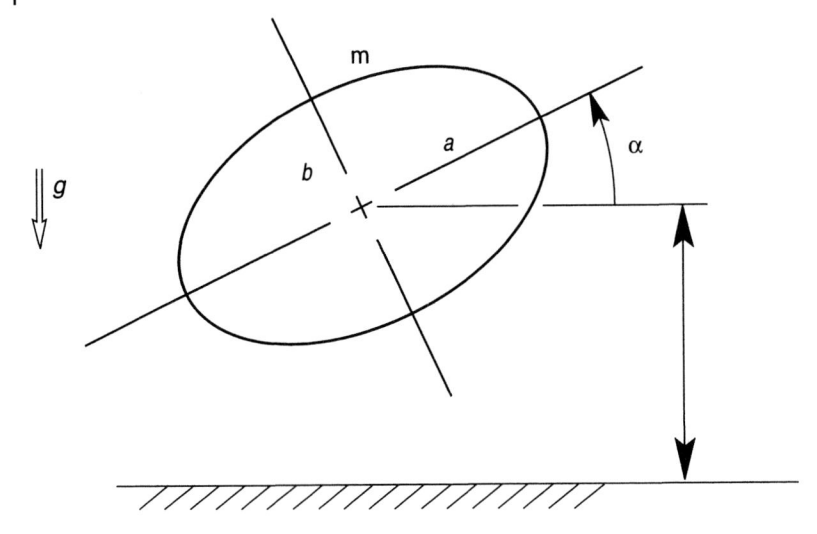

Fig. P6.29

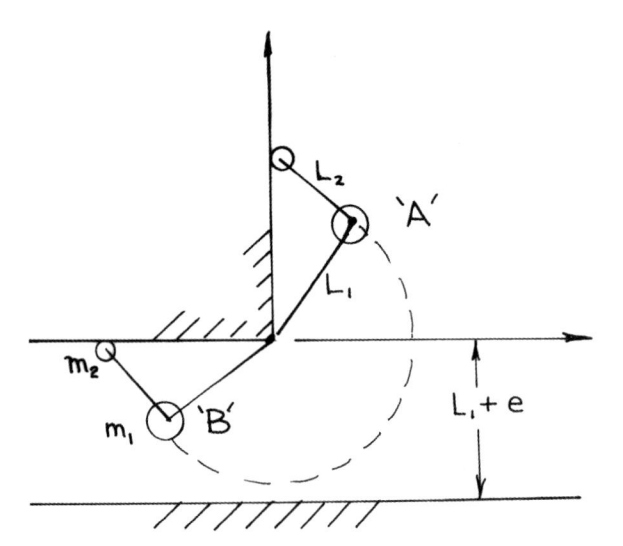

Fig. P6.31

6.30 Suppose the body in Fig. P6.29 is an oblate spheroid of radius a and the figure shows the cross section through the diameter of the perimeter circle. Suppose the material is steel and the half space is granite. Find the time of impact duration using the Hertz impact model, for $a = 4$ cm, $b = 2$ cm, $h_0 = 10$ cm, $\alpha = 0$.

6.31 (*Pick 'n Place Controller Design*) Consider the two-link serial link arm in Fig. P6.31. Assume the arm works in the plane. Design a PD controller to move the arm from position A to position B. Use the *Design Rules* in Section 6.6 for critical damping and resonance avoidance, to choose the control

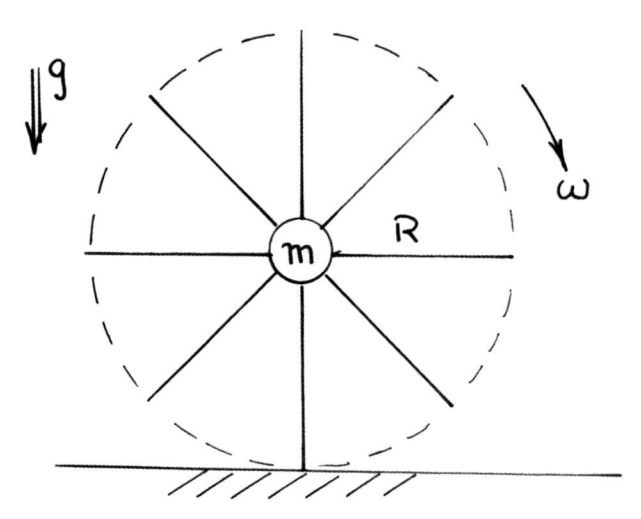

Fig. P6.33

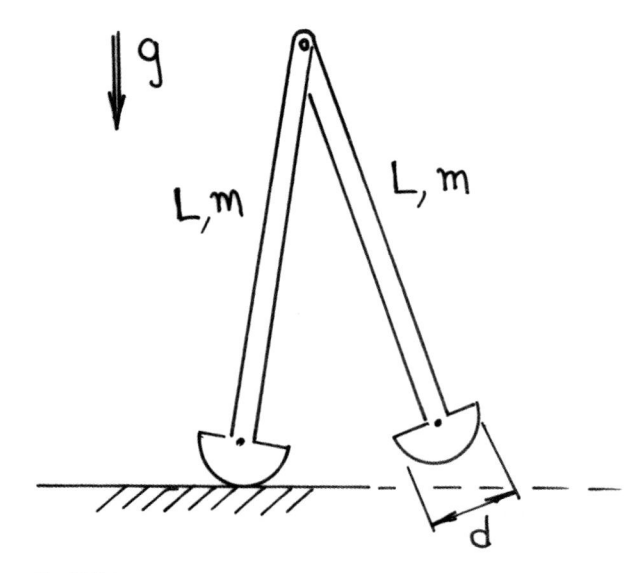

Fig. P6.34

gains for each joint actuator. Use MATLAB to verify that the arm moves to the desired position. Adjust the gains to minimize overshoot.

6.32 (*Reflected Inertia*) Consider the single degree of freedom link under control of a torque actuator motor with a two-stage speed reducer, with step down speed ratios n_1, n_2. The intermediate gear has an inertia I_G, and the pinion inertia is I_P. Use Lagrange's equations to find the equivalent inertia including the intermediate gear inertias when the dynamics are written in terms of the joint variable $\theta(t)$.

6.33 (*Impact Dynamics of a "Walking" Machine*) In the last several decades there has been interest in the dynamics of machines that walk without feedback or with minimal control. A discussion of research on walking machines may be found in the papers of McGeer (1990) and Collins et al. (2005). One of the simplest examples is the rimless wheel, discussed by McGeer (1990). In Fig. P6.33 is a sketch of an eight-spoke rimless wheel that "walks" by impacting each successive radial link with the ground. Assume an inelastic impact and utilize conservation of angular momentum to calculate the change in angular velocity before and after each impact.

6.34 (*Walking Machine Dynamics*) The two-link model of a passive walking machine is shown in Fig. P6.34. The "foot" is modeled by a circular rim that rolls without slipping. Model the legs as two uniform mass links of mass m. Use Lagrange's equation to derive the equations of motion between foot strike times. How is this problem similar to the double pendulum with inverted gravity forces? (Check the website of Prof. A. Ruina, Cornell University for references to this problem as well as videos of walking machines.)

7
Orbital and Satellite Dynamics

7.1
Introduction

In the age of the communication revolution and the information superhighway, it is often forgotten that satellite technology plays an essential role in this revolution. In the design of satellite systems (Fig. 7.1), the principles of dynamics are important in three phases; earth-to-orbit launch, orbit-to-orbit transfer, and attitude stability. In this chapter we present only an introduction to these problems. The reader who wants a more advanced discussion of these topics should consult such monographs as Wiesel (1989), Kane et al. (1983), and Rimrott (1988). In addition, there are many books on the physics and dynamics of the solar system that treat the dynamics of planets, moons, comets, and other natural objects in our solar system. (See, e.g., Burns and Mathews, 1986.)

Modern dynamics of orbiting bodies in our solar system began with the work of Nicolaus Copernicus (1473–1543), Galileo Galilei (1564–1642) and Johannes Kepler (1571–1630). Kepler succeeded the Dane, Tycho Brahe (1546–1601), as court mathematician in Prague in 1601. Kepler was fortunate to have inherited Tycho's astronautical observations, and in 1609 he published two laws of orbital motion: the first that the planets had elliptical paths around the sun, and the second that the radius vector swept out equal areas around the ellipse. Later in 1619 he added a third law relating the period of the orbits to the major diameter of the ellipse. As we all know, Sir Isaac Newton (1642–1727) proposed his famous laws of motion as well as the inverse square law of gravitational force with which he was able to derive Kepler's three laws. This legacy is the foundation of modern applied orbital dynamics.

As already discussed in the introduction to Chapter 1, the modern science of gravity was changed dramatically by the ideas of Albert Einstein (1879–1955), who proposed in his general theory of relativity (1905–1916) that the so-called "force of gravity" could be replaced by a curvature of space–time induced by gravitational masses. It is interesting to note that Einstein developed part of this theory in Prague, 1911–1912, where Kepler and Tycho Brahe had worked 300 years earlier. Although this radical revision of the theory of gravity changed physics, it has not

Applied Dynamics. With Applications to Multibody and Mechatronic Systems. Second Edition. Francis C. Moon
Copyright © 2008 WILEY-VCH Verlag GmbH & Co. KGaA, Weinheim
ISBN: 978-3-527-40751-4

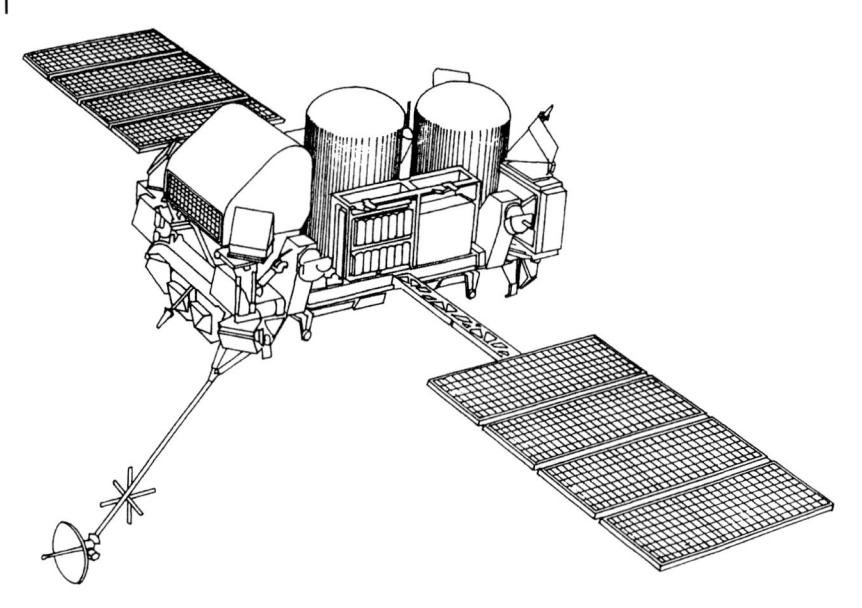

Fig. 7.1 Sketch of satellite with solar panels and antenna. (From Caprara, 1986, p. 173.)

altered significantly the methods of calculation of orbital dynamics based on the Newtonian concept of force. The modern theory of gravity, however, did replace the "action at a distance" nature of Newtonian theory. Modern theories presume that gravitational "forces" propagate at the speed of light, though to date, gravitational waves have yet to be measured in our solar system. The interested reader should consult the text *Gravitation*, by Misner, Thorne, and Wheeler (1973), or Zee (1989).

However, the equivalence of inertia and gravitational masses, originally proposed by Newton following the observation of Galileo, were confirmed in experiments by the Hungarian Baron Roland Eötvös in 1922 to five parts in a billion. This equivalence has formed one of the cornerstones of modern gravitational theory (see, e.g., Lederman, 1993, pp. 94–95).

In this chapter we review the Newtonian theory of orbital mechanics. The Lagrangian formulation is employed to develop the idea of an *equivalent potential energy*, using the ignorable angular variable. Orbit transfer is discussed next. Finally, rigid-body motions of satellites and their attitude or pointing stability will be introduced, as well as tethered satellite dynamics and control moment gyros.

7.2
Central-force Dynamics

We consider the dynamics of a mass under a vector force always directed toward a fixed center (Fig. 7.2). This assumption approximates the case of a small mass near

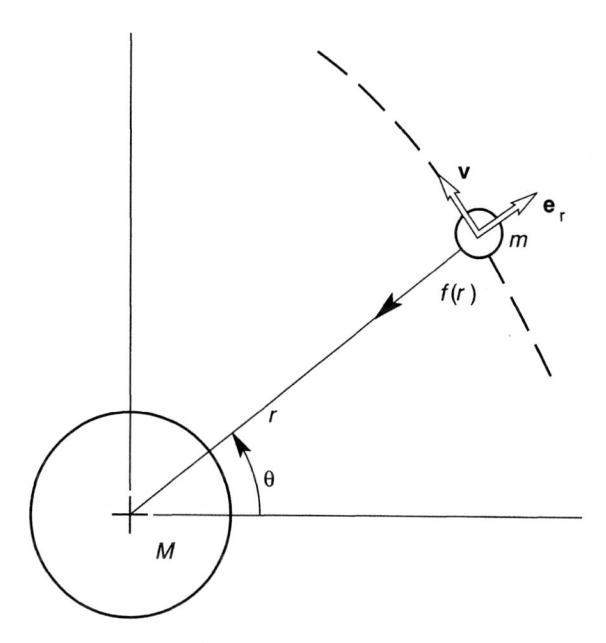

Fig. 7.2 Mass particle motion under a central-force attractor.

a large spherically symmetric gravitational mass. We further assume that the force **F** can be written in the form

$$\mathbf{F} = -f(r)\mathbf{e}_r \tag{7.1}$$

where the position vector is given by $\mathbf{r} = r\mathbf{e}_r$. Newton's law for this problem is given by

$$m\dot{\mathbf{v}} = -f(r)\mathbf{e}_r \tag{7.2}$$

Taking the moment ($\mathbf{r} \times [\ ]$) of both sides of this equation, we arrive at the result that the angular momentum **H** about the force center is conserved, i.e.,

$$\mathbf{r} \times \frac{d}{dt} m\mathbf{v} = \mathbf{r} \times f(r)\mathbf{e}_r = 0$$
$$\mathbf{H} = \mathbf{r} \times m\mathbf{v} = \text{constant} \tag{7.3}$$

The consequence of this result is that the vector direction of **H**, as well as its scalar magnitude, is fixed. Using cylindrical coordinates with the $\mathbf{e}_z$ axis aligned with **H**, this result states that

$$H_z = mrv_\theta = \text{constant} \tag{7.4}$$

We see that the motion of a mass under a central force is planar. The relation in (7.4), however, does not prescribe $v_r = \dot{r}$. To determine $\dot{r}$, we must make a further assumption about the central-force law $f(r)$. Central forces in nature occur in gravitational problems as well as in the motion of electrical charges [see, e.g., (4.80b)]. Electrical forces on charged particles may be repulsive or attractive, depending on the sign of the change. In the remainder of this chapter we focus on gravitational-force problems, which in Newton's theory are always attractive, and follow an inverse-square law, i.e.,

$$f(r) = \frac{GMm}{r^2} \qquad (7.5)$$

where G is the gravitational constant, and M is the mass of the attracting fixed-mass center.

The key to deriving Kepler's law of elliptic orbits lies in Newton's assumption of an inverse-square force law for gravitation between bodies. In the case of a small mass orbiting a large mass, we can use the central-force model. Also we can use the fact that the motion is planar and write Newton's law in polar coordinates (see (2.8)), where the position vector of the orbiting mass is given by $\mathbf{r} = r\mathbf{e}_r$, as shown in Fig. 7.2:

$$m\ddot{\mathbf{r}} = -\frac{mGM}{r^2}\mathbf{e}_r$$

or

$$\ddot{r} - r\dot{\theta}^2 = -\frac{GM}{r^2} \qquad (7.6a)$$

$$r\ddot{\theta} + 2\dot{r}\dot{\theta} = 0 \qquad (7.6b)$$

The second equation is just the conservation of angular momentum, which we can integrate to obtain the scalar form of (7.3),

$$r^2\dot{\theta} = h_0 \qquad (7.7)$$

It is easy to show that $h_0/2$ is the area swept out per unit time. Thus, (7.7) states that the area swept out by the radius is a constant, which is another of Kepler's laws. The constant is the angular momentum per unit mass of the orbiter. Substituting this expression into (7.6a), we obtain

$$\ddot{r} - \frac{h_0^2}{r^3} + \frac{GM}{r^2} = 0 \qquad (7.8)$$

This is a nonlinear, second-order differential equation. However, there is a dependent variable transformation that will yield a linear differential equation, namely, $u = 1/r$. In order to derive the equation of an ellipse and other conic sections, we also have to change the independent variable from time to the angle θ, i.e.,

$$\dot{r} = \frac{dr}{d\theta}\frac{d\theta}{dt} = -u'(r^2\dot{\theta})$$

or

$$\frac{d^2 r}{dt^2} = -h_0^2 u^2 \frac{d^2 u}{d\theta^2} \tag{7.9}$$

Substituting this expression into (7.8), the radial component of Newton's law becomes

$$u'' + u = \frac{GM}{h_0^2} \tag{7.10}$$

where the dependent variable is now the angle θ. This equation has trigonometric solutions that can be written

$$u(\theta) = \frac{1}{r} = \frac{GM}{h_0^2} + A\cos(\theta + \theta_0) \tag{7.11}$$

The constants A, θ_0 are determined by the initial radius and angle. This equation describes curves that belong to the family of *conic sections*; circles, ellipses, parabolas, and hyperbolas. These curves can be obtained from the intersection of a plane and a cone, as shown in Fig. 7.3.

Circular Orbits

When the orbit is a circle, a much more direct derivation of orbit parameters can be obtained using high school physics. In circular motion, the radial acceleration, $r\dot\theta^2 = v^2/r$, must balance the gravitational force per unit mass, i.e.,

$$m\frac{v^2}{r} = \frac{GMm}{r^2}$$

or

$$v = (GM/r)^{1/2} \tag{7.12}$$

Thus the orbiting velocity decreases as the diameter of the circle increases. We can also determine the period of the motion T, from the law of angular momentum. In one orbit, the area swept out is $A = \pi r^2$, so that

$$\frac{h_0}{2} = \frac{rv}{2} = \frac{A}{T}$$

or using (7.12)

$$T = \frac{2\pi r}{v} = 2\pi (r^3/GM)^{1/2} \tag{7.13}$$

Thus T^2 is proportional to r^3, which is Kepler's third law of orbital motion of the planets.

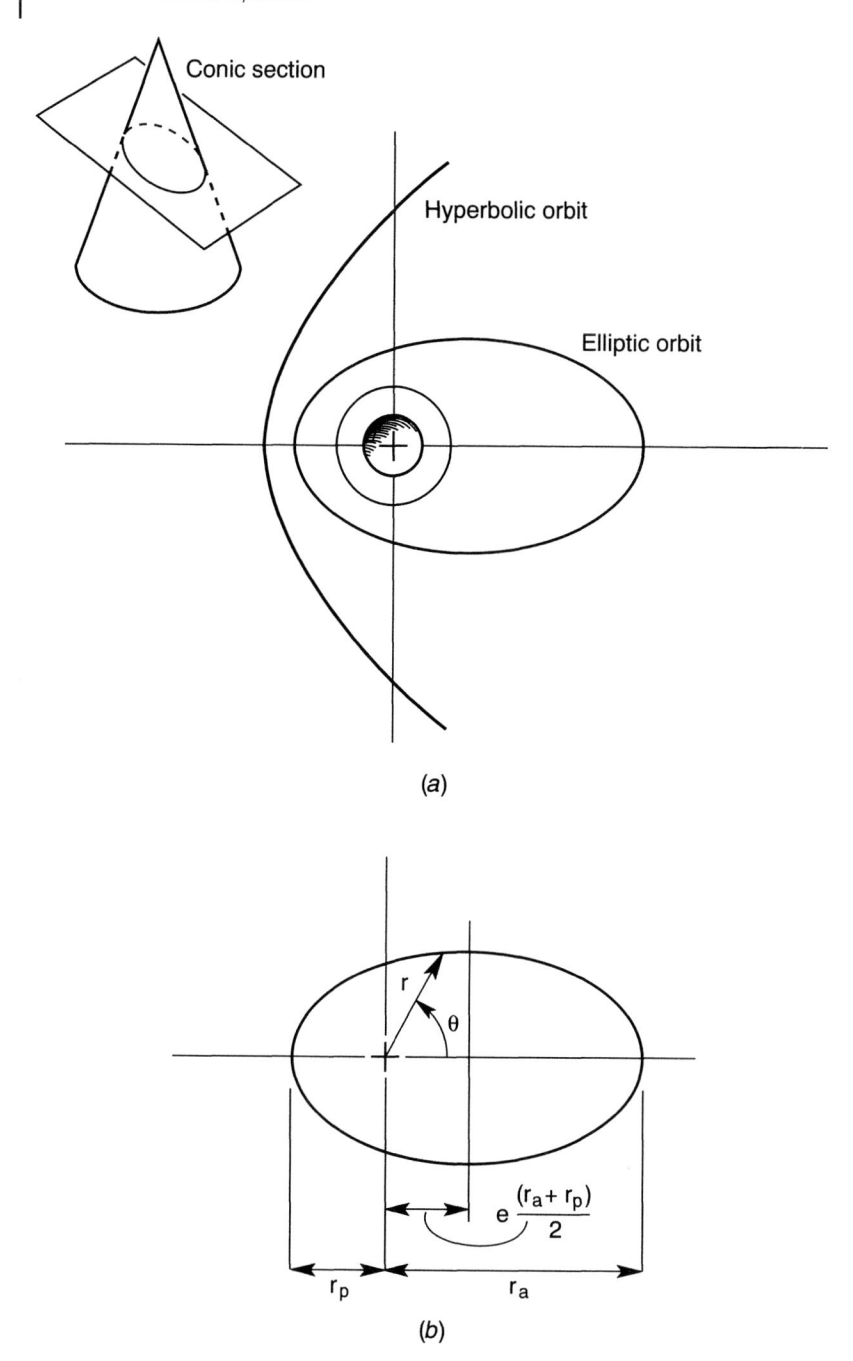

Fig. 7.3 (a) Conic sections, circular, elliptic, hyperbolic orbits under gravitational inverse-square law. (b) Geometric parameters of an elliptic orbit about a fixed gravitational attractor.

Example 7.1. For the case of a satellite in a circular orbit about the earth we wish to find the orbit velocity and period. To remember the gravitational constant GM, a clever trick is to set the gravitational force equal to its value on the surface of the Earth, i.e.,

$$\frac{GM}{r_e^2} m = mg$$

or

$$GM = gr_e^2 \tag{7.14}$$

For $r_e = 6.38 \cdot 10^6$ m, $GM = 3.99 \cdot 10^{14}$ m^3/s^2. Thus the velocity for a circular orbit using (7.12) is given by

$$v = (gr_e^2/r)^{1/2} \tag{7.15}$$

Close to the earth's surface $r = r_e$, and

$$v = (gr_e)^{1/2} = 7906 \text{ m/s} = 28{,}460 \text{ km/h } (17{,}790 \text{ mph})$$

The period of an orbit at the earth's surface is $T = 84$ min. An interesting exercise is to find the orbit radius for a geosynchronous orbit where $T = 24$ h. Thus, we solve for

$$r_{\text{geo}} = \left(\frac{GMT^2}{4\pi^2}\right)^{1/3}$$
$$= 42{,}260 \text{ km } (26{,}410 \text{ mi}) \tag{7.16}$$

Of course r_{geo} is measured from the earth's center and not the surface.

Elliptical Orbits

The general bounded solution of the orbital equation of motion (7.10) in polar coordinates (7.11) has the form of the equation for an ellipse (Fig. 7.3). One of the two foci of the ellipse is centered at the gravity center. The ellipse is defined by the constant ratio of the radial distance from the focus divided by the perpendicular distance PD from a fixed plane called a *directrix*; i.e. (see Homework Problems 7.1, 7.2, Fig. P7.1)

$$r/PD = \varepsilon$$

where

$\varepsilon = 0,$ for a circular orbit
$\varepsilon < 1,$ for an ellipse
$\varepsilon = 1,$ for a parabola
$\varepsilon > 1,$ for a hyperbola

The constant ε is called the *eccentricity* of the orbit. The eccentricity is related to the semimajor and minor axes of the ellipse a, b by the expressions

$$r_a = a(1 + \varepsilon)$$
$$r_p = a(1 - \varepsilon)$$
$$b = a\sqrt{(1 - \varepsilon^2)} \tag{7.17}$$

Here r_a is the *apogee* or the greatest distance from the focus of the elliptic orbit and r_p is called the *perigee*, or closest distance from the focus of the orbit.

Effective Potential of Orbiting Masses

The orbit problem can be recast in Lagrangian dynamics by expressing the gravitational force in terms of a potential function, i.e.,

$$\mathbf{F} = -\frac{mGM}{r^2} = -\frac{\partial V(r)}{\partial r}$$
$$V_{(r)} = -\frac{mGM}{r} \tag{7.18}$$

The Lagrangian then becomes

$$\mathcal{L} = T - V(r)$$
$$T = \frac{1}{2}m(\dot{r}^2 + r^2\dot{\theta}^2) \tag{7.19}$$

Before deriving the equations of motion, we can first assert that in the absence of any dissipation, the total energy and angular momentum are conserved, i.e.,

$$E = T + V \equiv me_0 = \text{constant}$$
$$2e_0 = \dot{r}^2 + r^2\dot{\theta}^2 - \frac{2GM}{r}$$
$$h_0 = r^2\dot{\theta} \tag{7.20}$$

or

$$2e_0 = \dot{r}^2 + \frac{h_0^2}{r^2} - \frac{2GM}{r} \tag{7.21}$$

The two terms, $V(r) + mh_0^2/2r^2$ can be defined as an effective potential $V^*(r)$. The effective potential is plotted in Fig. 7.4. The energy per unit mass e_0 is related to the value of r at the perigee r_p, where $\dot{r}_p = 0$:

$$2e_0 = \frac{h_0^2}{r_p^2} - \frac{2GM}{r_p} \tag{7.22}$$

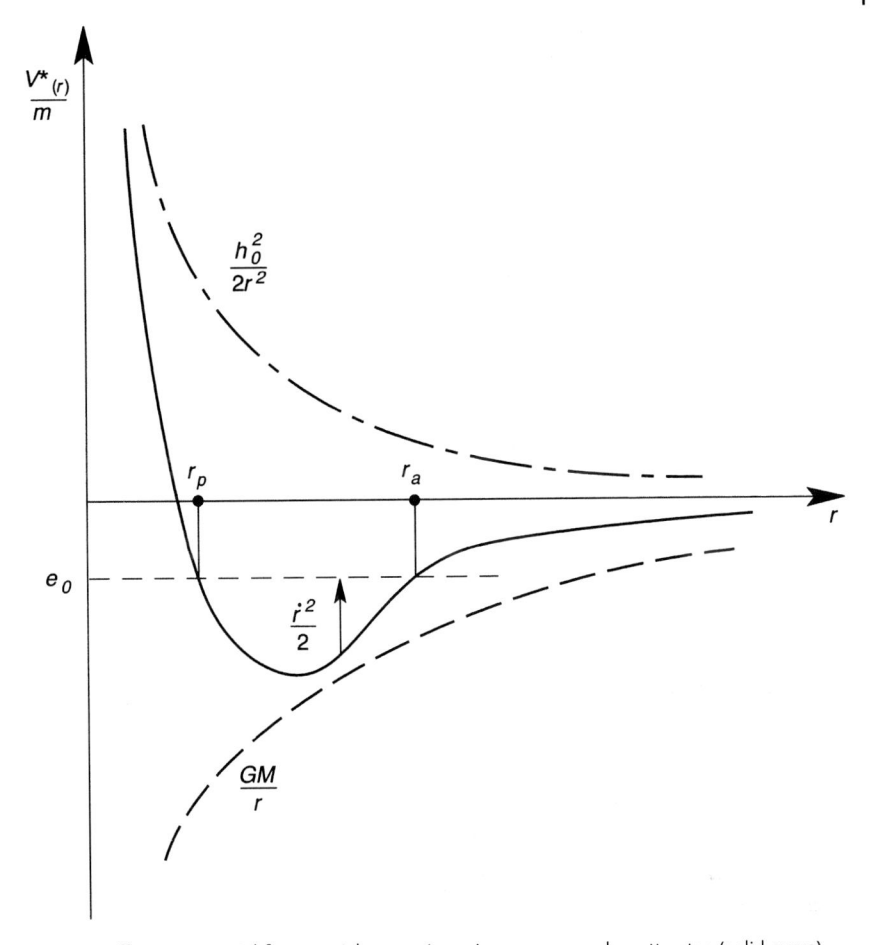

Fig. 7.4 Effective potential for a particle mass in an inverse-square law attractor (solid curve).

Lagrange's equations have the form

$$\frac{d}{dt}\frac{\partial T(q_k,\dot{q}_k)}{\partial \dot{q}_i} - \frac{\partial T}{\partial q_i} + \frac{\partial V(q_k)}{\partial q_i} = 0 \tag{7.23}$$

In the case of a single mass orbiting a fixed gravity center we have $q_1 = r$, $q_2 = \theta$. The two equations of motion, so derived are identical to the r, θ components of Newton's equations (7.6). From the conservation of angular momentum (7.7), we obtain a relationship between $\dot{\theta}, r$.

In general, when one has an ignorable coordinate, (i.e., $\partial \mathcal{L}/\partial \theta = 0$), one can define a new Lagrangian in the reduced coordinates (see Goldstein, 1980, p. 352). This new Lagrangian, $\mathcal{L}^*$, called a *Routhian*, now satisfies the equation

$$\frac{d}{dt}\frac{\partial \mathcal{L}^*}{\partial \dot{r}} - \frac{\partial \mathcal{L}^*}{\partial r} = 0 \tag{7.24}$$

where

$$\mathcal{L}^* = \mathcal{L} - p_\theta \dot{\theta}, \qquad p_\theta = \partial \mathcal{L}/\partial \dot{\theta} \qquad (7.25)$$

and

$$V^*(r) = m\left(\frac{h_0^2}{2r^2} - \frac{GM}{r}\right), \qquad T^* = \frac{m}{2}\dot{r}^2$$

Thus by using the conservation of angular momentum, we have reduced the problem to a single degree of freedom, one with an effective potential energy $V^*(r)$. This potential is shown in Fig. 7.4. The potential well results from the competing $1/r$ and $1/r^2$ singularities in V^*. The latter dominates near $r = 0$, while the gravitational potential term dominates as $r \to \infty$. The total energy per unit mass, e_0, satisfies the equation

$$m\dot{r}^2 = 2(me_0 - V^*) \geqslant 0 \qquad (7.26)$$

The effective potential energy is always less than me_0, so the orbit is confined to the shaded area in Fig. 7.4. This implies the following classification:

$e_0 < 0;$ bounded motion (elliptic orbits)

$e_0 = 0;$ minimum energy for escape (parabolic orbit)

$e_0 > 0;$ unbounded motion (hyperbolic orbit)

Example 7.2 Escape Velocity. The condition for escape from a bounded orbit around the earth is, $e_0 = 0$, according to Fig. 7.4. If we launch a satellite at an altitude A above the surface of the earth, with $\dot{r} = 0$, one can use (7.21), along with $v_{esc} = h_0/r$ to find the escape velocity, or

$$v_{esc}^2 = \frac{2GM}{R + A}$$

or

$$v_{esc} = \sqrt{2} v_{cir} \qquad (7.27)$$

Thus the difference between a tangential velocity for escape and that for a circular orbit is a factor of 1.41 or a 41% increase in velocity. The value of the escape velocity for earth when $A \ll R$ is found using the expression for GM:

$$GM = gR^2$$
$$v_{esc} = (2gR)^{1/2} \qquad (7.28)$$

This value decreases as the launch altitude A is increased relative to R. The escape velocities for a number of planetary bodies are given in Table 7.1.

Table 7.1 Planetary parameters.[a]

	Mean orbit radius (AV)	Mass (10^{24} kg)	Escape velocity (km/s)	Relative surface gravity	Rotation period (h, days)
Mercury	0.387	0.3303	4.25	0.38	58.65 d
Venus	0.723	4.870	10.4	0.88	243.01 d
Earth	1.000	5.976	11.2	1.0	23,9345 h
Mars	1.524	0.6421	5.02	0.38	24.6299 h
Jupiter	5.203	1900	59.6	2.65	9.841 h (equator)
Saturn	9.539	568.8	35.5	1.17	10.233 h (equator)
Uranus	19.182	86.87	21.3	1.05	17.24 h
Neptune	30.058	102.0	23.3	1.23	16.10 h
Pluto	39.44	0.013(?)	1.1(?)	0.16(?)	6.387 d

[a] Sources: J. A. Burns and M. S. Matthews, Eds. (1986), *Satellites*, Univ. Arizona Press, Tucson; *Encyclopedia Britannica*, 1964 ed.

Table 7.2 Solar system satellite parameters.

	Orbital semimajor axis (10^6 m)	Orbital period (days)
Moon (Earth)	384.4	27.3217
Phobos (Mars)	9.378	0.319
Deimos (Mars)	23.459	1.263
Metis (Jupiter)	127.96	0.2948
Ganymede (Jupiter)	1070	7.155
Atlas (Saturn)	137.64	0.602
Hyperion (Saturn)	1481.1	21.277
Triton (Neptune)	354.3	5.877 (Retrograde)

One can also show that the escape velocity for a vertical launch, i.e., $v_\theta = 0$, is given by the same value as for a tangential launch:

$$(v_r)_{esc} = \left(2\frac{GM}{r}\right)^{1/2} \tag{7.29}$$

Orbit-to-Orbit Transfer

We have seen that a number of orbit problems can be solved using the two principles of conservation of energy and angular momentum. Another important problem concerns the transfer between two circular coplanar orbits, as shown in Fig. 7.5. The problem was solved by Walter Hohmann in 1925 (see, e.g., Wiesel, 1989). The key idea to transfer from either the circular orbit to the ellipse or vice versa is to use an impulsive burn at the perigee and apogee points of the ellipse.

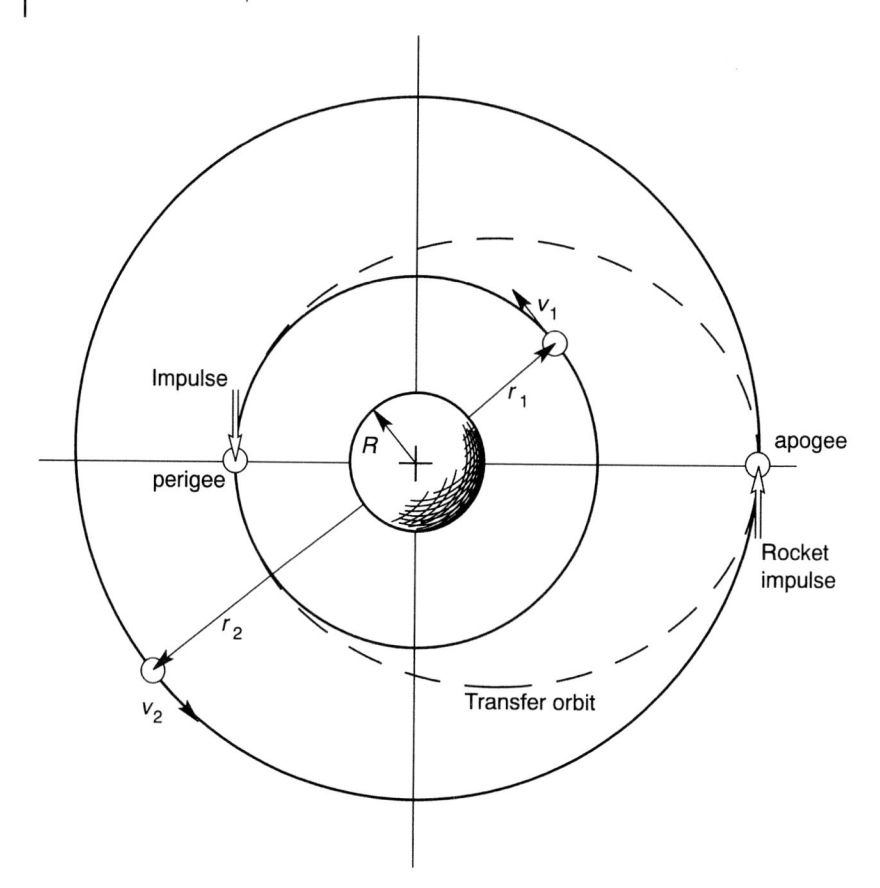

Fig. 7.5 Hohmann orbit-transfer geometry (Example 7.3).

The thrust capabilities of the satellite rockets are assumed to be large enough so that the burn time is very small compared to the orbit time $\sim 10^2$ min.

To describe the problem in mathematical terms, we define the following variables. The radii of the inner and outer circular orbits are r_1, r_2 respectively which are also the perigee and apogee, respectively, of the elliptic orbit. The velocities of the two circular orbits are then given by

$$v_1 = (gR^2/r_1)^{1/2}$$
$$v_2 = (gR^2/r_2)^{1/2} \tag{7.30}$$

where R is the radius of the earth or other mass center and g is the acceleration of gravity on the surface.

While the inner circular orbit and the ellipse share the same radius at the perigee, the tangential velocities are different. We denote the perigee velocity of the ellipse by v_p, and the apogee velocity by v_a. These two velocities are related by the conservation of angular momentum on the ellipse, i.e.,

$$r_1 v_p = r_2 v_a = h_0 \tag{7.31}$$

where h_0 is the specific angular momentum of the ellipse.

To determine the specific angular momentum of the elliptic orbit, we can use the conservation-of-energy principle;

$$v_r^2 + v_\theta^2 - \frac{2gR^2}{r} = 2e_0 \tag{7.32}$$

At the perigee and apogee points, $v_r = 0$. Also, $v_\theta = v_p$ or v_a can be replaced by h_0/r. Equating the energy at the two exchange points, we can determine h_0:

$$\frac{h_0^2}{r_1^2} - \frac{2gR^2}{r_1} = \frac{h_0^2}{r_2^2} - \frac{2gR^2}{r_2}$$

$$h_0^2 = 2v_1^2 r_1^2 [1-\beta][1-\beta^2]^{-1}$$

or

$$h_0 = \sqrt{2} v_1 r_1 [1+\beta]^{-1/2} \tag{7.33}$$

where $\beta = r_1/r_2$. Finally, we can calculate the specific impulse (impulse per unit mass)

$$(v_p - v_1) = \frac{h_0}{r_1} - v_1 = v_1\left[\sqrt{2}(1+\beta)^{-1/2} - 1\right]$$

$$(v_2 - v_a) = v_2 - \frac{h_0}{r_2} = v_2\left[1 - \sqrt{2}\beta^{1/2}(1+\beta)^{-1/2}\right] \tag{7.34}$$

Example 7.3 Hohmann Transfer. Find the orbit parameters and specific impulse for a transfer from an earth orbit of $r_1 = 6600$ km to an orbit of $r_2 = 7000$ km.

The radius of the earth is assumed to be $6.48 \cdot 10^6$ m, from which we can calculate v_1, v_2:

$$v_1 = 7.77 \text{ km/s}, \qquad v_2 = 7.55 \text{ km/s} \tag{7.35}$$

Using the ratio $\beta = r_1/r_2 = 0.943$, determine the specific angular momentum h_0 from which we can calculate the perigee and apogee velocities:

$$v_p = h_0/r_1 = \frac{\sqrt{2}v_1}{[1+\beta]^{1/2}} = 1.0146 v_1$$

and the specific impulse per unit mass is

$$(v_p - v_1) = 0.0146 v_1$$

Similarly,

$$v_a = h_0/r_2 = \frac{\sqrt{2}v_1\beta}{[1+\beta]^{1/2}} = 0.9852v_2$$

and

$$v_2 - v_a = 0.0148v_2 \tag{7.36}$$

7.3
Two-body Problems

In the previous sections, the gravity force center was assumed to be immovable, unperturbed by the orbiting mass. This assumption is a reasonable one in a two-body problem where one of the masses is orders of magnitude larger than the other, as in the earth and artificial satellites, or the sun and its smaller planets. In this section, we explore the consequences when both masses can affect the dynamics of the other. In our solar system, two examples are the earth–moon system, and the sun–Jupiter system. The restricted three-body problem is discussed in Homework Problems 7.24, 7.25, 7.26.

The geometry of the two-body problem is shown in Fig. 7.6. Each mass has a position vector r_1, r_2, from a fixed reference. The initial gravitational forces act along the vector difference $\mathbf{r} = \mathbf{r}_1 - \mathbf{r}_2$. The resulting equations of motion become

$$m_1\ddot{\mathbf{r}}_1 = -\frac{Gm_1m_2}{r^2}\mathbf{e}_r$$

$$m_2\ddot{\mathbf{r}}_2 = \frac{Gm_1m_2}{r^2}\mathbf{e}_r \tag{7.37}$$

The center of mass of the system is defined by $(m_1 + m_2)\mathbf{r}_c = m_1\mathbf{r}_1 + m_2\mathbf{r}_2$. It is easy to see that by adding the two equations of motion the mass center must have either a zero or fixed velocity relative to an inertial reference frame. The position vectors $\boldsymbol{\rho}_1$, $\boldsymbol{\rho}_2$ from the mass center are defined by

$$\mathbf{r}_1 = \mathbf{r}_c + \boldsymbol{\rho}_1, \qquad \mathbf{r}_2 = \mathbf{r}_c + \boldsymbol{\rho}_2 \tag{7.38}$$

These vectors must satisfy

$$m_1\boldsymbol{\rho}_1 + m_2\boldsymbol{\rho}_2 = 0 \tag{7.39}$$

Further, if we subtract the two equations of motion in the form, $\ddot{\mathbf{r}}_1 - \ddot{\mathbf{r}}_2 = \ddot{\mathbf{r}}$, it is easy to show that the difference vector $\mathbf{r}$ satisfies an equation of motion identical to that of a one-body problem

$$\ddot{\mathbf{r}} = -\frac{G(m_1 + m_2)}{r^2}\mathbf{e}_r \tag{7.40}$$

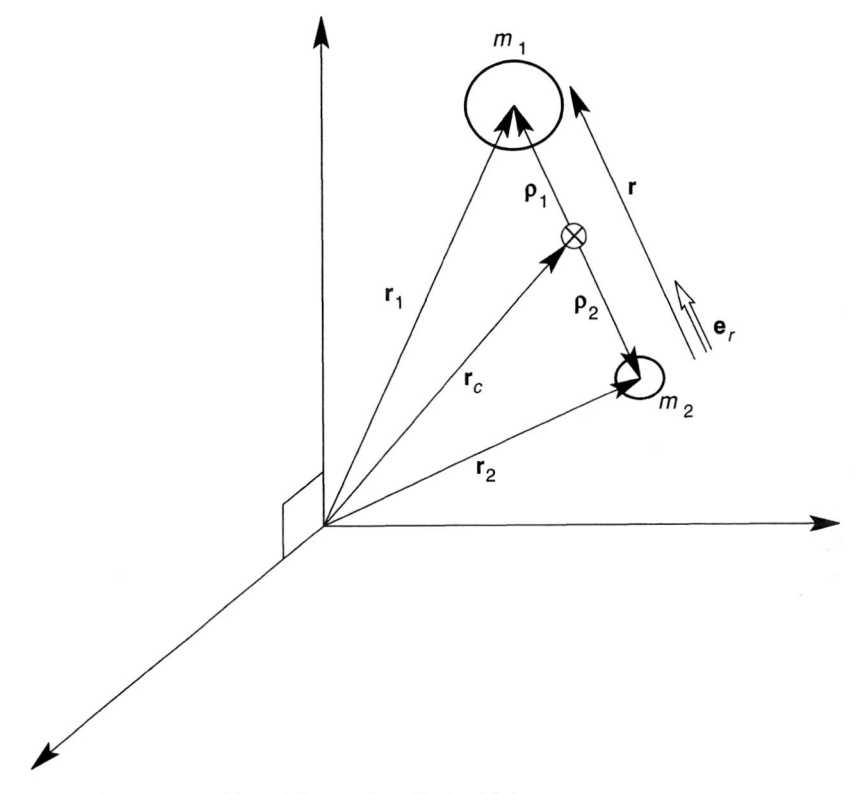

Fig. 7.6 Two-mass problem with mutual gravitational forces.

Using the definition of the position vectors $\boldsymbol{\rho}_1$, $\boldsymbol{\rho}_2$, this equation is also equivalent to

$$\ddot{\boldsymbol{\rho}}_1 = -\frac{Gm_2}{(1+(m_1/m_2))^2} \frac{\mathbf{e}_r}{\rho_1^2}$$

$$\ddot{\boldsymbol{\rho}}_2 = \frac{Gm_1}{(1+(m_2/m_1))^2} \frac{\mathbf{e}_r}{\rho_2^2} \qquad (7.41)$$

Thus, while each mass is coupled to the other throughout the motion of the center of mass, the dynamical equations only involve coordinates of one body. Thus, the results of the one-body problem are applicable to the two-body case, as illustrated in the following example.

Example 7.4 Circular Orbits of Two Bodies. To illustrate the application (7.40), assume that the distance between the two bodies is a constant, i.e., $\dot{r} = 0$. Then we are asked to find the motions of the two masses.

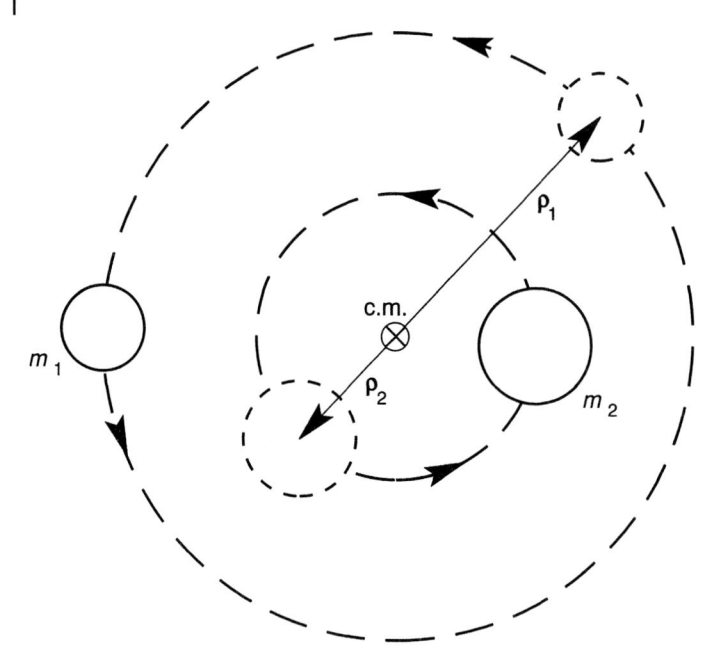

Fig. 7.7 Motion of two mutually attracting masses.

If $\dot{r} = 0$, then we have $\dot{\rho}_1 = \dot{\rho}_2 = 0$, or the masses orbit in circular paths relative to the mass center. Using the equation of motion for $\ddot{\mathbf{r}}$ in polar coordinates (2.8), it is easy to show that the angular velocity of $\mathbf{r}$ or $\mathbf{e}_r$ is given by

$$\Omega = \left(\frac{G(m_1 + m_2)}{r^3}\right)^{1/2} \qquad (7.42)$$

where $r = \rho_1 + \rho_2$. Using the vector equations for $\ddot{\boldsymbol{\rho}}_1$, $\ddot{\boldsymbol{\rho}}_2$ in polar coordinates, (7.41), we can also show that the two circular velocities are

$$v_1 = \left[\frac{Gm_2}{(1 + (m_1/m_2))^2 \rho_1}\right]^{1/2}$$
$$v_2 = \left[\frac{Gm_1}{(1 + (m_2/m_1))^2 \rho_2}\right]^{1/2} \qquad (7.43)$$

A sketch of the two orbits is shown in Fig. 7.7.

7.4
Gravity Force of Extended Bodies

The discussion thus far has assumed that each attracting particle is a point mass. The analysis of the gravitational dynamics of extended bodies is usually dealt with

7.4 Gravity Force of Extended Bodies

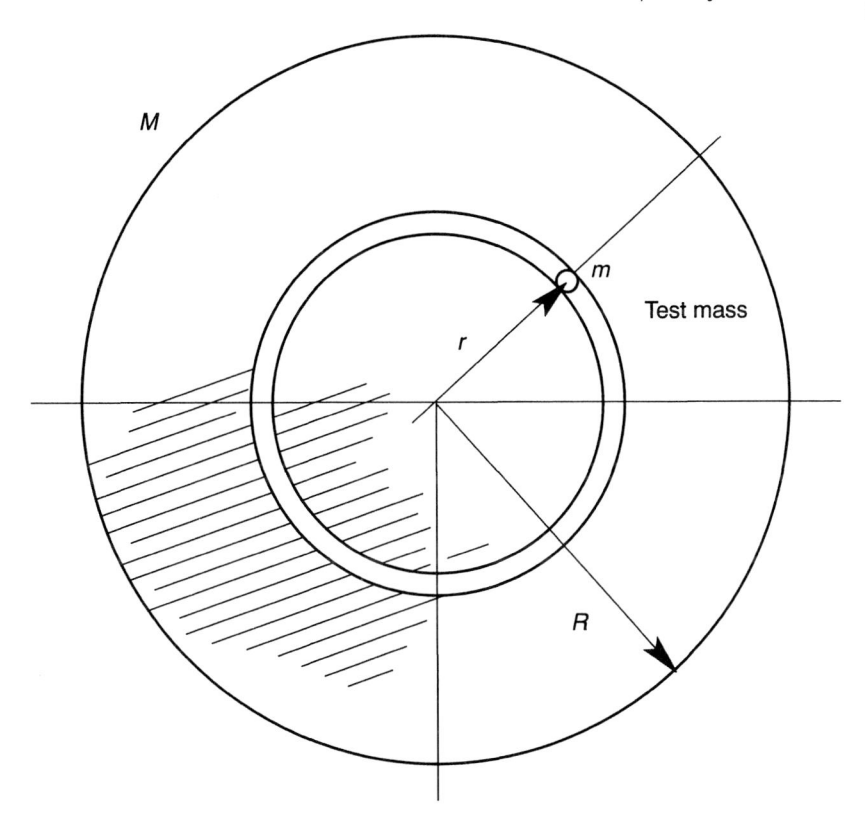

Fig. 7.8 Mass particle m inside a uniform distribution of mass density, ρ. Separation of the problem into an attracting sphere of radius, r, and a mass shell with inner and outer radii, r, R.

in advanced tests (see, e.g., Wiesel, 1989, or Meirovitch, 1970), however, a few remarks and an example are worth noting here. First it can be shown that a spherically symmetric mass distribution acts like a point-mass attractor when the other mass center or test mass is outside the sphere.

An interesting problem results when we consider the test mass to be inside the sphere of the extended mass center (Fig. 7.8). This can occur if the extended mass is gaseous or fluid, or in the case of a long tunnel in a solid-mass sphere. This problem has been treated in a classic text on potential theory (Kellogg, 1929). He divided the problem of calculating the force on the test mass by looking at two subproblems: the force of the mass shell that surrounds the test mass, and the force on the test mass due to the sphere bounded by the radius of the test mass. Kellogg's results are succinct enough to quote:

> ... a homogeneous shell attracts a particle at an exterior point as if the mass of the shell were concentrated at its center, and exercises no force on a particle in its interior.

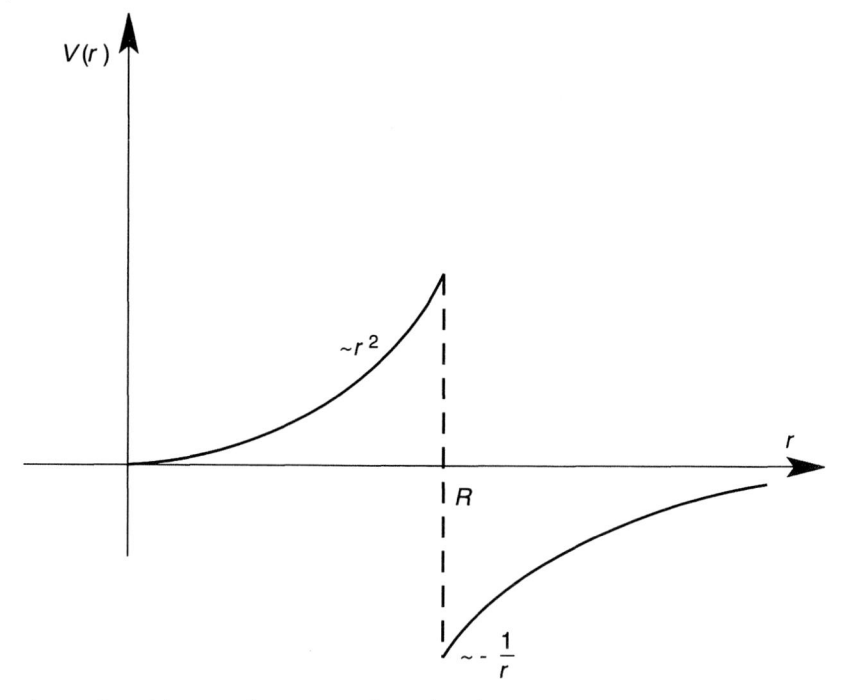

Fig. 7.9 Potential energy of a mass particle inside and outside of a homogeneous sphere of mass density, ρ.

To make this statement clear, consider a homogeneous sphere of radius R and mass density ρ. The mass is then, $M = \rho 4\pi R^3/3$, and the force on a test mass m is given by

$$F = -\frac{mMG}{r^2}; \quad r > R$$
$$= -m\frac{\rho G 4\pi r}{3}; \quad r < R \tag{7.44}$$

The force potential is shown in Fig. 7.9, and suffers a discontinuity at $r = R$, i.e.,

$$V(r) = m\rho\frac{G 2\pi}{3}r^2; \quad r < R \tag{7.45}$$

Two applications to gravity dynamics in tunnels in the earth are presented in two examples.

Example 7.5 The Gravity Grade Tunnel. Imagine a tunnel dug into the earth from city A to city B such that the path forms a straight line as shown in Fig. 7.10. Suppose an evacuated tube is built, and a vehicle is suspended using magnetic levitation such that there is no friction or aerodynamic losses. Then we are asked to find the time of travel between A and B and the maximum velocity at the midpoint.

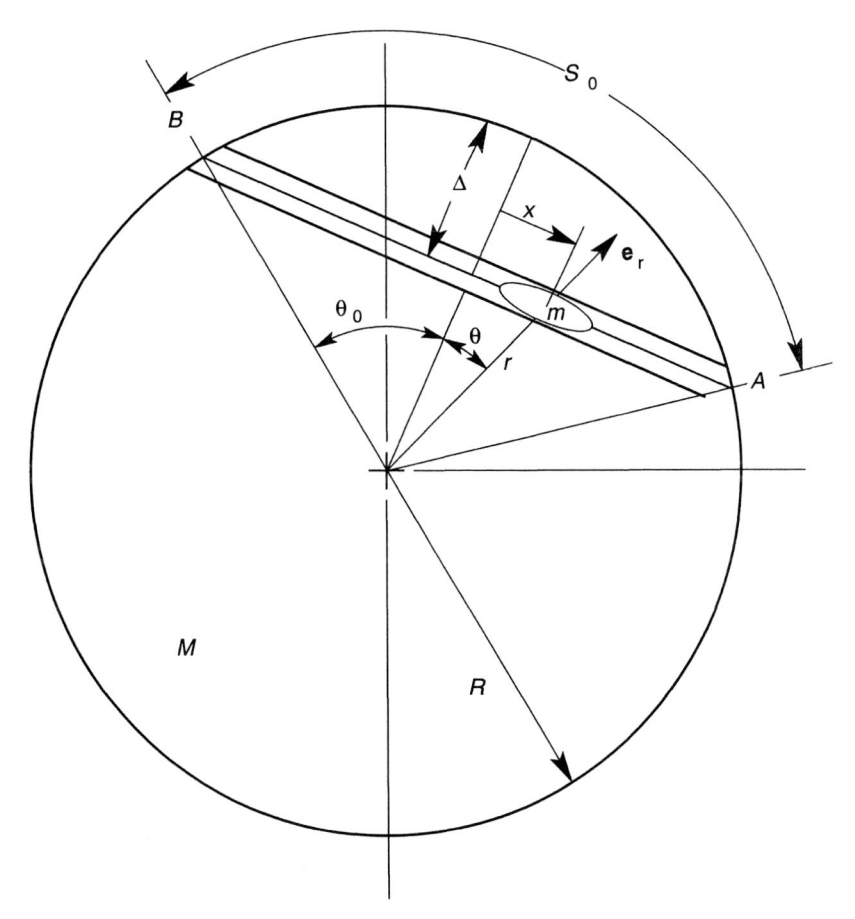

Fig. 7.10 Geometry of a vehicle mass in a gravity-grade tunnel in a homogeneous sphere (Example 7.5).

To analyze the problem, we draw a great circle between A, B (assumed to lie on a spherical earth). When the vehicle mass is below the surface, i.e., $r < R$, the gravitational force follows the linear law given in (7.44) and (7.45) and has a potential energy

$$V(r) = \frac{m\rho G 2\pi r^2}{3}$$
$$= mg\frac{r^2}{2R} \tag{7.46}$$

where we have used the identity $GM = gR^2$ and g is the acceleration of gravity at $r = R$. At the midpoint of the trip, the minimum value of r is given by

$$r_{min} = R\cos\theta_0 = R\cos\left(\frac{S_0}{2R}\right) \tag{7.47}$$

where S_0 is the trip length at the earth's surface. The maximum velocity is then determined from an energy balance:

$$\frac{1}{2}mv_{max}^2 = V(R) - V(r_{min})$$

or

$$v_{max}^2 = gR\left(1 - \cos^2\left(\frac{S_0}{2R}\right)\right) \tag{7.48}$$

To determine the time of the trip between A and B we can use D'Alembert's principle of virtual work (4.38) where the vehicle position x is used as a generalized coordinate:

$$\left(m\ddot{\mathbf{r}} + \frac{mg}{R}r\mathbf{e}_r\right) \cdot \frac{\partial \mathbf{r}}{\partial x} = 0 \tag{7.49}$$

The projection or Jacobian vector is found from the constraint

$$\mathbf{r} = x\mathbf{e}_x + (R - \Delta)\mathbf{e}_y$$

$$\frac{\partial \mathbf{r}}{\partial x} = \mathbf{e}_x$$

The resulting equation of motion, using $\mathbf{e}_r \cdot \mathbf{e}_x = \sin\theta$, is given by

$$\ddot{x} + \frac{g}{R}r\sin\theta = 0$$

or

$$\ddot{x} + \frac{g}{R}x = 0 \tag{7.50}$$

Thus the equation of motion is a harmonic oscillator similar to the small motion pendulum [see (1.7), (1.8)]. The period is given by $\tau = 2\pi(R/g)^{1/2}$, and it is independent of the distance between A and B.

For the earth $R = 6.4 \cdot 10^6$ m, and $\tau = 84.4$ min. However, the trip time between A and B is only $\tau/2$, so that the one-way trip time is 42.2 min.

This calculation was made by Robert Goddard, the rocket pioneer, in his design for a magnetically levitated tunnel train to go between several east coast cities in the United States. He published this in *Scientific American* in 1909 and applied for a patent much later in the 1940s.

Example 7.6 Design of Free Fall Tunnel. There exist several drop tunnels to perform experiments under weightlessness. One facility is at the NASA-Lewis Research Center in Cleveland. Suppose we are given a test time t_0. Then how deep will we have to drill the tunnel (Fig. 7.11) to achieve this specification without additional braking depth?

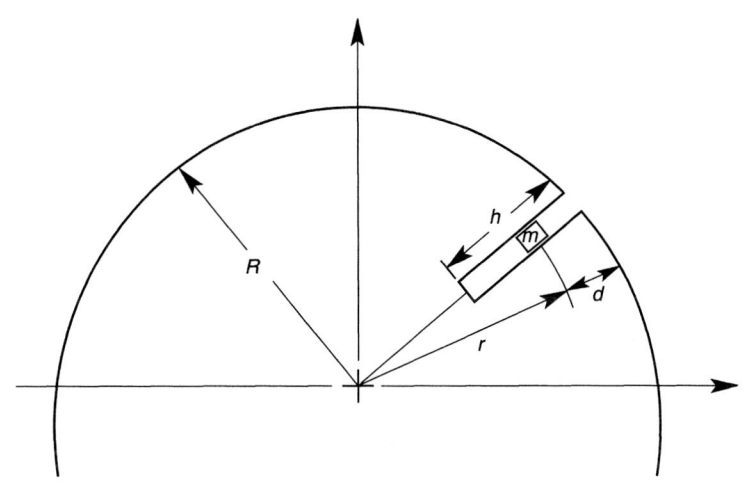

Fig. 7.11 Geometry in a free fall tunnel in a homogeneous gravitational sphere (Example 7.6).

For radial motion, Newton's law becomes

$$m\ddot{r} = -mg\frac{r}{R} \tag{7.51}$$

Starting at $r = R$ at $t = 0$, the solution is given by

$$r(t) = R\cos\sqrt{\frac{g}{R}}t \tag{7.52}$$

Thus the tunnel depth is given by

$$d = R - R\cos\sqrt{\frac{g}{R}}t_0 \tag{7.53}$$

For t_0 much smaller than the period $\tau = 84$ s

$$d \simeq gt_0^2/2 \tag{7.54}$$

Thus for $t_0 = 1.4$ s, $d = 9.8$-m deep or about the height of a three-story building. However, for $t_0 = 14$ s, $d = 980$ m, without additional depth for braking. Such a tunnel exists in an old mine in Hokkaido, Japan, with a depth of 2 km.

7.5
Rigid-body Satellite Dynamics

Gravitational Torque and Attitude Stability

This section describes two problems, one theoretical and the other practical. The first concerns the gravitational force on an extended nonspherically symmetric

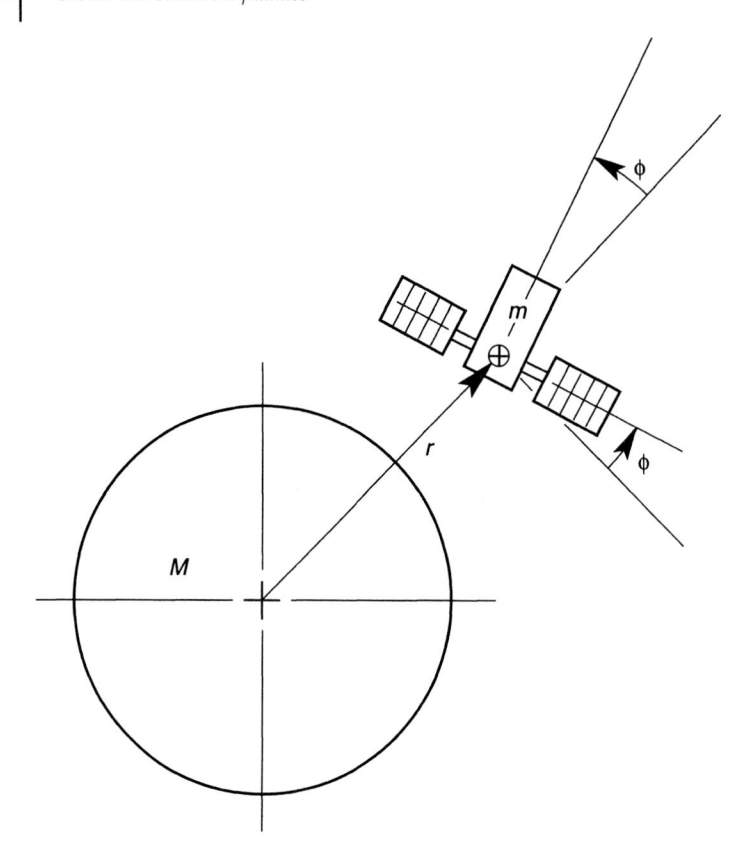

Fig. 7.12 Motion of a rigid body (satellite) about a gravitational mass attractor.

body (Fig. 7.12). Such problems lead to a gravitational torque in addition to an attractive force. The second problem of importance to the design of orbiting rigid bodies is to determine which orientations are stable with respect to the orbit plane. These subjects are discussed in more detail in advanced books on space dynamics, such as Rimrott (1988) and Wiesel (1989). For the student at the intermediate dynamics level, it is important to understand the basic phenomena of attitude stability through a couple of examples and to learn some of the mathematical difficulties.

The reader should be warned that our discussion of stability is based on linearized dynamics of a nonlinear system. In many nonlinear systems with dissipation there are theorems that validate linear stability theory. However, for nondissipative problems, sometimes called *Hamiltonian systems*, one must be careful to deduce long term stability predictions based on linearization, without looking at the nonlinear dynamics.

Another caution here is the assumption of decoupled rigid body dynamics from the orbital problem found even in the advanced books. For short-term stability, this assumption may be valid, but the weak coupling between the orbit dynamics and the attitude orientation might manifest itself over many orbit periods.

Finally, *gravity-gradient attitude stabilization* is only one of the tools available to the spacecraft engineer. One can also use active methods using pulsed thrusters or torques created by the earth's magnetic field. Also at low orbits, one must take into account high atmospheric drag and drag induced torques on the spacecraft.

With these caveats we examine a few simple examples of a two-mass, dumbbell satellite, then look at the results for an orbiting rod, and finally derive equations for the torque on a more complex rigid body satellite.

Center of Gravity

Consider first the planar motion of a two mass, rigid satellite in the plane of the circular orbit shown in Fig. 7.13. We make the assumptions that the orbit radius r is constant and is unperturbed by the angular motion $\phi(t)$. (In a more advanced analysis this assumption would be relaxed.) It is easy to see in this example that the gravitational force on the inner mass m_1, is greater than the force on the outer mass m_2. This means that the center of gravity is not located at the center of mass. Thus when $\phi \neq 0$, the net gravitational force, acting through the center of gravity, produces a torque or force moment about the center of mass.

In this example the center of gravity, r_G, is easily calculated for the alignment $\phi = 0$. The center of gravity is defined by the equation

$$\frac{(m_1 + m_2)GM}{r_G^2} = \frac{m_1 GM}{r_1^2} + \frac{m_2 GM}{r_2^2} \tag{7.55}$$

The center of mass, however, is defined by

$$(m_1 + m_2)r_c = m_1 r_1 + m_2 r_2 \tag{7.56}$$

If we define local coordinates $r_1 = r_c - \rho_1, r_2 = r_c + \rho_2$, the center-of-mass definition is equivalent to

$$m_1 \rho_1 = m_2 \rho_2 \tag{7.57}$$

Expanding $1/r_G^2$ to second order in ρ_1, ρ_2, it is straightforward to arrive at the expression

$$\left(\frac{r_G}{r_c}\right)^2 = 1 - 3\frac{\beta \rho_1 (\rho_1 + \rho_2)}{r_c^2} \tag{7.58}$$

where $\beta = m_1/(m_1 + m_2)$.

For equal masses $\beta = 1/2, \rho_1 = \rho_2 = L/2$,

$$\left(\frac{r_G}{r_c}\right)^2 = 1 - \frac{3}{4}\left(\frac{L}{r_c}\right)^2 \tag{7.59}$$

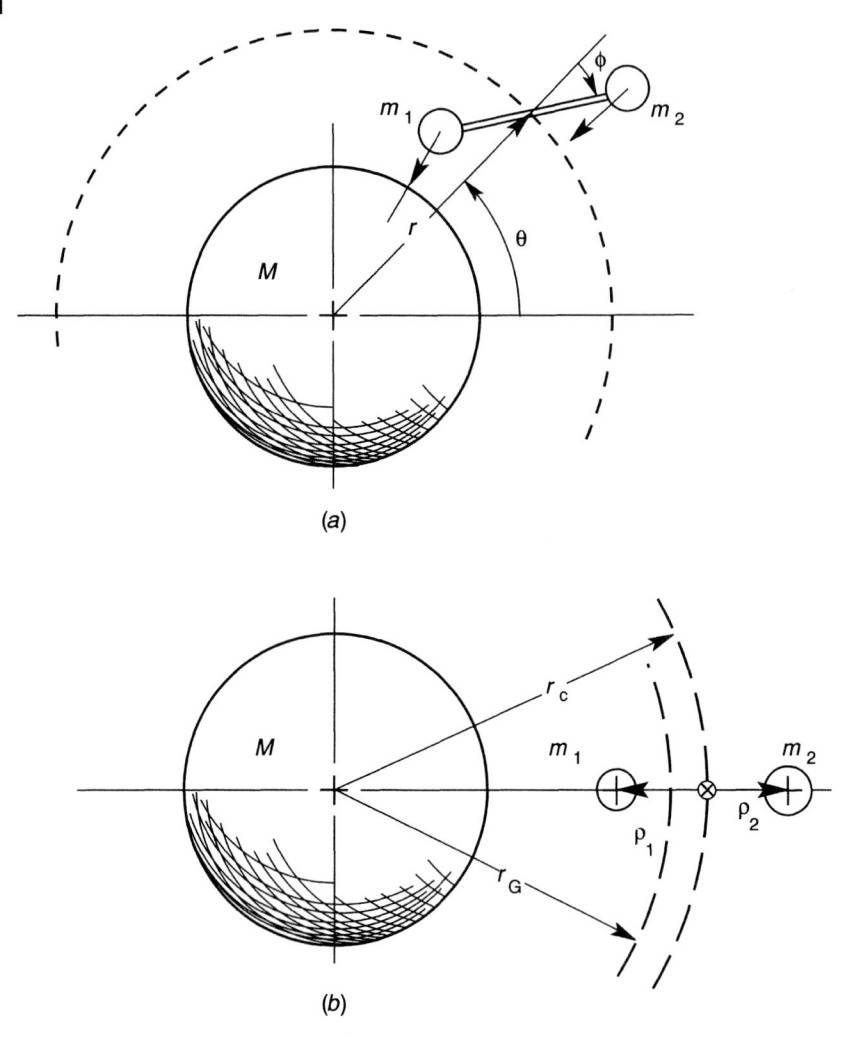

Fig. 7.13 Geometry of a dumbbell satellite in a circular orbit about a gravitational attractor.

For distributed mass in a satellite, the center of mass and center of gravity are defined by

$$m\mathbf{r}_c = \int \mathbf{r}\, dm$$

$$\frac{m}{r_G^3}\mathbf{r}_G = \int \frac{\mathbf{r}\, dm}{r^3} \qquad (7.60)$$

The example of a rod-shaped satellite of length L is given in Rimrott (1988) (see Fig. 7.14). When ϕ is 0, i.e., the rod is aligned with the radius,

$$r_G = r_c \left(1 - \frac{L^2}{4r_c^2}\right)^{1/4} \tag{7.61}$$

This produces a stable restoring moment when $\phi \neq 0$, but small.

However, when $\phi = \pi/2$, the center of gravity is behind the satellite-mass center and a slight disturbance will produce a destabilizing gravitational torque. In this case, the center of gravity is given by

$$r_G = r_c \left(1 + \frac{L^2}{4r_c^2}\right)^{1/4} \tag{7.62}$$

Stability of a Dumbbell Satellite

The equations of motion for a two-mass satellite in a circular orbit (Fig. 7.13a) can be derived using Lagrange's equations, noting that the motion of the satellite occurs in a moving reference frame of the circular orbit. The case of unequal masses is left as an exercise. For this example $m_1 = m_2 = m$; $\rho_1 = \rho_2 = L$. For a rigid body the kinetic energy is

$$T = \frac{1}{2}(2m)v_c^2 + \frac{1}{2}\mathbf{H} \cdot \boldsymbol{\omega}$$

or

$$T = m\left(\dot{r}_c^2 + r_c^2 \dot{\theta}^2\right) + mL^2(\dot{\theta} - \dot{\phi})^2$$

The potential energy follows from (7.18)

$$V = -\frac{mGM}{r_1} - \frac{mGM}{r_2} \tag{7.63}$$

where

$$r_1^2 = L^2 + r_c^2 - 2r_c L \cos \phi$$
$$r_2^2 = L^2 + r_c^2 + 2r_c L \cos \phi \tag{7.64}$$

Here we make the assumption that $\dot{r}_c = 0$, $\ddot{\theta} = 0$. However, in a full nonlinear analysis, we would include the change in the orbit parameters induced by the rigid-body motions. With these assumptions, ϕ becomes the only generalized coordinate and its equation of motion follows from Lagrange's equation

$$\frac{d}{dt}\frac{\partial T}{\partial \dot{\phi}} - \frac{\partial T}{\partial \phi} + \frac{\partial V}{\partial \phi} = 0$$

or

$$\ddot{\phi} + \frac{GMr_c}{2L}\sin\phi\left[\frac{1}{r_1^3} - \frac{1}{r_2^3}\right] = 0 \tag{7.65}$$

The bracketed term also depends on the angle ϕ, which for small L/r_c is given by

$$\left[\frac{1}{r_1^3} - \frac{1}{r_2^3}\right] = \frac{6L}{r_c^4}\cos\phi \tag{7.66}$$

Substituting this expression into (7.65), we obtain

$$\ddot{\phi} + \frac{3GM}{2r_c^3}\sin 2\phi = 0$$

For small motions about $\phi = 0$, we have

$$\ddot{\phi} + \frac{3GM}{r_c^3}\phi = 0$$

or

$$\ddot{\phi} + 3\Omega_0^2\phi = 0 \tag{7.67}$$

where $\Omega_0^2 = GM/r_c^3$ is the square of the frequency of the orbit. Thus when the satellite is aligned with the radius, its dynamics is neutrally stable and it oscillates with a frequency $\sqrt{3}$ times the orbital frequency.

When $\phi = \pi/2$, the equation of motion is similar to (7.67), but the sign of the gravitational torque is positive, which leads to an exponentially growing solution or instability.

Rimrott (1988) shows in his book that the same results are obtained for a rod-shaped rigid satellite.

Out of Plane Motion

It is left as an exercise to derive the linearized equation of motion for small motion of the rigid, rod shaped satellite out of the orbital plane. The results for out of plane libration, and $\phi = 0$, for a rod-shaped body are given in Rimrott (Fig. 7.14):

$$\ddot{\psi} + 4\Omega_0^2\psi = 0 \tag{7.68}$$

Thus, the libration frequency is twice the orbital frequency of the satellite.

Example 7.7 Stability of a Four-mass Cross Satellite. It is instructive to investigate the linear stability of the four-mass rigid cross satellite (Fig. 7.15). We already know that the m_1 mass pair by itself is stable and the m_2 mass pair with $m_1 = 0$ is unstable near $\phi = 0$. Therefore we determine the combination of $\{m_1, m_2, L_1, L_2\}$, for which the four-cross is stable. We will show that the equation that governs the planar motion is given by

$$I_3\ddot{\phi} + 3\Omega_0^2(I_2 - I_1)\cos\phi\sin\phi = 0 \tag{7.69}$$

where $\{I_1, I_2, I_3\}$ are the principal moments of inertia.

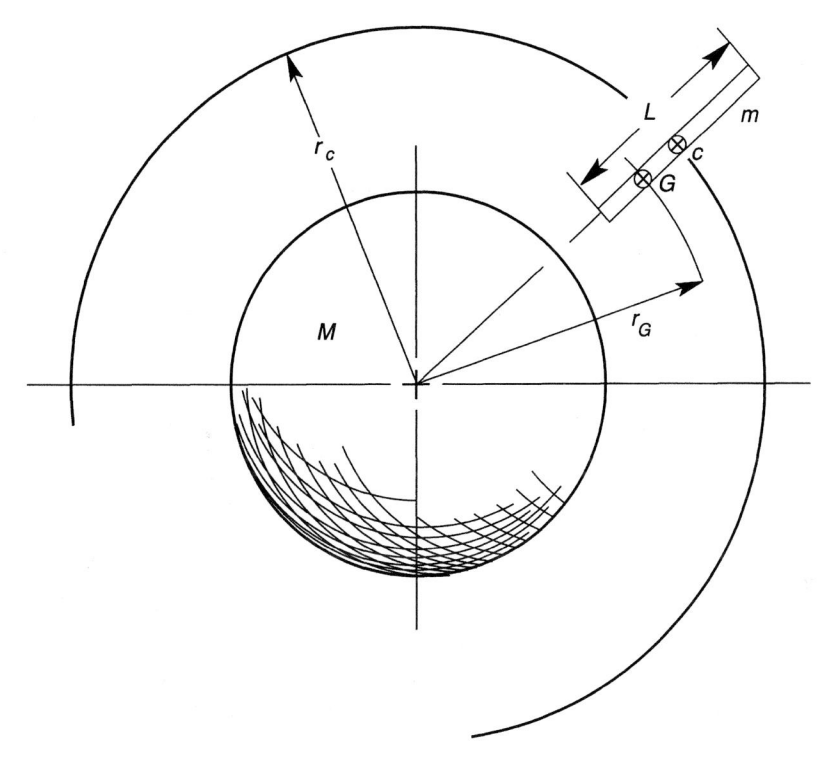

Fig. 7.14 Geometry of a rod-shaped rigid body in circular orbit.

In this example, we use the method of D'Alembert's virtual work, (2.61) or (4.38). Since there is only one degree of freedom, ϕ, this principle takes the form

$$\sum (m_i \ddot{\mathbf{r}}_i - \mathbf{F}_i) \cdot \boldsymbol{\beta}_i = 0$$

where $\boldsymbol{\beta}_i = \partial \mathbf{r}_i / \partial \phi$. In this example, the position vectors are given by

$$\mathbf{r}_1 = \mathbf{r}_c + L_2 \mathbf{e}_2, \qquad \mathbf{r}_2 = \mathbf{r}_c - L_2 \mathbf{e}_2$$
$$\mathbf{r}_3 = \mathbf{r}_c - L_1 \mathbf{e}_1, \qquad \mathbf{r}_4 = \mathbf{r}_c + L_1 \mathbf{e}_1$$

The gravity forces all take the form

$$\mathbf{F}_i = -\frac{m_i GM}{r_i^3} \mathbf{r}_i$$

Using the fact that $\partial \mathbf{e}_1 / \partial \phi = \mathbf{e}_2$ and $\partial \mathbf{e}_2 / \partial \phi = -\mathbf{e}_1$, the projection vectors are given by

$$\boldsymbol{\beta}_1 = -L_2 \mathbf{e}_1, \qquad \boldsymbol{\beta}_2 = L_2 \mathbf{e}_1$$
$$\boldsymbol{\beta}_3 = -L_1 \mathbf{e}_2, \qquad \boldsymbol{\beta}_4 = L_1 \mathbf{e}_2$$

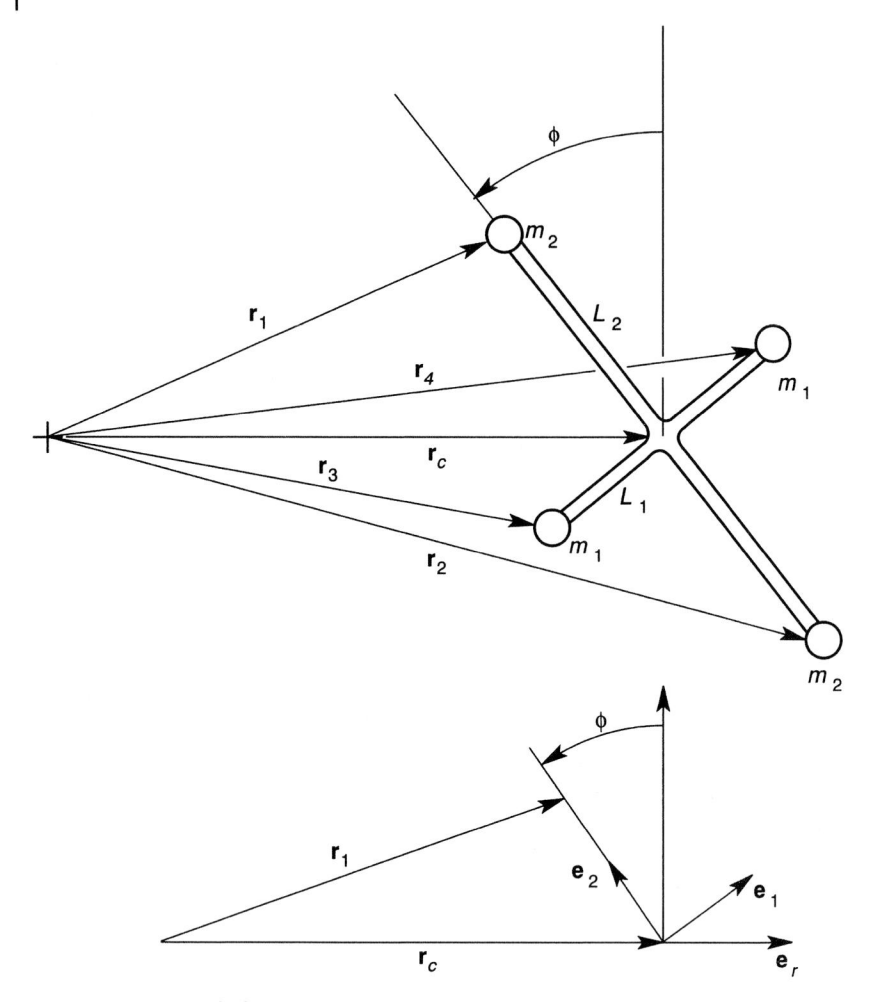

Fig. 7.15 Geometry of a four-mass, rigid body satellite in a circular orbit (Example 7.7).

Combining these equations, we arrive at an expression for the generalized force

$$\sum \mathbf{F}_i \cdot \boldsymbol{\beta}_i = GMr_c m_2 L_2 \cos\phi \left[\frac{1}{r_1^3} - \frac{1}{r_2^3}\right]$$

$$+ GMr_c m_1 L_1 \sin\phi \left[\frac{1}{r_4^3} - \frac{1}{r_3^3}\right]$$

Using approximations as in (7.66) for $L_i/r_c \ll 1$,

$$\sum \mathbf{F}_i \cdot \boldsymbol{\beta}_i = -\frac{3GM}{r_c^3} \cos\phi \sin\phi \{2m_1 L_1^2 - 2m_2 L_2^2\}$$

To compare this to the expression at the beginning of this example, we note that $I_2 = 2m_1 L_1^2$, $I_1 = 2m_2 L_2^2$, $\Omega_0^2 = Gm_0/r_c^3$.

Finally, to find the inertial term in D'Alembert's principle of virtual work, we use relations of the form

$$\dot{\mathbf{v}}_1 = \dot{\mathbf{v}}_c - L_2 \ddot{\phi} \mathbf{e}_1 - L_2 \dot{\phi}^2 \mathbf{e}_2$$

$$m_2 \dot{\mathbf{v}}_1 \cdot \boldsymbol{\beta}_1 = m_2 L_2^2 \ddot{\phi}$$

$$\sum m_i \dot{\mathbf{v}}_i \cdot \boldsymbol{\beta}_i = I_3 \ddot{\phi}$$

where $I_3 = 2m_1 L_1^2 + 2m_2 L_2^2$ is the second moment of inertia about the axis normal to the plane.

Thus we see that for small motions

$$\ddot{\phi} + \frac{3\Omega_0 (I_2 - I_1)}{I_3} \phi = 0 \tag{7.70}$$

and stability requires $I_2 - I_1 > 0$ or

$$m_1 L_1^2 > m_2 L_2^2$$

The general case of (7.69) for a distributed mass rigid body may be found in Meirovitch (1970), Rimrott (1988), or Wiesel (1989).

Dual Spin Satellite Dynamics

The use of gravitational moments to achieve pointing stability of orbiting satellites is only one of several technologies. We recall that a spinning rigid body can maintain orientation of the spin axis in a gimbal mounted gyro. To achieve both a nonspin platform for instruments and antennas, and gyro stabilization, engineers sometimes use a dual spin concept illustrated in Fig. 7.16. This device consists of a slowly rotating platform, orbiting the earth, and an angular momentum storage wheel with a relatively high rate of spin. This system can be viewed as a two-body or multi-rigid body system as in Chapter 6.

Dual spin satellites, sometimes called gyrostats (see also Rimrott, 1988, Chapter 11), have been used on dozens of spacecraft. According to Kinsey et al. (1996), a typical size is of the order of a ton with dimensions on the order of 2–3 meters in diameter and up to 8 meters in length.

In operation, the platform and rotor initially have identical spin. Then a motor places equal and opposite torques on the two bodies, such that the platform despins and the rotor spins up. However, there are known cases where a tumbling instability can occur, i.e., nutation, and precession can grow in a dynamic instability. This coning motion can cause vibrations that can either damage the equipment or prevent stable communication or operation of the satellite. One recent paper which analyzes these instabilities is by Kinsey et al. (1996) based on work at UCLA and Cornell University.

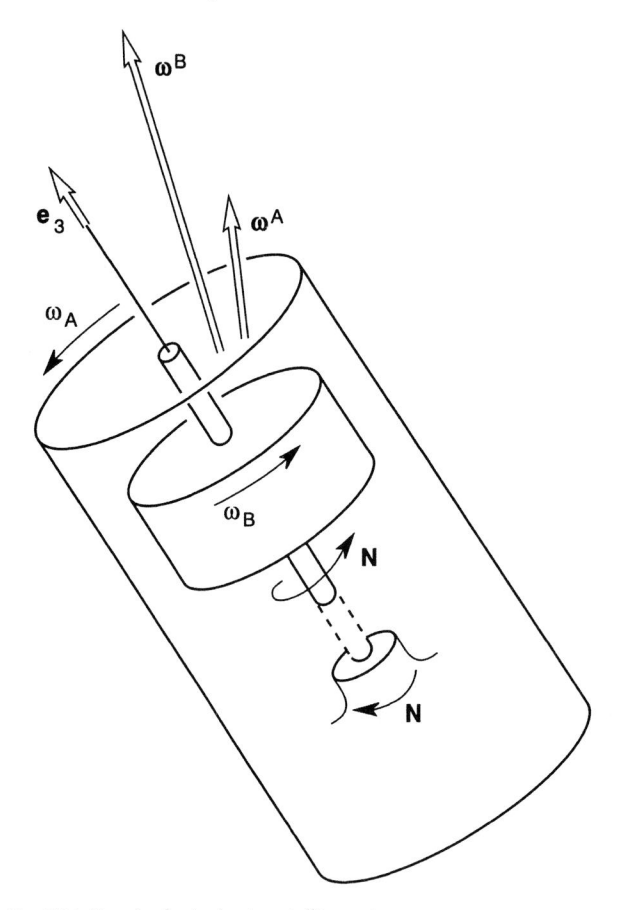

Fig. 7.16 Sketch of a dual-spin satellite system.

In this section we derive the equations of motion for two-symmetric rotors. The case of cross inertias and imbalance in the spin rotor can be found in Kinsey et al. (1996).

Because both bodies share a common axis of rotation we can write the two angular velocity vectors

$$\boldsymbol{\omega}^A = \omega_1 \mathbf{e}_1 + \omega_2 \mathbf{e}_2 + \omega_A \mathbf{e}_3$$

$$\boldsymbol{\omega}^B = \omega_1 \mathbf{e}_1 + \omega_2 \mathbf{e}_2 + \omega_B \mathbf{e}_3$$

where $\{\mathbf{e}_1, \mathbf{e}_2, \mathbf{e}_3\}$ are fixed to the rotor or body B. We consider the two motions to be pure rotation about the center of mass of the two bodies. The equation of motion can be derived in several ways, Newton–Euler, Lagrange's equation, or virtual power. In the case of virtual power, we can use a generalization due to Kane in which the generalized velocities are chosen to be the set $\{u_i\} = \{\omega_1, \omega_2, \omega_A, \omega_B\}$ and the principle takes the form

$$(\dot{\mathbf{H}}_A - \mathbf{M}_A^a) \cdot \boldsymbol{\gamma}_{Ai} + (\dot{\mathbf{H}}_B - \mathbf{M}_B^a) \cdot \boldsymbol{\gamma}_{Bi} = 0$$

where

$$\boldsymbol{\gamma}_{Ai} = \frac{\partial \boldsymbol{\omega}^A}{\partial u_i}, \qquad \boldsymbol{\gamma}_{Bi} = \frac{\partial \boldsymbol{\omega}^B}{\partial u_i}$$

The $\{u_i\}$ are sometimes called "quasi-coordinates" because they cannot be integrated to obtain generalized coordinates. Instead, one must use the kinematic equations, for example Euler angles in Chapter 5, to relate the $\{u_i\}$ to angles that give the global orientation. Whatever the method, the resulting equations become for equal and opposite torques $\mathbf{N} = N\mathbf{e}_3$,

$$(\dot{\mathbf{H}}_A + \dot{\mathbf{H}}_B) \cdot \mathbf{e}_1 = 0$$
$$(\dot{\mathbf{H}}_A + \dot{\mathbf{H}}_B) \cdot \mathbf{e}_2 = 0$$
$$\dot{\mathbf{H}}_A \cdot \mathbf{e}_3 + N = 0$$
$$\dot{\mathbf{H}}_B \cdot \mathbf{e}_3 - N = 0$$

where the last two equations can be replaced by

$$(\dot{\mathbf{H}}_A + \dot{\mathbf{H}}_B) \cdot \mathbf{e}_3 = 0$$
$$I_{33}^A \dot{\omega}_A = -N$$

These equations become (with $I_1 = I_{11}^A + I_{11}^B$)

$$I_1 \dot{\omega}_1 = -\omega_2 \left[I_{33}^B \omega_B + I_{33}^A \omega_A - I_1 \omega_B \right]$$
$$I_1 \dot{\omega}_2 = \omega_1 \left[I_{33}^B \omega_B + I_{33}^A \omega_A - I_1 \omega_B \right]$$
$$I_{33}^A \dot{\omega}_A = -N$$
$$I_{33}^B \dot{\omega}_B = N \tag{7.71}$$

If the bodies are initially at ω_0, and the torque N is constant, then the last two equations can be integrated to obtain

$$\omega_A = \omega_0 - \frac{N}{I_{33}^A} t$$
$$\omega_B = \omega_0 + \frac{N}{I_{33}^B} t$$

Thus the first two equations become coupled linear, first-order equations with time varying coefficients. The rates ω_1, ω_2 determine the extent of nutation and coning.

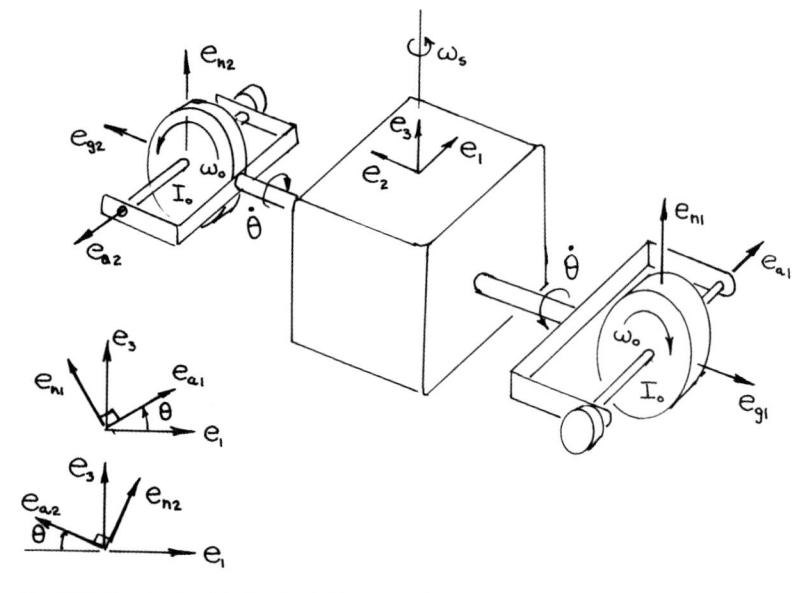

Fig. 7.17 Sketch of an idealized satellite, control-moment gyro (CMG) attitude stabilizer, with two paired single-axis gyros with coordinated nutation rotation speeds $\dot\theta$.

7.6
Control Moment Gyros

The control of spacecraft attitude is a major design problem in the creation of artificial satellites. As we saw in Section 7.4, the orientation of a satellite can become unstable in gravity gradients. There are also other disturbances in space as well as maneuvers that can change the orientation of antenna that are essential to satellite operations. Two attitude control techniques are momentum reaction wheels and the use of the Earth's magnetic field to create a magnetic control torque on the satellite. The use of control moment gyros (CMG) for attitude control has been studied since the early days of the space age, in the 1960s. In this section we present the introductory concepts of CMG so that the reader is able to read the literature on the subject. (See, e.g., Kurokawa, 2007; Carpenter and Peck, 2008.)

When reaction wheels are used to change the spacecraft angular momentum, energy is required to spin up or slow down the reaction wheels. In CMG technology, one stores angular momentum in pairs of flywheels and then transfers this angular momentum to the satellite through a control system with only a small expenditure of energy. We review the basic geometric configuration of a single gimbal, scissored-pair gyro system as illustrated in Fig. 7.17. There are many other configurations as well as double gimbal CMGs, and the reader is referred to more advanced references to learn about these more complex designs.

The basic idea of a CMG is to store angular momentum in two flywheels that are initially oriented in opposite directions so that no additional angular momentum is required of the overall satellite system. When the attitude control system requires some torque, the gimbal motors rotate the wheels in opposite directions such that a net angular momentum is created along an axis normal to the plane of the gimbal and initial spin axes. This change in angular momentum in the CMG sub assembly produces a torque on the CMG, T, and an equal and opposite torque $-T$ on the satellite structure to which it is attached. In order to achieve three-dimensional attitude control, several pairs of CMG are usually employed. These multi CMG systems often have singularities in the control operation that must be considered in the overall design.

First we define gimbal, spin and torque axes as follows (Fig. 7.17):

$$\{\mathbf{e}_{gi}, \mathbf{e}_{ai}, \mathbf{e}_{ni}\}$$

where

$$\mathbf{e}_{gi} \times \mathbf{e}_{ai} = \mathbf{e}_{ni} \quad \{i = 1, 2\}$$
$$\mathbf{e}_{g2} = -\mathbf{e}_{g1} \tag{7.72}$$

For simplicity we attach another reference system $\{\mathbf{e}_1, \mathbf{e}_2, \mathbf{e}_3\}$ along the principal axes of the satellite body. We assume that one of these axes is aligned with the two gimbal axes such that $\mathbf{e}_{g1} = -\mathbf{e}_2 = -\mathbf{e}_{g2}$. When the gimbal angles are zero, i.e. $\theta_1 = \theta_2 = 0$, the initial angular momentum vectors are aligned with the satellite axes such that $\mathbf{e}_{a1} = \mathbf{e}_1 = -\mathbf{e}_{a2}$.

The angular velocity vectors for the satellite body and two flywheels are:

$$\omega_s = \dot{\phi}\mathbf{e}_3$$
$$\omega_1 = \dot{\phi}\mathbf{e}_3 + \dot{\theta}\mathbf{e}_{g1} + \omega_0\mathbf{e}_{a1}$$
$$\omega_2 = \dot{\phi}\mathbf{e}_3 + \dot{\theta}\mathbf{e}_{g2} + \omega_0\mathbf{e}_{a2}$$
$$\mathbf{e}_{g2} = -\mathbf{e}_{g1} = \mathbf{e}_2 \tag{7.73}$$

For a scissored pair operation, the two gimbal angles are equal, i.e. $\theta_1 = \theta_2$.

The resulting angular momentum vectors relative to the rotor center of mass for the two paired flywheels are

$$\mathbf{H}_{1c} = (I_0\omega_0 + \dot{\phi}\sin\theta)\mathbf{e}_{a1} + I_1\dot{\theta}\mathbf{e}_{g1} + I_1\dot{\phi}\cos\theta\mathbf{e}_{n1}$$
$$\mathbf{H}_{2c} = (I_0\omega_0 + \dot{\phi}\sin\theta)\mathbf{e}_{a2} + I_1\dot{\theta}\mathbf{e}_{g2} + I_1\dot{\phi}\cos\theta\mathbf{e}_{n2} \tag{7.74}$$

For an open loop control problem, the gimbal angle $\theta(t)$ function is given and our goal is to determine the change in the angular velocity of the satellite. Since there are no external torques or force moments on the system, we assume that the combined angular momentum of the three bodies about the fixed center of mass of the system will remain constant in magnitude and direction. We also assume that the main satellite body will only rotate about its vertical axis, $\mathbf{e}_3$, and that this axis is one of the principle inertia axes of the satellite without the gimbal systems.

Asserting the equality of angular momentum before and after gimbal motion we obtain

$$\mathbf{H}_{10} + \mathbf{H}_{20} + \mathbf{H}_s = 0 \tag{7.75}$$

$$\mathbf{H}_s = I_{s3}\dot{\phi}\mathbf{e}_3 \tag{7.76a}$$

$$\mathbf{H}_{1c} + \mathbf{H}_{2c} = \{2I_0(\omega_0 + \dot{\phi}\sin\theta)\sin\theta + 2I_1\dot{\phi}\cos^2\theta\}\mathbf{e}_3$$

$$\mathbf{H}_{10} = \mathbf{H}_{1c} + m_g\mathbf{r}_{01} \times \mathbf{v}_{1c}$$

$$\mathbf{H}_{10} + \mathbf{H}_{20} = \mathbf{H}_{1c} + \mathbf{H}_{2c} + 2m_g L^2 \mathbf{e}_3 \tag{7.76b}$$

Here, m_g is the mass of one of the gyro assemblies and L is the perpendicular distance of its center of mass from the main satellite spin axis, $\mathbf{e}_3$. Combining the satellite and gyro mass moment of inertias into $\tilde{I}_{s3} = I_{s3} + m_g L^2$, and summing the angular momentum contributions from the three bodies, the conservation of angular momentum principle (7.75) yields a relationship between the spin rate of the satellite and the control angle of the scissored pair:

$$\tilde{I}_{s3}\dot{\phi} = -\{2I_0(\omega_0 + \dot{\phi}\sin\theta)\sin\theta + 2I_1\dot{\phi}\cos^2\theta\} \tag{7.77a}$$

or

$$\phi(t) = -2I_0\omega_0 \int_0^t \frac{\sin\theta(\tau)\,d\tau}{[\tilde{I}_{s3} + 2I_0\sin^2\theta + 2I_1\cos^2\theta]} \tag{7.77b}$$

To obtain the net gyroscopic torque T applied to the satellite structure, we differentiate the conservation of angular momentum expression (7.77a) above and set $T = \tilde{I}_{s3}\ddot{\phi}$, or

$$T = -[2I_0\omega_0\dot{\theta}\cos\theta + 4(I_0 - I_1)\dot{\phi}\dot{\theta}\sin\theta\cos\theta + 2(I_0\sin^2\theta + I_1\cos^2\theta)\ddot{\phi}] \tag{7.78}$$

When the gyro spin rate ω_0 is much larger than that of the gyro or satellite angular velocity rates, $\dot{\theta}$, $\dot{\phi}$, then the expression (7.78) can be simplified:

$$T \cong -2I_0\omega_0\dot{\theta}\cos\theta \tag{7.79}$$

In practice satellite systems are designed to carry several pairs of scissored CMGs. In this case the engineer has a more complex task of designing a complete 3D spatial attitude control system. When there are several CMGs in an array, there is the question of singularities in the geometry where the applied CMG gimbal angles cannot produce a desired toque on the satellite. For a discussion of the subject of CMG singularities the reader is referred to Kurokawa (2007).

Some technical details of CMGs can be found on the website of Honeywell Aerospace Electronic Systems. Search for the M50 Control Moment Gyroscope. The rotor speed is on the order of 5000 RPM and the stored angular momentum is around 25–75 Newton-meter-seconds. The output torque is on the order of 75 Nm and the mass is 28 kg.

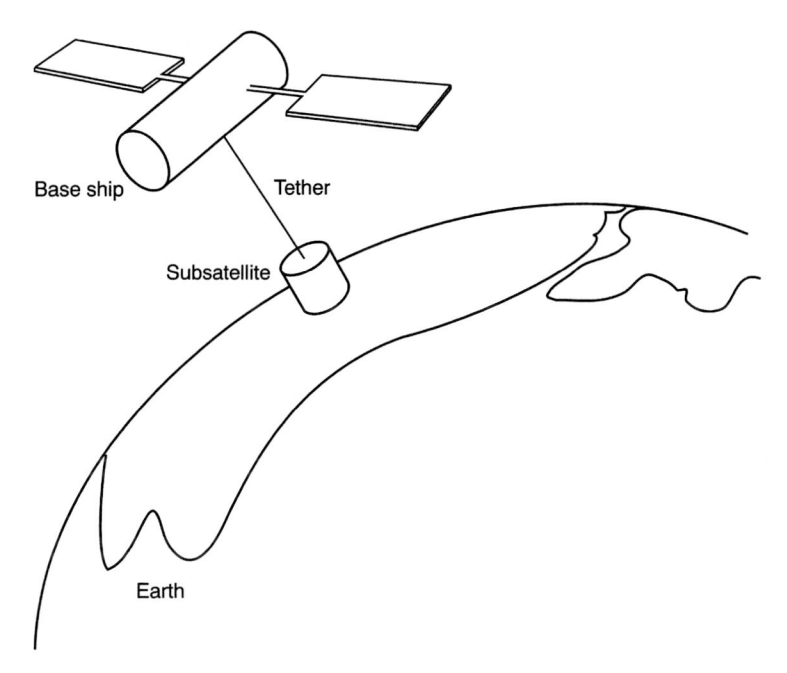

Fig. 7.18 Tethered satellite on a cable in orbit.

7.7
Tethered Satellites

Artificial earth satellites have become more complex since the launch of the first spherically shaped sputnik in 1957. Deploying antenna, folded solar panels, trusslike structures, gyro, and attitude stability wheels have all added dynamic complexities to satellite design (see Fig. 7.1). One of the latest configurations is the tethered satellite pair shown in (Fig. 7.18). Applications for tethered satellites include:

- Orbiting antenna and interferometer
- Upper-atmosphere probes
- Electric power source
- Refueling platform
- Artificial-gravity platform
- Microgravity platform

This section can only provide an introduction to the dynamics of tethered satellites. The interested reader should consult the excellent monograph by two Russian experts, Beletsky and Levin (1993). They trace the history of the tethered concept back to Tsidkovskii in 1895 and provide detailed analysis of modern tethered satellite design.

An early test of a tethered pair in orbit was the U.S. *Gemini II* tethered to an Agena rocket stage in 1966. The pair was set into a rotation rate 13.5 times the orbital rate, which produced an artificial gravity. More recently a U.S.–Italian space shuttle experiment deployed a satellite several kilometers from the shuttle tethered by a small-diameter cable that transmitted electrical power to the shuttle. The cable eventually failed, however. The cause of this failure was not known at the time of writing of this book. But it shows that designing multibody satellites is still a developing field today.

Dynamics is not the only problem the designer of a tethered pair must consider. According to Beletsky and Levin, other design concerns include:

- Tether-cable material strength
- Solar and vehicle attraction
- Tether bending stiffness
- Waves on the tether and tangling
- Cable deployment and retrieval
- Residual stresses
- Internal friction in multifiber or strained cables
- Electrostatic charge effects
- Plasma drag
- Electric currents in the tether
- Air drag
- Tether heating
- Meteor impact

Again the reader is guided to the work of Beletsky and Levin, who also include 215 references on tether research and design.

Our goal in this section is to show how simple dynamics tools can be used to analyze the most elementary motions of an orbiting tethered pair. In this task we necessarily make simplifying assumptions. In practice we must consider the tether elasticity. Two types of tether motions are related to tether deformation: extensional waves, and string or transverse waves. The string wave speed v_t is related to the tension in the tether, T, and the mass density per unit length γ:

$$v_t = (T/\gamma)^{1/2}$$

The extensional wave speed, v_e, is related to the composite cable Young's modulus, Y, and the cross-sectional area A;

$$v_e = (YA/\gamma)^{1/2}$$

According to Beletsky and Levin, a stainless-steel tether with a tension stress equal to 1/10 of the rupture stress would yield $v_e = 5.0$ km/s and $v_t = 0.16$ km/s. For a Kevlar composite cable, $v_e = 9.5$ km/s, $v_t = 0.44$ km/s. Thus we can see that motions associated with tether deformations occur on time scales much smaller than the orbital or libration periods for cable lengths of 1–10 km.

We consider next the simplified problem of a straight, inextensible-cable tethered pair. This problem is similar to those studied in Section 7.5, and in particular the

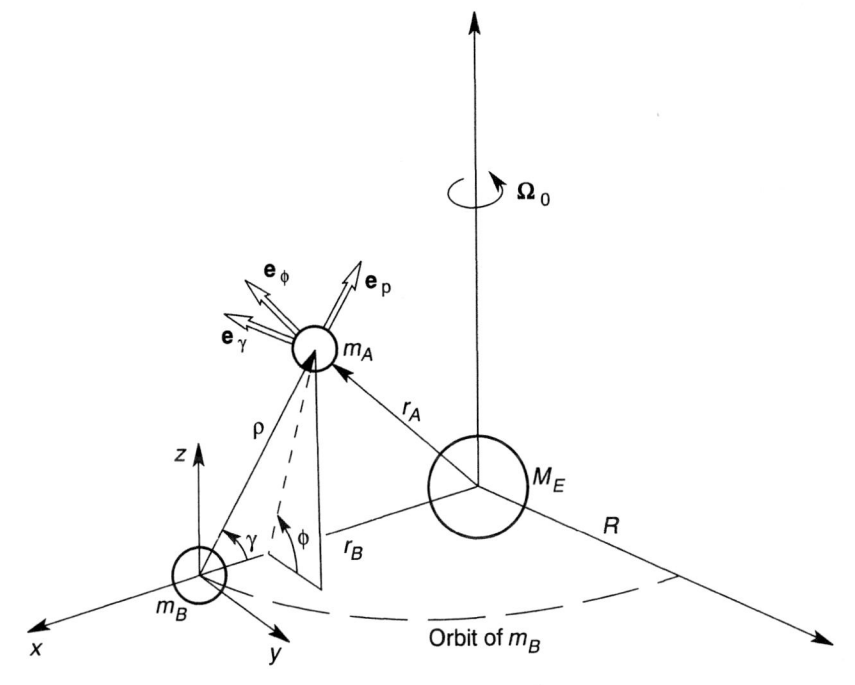

Fig. 7.19 Geometry of earth, base satellite m_B, and tethered satellite, m_A. Base satellite in a circular orbit.

dumbbell satellite. Further if we neglect the rigid-body orientation of the mother ship and the subsatellite, the tethered-pair dynamics is identical to rigid-body dynamics of an unequal mass dumbbell satellite as long as the cable tension remains positive. In this problem we neglect the gravitational force between m_A, m_B.

Circular Orbit of a Large Base Satellite

There are two further assumptions that can be made to obtain a first-order analysis of the dynamics of a tethered pair of masses. First if the masses of the two coupled satellites are comparable, then a reference system centered at the center of mass would be appropriate. But if one mass is very much larger than the subsatellite, as in the U.S.–Italian 1996 space shuttle experiment, then the center of mass is almost identical to that of the larger satellite. In this case we assume that the orbit of the large mass is not disturbed significantly by the smaller tethered mass.

Second as a matter of mathematical convenience, we examine only the case of a circular orbit of the base satellite. The case of an elliptical orbit may be found in Beletsky and Levin (1993). We assume that a base satellite of mass m_B orbits the earth E in the circular orbit shown in Fig. 7.19 such that the following balance law is valid

$$\ddot{\mathbf{r}}_B = -\frac{GM_E \mathbf{r}_B}{R^3} \tag{7.80}$$

where $\mathbf{r}_B = -R\mathbf{e}_x$. The mass m_B is in circular motion with rotation vector $\boldsymbol{\Omega}_0$ given by $\boldsymbol{\Omega}_0 = \Omega_0 \mathbf{e}_z$, $\Omega_0^2 = GM_E/R^3$. We further assume a deployed tethered subsatellite of mass m_A on an inextensible cable of length ρ. The equation of motion of the subsatellite is given by Newton's law

$$m_A \ddot{\mathbf{r}}_A = -\frac{m_A GM_E}{r_A^3}\mathbf{r}_A + \mathbf{T} \tag{7.81}$$

where $\mathbf{T}$ is the cable-tension vector.

Our goal in this analysis is to write this equation in rotating coordinates moving with the mother ship $\{\mathbf{e}_x, \mathbf{e}_y, \mathbf{e}_z\}$. It is also convenient to express the vectors in the spherical set of basic vectors $\{\mathbf{e}_\rho, \mathbf{e}_\phi, \mathbf{e}_\gamma\}$ shown in Fig. 7.19. To rewrite (7.81) in a moving frame of reference, we use the relation (3.18) and note that $\mathbf{r}_A = \mathbf{r}_B + \boldsymbol{\rho} = R\mathbf{e}_x + \rho \mathbf{e}_\rho$. Noting that $\dot{\boldsymbol{\Omega}}_0 = 0$, we obtain

$$\mathbf{a}_\rho + \boldsymbol{\Omega}_0 \times (\boldsymbol{\Omega}_0 \times \boldsymbol{\rho}) + 2\boldsymbol{\Omega}_0 \times \mathbf{v}_\rho = -\frac{GM_E}{r_A^3}\mathbf{r}_A - \ddot{\mathbf{r}}_B + \mathbf{T} \tag{7.82}$$

In this equation, $\mathbf{a}_\rho$, and $\mathbf{v}_\rho$ are measured with respect to the mother satellite reference system. The first two terms on the right represent the gravity gradient force $\mathbf{f}$. The term $\mathbf{T} = -T\mathbf{e}_\rho$ is the cable tension. Using the equation of motion for m_B (7.80), this force has the form

$$\mathbf{f} = GM_E\left(\frac{\mathbf{r}_B}{R^3} - \frac{\mathbf{r}_A}{r_A^3}\right)$$

In practical applications, $\rho \ll R$. In this case, Beletsky and Levin (1993) show that $\mathbf{f}$ is approximately given by

$$\mathbf{f} = -GM_E\left(\boldsymbol{\rho} - 3(\boldsymbol{\rho} \cdot \mathbf{r}_B)\frac{\mathbf{r}_B}{R^5}\right) \tag{7.83}$$

Using mixed-basis vectors, this takes the form

$$\mathbf{f} = -\frac{\rho GM_E}{R^3}(\mathbf{e}_\rho - 3\cos\gamma \mathbf{e}_x) \tag{7.84}$$

Using transformation matrices, we can show that the two sets of basis vectors are related by the matrix equation

$$\begin{Bmatrix} \mathbf{e}_x \\ \mathbf{e}_y \\ \mathbf{e}_z \end{Bmatrix} = \begin{bmatrix} -C\gamma & 0 & S\gamma \\ C\phi C\gamma & -S\phi & -C\phi S\gamma \\ S\phi S\gamma & C\phi & S\phi C\gamma \end{bmatrix} \begin{Bmatrix} \mathbf{e}_\rho \\ \mathbf{e}_\phi \\ \mathbf{e}_\gamma \end{Bmatrix} \tag{7.85}$$

The equations of motion based on (7.82) can be represented in spherical coordinates using

$$\boldsymbol{\rho} = \rho \mathbf{e}_\rho$$

$$\mathbf{v} = \dot{\rho}\mathbf{e}_\rho + \rho\dot{\phi}\sin\gamma\,\mathbf{e}_\phi + \rho\dot{\gamma}\mathbf{e}_\gamma$$

$$\mathbf{a}_\rho = (\ddot{\rho} - \rho\dot{\gamma}^2 - \rho\dot{\phi}^2\sin^2\gamma)\mathbf{e}_\rho$$

$$+ (\rho\ddot{\phi}\sin\gamma + 2\dot{\rho}\dot{\phi}\sin\gamma + 2\rho\dot{\gamma}\dot{\phi}\cos\gamma)\mathbf{e}_\phi$$

$$+ (\rho\ddot{\gamma} + 2\dot{\rho}\dot{\gamma} - \rho\dot{\phi}^2\sin\gamma\cos\gamma)\mathbf{e}_\gamma \tag{7.86}$$

We now have assembled all the equations to derive a set of three coupled differential equations for the motion of the tethered mass m_A with respect to the mother ship m_B. The unknowns in this problem are $\{\rho(t), T, \gamma(t), \phi(t)\}$. If we neglect the elasticity of the cable, then $\rho(t)$ is prescribed during deployment or reel-in. Once deployed, we can set $\dot{\rho} = \ddot{\rho} = 0$, in the preceding equations. The cable tension $T(t)$ will vary in time with the motion. However, we must check that $T > 0$ during integration of the equations. In the radial equilibrium state we have a balance of the gravity-gradient force, the tension, and the centripetal acceleration in (7.82)

$$T_0 = 2\rho\frac{GM_E}{R^3} + \Omega_0^2\rho \tag{7.87}$$

Thus we see that the cable tension is proportional to the deployed cable length.

It serves no pedagogical purpose to write out the equations of motion, given their complexity. There is an exercise in the Problems to use *MATHEMATICA* or other symbolic codes to write out the complete equations.

For small angle γ, one can expect pendulum-type motions of the kind we examined in Section 7.5 on the dumbbell satellite. Further details of tether dynamics may be found in Beletsky and Levin (1993).

Homework Problems

7.1 The orbit geometry of a central-force, inverse-square-force law includes the circle, ellipse, parabola, and hyperbola. These curves are often referred to as "conic sections" because they result from slicing a right circular cone with a plane. The geometry of an ellipse is shown in Fig. P7.1. The conic section is defined by the constant ratio of $r/PD = \varepsilon$, where ε is called the eccentricity. (The origin of r is called the *focus*, and the line from which PD is measured is called the *directrix*.) Show that $\varepsilon = 0$ for a circle; $\varepsilon > 1$ for a hyperbola.

7.2 For an elliptic orbit (Fig. P7.1) show that the apogee and perigee radii are given by $r_a = a(1 + \varepsilon)$, $r_p = a(1 - \varepsilon)$, where ε is the eccentricity (see Problem 7.1) and $2a$ is the major diameter of the ellipse.

7.3 One of Kepler's laws of orbital motion states that the area swept out per unit time is a constant. Use this fact to derive the relation between the period of the orbit T and the angular momentum per unit mass h:

$$T = 2\pi ab/h$$

(Note that πab is the area of the ellipse; see Fig. P7.1.)

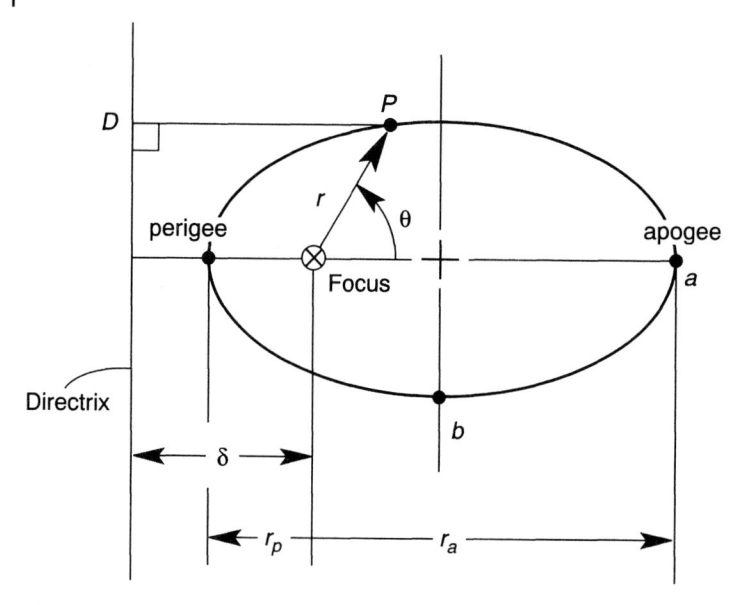

Fig. P7.1

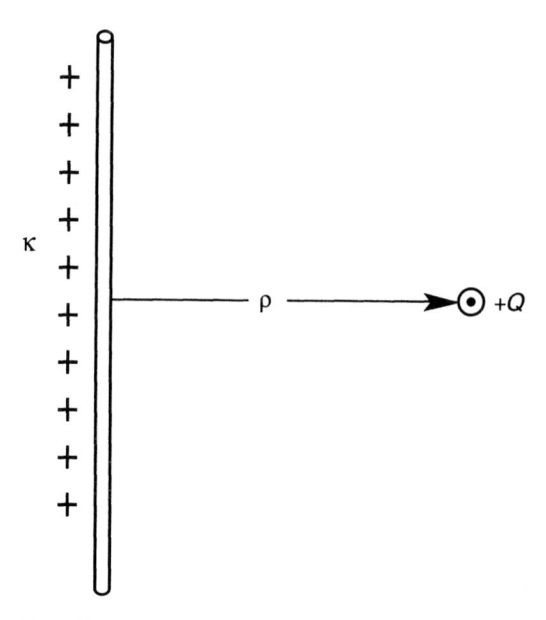

Fig. P7.4

7.4 Electric charges also experience an inverse-square-force law with the radial force between two point charges given by

$$F_r = \frac{1}{4\pi\varepsilon_0}\frac{Q_1 Q_2}{r^2}$$

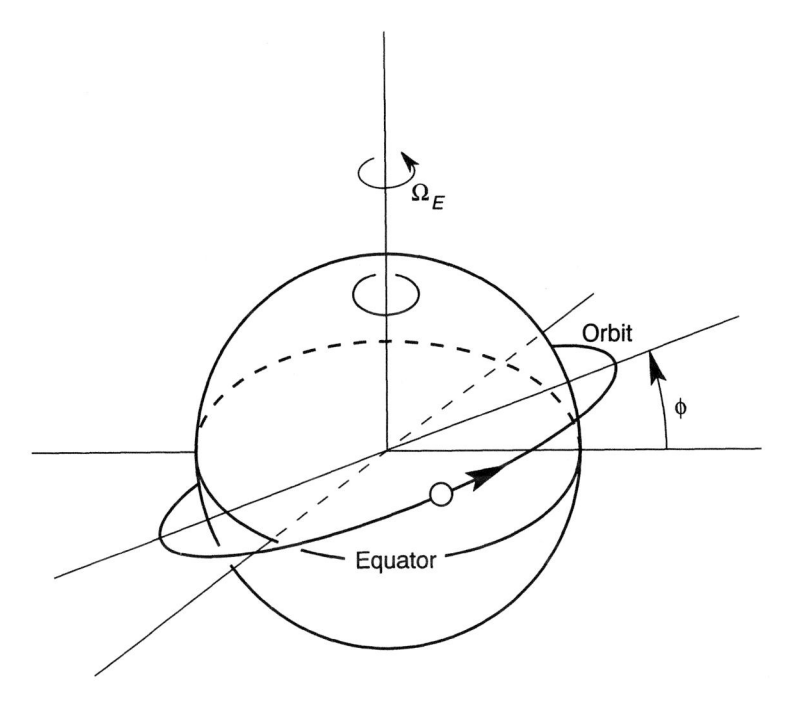

Fig. P7.6

Assuming that an infinite wire contains a distributed charge of κ coulombs per meter (C/m), derive the force law between a charge Q and the distributed charged wire (Fig. P7.4).

7.5 For the example of Problem 7.4, assume that the charged wire and point charge Q are of opposite charge. Examine the equations of motion of the point charge in a plane normal to the wire. Do the equations admit closed orbits?

7.6 Imagine a satellite in a low-altitude earth orbit with the plane of the orbit inclined to the equatorial plane at an angle ϕ (see Fig. P7.6). For small ϕ, show that the radial projection of the orbit onto the earth's surface (or ground-plane projection) is approximately sinusoidal. Calculate the distance between the points that the orbit shadow crosses the equator between two successive orbits.

7.7 The sun's diameter is approximately 109 times the diameter of the earth. Calculate the center of mass of the sun–Jupiter system as a ratio of the sun's radius. Given the masses and separation distance of the sun and Jupiter, find their mutual rotation orbit as a two-body problem, assuming circular orbits. (See Example 7.4.)

7.8 Suppose a small satellite is launched from the surface of the earth in an equatorial orbit with orbit insertion velocities $v_r = v_0 \sin \alpha$, $v_\theta = v_0 \cos \alpha$ relative to the surface of the earth (Fig. P7.8). Find the orbit parameters, including

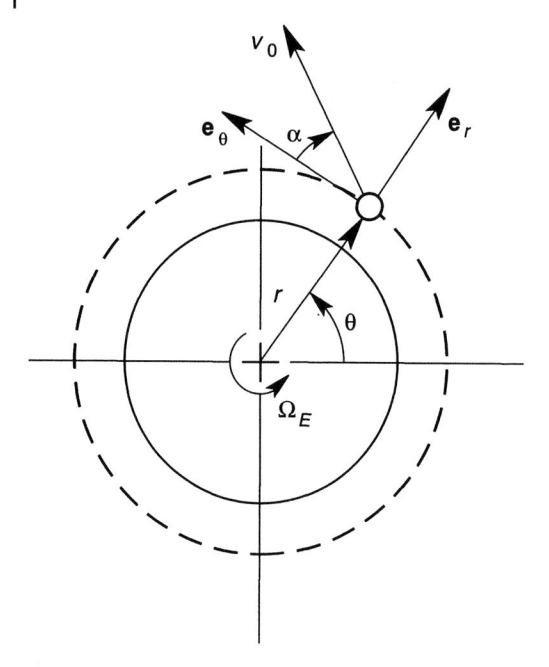

Fig. P7.8

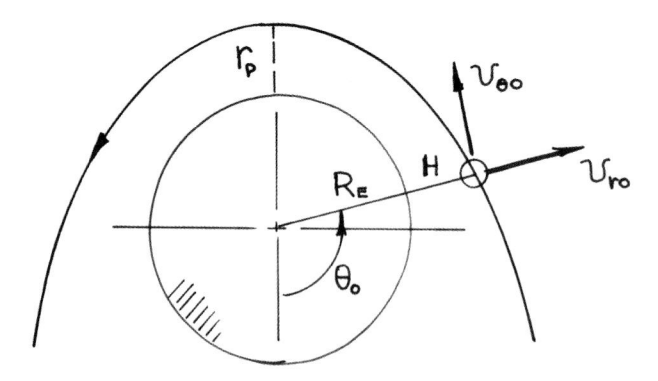

Fig. P7.9

the angle of the major axes relative to the original launch radius in a fixed frame of reference.

7.9 (*Hyperbolic Orbit*) A satellite is launched from the Earth at a height H, above the surface and inserted in a hyperbolic orbit with a radial velocity of v_{r0}, and a circumferential velocity of $v_{\theta 0}$ (Fig. P7.9). What is the condition on the initial velocity components for a hyperbolic orbit? Find the perigee of the hyperbolic orbit and the location of the perigee relative to the launch origin.

7.10 (*Transfer Orbits*) In Fig. P7.10 a satellite is in an elliptical orbit with apogee r_{a1}, and perigee r_{p1}, where the major axis length a and eccentricity ε are determined from $r_a = a(1 + \varepsilon), r_p = a(1 - \varepsilon)$. Suppose engineers want to

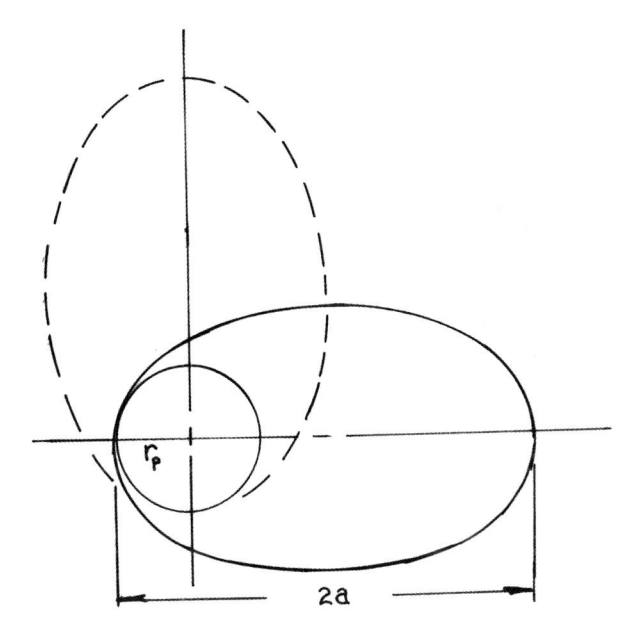

Fig. P7.10

transfer the satellite to another elliptical orbit at 90 degrees from the original orbit as shown in Fig. P7.10, where the new apogee is equal to the old perigee and the new orbit has the same eccentricity ε as the original. Use a circular transfer orbit and find the two impulse burns necessary, assuming that the burn times are very short compared to the orbit period.

7.11 Find an orbit around Mars that is stationary with the rotation of Mars.

7.12 Suppose a satellite orbits the Earth at a distance H above the surface and has an eccentricity equal to 0.7. What is the time to traverse the orbit from the apogee from 0 to 90 degrees in the circumferential direction?

7.13 (*Hohmann Transfer Orbit*) Find a Hohmann transfer orbit to go from a circular orbit above Mars at a radius equal to the diameter of Mars, to a circular orbit close to the surface of Mars.

7.14 (*Center of Gravity*) Use the definition of the center of gravity, (7.60), for a distributed mass satellite to derive the approximate expression (7.61) for a uniform rod of length L, orbiting in a circular orbit of radius r_c, and aligned with its long length in the radial direction.

7.15 (*Attitude Stability*) A satellite can be modeled as a cylindrical shell with mass m_1, and closed circular plate ends each of mass m_2 (Fig. P7.15). Use Eq. (7.70) in Example 7.7 to determine the stable and unstable axes for attitude stability under gravity gradients.

7.16 For numerical integration of the orbit Eqs. (7.6), using *MATLAB*, it is convenient to use nondimensional variables. Let the radial variable r be normalized with respect to the earth's radius r_e and the time be normalized by the period

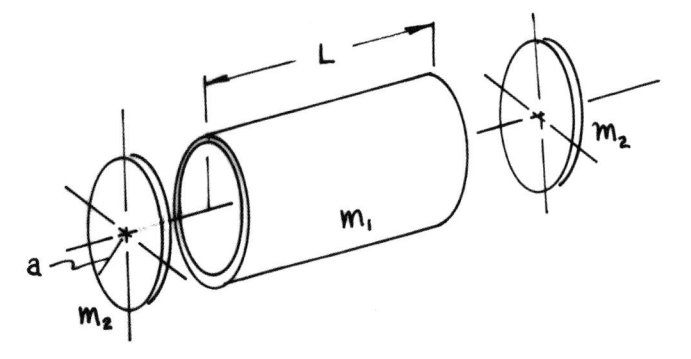

Fig. P7.15

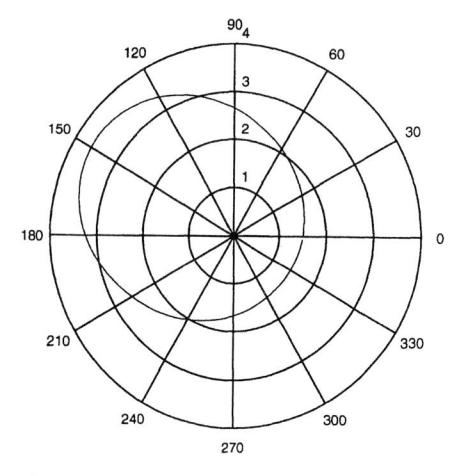

Fig. P7.17

of a circular orbit near the earth's surface, i.e., $\tau = 2\pi (r_e^3/GM)^{1/2}$ [see (7.13)]. Then show that the nondimensional equations of motion become

$$\ddot{r} = r\dot{\theta}^2 - 4\pi^2/r^2$$
$$\ddot{\theta} = -2\dot{\theta}\dot{r}/r$$

7.17 The following MATLAB program will integrate the orbit equations in Problem 7.16. In this code the two second-order polar equations of motion are rewritten as a set of first-order equations in time. The components of the state vector **x** are defined by $x_1 = r$, $x_2 = \dot{r}$, $x_3 = \theta$, $x_4 = \dot{\theta}$, where r, θ are defined in Problem 7.16. Examine the cases of a circular orbit for $\mathbf{x0} = [1, 0, 0, 2\pi]^T$ and an elliptical orbit for $\mathbf{x0} = [2, 0, 0, 2]^T$ with a time of integration given by $t0 = 0$, $tf = 4$. (*Note:* the code may take too large a time step in the case of a circular orbit.) Run the case $\mathbf{x0} = [1.5, 1.0, 0, 4]^T$ (Fig. P7.17). Why does this case give an inclined elliptic orbit?

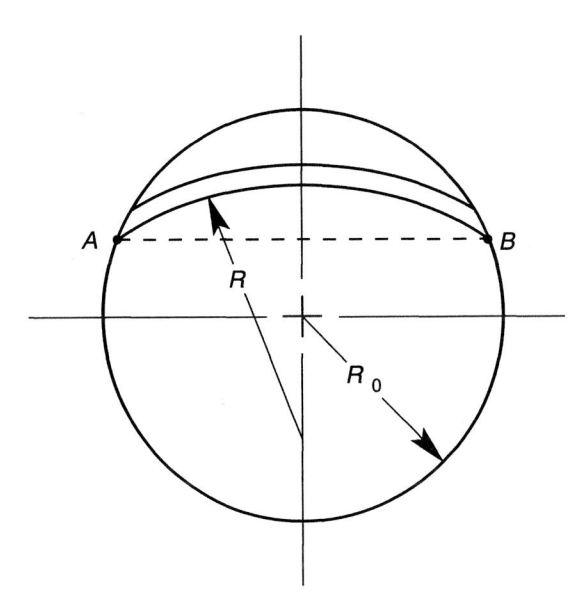

Fig. P7.18

```
MATLAB function orbit.m

function dx = orbit(t,x)
dx = zeros(4,1);
dx(1) = x(2);
dx(2) = x(1) .* x(4).^2 - 4 .* pi.^2 ./ x(1).^2;
dx(3) = x(4);
dx(4) = -2 .* x(4) .* x(2) ./ x(1);
MATLAB integration program Plotorbit.m
t0 = 0; tf = 4;
x0 = [1.5,0,0,4]';
[t,x] = ode45('orbit',t0,tf,x0);
polar(x(:,3),x(:,1))
```

7.18 Suppose the gravity-grade tunnel system described in Example 7.5, is a circular arc curve instead of a straight chord. Compare the time of travel between points A, and B on the surface of a homogeneous planet for the arc and the straight tunnel (see Fig. P7.18).

7.19 Examine the attitude equilibrium, stability, and vibration of a two-mass dumbbell satellite in a circular orbit with unequal masses, $m_1 \neq m_2$ (Fig. P7.19). Consider only planar motions.

7.20 Analyze the out-of-plane stability and vibrations of the two-mass dumbbell satellite with the axis of rotation in the plane of the circular orbit (Fig. P7.20). Show that the equation of motion is given by (7.68).

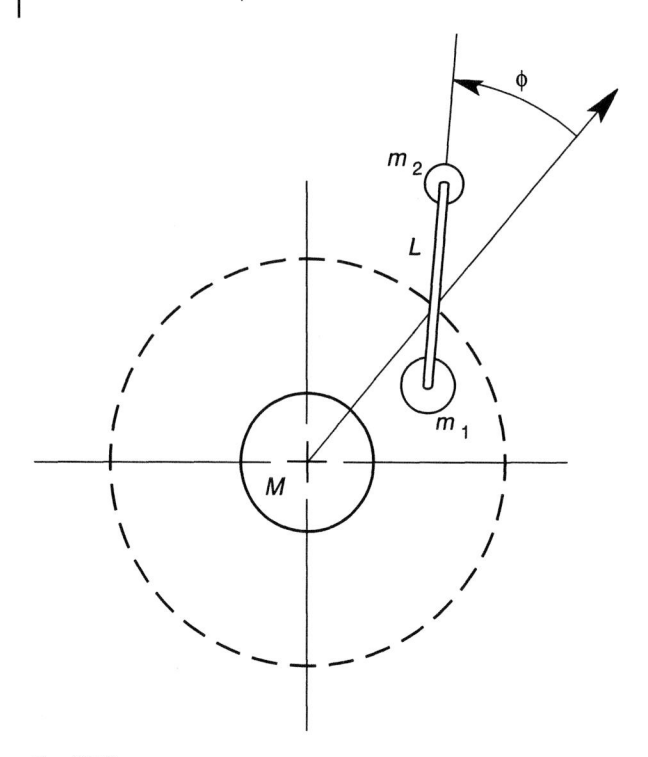

Fig. P7.19

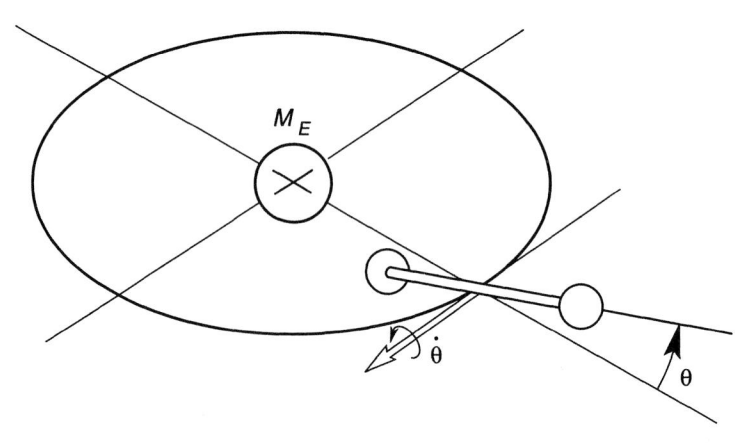

Fig. P7.20

7.21 Consider the three-mass orbiting satellite shown in Fig. P7.21. Assume that the satellite is in a circular orbit that remains unaffected by the satellite oscillations. Determine the planar equilibrium orientation of the satellite in orbit. Using the perturbation methods presented for the two-mass and four-mass satellite examples, determine the stable configurations and natural frequencies for planar oscillations.

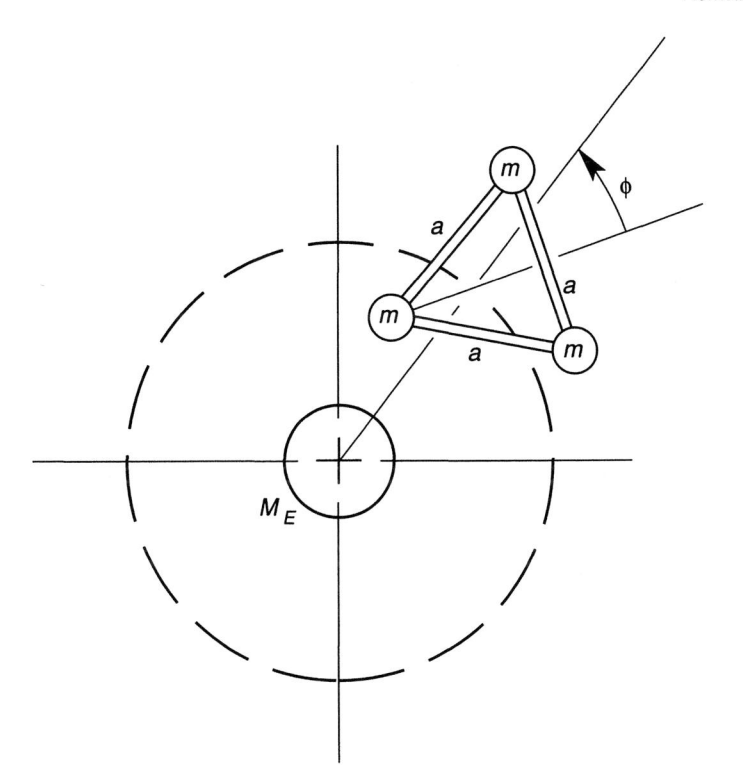

Fig. P7.21

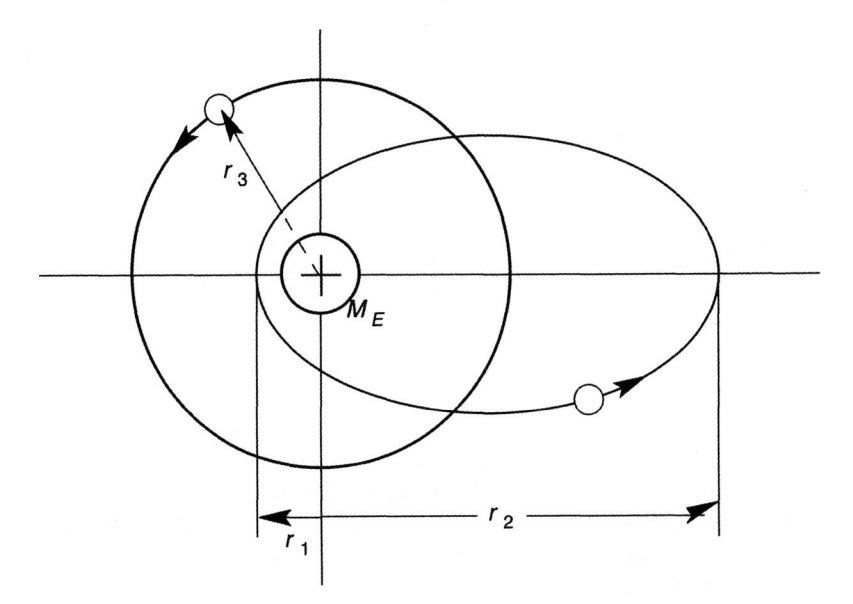

Fig. P7.22

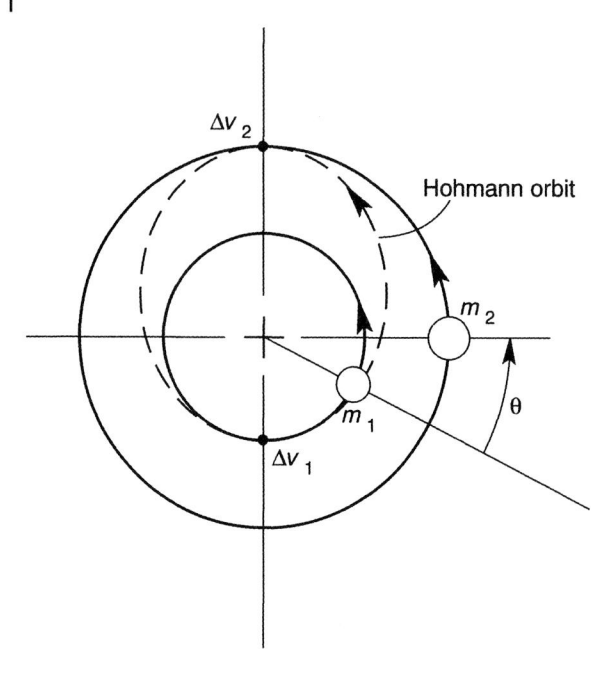

Fig. P7.23

7.22 A satellite is in an elliptical orbit about the earth where the apogee and perigee are known (Fig. P7.22). Find a transfer orbit to get into a circular orbit of radius r_3. Find expressions for the *two* Δv burns in terms of r_1, r_2, r_3, r_e, g, where r_e is the radius of the earth. Calculate for

$$r_e = 6400 \text{ km}$$
$$r_1 = 6600 \text{ km}, \qquad r_2 = 8000 \text{ km}$$
$$r_3 = 7000 \text{ km}$$

7.23 (*Rendezvous Transfer Orbit*) In Fig. P7.23, two satellites are in circular orbits r_1, r_2. Find a Hohmann transfer orbit and impulse rocket burns such that m_1 will rendezvous with m_2. (*Hints*: (1) Find the phase angle between m_1, and m_2 such that they both arrive at the same point on orbit, r_2. (2) Use the relation between the time of orbit and area of the elliptic orbit.) See Problems 7.1, 7.2, and 7.3.

7.24 A special case of the three-body gravitational dynamics problem is the motion of a small mass near the circular orbit of two larger masses m_1, m_2 shown in Fig. P7.24. In this problem assume that the small mass gravity forces have negligible effect on the co-rotating large masses. As a further assumption the small mass motions are restricted to the orbiting plane of the large masses. Examine the planar dynamics of m using a coordinate system located at the center of mass of m_1, m_2 and rotating with the two masses with angular ve-

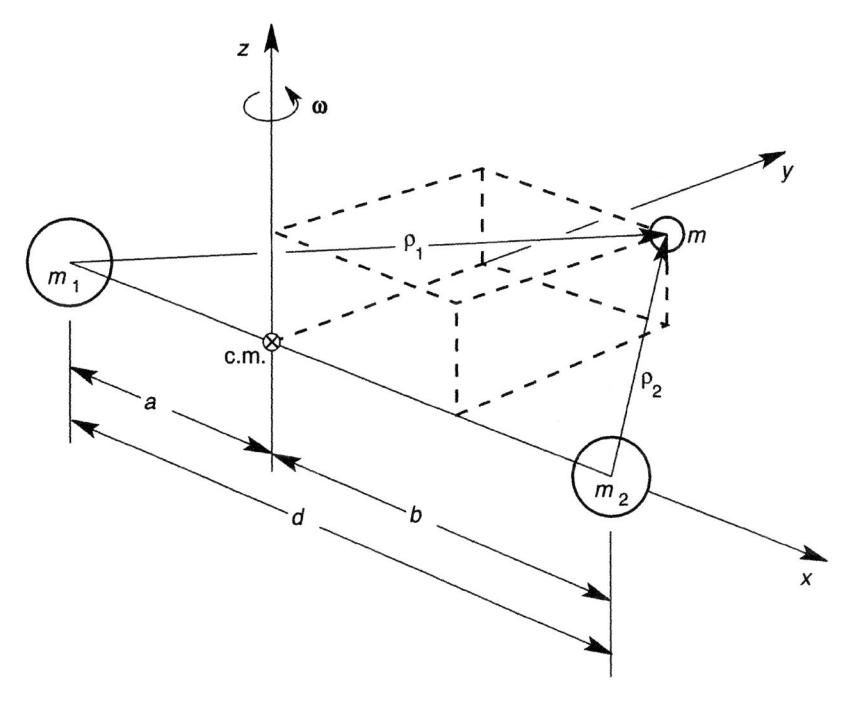

Fig. P7.24

locity given by $\omega^2 = G(M_1 + m_2)/d^3$, where d is the separation of the masses m_1, m_2. Using Lagrange's equations, show that the equations of planar motion are given by

$$\ddot{x} - 2\omega\dot{y} = -\frac{1}{m}\frac{\partial U}{\partial x}$$

$$\ddot{y} + 2\omega\dot{x} = -\frac{1}{m}\frac{\partial U}{\partial y}$$

where $U = V - \frac{1}{2}m\omega^2(x^2 + y^2)$ and V is the gravitational potential

$$V = -Gm\left(\frac{m_1}{\rho_1} + \frac{m_2}{\rho_2}\right)$$

where $\rho_1^2 = (x+a)^2 + y^2$, $\rho_2^2 = (x-b)^2 + y^2$. For a complete three-dimensional analysis, see Meirovitch (1970, Chapter 11).

7.25 Using the equations for the restricted three-body problem in Problem 7.24, write a MATLAB program to integrate the equations of motion and plot the orbits in the rotating x–y plane. Let m_1 be the earth and m_2, earth's moon. Choose initial conditions near the moon and examine how far a body can stray from the moon before the earth's gravitational field affects the moon–satellite orbits.

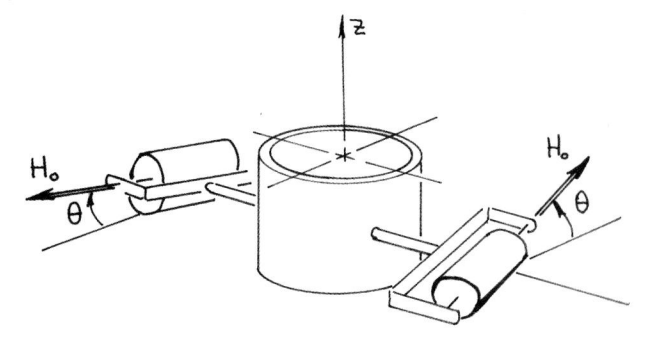

Fig. P7.28

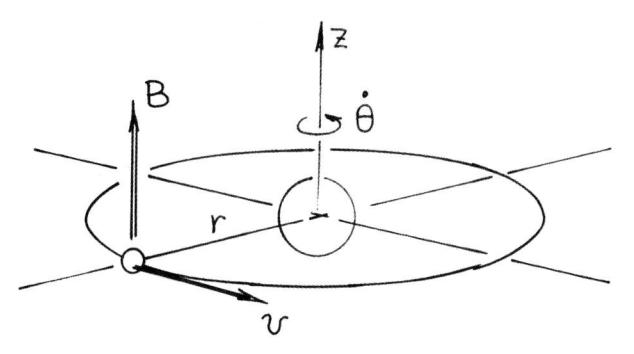

Fig. P7.29

7.26 In the restricted three-body problem of Problem 7.24, there are five equilibrium points known as the Lagrange points $\{L_1, L_2, L_3, L_4, L_5\}$. Set $\ddot{x} = \ddot{y} = \dot{x} = \dot{y} = 0$ in the equation of motion and show that three points L_1, L_2, L_3 lie on the axis connecting the two larger masses, and two other points L_4, L_5 lie off axis at points forming an equilateral triangle with m_1, m_2. For the earth–moon case, use *MATLAB* or other numerical tools to find $\{L_i\}$. (See also Meirovitch, 1970, Chapter 11.)

7.27 Use *MATHEMATICA* or other symbolic code to derive equations of motion for the tethered satellite described by (7.81)–(7.86). Linearize the equations about $\gamma = 0$ (see Fig. 7.18) and find the natural frequency of the tethered satellite.

7.28 (*Control Moment Gyro*) Consider a satellite shaped as a cylindrical shell with a scissored control moment gyro pair as in Fig. P7.28. Suppose we wish to slew the satellite about the z axis by π radians in a time t_0 and bring it to rest in this orientation. Find an open loop function $\theta(t)$ to gimbal the CMG pair such that there is no sudden angular acceleration of the satellite. (*Hint:* Use a gimbal function with zero velocity at the beginning and end of the maneuver.)

7.29 (*Charged Satellite*) Suppose a satellite accumulates an amount or charge Q in Coulombs. Assume that the satellite is initially in a circular orbit around the Earth and encounters a region of magnetic field B normal to the orbital plane

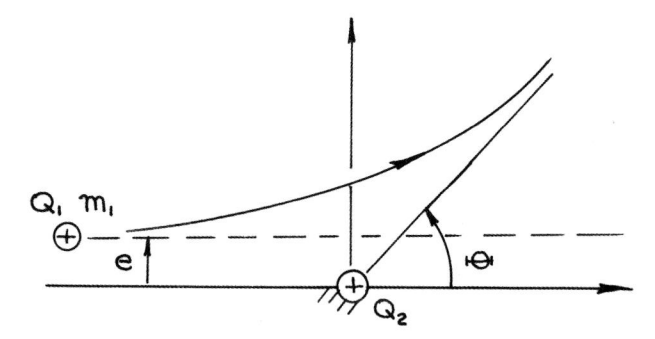

Fig. P7.30

(Fig. P7.29). If the electric force is $\mathbf{F} = Q\mathbf{v} \times \mathbf{B}$, determine the effect of the electric force on the satellite motion.

7.30 (*Scattering Dynamics*) In the physics of charged particles, same sign charges repel one another. In the diagram of Fig. P7.30, a charged particle with mass and charge, m_1, Q_1, approaches a fixed charge of strength Q_2 at an offset distance e. Use the equations of electric force to calculate the scattering angle as the moving charge moves well past the force center. Assume that the scales are large enough so that classical dynamics is valid. (See Goldstein et al., 2002, Chapter 3 for a solution.)

8
Electromechanical Dynamics: An Introduction to Mechatronics

8.1
Introduction and Applications

The design of modern machines involves dynamics, control, and intelligence. The so-called "smart machine" obtains information about both itself and its operating environment, processes this information into commands for control, and produces actuation forces and torques to enable the machine to move dynamically in space and time to accomplish some desirable goal. Examples of smart machines include VCRs, automatic cameras, cruise control and air-bag systems in automobiles, robots, flight control systems on aircraft, as well as most satellite systems. The control of machines goes back to water clocks in the Renaissance and speed controllers on steam engines in the nineteenth century. The study of "automatic controlled" machines is quite old. However, the modern computer and microchip era has introduced a greater capacity for processing information. Also modern electronics has enabled both sensing and information-processing hardware to be embedded into the machines, whereas in the past they were often separate systems, larger than the machines they were supposed to control.

The integration of dynamic and kinematic elements with sensing, actuation, and information processing has acquired the name mechatronics. The ultimate mechatronics device is the human and animal machine.

Two applications of mechatronics that span the extremes of scale are microelectromechanical systems (MEMS) and magnetically levitated (Mag-Lev) trains. MEMS devices have a size scale on the order of one micron (10^{-6} m) (Fig. 8.1), while Mag-Lev vehicles have a scale on the order of tens of meters (Fig. 8.2). (Since 2003 the German Mag-Lev system has been built in Shanghai, China.)

Although controlled machines often contain hydraulic or pneumatic actuators, electromagnetic sensors and actuators are increasingly becoming important components in the design of smart machines or structures. Yet, there are very few textbooks on dynamics that incorporate electromagnetic forces (see, e.g., Crandall et al., 1968, or Woodson and Melcher, 1968). In this chapter we define several types of electromagnetic forces and demonstrate their use in dynamic analysis using either the Newton–Euler or Lagrange formulation. The complete study of mechatron-

Applied Dynamics. With Applications to Multibody and Mechatronic Systems. Second Edition. Francis C. Moon
Copyright © 2008 WILEY-VCH Verlag GmbH & Co. KGaA, Weinheim
ISBN: 978-3-527-40751-4

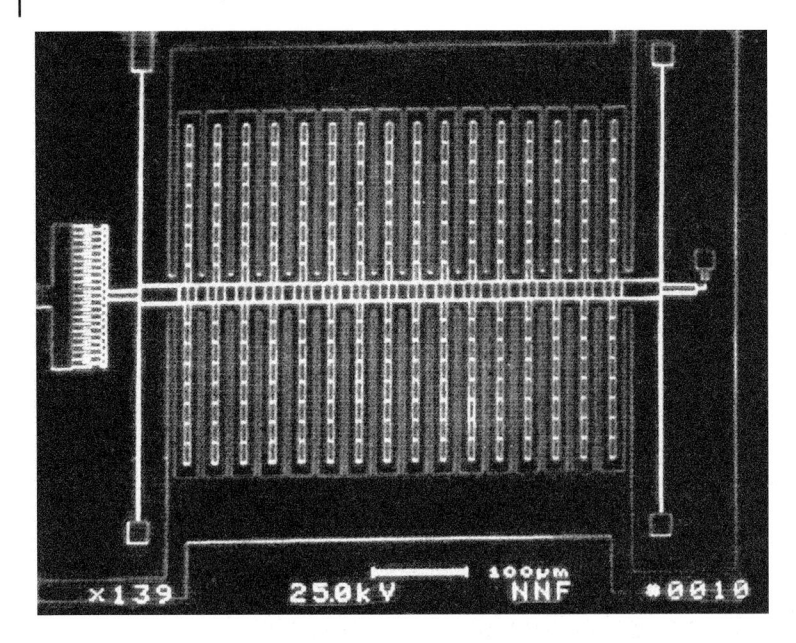

Fig. 8.1 Microelectromechanical actuator photo etched in silicon for a tunable microaccelerometer. (From S. G. Adams, *Design of Electrostatic Actuators to Tune the Effective Stiffness of Micro-Electromechanical Systems*, 1996.)

ics must include sensor theory, control theory, microelectronics, and information processing. We do not attempt to cover all of these elements. However, a principal component of mechatronic design is to be able to incorporate actuation forces into the dynamics. Thus, the goal of this chapter is to understand how to model several electromagnetic forces and to derive them along with the dynamical equations. We also see how an understanding of electromechanical dynamics can lead to some insights into the design of several application devices.

MEMS and NEMS Applications

Microelectromechanical devices made their entry into technology in the early 1990s with their use as accelerometers in automotive air bag sensor systems. Since then these micron-size mechanical systems have been used in other technologies such as optical switching and optical mirror systems for digital display systems as well as bio-medical applications. In the last decade, similar mechanical dynamical applications have appeared at scales approaching 10 nm that is the size of large molecules. Typically the mechanical components are etched out of silicon and are best represented as structural elements requiring analysis based on the theory of elasticity and structural mechanics. However in many devices, compliant mechanisms are employed in which the elastic deformation is localized at a "joint" or flexible hinge and a large part of the motion can be represented by rigid body dynamics.

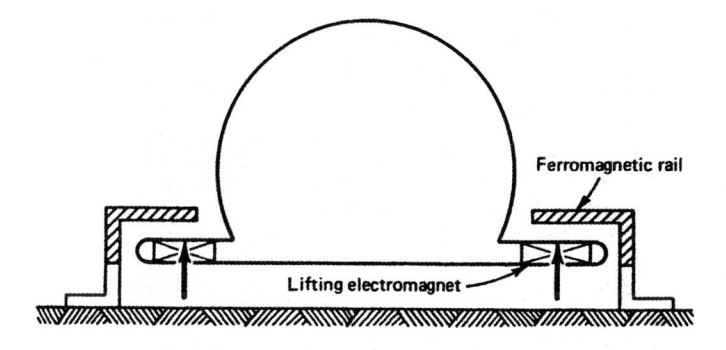

Fig. 8.2 German Mag-Lev 450-km/h passenger vehicle suspended by feedback control magnetic fields (Moon, 1994).

Other structures that are under study at the NEMS/MEMS level are string-like DNA strands, carbon nanotubes (C_{60}), graphene sheets made from unrolling carbon nanotubes, and hydrocarbon *molecular motors*. Two classes of molecular machines that are under study by chemists are molecular structures that link one another called *catenanes* and another called *rotaxanes* in which one chemical structure can slide along another chemical structure as in a prismatic joint. (See, e.g., Kelly, 2005.) Although computational simulation of these molecular structures is based on intermolecular forces, large calculation times have forced a form of coarse-graining in which simplified models are used that resemble coupled rigid and flexible bodies. Thus some of the concepts of classical dynamics are being adapted for use at the nano and molecular levels.

As scales approach atomic and molecular dimensions, one can ask whether classical mechanical models and electromagnetic forces are sufficient to explain experimental observations. The good news is that for the most part classic structural models such as the Euler–Bernoulli beam and the theory of elasticity are valid at the 10 nm scale and higher. However the forces acting on these classical structures

Table 8.1 Electromagnetic units.

Symbol	Physical quantity	MKS unit
B	Magnetic flux density	tesla
C	Capacitance	farad
D	Electric displacement	coulombs/meter^2
E	Electric-field intensity	volts/meter
H	Magnetic-field intensity	amperes/meter
I	Electric current	amperes or coulombs/s
J	Current density	amperes/meter^2
L	Inductance	henrys
M	Magnetization density	amperes/meter
m	Magnetic dipole	amperes/meter^2
P	Electric polarization density	coulombs/meter^2
Q	Electric charge	coulombs
q	Electric-charge density	coulombs/meter^3
R	Electrical resistance	ohms
V	Electrical voltage	volts
ε_0	Permittivity of vacuum	8.854×10^{-12} farad/meter
μ_0	Magnetic permeability of vacuum	$4\pi \times 10^{-7}$ henry/meter
τ	Electrical conductivity	(ohm · meter)$^{-1}$
ϕ	Magnetic flux linkages	weber-turns
Φ	Magnetic flux	webers
χ	Magnetic susceptibility	dimensionless

often involve quantum mechanics such as Casimir forces and quantum tunneling effects.

8.2
Electric and Magnetic Forces

There are many different types of electric and magnetic forces. However, many books on electromagnetics focus mainly on electric and magnetic fields and not very deeply on forces. Oddly, some of the more detailed discussions of electromagnetic forces are in old or out-of-print books such as Roters (1941) (on magnetic actuators) and Cady (1964) (piezoelectric forces).

In this chapter we encourage the student to review electromagnetic concepts and principles using a good introductory physics book. The basic objects upon which electromagnetic forces act are electrons, electric current, electric dipoles or polarization, and magnetic dipoles or magnetization. One should also review the concepts of electric and magnetic fields, as well as electric and magnetic circuit theory. In this book, we use the meter–kilogram–second–coulomb (MKSC) system of units. In this set of units, forces, energy, and power are expressed in newtons, joules and watts respectively. Electric units are expressed in amperes, volts, tesla, etc. (see Table 8.1).

8.2 Electric and Magnetic Forces

A special feature of electromechanical dynamics is the introduction of additional state variables that describe the state of the electric or magnetic circuit. It is often through these added state variables, such as a voltage across a capacitor or a magnetic field in an actuator, that control is effected. An understanding of the mutual interaction of the mechanical and electrical subsystems is the key to optimizing the design of electromechanical systems.

In electromagnetic systems, we can use either mass-based measures such as charge or dipole strength or force measures such as voltage or magnetic-field strength as state variables. In a continuum description we often use *charge density* q (C/m^3), *current density* **J** (A/m^2), *polarization density* **P** (C/m^2), and *magnetization density* **M** (A/m). The long-range forces on charges, currents, and dipoles are described with auxiliary variables **E**, **D**, **B**, **H**, called, respectively, the *electric field, electric displacement, magnetic-field density*, and *magnetic-field strength*. Thus, the body force acting on a volume element Δv with charge density q, is given by

$$\Delta \mathbf{F} = q \Delta v \mathbf{E}$$

or

$$\mathbf{f} = q\mathbf{E} \ (\text{N/m}^3) \tag{8.1}$$

Similarly, the body force per unit volume on a volume element carrying current, **J**, in a magnetic field, **B**, is given by

$$\mathbf{f} = \mathbf{J} \times \mathbf{B} \tag{8.2}$$

where the force, current, and magnetic field directions are related by **f** acting normal to the plane of **J** and **B**.

The total charge in a finite volume is denoted by Q, and the total force on Q in an electric field uniform throughout this volume is given by

$$\mathbf{F} = Q\mathbf{E} \tag{8.3}$$

Thus we can see that the unit for electric field, **E** is newtons/coulomb. Similarly, the total force on a wire element carrying current **I** (coulombs/s) of unit length is

$$\hat{\mathbf{f}} = \mathbf{I} \times \mathbf{B} \tag{8.4}$$

Here, however, the unit of $\hat{\mathbf{f}}$ is N/m. If **I** flows in a closed circuit as shown in Fig. 8.3, then the total force is given by the integral of **I** around the circuit

$$\mathbf{F} = \oint \mathbf{I} \times \mathbf{B} \, ds \tag{8.5}$$

We note in the figure that if **B** is uniform, then **F** in (8.5) is zero. Thus a nonzero force on a closed circuit must involve an inhomogeneous magnetic field. The elec-

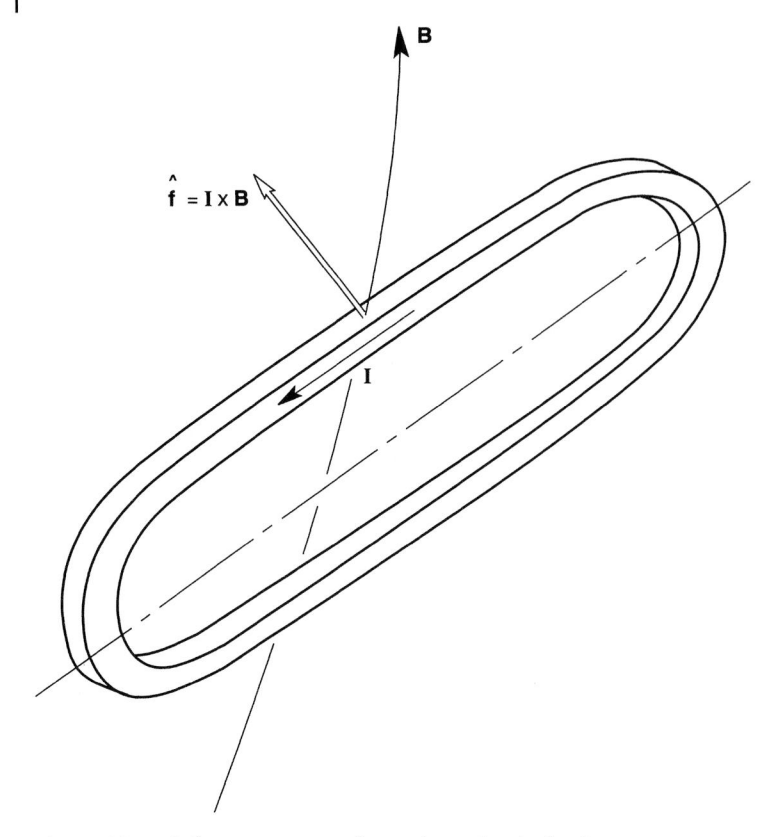

Fig. 8.3 Magnetic force on a current element in an electric circuit.

tric force between two charges is given in Chapter 4, (4.80b) as well as the force between parallel current filaments (4.80a).

Motion-induced Voltages

It is a well-known law of electromagnetics that the motion of an object in a magnetic field will produce an electric field and hence a voltage sometimes known as a back electromotive force or back emf (Lenz's law):

$$\mathbf{E} = \mathbf{v} \times \mathbf{B} \tag{8.6}$$

$$V_b = \int \mathbf{E} \cdot d\boldsymbol{\ell} \tag{8.7}$$

For a wire of length ℓ moving perpendicular to a magnetic field

$$V_b = -vB\ell = -\gamma v \tag{8.8}$$

If the wire carries current I, such that the magnetic force is in the same direction as $\mathbf{v}$, then

$$F = IB\ell = \gamma I \tag{8.9}$$

Thus the minus sign in V_b indicates that the induced voltage is opposite to the voltage that would produce the current I. Another basic fact is that the constant γ, in V_b, F are the same, which can also be shown using Lagrange's equations below.

Example 8.1 Voice Coil Actuator. A typical linear magnetic actuator is found in conventional acoustic speakers that can also be used as an actuator in other electromechanical systems. In this configuration, shown in Fig. 8.4, a radial magnetic field is produced by a permanent magnet and confined to a magnetic circuit by soft iron. In this figure we show only one turn, but in practice the number of turns is quite large, e.g., 10^3. To derive the equations of motion of the coupled system we assume that a voltage applied to the coil produces a current I. Using (8.5) the force on the wire is

$$\mathbf{F} = \oint \mathbf{I} \times \mathbf{B}\, d\ell = 2\pi r B_r I \mathbf{e}_x$$

However, once the coil begins to move with velocity $\dot{x}$, a back emf will be induced according to Lenz's law, (8.7):

$$V_b = \oint (\mathbf{v} \times \mathbf{B}) \cdot d\ell = 2\pi r B_r \dot{x}$$

Adding damping to the moving mass and resistance to the coil circuit, Newton's law and Kirchhoff's law become

$$m\ddot{x} + b\dot{x} = \gamma I$$

$$L\dot{I} + RI = -\gamma \dot{x} + V \tag{8.10}$$

where $\gamma = 2\pi r B_r$ for one turn and $\gamma = 2\pi r N \hat{B}$ for a multiturn coil with an average radial field $\hat{B}$. This is a linear, third-order system and as such we can determine the modes of the system by setting $V = 0$ and

$$\begin{vmatrix} x \\ I \end{vmatrix} = e^{st} \begin{bmatrix} x_0 \\ I_0 \end{bmatrix}$$

The eigenvalues of the system are found to be solutions of

$$(ms^2 + bs)(Ls + R) + \gamma^2 s = 0$$

When b, R are small, we find $s = 0$, $s = \pm i(\gamma^2/mL)^{1/2}$. The root $s = 0$ implies that x is constant, and the second two roots correspond to an oscillatory solution

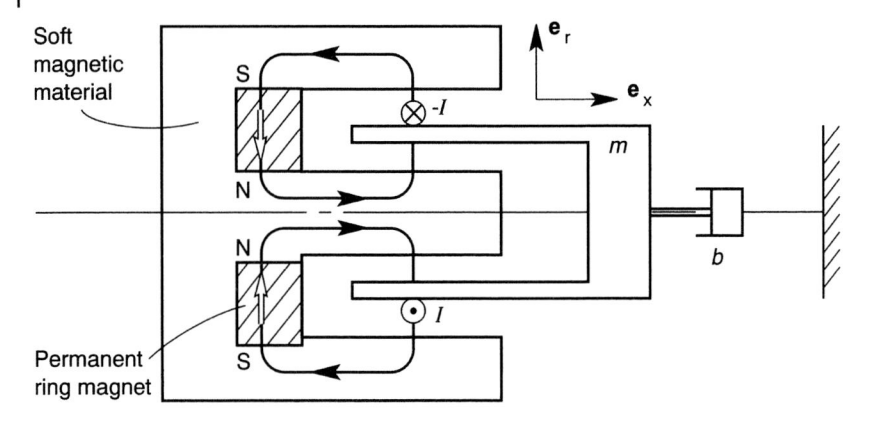

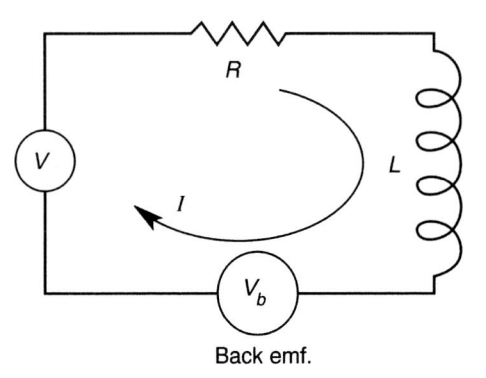

Fig. 8.4 Voice-coil actuator and its equivalent circuit showing a velocity-induced back-emf voltage source (Example 8.1).

of frequency $(\gamma^2/mL)^{1/2}$ radians per second (rad/s). Another interesting property of this system can be found by multiplying Newton's law by the velocity $\dot{x}$, and the circuit equation by the current I. Adding the resulting equations one obtains

$$\frac{d}{dt}\left(\frac{1}{2}mv^2 + \frac{1}{2}LI^2\right) = VI - RI^2 - b\dot{x}^2 \tag{8.11}$$

Thus the change in the kinetic plus magnetic energies is increased by the applied voltage and decreased by the dissipation losses in the resistor and the damper.

Example 8.2 Magnetic Accelerator. Imagine a one-meter-long wire carrying 100 A on a movable mass of $m = 1$ kg (Fig. 8.5). Suppose a nearby wire receives a sudden large current, of the opposite sense to the first, say 10,000 A, which is typical of a

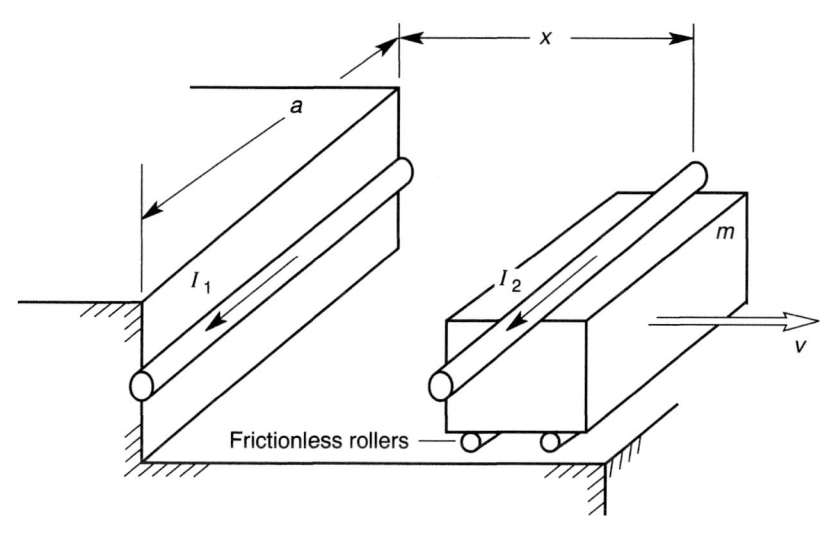

Fig. 8.5 Simple model of a magnetic-force accelerator.

small lightning strike. What is the velocity that the mass could attain? The equation of motion is given by

$$m\dot{v} = -\frac{\mu_0 I_1 I_2}{2\pi x} a \tag{8.12}$$

where a is the length of the wire.

If we multiply each side of the equation by $v = dx/dt$ and integrate with respect to time, we obtain a relation between the velocity (assuming a zero velocity start) and the initial and final distance between the wires (essentially this is a form of conservation of energy):

$$\frac{m}{2}v^2 = -\frac{\mu_0 I_1 I_2}{2\pi} a \ln \frac{x}{x_0} \tag{8.13}$$

or using $I_1 = 100$ A, $I_2 = -10{,}000$ A, $a = 1$, $m = 1$

$$v^2 = 0.4 \ln(x/x_0)$$

If $x/x_0 = 10$, then $v = 0.96$ m/s.

This example illustrates the basic principle in magnetic-forming devices and magnetic rail guns (see, e.g., Moon, 1984).

Electric and Magnetic Stresses

Another method to determine electric and magnetic forces in mechatronic systems is based on the concept of electric and magnetic stresses. (See Moon, 1994.) This idea, which had its origins in the works of Faraday and Maxwell, replaces electromagnetic body forces with surface stresses. Thus for example, in the case of the force on a ferromagnetic body in a magnetic field, the forces between the magnetization in the body and the external field are replaced with a *magnetic tension* $B_n^2/2\mu_0$ on the body surface with normal **n**. Similarly the force between a charged body in an electric field can be replaced by a *electric tensile stress* on the surface of normal **n**, $\varepsilon_0 E_n^2/2$. In both examples, the magnetic or electric field vectors are almost normal to the surface. When the total magnetic field is tangential to the surface, as in the case of a type I superconductor, the magnetic force can be calculated using a *magnetic pressure*, $B_s^2/2\mu_0$, where **s** is a unit vector, tangential to the surface and $\mathbf{B} = B_n \mathbf{n} + B_s \mathbf{s}$.

General expressions for magnetic and electric stresses are given by the following two surface integrals (see Stratton, 1941, p. 103):

$$F_m = \frac{1}{2\mu_0} \int_\Sigma \left[(B_n^2 - B_s^2)\mathbf{n} + 2 B_n B_s \mathbf{s} \right] dA \tag{8.14}$$

$$F_e = \frac{\varepsilon_0}{2} \int_\Sigma \left[(E_n^2 - E_s^2)\mathbf{n} + 2 E_n E_s \mathbf{s} \right] dA \tag{8.15}$$

Here the integrals are carried out on a surface surrounding the body in a medium for which the electric permittivity ε_0, and the magnetic permeability μ_0, are close to that of a vacuum.

As an example consider a MEMS cantilever beam that is one half of a capacitor circuit with a voltage V_0 across the gap d_0 that separates the two charged surfaces. The electric field in the gap between the two charged plates is $E = V_0/d_0$. To calculate the deflection of the elastic cantilevered beam one can apply a tensile stress along the beam with strength $t_{nn} = \varepsilon_0 V_0^2/(2 d_0^2)$. Examples of MEMS problems using this technique may be found in the book *Microsystems Mechanical Design*, edited by De Bona and Enikov (2006).

Forces at Nanoscale Dynamics

Classic electromagnetic forces are important concepts at macroscale mechanics. However at the order of 10^{-9} meters, one must also consider quantum mechanical effects. Several effects that become important in modeling dynamic systems at the nanometer scale include the following:

(i) Casimir force
(ii) Van der Waals forces
(iii) Leonard–Jones potentials
(iv) Quantum tunneling effects
(v) Photon pressure
(vi) Adhesion effects
(vii) Capillary forces

At the atomic and molecular scale the physics of motion are governed by *Schrödinger's* equation of quantum mechanics. However Schrödinger's equation uses force potentials that are derived from phenomena at the macroscale. Also physicists often use a semiclassical theory of quantum mechanics in which the $F = ma$ form of Newtonian mechanics is used, but with forces and force potentials derived from quantum theory. The first three effects above are discussed briefly below.

Casimir Effect

Both the van der Waals and the Leonard–Jones effects are associated with the interactions between electron shells in atoms and molecules when they approach one another. However the Casimir force is associated with the energy in a vacuum, an idea that defies traditional concepts of forces between mass carrying objects. This effect was discovered by two Dutch physicists H. B. G. Casimir and D. Poddar at Phillips Research Labs in 1948. In the classic Casimir effect an attractive force is calculated between two parallel conducting plates that carry no charge. In recent experiments the force has been measured between a large sphere and a parallel plate. The attractive force between two plates of area 1 cm^2, and separated by a micron (10^{-6} m) is calculated to be around 10^{-7} N, whereas for separation distances on the order of 10 nm (10^{-8} m) the Casimir attractive stress is of the order of the atmospheric pressure.

There are several ways to explain the Casimir effect and the reader can find definitions from web-based sources. From the viewpoint of quantum theory, although the vacuum field between the plates are zero, there are fluctuations of the electromagnetic field. Quantized electromagnetic waves are called *photons*. When these waves are contained by two conducting plates certain waves are reinforced while others are not. The radiation pressure on one side of the plate becomes greater than that on the inside and a net attractive force results. The theoretical value in the literature of the Casimir pressure is

$$\frac{F}{A} = \frac{\pi^2 hc}{240 r^4} \tag{8.16}$$

Here h is Planck's constant from quantum mechanics, c is the velocity of light and r is the separation distance of the two neutral conducting plates.

This theoretical value is modified in a real system because real materials reflect photon waves differently for different frequencies. Also the simple theory does not account for thermal fluctuations as well as geometry effects. Some references suggest the possibility of a repulsive Casimir force for certain geometries of the interacting bodies and the potential of Casimir levitation of two bodies at the nanometer scale.

Van der Waals and Leonard–Jones Forces

One of the problems in MEMS and NEMS devices is stiction between two components. One explanation for adhesion at these scales is the interaction of electron shells of atoms and molecules when surfaces get too close to one another. In the case of two neutral molecules, physicists have used an approximate expression for the force potential sometimes referred to as a "6–12" law of the form

$$V(r) = \kappa \left[\left(\frac{\sigma}{r}\right)^{12} - \left(\frac{\sigma}{r}\right)^{6} \right]$$

The constant σ is the distance at which the potential becomes zero and is the order of several Angstroms. The related force, calculated in the classical way, is

$$F = -\frac{\partial V}{\partial r} = 6\kappa \left[2\left(\frac{\sigma}{r}\right)^{13} - \left(\frac{\sigma}{r}\right)^{7} \right] / \sigma$$

The Leonard–Jones force is repulsive at a few atomic radius distances and attractive at distances larger than several atomic radii depending on the molecules. These forces are often used in molecular dynamics simulation codes.

At larger scales these intermolecular forces have gone under the name of van der Waals forces and like Casimir forces are attractive in nature. They have been found to be of importance in atomic force microscopy between the probe tip and nano fabricated mechanical structures. It is beyond the subject of this book to go into further details. These forces are mentioned briefly here as a caution to engineers and applied scientists who seek to apply dynamics modeling at very small scales. Below a micron and when electromagnetic forces are nominally zero, van der Waals and Casimir forces can lead to unanticipated mechanical phenomena that is not explained by macroscopic Newtonian and electromagnetic forces.

8.3
Electromechanical Material Properties

Some electromechanical devices are based on multicomponent machines such as servomotors, generators, solenoid actuators, interferometers, while others are based on electromechanical material properties. Devices in this category include a large class of sensors such as strain gauges, thermoelectric devices, piezoelectric, and magnetostrictive devices. In this section we review the basic electric and magnetic material properties and present an introduction to a few electromechanical material-based devices. In recent years electro-optical, especially solid state laser devices, have become important in mechatronic applications. We can only review a few of the more common material behavior-based properties here.

Dielectric and Magnetic Materials

Electromagnetic materials can not only support charge storage and charge transport but also involve electric and magnetic dipoles, sometimes called *electric polarization* or *magnetization*, respectively. A dipole can be represented by a positive and negative charge separated by a small distance, as shown schematically in Fig. 8.6a. When an electric field is placed across a dielectric material, a dipole field, **P**, is assumed to be induced, and the sum of the applied and induced fields are called the *electric displacement* **D**:

$$\mathbf{D} = \varepsilon_0 \mathbf{E} + \mathbf{P} \tag{8.17}$$

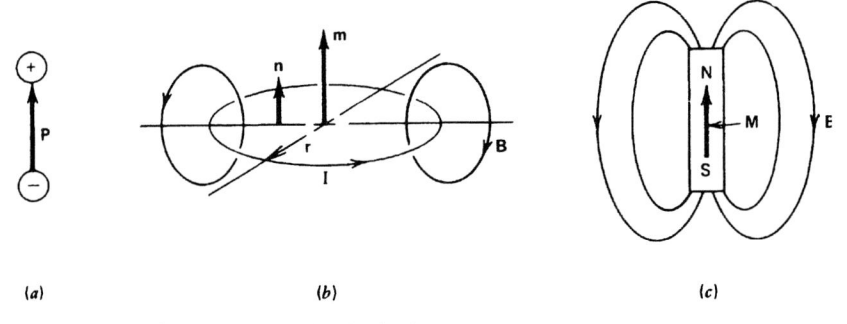

Fig. 8.6 Models for (a) an electric dipole; (b) electric current dipole; (c) a magnetic dipole (Moon, 1994).

In a nondielectric material, **D** and **E** are essentially the same fields, related by the constant ε_0 called the *electric permittivity* of vacuum. For a linear nonconducting, dielectric material, the electric polarization is related to the applied field, **E**:

$$\mathbf{P} = \varepsilon_0 \eta \mathbf{E} \quad \text{or} \quad \mathbf{D} = \varepsilon_0(1 + \eta)\mathbf{E} \tag{8.18}$$

Here η is called the *electric susceptibility*.

In a magnetic material, the magnetic dipole has two models, as shown in Figs. 8.6b and 8.6c. One model is based on the bar magnet in which positive and negative magnetic monopoles are imagined to exist at each end of the small magnet. In another model, the dipole field is assumed to be generated by small current loops or vortices. It is a peculiarity of magnetics that both models yield the same magnetic-field structure far from the bar magnet or the closed circuit. However, the isolated magnetic monopole is not believed to be a real physical object. The dipole model is considered a useful concept.

The volume dipole magnetization density in a material is denoted by **M** and in linear materials is assumed proportional to the applied field, **H**:

$$\mathbf{M} = \chi \mathbf{H} \quad \text{or} \quad \mathbf{B} = \mu_0(1 + \chi)\mathbf{H} \tag{8.19}$$

The constant χ is called the *magnetic susceptibility* and $\mu_r = (1 + \chi)$ is called the *relative permeability*. In nonferromagnetic metals such as aluminum or copper, $\chi \sim O(10^{-3})$, whereas for linear ferromagnetic silicon–iron $\chi \sim 10^{-3}$–10^5. Materials with a large μ_r, or χ, are often compounds or alloys of iron, nickel, or cobalt. Many ferromagnetic materials are nonlinear and hysteretic, i.e., the relation **M(H)** is not unique. An example of a nonlinear ferromagnetic material is shown in Fig. 8.7 for a neodymium–iron–boron material. If a strong magnetization exists when **H** is zero, i.e., $\mathbf{M}(0) \neq 0$, the material is called a *permanent* magnetic material. However, so-called permanent magnets can be demagnetized in a large applied field. Permanent magnets based on rare-earth and ferromagnetic elements make very strong magnetic-field sources of $B \sim 0.5$ tesla (T) in free space and $B \sim 1.0$ T in a closed magnetic circuit. These materials have made possible some very small motors and magnetomechanical actuators in mechatronic devices.

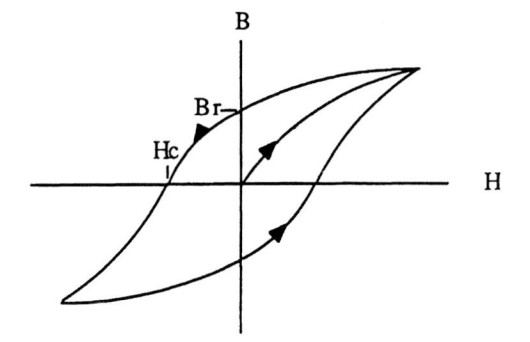

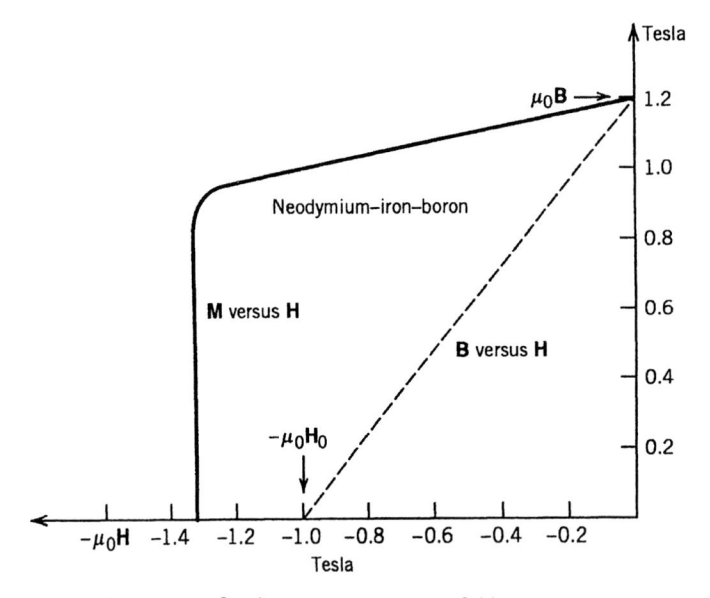

Fig. 8.7 (a) Magnetic flux density, B, vs. magnetic-field strength, H, for a typical ferromagnetic material. (b) B vs. H and M vs. H magnetization curve for commercial rare-earth magnet neodymium–iron–boron.

Both dielectric and ferromagnetic materials are sensitive to temperature. Thus, for example, rare-earth magnets can suffer spontaneous demagnetization above a critical Curie temperature, which may only be around 10^2 °C.

Recently, materials scientists have developed superconducting permanent magnets that can carry fields up to 5 T.

Resistors and Diodes

Classic resistive electric materials are assumed to be linear in the relation between current density, **J**, and applied electric field, **E**:

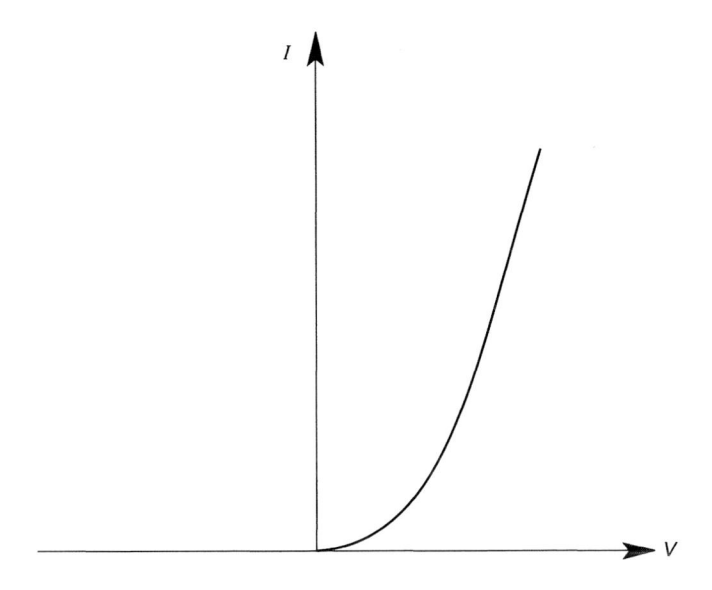

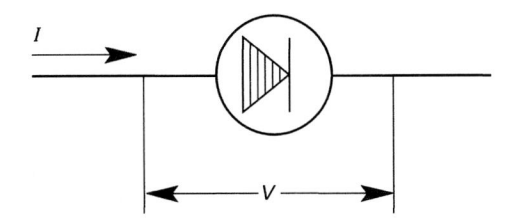

Fig. 8.8 Current–voltage properties of an ideal diode.

$$\mathbf{E} = \rho \mathbf{J} \tag{8.20}$$

where ρ is called the *resistivity*. The inverse, $\sigma = 1/\rho$, is called the *electrical conductivity*. For a lumped-material system we often write a relation between the electric voltage, V, and the current, I:

$$V = RI \tag{8.21}$$

For a cylinder of length ℓ and cross-sectional area A,

$$V = E\ell, \quad I = JA$$

$$R = \rho \ell / A \tag{8.22}$$

Materials that restrict the flow of electricity have values of R in the range of 1–10^6 ohms (Ω). Materials that easily permit the flow of charge, like aluminum and copper are called normal conductors and have resistivity ρ in the range 10^{-8} $\Omega \cdot$m. These properties also depend on temperature.

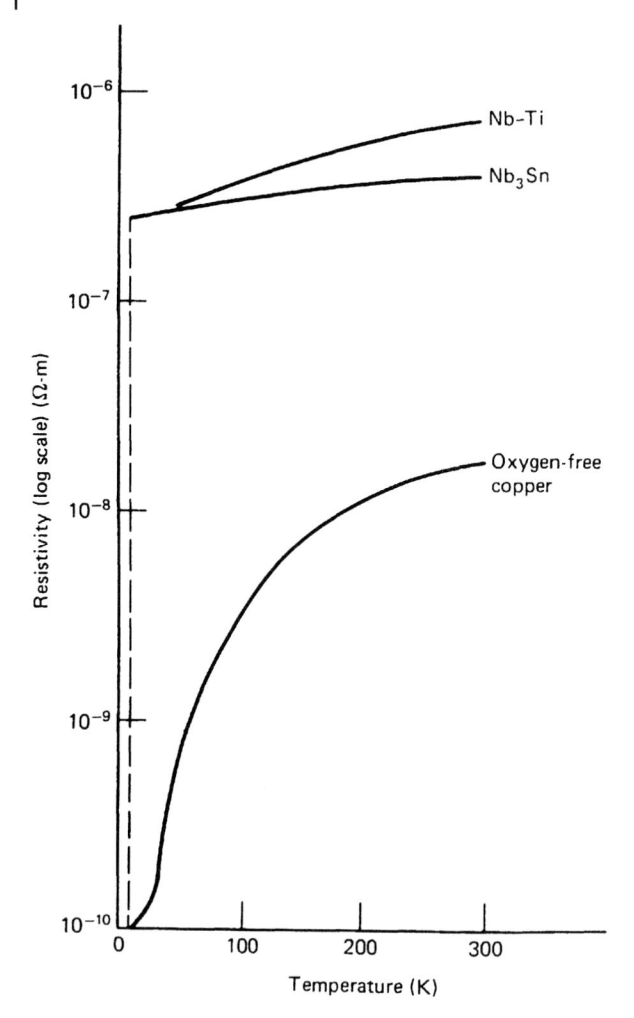

Fig. 8.9 Electrical resistivity of low-temperature ($T < 20$ K) superconducting wires and oxygen-free appear.

Diodes, Semiconductors, and Superconductors

Many materials that support charge transport do not fall into the simple categories of linear normal conductor or resistor. Exceptions include nonlinear devices such as diodes or semiconductors and superconductors. Diodes are often bilayer composite semiconductor devices that permit conduction for a negative voltage and restrict conduction for positive voltage. An example of a $V(I)$ curve for a diode is shown in Fig. 8.8. Many other multilayer electric composites are used in modern electronics, including numerous transistor devices. (See, e.g., Fig. 8.14.)

Another nonlinear material is the superconductor. This material is characterized by the disappearance of resistance at a nonzero temperature, usually well be-

Fig. 8.10 Photograph of a magnetically levitated rare-earth magnet with a turbine disc above a high-temperature superconducting material at 78 K (YBa$_2$Cu$_3$O$_7$).

low the freezing of water (see Fig. 8.9). However, a new class of materials called high-temperature superconductors has zero resistance at the temperature of liquid nitrogen ($T = 78$ K). These materials include a compound yttrium–barium–copper–oxide (YBCO) and are brittle, ceramic like materials. Metal-like superconductors include niobium–titanium, which require a temperature near liquid helium ($T = 4.2$ K).

These materials not only exhibit perfect conductivity (zero resistance) but can also trap magnetic flux and are useful for levitation devices, superconducting permanent magnets, and flux-shielding devices (see Fig. 8.10).

Levitation and Suspension Forces in High-Temperature Superconductors

As new electromagnetic materials and technologies are developed such as superconductors and nanotechnology, there have appeared new types of electromagnetic forces that are not easy to derive from the basic principles. One of these forces is the interaction of permanent magnets and high-temperature superconductors (HTSC). These magnetic forces have been used for levitation in magnetic bearings and are exceptions to Earnshaw's stability theorem (Moon, 1994). To characterize these forces one often has to rely on empirical studies as illustrated in Figs. 8.11a and 8.11b. Fig. 8.11a shows the repulsive force between a pellet of yttrium–barium–copper–oxide, or YBCO, and a rare-earth permanent magnet. The YBCO is immersed in liquid nitrogen (78 K). The force between the magnet and YBCO is strongly hysteretic, depending on whether the test magnet is approaching or withdrawing from the surface of the superconductor. In fact one can see that there is a regime where the force is actually tensile or capable of suspending the test magnet below the superconducting pellet that has in fact been observed.

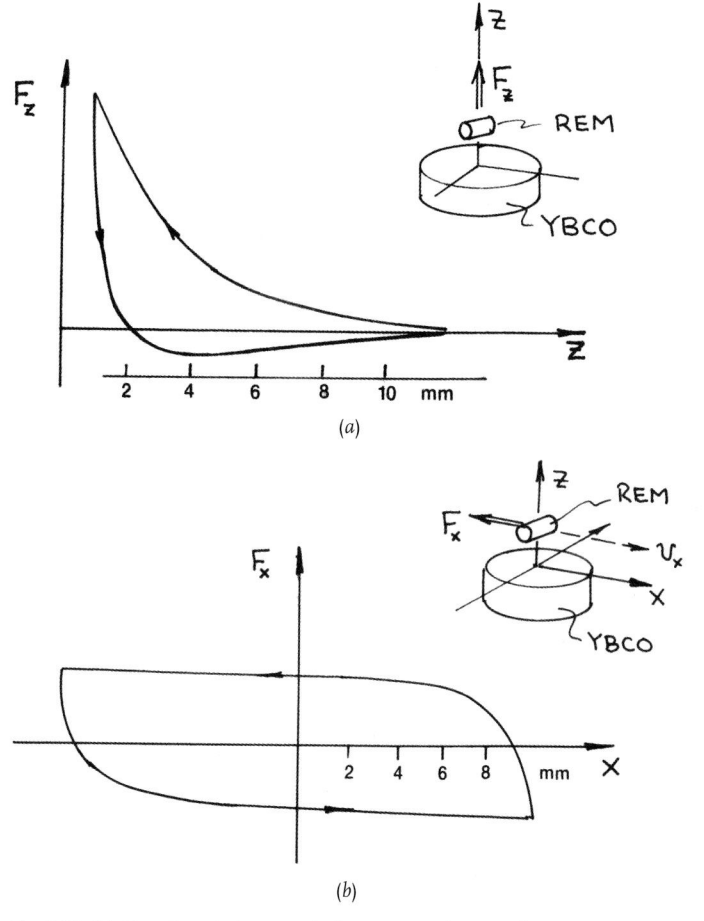

Fig. 8.11 (a) Plot of vertical magnetic force versus distance for a small permanent rare-earth magnet near a high temperature superconductor (YBCO) showing hysteretic behavior. (b) Plot of lateral magnetic "friction" versus distance for a permanent rare-earth magnet near a high temperature superconductor showing hysteretic behavior (Moon, 1994).

Another strange feature of HTSC-magnet forces is the existence of a noncontacting magnetic friction force as illustrated in Fig. 8.11b. Here the test magnet is moved laterally along the surface of the HTSC and a drag force is measured. The force reverses direction when the magnet direction is reversed.

These magnetic forces have been modeled using the basic physics of superconducting electric currents, and involve pinning centers that tend to prevent the circulating superconducting current loops from moving as the applied magnetic field is changed. However the calculations are not simple and for the present experimental data as in Fig. 8.11 are used to design superconducting bearing and other HTSC force devices such as flywheels and cryo-cooler turbines.

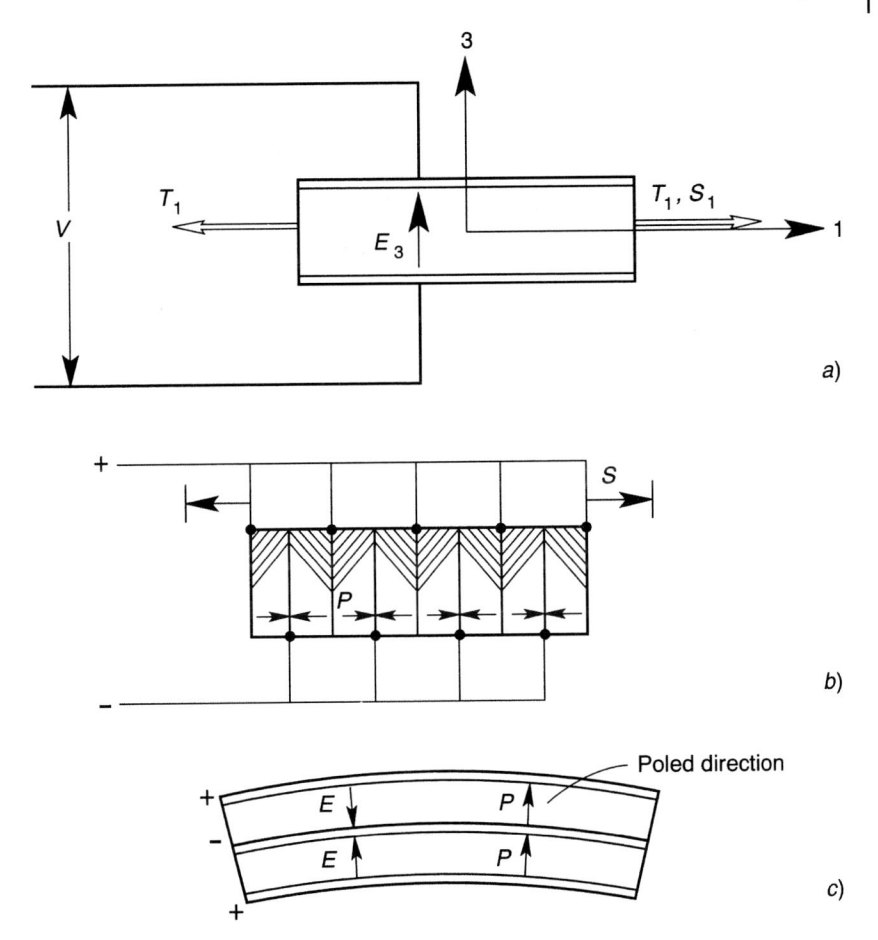

Fig. 8.12 (a) Vector components of stress T_1, elastic strain S_1, and electric field E_3 in a piezoelectric actuator-sensor. (b) Piezoelectric stack actuator. (c) Piezoelectric bender actuator.

Piezoelectric Materials

Piezoelectric materials are dielectrics whose electrical properties depend on the applied mechanical stresses and strains and whose stress–strain behavior depends on the applied electric field or voltage. These materials are widely used in so-called "smart structures" for both sensors and actuation. Several configurations are shown in Fig. 8.12. These materials behave anisotropically in their stress–strain–voltage properties. We do not describe the general theory here. One of the most popular materials is lead zirconate titanate or PZT. For the simple configuration in Fig. 8.12a, a voltage V is applied to the 3 direction and a stress T_1 is applied in the horizontal direction. We assume that the strain $S_2 = 0$ and that the stress in the vertical direction is zero (i.e., $T_3 = 0$). For these conditions a linear relation

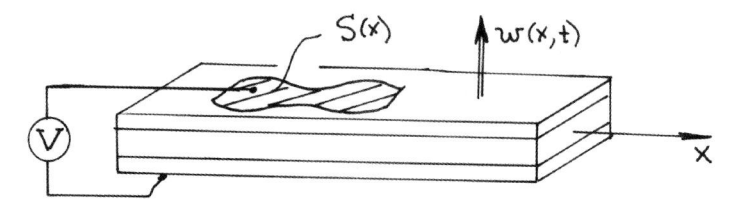

Fig. 8.13 Composite piezoelectric beam structure with variable conductor area $S(x)$ and applied voltage V.

between stress T_1, voltage $V = E_3\Delta$, strain S_1, and electric displacement D_3 can be found:

$$T_1 = CS_1 - e_{13}E_3$$
$$D_3 = e_{13}S_1 + \varepsilon_3 E_3 \qquad (8.23)$$

Thus, a voltage can produce a transverse mechanical displacement that can serve as an electric piston. Piezo materials come in three forms:

- Piezoceramic (e.g., PZT or BaTiO$_3$)
- Piezofiber composites
- Piezopolymer films (e.g., polyvinyl difluoride or PVDF)

Piezoceramics are manufactured in so-called "piezostacks," shown in Fig. 8.12b, which can produce axial displacement of the order of 400 μm, and piezobimorphs or benders (Fig. 8.12c), which can produce bending displacements. Piezopolymer film is easily attached to platelike structures, and can act as a sensor of mechanical motion, but PVDF produces very small forces, which are only suitable for small devices.

Piezoelectric Beam

Piezoelectric beams can be fabricated using several combinations of piezo and structural layers (Fig. 8.13). In the research paper of Lee and Moon (1989), the equation for a three layer beam with two piezo outer layers and a central structural layer is found to have the following form when the outer piezo layers are of opposite polarization:

$$D\frac{\partial^4 w}{\partial x^4} + \rho A \frac{\partial^2 w}{\partial t^2} = -2e_{31}z_0 \frac{\partial^2 S(x)V_3}{\partial x^2}$$
$$z_0 = (h_s + h_p)/2 \qquad (8.24)$$

In this equation the constant D is the beam stiffness constant, e_{31} is the piezo material constant, z_0 is an average of the piezo and structural thicknesses, $S(x)$ is the electrode area pattern, and V_3 is the transverse voltage across the composite plate. The effective voltage $S(x)V_3$ can be made to vary along the length of the beam by creating an electrode pattern on the top and bottom of the plate. If the voltage V_3 is uniform along the plate, $S(x) = 1$, then a concentrated bending moment is placed at the end of the beam. For details the reader is referred to the research paper of Lee and Moon (1989) or the textbook by Miu (1993).

8.4
Dynamic Principles of Electromagnetics

The basic principles of electromagnetics are embodied in a set of partial differential equations called *Maxwell's*[1] *equations*. They govern the spatial and temporal dynamics of continuous charge, current, polarization and magnetization densities as well as the electric- and magnetic-field quantities defined earlier. These equations are used to describe phenomena such as wave propagation in space, detailed field distributions in and around magnets, and capacitor elements. However, in many applications the charges, currents, and magnetic fields are confined to material geometries with high aspect ratios that we call *circuits*. Technical systems with electric and magnetic circuits constitute an important class of devices in mechatronics, including wire and lumped-element circuits, relays, magnetic actuators, and some piezoelectric and magnetostrictive devices. Thus, we describe the principles of electromechanics for circuit systems in the remainder of this chapter.

One of the consequences of the localization of electric and magnetic fields in circuits is that the dynamic principles can be written in the form of ordinary differential equations similar to Newton's laws or Lagrange's equations. However, whereas the electromagnetic forces have clear definitions in a continuum field problem [as in (8.2) and (8.3)], in circuits the determination of these forces is not so direct and often an energy method is employed. In the following, we see how such forces can be related to electric- and magnetic-energy functions and incorporated in Lagrange's equations.

Circuit Elements

The most primitive circuit elements are resistors, capacitors, and inductors. Their respective graphical symbols are shown in Fig. 8.14. Intermediate-level circuit elements include diodes and transistors. Higher level circuit devices include many kinds of amplifiers, switches, and logic circuits. These so-called integrated-circuit devices often contain tens and even hundreds of the primitive and intermediate-level elements. The dynamical behavior of the primitive elements can be formulated using the basic principles of electromagnetics. However, intermediate-level devices are often so complex that their behavior is only understood with empirical properties relating input variables to output variables.

---

[1] Named after British physicist James Clerk Maxwell (1831–1879).

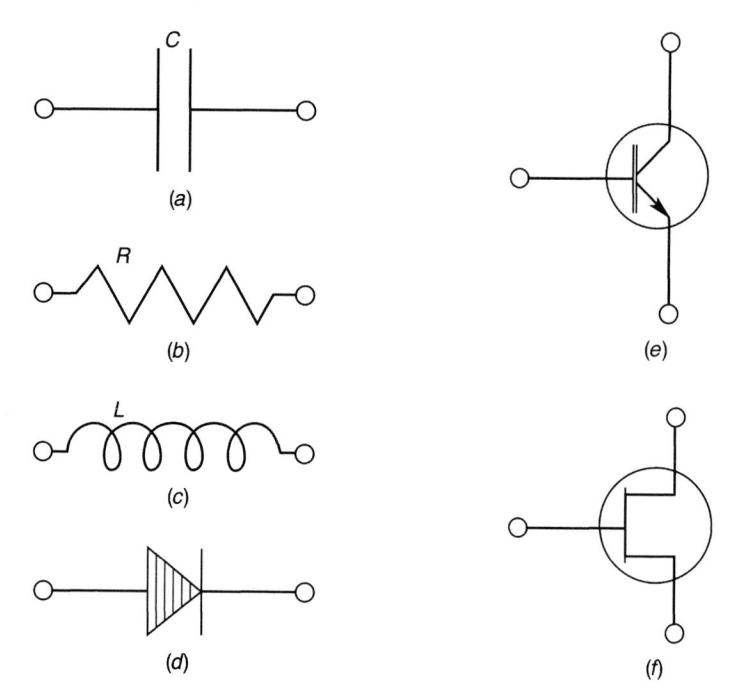

Fig. 8.14 Basic-electric circuit symbols for lumped electromagnetic components. (*a*) Capacitor. (*b*) Resistor. (*c*) Inductor. (*d*) Ideal diode. (*e*) Transistor. (*f*) Field-effect transistor (FET).

Dynamical Equations of Electric and Magnetic Circuits

The dynamical equations of lumped electromagnetic systems are similar in structure to Newton's laws for mechanical systems. In simple one-degree-of-freedom mechanics we have a kinematic relation and a dynamical law, as shown below:

$$\frac{dx}{dt} = v \quad \text{(Kinematic relation)}$$

$$\frac{d}{dt}(mv) = F(x, v, t) \quad \text{(Dynamical law)} \tag{8.25}$$

Electric and magnetic circuits have an analogous structure. Consider, for example, the circuit shown in Fig. 8.15. The capacitor C, stores electric energy and is analogous to the elastic spring. The voltage V across the capacitor is related to the electric field, **E**, and the resulting stored charge, Q, is related to the voltage by the equation

$$V = \int \mathbf{E} \cdot d\boldsymbol{\ell}$$

$$Q = CV \tag{8.26}$$

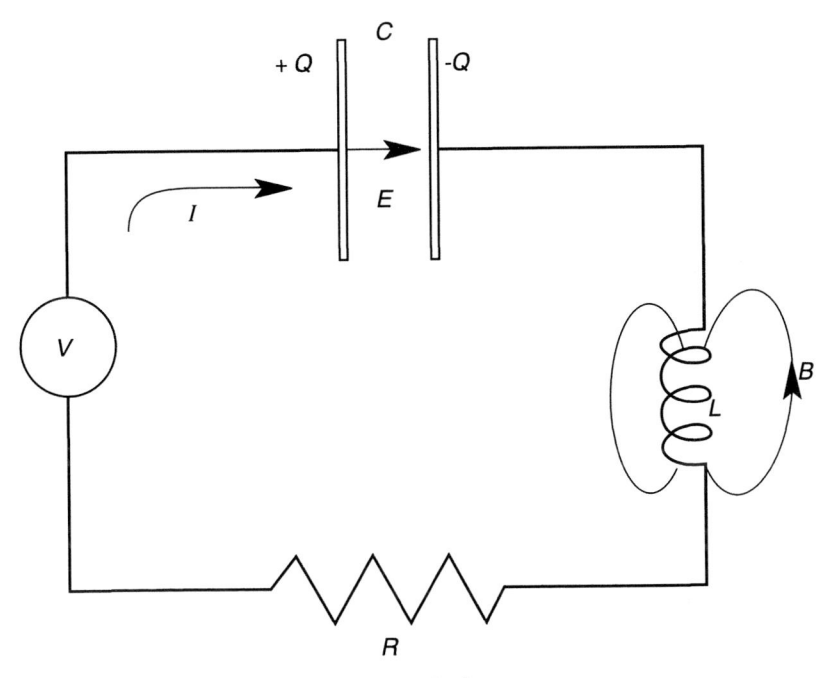

Fig. 8.15 Classic *LCR* electric circuit with applied voltage. Analog to a spring–mass–damper mechanical oscillator.

The inductor, L, is usually a multiturn-coil-wound device that stores magnetic energy and is analogous to the mechanical inertia or kinetic energy storage.

Magnetic flux Φ is defined as the integral of the normal component of the magnetic field over an area

$$\Phi = \int_A \mathbf{B} \cdot \mathbf{n}\, da \tag{8.27}$$

Magnetic flux linkages, ϕ, in a multiturn coil with N turns is defined by $\phi = N\Phi$. The basic laws for an electric circuit take the form

$$\frac{dQ}{dt} = I \quad \text{(Conservation of charge)} \tag{8.28}$$

$$\frac{d\phi}{dt} = V(\phi, I, t) \quad \text{(Faraday–Henry law of flux change)}$$

Clearly, there are one too many state variables. If we apply Ampere's law to the inductive device, we can often relate the magnetic flux linkages in the magnetic-circuit to the current in the electric circuit, i.e.,

$$\phi = f(I) \tag{8.29}$$

In the case of a linear system

$$\phi = LI \tag{8.30}$$

where L is called the *inductance* and in measured in henrys. When the magnetic circuit is coupled to a mechanical circuit, the inductance is sometimes a nonlinear function of the mechanical displacement.

To close the system, the voltage across the inductor must be related to the voltages in the rest of the circuit, e.g., across the resistive, capacitive, and applied-voltage device. In electromechanical systems the force on the mechanical system must relate to the electromechanical state variables. For the one-dimensional system shown in Fig. 8.16a the closed set of equations takes the form

$$\frac{dx}{dt} = v$$
$$\frac{dmv}{dt} = F(x, v, Q, I, t)$$
$$\frac{dQ}{dt} = I$$
$$\frac{d(L(x)I)}{dt} = V(Q, I, x, t) \tag{8.31}$$

This is the basic form of the equations of motion for an electro-mechanical or mechatronic system.

Kirchhoff's Circuit Laws

Multiloop circuits with passive elements containing many inductors and capacitors, are governed by two basic principles, called *Kirchhoff's laws*;

1. The sum of all the currents entering a circuit node must equal zero.
2. The sum of all the voltage drops around each circuit loop must equal the applied voltage sources.

In the following section we show how these laws can be derived from Lagrange's equations by identifying a magnetic-energy storage function and an electric-energy storage function.

8.5
Lagrange's Equations for Magnetic and Electric Systems

The direct formation of equations of motion using Newton's law and the circuit laws requires explicit knowledge of electromagnetic forces and motion-induced voltages. In Chapter 4 we saw how equations of motion could be derived from potential- and kinetic-energy functions. In electro-mechanical systems we have to identify a set of generalized displacement and velocity variables. For electric and magnetic systems the choice of these variables is guided by the general-

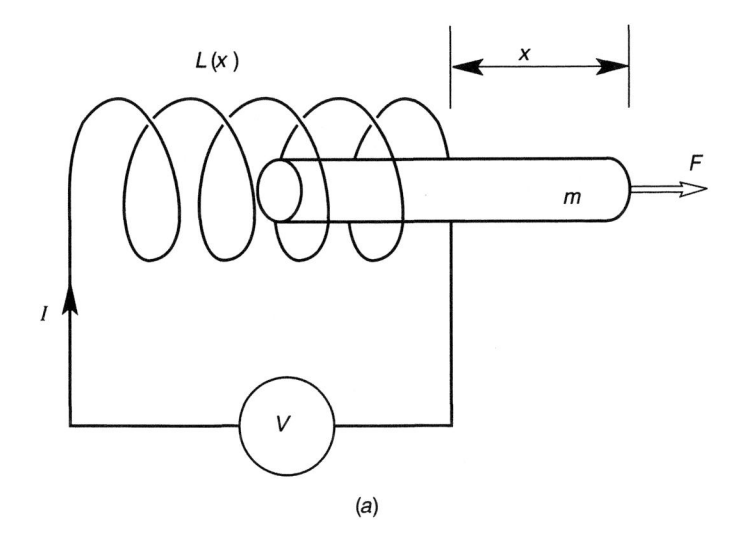

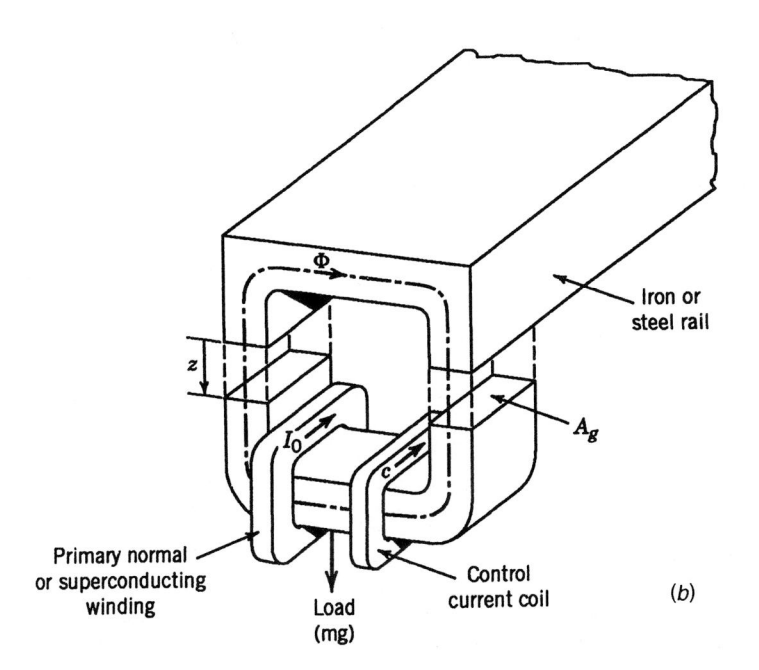

Fig. 8.16 (a) Ferromagnetic, linear magnetic actuator.
(b) Geometric arrangement of an electromagnet and guideway for a feedback-stabilized Mag-Lev suspension.

ization of the principal of virtual work to electromechanical problems (see also Crandall et al., 1968).

In order to formulate a variational principle for magnetomechanical systems, we must first outline the concept of a magnetic energy function. Consider the system in Fig. 8.16a, in which a voltage source $V(t)$ pumps current into multiturn coil wrapped around a ferromagnetic core.

To get an energy principle we start with the dynamic equation for the circuit, assuming that there is no resistance and no mechanical motion. Kirchhoff's law of voltages requires

$$V = \frac{d\phi}{dt}$$

where V is the voltage source and ϕ is the magnetic flux linkage through the coil. Multiplying the equation by the current I and integrating with respect to time we obtain

$$\int IV\,dt = \int I\,d\phi$$

We can view the left-hand side as the energy input (IV has units of power) and the integral on the right as a change in stored magnetic energy $W(\phi)$, in which we view the current as a function of the flux linkages, i.e.,

$$W(\phi) = \int_0^\phi I(\phi')\,d\phi'$$

Now consider a static problem for a mechanical system under magnetic forces produced by an electromagnet. Such problems occur in magnetic-bearing support of machine rotors or in magnetic levitation of vehicles (see Fig. 8.16b). We assume that there are n mechanical generalized displacements $\{q_k;\ k = 1, 2, \ldots, N\}$ and M magnetic fluxes $\{\phi_j;\ j = 1, 2, \ldots, M\}$. The generalized mechanical forces are assumed to be derivable from an elastic energy function $\mathcal{V}(q_k)$. For the present discussion we assume that the M circuits contain electromagnets, for which we write a magnetic-energy function, W, that depends not only on the magnetic fluxes in the electro-magnets but also on the mechanical displacements, i.e., $W = W(\phi_j, q_k)$.

The principle of virtual work for electromechanical systems states:

> The change in both mechanical- and magnetic-energy functions under small changes of the independent variables is equal to the change of work done on the system by external forces and voltages.

When electric currents provide energy input, as in magnetic bearings, the work done in a small time, dt, is given by the product of current, voltages and time:

$$\sum I_j V_j\,dt \tag{8.32}$$

Here V_j is the voltage that drives the current I_j in the jth circuit. In a circuit with only an electromagnet, Faraday's law relates these voltages to the change in flux linkage, i.e.,

$$V_j = \frac{d\phi_j}{dt} \tag{8.33}$$

so that

$$\sum I_j V_j \, dt = \sum I_j \, d\phi_j \tag{8.34}$$

Thus in the principle of virtual work we vary the magnetic fluxes under fixed currents. Such a variation is denoted by $\delta\phi_j$. The principle of virtual work for magnetomechanical systems is then

$$\sum I_j \delta\phi_j = \delta W + \delta V \tag{8.35}$$

The right-hand side is expanded as follows, noting the functional dependence, $W(\phi_j, q_k), V(q_k)$:

$$\delta W + \delta V = \sum \left(\frac{\partial W}{\partial q_k} + \frac{\partial V}{\partial q_k} \right) \delta q_k + \sum \frac{\partial W}{\partial \phi_j} \delta\phi_j \tag{8.36}$$

Because we have assumed that the small variations δq_k and $\delta\phi_j$ are independent, the terms corresponding to each variation on either side of (8.25) must be equal, i.e.,

$$I_j = \frac{\partial W}{\partial \phi_j}$$

$$\frac{\partial W(\phi_j, q_k)}{\partial q_k} = -\frac{\partial V}{\partial q_k} \tag{8.37}$$

Looking at a single-degree-of-freedom problem, as in Fig. 8.16, the second equation (8.37) represents a balance of mechanical and magnetic forces. Thus the magnetic force must be given by

$$F_k^m = -\frac{\partial W(\phi_j, q_k)}{\partial q_k} \tag{8.38}$$

This formula is a very important result. It says that magnetic forces can be determined by observing the change in the magnetic energy of the circuit due to mechanical displacement q_k. This can be made even more explicit if we consider the form of the magnetic energy function for the inductor element shown in Fig. 8.16. From (8.30) the magnetic flux linkage ϕ is related to the current by the relation

$$\phi = LI \tag{8.39}$$

where L is the inductance. For one mechanical degree of freedom q, the magnetic energy function can be shown to be

$$W(\phi, q) = \frac{1}{2} \frac{\phi^2}{L(q)} \tag{8.40}$$

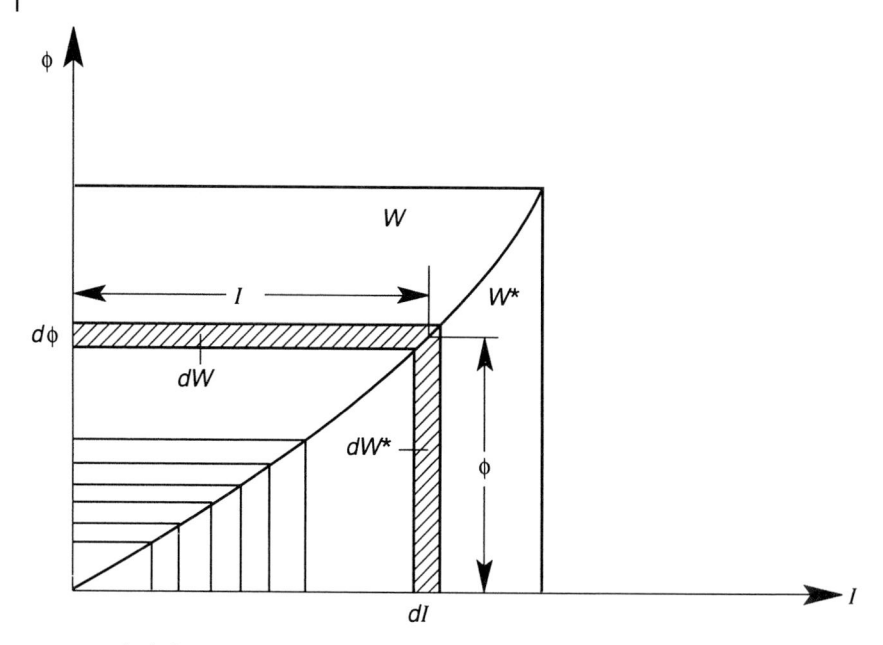

Fig. 8.17 Flux linkage ϕ vs. current I, showing the differential magnetic energy dW and differential comagnetic energy dW^*.

Thus if W depends on the displacements, it must be through the dependence of the inductance $L(q)$ on the displacement. In practical terms, the magnetic force of the electromagnet on the mechanical circuit can be determined simply by measuring the change of inductance due to displacement. The magnetic force is then

$$F^m = \frac{\phi^2}{2L^2} \frac{dL(q)}{dq} \tag{8.41}$$

In applying the force formula (8.38), we note that since ϕ, q are assumed to be independent, ϕ is held fixed when we take the derivative with respect to q. We note that if q is in an angle, F^m has units of torque.

When there are several flux paths and more than one current circuit, the flux ϕ_i could depend on more than one current, i.e.,

$$\phi_i = \sum L_{ij} I_j \tag{8.42}$$

and

$$W = \frac{1}{2} \sum \sum L_{ij} I_i I_j$$

Here the diagonal terms of the inductance matrix, L_{ii}, are called *self-inductances*, and the off diagonal terms, L_{ij}, are called *mutual inductances*. In many systems it is more convenient to choose the currents $\{I_j\}$ as the generalized variables instead of the fluxes. This requires defining the magnetic energy in terms of currents instead of fluxes. Referring to Fig. 8.17, if the current and flux are uniquely related,

$$I = g(\phi) \tag{8.43}$$

then W can be interpreted as the area under the curve $g(\phi)$. The difference between the area $I\phi$ and $W(\phi)$ defines an area function $W^*(I)$,

$$W^*(I) = I\phi - W(\phi) \tag{8.44}$$

Thus, if W is changed by a mechanical displacement, the change is also reflected in W^*, i.e.,

$$dW^* = \phi\, dI + I\, d\phi - dW$$

or

$$\phi = \frac{\partial W^*(I, q)}{\partial I}$$

$$I = \frac{\partial W(\phi, q)}{\partial \phi}$$

$$F^m = \frac{\partial W^*(I, q)}{\partial q} = -\frac{\partial W(\phi, q)}{\partial q} \tag{8.45}$$

W^* is called the *comagnetic energy function* (see also Crandall et al., 1968). For a simple electromagnet it is easy to show that this coenergy is given by

$$W^* = \frac{1}{2} L(q) I^2 \tag{8.46}$$

and the magnetic force is given by

$$F^m = \frac{\partial W^*(I, q)}{\partial q} = \frac{I^2}{2} \frac{dL(q)}{dq} \tag{8.47}$$

It is important to note the sign change in (8.45). Using the flux as a state variable leads to a minus sign in the force law, and using current implies the positive sign. Thus the choice of magnetic variables is important, though either choice will lead to the same answer provided one is consistent in the force laws.

For dynamic problems the principle of virtual work is extended in D'Alembert's principle, as discussed in Chapter 4. In electromechanical dynamics problems we assume that the kinetic-energy function $T(q_k, \dot{q}_k)$ is independent of the electric and magnetic variables, such as the fluxes or currents. Using the results just derived for a static magnetomechanical problem and the results of Chapter 4, the Lagrangian for a magnetomechanical problem can be shown to take the following form when we choose the fluxes as state variables

$$\mathcal{L} = T(q_k, \dot{q}_k) - \mathcal{V}(q_k) - W(\phi_j, q_k) \tag{8.48}$$

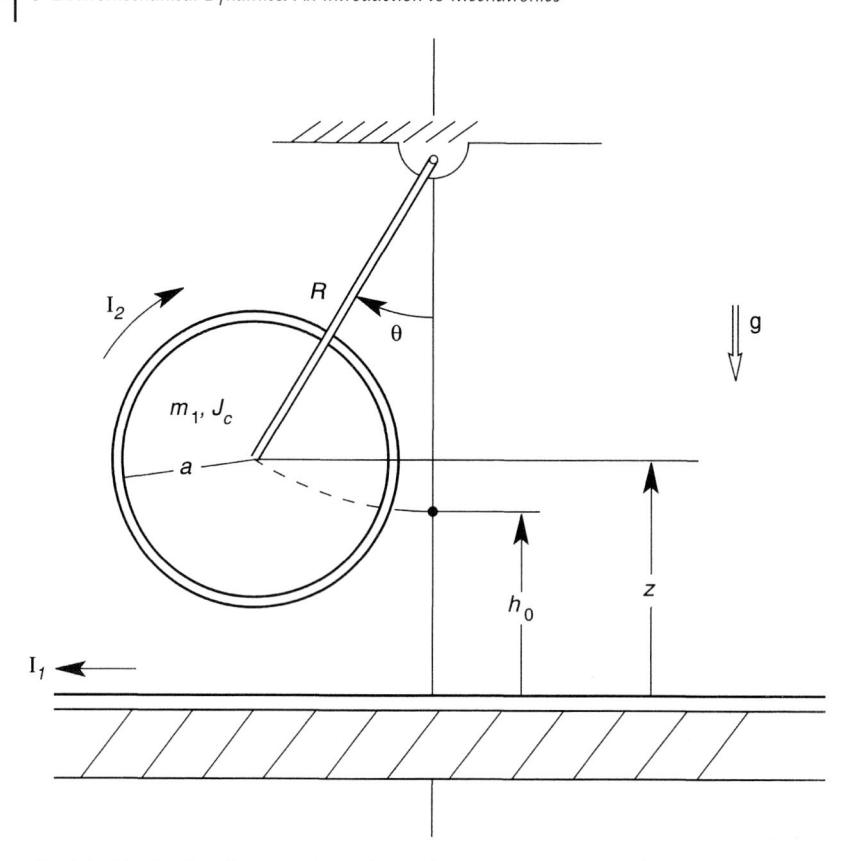

Fig. 8.18 Sketch of an electromechanical variable-frequency seismograph (Example 8.3).

Applying the variations of the mechanical displacements $\{q_k\}$ with the magnetic fluxes held fixed generates the equations of motion for the mechanical subsystem

$$\frac{d}{dt}\frac{\partial T}{\partial \dot{q}_k} - \frac{\partial T}{\partial q_k} + \frac{\partial \mathcal{V}}{\partial q_k} + \frac{\partial W}{\partial q_k} = 0; \quad k = 1, 2, \ldots, N \quad (8.49)$$

When it is convenient to use currents, we replace the term $(\partial W/\partial q_k)$ with $-(\partial W^*/\partial q_k)$ in Lagrange's equations as in (8.45).

However, the preceding equations do not generate the circuit equations. While we have chosen the fluxes as generalized circuit displacements, we have not identified generalized circuit velocities. To do this we now focus the discussion on an electromechanical problem with a charge storage device (a capacitor) instead of a flux storage inductor, after the following example.

Example 8.3 Variable Frequency Seismograph. To illustrate the magnetic-energy method for magnetomechanical dynamics, consider the pendulum shown in Fig. 8.18. Pendulum systems are often used in seismographs. In the device shown two current filaments are used to change the natural frequency of the pendulum. In this system we have two circuits with currents I_1, I_2. In such cases we need only consider that part of the magnetic energy that depends on the motion θ. The magnetic coenergy can be shown to be given by the mutual inductance $L(\theta)$:

$$W^*(I_1, I_2, \theta) = I_1 I_2 L(\theta)$$

where $L(\theta)$ represents the magnetic-flux threading circuit 2 due to a unit current in circuit 1. The magnetic field due to unit current in 1 is given by $B = \mu_0/2\pi z$. Integration of this flux in the circular area can be shown to be

$$L(z) = \mu_0 \left[z - (z^2 - a^2)^{1/2} \right] \tag{8.50}$$

The constraint between z and θ is

$$z = h_0 + R(1 - \cos\theta)$$

and the Lagrangian is

$$\mathcal{L} = \frac{1}{2}(J_c + R^2 m)\dot{\theta}^2 + I_1 I_2 L(z(\theta)) - mgz(\theta) \tag{8.51}$$

where J_c is the moment of inertia of the cylinder mass about the center of mass.
Lagrange's equation can then be found to be

$$(J_c + mR^2)\ddot{\theta} + mgR\sin\theta = I_1 I_2 \frac{dL}{dz} R\sin\theta$$

and

$$\frac{dL}{dz} = \mu_0 \left[1 - \left(1 - \frac{a^2}{z^2}\right)^{-1/2} \right] \tag{8.52}$$

To maximize the magnetic force, we assume that the gap Δ is small compared with the coil radius a:

$$h_0 = a + \Delta, \qquad \Delta/a \ll 1$$

and

$$\left.\frac{dL}{dz}\right|_{\theta=0} \simeq -\mu_0 \left(\frac{a}{2\Delta}\right)^{1/2} \tag{8.53}$$

For small motions, the linearized equation of motion becomes

$$J_0\ddot{\theta} + \left(mgR + \mu_0 I_1 I_2 R \left(\frac{a}{2\Delta}\right)^{1/2} \right)\theta = 0 \tag{8.54}$$

Thus by varying the relative sign of $I_1 I_2$, the natural frequency can be increased ($I_1 I_2 > 0$) or decreased ($I_1 I_2 < 0$).

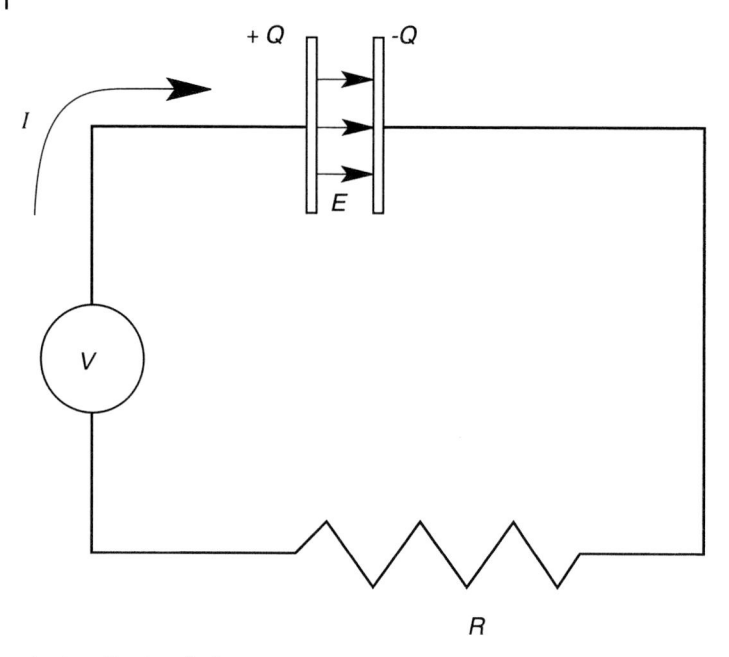

Fig. 8.19 Circuit with electric energy storage (capacitor C), electric-energy dissipator (resistor R), and electric energy current source V.

Lagrange's Equations for Electric Field Systems

To motivate the choice of electromagnetic dynamical state variables consider the electric circuit in Fig. 8.19, with a series-connected voltage source, capacitor, and resistor. The capacitor is a device that contains two platelike conductors separated by a dielectric material. When a voltage is applied, positive charge piles up on one plate and negative charge on the other so that an electric field, E, exists in the gap between the plates. The integration of the electric field across the gap defines a voltage on the capacitor proportional to the total positive charge, Q,

$$V_c = \int_1^2 E\, dx = \frac{Q}{C} \tag{8.55}$$

C is called the capacitance.

If a charge source device puts out a certain rate of charge, Q, or current, I, then the charge continuity principle equates I and dQ/dt as discussed in Section 8.2. The charge source, sometimes called a *signal generator* or *power supply*, is usually characterized by a voltage time history, $V(t)$. The dynamic principle for an electric circuit states that the applied voltage V must equal the sum of voltage drops across the capacitor and in the case of Fig. 8.16, the resistor, RI. (This is one of Kirchhoff's circuit laws. See pages 457–458.) Thus the equations of motion for the R–C circuit are

$$\frac{dQ}{dt} = I$$

$$0 = -RI - \frac{Q}{C} + V(t) \tag{8.56}$$

Comparing this equation with that for the mechanical system, (8.25) we are tempted to choose the charge, Q, as a generalized "position" state variable, and the current I as the generalized "velocity" state variable.

However, in this analogy, we can see that there is no analog to momentum or kinetic energy. If we add an inductor to the circuit, we must add a voltage drop across the inductor $d(LI)/dt$ where we write the flux linkages in terms of the inductance L and current, (8.30). Then the circuit equation has a form more analogous to the mechanical system (8.25)

$$\frac{dQ}{dt} = I$$

$$\frac{dLI}{dt} = -RI - \frac{Q}{C} + V(t) \tag{8.57}$$

Thus, the capacitor, which stores electric charge, acts as a spring or potential energy function, while the inductor, which stores magnetic flux, acts as a momentum or kinetic energy device. The resistor in (8.57) is analogous to a linear damper.

It is quite natural, then, to propose an *electromagnetic Lagrangian* $\mathcal{L}_{em}$ with Q as a generalized position variable from which we can derive the circuit equation (8.57),

$$\mathcal{L}_{em} = \frac{1}{2}LI^2 - \frac{1}{2}\frac{Q^2}{C}$$

$$\frac{d}{dt}\frac{\partial \mathcal{L}_{em}}{\partial \dot{Q}} - \frac{\partial \mathcal{L}_{em}}{\partial Q} + \frac{\partial \mathcal{R}}{\partial \dot{Q}} = \mathcal{Q} \tag{8.58}$$

where $\mathcal{R} = \frac{1}{2}RI^2 = \frac{1}{2}R\dot{Q}^2$ is the electromagnetic Rayleigh dissipation function similar to the mechanical analog in Chapter 4, and $\mathcal{Q}$ is the generalized force that, in the case of (8.57), is the voltage $V(t)$. It is easy to see that application of (8.58) to the circuit will yield the Kirchhoff circuit law (8.57).

The first term in $\mathcal{L}_{em}$ is just the magnetic coenergy $W^*(Q) = L\dot{Q}^2/2$ while the second term can be interpreted as an electric energy function, $W_e(Q)$. As in the case of the magnetic energies, one can define an electric coenergy $W_m^*(V)$ which is a function of the voltage across the capacitor, $V = Q/C$, instead of a function of the charge Q;

$$W_e^*(V) = QV - W_e(Q)$$

Use of a Lagrangian formulation for a simple electric circuit would be overkill if it cannot be coupled with the mechanical system. It can be shown that for the coupled system in Fig. 8.20, we can construct a combined mechanical and electromagnetic Lagrangian that will yield both Newton's laws and Kirchhoff's laws. In addition,

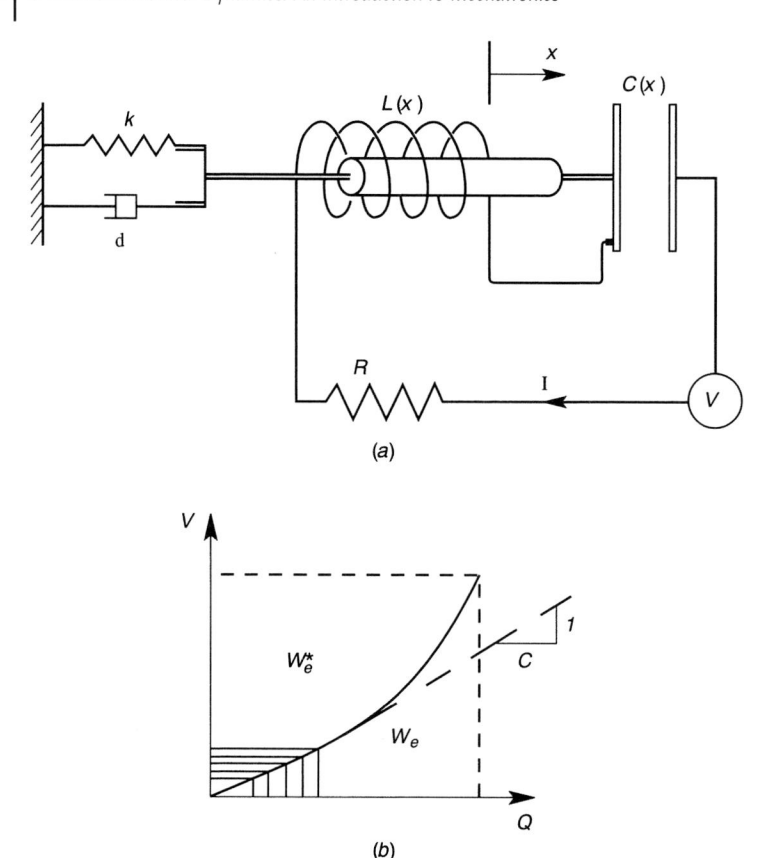

Fig. 8.20 (a) Coupled electromechanical system: a spring–mass–damper oscillator with magnetic- and electric-force actuators. (b) Electric field vs. charge on the variable capacitor.

this formulation produces the proper form of the electric and magnetic forces, as well as the motion-induced voltages on the circuit.

To carry this out in detail, we choose a mechanical variable $x(t)$ and a charge $Q(t)$ on the capacitor as generalized position variables, and write

$$\mathcal{L} = \frac{1}{2}m\dot{x}^2 - \frac{1}{2}kx^2 + \frac{1}{2}L\dot{Q}^2 - \frac{1}{2}\frac{Q^2}{C}$$

$$\mathcal{R} = \frac{1}{2}\delta\dot{x}^2 + \frac{1}{2}R\dot{Q}^2 \tag{8.59}$$

$$\frac{d}{dt}\frac{\partial \mathcal{L}}{\partial \dot{x}} - \frac{\partial \mathcal{L}}{\partial x} + \frac{\partial \mathcal{R}}{\partial \dot{x}} = \mathcal{Q}_x(t)$$

$$\frac{d}{dt}\frac{\partial \mathcal{L}}{\partial \dot{Q}} - \frac{\partial \mathcal{L}}{\partial Q} + \frac{\partial \mathcal{R}}{\partial \dot{Q}} = \mathcal{Q}_Q(t) \tag{8.60}$$

Carrying this calculation one more step in detail yields the equations of motion:

$$\frac{d}{dt}(m\dot{x}) + kx + \delta\dot{x} = F(t) + \frac{I^2}{2}\frac{dL}{dx} - \frac{Q^2}{2}\frac{d}{dx}\left(\frac{1}{C}\right)$$

$$\frac{d}{dt}(LI) + \frac{Q}{C} + RI = V \tag{8.61}$$

Thus we see that the magnetic force is proportional to the change in inductance with mechanical displacement and the electric force is proportional to the change of capacitance with displacement. As in the case of the magnetic force formula (8.45), the electric force in the capacitor-based actuator can be calculated from either $W_e^*(V, x)$ or $W_e(Q, x)$. Using $Q = C(x)V$, and (8.61), one can easily see that

$$F_e = -\frac{Q^2}{2}\frac{d}{dx}\left(\frac{1}{C(x)}\right) = -\frac{\partial W_e(Q, x)}{\partial x} = \frac{\partial W_e^*(V, x)}{\partial x} \tag{8.62}$$

where

$$W_e^* = \frac{1}{2}C(x)V^2, \quad W_e = \frac{1}{2}Q^2/C(x)$$

The total Lagrangian for the electromechanical system with change as a generalized coordinate becomes

$$\mathcal{L} = T - V + W_m^*(\dot{Q}, x) - W_e(Q, x)$$

Hidden in the equation is the *back emf* or motion-induced voltage in the circuit. If we expand the left-hand side of (8.61), recognizing that $L(x(t))$ is an implicit function of time, we obtain

$$\frac{d}{dt}LI = L\frac{dI}{dt} + I\frac{dL}{dx}\dot{x}$$

The second term represents the velocity-induced voltage on the circuit, sometimes called a "back emf."

These equations can be extended to multiple mechanical degrees of freedom and multiple circuits, and is given as an exercise in the Problems. For the novitiate to electromechanical systems, however, it is important to first understand the examples given in Section 8.6.

Example 8.4 Linear Electric Actuator. As an application of the capacitance method of calculating electric forces consider the nested cylinders with radii a, b, shown in Fig. 8.21. This configuration could be used in a MEMS device to produce an out-of-plane motion. The capacitance between two nested cylinders with a vacuum (or air) gap can be calculated by integrating the electric energy density $\varepsilon_0 E^2/2$ between the cylinders. The stored coenergy in a capacitor is

$$W^* = \frac{1}{2}CV^2 \tag{8.63}$$

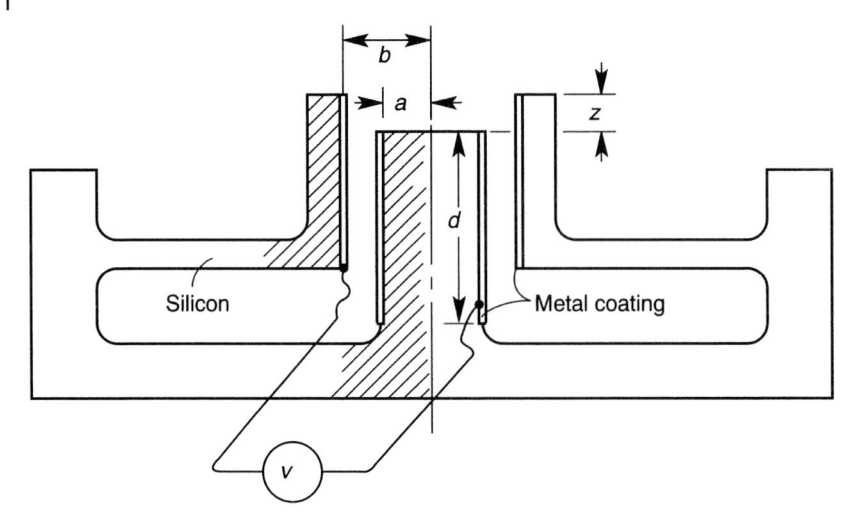

Fig. 8.21 Sketch of a MEMS electric force, nonplanar actuator.

where V is the voltage across the two cylinders. The capacitance is given by

$$C = \frac{2\pi \varepsilon_0 \ell}{\ln(b/a)} \tag{8.64}$$

where ℓ is the overlap length, i.e., $\ell = d - z$. The relation between the force and electric coenergy is [see (8.62)]

$$F_z = \frac{\partial W^*}{\partial z} = -\frac{V^2 \pi \varepsilon_0}{\ln(b/a)} \tag{8.65}$$

Thus the force tends to pull the inner cylinder into the outer cylinder. Note that the force is quadratic in the voltage and independent of the displacement z.

8.6
Applications

Electrodynamic Accelerators: Mass Drivers

Rotating electromechanical devices were developed in the late nineteenth century. Linear motion electromechanical devices have received significant development in the late twentieth century. These applications include voice coil actuators, magnetic hammers, linear induction and synchronous motors, and electromagnetic launchers or mass drivers, which have been proposed as spacecraft launch systems for the moon or Mars.

As an example of the application of Lagrange's equations to electromechanics, we consider the acceleration of a passive conducting body by a neighboring coil

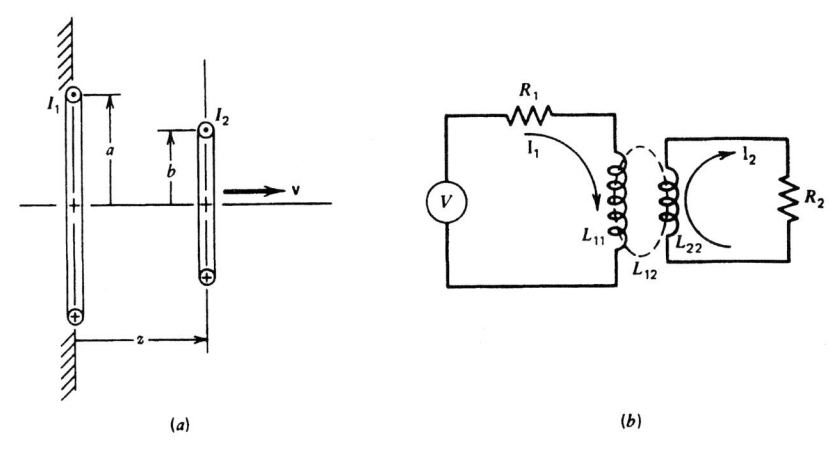

Fig. 8.22 Two-coil magnetic accelerator. (*a*) Geometry of physical system. (*b*) Equivalent circuit of driver and accelerating coils.

with active currents, as shown in Fig. 8.22. This simple problem incorporates the basic physics of a *magnetic hammer*. The driven coil can be an aluminum plate that is thrown against a die and formed into a shape. Here we want to transfer magnetic energy into kinetic energy.

The two-coil accelerator is modeled as two circuits shown in Fig. 8.18. The driving circuit is assumed to have a voltage power supply, a resistance R, and self-inductance L_{11}. The flux generated by the flow of current $I_1(t)$ also links the passive coil circuit. The moving or passive coil is assumed to have a resistance R_2 and self-inductance L_{22}. The part of the flux generated by $I_2(t)$ is assumed to link the driven coil. The colinkage of fluxes is described by a mutual inductance $L_{12}(x)$, which depends on the relative distance $x(t)$ between the two coils. The mutual inductance appears as a coefficient in the magnetic coenergy function.

$$W^* = \frac{1}{2}\left(L_{11}I_1^2 + 2L_{12}I_1 I_2 + L_{22}I_2^2\right) \tag{8.66}$$

To obtain the equations of motion for the moving coil and the two circuits, we choose $x(t)$ as a generalized position variable, and $I_1(t)$, $I_2(t)$ as generalized velocity variables. Since we do not have any assumed capacitance devices, we can use I_1, I_2 instead of $\dot{Q}_1$, $\dot{Q}_2$ in the equations. There are also no assumed gravity or elastic forces. The Lagrangian for the system is just the kinetic and magnetic energy functions:

$$\mathcal{L} = T + W^*(I_1, I_2, x)$$

where

$$T = \frac{1}{2}m\dot{x}^2 \tag{8.67}$$

To account for the resistance or energy dissipation, we write down a Rayleigh dissipation function

$$\mathcal{R} = \frac{1}{2} R_1 I_1^2 + \frac{1}{2} R_2 I_2^2 \tag{8.68}$$

The general form of the equations of motion are

$$\frac{d}{dt}\frac{\partial \mathcal{L}}{\partial \dot{x}} - \frac{\partial \mathcal{L}}{\partial x} + \frac{\partial \mathcal{R}}{\partial \dot{x}} = 0$$

$$\frac{d}{dt}\frac{\partial \mathcal{L}}{\partial I_1} + \frac{\partial \mathcal{R}}{\partial I_1} = V$$

$$\frac{d}{dt}\frac{\partial \mathcal{L}}{\partial I_2} - \frac{\partial \mathcal{R}}{\partial I_2} = 0 \tag{8.69}$$

In the second equation, V represents the generalized force corresponding to the generalized velocity I_1. This term was obtained using the principle developed in Chapter 4 by writing the power input to the system which is $V I_1 = V \dot{Q}_1$. The coefficient of $\dot{Q}$ is the generalized force in the second equation.

The specific form of the equations can be written provided that we recognize that $L_{12}(x(t))$ is an explicit function of $x(t)$ and an implicit function of time:

$$m\ddot{x} = I_1 I_2 L'_{12}(x)$$

$$L_{11}\dot{I}_1 + L_{12}\dot{I}_2 + R_1 I_1 + I_2 L'_{12}\dot{x} = V(t)$$

$$L_{12}\dot{I}_1 + L_{22}\dot{I}_2 + R_2 I_2 + I_1 L'_{12}\dot{x} = 0 \tag{8.70}$$

The equation of $L_{12}(x)$ can be obtained from tables of inductance functions in handbooks or can be calculated numerically. A typical mutual inductance gradient $L'_{12}(x)$ is shown in Fig. 8.23. We can see that this function is highly nonlinear. Also the terms $I_2\dot{x}$ and $I_1\dot{x}$ are nonlinear terms in the second and third equations in (8.70).

As a special case, we assume that the current $I_1(t)$ is specified instead of the voltage $V(t)$. This eliminates one of the equations, and we can write the equations as a set of first-order differential equations suitable for a numerical integration routine. (The student is encouraged to try this using the integration routine MATLAB ode23 or ode45 or similar integration software.)

$$\frac{dx}{dt} = v$$

$$\frac{m dv}{dt} = I_1 I_2 L'_{12}(x)$$

$$L_{22}\frac{dI_2}{dt} = -R_2 I_2 - L_{12}\frac{dI_1}{dt} - L_{12} I_1 v \tag{8.71}$$

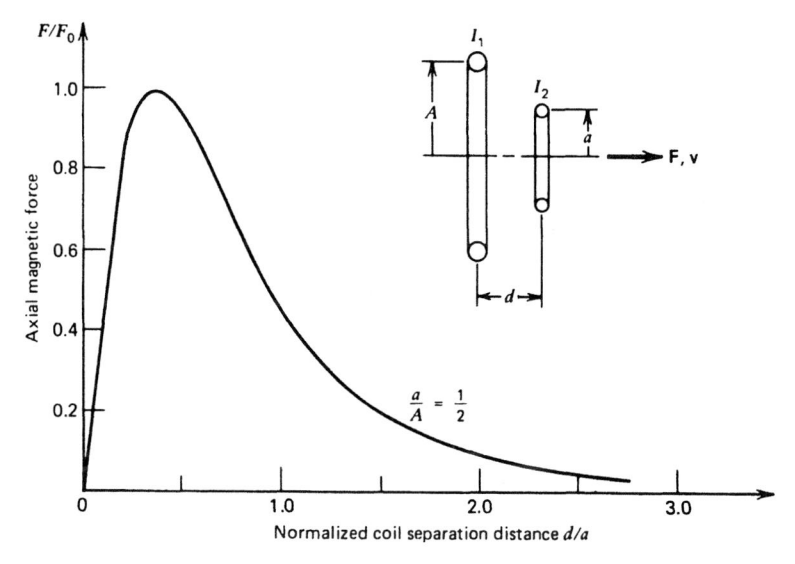

Fig. 8.23 Normalized magnetic force between two coaxial circular coils vs. axial distance.

As a special case, we assume a pulsed half-sine wave in the drive coil:

$$I_1 = I_0 \sin \frac{\pi t}{\tau_0}, \quad 0 \leqslant t \leqslant \tau_0, \quad I_1 = 0, \quad t > \tau_0 \tag{8.72}$$

The time history of the coil velocity $v(t)$ and the induced current $I_2(t)$ is shown in Fig. 8.24. It is of interest to note the current reversal at the end of the drive-coil pulse. This current reversal decreases the maximum attainable terminal mass velocity.

The change in velocity can be calculated using impulse methods discussed in Chapter 2:

$$\Delta v = \frac{1}{m} \int_0^{\tau_0} I_1 I_2 L'_{12} \, dt \tag{8.73}$$

We can see here that if I_2 becomes negative, the terminal velocity will be reduced.

Electromagnetic Launcher Dynamics

Another magnetic accelerator is shown in Fig. 8.25, known as an electromagnetic rail launcher (EML) or *magnetic railgun*. Electric currents on the order of 10^4 to 10^6 amperes are passed though a conducting rail across a moving armature and back down a parallel rail. The current is usually discharged from a capacitance energy storage bank or from a flywheel into a magnetic generator. The closed circuit produces a large magnetic field between the rails that propels the armature to very high speeds of from 0.5 to 5 kilometers per second. The length of the rails is on the order from 1 to 3 meters, and the acceleration of the armature is extremely large.

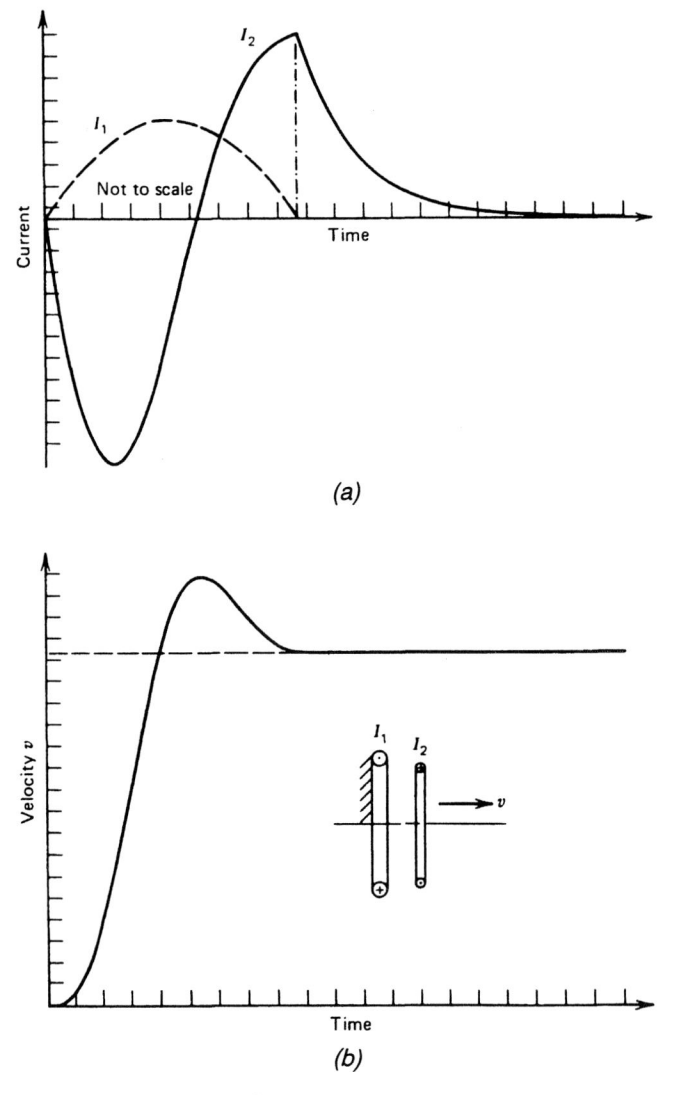

Fig. 8.24 Dynamic response of an accelerated coil (Fig. 8.23) due to half-sine pulsed current in the driver coil: (*a*) Induced current in accelerated coil vs. time. (*b*) Axial velocity of accelerated coil vs. time.

These EML devices have and are being built for both military and space launch applications. They have also been used for experimental work in the impact of materials.

The magnetic force in the EML can be calculated in several ways using either direct **I** × **B** vector methods, magnetic pressure (8.14), or the Lagrangian method in

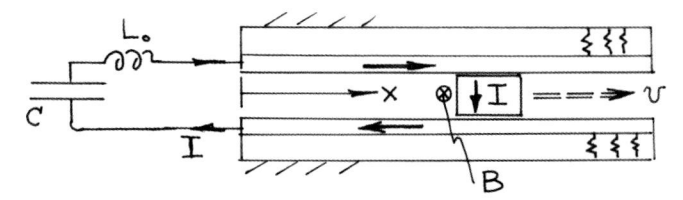

Fig. 8.25 Sketch of electromagnetic launcher showing energy storage capacitor, conducting rails, armature and elastic support structure. (See Johnson and Moon, 2006).

which the force on the armature is related to the gradient of the magnetic energy stored in the rail and hence related to the gradient of the rail inductance, i.e.,

$$F_{\text{mag}} = \frac{\partial W_{\text{mag}}}{\partial x} = \frac{1}{2} I^2 \frac{dL(x)}{dx} \tag{8.74}$$

For parallel rails with uniform geometry along the rail, $L(x)$ is a linear function of x, i.e. $L(x) = L_0 + L'x$, and $F_{\text{mag}} = I^2 L'/2$. EML devices are usually characterized by the constant L'.

EML systems are examples of dynamics under very extreme conditions of high currents, magnetic fields as well as velocities and accelerations. One of the consequences of such extreme conditions is melting, gouging and wear along the rail. Although EML technology has been known since the 1970s, only recently have problems related to wear and rail damage been addressed. Another issue in these devices is the dynamics of the rails themselves. Anthony Johnson (now at General Atomics) and the Author have investigated elastic waves generated by the moving armature and the effect of these waves on the contact pressure that controls the contact resistance as well as the wear and friction mechanics. (See, e.g., Johnson and Moon, 2006.)

Using the Lagrange method for both the circuit and the motion of the armature, the coupled equations of motion can be shown to take the following form when the rails are connected to a charged capacitance bank of capacitance C_0 (Johnson, 2008). At the initial time $t = 0$, the velocity $v = 0$ and the charge has a starting value Q_0.

$$\dot{x} = v$$
$$m\dot{v} = \frac{1}{2} I^2 L' - f(v)$$
$$\dot{Q} = I$$
$$(L_0 + L'x)\dot{I} + (R_0 + R_1 x)I + \frac{Q}{C_0} = 0 \tag{8.75}$$

In the second equation, $f(v)$ is the friction force between the sliding armature and the rails. A typical value for $dL/dx = 0.3 \cdot 10^{-6}$ Henries per meter, for a laboratory EML device. A typical current pulse time for small capacitance banks is on the order of several milliseconds.

Vibrations of a Levitated Superconducting Coil

One of the exciting transportation technologies to emerge in the late Twentieth Century is magnetically levitated vehicles and trains, or Mag-Lev (see Fig. 8.2). In this example we discuss the vibration dynamics of a Mag-Lev vehicle that carries superconducting magnets (see also Moon, 1994). The levitation forces are generated when a current-carrying superconducting coil moves over a normal conductor. In the example here, we assume that the coil moves over a thin conducting sheet, as shown in Fig. 8.26. The steady design height of the coil above the guideway is h_0 and the perturbed heave displacement is $u(t)$ so that the total coil–track gap is $h(t)$,

$$h(t) = h_0 + u(t) \tag{8.76}$$

Our goal here is to find the natural frequency of the vehicle when it is disturbed. We show the direct method, which is based on Newton's law and the perturbation of the magnetic-force function.

In the direct method the magnetic lift force, $F(h)$, is approximated by the first two terms in a Taylor series:

$$F(h) = F(h_0) + \left.\frac{\partial F}{\partial h}\right|_{h=h_0} (h - h_0) \tag{8.77}$$

The equation of motion for the one-degree-of-freedom problem is

$$m\ddot{h} = F(h) - mg$$

or

$$\ddot{u} - \frac{1}{m}\frac{\partial F}{\partial h} u = 0 \tag{8.78}$$

where the derivative is evaluated at $h = h_0$. As an example, we consider a moving current element of length b and carrying I Ampere-turns of current. The levitation force, h as the form

$$F = \mu_0 I^2 f(b/h) \tag{8.79}$$

At equilibrium the lift force should balance the gravitation force:

$$\mu_0 I^2 f(b/h_0) = mg \tag{8.80}$$

It is straightforward to show that

$$\ddot{u} - g\frac{\partial \log f}{\partial h} u = 0 \tag{8.81}$$

If we assume that the perturbed motion is sinusoidal,

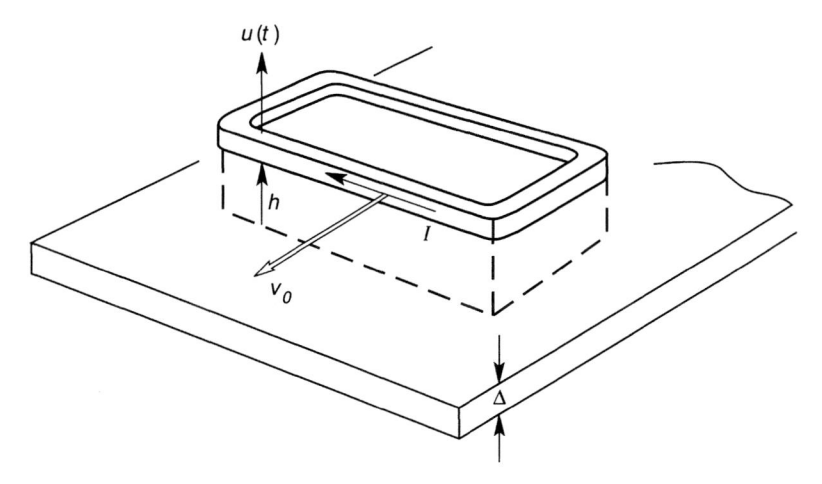

Fig. 8.26 Levitated superconducting coil moving above a thin planar conducting sheet.

$$u(t) = A\cos(\omega t + \varphi_0) \tag{8.82}$$

then the natural frequency takes the form

$$\omega^2 = -g\frac{\partial \log f}{\partial h} \tag{8.83}$$

or that the natural frequency of a levitated vehicle is *independent of the mass*. For example, the steady levitated force of a moving wire over a thin conducting sheet with velocity v_0, is given by (see, e.g., Moon, 1994)

$$F = \frac{\mu_0 I^2}{4\pi} \frac{v_0^2}{v_0^2 + w^2}\left(\frac{b}{h}\right) \tag{8.84}$$

where $w = 2/(\mu_0 \sigma \Delta)$ is a characteristic velocity, σ is the sheet conductivity, and Δ is the sheet thickness. Using (8.83), the natural frequency is

$$\omega = \left(\frac{g}{h_0}\right)^{\frac{1}{2}} \tag{8.85}$$

This means that the natural frequency of a levitated coil over a sheet guideway is the same as a pendulum with length equal to the gap h_0.

Microelectromechanical Accelerometer

Microelectromechanical systems (MEMS) involve the etching of 1–10 μm scale mechanical devices in silicon using photolithography technologies. This example is taken from the doctoral thesis of S. G. Adams (1996) at Cornell University (see also Adams et al., 1995). The object of this study was to design and fabricate a miniature MEMS accelerometer with a tunable natural frequency. The small size of the

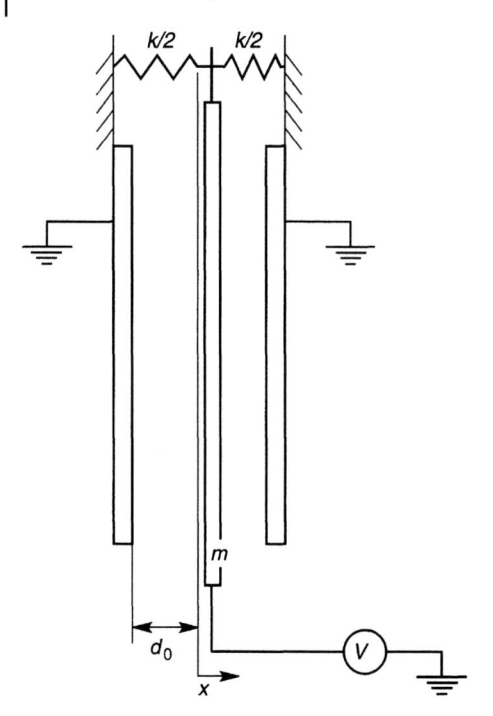

Fig. 8.27 Model for a MEMS electric force actuator (see also Fig. 1.8).

device makes it suitable for embedding in a machine or structural component for use as an accelerometer.

A photomicrograph of the device is shown in Fig. 8.1.

The platelike arms serve two functions: (1) to provide a plate electrode for a capacitor or electric-energy storage, and (2) to provide elastic or mechanical stiffness. A simplified model of the device is shown in Fig. 8.27. The stator plates of the capacitor are grounded by a voltage V relative to the grounded stator. Although the flexible structure has many modes of vibration, we consider a model with only one degree of freedom denoted by $x(t)$ that describes the lateral motion of the center plate m in Fig. 8.27.

According to the electric-energy formulation of electromechanics (8.62), the electric force between the positive charge on the moving plate and the negative charges on the stator plates of the capacitor is given by

$$F^e = \frac{\partial W^*(V, x)}{\partial x} \tag{8.86}$$

and the equation of motion of the moving plate is given by

$$m\ddot{x} + kx = \frac{\partial W^*}{\partial x} \tag{8.87}$$

The electric energy stored in a capacitor volume is given by

$$W^* = \int \frac{1}{2}\varepsilon_0 E^2 \, dv \tag{8.88}$$

where E is the electric field and ε is the electric permittivity (see, e.g., Jackson, 1962). The electric field is related to the voltage by

$$V = \int \mathbf{E} \cdot d\boldsymbol{\ell} = Ed \tag{8.89}$$

where d is the gap between the plates. If the area of the plate is denoted by A, then the expression for $W^*(V, x)$ is given by

$$W^* = \frac{\varepsilon_0}{2} V^2 A \left[\frac{1}{d_0 - x} + \frac{1}{d_0 + x} \right]$$

or

$$W^* = \frac{\varepsilon_0}{2} V^2 A \frac{d_0}{d_0^2 - x^2} \tag{8.90}$$

where ε_0 is the permittivity of vacuum, and d_0 is the initial gap between the plates. The electric force that results is a nonlinear function of the displacement x and a quadratic nonlinear function of the voltage V:

$$F^e = \varepsilon_0 \frac{V^2 A}{d_0} \frac{x}{(1 - x^2/d^2)^2} \tag{8.91}$$

For small motions compared to the gap we can expand the force in a Taylor series and obtain a negative electric stiffness, κ:

$$m\ddot{x} + kx = \kappa x \tag{8.92}$$

where

$$\kappa = \varepsilon_0 V^2 A / d_0 = c_0 V^2$$

When the mechanical stiffness is greater than the negative electric stiffness, $k > \kappa$, there exists a sinusoidal notion $x(t) = B \cos \omega t$, where the natural frequency can be tuned, depending on the square of the applied voltage, i.e.,

$$\omega^2 = (k - c_0 V^2)/m \tag{8.93}$$

This behavior was obtained for a MEMS device similar to that shown in Fig. 8.1, and experimental results are shown in Fig. 8.29 (Adams, 1996).

Casimir Force Instability

In Fig. 8.28 is shown a variation of an experiment to measure the Casimir force at submicron scales. A torsional device is fabricated with two paddles or small plates

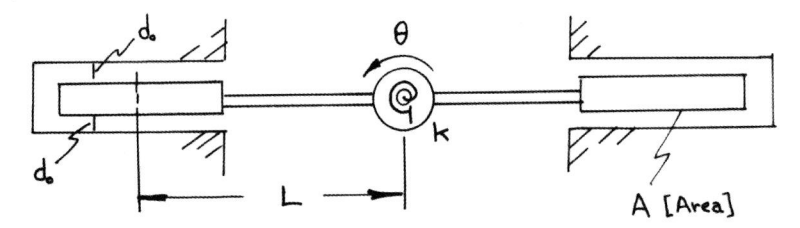

Fig. 8.28 Sketch of a nano-scale torsional experiment to measure the Casimir force effect.

on either side that are coated with a conductor. The plates are placed into two cavities in which the plates can move. The problem here is to calculate at what gap the torsional structure will become unstable due to the Casimir attraction of the plates to the walls of the cavities. Neglecting thermal effects we assume that each of the walls will attract the paddles. We also neglect the tilt in calculating the force on one paddle when displaced from its center position. Using Eq. (8.16), the torque and related force on each paddle is then (see Guo and Zhao, 2004; Lin and Zhao, 2005)

$$T = 2FL \tag{8.94}$$

$$F = \frac{\pi^2 hcA}{240}\left[\frac{1}{(d_0 - L\theta)^4} - \frac{1}{(d_0 + L\theta)^4}\right]$$

$$F = \frac{\pi^2 hcAL}{30 d_0^5}\theta \tag{8.95}$$

The torque is destabilizing, increasing with the angle θ. The Casimir torque will be resisted by the structural torque such that the equation of angular motion becomes

$$I_0 \ddot{\theta} + (k - \kappa)\theta = 0 \tag{8.96}$$

where

$$\kappa = \frac{\pi^2 hcA}{15 L^3}\left(\frac{L}{d_0}\right)^5$$

This equation will yield a harmonic solution as long as the bracketed term in the equation of motion is positive. The natural frequency is a function of the geometric parameter $(L/d_0)^5$, i.e.

$$\theta(t) = \vartheta_0 \cos(\omega t + \varphi_0)$$
$$\omega^2 = \omega_0^2 (1 - \kappa/k)$$
$$\omega_0^2 = k/I_0 \tag{8.97}$$

Thus by choosing an appropriate gap to length ratio d_0/L, one can tune the natural frequency below the structural natural frequency similar to results for a MEMS capacitor in Fig. 8.29. A critical value of the tuning parameter d_0/L exists when the

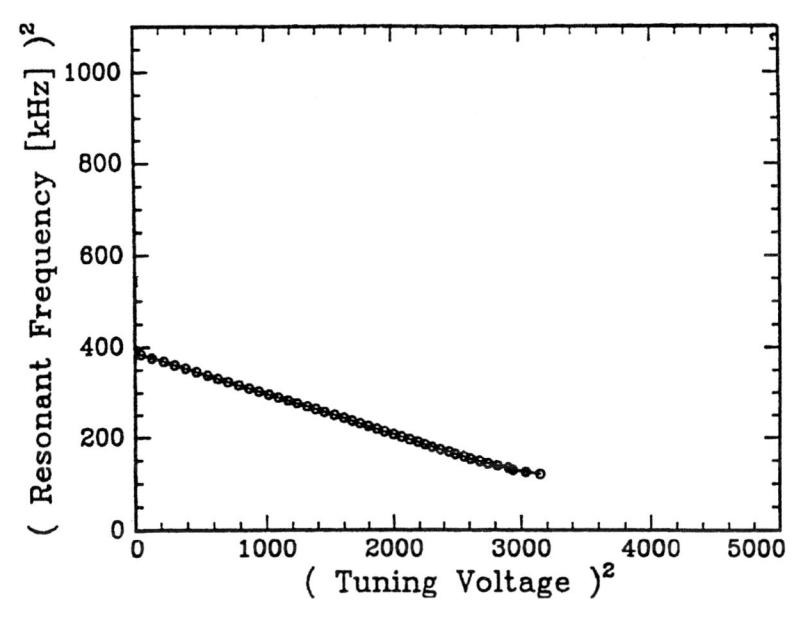

Fig. 8.29 Natural frequency squared vs. voltage squared for a MEMS actuator (see Fig. 8.1, Eq. (8.93)) (Adams, 1996).

natural frequency approaches zero:

$$\omega \to 0$$

$$d_0/L \to (d_0/L)_{crit}$$

$$\left(\frac{d_0}{L}\right)_{crit} = \left[\frac{\pi^2 hcA}{15L^3 k}\right]^{1/5} \tag{8.98}$$

A similar detuning technique using an electric field has been used in a larger scale MEMS oscillator (see Figs. 8.27, 8.29). This instability is known in pure structural mechanics as *buckling*. Near or beyond the buckling parameter the linear model is not valid and a nonlinear equation of motion must be used. Also instead of a cavity, experimental physicists have used spherical surfaces with a large radius of curvature.

8.7
Control Dynamics

The introduction of control in the dynamics of machines implies that the engineer wants to design the dynamical behavior to fit some specifications. Among the control objectives control designers often use are:

- Maintain position, velocity, or rpm
- Damp out unwanted motions

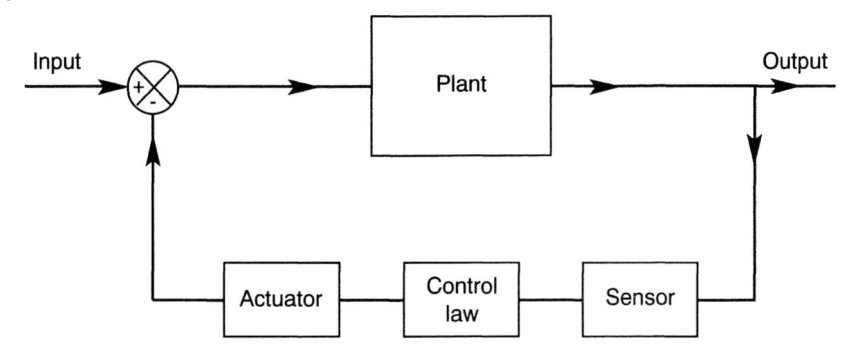

Fig. 8.30 Input–output model of a mechanical system (plant) with a feedback controller (sensor-control electronics–actuators).

- Keep the device close to a given trajectory
- Stabilize an otherwise unstable motion
- Optimize for minimum time
- Optimize for minimum energy or fuel

The methods to achieve these objectives are different from those used in traditional dynamic analysis. That is why dynamics and control are usually taught in different courses and textbooks. But in the new age of mechatronics, knowledge of both systems of analysis in control and dynamics are beginning to merge. This section can only give an example of how control and dynamics interact.

The conventional view of a dynamical system from the control analyst is captured in Fig. 8.30. The uncontrolled machine is represented by the box called *plant*, while the control engineer's contribution is contained in the box called *control*. This representation is not always unique. The control box can also be split into a sensor system, control law, and actuator power supply. An example of a controlled machine is the two link robot arm shown in Fig. 8.31, shown with a linear magnetic actuator. In practice there would be another actuator at the rotary joint, but for simplicity we have replaced the actuator with an elastic spring. Thus the *plant* in this case consists of a two-degree-of-freedom mechanical system and a double-loop electric circuit. The two systems are connected with a magnetic force and motion-induced back emfs in the circuits V_{b1}, V_{b2}, similar to (8.10) in Example 8.1.

A simple control objective here would be: given a disturbance of the "arm," damp out the vibration in a shorter time than natural viscous or aerodynamic damping. To meet this goal a classic control engineer would measure some or all of the state variables in the plant $\{x, \dot{x}, \theta, \dot{\theta}, I_1, I_2\}$ and introduce a control voltage in the circuit V_c that is proportional to some or all of these variables. Such a method is known as *linear, analog, state feedback control*. Thus, if **z** is an N-dimensional state vector, then

$$V_c = \mathbf{G} \cdot \mathbf{z}$$

where **G** is an N-dimensional control gain vector.

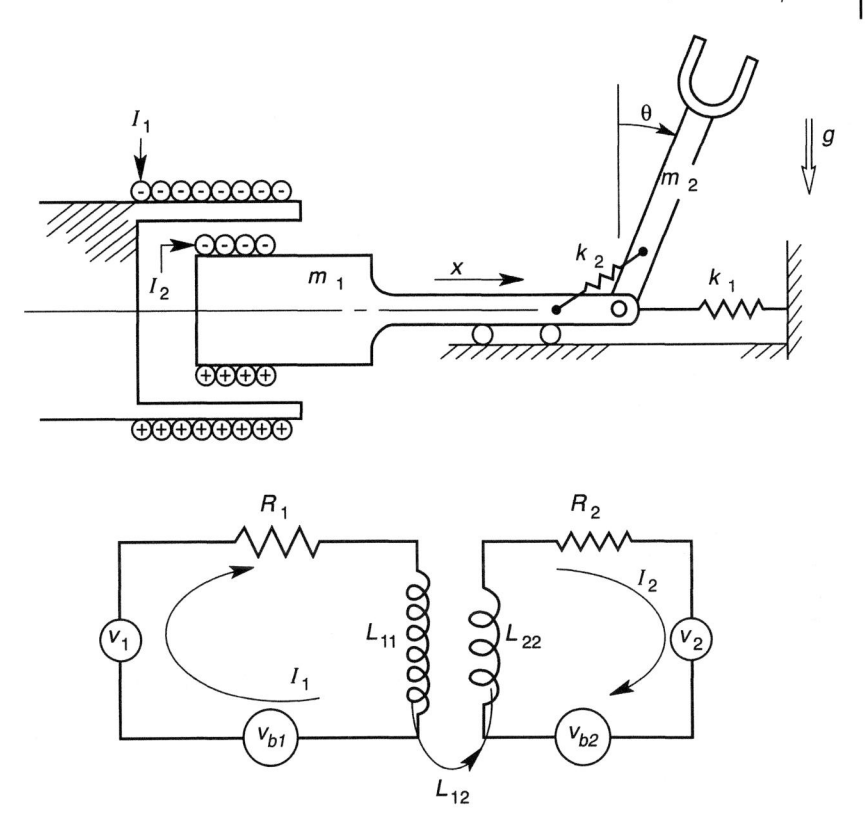

Fig. 8.31 Two-link arm with linear magnetic actuator.

The procedure for analysis involves several steps:

- Develop physical model for the magnetic force and back emf
- Derive dynamical equations of motion, including the electric drive circuit
- Decide on control strategy, e.g., analog, digital, adaptive
- Optimize the control gains to meet specifications
- Check stability of the closed-loop system

Obviously the focus of this book is on the first two tasks. However, to give an example of the procedure, we write the overall equation for the system in Fig. 8.31. The dynamics of the system are similar to that in Example 4.7, except we have added a torsional spring. The equation takes the form of a multibody system (ρ_2 locates the mass center for m_2).

$$\begin{bmatrix} (m_1+m_2) & m_2\rho_2\cos\theta \\ m_2\rho_2\cos\theta & J_0 \end{bmatrix} \begin{bmatrix} \ddot{x} \\ \ddot{\theta} \end{bmatrix} + \begin{bmatrix} k_1 & 0 \\ 0 & k_2 \end{bmatrix} \begin{bmatrix} x \\ \theta \end{bmatrix} - \begin{bmatrix} m_2\rho_2\sin\theta \\ 0 \end{bmatrix} \dot{\theta}^2$$
$$+ m_2 g \begin{bmatrix} 0 \\ -\rho_2\sin\theta \end{bmatrix} = \begin{bmatrix} I_1 I_2 \frac{\partial L_{12}}{\partial x} \\ 0 \end{bmatrix} \qquad (8.99)$$

$$L_{11}\dot{I}_1 + R_1 I_1 + I_2 \dot{x}\frac{dL_{12}}{dx} = V_c(t)$$

These equations were derived using a magnetic energy assumption in Lagrange's equations (8.49):

$$W^* = \frac{1}{2}\left(L_{11}I_1^2 + 2L_{12}(x)I_2 I_1 + L_{22}I_2^2\right)$$

and I_2 = constant. For simplicity, we assume

$$L_{12} = L_0 - \lambda x \tag{8.100}$$

To introduce control we consider two problems: in the first case, the position $\theta = 0$ is stable ($k_2 > m_2 g \rho_2$) and we wish to introduce active damping; in the second case, $k_2 < m_2 g \rho_2$ and the $\theta = 0$ position is statically unstable. In this case, control is used to stabilize the system about $\theta = 0$. In either problem, the system as written is nonlinear in the terms $\dot{\theta}^2$, $\sin\theta$. In many dynamical systems stability can be established for the linearized system. There are exceptions, however, where nonlinear terms are essential (see, e.g., Guckenheimer and Holmes, 1983). We make the classic assumption and linearize the system about $\theta = 0$.

Next we introduce the control law. Many modern technologies use digital control methods. We assume analog control and assume that we can produce a control voltage proportional to the state variable, i.e.,

$$V_c = G_1 x + G_2 \dot{x} + G_3 \theta + G_4 \dot{\theta} + G_5 I_1$$

Other control laws are possible, including terms proportional to the integral of some of the state variables. The gains $\{G_i\}$ involve both sensor and power-supply parameters.

The job of the control designer is to choose a control law (e.g., pick the gains $\{G_i\}$) that will produce damping in the system without driving the system unstable. For an introduction to some of these methods, review an introductory control test such as Ogata (1997).

In many magnetic actuators the voltage induced by the self-inductance $L_{11}\dot{I}_1$ is small compared to $R_1 I_1$ for low enough frequencies. Thus, in this approximation we can reduce the order of the system by replacing the current I_1 in the force relation by

$$R_1 I_1 = V_c + I_2 \lambda \dot{x} \tag{8.101}$$

If we try a control law $V_c = G_2 \dot{x} + G_4 \dot{\theta}$, these linearized equations become

$$\begin{bmatrix} (m_1 + m_2) & m_2 \rho_2 \\ m_2 \rho_2 & J_0 \end{bmatrix}\begin{bmatrix} \ddot{x} \\ \ddot{\theta} \end{bmatrix} + \frac{\lambda I_2}{R_1}\begin{bmatrix} (G_2 + \lambda I_2) & G_4 \\ 0 & 0 \end{bmatrix}\begin{bmatrix} \dot{x} \\ \dot{\theta} \end{bmatrix}$$
$$+ \begin{bmatrix} k_1 & 0 \\ 0 & k_2 - m_2 g \rho_2 \end{bmatrix}\begin{bmatrix} x \\ \theta \end{bmatrix} = 0 \tag{8.102}$$

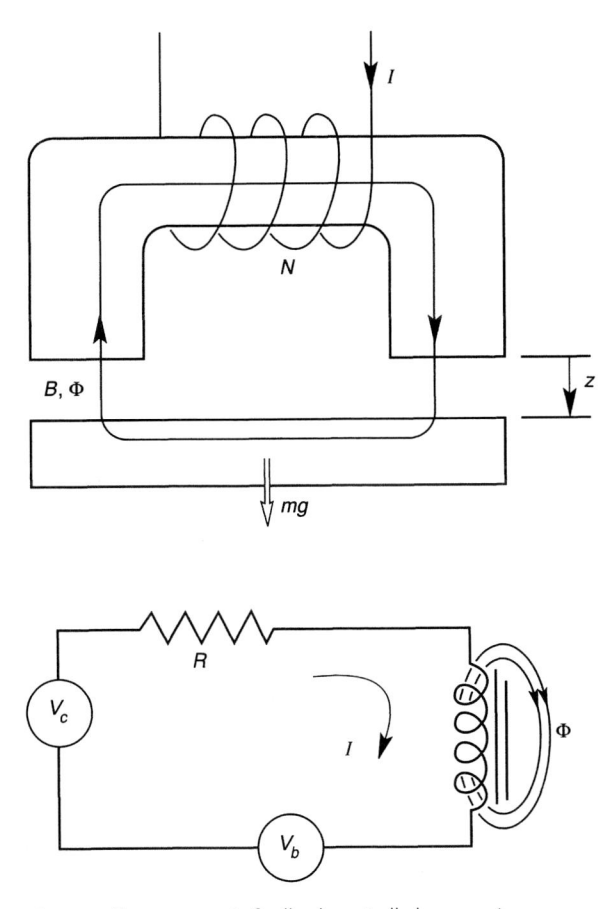

Fig. 8.32 Electromagnetic feedback controlled suspension system or magnetic bearing (Example 8.5).

First we see that the induced voltage due to Lenz's law introduces a "natural" damping $I_2^2 \lambda^2 / R_1$. Second the control law has introduced "active" damping through the gains G_2, G_4. It should be noted that in neglecting $L_{11} \dot{i}_1$ we must be careful, since at high frequencies this term can lead to instabilities in the control system.

The use of the proportional gains G_1, G_3 will introduce control stiffness terms into the system. An example is given below.

Example 8.5 Active Controlled Magnetic Levitation. There are two types of magnetic levitation. Passive diamagnetic, superconducting and eddy current systems do not require active control for stability. However, electromagnetic levitation systems (EML) are inherently unstable without control as a consequence of a basic theorem of electromagnetics called *Earnshaw's theorem* (see Moon, 1994). A typical EML system is shown in Fig. 8.32. The engineer is asked to find a control law that will make the suspended mass stable, i.e., the system will exhibit positive stiffness and damping. The equations of motion of the suspended mass and the electromagnet circuit are

$$m\ddot{z} = mg + F_z$$
$$N\dot{\Phi} = -RI + V_0 + V_c \tag{8.103}$$

where V_0, V_c are *dc* and control voltages, respectively, and N is the number of turns in the circuit.

The magnetic flux Φ in the gap can be related to current through an application of Ampere's law:

$$\oint \mathbf{H} \cdot d\ell = NI$$

In the air gap $B = \mu_0 H$, $\Phi = BA$ where A is the area of the magnet pole. These equations result in

$$\Phi = NI/\mathcal{R}$$
$$\mathcal{R} = \mathcal{R}_{fe} + \frac{2z}{\mu_0 A} \tag{8.104}$$

$\mathcal{R}$ is called the magnetic reluctance and is the sum of the reluctance in the iron $\mathcal{R}_{fe}$ and the gap. The magnetic force can be derived from the concept of *magnetic stress*

$$F = \int \frac{B^2}{2\mu_0} dA \tag{8.105}$$

where we integrate across the two gaps. This method leads to the relation

$$F_z = -\frac{\Phi^2}{\mu_0 A} \tag{8.106}$$

In Eq. (8.103) V_0 provides the gravity equilibrating force where

$$V_0 = RI_0$$
$$mg = \frac{(NI_0)^2}{\mu_0 A \mathcal{R}_0^2} \tag{8.107}$$

where $z = z_0$ is the equilibrium position. However, Earnshaw's theorem tells us that with $V_c = 0$, the perturbed system will have negative magnetic stiffness and will be unstable. To introduce control we write a linearized set of equations and add a state-variable control:

$$z = z_0 + h(t)$$
$$I = I_0 + c(t)$$

$$\mathcal{R} = \mathcal{R}_0 + \mathcal{R}_1 h(t) \tag{8.108}$$

We also expand the nonlinear force expression

$$\Phi = \frac{NI_0}{\mathcal{R}_0}\left(1 + \frac{c}{I_0} - \frac{\mathcal{R}_1}{\mathcal{R}_0}h\right)$$

$$F_z = -mg\left(1 + \frac{2c}{I_0} - \frac{2\mathcal{R}_1}{\mathcal{R}_0}h\right) \tag{8.109}$$

The coupled linearized equations then take the form

$$\ddot{h} - 2g\frac{\mathcal{R}_1}{\mathcal{R}_0}h = -\frac{2gc}{I_0}$$

$$\frac{N^2}{\mathcal{R}_0}\dot{c} + Rc = \frac{N^2 I_0}{\mathcal{R}_0^2}\mathcal{R}_1\dot{h} + V_c \tag{8.110}$$

We now introduce the following control law

$$V_c = G_1 h + G_2 \dot{h} + G_3 \ddot{h} \tag{8.111}$$

(The use of acceleration feedback $G_3 \ddot{h}$ is equivalent to using the state variable I.)
These equations can be rewritten in simplified notation

$$\ddot{h} - \alpha^2 h = -\beta c$$

$$\dot{c} + \gamma c = \Gamma_1 h + (\Gamma_2 + \delta)\dot{h} + \Gamma_3 \ddot{h} \tag{8.112}$$

where $\{\Gamma_1, \Gamma_2, \Gamma_3\}$ are the control gains. Note that with $c = 0$, the gap variable $h(t)$ grows exponentially. This system of equations can be solved using the Laplace transform, or more simply by looking for a solution of the form

$$\begin{bmatrix} h(t) \\ c(t) \end{bmatrix} = e^{st} \begin{bmatrix} \bar{h} \\ \bar{c} \end{bmatrix}$$

To achieve stability, we choose $\Gamma_1, \Gamma_2, \Gamma_3$ such that the control force $-\beta\bar{c}$ appears as if it were a restoring spring and damper:

$$-\beta\bar{c} = -k\bar{h} - bs\bar{h} \tag{8.113}$$

where k is similar to a spring constant and b is a damping constant. This choice is achieved by setting the gains to the following values:

$$\Gamma_1 = k\gamma/\beta$$

$$\Gamma_2 = -\delta + (k + \gamma b)/\beta$$

$$\Gamma_3 = b/\beta \tag{8.114}$$

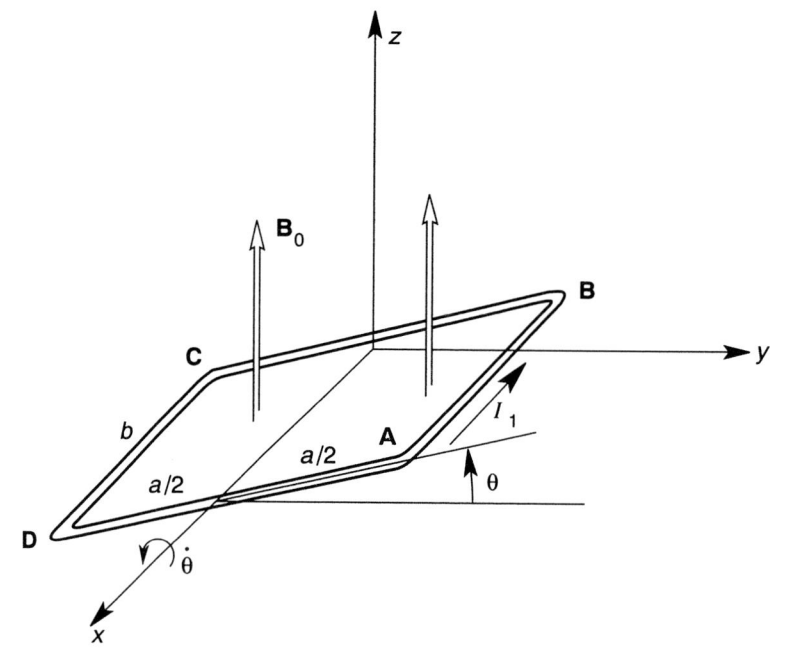

Fig. P8.1

For stability we require that $k > \alpha^2$, i.e. the control stiffness k must be greater than the negative magnetic stiffness α^2. In terms of our original physical variables, we require

$$G_2 > \frac{V_0 \mathcal{R}_1}{\mathcal{R}_0} \tag{8.115}$$

There are many other control schemes that could achieve the same objective, including so-called digital control methods. There are also control methods that can account for the nonlinear nature of the problem.

Homework Problems

8.1 Consider a rectangular coil with current I_1 in a uniform magnetic field B_0 as shown in Fig. P8.1.
   (a) Calculate the torque about the x axis. (Note that the current elements AD, BC do not create a torque about the x axis.)
   (b) Show that the back emf in the coil due to its motion is given by $2ab\dot\theta B_0 \sin\theta$. Also show that this voltage can be calculated by using either the rate of change of magnetic flux normal to the plane of the coil or by using the basic formula for the back emf, (8.6), (8.7).

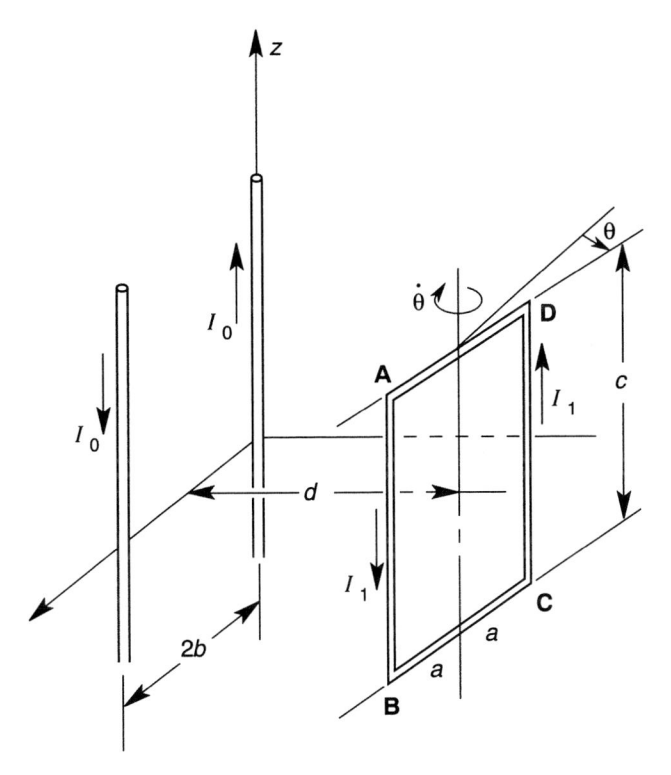

Fig. P8.4

8.2 In the Problem '8.1, assume that the current I_1 is held fixed. Derive an equation of motion near $\theta = 0$. Find the natural frequency. What is the magnetic torsional stiffness?

8.3 In Problem 8.2, assume that the current I_1 is produced in a circuit with constant voltage V_0 and resistance R. Derive the coupled equations of motion for $\theta(t)$, $I_1(t)$. Neglect the self-inductance. For $\theta \simeq 0$, find a solution for the free vibration.

8.4 Consider two long current filament wires carrying equal and opposite currents I_0, as shown in Fig. P8.4. Nearby a rectangular coil with current I_1 is constrained to rotate about an axis parallel to the I_0 pair. Derive an expression for the torque on the coil $ABCD$ due to the long current filament pair. (*Hint*: The magnetic forces on the elements AD, BC do not produce torques about the axis.)

8.5 Suppose the current I_1 in Problem 8.4 is produced by a circuit with applied voltage V and resistance R. Derive the coupled equations of motion for $\theta(t)$, $I(t)$. (Neglect the self-inductance.)

8.6 A magnetic actuator is shown in Fig. P8.6. The inductance is assumed to vary as a linear function of the position of the ferromagnetic plunger $x(t)$, i.e., $L = L_0 + L_1 x$. If the magnetic energy is given by $W = (1/2)LI^2$, derive the equations of motion for both the position of the actuator mass and the

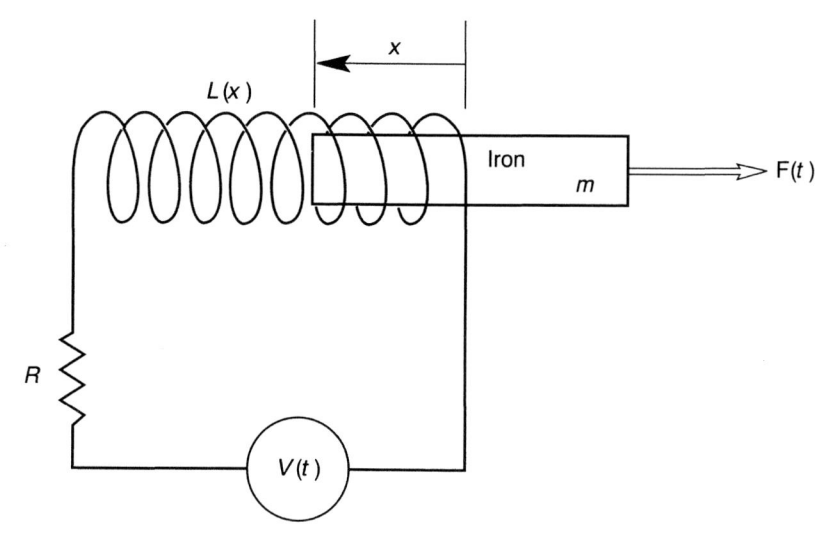

8 Electromechanical Dynamics: An Introduction to Mechatronics

Fig. P8.6

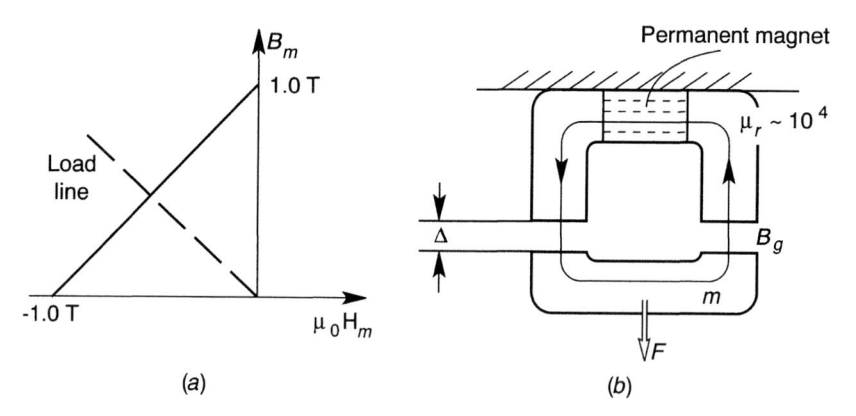

Fig. P8.8 (a) (b)

current in the circuit. Assume that the voltage $V(t)$ is a known function of time.

8.7 In Problem 8.6, suppose $V(t)$ is chosen to make the current constant. Suppose $m = 100$ g, $L_1 = 2 \cdot 10^{-4}$ H/m, $I = 100$ A. Find the acceleration of the mass.

8.8 (*Lift Force of a Permanent Magnet*) Permanent magnets are used in many magnetomechanical devices, yet the calculation of the pulling or lifting force is not easily obtained from the magnet characteristics. In this problem use the magnetization curve in Fig. P8.8 for neodymium–iron–boron, which is a popular permanent magnet material. Now consider the magnetic circuit shown in Fig. P8.8. A soft ferromagnetic material (e.g., silicon–iron) guides the magnetic flux so that it enters the mass m. The magnetic energy is approximately

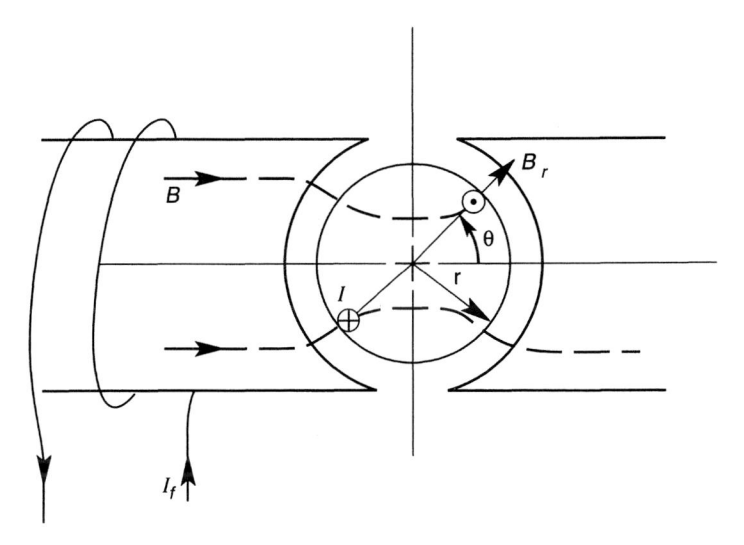

Fig. P8.9

given by $W_m = (1/2\mu_0)B_g^2 A_g 2\Delta$, where B_g is the magnetic-field density in the gap and A_g is the gap area. Then since there is no current encircling the magnetic circuit, Ampere's law can be shown to relate H_m in the permanent magnet to the field in the gap by

$$\mu_0 H_m l_m + 2\Delta B_g = 0$$

This relation is known as the *load line*. By the conservation of magnetic flux, we also have $B_g A_g = B_m A_m$. In this problem assume that $A_g = A_m$.

For $A_m = 1 \text{ cm}^2$, $B_r = 1.0$ T, $\mu_0 H_c = 1.0$ T, $\Delta = 1$ mm. Find the force in the mass. (*Hint*: Use the load line together with the demagnetization law in Fig. P8.8 to eliminate H_m.)

8.9 A model for a direct-current (dc) motor is shown in Fig. P8.9. A geometric device, known as a commutator switches the current in the armature coil when $\theta > \pi/2$, so that the torque is given by $T = IB_r \ell_a 2r$, where ℓ_a is the length of the armature coil. In addition, the moving coil element in a magnetic fluid generates a back emf given by $V_b = -2r\dot{\theta} B_r \ell_a$. If the gap field B_r is held fixed, the torque and back emf for a multiturn coil may be expressed by $T = \alpha I$, $V_b = -\alpha \dot{\theta}$. Show that the equations of motion for the dc motor are given by

$$J\ddot{\theta} + b\dot{\theta} = \alpha I + T_L$$
$$L\dot{I} + RI + \alpha \dot{\theta} = V(t)$$

where $V(t)$ is the armature drive voltage; L and R, the armature coil inductance and resistance; b is the viscous torque; and T_L the load torque on the rotor.

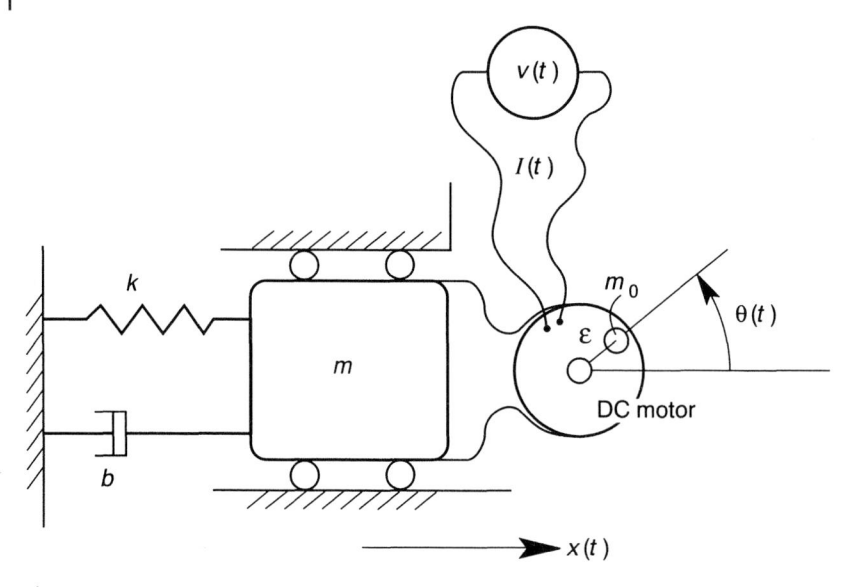

Fig. P8.10

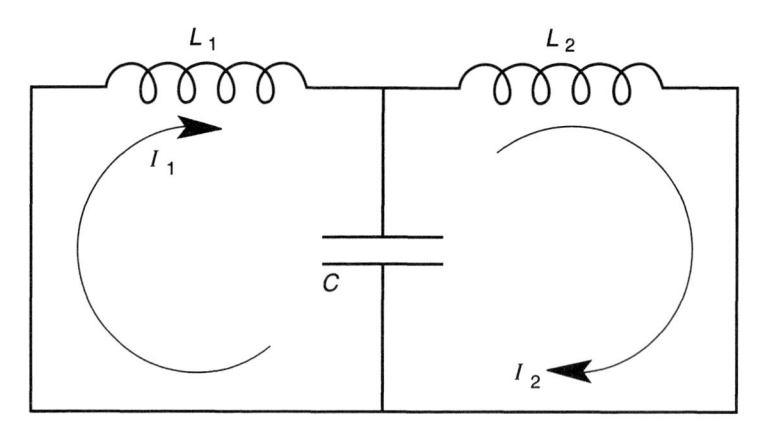

Fig. P8.11

8.10 (*Energy-limited Vibration Excitation*) A dc motor drives an excentric mass m_0 as it rotates to produce a sinusoidal excitation on the linear spring–mass–damper system shown in Fig. P8.10. Derive the coupled equations of motion for $\theta(t)$, $x(t)$, $I(t)$ when the armature circuit for $I(t)$ is driven by a constant voltage V_0. (*Hint:* Find the torque on the rotating mass due to the vibration $x(t)$.) Under what conditions will $\Omega = \dot{\theta}$ be approximately constant?

8.11 Consider the circuit in Fig. P8.11. Choose the currents in each leg I_1, I_2 as generalized velocities. Write the Lagrangian for the coupled loops. Show that the equations of motion are analogous to two masses coupled by a spring. (*Hint:* Show that $\mathcal{L} = (1/2)(L_1 \dot{Q}_1^2 + L_2 \dot{Q}_2^2) - (Q_1 - Q_2)^2/C$.)

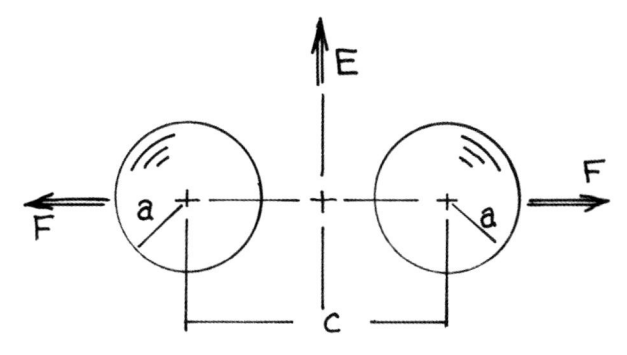

Fig. P8.13

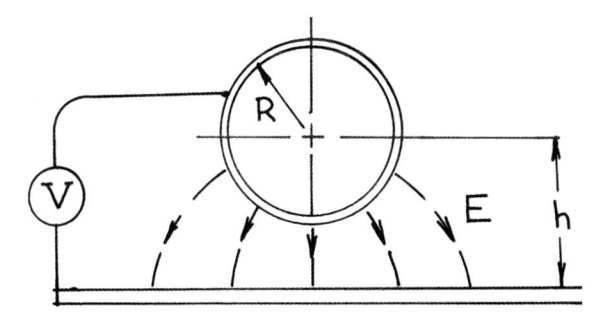

Fig. P8.14

8.12 *(Forces on Electric Dipoles)* The force on an electric dipole $\mathbf{p}_1$ is given as the gradient of the electric field $\mathbf{E}_2$, where $\mathbf{E}_2$ is produced by another set of charges, i.e. $\mathbf{F}_1 = (\mathbf{p}_1 \cdot \nabla)\mathbf{E}_2$. Assume that $\mathbf{E}_2$ is produced by another dipole of strength $\mathbf{p}_2$, and find an expression for the force between two planar electric dipoles. Assume that the electric field $\mathbf{E}_2$ is given by $\mathbf{E}_2 = -\nabla V$, where $V = \mathbf{p}_2 \cdot \mathbf{r}/(4\pi\varepsilon r^3)$ (Smythe, 1968, p. 5). At what orientation of the dipoles is the electric force at a maximum?

8.13 *(Spherical Conductors)* A spherical conductor when placed in a uniform electric field will expel the electric field inside the sphere (Fig. P8.13). When two conducting spheres are placed in a uniform electric field, they will have an induced repulsion force similar to that of two electric dipoles. An approximate expression for this force between the spheres may be found in the classic text of Smythe (1968, p. 225):

$$F = 12\pi\varepsilon E^2 \frac{a^6}{c^4}\left[1 - 2\frac{a^3}{c^3} - 8\frac{a^5}{c^5} + \cdots\right]$$

Find an expression for the electric energy function of the two spheres in a uniform electric field.

8.14 *(Force Between a Charged Cylinder and a Grounded Plane)* A long cylinder is connected to a voltage source V with its axis parallel to a grounded plane

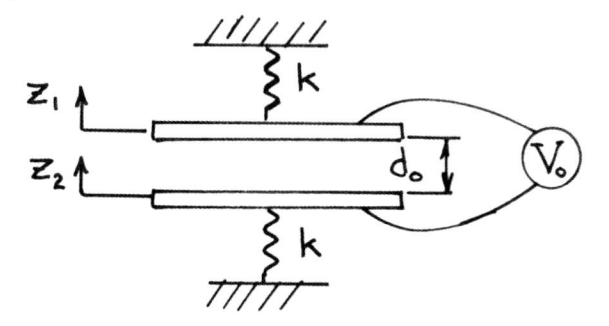

Fig. P8.15

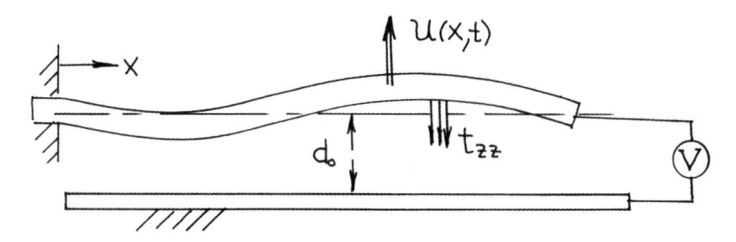

Fig. P8.16

as shown in Fig. P8.14. The expression for the electric capacitance of this configuration may be found in Smythe (1968) as

$$C = 2\pi\varepsilon\left[\cosh^{-1}\left(\frac{h}{R}\right)\right]^{-1}$$

where R is the radius of the cylinder and h is the distance of the center of the cylinder from the grounded plane. Find an expression for the force between the cylinder and the plane. Can you show that this is similar to the force between two charged cylinders at a separation distance $2h$?

8.15 (*Electric Force Attraction of Two Elastic Plates*) In a problem similar to the example in Section 8.6, two plates of a capacitor are attached to elastic structures with a spring constant k as shown in Fig. P8.15. Using a mutual force between each plate based on the electric stress $t_{zz} = \varepsilon_0 V_0^2/(2d_0^2)$, replace the gap parameter with $(d_0 - z_2 - z_1)$ and investigate the stability and dynamics of the two coupled oscillators. (See De Bona and Enikov, 2006 for a similar problem.)

8.16 (*MEMS Cantilever with Distributed Electric Stress*) Consider the simple cantilever beam in Fig. P8.16 that serves as one plate of a capacitor with a fixed voltage V_0 across the plates. Use the Euler–Bernoulli beam equation with distributed electric stress $t_{zz} = \varepsilon_0 V_0^2/2(d_0 - u(x))^2$ to examine the bending and stability of the cantilever structure. [*Hint:* Assume that the deflection of the

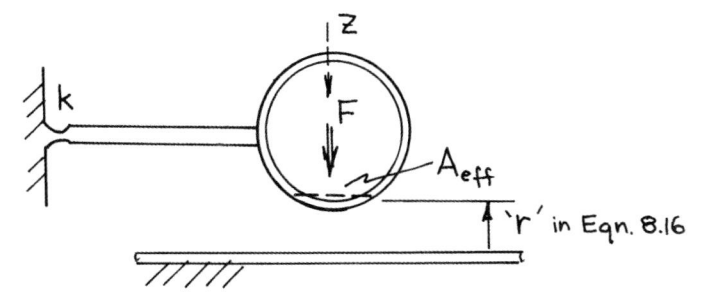

Fig. P8.17

beam $u(x)$ is much smaller that the gap d_0 and use a linearized force on the beam. The classic beam equation has the form

$$D\frac{\partial^4 u}{\partial x^4} + \rho A \frac{\partial^2 u}{\partial t^2} = f_z]$$

8.17 (*Casimir Nano Forces*) A sphere with a conducting surface is brought in close proximity to another conducting surface on a levered structure such that there exists a Casimir force (8.16) between them (Fig. P8.17). Assume that the levered structure has a torsional inertia J and rotational spring constant k. Derive an equation of motion for the motion of the torsional lever assuming that the effective Casimir area is A_{eff}.

8.18 Two plates of a capacitor, one rectangular and one triangular shaped, separated by a gap, D, act as an electric actuator for a microelectromechanical mechanism (Fig. P8.18). Assume that the capacitance is proportional to the overlap area $A(s)$, where $s = s_0 + x(t)$ and $C = \varepsilon_0 A(s)/\Delta$. Using an electric-energy function $W_e = Q^2/2C(s)$, derive the equations of motion using Lagrange's equations for the generalized variables $x(t), Q(t)$. Assume that the capacitor circuit has a resistance R and an applied voltage $V(t)$. (*Hint*: Show that the electric actuation force in the x-direction is given by $F_e = V^2 \varepsilon_0 (s_0 + x)/\Delta\sqrt{3}$.)

8.19 A concentrated mass, with charge Q, is supported by a linear spring. The charge is brought close to a grounded conducting surface. The induced charge distribution on the surface produces a field near Q that is equivalent to an image charge below the surface (Fig. P8.19). If the unstretched spring length is z_0, derive an equation of motion for the motion of the mass with charge. For what values of the spring constant is the system statically stable?

8.20 The electric force between two charges is an inverse-square law. Show that the force between two parallel-line distributions of charge with Q_1, Q_2 c/m vary with the inverse of radial separation r. [Answer: $f = Q_1 Q_2/2\pi\varepsilon_0 r$.] (*Hint*: Use Gauss' law, $\int \varepsilon_0 \mathbf{E} \cdot \mathbf{n}\, dA = Q$, to show that the electric field of a line charge distribution decreases with the radial distance.)

8.21 Consider the micromirror tilting device for an optical scanner shown in Fig. P8.21. The flat metallic conductor is grounded and two parallel conduc-

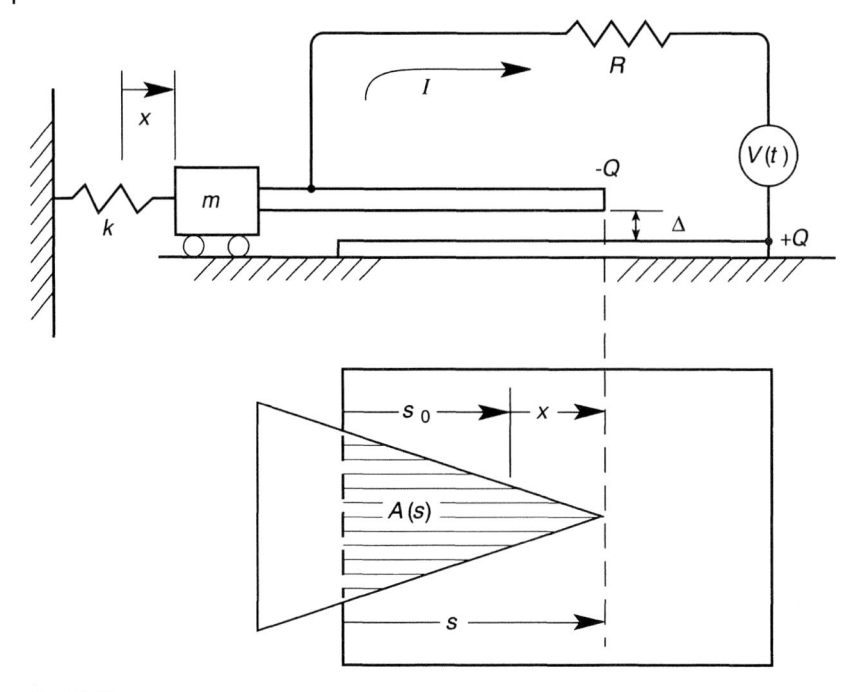

8 Electromechanical Dynamics: An Introduction to Mechatronics

Fig. P8.18

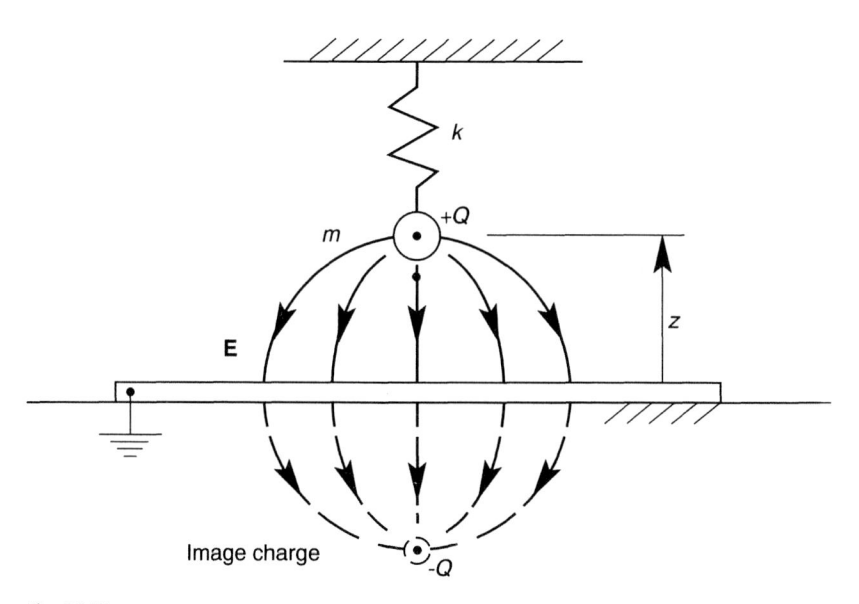

Fig. P8.19

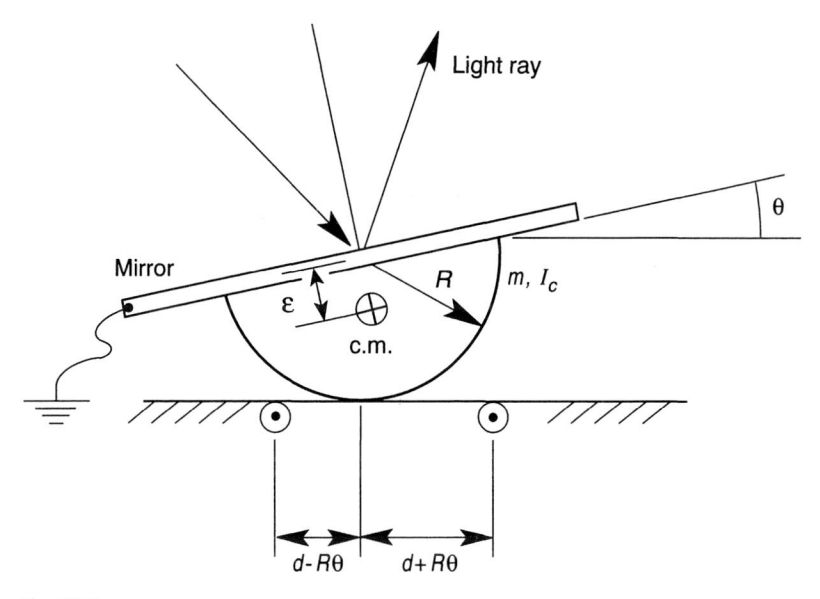

Fig. P8.21

tors carry voltages V_1, V_2 and act like distributed line charges. Add a torsional elastic spring to the mirror and derive an approximate dynamic model for the rocking motion of the mirror. Assume rolling without slip. Calculate approximate electric forces by assuming that the grounded surface is an infinite plane and use the theory of image charges as in Problem 8.19.

8.22 A MEMS device contains four cantilever beams oriented at right angles to one another, as shown in Fig. P8.22. The motions of the tips of the beams are assumed to be small. Suppose we bias the tips of the beams with charges of equal strength, Q. When $Q = 0$, assume that each of the beams can be characterized dynamically by a common modal mass m, and modal stiffness k corresponding to the lowest vibration mode. Use Lagrange's equations to derive coupled equations of motion assuming that the charges are held constant.

8.23 In Problem 8.22 determine the coupled-vibration modes. Find the frequency spectrum as a function of the charge, Q.

8.24 Suppose the charges in Problems 8.22 and 8.23 alternate positive and negative around the torus. How do the equations of motion change? How does the frequency spectrum change with the charge, Q?

8.25 (*Dynamics of Electromagnetic Levitation*) A current-carrying wire is shown moving past an electrically conducting plate in Fig. P8.25a. There is an induced lift force F_L, as well as an induced eddy current drag force F_D. It can be shown that the expressions for these magnetic forces are given by (Moon, 1984, 1994)

$$F_L = \frac{\mu_0 I^2}{4\pi h} \frac{v^2}{v^2 + h^2}; \qquad F_D = \frac{w}{v} F_L$$

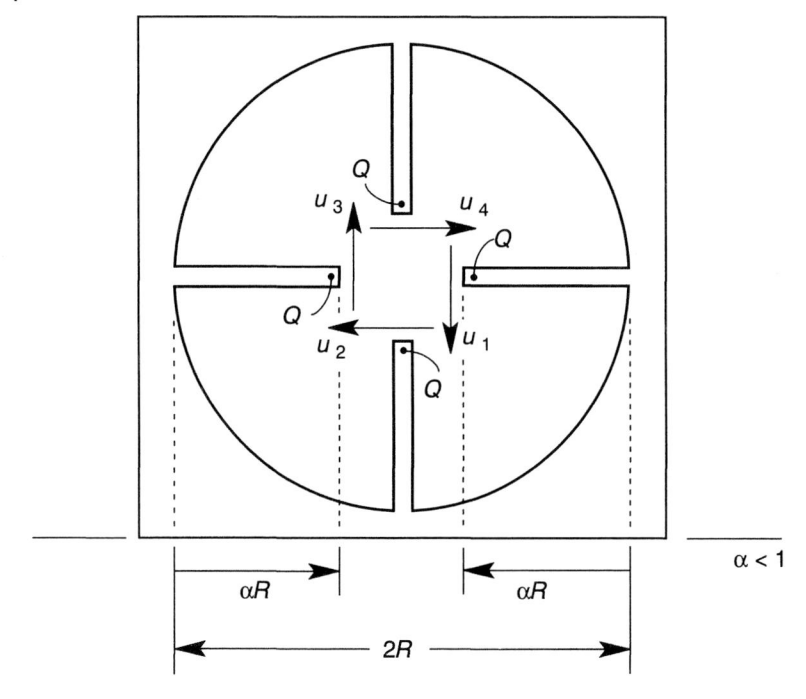

Fig. P8.22

where
$$w = \frac{1}{\mu_0 \sigma \Delta}$$

Here Δ is the track thickness, and we can assume that the product of permeability and conductivity $\mu_0 \sigma = 44$, for aluminum.

Suppose a model of a Mag-Lev vehicle contains a rectangular current-carrying coil, shown in Fig. P8.25b, is moved over a conducting plate with a speed v. Derive coupled equations of motion for heave (vertical motion) and pitching motions.

8.26 (*Mag-Lev Dynamics*) In Problem 8.25, suppose that the model is pulled along the conducting track in Fig. P8.25b with a constant thrust F_0, instead of a constant speed v. As the vehicle vibrates in the vertical direction, $h = h_0 + z(t)$, the drag will change and the horizontal velocity v will change, i.e. $v = v_0 + \phi(t)$, and the vehicle will exhibit hunting vibrations. Neglect the pitching motions and derive equations of motion for the vertical and hunting vibrations.

8.27 (*Magnetic Forces and Potentials*) A small ferromagnetic sphere is placed next to a current carrying wire with steady current I as shown in Fig. P8.27. If the material is a so-called "soft" ferromagnetic sphere, the magnetic field of the wire will induce magnetization in the material and the sphere will be

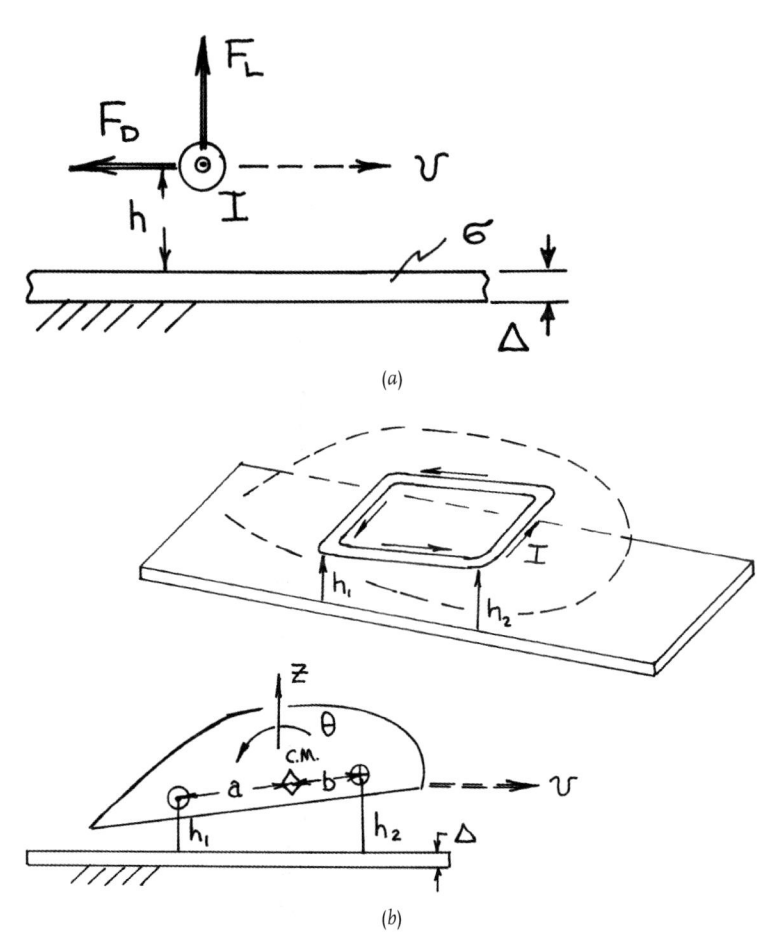

Fig. P8.25

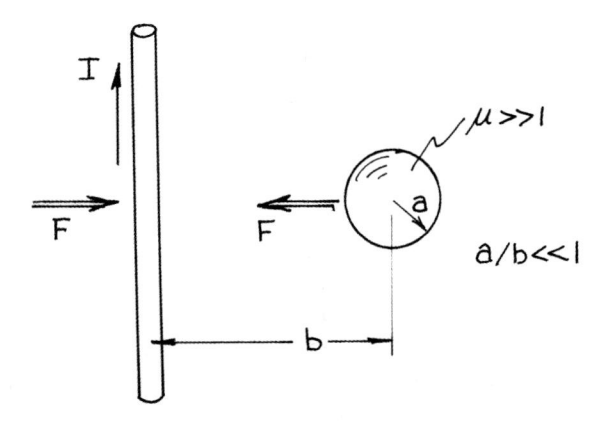

Fig. P8.27

attracted to the wire. For a strong ferromagnetic material, such as soft iron, the magnetic force is approximately given by (see Moon, 1984)

$$F = \frac{\mu_0 I^2}{\pi}\left(\frac{a}{b}\right)^3$$

where a is the radius of the sphere and b is the distance of the center of the sphere from the wire. Find an expression for the magnetic energy of interaction between the magnetic sphere and the current-carrying wire.

8.28 (*Electromagnetic Launcher*) A magnetic rail launcher is shown in Fig. 8.25. Assume that the inductance gradient is designed to increase with the distance along the rail. $L' = \lambda_0 + \lambda_1(x/L)$, where L is the length of the launcher. Also assume that the current pulse is a half sine function of time before it shuts off. Find an equation of motion to model the dynamics of the armature and find the velocity at $x = L$, as a function of the peak current I_0, and the half sine pulse time. [*Hint*: Use (8.75a,b) and integrate with MATLAB.]

9
Introduction to Nonlinear and Chaotic Dynamics

9.1
Introduction

Sources of Nonlinear Dynamics and Chaos

The goal of this chapter is to describe the range of dynamic motions possible in the particle and rigid body systems discussed in this book. In contrast to structural dynamics, where motions are usually small and linear differential equations result, rigid body and multibody problems often result in nonlinear differential equations of motion that generate a wider class of dynamics problems.

Sources of nonlinearities in mechanical and electromagnetic systems are listed below. Chaotic dynamics are possible whenever systems have strong nonlinear elements such as:

- Kinematic nonlinear accelerations, as in serial-link mechanisms
- Nonlinear damping or friction
- Backlash, play, or limiter elements
- Fluid-related forces
- Nonlinear feedback control forces such as saturation
- Nonlinear resistors, inductors, or capacitive elements, especially ferromagnetic elements with hysteresis
- Diodes, transistors
- Nonlinear optical sensors
- Electric and magnetic forces

Mechanical systems that are prone to nonlinear behavior, including chaos, are:

- Vibration of buckled structures
- Gear transmission elements with play or backlash
- Systems with sliding dry friction
- Gyroscopic systems
- Fluid-structure dynamics
- Magnetomechanical actuators

Applied Dynamics. With Applications to Multibody and Mechatronic Systems. Second Edition. Francis C. Moon
Copyright © 2008 WILEY-VCH Verlag GmbH & Co. KGaA, Weinheim
ISBN: 978-3-527-40751-4

- Multilink systems, such as robots

Dynamics is the study of how systems change in time. For most of this book, we have focused on the derivation of equations of motion of systems that are governed by Newton's laws, and in Chapter 8 on systems that are governed by the laws of electromagnetism. In this chapter we review the time history behavior of a few nonlinear systems governed by these laws. There are three types of mathematical models in dynamics:

1. Partial differential equations in space and time
2. Differential equations in time
3. Difference equations or maps that describe a sequence of events

The solution of partial differential equations is beyond the scope of this book but is of great importance to problems in elasticity, fluid mechanics, heat transfer, and electromagnetics. However, there is a large class of problems for which we can integrate over space and obtain a set of equations in time alone; the problems of particle and rigid-body dynamics fall into this category. These equations often take the form of

$$\dot{\mathbf{x}} = \mathbf{f}(\mathbf{x}, t) \tag{9.1}$$

There is another class of problems for which we can integrate between events in time; e.g., between the impacts of a ball on a rigid surface. For these problems it is sometimes possible to write the equations of motion exactly as a set of difference equations:

$$\mathbf{x}_{n+1} = \mathbf{F}(\mathbf{x}_n) \tag{9.2}$$

In the last two decades of the twentieth century, there has been a revolution in the understanding of how systems behave in time. This revolution has been named *chaos theory*, but the proper description of this theory is *nonlinear dynamical systems*. [See, e.g., Moon (1992); Strogatz (1994)]. As we explain in this chapter, it is the nonlinear nature of the equations of motion of either (9.1) or (9.2) that is responsible for the complex behavior seen in even simple systems. The variety of nonlinear motions are summarized in Table 9.1.

Chaotic Dynamics

Until the late 1970s, most physicists and engineers believed that the classical laws of dynamics stemming from Newton led to predictable dynamical behavior. However, at the beginning of the twentieth century Henri Poincaré showed that the three-body problem in celestial mechanics could result in unpredictable behavior sensitive to small changes in initial conditions. To quote from his essay on Science and Method in 1908,

> It may happen that small differences in the initial conditions, produce very great ones in the final phenomena. A small error in the former will produce an enormous error in the latter. Prediction becomes impossible.

Table 9.1 Classification of nonlinear motions.

Periodic motion	The system returns to its state after a finite time T. The trajectory in phase space is a closed curve.
Subharmonic motion	In addition to a periodic solution of period T, the system exhibits solutions with multiple primary periods, nT. The trajectory in phase space is a closed curve.
Quasi-periodic motions	These are motions that exhibit two or more periodic components of incommensurate period or frequency, i.e., T_1/T_2 is not a rational number. The trajectory covers a torus in phase space.
Chaotic motions	Nonclosed orbit trajectories are generated by the solution of a deterministic set of ordinary differential equations. The Fourier spectrum has a continuous component. In dissipative systems, the trajectory fills a fractal object in phase space. Solutions are sensitive to initial conditions. Nearby trajectories diverge on average (positive Lyapunov exponent).

This was largely ignored by practitioners of dynamics until the 1960s when Edward Lorenz of MIT observed strange dynamics in the computer simulation of a model of atmospheric convection. Finally in the late 1970s, evidence from applied mathematics, coupled with experiments from physics and engineering, began to support the idea of unpredictable deterministic dynamics, now called *chaos*. Classical nonlinear dynamics admitted steady nonlinear periodic motions as well as quasi-periodic motions (the sum of two or more incommensurate periodic motions). Now chaos theory added a new type of motion, the *strange attractor*. The hallmarks of chaotic dynamics or strange attractors include:

- Sensitive dependence on initial conditions
- Broad spectrum of frequencies in the Fourier transform
- Fractal structure in the Poincaré section in phase space
- Transient bursts of irregular motion
- Pattern of increasingly complex dynamics as some parameter is varied, e.g., period doubling

One measure of steady-state chaos and the sensitivity to initial conditions is the average exponential divergence of nearby trajectories in the phase space. This measure is called the Lyapunov exponent Λ. If $\Lambda > 0$ then the system is said to be chaotic. An introduction to Lyapunov exponents may be found in Abarbanel (1996) or Moon (1992).

9.2
Nonlinear Resonance

Resonance in a linear oscillator was discussed briefly in Chapter 1 for the system with linear damping and linear stiffness (1.12). When a lightly damped linear system experiences a sinusoidal force of frequency ω, the steady-state response exhibits an amplification near the undamped natural frequency, as shown

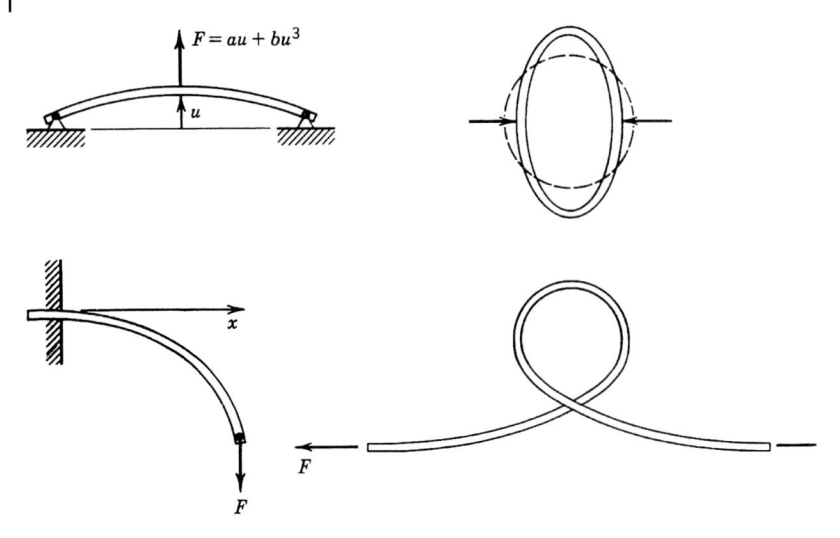

Fig. 9.1 Nonlinear force-displacement elements in mechanical systems.

in Fig. 1.18. However, many elastic systems exhibit a strong nonlinear relation between an applied force and the displacement. Several examples are shown in Fig. 9.1.

The one-dimensional dynamics of a driven mass under a nonlinear spring with a cubic nonlinearity and linear damping takes the form

$$m\ddot{x} = -b\dot{x} - kx - \kappa x^3 + f(t) \tag{9.3}$$

A classic nonlinear problem is the periodically forced motion where $f(t) = f_0 \cos \omega t$. We can divide by the mass to obtain the standard cubic nonlinear oscillator named after the German engineer G. Duffing who first studied it in 1918.

$$\ddot{x} + 2\gamma \dot{x} + \omega_0^2 x + \beta x^3 = \alpha_0 \cos \omega t \tag{9.4}$$

There are numerous books on nonlinear oscillations that show how to analyze this equation, such as Minorsky (1962), Guckenheimer and Holmes (1983), and Nayfeh and Balachandran (1995). We summarize the results here.

1. *The Linear Oscillator:* $\kappa = \beta = 0$. When $f_0 = \alpha_0 = 0$, in (9.4) the free oscillations are those of a damped oscillator where

$$x(t) = A_0 e^{-\gamma t} \cos\{[\omega_0^2 - \gamma^2]^{1/2} t + \phi_0\} \tag{9.5}$$

The damped natural frequency $\omega_d = [\omega_0^2 - \gamma^2]^{1/2}$ is independent of the initial amplitude.

2. *The Unforced Nonlinear Oscillator with Zero Damping.* In this case we have

$$\ddot{x} + \omega_0^2 x + \beta x^3 = 0 \tag{9.6}$$

9.2 Nonlinear Resonance

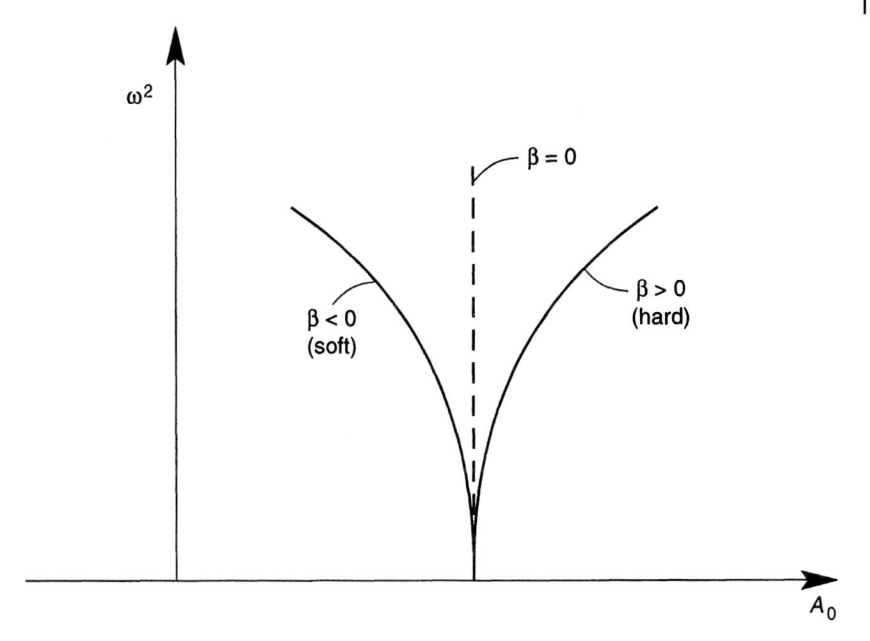

Fig. 9.2 (a) Natural frequency vs. amplitude for hard and soft springs.

Although exact solutions to the equations of nonlinear vibrations are not generally available, there are numerous perturbation analyses that yield the natural frequency of the cubic spring oscillator as a function of vibration amplitude. If $x(t)$ is approximately given by

$$x(t) = A_0 \cos \omega t + A_1 \cos 3\omega t + \cdots$$

perturbation analysis reveals the relation

$$\omega^2 = \omega_0^2 + \frac{3}{4}\beta A_0^2 \tag{9.7}$$

For a *hard spring* ($\beta > 0$), the natural frequency increases with amplitude A_0, while for a *soft spring* ($\beta < 0$), the vibration frequency decreases with amplitude A_0 (see Fig. 9.2). These results, however, are not valid for arbitrarily large A_0.

The cubic spring oscillator is a member of a larger class of problems where the restoring spring force can be derived from a potential function $\mathcal{V}(x)$, i.e.,

$$m\ddot{x} = -\frac{d\mathcal{V}}{dx}$$

For this case we have conservations of energy,

$$\frac{1}{2}m\dot{x}^2 + \mathcal{V}(x) = E_0$$

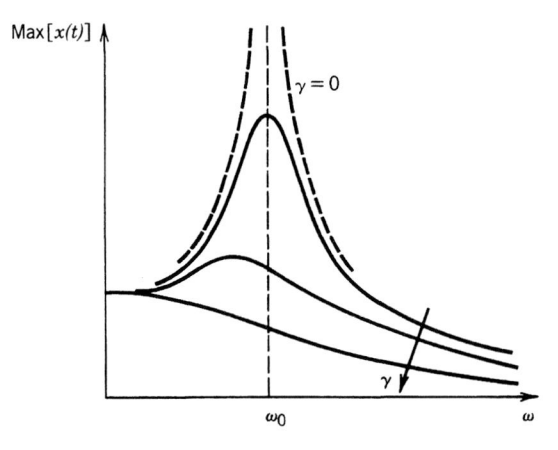

Fig. 9.2 (*b*) Linear resonance curve.

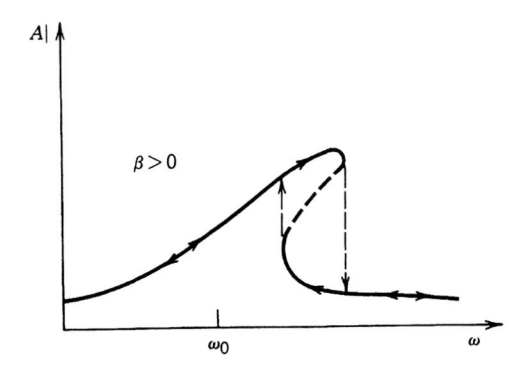

Fig. 9.2 (*c*) Nonlinear resonance curve.

If we assume that the motion in the phase plane $\{x, \dot{x}\}$ is bounded and periodic, then an integral expression can be found for the vibration period τ:

$$\tau = \sqrt{2m} \int_{-A_0}^{A_0} \frac{dx}{(E_0 - \mathcal{V}(x))^{1/2}} \qquad (9.8)$$

Here we have assumed that the motion is symmetric, i.e., when $\dot{x} = 0$, $x = \pm A_0$. The energy is found from the conservation law, $\mathcal{V}(A_0) = E_0$, when $\dot{x} = 0$. In general, this integral must be evaluated using either approximate, perturbation, or numerical techniques.

The change of frequency with amplitude (9.7), is shown in Fig. 9.2 for the Duffing oscillator. Nonlinear oscillators generally exhibit a continuous *spectrum* of natural frequencies at which forced resonance is possible, in contrast to linear systems for which forced resonance is possible only at discrete frequencies.

3. *The Forced-periodic Duffing Oscillator* (9.4). This case is more complicated than the others in that are several possible types of motion:

- Periodic motion with frequency ω
- Periodic motion with subharmonic frequency ω/n
- Chaotic motion

When the damping, forcing amplitude and nonlinearity are all small, then a classic result is the case of *nonlinear resonance* shown in Fig. 9.2c. Fig. 9.2c shows that the steady-state motion for the hard spring ($\beta > 0$) is *hysteretic*. That is, the amplitude vs. frequency curves are different, depending on whether the driving frequency is increasing or decreasing. The sharp jumps in amplitude, shown as dashed lines in Fig. 9.2c are characteristic of forced nonlinear oscillators.

9.3
The Undamped Pendulum: Phase-Plane Motions

The circular motion of a mass under the force of gravity is a classic problem in nonlinear dynamics (Fig. 9.3). The equation of motion can be derived by either using the law of angular momentum [see (1.6), (1.7)] or using a Lagrangian

$$\mathcal{L} = \frac{1}{2}mL^2\dot\theta^2 - mgL(1 - \cos\theta)$$

where g is the gravitational constant. The resulting equation of motion is

$$\ddot\theta + (g/L)\sin\theta = 0 \tag{9.9}$$

A similar equation of motion results for any rigid body in rotation about a fixed axis normal to the direction of gravity (Fig. 9.3b). The Newton–Euler equation of motion is given by

$$\dot{\mathbf{H}}_0 = \mathbf{M}_0$$

where

$$\mathbf{H}_0 = \int \mathbf{r} \times (\boldsymbol{\omega} \times \mathbf{r})\, dm = \dot\theta \int r^2\, dm\, \mathbf{e}_z$$

$$\mathbf{M}_0 = \int \mathbf{r} \times \mathbf{g}\, dm = \int \mathbf{r}\, dm \times \mathbf{g}$$

Here r is the distance from the axis of rotation to the mass element dm. The first moment of mass in $\mathbf{M}_0$ is simply the vector position of the center of mass, $\mathbf{r}_c$ times the mass. The second moment of mass is related to the radius of gyration, r_G, i.e.,

$$\mathbf{H}_0 = \dot\theta r_G^2 m \mathbf{e}_z$$

$$\mathbf{M}_0 = m\mathbf{r}_c \times \mathbf{g}$$

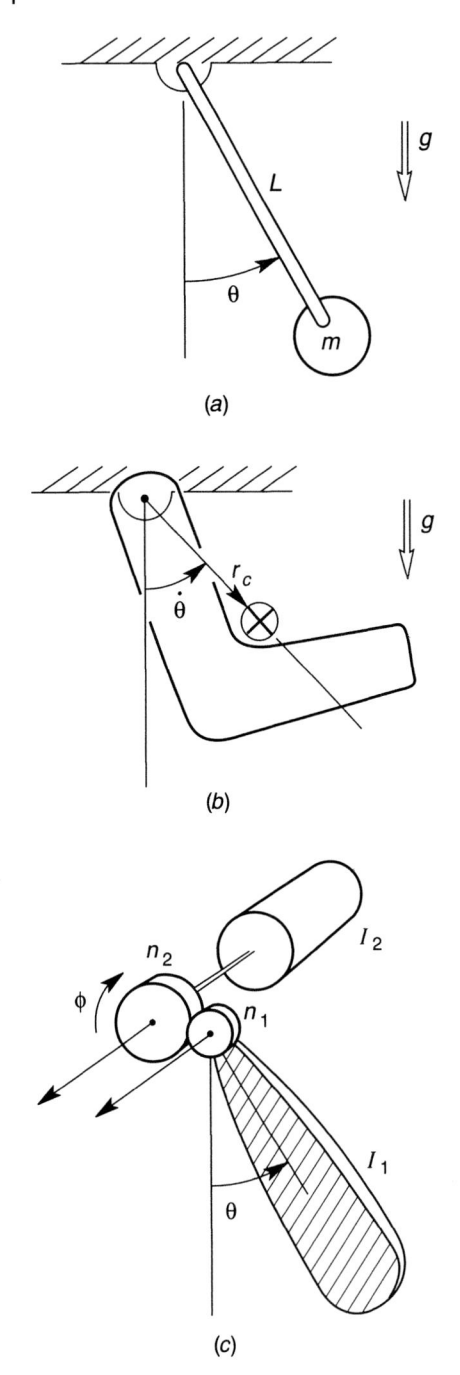

Fig. 9.3 (a) Mass-particle pendulum. (b) Rigid body pendulum. (c) Multibody pendulum system.

9.3 The Undamped Pendulum: Phase-Plane Motions

If θ is measured from the vertical or gravity direction, as in Fig. 9.3, then the scalar equation of motion becomes

$$r_G^2 m\ddot{\theta} + r_c mg \sin\theta = 0 \qquad (9.10)$$

or

$$\ddot{\theta} = -\frac{gr_c}{r_G^2}\sin\theta$$

which is similar to (9.9) if we let $L = r_G^2/r_c$.

In nonlinear systems, it is useful to rewrite this second-order equation as a set of two first-order differential equations;

$$\dot{\theta} = \omega$$
$$\dot{\omega} = -\omega_0^2 \sin\theta \qquad (9.11)$$

where $\omega_0^2 \equiv g/L$, or $\omega_0 = gr_c/r_G^2$. The motions are then described by curves in the $\{\theta, \omega\}$ plane or phase plane. Equilibrium points in this plane are defined by $\dot{\theta} = 0$, $\dot{\omega} = 0$ or

$$\omega_e = 0$$
$$\sin\theta_e = 0$$

Three equilibrium points in $0 \leq \theta \leq 2\pi$ are found: $\{0, 0\}$, $\{-\pi, 0\}$, $\{\pi, 0\}$. Motions about the origin are found to be closed curves, and the equilibrium point is called a *center*. The other two points at $\theta = \pm\pi$ are known as *saddle points*. At a saddle point there are two trajectories that approach the point and two that move away from the point. Motions near the saddle are unstable, i.e., they move away from the point for increasing time (see Fig. 9.4).

The family of possible motions can be found by integrating the energy equation

$$\frac{\dot{\theta}^2}{2} + \frac{g}{L}(1 - \cos\theta) = e_0 \qquad (9.12)$$

The resulting motions are represented by elliptic integrals. A sketch of these motions in the phase plane $\{\theta, \dot{\theta}\}$ shows three types of motions (Fig. 9.4a):

- Closed periodic orbits inside the separatrix
- Infinite time motions between one saddle point at $\theta = -\pi$ and the other at $\theta = \pi$, i.e., on the separatrix
- Open orbits outside the separatrix representing complete circular motions

The dynamics of the pendulum can also be represented by a cylindrical phase space, as shown in Fig. 9.4b. In this space, the two saddles at $\theta = \pm\pi$ become one. [See also Figs. 1.1, 1.15, 1.16.]

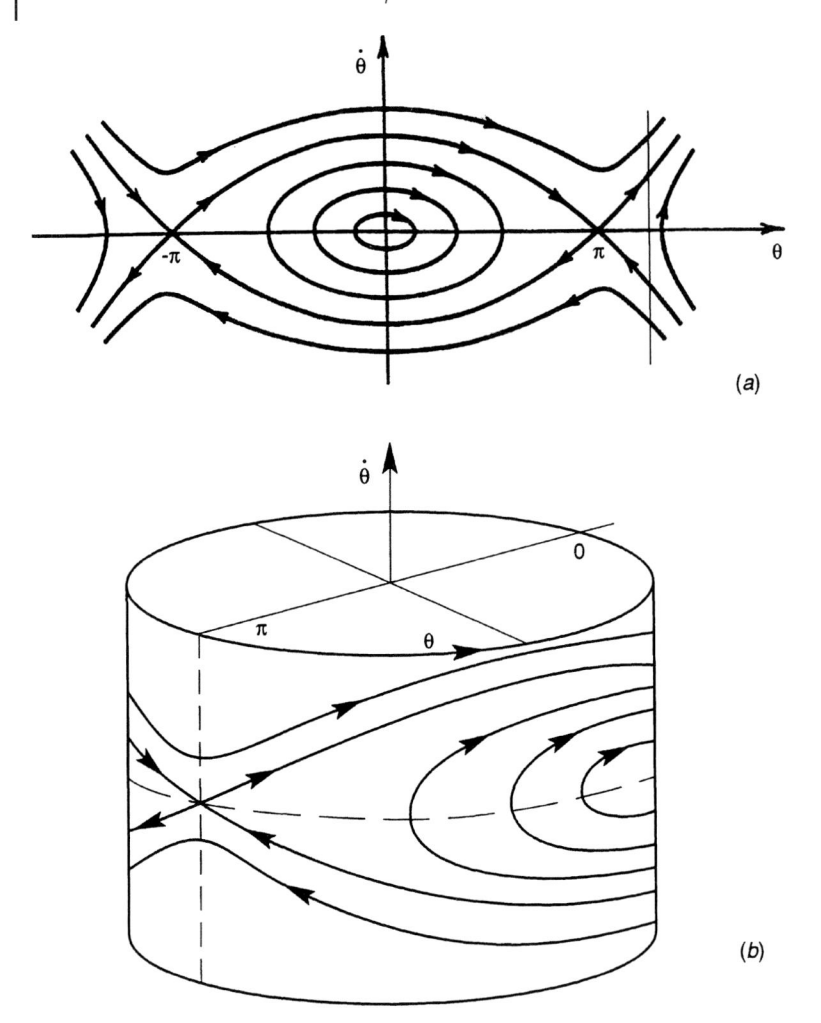

Fig. 9.4 (a) Phase plane for free pendulum motions showing center and saddle points. (b) Cylindrical phase space for free pendulums.

Example 9.1. A pendulum system consists of two rigid bodies in rotation about parallel axes, as shown in Fig. 9.3c. Body 1, which rotates about an axis, is at a distance r_c from its center of mass, and body 2 rotates about an axis that contains its center of mass. The two motions are constrained by a gear system with radii r_1, r_2 and a teeth ratio of n_1/n_2. Are the dynamics similar to the simple pendulum, and if so, what is the effective small angle frequency ω_0? How does the frequency depend on the amplitude for moderate angles $\theta < \pi/2$?

To solve this problem we calculate the kinetic energy function and use the velocity constraint

$$\dot{\phi} = \dot{\theta} r_1/r_2 = \dot{\theta} n_1/n_2$$

$$T = \frac{1}{2} I_1 \dot{\theta}^2 + \frac{1}{2} I_2 \dot{\phi}^2$$

$$T = \frac{1}{2}(I_1 + I_2 n_1^2/n_2^2)\dot{\theta}^2$$

If we use Lagrange's equation, the potential energy will be

$$V = m_1 g r_c (1 - \cos\theta)$$

Thus the Lagrangian is similar in form to both the single-particle or rigid-body pendulum (9.10), provided we use the effective inertia in place of the moment of inertia I_1, i.e.,

$$(I_1 + I_2 n_1^2/n_2^2) = r_G^2 m_1$$

Then it is easy to see that the small motion natural frequency is given by

$$\omega_0^2 = \frac{m_1 r_c g}{(I_1 + I_2 n_1^2/n_2^2)}$$

For moderate angles, the sine function can be expanded in a Taylor series. The approximate equation of motion becomes

$$\ddot{\theta} + \omega_0^2 \left(\theta - \frac{1}{6}\theta^3\right) = 0.$$

Compare this to (9.6) and use (9.7). One finds $\beta = -\omega_0^2/6$, and

$$\omega^2 = \omega_0^2(1 - A_0^2/8).$$

Thus the period increases with amplitude.

9.4
Self-Excited Oscillations: Limit Cycles

There are many dynamical systems in which a steady energy source can be converted into oscillatory motion. Examples include:

- Relative motion between two solids
- Fluid flow around solid and elastic objects
- Wind blowing over water
- Rolling objects, e.g., wheel shimmy
- Active electronic circuits with feedback
- Computer controlled systems
- Thermal fluid or thermoelastic systems
- Biochemical and chemical reactions

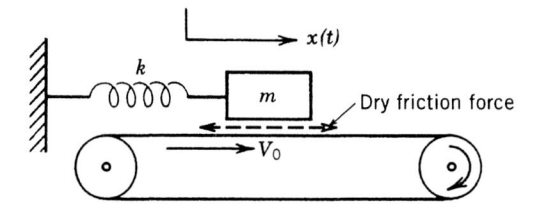

Fig. 9.5 (a) Sketch of a friction-driven self-excited oscillator.

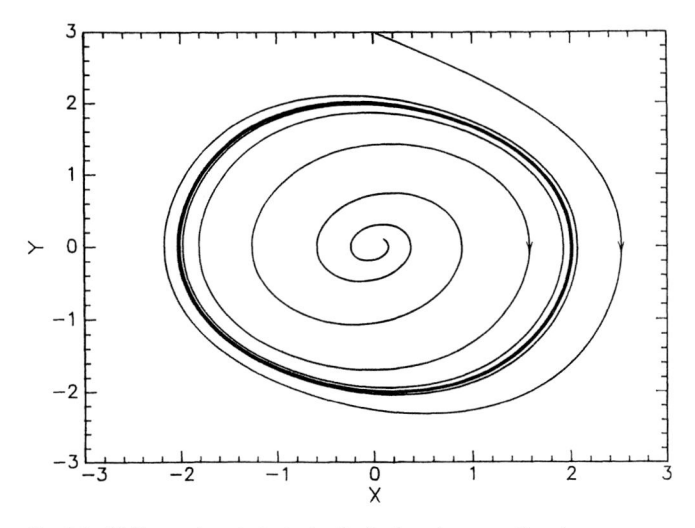

Fig. 9.5 (b) Phase-plane trajectories for limit cycles x vs. $Y = \dot{x}$.

When the energy input is balanced by an energy-dissipation mechanism, the motion can limit itself onto either a periodic, a quasi-periodic, or chaotic attractor. An example of such a limit cycle motion in the phase plane is shown in Fig. 9.5b. Near an equilibrium point at the origin, the unstable motion spirals outward and approaches the limit cycle asymptotically. For a stable limit cycle, initial conditions outside the periodic closed orbit spiral inward onto the periodic orbit.

A classic mechanical self-excited oscillator is a solid block moving over a belt undergoing stick-slip dry-friction-induced motions shown in Fig. 9.5a. Another is the fluid-excited motions or flutter around an airfoil whose angle of attack depends on an elastic restoring force. Nonlinear, self-excited circuits have been known from the time of the vacuum tube, which has a nonlinear voltage–current relation. Lord Rayleigh in 1896 and 8, Van der Pol in 1927 studied the following differential equation which describes the dynamics of a limit cycle

$$\ddot{x} - \gamma\dot{x}(1 - \beta x^2) + \omega_0^2 x = 0 \tag{9.13}$$

In this model, the small motions ($\beta x^2 \ll 1$) are represented by a negative damping or positive energy input $\gamma\dot{x}$. The linear motion is an unstable spiral, as shown

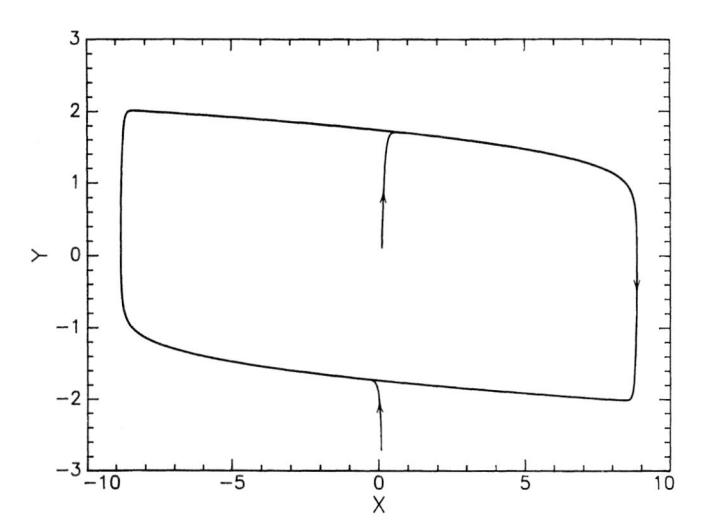

Fig. 9.5 (c) Relaxation oscillations.

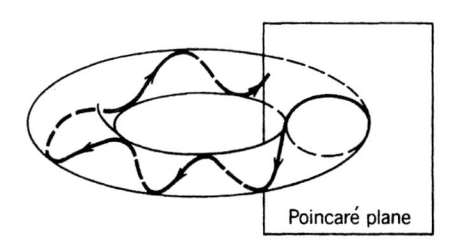

Fig. 9.6 Sketch of quasi-periodic orbit in phase space around a torus.

in the phase-plane plot in Fig. 9.5b. However, as the amplitude increases, the non-linear damping terms take energy out of the motion, resulting in a bounded limit-cycle oscillation.

When the damping term is small, the limit-cycle oscillation is nearly sinusoidal with frequency ω_0. However, when γ is large, the motion takes the form of a relaxation oscillation shown in Fig. 9.5c.

If sinusoidal forcing is added to the right-hand side of (9.13) (i.e., $f(t) = f_0 \cos \omega_1 t$) the motion can exhibit quasi-periodic oscillations of the form

$$x(t) = A_1 \cos \omega_1 t + A_2 \cos \omega_2 t$$

where ω_2 is close to ω_0. When ω_1 and ω_2 are incommensurate (i.e., ω_1/ω_2 is an irrational number), the motion is said to be *quasi-periodic*. We can envision this motion as taking place around the surface of a torus in phase space, as shown in Fig. 9.6.

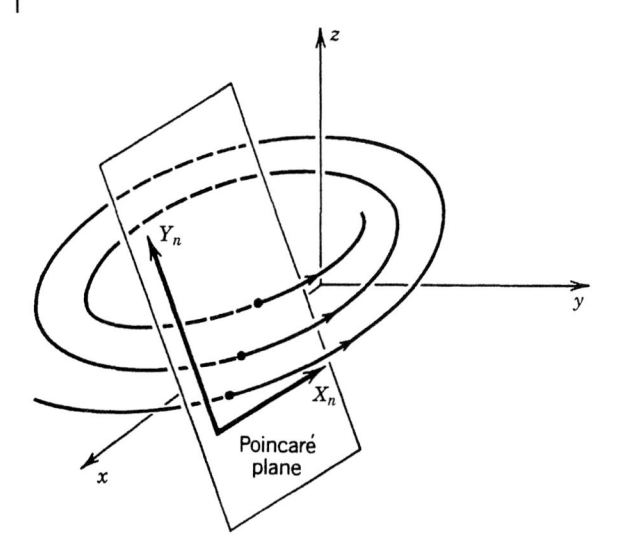

Fig. 9.7 Sketch of two-dimensional Poincaré section in a three-dimensional phase space.

9.5
Flows and Maps: Poincaré Sections

The dynamics of a system governed by Newton's laws or the laws of lumped circuit elements can be represented as a trajectory in phase space.

Consider an electromechanical system governed by the equations

$$m\ddot{x} + g(x, \dot{x}) + f(x) = \beta I^2$$
$$L\frac{dI}{dt} + RI + \beta \dot{x} I = V_0 \quad (9.14)$$

We define the state vector as $\mathbf{s} = [x, v, I]^T$, where $v = \dot{x}$. Then the equations can be written in the form of a set of first-order differential equations

$$\dot{x} = v$$
$$\dot{v} = -\frac{1}{m}\left[g(x, v) + f(x) - \beta I^2\right]$$
$$\dot{I} = -\frac{1}{L}[RI + \beta v I - V_0]$$

or

$$\dot{\mathbf{s}} = \mathbf{F}(\mathbf{s})$$

We can view the dynamics as a trajectory in a three-dimensional phase space (Fig. 9.7) $\{x, y = v, z = I\}$. When initial conditions are varied, a set of nonoverlapping trajectories is drawn in the phase space, which is similar to a bundle of particles moving in a fluid. Thus the solutions to a dynamical system are sometimes referred to as a *flow*. At the turn of the twentieth century, Henri Poincaré (1854–1912) developed a technique to analyze the motions of a flow as a set of difference equations. This technique, illustrated in Fig. 9.7, is now called a Poincaré map. To obtain a Poincaré map we construct a surface that intersects the trajectories, e.g., we can choose a plane $ax + by + cz = d$. On this plane we define coordinates $\{X_n, Y_n\}$. The index n marks the cycle or sequential penetrations of the continuous time history trajectory with the Poincaré surface. We can see that a three-dimensional flow (third-order ordinary differential equations) generates a set of two first-order difference equations

$$X_{n+1} = F(X_n, Y_n)$$

$$Y_{n+1} = G(X_n, Y_n)$$

These equations take a contiguous set of points at time t_n and map them into a distorted set of points at t_{n+1}.

The advantage of looking at the difference equation generated by the Poincaré section rather than the original differential equations is that we can more easily see how bundles of trajectories change in time. This is done by asking how an area in the $[X_n, Y_n]$ plane is transformed in one iteration. Consider, for example, the following set of equations

$$X_{n+1} = 1 - \alpha X_n^2 + Y_n$$

$$Y_{n+1} = \beta X_n \tag{9.15}$$

This is called the Hénon map proposed by the French astronomer M. Hénon in 1976. When $\beta < 1$, the map contracts areas in the $X = Y$ plane. It also stretches and bends areas in the Poincaré phase plane, as illustrated in Fig. 9.8. This stretching, contraction, and bending or folding of areas in phase space is what leads to chaos and unpredictability in nonlinear systems. Maps that stretch and fold like the Hénon map are called *horseshoe maps*. After many iterations, the original compact area of neighboring initial conditions is stretched and folded over and over until it has a fractal structure, as shown in Fig. 9.8b.

Fractals and Poincaré Maps

One of the most singular characteristics of chaotic dynamics is the mazelike, multisheeted structure of the Poincaré section of the phase-space trajectories. When this pattern of points has a similar pattern on finer and finer length scales, the term *fractal* is often applied. There are several measures of a fractal set. One of the most intuitive is the *fractal dimension* or *box counting dimension d*. Fractal dimension is a concept applied to a distribution in an n-dimensional space. For example, in three dimensions the number of cubes of size, ε, $N(\varepsilon)$, necessary to cover a uniform distribution of points, scales as $N(\varepsilon) \sim \varepsilon^{-3}$. But in a fractal distribution of points

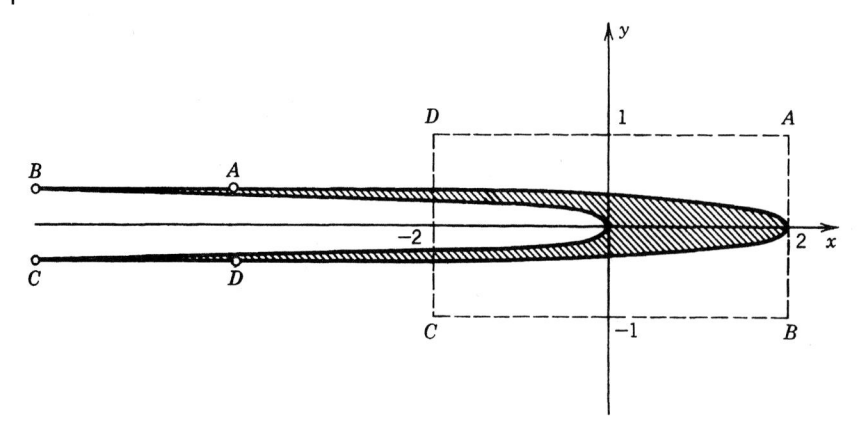

Fig. 9.8 (*a*) Stretching and folding operations of the Hénon map.

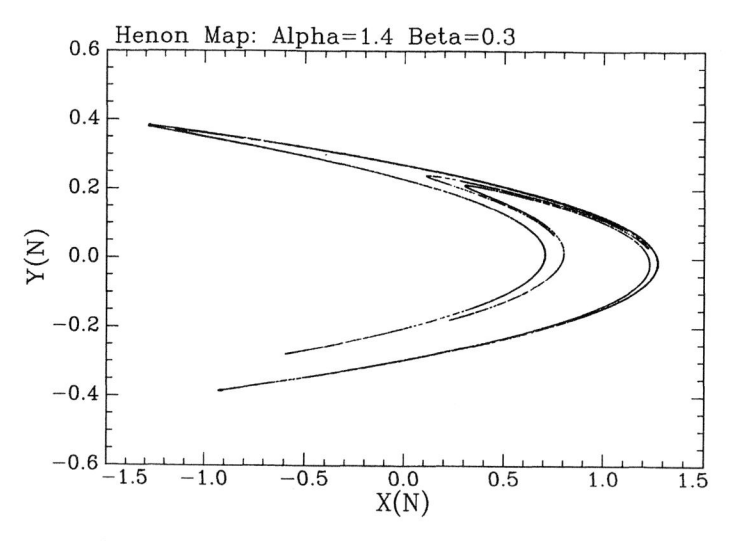

Fig. 9.8 (*b*) Iterations of the Hénon map showing fractal structure.

there are gaps in the pattern at finer and finer scales. Thus the box counting scaling may result in a noninteger dependence of $N(\varepsilon)$ on ε, i.e.

$$N(\varepsilon) \sim \varepsilon^{-d}$$

or

$$d = \lim_{\varepsilon \to 0} \frac{\log N(\varepsilon)}{\log(1/\varepsilon)} \qquad (9.16)$$

A basic set of points with fractal dimension is the *Cantor sets*. This set is created by an iterative process. Starting with a uniform distribution and a line, say from zero to one, the middle third of points is thrown out, with two remaining sets on

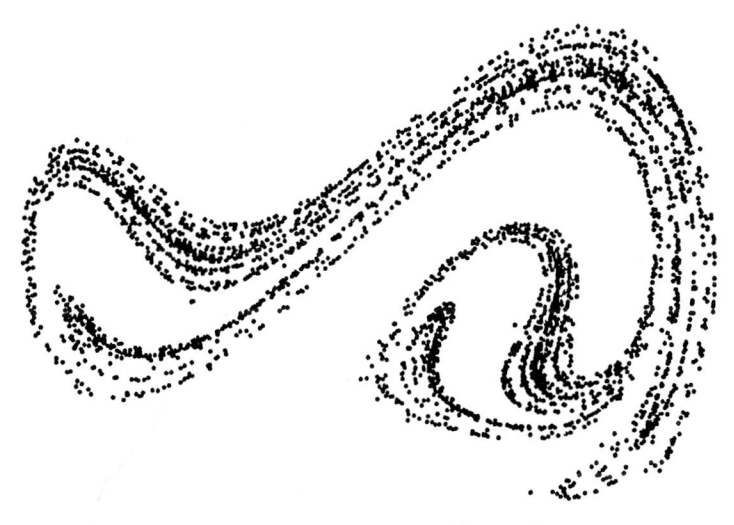

Fig. 9.8 (c) Fractal structure of the Poincaré map of a mass in a two-well potential (see Moon, 1992).

the line. Then the middle third of the two remaining sets is thrown out and the process is repeated. The limiting set of points, called a *Cantor dust* has a fractal dimension of

$$d = \frac{\log 2}{\log 3}$$

This is obtained from the formula for d where at the nth iteration $N(\varepsilon) = 2^n$, and $\varepsilon = (1/3)^n$. Noninteger dimensions of Poincaré sections of a trajectory in phase space is characteristic of strange attractors. For example, the motion of a particle in a two-well potential can be described by the equation of motion (Guckenheimer and Holmes, 1983)

$$\ddot{x} + \gamma \dot{x} - \frac{1}{2}(x - x^3) = f \cos \omega t$$

The motion becomes chaotic for the parameter values $\gamma = 0.115$, $f = 0.23$, $\omega = 0.8333$ (Fig. 9.8c). For this trajectory, the calculated fractal dimension of the Poincaré section is found to be $d = 2.32$ (see Moon, 1992). The fractal dimension of the Hénon map described above is $d = 1.26$. The fractal dimension and the Lyapunov exponent measures of chaos can be related in most cases. The reader is referred to advanced books on chaos theory such as Abarbanel (1996) for more details.

One-dimensional Maps: Period Doubling and Chaos

In many dynamics problems, the motions may be described by a first-order difference equation of the form

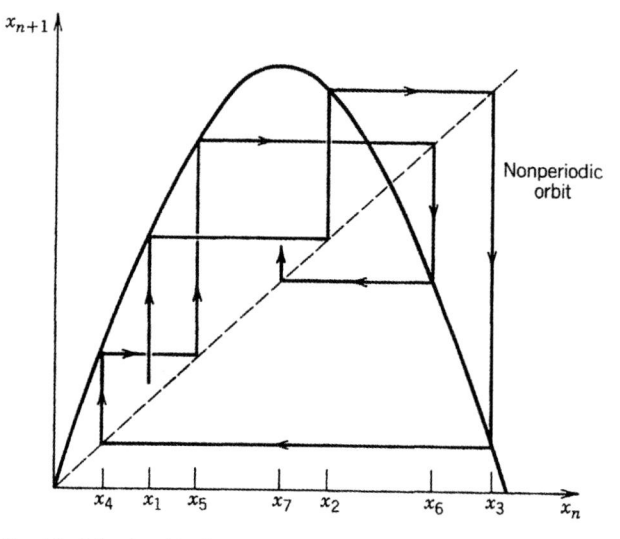

Fig. 9.9 Cobweb orbits for a return map with an extremum.

$$X_{n+1} = F(X_n) \tag{9.17}$$

For example, in the Hénon map (9.15), when $\beta = 0$, $F(X_n)$ is a quadratic function. The simplest equation of this form is the linear equation:

$$X_{n+1} = \lambda X_n$$

Unlike nonlinear equations, this equation can be solved explicitly by trying a solution of the form $X_n = A\lambda^n$. This solution is stable (i.e., $|x_n| \to 0$ as $n \to \infty$) if $|\lambda| < 1$, and unstable if $|\lambda| > 1$. This equation is sometimes used as a model for population growth in biology and chemistry. A population model that accounts for limited resources is the so-called logistic equation or quadratic map:

$$X_{n+1} = \lambda X_n (1 - X_n) \tag{9.18}$$

The solutions to this equation (sometimes called *orbits*) can be visualized in the *cobweb diagram* shown in Fig. 9.9. Fixed points are defined as iterations that return to themselves:

$$X_{n+1} = X_n = \lambda X_n (1 - X_n) \tag{9.19}$$

The origin is one fixed point, while the second is given by the intersection of the parabola with the identity line,

$$X = (\lambda - 1)/\lambda \tag{9.20}$$

provided that $\lambda > 1$, or the slope at the origin is greater than one.

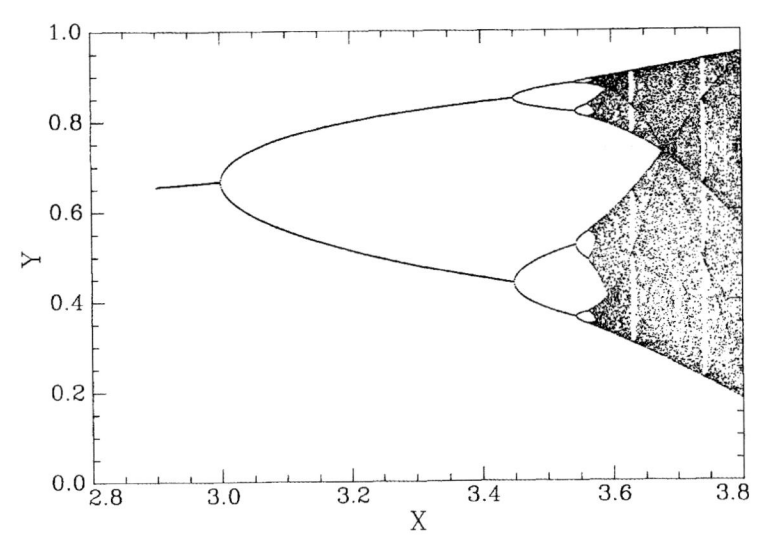

Fig. 9.10 Bifurcation diagram for the quadratic map showing period doubling and chaotic regimes. [$X = \lambda$, $Y = X_{n+1}$ in (9.18).]

We can also find period-two fixed points by solving

$$X_{n+2} = \lambda X_{n+1}(1 - X_{n+1}) = X_n$$

or

$$\lambda\lambda X_n(1 - X_n)(1 - \lambda X_n(1 - X_n)) = X_n \qquad (9.21)$$

These period-two fixed-point solutions only occur for certain values of the control parameter λ. There are, in fact, sequences of period-2^N orbits (called subharmonics in differential equations). The ranges for different subharmonic solutions are shown in Fig. 9.10, called a *bifurcation diagram*. We can see that the higher period two orbits have a narrower range of λ than the lower period two orbits. Mitchel Feigenbaum (1978) discovered a relation between the critical values of λ where period-doubling orbits change period:

$$\lim_{n \to \infty} \frac{\lambda_n - \lambda_{n-1}}{\lambda_{n+1} - \lambda_n} = 4.6692\ldots \qquad (9.22)$$

This relation is important because it is not only valid for the quadratic map (9.18) but is valid for a wide class of one-hump maps.

In the case of the logistic map, when $\lambda = \lambda_c = 3.57\ldots$, the periodic orbits are no longer stable, and a nonrecurring, yet bounded orbit appears that has been called a *chaotic* orbit.

The following *MATLAB* program will generate a bifurcation diagram similar to Fig. 9.10 for (9.18).

```
Comment % Set initial conditions.
x = .1;
y = .1;
a = 2.1;aa = 0;aa = [];
z = [x,y];zz1 = [0 0 0 0 0] ' ;zz1 = [];
Comment % Loop for different values of λ = a.
for j = 1:50;
a = a + 0.03 ;
z = [.1 .1];
Comment % Iterate Logistic Map.
for k = 1:40 ;
x = y;
y = a.*y .*y(1-y) ;
z = [z ;x,y];
end
aa = [aa;a] ;

zz = z (36:40,1) ;
zz1 = [zz1 zz];
end
plot(aa,zz1, '+b')
```

Example 9.2 Magnetic Levitation Chaos. To illustrate how a discrete-time map one can be obtained from a differential equation, we examine the problem of magnetically suspending a mass moving in a viscous fluid under gravity, as shown in Fig. 9.11. We assume that the viscous drag force is linearly proportional to the velocity. Without other forces, the mass would fall under gravity. To suspend the mass, we apply a magnetic impulse at discrete times $\{\ldots t_n, t_{n+1} \ldots\}$. To derive an equation of motion requires two steps:

1. Determine the motion between impulses by solving the linear differential equation.
2. Apply a rule to change the velocity during the magnetic impulse.

The equation of motion between impulses is

$$m\frac{dv}{dt} = mg - bv$$

or

$$\dot{v} + \gamma v = g \tag{9.23}$$

where $\gamma = b/m$. The time between impulses is τ, and we solve for the velocity between $n\tau < t < (n+1)\tau$:

$$v(t) = g\gamma^{-1} + Be^{-\gamma(t-n\tau)} \tag{9.24}$$

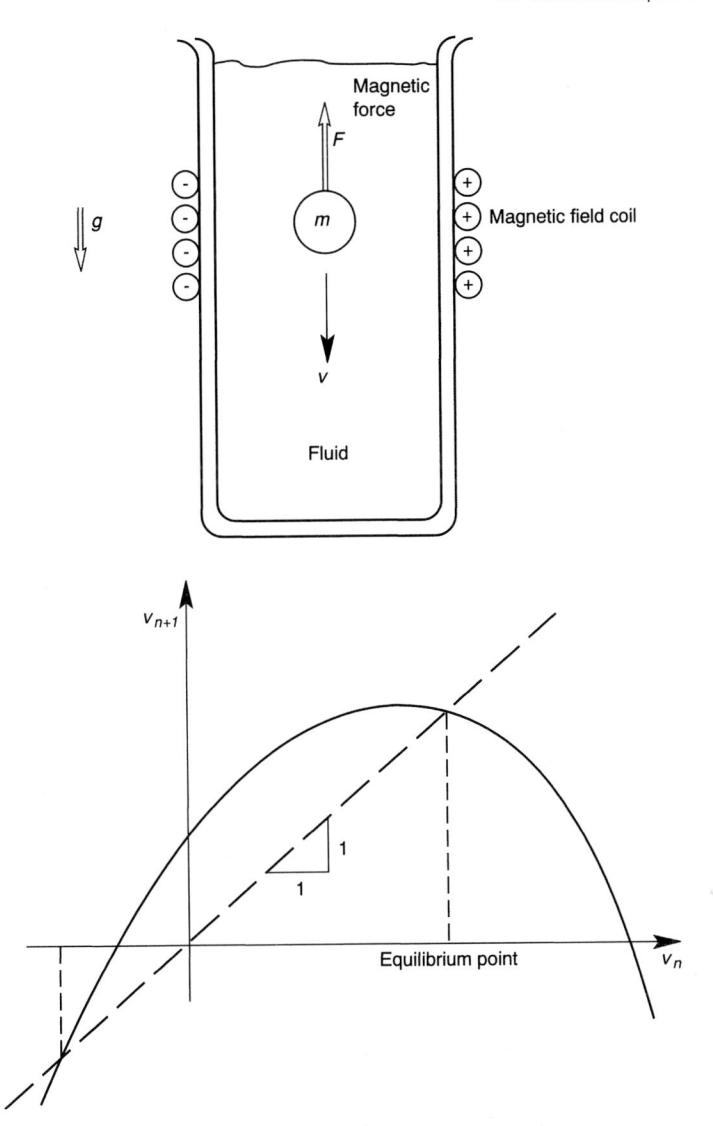

Fig. 9.11 (a) Magnetic-levitation device consisting of a ferromagnetic sphere in a viscous fluid. (b) Return map for the levitation feedback system in part (a).

The constant B is determined from the initial condition

$$v(n\tau + \varepsilon) = v_n; \qquad \varepsilon \to 0$$

or

$$B = v_n - g\gamma^{-1} \tag{9.25}$$

Thus at the end of the free-fall regime,

$$v^-_{n+1} = g\gamma^{-1} + (v_n - g\gamma^{-1})e^{-\gamma\tau} \tag{9.26}$$

Next we must make an assumption about the magnetic-impulse-induced velocity change, i.e., after the impulse we assume

$$v_{n+1} = v^-_{n+1} + F \tag{9.27}$$

We arbitrarily choose a nonlinear feedback law that relates the magnetic impulse F to the square of the velocity after the previous impulse at $t = n\tau$;

$$F = -\Gamma v_n^2 \tag{9.28}$$

If we rescale the gain $\Gamma = \beta e^{-\gamma\tau}$, we can now write a relation between the velocity after impulse at $t = n\tau$ to the velocity after impulse at $t = (n+1)\tau$, i.e., we obtain a first-order difference equation or iterated map:

$$v_{n+1} = g\gamma^{-1}(1 - e^{-\gamma\tau}) + e^{-\gamma\tau}(v_n - \beta v_n^2) \tag{9.29}$$

This map is shown in Fig. 9.11b, and is similar to the logistic or quadratic map (9.18). In this physical problem we can use either the control gain β as the variable parameter, or we could change the time between impulses τ. In either case, we can ask several questions about this map.

1. What are the possible equilibrium states?
2. Is the system stable or unstable near equilibrium?
3. What are the critical values of β or τ for which subharmonic motions can be seen?
4. At what values of the parameters does the system exhibit chaotic dynamics?

Example 9.3 The Kicked Rotor. As another example of how to obtain an explicit Poincaré map or discrete-time equations from continuous time dynamics, consider the rotor in Fig. 9.12a. We assume that the rotor is subject to impulse forces always aligned with the vertical axis. The impulse force F_0 creates an impulse moment $M_0 = F_0 R \sin\theta$ that increases the angular velocity at the discrete times $t = n\tau$. $I[\omega(t = n\tau + \varepsilon) - \omega(t = n\tau - \varepsilon)] = F_0 R \sin\theta(n\tau)$, where ε is much smaller than τ. We also assume that a constant torque is applied, $c\omega_0$, as well as viscous damping $c\omega$. The resulting equations of motion are

$$\dot\theta = \omega$$
$$I\dot\omega + c\omega = c\omega_0 + F_0 R \sin\theta \sum \delta(t - n\tau) \tag{9.30}$$

Between impulses, $n\tau < t < (n+1)\tau$, the system is linear and exhibits damped motions

$$\omega = Ae^{-ct/I} + \omega_0$$

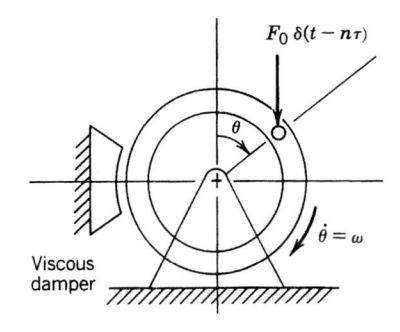

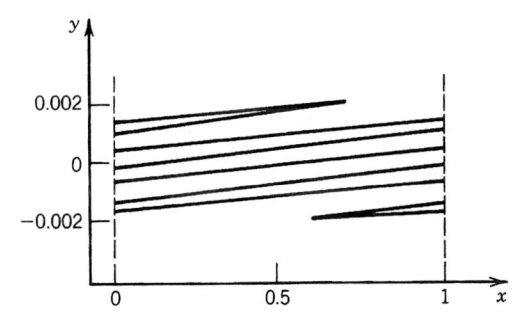

Fig. 9.12 (a) The kicked rotor with viscous damping and periodically excited torque. (b) Iterated Poincaré map showing a strange attractor: x represents the angular rotation (mod 1), and y represents the angular velocity.

The integral of this expression will give $\theta(t)$ between impulses. In this problem, the state variables are $\{\theta, \omega\}$. We define the discrete-valued variables $\{\theta_n, \omega_n\}$ to be the value of the state variables just before the impulse force at $n\tau$, i.e.,

$$\theta_n = \theta(t = n\tau - \varepsilon), \quad \varepsilon > 0, \quad \varepsilon \ll \tau$$
$$\omega_n = \omega(t = n\tau - \varepsilon)$$

To obtain a Poincaré map we try to find a relation for $\{\theta_{n+1}, \omega_{n+1}\}$ in terms of $\{\theta_n, \omega_n\}$. This is done by using the preceding solution and the impulse conditions to relate $\omega(n\tau + \varepsilon)$ and $\omega(n\tau - \varepsilon) = \omega_n$. We also use the fact that the angles are continuous at the impulse events, i.e., $\theta(n\tau + \varepsilon) = \theta(n\tau - \varepsilon) = \theta_n$. With these observations, we can show that the exact Poincaré map is given by

$$\omega_{n+1} = \frac{c\tau}{I}\omega_0 + \omega_n - \frac{c}{I}(\theta_{n+1} - \theta_n) + \frac{1}{I}F_0 R \sin\theta_n$$
$$\theta_{n+1} = \omega_0\tau + \theta_n + \frac{I}{c}\left(1 - e^{-c\tau/I}\right)\left(\omega_n + \frac{1}{I}F_0 R \sin\theta_n - \omega_0\right) \quad (9.31)$$

These equations were originally derived by the Soviet physicist George Zaslavsky in 1978 to model the nonlinear interaction between two oscillators in plasma physics (see Moon, 1992). This two-dimensional map is often nondimensionalized, using the definition:

$$x_n = \frac{\theta_n}{2\pi} \pmod 1$$

$$y_n = \frac{\omega_n - \omega_0}{\omega_0}$$

The mod 1 means that since the motion is circular, we only plot angles between $0 \leqslant \theta \leqslant 2\pi$ and $0 \leqslant x \leqslant 1$. Thus, if $\theta = 2\pi + \phi$, $\phi < 2\pi$, by mod 1 we plot $\phi/2\pi$. In the new variables, the map is written

$$y_{n+1} = e^{-\Gamma}(y_n + \beta \sin 2\pi x_n)$$

$$x_{n+1} = \left\{ x_n + \frac{\Omega}{2\pi} + \frac{\Omega}{2\pi\Gamma}(1 - e^{-\Gamma})y_n + \frac{K}{\Gamma}(1 - e^{-\Gamma}) \sin 2\pi x_n \right\} \quad (9.32)$$

where the brackets { } indicate the use of only the fractional part or mod 1. Also $\beta = F_0 R/I\omega_0$, $K = \beta\Omega/2\pi$, $\Gamma = c\tau/I$, $\Omega = \omega_0\tau$. Here y_n measures the departure of the speed from the unperturbed equilibrium speed $\omega = \omega_0$. This set of difference equations is much faster to iterate in time than numerically integrating the original differential equations (9.30). What is remarkable about such nonlinear maps is the complexity of motions that they exhibit. Although the impulses are periodic in time, the output dynamics may be periodic, quasi-periodic, or chaotic. For example, it has been found that this map may exhibit chaos when

$$1 < \frac{\Gamma}{1 - e^{-\Gamma}} < K$$

The student might try the following parameters: $\Gamma = 5$, $\beta = 0.3$, $\Omega = 100$, $K = 9$. (See Fig. 9.12b.)

9.6
Complex Dynamics in Rigid Body Applications

Since the discovery in the late 1970s that many complex motions are deterministic chaos, many physical systems have been found to exhibit chaotic dynamics. A survey of many of these problems in mechanical, electrical, optical, and biological systems may be found in Moon (1992). In this final section we discuss a few examples from rigid body dynamics. The examples were chosen to be similar to some of the examples discussed in the earlier chapters. Our goal in this chapter is to show the variety of motions and complexity in the dynamic response of rigid body and mechatronic systems beyond their linear dynamic behavior. We also wish to emphasize that nonlinear dynamics is not a closed area. There are many multibody and mechatronic systems for which we do not know the complete range of dynamic behavior. Examples include many rolling problems; dynamics with friction and fluid forces; human and animal-gait dynamics; walking machines; and

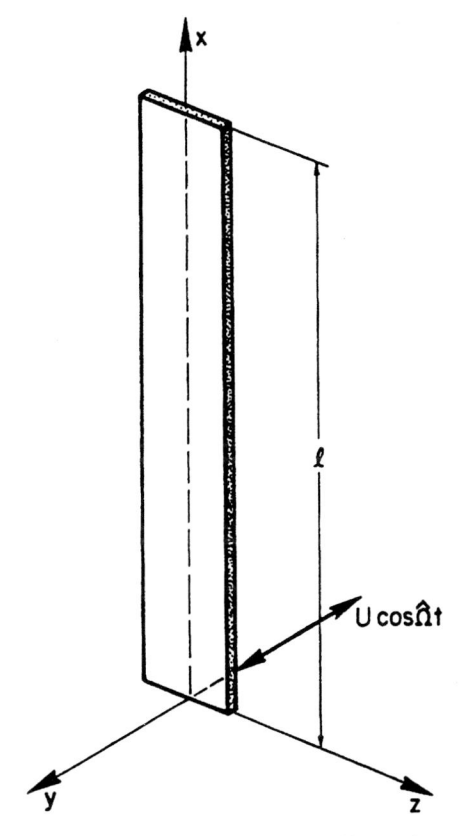

Fig. 9.13 (*a*) Continuous cantilevered beam showing torsional and bending deformations.

robots to name a few. The student interested in new research areas will find a rich mine of interesting dynamics research in these problems.

The nonlinear dynamics in the following examples have been explored by analytical, numerical and experimental observations:

- Torsional-bending vibrations (Chapter 4)
- Gear rattling (Chapter 5)
- Multipendula kinetic sculptures (Chapter 6)
- Tumbling of Hyperion (Chapter 7)
- Printer actuator chaos (Chapter 8)

Bending–Torsional Oscillations

The lateral bending of a cantilever beam is a basic paradigm of structural dynamics. In conventional structural modeling, the bending vibrations are assumed to be uncoupled from the torsional deformations of the torsion beam (Fig. 9.13a). However, it has recently been discovered that in thin, plate-like beams, as in Fig. 9.13,

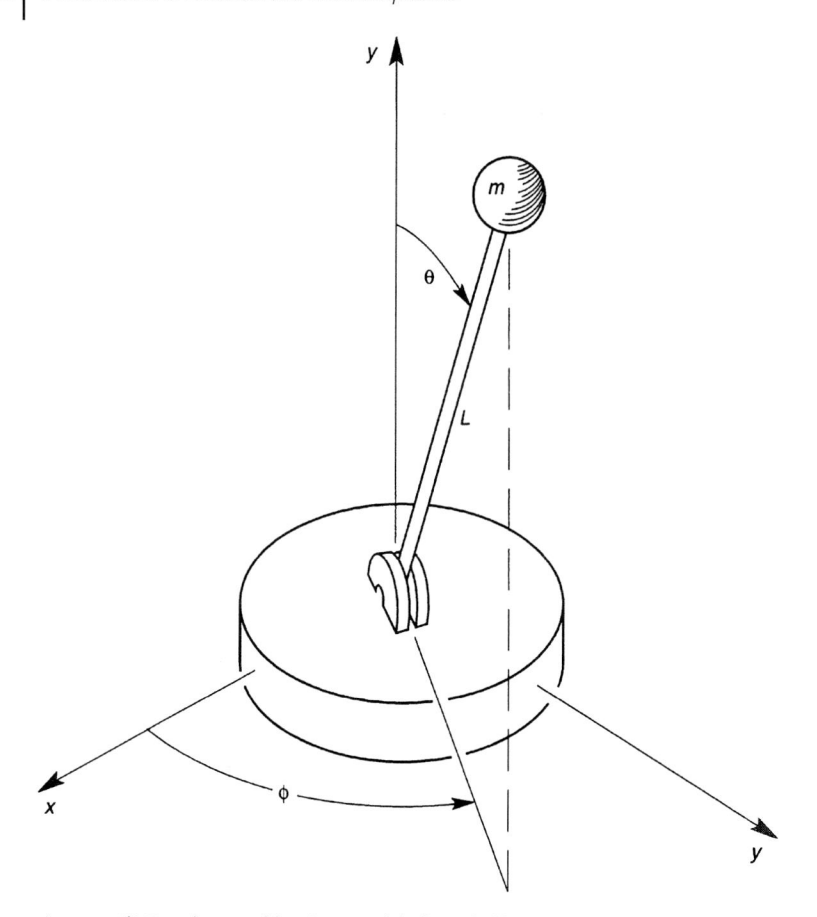

Fig. 9.13 (*b*) Two-degree-of-freedom model of coupled bending–torsional motions in part (*a*).

not only do the lateral and torsional modes couple, but this interaction can exhibit chaotic vibration under periodic excitation (Cusumano and Moon, 1995). The analysis of continuous elastic structures is beyond the scope of this book; however, a two-degree-of-freedom model can be used to illustrate the nature of the torsion–lateral coupling, as shown in Fig. 9.13*b*. In the model, the lateral bending vibrations are captured by the angular motion of the particle on a rod that is constrained to vibrate on a horizontal axis with a restoring spring, k_B. The torsional vibration is modeled as a rotary inertia, I_T, rotating about the vertical axes with a spring constant, k_T. The uncoupled linear lateral and torsional frequencies are found to be

$$\omega_B^2 = k_B/mL^2$$
$$\omega_T^2 = k_T/I_T$$

To derive the nonlinear coupled equations of motion we can use Lagrange's equation with generalized variables $\theta(t)$, $\phi(t)$. The kinetic and potential energies may be found to be

$$T = \frac{1}{2}m[\dot{x}^2 + \dot{\theta}^2 L^2 + \dot{\phi}^2 L^2 \sin^2\theta$$
$$\quad - 2\dot{x}\dot{\phi}L\sin\theta\sin\phi + 2\dot{x}\dot{\theta}L\cos\theta\cos\phi] + \frac{1}{2}I_T\dot{\theta}^2$$
$$V = \frac{1}{2}k_B\theta^2 + \frac{1}{2}k_T\phi^2 \quad (9.33)$$

In this model we neglect the effect of gravity. This is valid if k_B is large enough. Also we assume that the rectilinear base motion $\dot{x}(t)$ is given, i.e., $x(t) = A\cos\omega_0 t$, and $\dot{x} = -A\omega_0\sin\omega_0 t$.

Carrying out the derivatives in Lagrange's equations we can show that the following equations result:

$$\ddot{\theta} + \omega_B^2\theta - \dot{\phi}^2\sin\theta\cos\theta = -\frac{\ddot{x}}{L}\cos\theta\cos\phi$$
$$[\eta + \sin^2\theta]\ddot{\phi} + \omega_T^2\eta\phi + 2\dot{\phi}\dot{\theta}\sin\theta\cos\theta = \frac{\ddot{x}}{L}\sin\theta\sin\phi \quad (9.34)$$

where $\eta = I_T/mL^2$.

We can see that the coupling is due to the centripetal acceleration $\dot{\phi}^2$ in the first equation and the Coriolis acceleration $2\dot{\phi}\dot{\theta}$ in the second equation.

Both analysis and experiments show that there exists a nonlinear oscillation in which θ remains either positive or negative, and the torsional motion balances the centripetal acceleration. The frequency of this mode, however, depends on the amplitude of the oscillation, as is typical in nonlinear oscillation (Fig. 9.14). When $x(t) = A\cos\omega_0 t$, it has also been observed that chaotic oscillations can occur (see Cusumano and Moon, 1995; Pak, 1999).

Gear-rattling Chaos

In Chapter 1 we discussed how noise generated in mechanical systems can be a failure mode. One of the recent discoveries of chaos theory is that gear transmission systems can be a source of deterministic noise or chaos in multibody systems (Pfeiffer, 1994). Kinematic mechanisms are generally input–output devices that convert one motion into another, e.g., linear to rotary motion or motion at one speed into motion at another speed, as in a gear pair. However, in real mechanical devices there exist departures from the ideal mechanism due to gaps, play, backlash, elasticity, and inelasticity. In these less than ideal mechanisms dynamics enters and the output is no longer kinematically determined by the input.

As an example, consider the two spur gears in Fig. 9.15a. Without gaps or play they have a speed ratio ω_1/ω_2 equal to d_2/d_1. However, when the gear pitches do not exactly match due to play, then the kinematic problem becomes a dynamic one. This problem is analogous to the two-body rectilinear-motion problem shown in

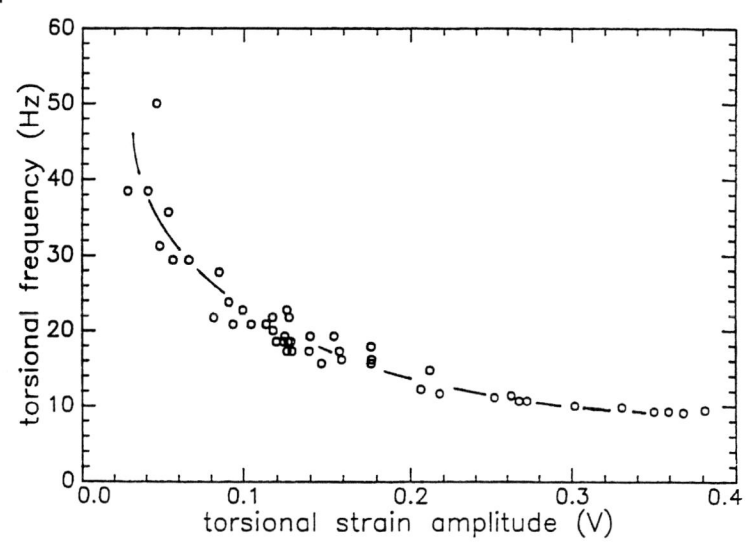

Fig. 9.14 Natural frequency vs. amplitude for coupled bending–torsional oscillations (see Fig. 9.13). (From Cusumano and Moon, 1995.)

Fig. 9.15b. When the inner mass is not ill contact with the outer mass, it conserves linear momentum. However, at impact it can either gain or lose linear momentum to the outer mass. Impact problems of this kind have been solved by constructing a Poincaré map similar to the previous examples. The Poincaré map is constructed when the relative motion is zero, $x = 0$, and the pair $\{v_n = \dot{x}(t_n), \omega t_n; \mod 2\pi\}$ is plotted where t_n is the time of contact of the nth impact. An example of such a Poincaré map is shown in Fig. 9.15c, which is from the work of Li et al. (1990). The fractal nature of the Poincaré map is evidence for deterministic chaos, i.e., the input frequency, ω will generate a broad spectrum of output frequencies which we interpret as structure-borne noise. A phase-plane trajectory of a chaotic motion is shown in Fig. 9.16.

This system has also been shown to exhibit period doubling or a subharmonic route of bifurcations to chaos.

Professor F. Pfeiffer and his laboratory at the Technical University of Munich have pioneered in the application of Poincaré maps and other nonlinear analysis methods to predicting deterministic noise in automotive and other gear transmission systems.

Kinetic Sculpture

Sculptural art of the twentieth century has taken on a dynamic quality, especially in the works of Alexander Calder, Jean Tinguely, and George Rickey. Often these works can be seen in public spaces such as museums and airports. Part of the fascination of kinetic sculpture is the changing variety of patterns. In fact, many

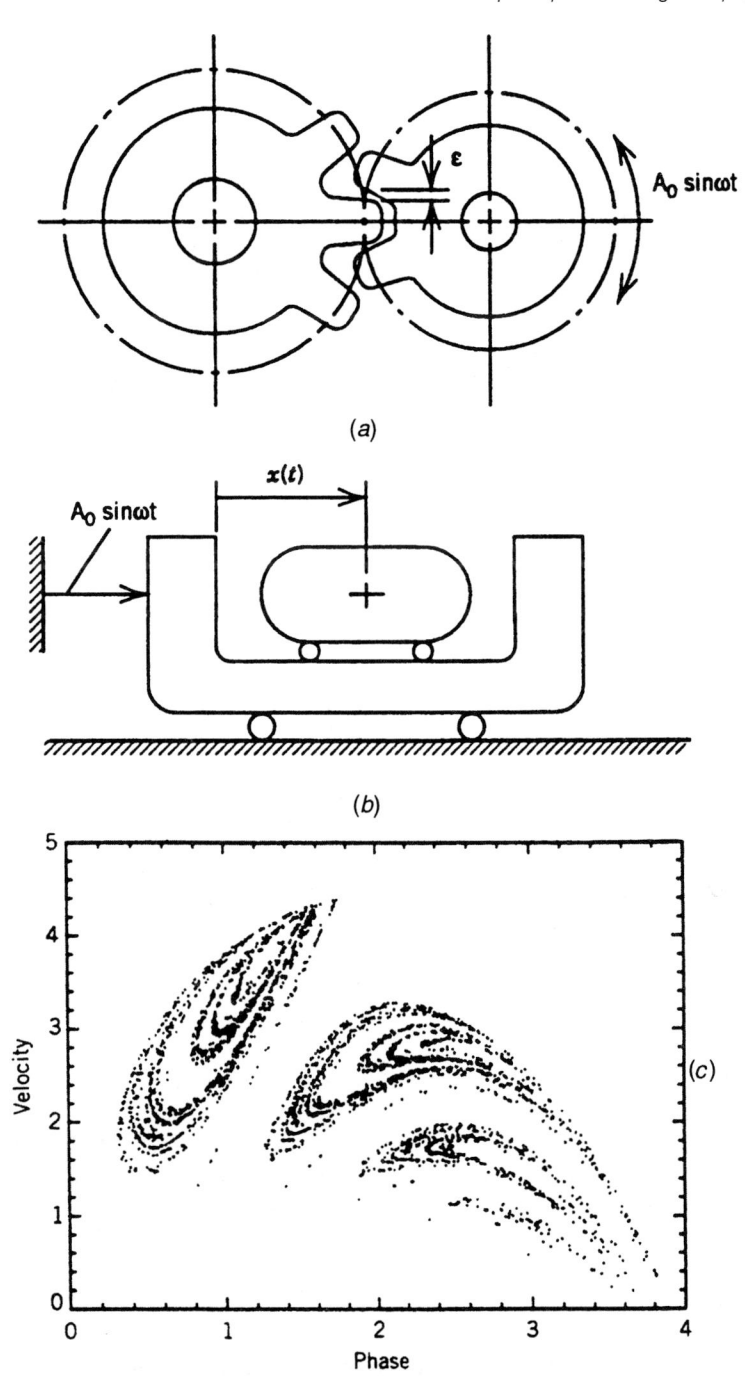

Fig. 9.15 (a) Sketch of enmeshed gears with play ε.
(b) Rectilinear motion of a mass moving between two vibrating constraints. (c) Poincaré map for gear-rattling chaos.

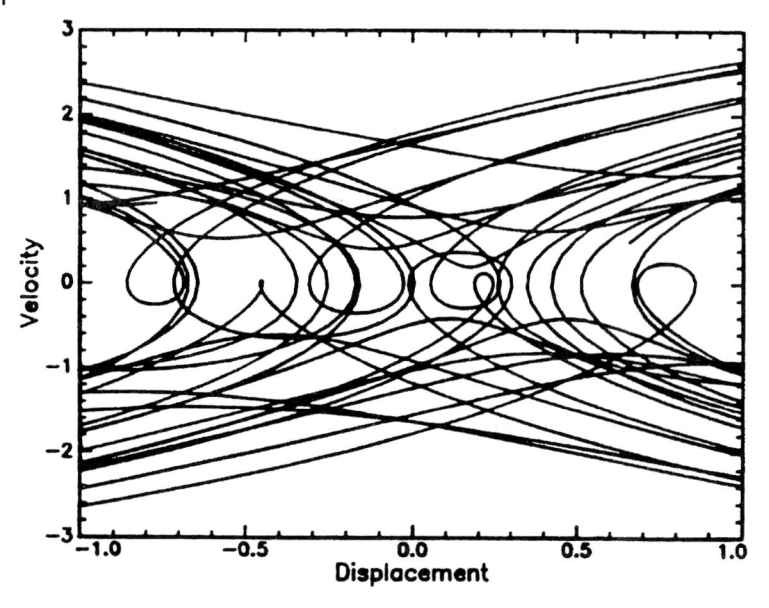

Fig. 9.16 Chaotic rattling oscillations in the phase plane for systems in Fig. 9.15b. (From Li et al., 1990.)

kinetic sculptors recognize the need to design in unpredictable random or chaotic dynamics if their works are to be successful. Both Calder and Rickey use the dynamics of the pendulum and coupled pendulums. Mobiles inspired by Calder are a staple of modern art and gift shops.

A chaotic toy that I have seen on many a dynamicist's desk is a two- or three-arm coupled pendulum device. Professor N. Rott of Stanford University some time ago published a paper describing the mechanics and construction of these toys (Rott, 1970). The analysis is based on nonlinear resonance and, in particular, the 2:1 resonance. For example, in the two-link device shown in Fig. 9.17, two angles describe the configuration: α, γ.

For small motions about the stable equilibrium position, Rott derives equations of motion of the form

$$\ddot{\alpha} + \omega_1^2 \alpha = F(\alpha, \gamma, \dot{\gamma}, \ddot{\gamma})$$
$$\ddot{\gamma} + \omega_2^2 \gamma = G(\alpha, \gamma, \dot{\alpha}, \ddot{\alpha}) \qquad (9.35)$$

where F, G are nonlinear coupling terms and the constants ω_1^2, ω_2^2 are related to the geometry. Neglecting the coupling, there are two oscillators with frequencies ω_1, ω_2. Rott adjusts the geometry of his pendulums so that $\omega_1/\omega_2 = 1:2$, and optimizes the design so that the relative amplitudes of the coupled device are equal. This allows energy to flow easily from one pendulum to the other and can result in some transient chaotic behavior. He also has designs for two other devices, one of which is a three-pendulum device that looks like a puppet (Fig. 9.17c). Needless to

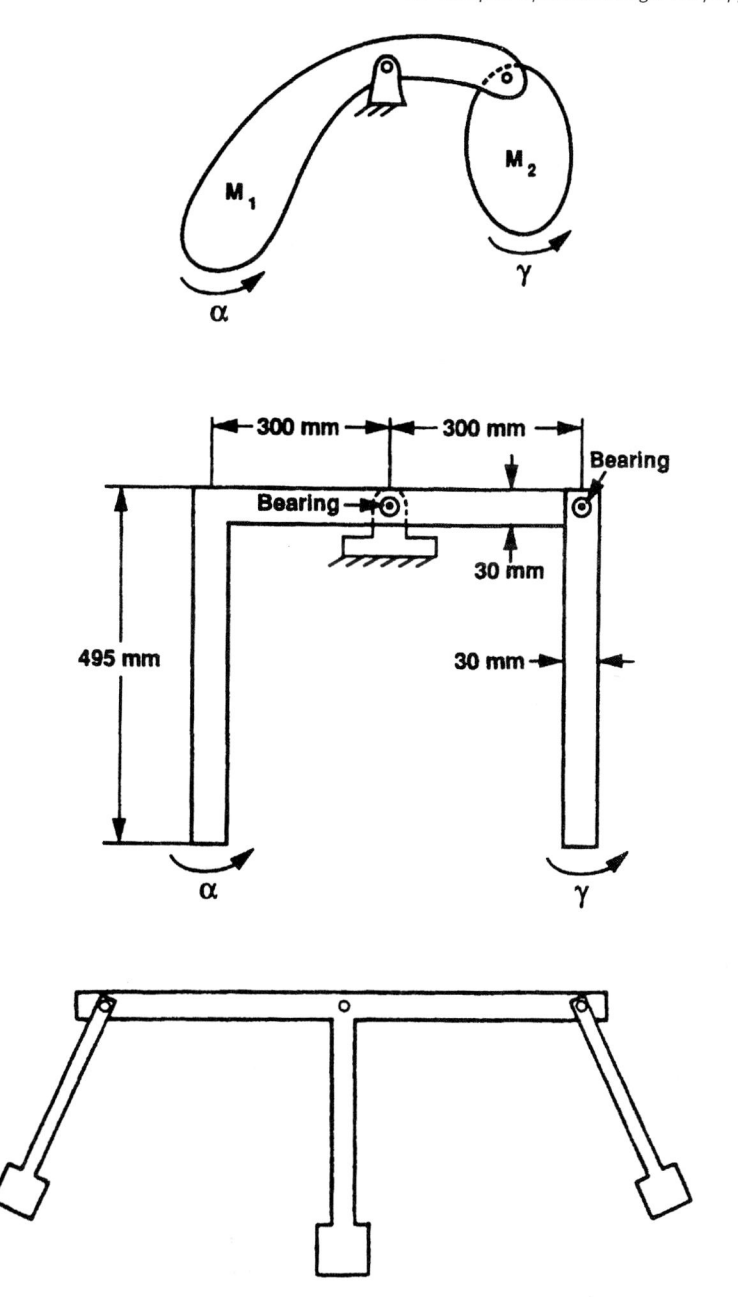

Fig. 9.17 Multibody pendulum systems for kinetic sculpture and toys. (After Rott, 1972.)

say, the choice of good bearings in these devices is essential. Professor Rott recently (ca. 1992) marketed an executive toy called "Pendemonium" based on his principles of nonlinear resonances. This example illustrates how knowledge of the possibilities of nonlinear dynamics can inspire design choices without detailed solution of the nonlinear equations of motion.

Chaotic Tumbling of Hyperion

Given the success of the Newtonian model of the dynamics of our solar system, we have come to accept the predictable nature of Newton's orbital dynamics. The predictable time history of our planet around the sun has been used to measure our own history of our world. Now, three centuries after publication of Newton's *Principia*, some are challenging the notion of absolute predictability in the motions of a few of the objects in our solar system such as comets and moons of planets. One example is the apparent chaotic motion of one of Saturn's moons.

The NASA mission of *Voyager 2* transmitted pictures of an irregularly shaped satellite of Saturn called Hyperion (Fig. 9.18). The pioneering work of J. Wisdom of M.I.T. (Wisdom et al., 1984) showed how this nonsymmetric celestial object could exhibit chaotic tumbling in its elliptical orbit around Saturn. Later work may be found in Black et al. (1995) and Thomas et al. (1995).

It is well known that an elongated satellite such as a dumbbell-shaped object orbiting in a circular orbit could exhibit oscillating planar rotary motions about an axis through the center of mass and normal to the plane of the orbit [see (7.67) and (7.68)].

When the satellite is asymmetric with three different moments of inertia, $A < B < C$, Wisdom et al. (1984) show that the planar dynamics are described by

$$\frac{d^2\theta}{dt^2} + \frac{\omega_0^2}{2r^3} \sin 2(\theta - f) = 0 \tag{9.36}$$

where time is normalized by the orbital period $T = 2\pi$ and where $r(t)$ and $f(t)$ are 2π periodic. Here $\omega_0^2 = 3(B - A)/C$, r is the radius to the center of mass, and $\theta(t)$ measures the orientation of the satellite.

This equation is similar to that of a parametrically forced pendulum that has been found to exhibit chaotic dynamics. Wisdom et al. (1984) show that these planar motions can become unstable with the possibility of three-dimensional tumbling of the satellite in its orbit around Saturn. Imagine living on such a world where the Saturn rise and set are unpredictable and where definitions of east and west, defined on earth by the fixed axis of rotation, would be hard to determine by intuition.

In Wisdom's model, $f(t)$ and $r(t)$ are periodic in time and are found from the elliptic orbit of the satellite about Saturn using an eccentricity of the orbit, $e = 0.1$. Equation (9.36) is similar to a forced pendulum. As an exercise, you should assume that r is not time-dependent, but that $\omega_0 = 0.2$, $f = \theta_0 \cos t$, and numerically integrate Eq. (9.36). To obtain a Poincaré map, plot $\{\theta, \dot{\theta}\}$ when $f(t) = 0$. Choose several values of $\theta_0 < \pi/2$ and several initial conditions. The results can be compared with figures 1 and 2 of Wisdom et al. (1984).

9.6 Complex Dynamics in Rigid Body Applications | 533

100 km

(a)

(b)

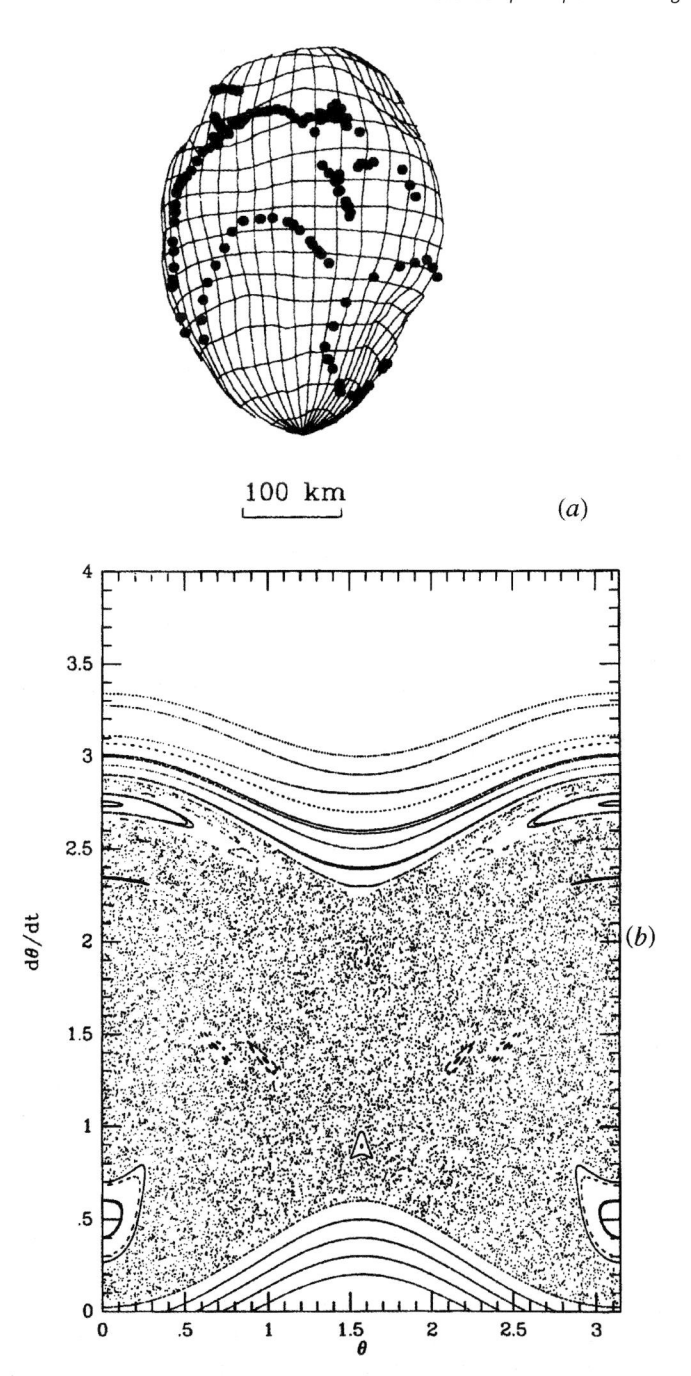

Fig. 9.18 (a) Computer generated shape of Saturn's moon Hyperion. (From Black et al., 1995.) (b) Poincaré map of model dynamics of Hyperion showing chaotic orbit.

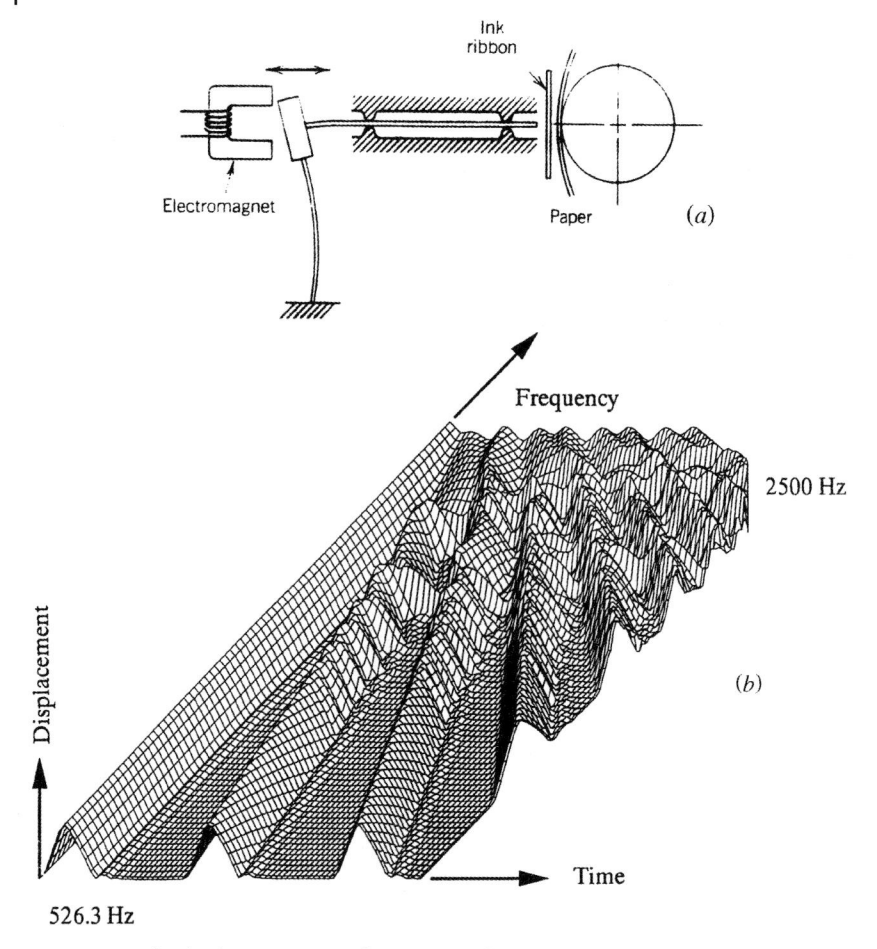

Fig. 9.19 (*a*) Sketch of a pin-actuator of a printer mechanism. (*b*) Displacement of a printer actuator as a function of time and different frequencies showing loss of predictable output. (From Hendriks, 1983, copyright 1983 by IBM Corporation, reprinted with permission.)

Impact-print Hammer

Impact-type problems have emerged as an obvious class of mechanical examples of chaotic vibrations. A practical realization of impact-induced chaotic vibrations is the impact-print hammer experiment studied by Hendriks (1983) at IBM (Fig. 9.19a). In this printing device, a hammer head is accelerated by a magnetic-force actuator and the kinetic energy is absorbed in pushing ink from a ribbon onto paper. Hendriks uses an empirical law for the impact force vs. relative displacement after impact; u is equal to the ratio of displacement to ribbon–paper thickness:

9.6 Complex Dynamics in Rigid Body Applications

$$F = \begin{cases} -AE_p u^{2.7}, & \dot{u} > 0 \\ -AE_p \beta u^{11}, & \dot{u} < 0 \end{cases} \tag{9.37}$$

where A is the area of hammer-ribbon contact, E_p acts like a ribbon–paper stiffness, and β is a constant that depends on the maximum displacement. It is clear that this force is extremely nonlinear.

When the print hammer is excited by a periodic voltage, it will respond periodically as long as the frequency is low. But as the frequency is increased, the hammer has little time to damp or settle out, and the impact history becomes chaotic (see Fig. 9.19b). Thus, chaotic vibrations can restrict the speed at which the printer can work. One solution that has been explored is adding feedback control, but the increased cost has discouraged technical implementation of this option.

MATLAB Example: Spinner Toy – The Wild Mouse Ride

In homework problems in Chapters 5 and 6 we introduced the problem of a spinning body with offset center of mass on an oscillating base (Figs. P5.33, P6.16). This problem is a model for a child's folk toy as well as a simplified model of an amusement ride called the "Wild Mouse" in which a passenger car is shaken side to side by an oscillating track (Chapter 6, homework problem 6.16). The form of the equation of motion is a second-order nonlinear ordinary differential equation (ode) with a sinusoidal parametric term and linear damping:

$$\ddot{\theta} + \gamma \dot{\theta} + \alpha \sin \omega t \sin \theta = 0 \tag{9.38}$$

In this problem, the frequency ω can be set to unity if one scales the damping γ and the driving amplitude α. This equation can be set into the form of a set of coupled first-order odes:

$$x_1 \equiv \theta, \qquad x_2 \equiv \dot{\theta} \tag{9.39}$$

$$\dot{x}_1 = x_2$$

$$\dot{x}_2 = -\gamma x_2 - \alpha(\sin t)\sin x_1 \tag{9.40}$$

The equilibrium points in the phase plane x_1–x_2 are on the $x_2 = 0$ axis and at points where $\sin x_1 = 0$, or where $x_1 = 0, \pi, -\pi$.

The $\sin x_1$ term is similar to that in a damped pendulum (Chapter 1), with an oscillating gravity term: that is for half the time cycle ($0 < t < \pi$) the motion acts as a damped pendulum, while in the next half cycle ($\pi < t < 2\pi$) the force seems to act as a saddle point or inverted pendulum, moving the pendulum away from the position $x_1 = 0$.

This motion can also be viewed as occurring in a three-dimensional state space with $\sin t$ replaced by $\sin x_3$, and adding another first-order equation differential equation $dx_3/dt = 1$. This artifact creates a three-dimensional autonomous system of odes in a cylindrical state space in which the variable x_3 is restricted to the range

($0 \leqslant t < 2\pi$). [See, e.g., Guckenheimer and Holmes (1983) for the mathematical theory.]

However the form (9.40) is suited for a numerical solution using the code MATLAB, in which one defines a MATLAB function with the right-hand side of (9.40) and calls for an integration subroutine such as "ode45" or "ode23." An example of a standard MATLAB ode integration code is given below in MATLAB Version 7.3. (See also the MATLAB primer by Pratap, 2004.) In this problem one can explore the parameter space [α, γ] to look for regular, periodic solutions and regions of chaotic solutions (Figs. 9.20a, 9.20b).

```
%% Define Function for the ODE solver
function dx = spinnertoy (t,x)

    alpha = 0.2 % Alpha driving variable (normalized)
    gamma = 0.01; % Damping Constant
    dx = zeros(2,1);

    dx(1) = x(2);
    dx(2) = alpha*sin(w*t)*sin(x(1)) - gamma*x(2);
% End of Function Definition.

% Set the time parameters t0 and tf.
t0 = 0;
tf = 60;

% Set the initial state vector
x0= [pi()/3; 0];

% Solve the ODE and Plot: chaotic case:
[t, x] = ode45(@spinnertoy, [t0 tf],x0);

figure(1);
plot(x(:,1),x(:,2));
xlabel('Angular Displacement [rad]');
ylabel('Angular Velocity [rad/s]');
title('Angular Velocity versus Angular Displacement for
Spinnertoy');
grid on;
```

In Fig. 9.20a, the motion starts at an initial angle and then evolves to an orbit near $\pi/2$ radians. In Fig. 9.20b, the dynamics does not settle into a limiting orbit but spins around the center exceeding 2π but spins in a nonregular time pattern. In order to classify this motion as chaotic, one would have to perform one or two tests on a long time series of the motion such as calculating the Lyapunov Exponent or taking a Poincaré map and looking for fractal structure in the sequence of points in the x_1–x_2 phase plane. It is possible to take a Poincaré map using the MATLAB function OPTION and EVENTS. One samples the position and velocity variables

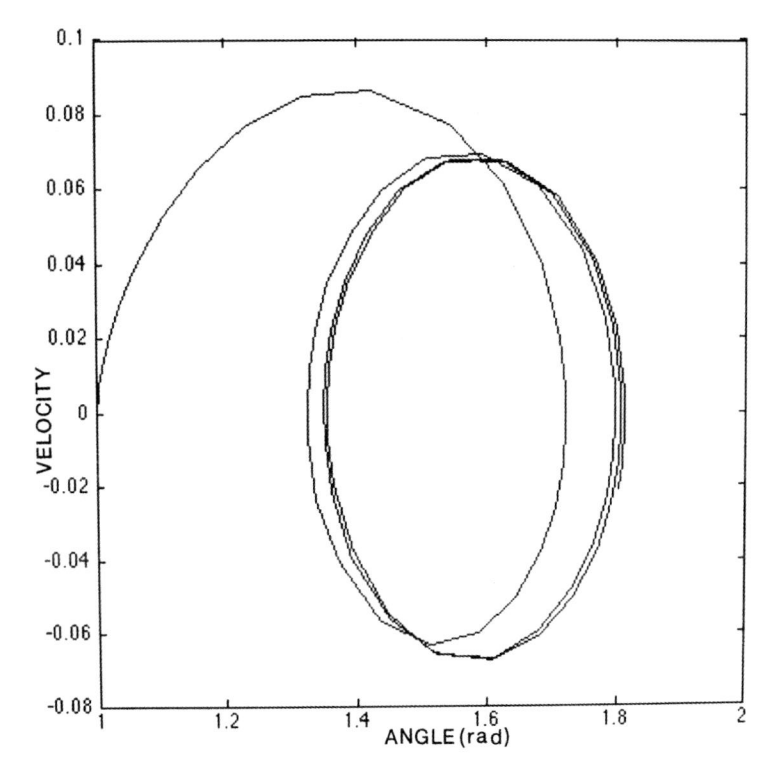

Fig. 9.20 (a) MATLAB integrated solution of "spinnertoy" nonlinear equations (9.40) for parameters leading to a long time periodic solution.

at 2π multiples of the time variable and plotting the points in the phase plane as in Fig. 9.yy. The reader is referred to the MATLAB HELP to use these functions.

Homework Problems

9.1 Sometimes the roll dynamics of a ship can be modeled by a particle of mass m moving in a potential field with $V(x) = ax^2 - bx^3$; $a, b > 0$.
   (a) Find the fixed points of the motion and establish the local stability.
   (b) Sketch the flow lines for different initial conditions when there is no damping.
   (c) Sketch the flow lines when there is small linear damping. (See, for example, the work of Thompson et al., 1990.)

9.2 The following equation can be derived from the dynamics of an electron or proton in a circular accelerator (e.g., see Helleman, 1980):

$$y_{n+1} + by_{n-1} = 2cy_n + 2y_n^2, \quad |b| \leqslant 1$$

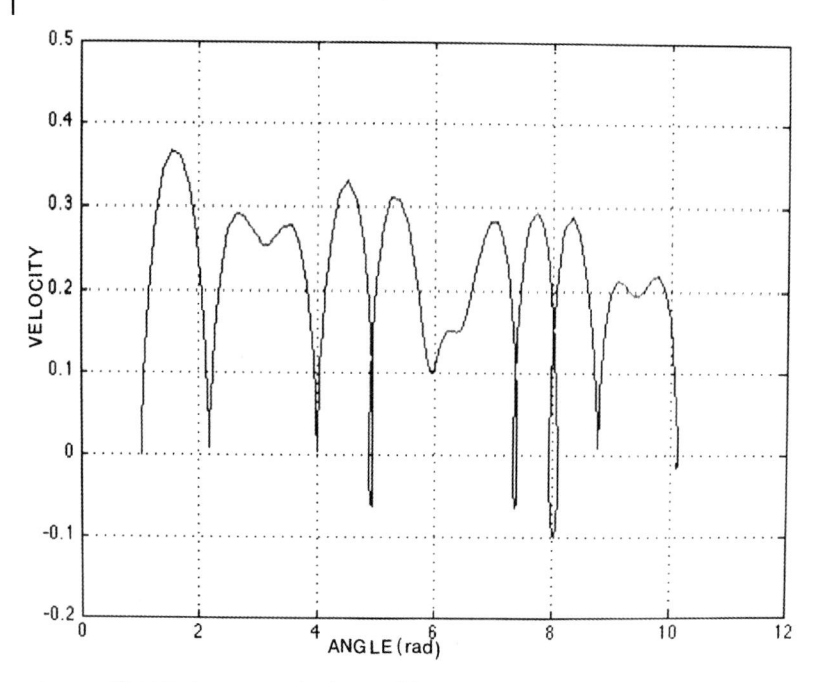

Fig. 9.20 (*b*) MATLAB integrated solution of the "spinnertoy" nonlinear equations (9.40) for parameters leading to a long time chaotic solution.

(a) Rewrite this equation as a set of first-order difference equations.
(b) For $b = 1$, show that this equation is equivalent to a special Hénon map (9.15).
(c) For $b = 0$, show that this equation reduces to the logistic map (9.18).

9.3 Suppose that the output of a dynamical system has two frequency components, that is,

$$x(t) = A \cos \omega_1 t + B \cos(\omega_2 t + \phi)$$

(a) If you take a Poincaré map on the phase $\omega_1 t$, show that the map $(x, y = \dot{x})$ is an ellipse when ω_1/ω_2 is incommensurate.
(b) Describe the map dynamics if $\omega_1/\omega_2 = p/q$, where p and q are integers.
(c) What does the map look like in the phase plane (x, y) if there is a third sinusoid of frequency ω_3?

9.4 (*Pseudo-Phase Plane*) The equation for a linear harmonic oscillator (spring and mass or inductor and capacitor circuit) is given by $\ddot{x} + x = 0$. The solutions for this equation can be represented in the phase plane by an ellipse written in parametric form:

$$x = A \sin t, \qquad y = \dot{x} = A \cos t$$

Derive an expression for the solution $x = A \sin t$ in terms of pseudo-phase-plane variables (x, x'), where $x' = x(t+T)$. Plot this expression for different values of A.

[Answer: $(x' - x \cos T)^2 = (A^2 - x^2) \sin^2 T$]

9.5 Investigate the properties of the cubic map (see Holmes, 1979)

$$x_{n+1} = y_n$$
$$y_{n+1} = -bx_n + dy_n - y_n^3$$

Find the fixed points and determine their stability as a function of the parameters b, d. Iterate this map for $b = 0.2$, $d = 2.5, 2.65, 2.77$. Can you find a strange attractor?

9.6 Consider an inverted pendulum: a spherical mass at the end of a massless rod of length L. The pendulum is constrained by two rigid walls on each side. At equilibrium the pendulum mass will rest on one of the two walls. Assume that the rest angles are small and show that for undamped free vibrations the dynamics are governed by (see Shaw and Rand, 1989)

$$\ddot{x} - x = 0, \quad |x| < 1$$
$$x \to -x, \quad |x| = 1$$

Show that a saddle point exists at the origin of the phase space $(x, \dot{x})$. Sketch a few trajectories.

9.7 Consider the two-degree-of-freedom system of a linear spring–mass oscillator confined to the diameter of a circular disc with spring constant, k_r. Assume also that the disc can rotate about its axis with a linear torsional spring restraint, k_ϕ. Neglecting gravity and dissipation, show that the equations of motion take the form

$$m\ddot{r} - mr\dot{\varphi}^2 + k_r r = 0$$
$$J\ddot{\varphi} + m(r^2 \ddot{\varphi} + 2r\dot{r}\dot{\varphi}) + k_\phi \varphi = 0$$

(*Hint*: Note that the kinetic energy is given by $\frac{1}{2}m(\dot{r}^2 + r\dot{\varphi}^2) + \frac{1}{2}J\dot{\varphi}^2$.) Show that energy is conserved in this problem.

9.8 (*The Kicked Rotor*) Equations for the kicked rotor (9.31) were derived for a damped system. Derive the two-dimensional map for zero damping ($c = 0$). Iterate this map on a small computer. Do you expect fractal structure?

9.9 (*Tumbling of Hyperion*) The rigid body tumbling of the irregularly shaped satellite of Saturn called Hyperion has been modeled by (9.36) by Wisdom et al. (1984), where the perturbing functions $f(t)$ and $r(t)$ are periodic in time. In the original paper, $f(t)$ and $r(t)$ were found from the elliptic orbit of the satellite using an eccentricity of the orbit of $e = 0.1$. However, (9.36) is analogous to a periodically forced pendulum. As an approximation, assume that $r = 1$ (i.e., not time-dependent), $\omega_0 = 0.2$, and $f = \theta_0 \cos t$. Numerically integrate these equations on a computer and plot θ vs. $\dot{\theta}$ when $f(t) = 0$ (i.e., a Poincaré map). Choose several values of $\theta_0 < \pi/2$ and several different initial conditions and compare your results with figures 1 and 2 of Wisdom et al. (1984).

9.10 A magnetic compass needle is assumed to be pivoted at its center and subjected to a rotating magnetic field of intensity B_0. Assume that the needle carries a magnetic moment **M** and that the torque about the axis is given by the cross product of the moment **M** and the magnetic field. Derive the equation of motion (Croquette and Poitou, 1981)

$$J\ddot{\theta} = -\mu\big[\sin(\theta - \omega t) + \sin(\theta + \omega t)\big] - \gamma\dot{\theta}$$

When $\gamma = 0$, show either analytically or computationally that either clockwise or counterclockwise motions are solutions.

9.11 (*Period Doubling*) Use a small computer to enumerate the critical values of λ in the logistic equation (9.18), and show that the sequences of values of λ_n approach the universal number (9.22).

9.12 Another first-order iterated map that exhibits period doubling is the sine map

$$x_{n+1} = \lambda \sin \pi x_n, \quad 0 \leqslant x_n < 1$$

With a small computer, show that this map exhibits a period-doubling sequence. Also show (numerically) that the period-two map, $x_{n+2} = \lambda \times \sin \pi [\lambda \sin \pi x_n]$, has a double hump similar to the quadratic or logistic map (9.18).

9.13 Consider the construction of a Cantor set that starts with a uniformly dense distribution of points on a line and begins by throwing out the middle β percent of the set. Iteration of this process results in a Cantor-type fractal set of points. Show that the box-counting or capacity fractal dimension (9.16) is given by

$$d = \frac{\ln 2}{\ln[2/(1-\beta)]}$$

9.14 Define a fractal-creating operation that takes a line element of length L and replaces it by eight equal segments of length $L/4$, as shown in Fig. P9.14. Draw at least four iterations on a large piece of graph paper. Use the four sides of a square as the initial line elements.

9.15 (*Sierpinski Carpet*) In this construction of a two-dimensional fractal set, start with a square that is divided into nine equal squares. Then the central square is removed, leaving eight. Repeat this algorithm, dividing each of the eight into nine pieces and removing the central square. Sketch several iterations of this process on a large piece of graph paper. Show that the box-counting or fractal dimension (9.16) is $d = \log 8 / \log 3$.

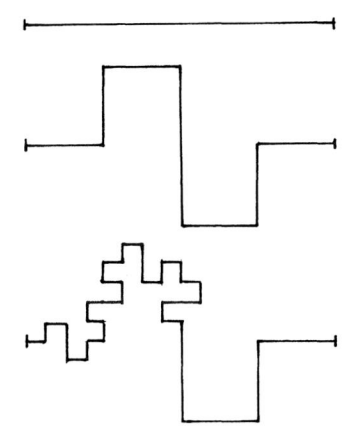

Fig. P9.14

9.16 Nonlinear lattice models have been used to model many systems, ranging from the dynamics of a long string of railroad cars to the dynamics of macromolecules, such as DNA. The equations can often be written in terms of a nonlinear potential function $V(r_n)$, where $r_n = x_{n+1} - x_n$ is the relative displacement between neighboring cells. Derive the equation of motion for general $V(r_n)$.

(*Solitons*) One nonlinear potential lattice model uses an exponential force potential (Toda, 1989) $V(r) = (a/b)e^{-br} + ar$ $(ab > 0)$. A lattice with this potential is known to admit so-called *solitary wave* or *soliton* solutions, where a given deformation pattern can propagate without distortion and where two such waves can interact and preserve their identity similar to linear-wave systems. Use MATLAB to plot the solitary wave solution for the Toda lattice given by

$$x_n = \frac{1}{b} \ln \frac{1 + e^{2(\kappa n - \kappa \pm \beta t)}}{1 + e^{2(\kappa n \pm \beta t)}} + \text{constant}$$

Here $\beta = (ab/m)^{1/2} \sinh \kappa$, and the width of the wave is proportional to $1/\kappa$. Also show that for the Toda lattice the solitary wave has a speed $C = \beta/\kappa$.

9.17 (*Atomic Force Microscope Dynamics*) In Chapter 8 we reviewed some of the nanoscale forces that are using in the dynamic modeling at the near molecular level. One of these is the so-called Leonard–Jones potential that accounts for forces between electron shells. This force has an attractive part at long distance and a repulsive part at short distance as shown in Section 8.2:

$$F = -\frac{\partial V}{\partial r} = 6\kappa \left[2\left(\frac{\sigma}{r}\right)^{13} - \left(\frac{\sigma}{r}\right)^{7} \right] / \sigma$$

One of the important modern tools of nanoscale science is the atomic field microscope. It consists of a cantilever with a sharp tip that scans the sur-

Fig. P9.17

$k(x) = k_o + k_1 x^2$

$r = (r_o - x)$

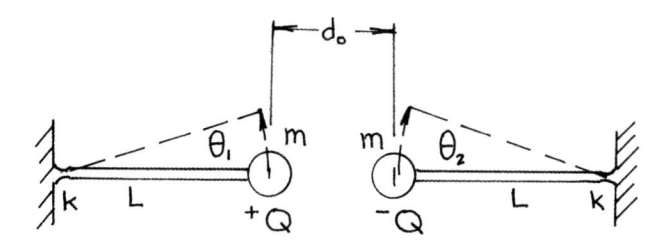

Fig. P9.19

face and maps the signals onto a two-dimensional picture of atomic level surface features. At close enough distances from surface, the Leonard–Jones force can bend the cantilever tip and change its dynamics. Use the attractive part of the L–J potential $[r^{-7}]$ to construct a mathematical model of the AFM cantilever with a lumped mass approximation as illustrated in Fig. P9.17. Assume the cantilever force can be modeling with a linear and cubic nonlinear spring. Add some small damping to the problem. Explore under what conditions there will exist a saddle point in the dynamics. Analyze the dynamics with MATLAB for both forced sinusoidal motion as well as impulse motion. [A more detailed analysis of this problem may be found in the papers of Hornstein and Gottlieb, 2008 and Couturier et al., 2002.]

9.18 (*MEMS Tuned Oscillator Dynamics*) In Chapter 8, Section 8.6, we examined the stability and linear dynamics of an oscillating mass between two capacitor plates with attractive electric forces. Add cubic as well as linear stiffness to this problem and examine the phase plane dynamics. Under what conditions will there be saddle points in the nonlinear problem? Analyze the transient dynamics with MATLAB.

9.19 (*Vibration of Electric Charge-coupled Oscillators*) Each of two MEMS scale cantilevers are modeled as a lumped mass electric pendulum with a weak compliant hinge of stiffness k (Fig. P9.19). Suppose the masses on the ends carry equal and opposite charges such that an attractive electric force exists between the two masses. Derive equations of motion for the two degree of freedom system using Lagrange's equations. Consider the anti-symmetric case where the rotation of one cantilever is opposite the other, i.e., $\theta_1 = -\theta_2$. Use a MATLAB simulation to determine if the natural frequency for this mode decreases or increases with amplitude as in Fig. 9.2. For the ambitious, use an analytic perturbation method such as described in Nayfeh and Balachandran (1995),

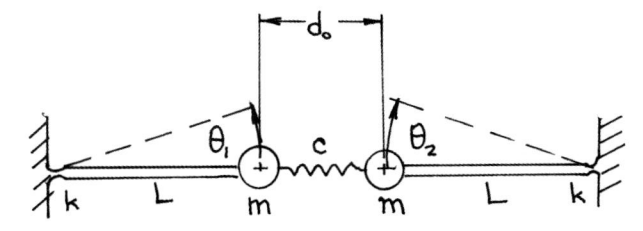

Fig. P9.20

to find out if the anti-symmetric vibration mode is "softening" or "hardening" nonlinearity.

9.20 In Problem 9.19, replace the electric charge force with a linear spring with an initial pretension Δ, and spring constant c (Fig. P9.20). Derive an equation of motion for the anti-symmetric mode and compare the results with that for the case of attracting charges. Is this problem a hardening or a softening nonlinearity?

Appendix A
Second Moments of Mass for Selected Geometric Objects

Table A1

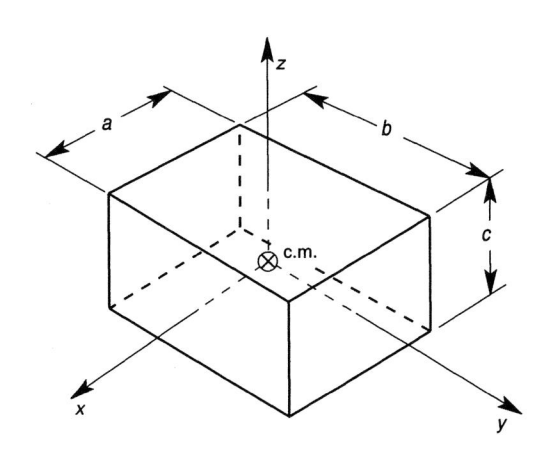

Rectangular parallelepiped

$I_{xx} = \frac{1}{12}m(b^2 + c^2)$

$I_{yy} = \frac{1}{12}m(a^2 + c^2)$

$I_{zz} = \frac{1}{12}m(a^2 + b^2)$

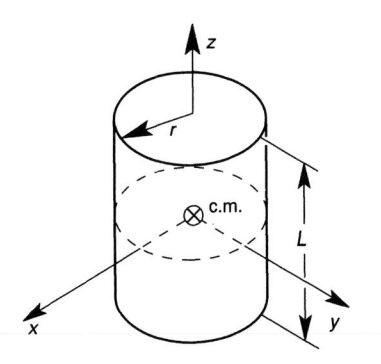

Circular solid cylinder

$I_{xx} = I_{yy} = \frac{1}{12}m(L^2 + 3r^2)$

$I_{zz} = \frac{1}{2}mr^2$

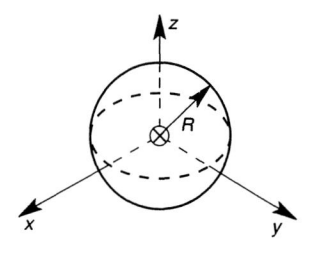

Appendix A. Second Moments of Mass for Selected Geometric Objects

Sphere

$$I_{xx} = I_{yy} = I_{zz} = \tfrac{2}{5}mR^2$$

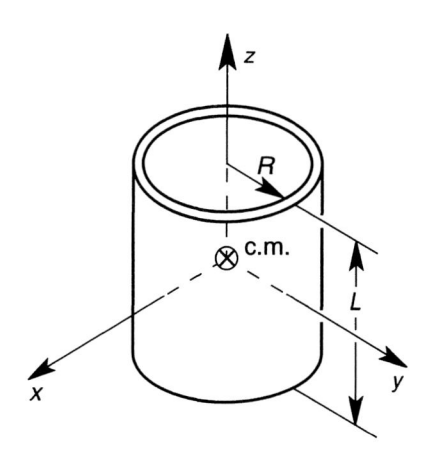

Circular shell

$$I_{xx} = I_{yy} = \tfrac{m}{12}(L^2 + 6R^2)$$
$$I_{zz} = mR^2$$

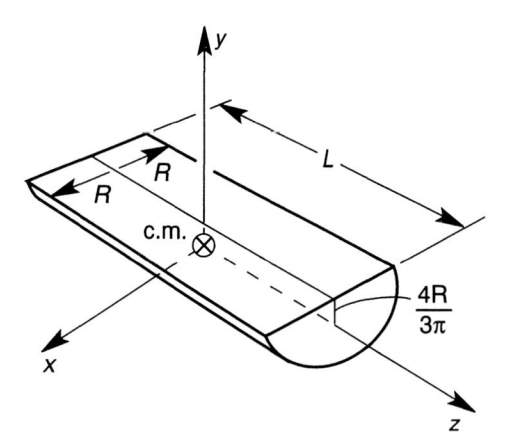

Half cylinder

$$I_{xx} = \tfrac{(9\pi^2 - 64)}{36\pi^2}mR^2 + \tfrac{1}{12}mL^2$$
$$I_{yy} = \tfrac{1}{12}m(3R^2 + L^2)$$
$$I_{zz} = \tfrac{(9\pi^2 - 32)}{18\pi^2}mR^2$$

Thin circular rod

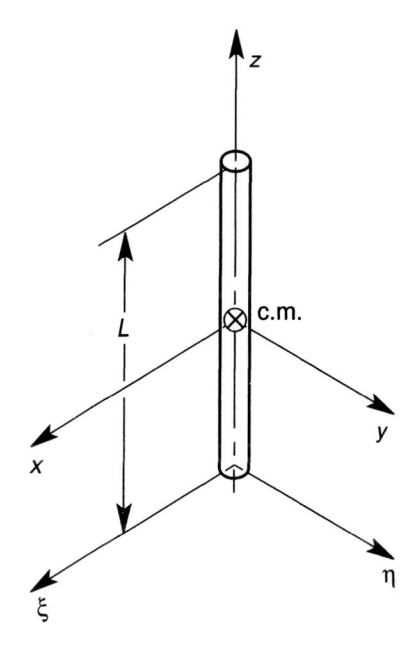

$I_{xx} = I_{yy} = \frac{1}{12}mL^2$

$I_{zz} = 0$

$I_{\xi\xi} = I_{\eta\eta} = \frac{1}{3}mL^2$

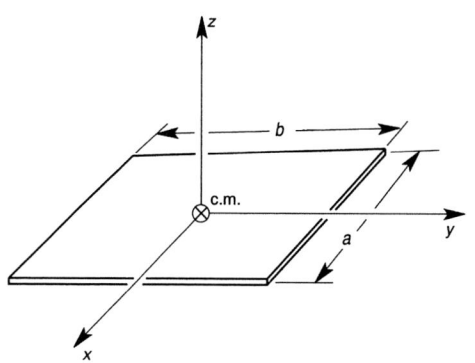

Thin rectangular plate

$I_{xx} = \frac{1}{12}mb^2$

$I_{yy} = \frac{1}{12}ma^2$

$I_{zz} = \frac{1}{12}m(a^2 + b^2)$

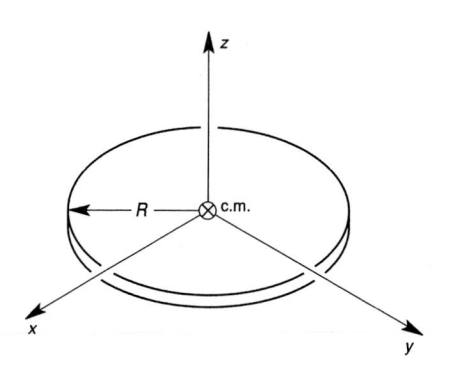

Thin circular disc

$I_{xx} = I_{yy} = \frac{1}{4}mR^2$

$I_{zz} = \frac{1}{2}mR^2$

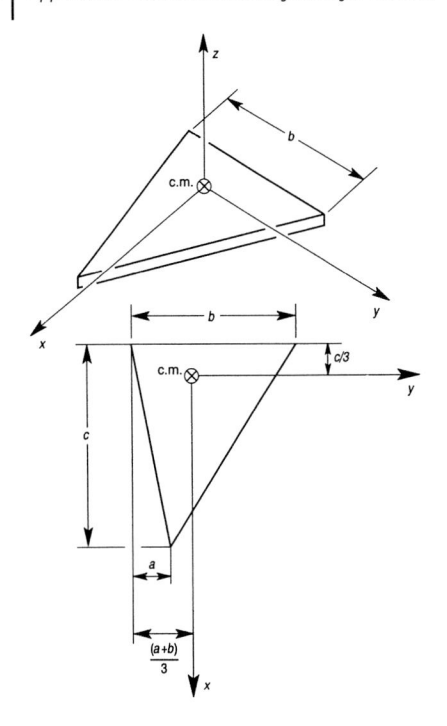

Thin triangular plate[a]

$I_{xx} = \frac{1}{18}m(a^2 + b^2 - ab)$

$I_{yy} = \frac{1}{18}mc^2$

$I_{zz} = \frac{1}{18}m(a^2 + b^2 + c^2 - ab)$

$I_{xy} = \frac{1}{36}mc(2a - b)$

a) Source: After J. H. Ginsberg and J. Gervin, *Dynamics*, Wiley, New York, 1977.

Appendix B
Commercial Dynamic Analysis and Simulation Software Codes

In the last two decades, a large number of analysis codes have become available to the dynamicist and design engineer. These new tools range from symbolic computation codes for deriving equations of motion, such as *MATHEMATICA* or *MAPLE*; numerical integration and calculation codes, such as MATLAB; to sophisticated "full service" codes, which include geometric modeling, dynamic equation formulation and analysis, and postprocessing simulation and animation. For the first time ever, engineers can easily construct three-dimensional dynamic models, calculate dynamic forces, and animate motions for complex multibody machines. However, like many commercial products, caveat emptor (let the buyer beware).

There are several hidden sources of potential problems that the user should be aware of:

1. In many codes, the method used to construct/derive the equations of motion is not described or is proprietary.
2. In several codes, the method of numerical integration of the equations of motion is not given or is not explicitly revealed to the user.
3. The method used to enforce constraints such as contact or rolling is not explicitly described in the user manuals.
4. The contact physics of impact and friction that are used in the simulation are often not described. For example, you would like to know whether Newton or Poisson impact laws are used or whether Coulomb or more modern state-variable friction laws are used.
5. The most critical feature that most commercial multibody codes lack is documented code verification, either with known analytical or numerical solutions or with careful laboratory experiments.

Unlike early finite-element structural/elasticity codes, which received extensive scrutiny in the published research literature, these new codes have been largely developed in the proprietary environment of private companies, though most had academic precursors. Thus, although many of these codes present beautiful color cartoon animation that looks real, the user should conduct his or her own verification tests before basing critical design decisions on the output of these exciting new design tools.

Applied Dynamics. With Applications to Multibody and Mechatronic Systems. Second Edition. Francis C. Moon
Copyright © 2008 WILEY-VCH Verlag GmbH & Co. KGaA, Weinheim
ISBN: 978-3-527-40751-4

Below we list some of the widely available software codes for dynamic analysis of rigid bodies. We realize that there are many excellent products that we have not listed, and there will no doubt appear many others after the publication of this book. These products are listed to encourage the student to test these new tools against some of the examples in this book.

Since the first edition of this book appeared in 1998, the number of multibody and special purpose dynamics software packages has increased. At the same time earlier software has been repackaged and bought by different companies. In addition, there has been an explosion of interest in extremely fast multibody physics solvers for entertainment games. These so-called SDK or software development kits, often deal with collisions and friction of objects in a way that looks "real" but is never tested against experimental data. Also multibody codes are now seen as a possible course-graining solution for million degree of freedom molecular dynamics simulations.

The author has not used all of these codes and cannot recommend them with any experience. In the past he has worked with WorkingModel (2D), DADS and SolidWorks/COSMOS Motion as well as with MATLAB.

It is expected that during the next several years new codes will appear and code distributors and developers will continue to change. As with any computer code, the motto is "caveat emptor" or "buyer beware." It is rare to find anyone who actually compares code output with experiments, but it would certainly be prudent to at least test a new code against a known analytical solution or a simulation for which one can trace the dynamics equations of motion as in a MATLAB integrator solution. In all codes, constraints, fiction and collisions (impact) are the most likely sources of error.

MAPLE

General-purpose symbolic programming code with numerical integration and graphics capabilities.

Source: Waterloo Maple Software, Waterloo, Ontario, Canada.
            Available for IBM/PC and Apple personal-computer systems.
Reference: A. Heck, *Introduction to Maple*, Springer-Verlag, New York.

MATHEMATICA

General-purpose symbolic programming code with numerical integration and graphics capabilities.

Source: Wolfram Research, Champaign, Illinois.
            Available for IBM/PC and Apple personal-computer systems.
Reference: S. Wolfram, *Mathematica*, Addison-Wesley, Reading, MA, 1991.

Auto-Lev

Uses symbolic programming (computer algebra) to derive equations of motion for multibody systems using Kane's method. Outputs a Fortran code for integration.

Appendix B. Commercial Dynamic Analysis and Simulation Software Codes | 551

Source: Prof. T. Kane, Stanford University and D. Levinson, Palo Alto, California.
Available for IBM/PC and Apple personal-computer systems.
Reference: Based on T. Kane, D. A. Levinson, *Dynamics Theory and Applications*, McGraw-Hill, New York, 1985.

Update 2008: This symbolic manipulator code is popular with academic dynamicists and a new version Autolev 4.x was released in March 2005.

NEWEUL

Uses symbolic programming to derive equations of motion for multibody systems. Based on Newton–Euler and virtual power (Jourdain/Kane) methods.

Source: Prof. W. Schiehlen, University of Stuttgart, Stuttgart, Germany.
Reference: W. Schiehlen, *Technische Dynamik*, Teubreug Stuttgart, 1990.

Update 2008: NEWEUL is maintained by the Universität Stuttgart, Institute für Technische und Numerische Mechanik in Germany. It provides both symbolic derivation of equations of motion as well as simulation of multibody dynamics. As of 2008, a demo version of the code for 4 degrees of freedom was available on the web for WINDOWS or LINUX. A manual was also available in PDF format. The code is advertised to handle serial, branched an closed chain problems in vehicle dynamics, mechanisms, satellite dynamics and biomechanics. This code is generally directed toward university and research users. A description can be found on the web at www.mechb_uni-stuttgart.de/research/neweul/neweul_de.php.

MATLAB

Acronym for "Matrix Laboratory." A general-purpose numerical code for performing operations on matrices, vectors, arrays of data. Includes subroutines to solve coupled differential equations. Also contains symbolic programming capability through *MAPLE*.

Source: The Math Works, Inc., 24 Pine Parkway, Natick, MA 01760.
Available for IBM/PC and Apple personal-computer systems.
Reference: R. Pratap, *Getting Started with MATLAB7*, Oxford Univ. Press, New York, 2004.

ADAMS

This software acronym stands for "Automatic Dynamic Analysis of Mechanical Systems." This professional, general-purpose, multibody code is one of the most widely used in industry. The code is based on a form of Lagrange's equations. Besides rigid-body problems, there are options for human-body modeling, vehicle dynamics, and flexible-body dynamics using a finite-element interface. The computer-assisted design (CAD) interface allows preprocessing geometric model development and postprocessing animation. The article by Ryan referenced below claims the code has been experimentally verified for some vehicle dynamics problems.

Source: Originally developed by Professor M. A. Chace and others at the University of Michigan; now marketed by a commercial company; Mechanical Dynamics, 2301 Commonwealth Boulevard, Ann Arbor, MI 48105.

Reference: See article by R. R. Ryan in W. Schiehlen, Ed., *Multibody Systems Handbook*, Springer-Verlag, Berlin, 1990, pp. 361–402.

DADS

This acronym stands for "Dynamic Analysis and Design System." Like ADAMS, this general-purpose, professional multibody code is widely used in industry. It includes a solid modeling processor, integration package, and animation postprocessor. The code is based on direct Newton–Euler formulation with explicit constraints, which leads to a large number of equations that are linear in the increments of the state variables. DADS is available on all major workstations and PCs.

Source: Computer Aided Design Software, Inc., 2651 Crosspark Road, Coralville, IA 52241.

Reference: The original code came with a text written by Prof. E. J. Haug of the University of Iowa, which is based on his text, *Intermediate Dynamics*, Prentice Hall, Englewood Cliff's, NJ, 1992. The text outlines the computational theory developed by Haug, which shuns the use of a minimal set of generalized coordinates, as in Lagrange's equations, Kane's method, or D'Alembert's method.

Update 2008: A new version (LMS DADS 9.6) was released by LMS International. This company markets design tools for automotive, aerospace and manufacturing applications such as virtual prototyping.

DYNAMECHS

DYNAMECHS is an acronym for "Dynamics of Mechanisms." This code was developed in 1991 by Scott McMillan, Electrical Engineering, Ohio State University for mechanism and robot simulation that involves serial and prismatic linked rigid bodies. It is implemented in C++. Currently, the web lists a Version 4.0. This package also includes sub codes to calculate hydrodynamic forces on underwater vehicles.

Pro/MECHANICA MOTION

This general-purpose professional multibody code is one of a family of design codes under the name of Pro/MECHANICA. This specific code performs three-dimensional rigid-body simulation and, like ADAMS and DADS, includes solid modeling preprocessing and postprocessing. This code uses a modified Kane's method. The program runs on all major UNIX, Windows NT, and Windows 95 platforms.

Source: Parametric Technology Corporation, 128 Technology Drive, Waltham, MA 02154.

Update 2008: ProEngineer/ProMechanica (Parametric Technology Corp). ProEngineer is a large suite of professional and student CAD/CAM software codes. They run on WINDOWS, LINUX and UNIX. ProMechanica is a sub suite of ProEngineer and will handle both rigid body and structural dynamics problems.

SolidWorks (CosmosWorksMotion)

This suite of CAD software will handle the design, assembly, dynamics and animation of a collection of rigid bodies, especially mechanisms. The student version is easy to learn and has been used at Cornell University by mechanical engineering sophomores. The dynamics solver component is CosmosWorksMotion. Problems are assembled in SolidWorks and then transferred to CosmosWorksMotion. This software is sold by SolidWorks Corporation.

Working Model

This general-purpose multibody code comes in two- and three-dimensional versions. Originally developed for university teaching, this code is quickly gaining popularity because of its price relative to the industry-oriented codes and its fast learning curve. Like ADAMS, DADS, and Pro/MECHANICA, it has solid modeling capabilities to help set up the geometry and animate the simulation results. Runs on Windows 95 and Windows NT PC's with a Pentium processor.

Source: Knowledge Revolution, 66 Bovet Road, San Mateo, CA 94402.

Update 2008: Working Model 2D is now marketed by Design Simulation Technologies, Inc.

Video Game Dynamic Animation Codes

As discussed in Chapter 1, the "techno-tainment" industry or computer games developers have pushed physics-based dynamic simulation to very high speeds using approximate or short-cut solutions to difficult problems of impact, friction and other geometric constraints. A recent book describing some of these codes for game programming is *Physics Based Animation*, by Kenny Erieben. Three video game dynamics codes are listed below.

BOX2D (Physics Engine). This planar dynamics code was developed by one of the Author's former Cornell students, Erin Catto, PhD. This code treats dynamics and collisions of rigid objects using impulse-momentum methods. As of 2008, the code could be downloaded from the web. The applications are for computer games.

ODE (Open Dynamics Engine). This code is an "Open Source" rigid body dynamics simulator based on C/C++. The principal author is Russell Smith. Many others have added on to this code through a WIKI website. As of 2008 Russell Smith had a power point slide presentation on the web (dated 2004) that provided an overview of the code (www.ode.org/slides/parc/index.html).

PhysX (Open Source Solver). This software is available from Ageia who make hard-wired software chips. Users can usually get the PhysX software free and buy a hard wired accelerator chip to speed up the performance even faster. According to a Wikipedia website entry, PhysX was used in SONY's PlayStation 3 video game console.

Multi-Physics Codes: Electric and Magnetic Fields, MEMS Design

In addition to dynamic equation generators and integration solvers, many codes have appeared to address other spatial physics modeling needs such as heat transfer and electric field calculations. Listed below are three software packages that treat electric and magnetic field problems.

ComSol

COMSOL offers a suite of physics-based field codes such as electric field solvers that are especially useful for MEMS applications. There is a wide range of users of these codes who periodically hold a conference; e.g., one was held in 2008.

CoventorWare ANALYZER

This software contains a package of 3D field solvers for structural and electrical problems especially suited for MEMS applications. The suite also handles problems in microfluidics and piezoelastic structures. The company Coventor also has a suite of codes for MEMS manufacturing process design, CoventorWare DESIGNER.

Vector Fields Software for Electromagnetics

For mechatronics applications, this code has general-purpose two and three dimension capabilities. The ELEKTRA/TR code in this family calculates transient three-dimensional eddy currents in magnetic and electric materials. The code includes solid modeling preprocessor and visualization of the dynamic magnetic and electric fields.

Source: Vector Fields Ltd., 24 Bankside, Kidlington Oxford, OX5 1JE, England.

References

1 ABARBANEL, H. (1996). *Analysis of Observed Chaotic Data*, Springer-Verlag, New York.

2 ADAMS, S. G. (1995). "Capacitance Based Tunable Micromechanical Resonators," *Transducers '95, Proc. 8th Int. Conf. Solid-State Sensors and Actuators*, Stockholm, Sweden, 438–441.

3 ADAMS, S. G. (1996). *Design of Electrostatic Actuators to Time the Effective Stiffness of Micro-Electromechanical Systems*, Ph.D. Dissertation, Cornell University, Ithaca, NY.

4 AMIROUCHE, F. M. L. (1992). *Computational Methods in Multibody Dynamics*, Prentice Hall, Englewood Cliffs, NJ.

5 AMIROUCHE, F. (2006). *Fundamentals of Multibody Dynamics: Theory and Applications*, Springer.

6 ANGELES, J. (1997). *Fundamentals of Robotic Mechanical Systems*, Springer-Verlag, New York.

7 APPELL, P. (1899). "Sur une forme nouvelle des equations de lu dynamique." *C. R. Acad. Sci. Paris*, 459–460 (cited in Neimark and Fufaev, 1972).

8 ARTOBOLEVSKY, I. (Ed.) (1945). *Scientific Contributions of P. L. Chebyshev, Second Part: Theory of Mechanisms* (in Russian), Academy of Science USSR, Moscow.

9 ARTOBOLEVSKY, I. I. (1979). *Mechanisms in Modern Engineering Design, Volume I, Lever Mechanisms*, MIR Publishers, Moscow.

10 ASADA, H., SLOTINE, J.-J. F. (1986). *Robot Analysis and Control*, Wiley, New York.

11 BEATTY, M. F. (1986). *Principles of Engineering Mechanics*, Plenum Press, New York.

12 BELETSKY, V. V., LEVIN, E. M. (1993). *Dynamics of Space Tether Systems*, American Astronautical Society, San Diego.

13 BLACK, G. J., NICHOLSON, P. D., THOMAS, P. C. (1995). "Hyperion: Rotational Dynamics," *Icarus* 117, 149–161.

14 BOTTEMA, O., ROTH, B. (1979). *Theoretical Kinematics*, North-Holland, Amsterdam and New York.

15 BRACH, R. M. (1991). *Mechanical Impact Dynamics: Rigid Body Collisions*, Wiley, New York.

16 Brown, D., Peck, M. (2008). "Control Moment Gyros as Space-Robotics Actuators," *AIAA Guidance, Navigation and Control Conference Proceedings*.

17 BURMESTER, L. (1888). *Lehrbuch der Kinematik*, Leipzig.

18 BURNS, J. A., MATHEWS, M. S. (Eds.) (1986). *Satellites*, University of Arizona Press, Tucson.

19 CADY, W. G. (1964). *Piezoelectricity*, Dover, New York.

20 CAPRARA, G. (1986). *The Complete Encyclopedia of Space Satellites*, Portland House, New York.

21 CARPENTER, M., PECK, M. (2008). "Dynamics of a High Agility, Low Power Imaging Payload," *IEEE Trans. Robotics*.

22 CECCARELLI, M. (2004). *Fundamentals of Mechanics of Robotic Manipulation*, Springer.

23 CHEBYSHEV, P. L. (1899–1907). *Oeuvres de P. L. Tchebychef*, 2 volumes,

St. Petersburg. See also Artobolevsky (1945).
24. COLLINS, S., RUINA, A., TEDRAKE, R., WISSE, M. (2005). "Efficient Bipedal Robots Based on Passive-Dynamic Walkers," *Science* 307, 1082–1085.
25. CRAIG, J. J. (1986). *Introduction to Robotics: Mechanics and Control*, Addison-Wesley, Reading, MA.
26. CRAIG, J. J. (2005). *Introduction to Robotics*, 3rd edition, Pearson Prentice Hall, New Jersey.
27. CRANDALL, S. H., KARNOPP, D. C., KURTZ JR., E. F., PRIDMORE-BROWN, D. C. (1968). *Dynamics of Mechanical and Electromechanical Systems*, McGraw-Hill, New York.
28. CROQUETTE, V., POITOU, C. (1981). "Cascade of Period Doubling Bifurcations and Large Stochasticity in the Motions of a Compass," *J. Phys. (Paris) Lett.* 42, L537–L539.
29. CUSUMANO, J. P., MOON, F. C. (1995). "Chaotic Non-Planar Vibrations of the Thin Elastics, Part II: Derivation and Analysis of a Low Dimensional Model," *J. Sound Vibration* 179(2), 209–226.
30. DA VINCI, L. (1493). *Codex Madrid or Tratado de Estatica y Mechanica en Italiano*. Facsimile edition: *The Madrid Codices*, National Library Madrid, No. 8937, translated by Ladilao Reti, McGraw-Hill Book Co. See KMODDL for on-line copy.
31. DANKOWICZ, H. (2004). *Multibody Mechanics and Visualization*, Springer.
32. DE BONA, F., ENIKOV, E. T. (2006). *Microsystems Mechanical Design*, Springer, Vienna.
33. DEN HARTOG, J. P. (1948). *Mechanics*, McGraw-Hill, New York.
34. DREXLER, K. E. (1986). *Engines of Creation*, Anchor Books, New York.
35. DREXLER, K. E. (1992). *Nanosystems: Molecular Machinery, Manufacturing and Computation*, J. Wiley & Sons, New York.
36. DUGAS, R. (1988). *A History of Mechanics*, Dover, New York.
37. ERDMAN, A. G. (Ed.) (1993). *Modern Kinematics: Developments in the Last Forty Years*, Wiley, New York.
38. ERDMAN, A. G., SANDOR, G. N., KOTA, S. (2001). *Mechanism Design: Analysis and Synthesis*, 4th edition, Prentice-Hall, New Jersey.
39. ERIEBEN, K. (2005). *Physics-Based Animations*, Charles Rivers Media.
40. FERGUSON, E. S. (1962). "Kinematics of Mechanisms from the Time of Watt," *United States National Museum Bull.* 228, 185–230.
41. FROST, N. E., MARSH, K. J., POOH, L. D. (1974). *Metal Fatigue*, Clarendon Press, Oxford.
42. GALILEI, G. (1600). "Le Meccaniche," in: *Galileo Galilei on Motion and on Mechanics*, translated by Stillman Drake, The University of Wisconsin Press, Madison, 1960.
43. GALILEI, G. (1648). *Two New Sciences*.
44. GINSBERG, J. H. (1995). *Advanced Engineering Dynamics*, Cambridge Press, Cambridge, MA.
45. GINSBERG, J. H., GERVIN, J. (1977). *Dynamics*, Wiley, New York.
46. GOLDSMITH, W. (1960). *Impact*, Edward Arnold Publishing, London.
47. GOLDSTEIN, H. (1980). *Classical Mechanics*, Addison-Wesley, Reading, MA.
48. GOLDSTEIN, H., POOLE, C., SAFKO, J. (2002). *Classical Mechanics*, Pearson Education.
49. GREENWOOD, D. T. (1988). *Principles of Dynamics*, Prentice-Hall, Englewood Cliffs, NJ.
50. GRÜBLER, M. (1883). "Allemeine Eigenschaften der zwangläufigen ebenen kinematische Ketten," *Der Civilingeniieur Neuen Band* 29, Leipzig.
51. GRÜBLER, M. (1917). *Getriebelehre: Ein Theorie des Zwanglaufes und ebenen Mechanismen*, Verlag von Julius Springer, Berlin.
52. GUCKENHEIMER, J., HOLMES, P. (1983). *Nonlinear Oscillations, Dynamical Systems, and Bifurcations of Vector Fields*, Springer-Verlag, New York.
53. GUO, J. G., ZHAO, Y. P. (2004). "Influence of van der Waals and Casimir Forces on Electrostatic Torsional Actuators," *J. Microelectromech. Syst.* 13, 1027–1035.

54 HAHN, J. K. (**1988**). "Realistic Animation of Rigid Bodies," *Computer Graphics* 22(4), 299.

55 HART, I. B. (**1925**). *The Mechanical Investigations of Leonardo da Vinci*, Chapman and Hall Ltd., London.

56 HARTENBERG, R. S., DENAVIT, J. (**1964**). *Kinematic Synthesis of Linkages*, McGraw-Hill Book Co., New York, p. 75.

57 HAUG, E. J. (**1992**). *Intermediate Dynamics*, Prentice-Hall, Englewood Cliffs, NJ.

58 HAWKING, S. (**1988**). *A Brief History of Time*, Bantum Books, New York.

59 HECK, A. (**199?**). *Introduction to Maple*, Springer-Verlag, New York.

60 HELLEMAN, R. H. G. (**1980**). "Self-Generated Chaotic Behavior in Nonlinear Mechanics," *Fund. Probl. Stat. Mech.* 5, 165–233. Reprinted in: Cvitanovic, B. P. (Ed.) (**1984**), *Universality in Chaos*, Adam Hilger Publishing Ltd., Bristol, 420–488.

61 HENDRICKS, F. (**1983**). "Bounce and Chaotic Motion in Print Hammers," *IBM J. Res. Dev.* 27(3), 273–280.

62 HOLMES, P. J. (**1979**). "A Nonlinear Oscillator with a Strange Attractor," *Philos. Trans. R. Soc. London A* 292, 419–448.

63 HORNSTEIN, GOTTLIEB, O. (**2008**). "Nonlinear Dynamics, Stability and Control of the Scan Process in Noncontacting Atomic Force Microscopy," in: *Nonlinear Dynamics*, Springer.

64 HUSTON, R. L. (**1990**). *Multibody Dynamics*, Butterworths-Heinemann, Boston.

65 HUTCHINGS, F. R., UNTERWEISER, P. M. (**1981**). *Failure Analysis*, American Soc. Metals, Metals Park, OH.

66 HUYGENS, C. (**1658**). *Horologium*. See also English translation of Richard J. Blackwell, *Christiaan Huygens' The Pendulum Clock or Geometrical Demonstrations Concerning the Motion of Pendula as Applied to Clocks*.

67 JACKSON, J. D. (**1962**). *Classical Electrodynamics*, Wiley, New York.

68 JARDINE, L. (**2004**). *The Curious Life of Robert Hooke*, HarperCollins, New York.

69 JOHNSON, A. J. (**2008**). *Elastic Dynamics of Sliding Electrical Contacts under Extreme Conditions*, Doctoral Dissertation, Cornell University, Ithaca, NY.

70 JOHNSON, A. J., MOON, F. C. (**2006**). "Elastic Waves and Solid Armature Contact Pressure in Electromagnetic Launchers," *IEEE Trans. Magn.* 42(3), 422–429.

71 JOHNSON, A. J., MOON, F. C. (**2007**). "Elastic Waves in Electromagnetic Launchers," *IEEE Trans. Magn.* 43(1), 141–144.

72 JOURDAIN, P. F. B. (**1909**). "Note on an Analogue of Gauss Principle of Least Constraint," *Quart. J. Pure Appl. Math.* 8L.

73 KANE, T. R. (**1961**). "Dynamics of Nonholonomic Systems," *Trans. ASME, J. Appl. Mech.*, 574–578.

74 KANE, T. R. (**1983**). "Formulation of Dynamical Equations of Motion," *Am. J. Phys.* 51(11), 974–977.

75 KANE, T. R., LEVINSON, D. A. (**1983**). "Multibody Dynamics," *J. Appl. Mech.* 50, 1071–1078.

76 KANE, T., LEVINSON, D. A. (**1985**). *Dynamics: Theory and Applications*, McGraw-Hill, New York.

77 KANE, T., LIKINS, P. W., LEVINSON, D. A. (**1983**). *Space Craft Dynamics*, McGraw-Hill, New York.

78 KAY, E. R., LEIGH, D. A. (**2005**). "Hydrogen Bond-Assembled Synthetic Molecular Motors," in: KELLY, T. R. (Ed.), *Molecular Machines*, Springer, Berlin.

79 KAY, E. R., LEIGH, D. A., ZERBETTO, F. (**2007**). "Synthetic Molecular Motors and Mechanical Machines," *Rev. Angew. Chem. Int. Ed.* 46, 72–191.

80 KELLOGG, O. D. (**1929**). *Foundations of Potential Theory*, Springer-Verlag, Berlin.

81 KELLY, T. R. (Ed.) (**2005**). *Molecular Machines*, Springer, Berlin.

82 KINSEY, R. J., MINGORI, D. L., RAND, R. H. (**1996**). "Nonlinear Control of Dual-Spin Spacecraft," *J. Guidance, Control, and Dynamics* 19(1), 60–67.

83 KMODDL (**2004**). "Kinematic Models for Design Digital Library," http://kmoddl.library.cornell.edu.

84 Kuo, B., Golnaraghi, F. (2002). *Automatic Control Systems*, Wiley, New York.

85 Kurokawa, H. (2007). "Survey of Theory and Steering Laws of Single-Gimbal Control Moment Gyros," *J. Guidance, Control, and Dynamics* 30(5), 1331.

86 Lagrange, J. L. (1788). *Traité de Méchanique Analytique*, Paris.

87 Lederman, L. (1993). *The God Particle*, Bantum Doubleday, New York.

88 Lee, C. K., Moon, F. C. (1989). "Laminated Piezopolymer Plates for Bending Sensors and Actuators," *J. Acoust. Soc. Am.* 85(6).

89 Lesser, M. (1995). *The Analysis of Complex Nonlinear Mechanical Systems: A Computer Algebra Assisted Approach*, World Scientific, Singapore.

90 Li, G.-X., Rand, R. H., Moon, F. C. (1990). "Bifurcations and Chaos in a Forced Zero-Stiffness Impact Oscillator," *Int. J. Non-Linear Mech.* 25(4), 417–432.

91 Lin, W.-H., Zhao, Y.-P. (2005). "Nonlinear Behavior for Nanoscale Electrostatic Actuators with Casimir Force," *Chaos Solitons Fractals* 23, 1777–1785.

92 Lipson, H. (2005). "Evolutionary Synthesis of Kinematic Mechanism," *J. Comput. Aided Design*. See also Zykov et al. (2005).

93 Lobontiu, N. (2007). *Dynamics of Microelectromechanical Systems*, Springer.

94 McCarthy, J. M. (1990). *An Introduction to Theoretical Kinematics*, MIT Press, Cambridge, MA.

95 McGeer, T. (1990). "Passive Dynamic Walking," *Int. J. Robotics Res.* 9, 62–82.

96 Meirovitch, L. (1970). *Methods of Analytical Dynamics*, McGraw-Hill, New York.

97 Menschik, A. (1987). *Biometrie*, Springer-Verlag, Berlin.

98 Minorsky, N. (1962). *Nonlinear Oscillations*, Van Nostrand, Princeton, NJ.

99 Misner, C. W., Thorne, K. S., Wheeler, J. A. (1973). *Gravitation*, W. H. Freeman, San Francisco.

100 Miu, D. K. (1993). *Mechatronics*, Springer-Verlag, New York.

101 Moon, F. C. (1984). *Magneto-Solid Mechanics*, Wiley, New York.

102 Moon, F. C. (1987). *Chaotic Vibrations*, Wiley, New York.

103 Moon, F. C. (1992). *Chaotic and Fractal Dynamics*, Wiley, New York.

104 Moon, F. C. (1994). *Superconducting Levitation*, Wiley, New York.

105 Moon, F. C. (2002a). "Modeling Electromechanical Systems," in: Bishop, R. H. (Ed.), *The Mechatronics Handbook*, CRC Press, New York.

106 Moon, F. C. (2002b). "Nonlinear Dynamics," in: *Encyclopedia of Physical Science and Technology*, 3rd edition, Vol. 10, 523–535.

107 Moon, F. C. (2003a). "Franz Reuleaux: Contributions to 19th Century Kinematics and History of Machines," *Appl. Mech. Rev.* 56(2), 261–285.

108 Moon, F. C. (2003b). "Robert Willis and Franz Reuleaux: Pioneers in the Theory of Machines," *Not. Rec. R. Soc.* 57(2), 209–230.

109 Moon, F. C. (2004). "The Reuleaux Models: Creating an International Digital Library of Kinematics History," in: Ceccarelli, M. (Ed.), *International Symposium on History of Machines and Mechanisms, Proc. HMM2004*, Kluwer Academic, Dordrecht.

110 Moon, F. C. (2007). *The Machines of Leonardo da Vinci and Franz Reuleaux*, Springer.

111 Moon, F. C., Broschart, T. (1991). "Chaotic Sources of Noise in Machine Acoustics," *Arch. Appl. Mech.* 61, 438–448.

112 Moon, F. C., Stiefel, P. D. (2006). "Coexisting Chaotic and Periodic Dynamics in Clock Escapements," *Philos. Trans. R. Soc. A* 364, 2539–2563.

113 Nayfeh, A. H., Balachandran, B. (1995). *Applied Nonlinear Dynamics*, Wiley, New York.

114 Nayfeh, A. H., Mook, D. T. (1979). *Nonlinear Oscillations*, Wiley, New York.

115 Neimark, Ju. I., Fufaev, N. A. (1972). *Dynamics of Nonholonomic Systems*, translated from Russian by J. R. Barbour, American Mathematical Society, Providence, RI.

116 NEWTON, I. (1686). *Principia*. Translated by Motte (1729), revised by Cajori (1962), UC Press, Berkeley, CA.

117 OGATA, K. (1997). *Modern Control Engineering*, Prentice-Hall, Upper Saddle River, NJ.

118 O'REILLY, O. M. (2000). *Engineering Dynamics: A Primer*, Springer.

119 PAK, C. H. (1999). *Nonlinear Normal Mode Dynamics*, INHA University Press, Inchon.

120 PAPASTRAVRIDIS, J. G. (1992). "On Jourdains Principle," *Int. J. Eng. Sci.* 30(2), 135–140.

121 PAUL, B. (1970). *Kinematics and Dynamics of Planar Machinery*, Prentice-Hall, Englewood Cliffs, NJ.

122 PAUL, B. (1979). *Kinematics and Dynamics of Planar Machinery*, Prentice Hall, New Jersey.

123 PAUL, R. P. (1981). *Robot Manipulators: Mathematics, Programming and Control*, MIT Press, Cambridge, MA.

124 PEAUCELLIER, C.-N. (1864/1873). "Note sur une question de geométrie de compass," *Nouvelles Ann. Math. Sér. 2* 12, 71–78.

125 PELESKO, J. A., BERNSTEIN (2003). *Modeling MEMS and NEMS*, Chapman & Hall/CRC, New York.

126 PFEIFFER, F. (1992). *Einführung in die Dynamik*, B. G. Teubner, Stuttgart.

127 PFEIFFER, F. (1994). "Unsteady Processes in Machines," *Chaos* 4(4), 693–705.

128 PFEIFFER, F., GLOCKER, C. (1996). *Multibody Dynamics with Unilateral Contacts*, Wiley, New York.

129 PHILLIPS, J. (1984). *Freedom in Machinery: Introducing Screw Theory*, Cambridge Univ. Press, Cambridge.

130 PRATAP, R. (1996). *MATLAB – A Quick Introduction for Scientists and Engineers*, Saunders, Philadelphia.

131 PRATAP, R. (2004). *Getting Started With MATLAB7*, Oxford Univ. Press, New York.

132 RAIBERT, M. H. (1986). *Legged Robots That Balance*, MIT Press, Cambridge, MA.

133 REULEAUX, F. (1876). *Reuleaux Kinematics of Machinery, Outline of a Theorem of Machines*, translated and edited by A. B. W. KENNEDY, Macmillan, London.

134 REULEAUX, F. (1893). *The Constructor: A Handbook of Machine Design*, 4th edition, translated by H. H. SUPLEE, Philadelphia.

135 ROBERSON, R. E., SCHWERTASSEK, R. (1988). *Dynamics of Multibody Systems*, Springer-Verlag, Berlin.

136 RIMROTT, F. P. J. (1988). *Introductory Attitude Dynamics*, Springer-Verlag, New York.

137 ROSHEIM, M. E. (1994). *Robot Evolution: The Development of Anthrobotics*, Wiley, New York.

138 ROSHEIM, M. E. (2006). *Leonardo's Lost Robots*, Springer Verlag, Berlin.

139 ROTERS, H. C. (1941). *Electromagnetic Devices*, Wiley, New York.

140 ROTT, N. (1970). "A Multiple Pendulum for the Demonstration of Nonlinear Coupling," *J. Appl. Math. Phys.* 21, 570–572.

141 SCHIEHLEN, W. (1986). *Technische Dynamik*, B. G. Teubner, Stuttgart.

142 SCHIEHLEN, W. (Ed.) (1990). *Multibody Systems Handbook*, Springer-Verlag, New York.

143 SCHIEHLEN, W., AMBROSIO, J. (Eds.) (1997–2008). *Multibody System Dynamics*, Springer.

144 SHABANA, A. A. (1989). *Dynamics of Multibody Systems*, Wiley, New York.

145 SHAMES, I. H. (1966). *Engineering Mechanics, Vol. II, Dynamics*, Prentice-Hall, New Jersey.

146 SHAW, S. W., RAND, R. H. (1989). "The Transition to Chaos in a Simple Mechanical System," *Int. J. Non-Linear Mech.* 24(1), 41–56.

147 SHIGLEY, J. E. (1969). *Kinematic Analysis of Mechanisms*, 2nd edition, McGraw-Hill, New York.

148 SMYTHE, W. R. (1968). *Static and Dynamic Electricity*, 3rd edition, McGraw-Hill Book Co., New York.

149 STRATTON, J. A. (1941). *Electromagnetic Theory*, McGraw-Hill, New York.

150 STROGATZ, S. H. (1994). *Nonlinear Dynamics and Chaos*, Addison-Wesley, Reading, MA.

151 STROGATZ, S., ABRAMS, D. M., MCROBIE, A., ECKHARDT, B., OTT, E. (2005).

"Crowd Synchrony on the Millennium Bridge," *Nature* **438**(3), 43–44.
152. Szabó, L. (1987). *Geschichte der mechanischen Prinzipien: und ihrer wichtigsten Anwendungen*, 3rd edition (in German), Birkhaüser Verlag, Basel.
153. Tao, D. C. (1967). *Fundamentals of Applied Kinematics*, Addison-Wesley, Reading, MA.
154. Thomas, P. C., Black, G. J., Nicholson, P. D. (1995). "Hyperion: Rotation, Shape, and Geology from Voyager Images," *Icarus* 117, 128–148.
155. Thompson, J. M. T., Rainey, R. C. T., Soliman, M. S. (1990). "Ship Stability Criteria Based on Chaotic Transients from Incursive Fractals," *Philos. Trans. R. Soc. London A* 332, 149–167.
156. Thorne, K. S. (1994). *Black Holes and Time Warps*, Norton, New York.
157. Toda, M. (1989). *Theory of Nonlinear Lattices*, 2nd edition, Springer-Verlag.
158. Truesdell, C. (1968). *Essays in the History of Mechanics*, Springer-Verlag, New York (Chapter V).
159. Weinberg, S. (1993). *Dreams of a Final Theory*, Random House, New York.
160. Wiesel, W. E. (1989). *Spaceflight Dynamics*, McGraw-Hill, New York.
161. Wisdom, J., Peale, S. J., Mignard, F. (1984). "The Chaotic Rotation of Hyperion," *Icarus* 58, 137–152.
162. Wittenburg, J. (1977). *Dynamics of Systems of Rigid Bodies*, B. G. Teubner, Stuttgart.
163. Wittenburg, J. (2007). *Dynamics of Multibody Systems*, 2nd edition, Springer.
164. Willis, R. (1841/1870). *Principles of Mechanisms*, London, England.
165. Wolfram, S. (1991). *Mathematica*, Addison-Wesley, Reading, MA.
166. Woodson, H. H., Melcher, J. R. (1968). *Electromechanical Dynamics, Parts I, II, III*, Wiley, New York.
167. Zee, A. (1989). *An Old Man's Toy: Gravity at Work and Play in Einstein's Universe*, Macmillan, New York.
168. Zykov, V., Mytilinaios, E., Adams, B., Lipson, H. (2005). "Self Reproducing Machines," *Nature* **435**(7038), 163–164. See also Lipson (2005).

Index

a

Acceleration 44, 85, 170, 178
　　– angular 215
　　– centripetal 86
　　– Coriolis 86, 527
　　– radial 31
　　– relative to moving coordinates 85
ADAMS 14, 80, 309, 551
Airbus-380 1
Angular acceleration 215
Angular momentum 19, 26, 30, 52, 54, 56, 60, 62, 73, 230, 333, 385
　　– about center of mass 52
　　– conservation of 54
Angular velocity 56, 80, 84, 316, 332
　　– matrix 83
Apogee 320
Aristotle 37
Artobolevsky 6
Atomic force microscope 541
Attitude stability 403, 425
Auto-Lev 550

b

Back emf 440, 469, 488
Ball joint 109
Barium titanate 454
Bending-torsional oscillations 525
Bernoulli, Jean 189
Bernoulli family 36, 189
Bifurcation diagram 15, 519
Biomechanical dynamics 32
　　– knee 135
　　– centrodes 123
Bobsled dynamics 303
Brachistochrone 189
Brahe, Tycho 35

c

Calder, Alexander 528, 530

Cantilevered beam 525
Cantor sets 516, 540
Capacitance 162
Capacitor 444, 456, 466
Casimir effect 445, 495
Cartesian coordinates 58
Catenary 189
Center
　　– of gravity 405, 425
　　– of mass 51, 52, 63
Central-force motion 28, 384
Centripetal acceleration 86
Centro 118
Centrodes 118, 121
　　– biomechanics 123
　　– Reuleaux's method 124
　　– rolling 135
Chain
　　– 4-bar closed 8
　　– closed 7, 308
　　– open 7
Chaos
　　– magnetic levitation 520
　　– gears 529
　　– theory 3, 502
Chaotic
　　– bouncing 365
　　– dynamics 501, 512, 532
　　– tumbling, Hyperion 532
Charge 31, 59
　　– conservation of 457
　　– density 439
Charged satellite 432
Chasle's Theorem 91, 119
Chebyshev 108, 114, 116
Circuit elements 455
Circuits 456
Circular orbits 387
Cobweb diagram 518
Codex Madrid 9, 105, 107

Applied Dynamics. With Applications to Multibody and Mechatronic Systems. Second Edition. Francis C. Moon
Copyright © 2008 WILEY-VCH Verlag GmbH & Co. KGaA, Weinheim
ISBN: 978-3-527-40751-4

Coefficient of restitution 352
Comagnetic energy 463
Compliant mechanism 108
 – rotary 59
Compound mechanism 115
Computer graphics 96
ComSol Code 554
Conic sections 388
Conservation laws 58
 – angular momentum 29, 59, 75, 386, 392
 – charge 457
 – energy 58, 59, 161, 245
 – linear momentum 59
Conservative forces 159, 166
Constraint force 65, 185, 325
Constraints 44, 57, 64, 88, 140, 175
 – holonomic 164
 – Jacobian 44, 88, 140
 – nonholonomic 5, 170, 175, 182
 – rolling 184
 – sliding knife-edge 279
Control
 – dynamics 32, 481
 – stiffness 346
 – robotic machines 346
 – law 487
 – moment gyros (CMG) 414, 432
Coordinate frames 56, 84, 315
Coordinate systems
 – Cartesian 42
 – cylindrical 42
 – path coordinates 170
Copernicus, Nicolaus 383
Coriolis, Gaspard-Gustave de 44
Coriolis Acceleration 86
Coulomb Friction 290
Cross products of inertia 62, 227
Cyclic coordinates 167
Cycloid 118, 190, 285
Cylindrical coordinates 42, 43

d

da Vinci, Leonardo 9, 36, 104, 106, 107
DADS 14, 80, 309, 552
D'Alembert, Jean Le Rond 36, 64, 143
D'Alembert's Principle 4, 64, 139, 143, 145
DC motor 491
Degrees of freedom 44, 88, 106, 113, 117, 140, 316
Denavit–Hartenberg Notation 341
Dielectric forces 446
Difference equation 515

Digital micro-mirror 12
Diodes 448, 456
Directrix 389
Dissipation function 162
Dressler, K. E. 13
Duffing, G. 504
Duffing Oscillator 506
Dynamics equations
 – for electric and magnetic circuits 458, 466
Dual spin satellite 411
Dynamic failures 15
Dynamic instabilities 15, 16

e

Earnshaw's Theorem 485
Effective potential 390
Einstein, Albert 37
Electrical conductivity 449
Electric
 – charges 422
 – displacement 439, 446
 – energy 59
 – field 31, 439
 – force 161, 196, 438
 – permittivity 444, 447
 – stiffness 479
 – stress 443, 494
 – susceptibility 447
Electro-actuator, comb drive 11
Electromagnetic
 – Lagrangian 467
 – rail launcher 470, 473, 500
 – levitation 497, 10, 437, 476, 485
 – units 438
Electromechanical dynamics 32, 184, 435, 514
Elliptic orbits 28, 388, 392
Energy
 – conservation of 59, 161
 – methods 139, 59
 – principle 57, 69, 73
Epicycloid 106
Equations of motion 24, 58, 178, 319, 321
 – for a multibody system 348
Equilibrium 47, 68
Escape velocity 392
Euler, L. 35, 61, 119, 190
Euler Angles 216, 219, 220, 223
Euler–Bernoulli Beam 434
Euler–Lagrange Equation 190, 191
Euler Parameters 95
Euler's Equations 63, 234, 243
 – modified 235, 240

Euler's Theorem   61, 81, 91
Extremum problems   188, 196

f
Failures, dynamic   15
Faraday–Henry Law   457
Fatigue failure   15
Finite
  – motions   91, 216
  – rotation   91
  – rotation transformation   97
  – transformation matrix   277
First moment of mass   52, 250
Flutter   16
Flywheel   240
Force
  – active   147, 334
  – conservative   159
  – electric dipoles   493
  – electric and magnetic   161
  – gravity   398
  – nanoscale   444
  – permanent magnet   490
  – wire element carrying current   161
Force-moment   26, 56, 231
Four-bar linkage   110, 112, 113, 114
  – see also Mechanisms
  – spherical   6
  – crossed   121
  – Jacobian   134
  – double-slider   119
Fractals   515
Fractal dimension   515
  – box counting   515
Free wall tunnel   402
Free-body diagram   68
Frequency   24

g
Galilei, Galileo   21, 35, 189
Gear-rattling chaos   527
Gear transmission   106
Generalized
  – coordinates   45, 57, 141, 316
  – forces   155, 250, 176, 178
  – inertia force   176, 178
  – momentum   153, 166
  – speeds   176
  – velocities   66, 164, 176
Geosynchronous orbit   389
Gradient operator   58
Graph Theory   7, 309
Gravitational torque   407
Gravitons   38

Gravity   21, 159
Gravity force   58, 159, 386
  – extended bodies   398
Gravity-grade tunnel   400
Gravity gradient stabilization   405
Grübler, M.   114, 116
Grübler's Mobility Criterion   113, 368
Gyro sensor   56, 297
Gyropendulum   264, 302
Gyroscopic
  – dynamics   26, 56, 414
  – effects   26, 28, 64, 215, 237, 242

h
Hamilton, William Rowan   36, 139
Hamilton's Principle   188, 192
Hard spring oscillator   505
Henon map   515, 516
Hertz impact   362
History of dynamics   35
Hohmann, Walter   393
Hohmann Orbit Transfer   394, 395, 425
Holonomic constraints   65, 164, 170
  – see also Constraints
Homogeneous
  – solution   24
  – transformation   100, 216
Horseshoe map   515
Hubble telescope   213, 286
Hyperbolic orbit   288, 392, 424
Hyperion   539
Hypocycloid   121
Hypotrochoid   121
Hysteretic effects   447, 452

i
Ignorable coordinate   167, 391
Impact   351
  – central   374
  – chaos   364
  – elastic theory   362
  – planar   356, 359
  – print hammer   534
Impulse
  – angular   246
  – impact force   246
Incidence matrix   310, 313
Inductance   458, 461
  – mutual   462
  – self   462
Inertia matrix   227
Instant center   118, 121
Internal
  – combustion engine   212

– forces 147
Inverse problems, robotics 339, 386
Inverse square law 29, 421
Involute curve 106

j
Jacobi, K. G. J. 44
Jacobian 44, 66, 140, 88, 316
 – matrix 44, 46, 50, 142, 318
 – vector 58, 149
Joints 108
 – ball 109
 – prismatic 111
 – revolute 111
 – screw 109
 – spherical 109
Jourdain, Philip E. B. 37, 170
Jourdain's Principle 4, 67

k
Kane, Thomas R. 4, 67, 170, 175, 279
Kane's Equations 67, 170, 176
Kellogg 399
Kepler, Johannes 28, 35
Kepler's Law 386, 421
Kicked rotor 522, 539
Kinematic chain 111
Kinematic mechanisms 103
 – *see also* Mechanisms
 – pairs 307, 108, 109
 – relations 219, 220, 250
Kinematics 41, 77, 315
 – connected rigid bodies 316
 – knee 135
 – mountain bike 135
 – planetary gear 133
 – rigid body 214
 – robot arm 285
Kinetic
 – energy 57, 59, 156, 248, 320
 – sculpture 528, 531
Kirchhoff's Circuit Laws 441, 458, 460
KMODDL (Kinematic Models for Design Digital Library) 6, 36, 42, 78, 108
Kutzbach Criterion 117

l
Lagrange, Joseph Louis 139, 151, 190
Lagrange Multiplier 182, 186, 206, 282
Lagrange Points 432
Lagrange's equations 139, 151, 154, 161, 183, 188, 391
 – electric field system 466
 – magnetic systems 466, 458

– multi-body systems 319
– nonholonomic problems 278, 282
– rigid body 248
Lagrangian 161, 166, 390, 517
 – electromechanical system 469
Least action, Principle of 192
Leibniz, G. W. 35, 379
Lenz's Law 440, 441
Leonard–Jones forces 445, 446
Levitation
 – active controlled magnetic 485
 – rare-earth magnet 451
 – superconductor coil 476
 – forces 451
Limit cycles 511
Linear
 – electric actuator 469
 – momentum 19, 52, 54
 – oscillator 504
Linkages 106
Logistic equation 518
Lorenz, Edward 503
Lyapunov Exponent 503
Lyapunov Function 349
Lyapunov Stability Theorem 348
 – for robotic systems 347

m
Machine noise 18, 354
MACSYMA 14, 111, 214
Mag-Lev, *see also* Levitation
Magnetic accelerator 442
Magnetic actuator 489
Magnetic couple 31
Magnetic bearings, superconducting 240
Magnetic energy 461, 467
Magnetic field 31, 439
Magnetic field density 439
Magnetic flux 457
 – linkage 457
Magnetic force 161, 438, 452, 462, 486, 498
Magnetic hammer 471
Magnetic levitation, *see* Levitation
Magnetic levitation 1, 9, 16
 – chaos 520
Magnetic materials 446
Magnetic permeability 444
Magnetic pressure 444
 – friction 452
 – tension 444
Magnetic rail guns 473
Magnetic reluctance 486
Magnetic stress 443, 486
Magnetic susceptibility 447

Magnetic thrust bearing   240
Magnetization density   439
Magnetization curve   448
MAPLE   14, 70, 111, 214, 326, 550
Maps   514
   – one-dimensional   517
Mass drivers   470
Mass matrix   348
Mass moments   224
MATHEMATICA   14, 70, 72, 111, 179, 214, 309, 326, 421, 550
MATLAB   14, 69, 97, 131, 158, 203, 261, 265, 277, 288, 304, 367, 426, 519, 535, 551
Matrix
   – angular velocity   83
   – finite rotation   91
   – identity   226
   – mass   348
   – orthogonal   92
   – 4 × 4 transformation   92, 98
Maxwell, James Clerk   36, 443
Maxwell's Equations   455
Mechanisms   78, 103
   – Artobolevsky   6
   – cam   108, 252
   – double-slider   119, 154
   – dwell   108
   – eight-bar mechanism   116
   – escapement   106
   – four-bar linkage   43, 304
   – four-bar spherical   6
   – Hooke's Joint   204
   – parallel   108
   – Peaucellier straight-line   108, 117
   – ratchet   106
   – Sterling engine rhombus   135
   – six-bar compound   115
   – slider crank   110
Mechatronics   9, 435
MEMS   1, 9, 32, 59, 104, 108, 192, 436, 497
   – cantilever beam   444, 494
MEMS micromirror   12, 495
   – slider-crank   206
   – rhombus drive   138
Microelectromechanical
   – accelerometer   11, 436, 477, 542
   – systems, see MEMS
MIT robot arm   46, 369
Mobility   113
Molecular motors   437
Moment of force   55
Moment of inertia   56, 60, 62, 227, 285
Momentum
   – angular   19, 26, 52, 60, 230, 333
   – generalized   166
   – linear   19, 52, 54
Motion-induced voltages, see Back emf
Multibody codes   309, 549
   – ADAMS   551
   – Auto-Lev   550
   – DADS   552
   – Working Model   553
   – Video game codes   553
Multibody dynamics   4, 7, 305

n

Nanomachines   13
Nanotubes   437
Natural frequency   23
NEMS   32, 436
NEWEUL   551
Newton, Isaac   28, 35, 37
Newton–Euler Equations   20, 104, 224, 258, 272, 304
Newton Law of Impact   352
Newton's Law   29, 50, 54, 67
   – gravity   29
   – third law   55
Newton's *Principia*   1, 2, 35, 189, 532
Nonholonomic problem   33
Nonlinear
   – dynamics   501
   – oscillator   504
   – resonance   503
Numerical integration, equations of motion   277, 330

o

Optical velocimeter   14
Orbital dynamics   32, 383, 387
Orbits   388
Orbit-to-orbit transfer   393
Orthogonal matrix   92
Oscillations, self-excited   511

p

Parallel axes theorem   228, 267
Partial velocities   175, 176
Particular solution   24
Passive walking machine   107
Path coordinates   170
PD control   346, 348
Peaucellier   108, 116
Peaucellier Mechanism   117
"Pendemonium"   531
Pendulum   3, 22, 66, 168, 177, 290
   – double   203
   – double-slider   256

– electric   204
– gyro   264
– rolling   204, 297
– spherical   204, 208
Perigee   390
Period doubling   517, 540
Periodic motion   503, 512
Permanent magnetic material   447
Perturbation theory   244, 274
Phase plane   23, 24, 27, 349
Piezoelectric   454
– bender actuator   453
– stack actuator   454
Piezoelectric materials
– BaTiO$_3$ 453
– PVDF   454
– PZT   453
Planetary
– gears   133
– parameters   393
Poincaré, Henri   502
Poincaré map   15, 365, 514, 517, 523, 528, 538
– gear rattling   529
Poinsot's Ellipsoid   119, 245
Polar coordinates   29, 42, 145
Polbahnen   119
Polyvinyl difluoride (PVDF)   454
Position vector   42, 51
Potential energy   58, 159
Potential function   58, 70, 166
Precession   26, 27
Principal
– axes   63, 227, 234
– inertias   62, 227
Prismatic joint   111, 307
Projection vectors   65, 150, 322, 333
Pro/MECHANICA MOTION   552
Pseudo phase plane   538
PZT   454

q
Quantum mechanics   37
Quasi-periodic motions   503, 512, 513

r
Rare-earth magnet   448
Rayleigh Dissipation Function   162
– electromagnetic   467
Rectangular coordinates   43
Relative permeability   447
Resistivity   449
Resistors   448, 456
Resonance   24

Reuleaux, Franz   5, 36, 41, 77, 104, 113, 154
Reuleaux Triangle   108, 285
Reuleaux Method of Centrodes   124
Revolute joints   111, 307
Rickey, George   528, 530
Rigid body   60
– collisions, impact   353
– general motion of   211
– moment-free dynamics   243
– motion about a fixed point   236
– satellite dynamics   403
Robot arm
– three link parallelogram   208
Robot manipulator arm   5, 46, 209
– kinematics   285, 368
– dynamics   369
Robotic manipulators, serial-link   33, 326
Robotics   7, 305
– reflected inertia   381
Rolling   118
– centrodes   135
– disc, stability of   274
– wheel   185, 186
– constraint   184, 266, 270
Rotation   61
– instantaneous center of   118
– induced stiffness   254
– matrix   219
– rate vector   61, 95
– transformation   93
Rott, N.   530, 531
Routhian   391

s
Saddle point   509
Satellite   33, 383, 403
– deployment of solar panel   86
– dumbbell   407
– four-mass cross   408
– Hubble   213, 286
– stability of dumbbell   407
– tethered   417
Scalar triple product   250
Scattering dynamics   433
Screw
– joint   307
– transformation   102
Second moments of mass   62, 227
Seismograph   464
Separatrix   509
Ship, roll dynamics   537
Sierpinski carpet   540
Simulation and animation software   549
Singular points   113, 367

Slider crank mechanism   6, 9, 110, 205, 211
Soft spring   505
Solar panel   127, 128, 384
Solar system
  – parameters   393
  – satellite   393
SOLID WORKS   14, 80, 553
Solitons   541
Spacecraft   127
  – Hubble telescope   213, 286
Stability
  – dumbbell satellite   407
  – rigid body rotation   245
  – rolling disc   274
  – rotating pendulum   261
  – satellite attitude   403
  – vehicle skid   279
State feedback control   482
State vector   25
Straight-line mechanism   108
Strange attractor   503
Subharmonic oscillations   503
Superconductors   450
  – high-temperature   451
Superconducting magnetic bearings   240
System identification   15

t
Tait–Bryan Angles   220
Tangent vectors   175, 178
TGV   1
Tinguely, Jean   528
Three-body gravitational dynamics   430
Three-link planar manipulator   125, 335, 374
Transfer orbit   424, 430
Transformation
  – matrix, homogeneous   100
  – matrices   93, 98, 100, 216, 420
Transistor   456
Triple vector product   62, 226
Trochoid   118
Two-body problems   396
Two-link manipulator arm   339, 367

u
Unimodal constraints   305
Universal joint   6, 12

v
Van der Pol   512
Van der Waals Forces   445
Variational methods   139, 188
Vehicle dynamics   279, 283
Vehicle skid   229
  – stability of   283
Velocity   43, 45, 170, 317
  – generalized   176
  – partial   175, 176, 178
  – virtual   175
Vibration
  – absorber   193
  – period   169
Video games codes   8
Virtual displacement   48, 64
Virtual power   4, 67, 139, 153, 169, 179, 272, 281, 331
  – connected rigid bodies   322
  – rigid body   260, 268
  – system of particles   174
  – virtual velocities   175
Virtual work   4, 48, 50, 64, 143
  – electromechanical systems   460
Vitesse virtuelle   189
  – *see also* Virtual velocities
Voice coil actuator   441, 442
Voltage, induced   440
Voyager 2   532

w
Walking machine dynamics   382
Watt, J.   113
Wild Mouse   372, 535
Willis, R.   36
Working Model   78, 309, 553
Workspace, robot arm   342

y
YBCO   451